D0145971

WITHDRAWN

Molecular and Cellular Exercise Physiology

Frank C. Mooren, MD
Institute of Sports Medicine, University Hospital Muenster
Klaus Völker, MD
Institute of Sports Medicine, University Hospital Muenster
Editors

Human Kinetics

Library of Congress Cataloging-in-Publication-Data

Molecular and cellular exercise physiology / Frank Mooren, Klaus Völker, editors.
p. ; cm.
Includes bibliographical references and index.
ISBN 0-7360-4518-X (hard cover)
1. Exercise--Physiological aspects. 2. Cell physiology. 3. Molecular biology.
[DNLM: 1. Cell Physiology. 2. Cellular Structures--cytology. 3. Exercise--physiology. QH 631 M7185 2005] I. Mooren, Frank, 1964- II. Völker, Klaus, 1968-
QP301.M57 2005
612'.044--dc22

2004010862

ISBN: 0-7360-4518-X

Acquisitions Editor: Michael S. Bahrke, PhD; **Managing Editor:** Amanda S. Ewing; **Assistant Editors:** Amanda Eastin and Anne Cole; **Copyeditor:** Joyce Sexton; **Proofreader:** Joanna Hatzopoulos Portman; **Indexer:** Frank C. Mooren; **Permission Manager:** Dalene Reeder; **Graphic Designer:** Fred Starbird; **Graphic Artist:** Denise Lowry; **Photo Manager:** Kareema McLendon; **Cover Designer:** Robert Reuther; **Art Manager:** Kelly Hendren; **Illustrator:** ICC; **Printer:** Sheridan Books

Printed in the United States of America 10 9 8 7 6 5 4 3 2 1

Human Kinetics
Web site: www.HumanKinetics.com

United States: Human Kinetics, P.O. Box 5076, Champaign, IL 61825-5076
800-747-4457
e-mail: humank@hkusa.com

Canada: Human Kinetics, 475 Devonshire Road Unit 100, Windsor, ON N8Y 2L5
800-465-7301 (in Canada only)
e-mail: orders@hkcanada.com

Europe: Human Kinetics, 107 Bradford Road, Stanningley, Leeds LS28 6AT, United Kingdom
+44 (0) 113 255 5665
e-mail: hk@hkeurope.com

Australia: Human Kinetics, 57A Price Avenue, Lower Mitcham, South Australia 5062
08 8277 1555
e-mail: liaw@hkaustralia.com

New Zealand: Human Kinetics, Division of Sports Distributors NZ Ltd., P.O. Box 300 226 Albany, North Shore City, Auckland
0064 9 448 1207
e-mail: blairc@hknewz.com

Contents

Contributors

Paul A. Adlard, PhD
Institute for Brain Aging and Dementia
University of California-Irvine
Irvine, California

Nicole C. Berchtold, PhD
Institute for Brain Aging and Dementia
University of California-Irvine
Irvine, California

Arend Bonen, PhD
Department of Human Biology and Nutritional Sciences
University of Guelph
Guelph, Ontario, Canada

Dieter Böning, MD
Institute of Sports Medicine
University Hospital Benjamin Franklin
Free University of Berlin
Berlin, Germany

Claude Bouchard, PhD
Pennington Biomedical Research Center
Human Genomics Laboratory
Baton Rouge, Louisiana

Gale B. Carey, PhD
Department of Animal and Nutritional Sciences
University of New Hampshire
Durham, New Hampshire

Carl W. Cotman, PhD
Institute for Brain Aging and Dementia
Department of Neurobiology and Behavior
University of California-Irvine
Irvine, California

David J. Dyck, PhD
Department of Human Biology and Nutritional Sciences
University of Guelph
Guelph, Ontario, Canada

Øyvind Ellingsen, PhD, MD
Department of Circulation and Medical Imaging/Department of Cardiology
St. Olavs Hospital, Norwegian University of Science and Technology
Trondheim, Norway

Elvira Fehrenbach, PhD
Department of Transfusion Medicine
University of Tübingen
Tübingen, Germany

Sataro Goto, PhD
Department of Biochemistry, Faculty of Pharmaceutical Sciences
Toho University
Funabashi, Chiba, Japan

John A. Hawley, PhD
Exercise Metabolism Group, School of Medical Sciences, Faculty of Health Sciences
RMIT University
Bundoora, Victoria, Australia

Laurie Hoffman-Goetz, PhD
Department of Health Studies and Gerontology
Faculty of Applied Health Sciences
University of Waterloo
Waterloo, Ontario, Canada

Vuokko M. Kovanen, PhD, MD
Department of Health Sciences
University of Jyväskylä
Jyväskylä, Finland

Jean-Marc Lavoie, PhD
Department of Kinesiology
University of Montréal
Montréal, Quebéc, Canada

Jan Pål Loennechen, PhD, MD
Department of Circulation and Medical Imaging/Department of Cardiology
St. Olavs Hospital, Norwegian University of Science and Technology
Trondheim, Norway

Norbert Maassen, PhD
Center of Sports Physiology and Sports Medicine
Medical School
Hannover, Germany

Heimo Mairbäurl, PhD
Department of Internal Medicine VII, Sports Medicine
University of Heidelberg
Heidelberg, Germany

Andreas M. Niess, MD
Medical Clinic and Policlinic/Department of Sports Medicine
University of Tübingen
Tübingen, Germany

Harshna Patel, medical student
Department of Health Studies and Gerontology
Faculty of Applied Health Sciences
University of Waterloo
Waterloo, Ontario, Canada

Bente Klarlund Pedersen, MD, DMSc
The Copenhagen Muscle Research Centre and the Department of Infectious Diseases, Rigshospitalet
University of Copenhagen
Copenhagen, Denmark

Victoria M. Perreau, PhD
Institute for Brain Aging and Dementia
University of California-Irvine
Irvine, California

Dirk Pette, MD, PhD
Department of Biology
University of Konstanz
Konstanz, Germany

Joe Quadrilatero, PhD student
Department of Health Studies and Gerontology
Faculty of Applied Health Sciences
University of Waterloo
Waterloo, Ontario, Canada

Zsolt Radák, PhD
Laboratory of Exercise Physiology, School of Sport Science
Semmelweis University
Budapest, Hungary

Tuomo Rankinen, PhD
Pennington Biomedical Research Center
Human Genomics Laboratory
Baton Rouge, Louisiana

Walter F.J. Schmidt, PhD
Department of Sports Medicine/Sports Physiology
University of Bayreuth
Bayreuth, Germany

Axel Schulz, MD
Johanna Etienne Clinic
Orthopaedic Department
Neuss, Germany

Hansjörg Teschemacher, MD
Rudolf-Buchheim-Institute for Pharmacology
Justus-Liebig-University
Giessen, Germany

Ulrik Wisløff, PhD
Department of Circulation and Medical Imaging/Department of Cardiology
St. Olavs Hospital, Norwegian University of Science and Technology
Trondheim, Norway

Juleen R. Zierath, PhD
Department of Surgical Sciences, Section for Integrative Physiology
Karolinska Institutet
Stockholm, Sweden

Preface

The last century has revolutionized our knowledge and understanding in the biological sciences. The frontiers between traditionally separated disciplines have disappeared and have been replaced by a more integrated approach with an increasing cellular and subcellular perspective. This approach has opened our view about both function and regulation of the body as well as the pathogenesis and pathophysiology of many diseases.

The development of new experimental techniques has permitted fascinating insights into the structure, function, and regulation of living cells, the smallest functional units of the body. Electrophysiological methods have allowed study of the plasma membrane, the natural barrier between the intracellular and extracellular space that is in control of fluid and substrate movements. Using novel approaches in biochemistry, molecular biology, expression studies, and stereochemistry, it became possible to characterize membrane-coupled and cytosolic receptors, which serve the communication of cells with their environment and with other cells. Fluorescence spectroscopy and the use of antisense oligonucleotides revealed the complexity of the intracellular signaling networks, which transduce information into a cellular response, connect the different cellular organelles, and control gene expression and regulation. Novel gene transfer techniques and transgenic animal models have helped us to understand the translation of genetic information into proteins. Although the human genetic code has now been sequenced, much work to elucidate its meaning is still underway. This includes investigation of the regulation and function of a single gene (functional genomics) and the study of how proteins are processed and regulated (proteomics).

The same advances that have made possible enormous progress in human biology have affected our knowledge in exercise physiology. Traditionally the focus of exercise physiology has been the response of the body and its organs and tissues to the physiological demands of physical activity. This perspective was broadened when enzyme biochemistry became available and metabolic processes and their adaptation to exercise were better understood. The recent advances in molecular techniques have further extended the field of exercise physiology and have permitted researchers to address the mechanisms involved on a subcellular and molecular level.

This development has not yet been reflected in the classical textbooks that cover exercise physiology. The present book therefore offers this novel approach as the first comprehensive text about the effects of physical activity on the cellular and molecular level. It focuses on the molecular and cellular basis of exercise adaptations and aims to give the reader an impression of how exercise can affect and modulate cellular homeostasis.

This approach allowed us to take an in-depth look at the basic and fundamental mechanisms associated with changes in stroke volume, blood gas homeostasis, pH alterations, blood pressure, and osmotic changes in response to exercise and the ways in which they are translated into specific cellular and subcellular changes. This includes the correlation of "macro-parameters" like heart volume, cardiac contractility, and cardiac muscle hypertrophy to "micro-parameters" such as growth factor up-regulation, intracellular calcium gradients, and production of transcription factors. Other examples of key questions include the following: How does exercise improve glucose metabolism on the cellular level? Why does the vascular smooth muscle cell produce less tone in response to the same stimulus after endurance training? How can exercise training modulate the activity of killer cells? These questions are dealt with in comprehensive and detailed chapters that address the basic cellular and molecular mechanisms involved in exercise. Such an approach seems to be necessary for an integral and comprehensive understanding of the training process and can stimulate the search for novel markers of performance status.

This, however, is not the only goal of the present monograph. Presenting the mechanisms of exercise adaptation should highlight the potential of physical training in the prevention and therapy of chronic diseases. The incidence of chronic diseases such as diabetes and cardiovascular diseases has increased dramatically, with enormous effects on public health systems. As indicated by several epidemiological studies, physical inactivity is a common risk factor for such multi-etiologic diseases as arteriosclerosis, hypertension, impaired glucose tolerance, obesity ("metabolic syndrome" or "syndrome X"), coronary heart disease, colon cancer, and type 2 diabetes. Therefore chapters are dedicated to the cellular and molecular causes of these diseases and the ways in which they are affected by exercise training. These discussions will permit a better understanding of the importance of primary prevention strategies. Moreover, it is important to emphasize that physical training is a "powerful drug" in treating these sedentary lifestyle-associated diseases (secondary and tertiary prevention).

All in all, extended knowledge of cellular and molecular exercise physiology should broaden our minds and may help to develop and improve traditional training regimes both in performance exercise and in prevention and rehabilitation programs.

The book is therefore targeted at a wider audience. It should serve as basic reading for graduate and postgraduate students in exercise physiology, sport, and nutrition as well as students of human biology and physiology with an interest in physical activity. In addition it should be a reference text for exercise physiologists, sports medicine specialists, sport nutritionists, physiologists, exercise physiology researchers, and finally physicians and clinicians with interests in preventive care medicine.

The book is organized into a general section (part I) and a specific section (part II). This division is useful because it allows the uninitiated reader to develop an understanding of the field in a step-by-step manner but will also give experienced exercise physiologists the opportunity to deepen their knowledge in specific fields. Part I contains a systematic overview of modern cell and molecular biology in relation to exercise physiology. Each chapter in part I begins with an introduction to the particular area such as cellular architecture, cellular life span, transfer of genetic information into a protein response, or cellular signaling and then provides a detailed description of the effects of exercise on that specific aspect of the organism.

The second part of the book concentrates on specific cell types such as the cardiac myocyte, the skeletal muscle cell, and the different types of blood cells. After a description of their characteristic cellular and molecular features, the effects of exercise on these cell types are presented. These chapters have been limited to cell types that have apparent relevance to exercise and for which sufficient data in the literature are available.

Tables and figures are used abundantly for a better presentation and to aid understanding of the mechanisms and concepts developed in the text. Each chapter uses an extended list of references in order to facilitate literature searches and to encourage additional reading. Finally, abbreviations are defined in a list at the end of the book.

It is the hope of the editors that this textbook will reach a wide readership and may stimulate further investigations in this relatively new and exciting research area.

Part I
Molecular Exercise Physiology

Chapter 1

The Cell

Frank C. Mooren

Cellular Architecture

The cell represents the smallest unit of a living organism and exists either as a free and migrating cell, like blood cells, or as a cell embedded within a tissue complex. Despite many contacts with the extracellular matrix or neighboring cells, the single cell functions independently and autonomously within a certain range. It is surrounded by a cell membrane that is responsible for perception of hormonal information and for controlling the exchange of ions, substrates, and other substances. The cell interior is organized and structured by the cytoskeleton and consists of the cytosol and various membrane-bound organelles. Such a compartmentalization enables the cell to carry out numerous functions and processes synchronously—for instance, energy generation and dissipation, protein synthesis and protein degradation.

Membranes

Membranes coat the entire cell as well as intracellular organelles. This structural homogeneity of cellular membranes is a prerequisite for vesicular exchange across the cell membrane and between cellular organelles. All membranes consist of lipids and proteins as their major components. The relative amounts of proteins and lipids can vary significantly between different cell types and between different organelles. Likewise, the protein content of neuronal myelin membranes is about 20% dry weight but increases to about 80% in the mitochondrial membrane. The membrane lipids belong to different lipid classes such as glycerolphospholipids, phosphosphingolipids, glycosphingolipids, and sterols. The major membrane lipids are phospholipids, which form a bilayer matrix with the polar regions facing the hydrophil extracellular fluid and the cytosol. The complex lipids like phosphosphingolipids are involved in cellular signaling processes or in identifying cells like glycosphingolipids that carry blood group antigens in the erythrocyte membrane. Cholesterol, another complex lipid, is by far the most commonly found sterol in membranes. While its content is high in the plasma membrane, it is nearly absent in the mitochondrial membrane. The reason for this diversity of the chemically distinct lipids in any given membrane is largely unknown.

Membrane proteins can be divided into two classes: integral and peripheral proteins. This classification denotes the degree of treatment required to release the proteins from the membrane. Peripheral proteins that usually associate with lipid bilayer in a noncovalently bound form can be released from the membrane by relatively gentle methods like washing of the membranes in buffers of different ionic strength and pH or in the presence of divalent cation chelators such as EDTA. In contrast, integral membrane proteins are partially embedded in the bilayer and can be released from the bilayer only through destruction of its structural integrity using detergents or organic solvents. However, this distinction between peripheral and integral membrane proteins emphasizes the relative strength of attachment but does not clearly define the mode of attachment to the bilayer. The attachment might depend on electrostatic or hydrophobic interactions, or the protein might be covalently bound to fatty acids or phospholipids that serve as an anchor within the lipid bilayer.

Integral proteins, which span the entire membrane, are called transmembrane proteins. They may have a single transmembrane segment or multiple transmembrane segments and are amphiphilic. The portions of the proteins that face the polar solvents are enriched in amino acid residues with polar and ionized side chains, whereas the parts of the molecule that are in contact with the lipid bilayer contain primarily nonpolar amino acid residues.

The concept of the plasma membrane as a fluid mosaic model describes the plasma membrane as a highly flexible and mobile structure. For more than three decades it has been known that all biomembranes display a lateral and transversal asymmetry for both constituents, proteins and lipids, suggesting that the mobility of molecules is still somewhat restricted (Bretscher 1972). The transversal asymmetry of membrane proteins depends on the way in which the protein is originally inserted into the membrane. Generally the translocation of proteins across the bilayer is negligible. In contrast, the lateral mobility of proteins is much higher, allowing the enhancement of receptor protein density in specialized areas of the cell or in contact zones to other cells or membranes.

Likewise, transversal lipid asymmetry can be found in the plasma membrane. While the choline-containing phospholipids, phosphatidylcholine and sphingomyelin, are primarily located in the cell's outer leaflet, the aminophospholipids phosphatidylserine and phosphatidylethanolamine are primarily exposed to the intracellular environment. This property can be used for diagnostic purposes, for example to determine apoptotic cells. In apoptotic cells, the membrane phospholipid phosphatidylserine is translocated from the inner to the outer leaflet of the plasma membrane. The dissipation of phospholipid asymmetry is facilitated by calcium-dependent and Mg-ATP-independent activation of scramblase. The asymmetry of plasma membrane lipids under physiological conditions is maintained through the activation of two active transporters, the phospholipid translocase (floppase) and the aminophospholipid translocase (flipase) (Diaz & Schroit 1996; Tang et al. 1996; Bitbol & Devaux 1988). In addition, there is evidence for a lateral inhomogeneity of the lipid composition of the plasma membranes resulting in certain lipid microdomains. Furthermore, the cytoskeleton plays an important role in transversal and lateral distributions of membrane components. This is true not only for the membrane vesicle movements, which involve endocytosis, exocytosis, or phagocytosis, but also for cytoskeletal interactions with the membrane, which play a role in determination of asymmetry and stabilization of microdomains as well as plasma membrane stability (Manno et al. 2002).

Cytoskeleton

The cytoskeleton is responsible for the cell's shape and structure. The cytoskeleton is not a solid and static framework. Instead, it is highly dynamic and is involved in a variety of processes such as cellular movement and locomotion, cell division, phagocytosis, volume regulation, and cell signaling. Furthermore, it is involved in the organization of vesicular transport inside the cell and in directing vesicular endocytotic and exocytotic processes.

The cytoskeleton consists of three major types of filamentous proteins: actin filaments or microfilaments, intermediate filaments, and microtubules (Frixione 2000). Actin filaments are small (about 8 nm thick) filaments of polymerized actin, a 43-kDa globular protein. Polymerization is dependent on ATP and cations, most likely potassium and magnesium. There are a number of actin-associated proteins that are important for the structure and function of the actin network (table 1.1); for example, spectrine can serve as a cross-linker of actin filaments. Moreover, it connects the actin cytoskeleton to several other membrane proteins and ion transport systems. The latter function has also been established for ankyrin, another actin-associated protein. Interactions between ankyrin and the sodium pump, sodium channels as well as H^+-K^+ ATPases have been shown (Denker & Barber 2002).

The actin cytoskeleton is primarily located in the cell periphery. In polarized cells like epithelial cells, the structure of the actin meshwork differs between the basolateral and the apical part. Moreover, the actin cytoskeleton builds up the core of cellular extensions like microvilli.

The structure of the actin cytoskeleton is highly dynamic and can rapidly change upon cellular activation or stress. Therefore the actin filament turnover is linked to intracellular signaling processes adapting the cytoskeletal network to the physiological demands, for example during cell activation or motility (Mitchison & Cramer 1996). Important effectors on the actin cytoskeleton are the PI 3-kinase and small GTPases including Rho, Rac, and CDC-42. Each of these has specific inter-

Table 1.1 Classification of Actin-Associated Proteins

Actin-associated proteins	Name	Function
Actin monomer binding proteins	Profilin, cofilin, thymosin	Bind monomeric G actin; regulate the availability of polymerizable actin; actin disassembly
Actin filament capping proteins	Gelsolin, villin, leufactin	Regulate monomer addition at the filament end
Actin filament nucleating proteins	ARP2/3	Generate new filaments
Actin filament binding proteins		
Cross-linking proteins	Spectrin, fimbrin	Connect filaments to form a 3-dimensional network
Membrane anchors	Talin, ankyrin, ezrin	Connect filaments to the plasma membrane
Motor proteins	Myosin I, II, V, VI	Drive vesicular transport

actions with particular actin-associated proteins inducing the formation of filopodia, lamellipodia, and stress fibers (Nunoi et al. 2001).

Microtubules are polymers made of tubulin, a well-characterized heterodimeric protein. Microtubules originate from the centrosome, an area in the cytoplasm characterized by the presence of centriols, and require cations and GTP for polymerization. The microtubule network is located below the actin cytoskeleton spanning the distance from the plasma membrane to the nucleus. Because microtubules are polarized, their parallel arrangement within the cell defines the polarity of the cell. The microtubule cytoskeleton can rapidly assemble into a variety of distinct configurations in order to carry out different tasks. During mitosis, microtubules build up the mitotic spindle responsible for the chromosome's movement. During interphase, microtubules are responsible for the arrangement of cellular organelles and play a role in the movement of vesicles between organelles and the plasma membrane (Allan et al. 2002).

Similar to the situation with the actin cytoskeleton, there exist also a number of microtubule-associated proteins (MAPs) that serve in stabilization and cross-linking of microtubules. Furthermore, MAPs convey signaling molecules such as kinases and phosphatases to the microtubule cytoskeleton (Gundersen & Cook 1999). Motorproteins of the kinesin and dynein complexes are involved in the directed movement of vesicles. While dynein 1 (CD1) drives the cargo toward the minus ends of microtubules, most members of the kinesin superfamily drive the vesicle movement in the opposite direction (Allan et al. 2002; King 2000).

Finally, another group of cytoskeletal filaments, the intermediate filaments, should be mentioned. This group is more heterogeneous than those discussed so far. Encoded by more than 50 known genes, intermediate filaments consist of associated tetramers of fibrous proteins and are divided into six major classes (Fuchs & Weber 1994). One group encompasses the ceratines, which can be found predominantly in epithelial cells. Another group contains vimentin, found predominantly in mesenchymal cells, and desmin, expressed in all muscle cells. Intermediate filaments can be found in cell–cell contact zones and around the nuclear pores. A major function of several intermediate filaments is to stabilize cellular architecture against mechanical forces (Coulombe et al. 2000).

Cell Organelles

The cytosol contains a number of membrane-bound structures called cell organelles. Because of their different sizes and densities, organelles can be isolated after cellular homogenization by density gradient centrifugation. Organelles fulfill different functions in cellular metabolism and enable the cell to separate the multiple biochemical reactions, for example synthetic and degradation pathways, or energy-producing and -consuming reactions (figure 1.1).

Nucleus

The cell nucleus is usually the largest cell organelle and is composed of two membranes, forming the nuclear envelope, which is tethered to the endoplasmic reticulum (figure 1.1*a*). The nucleus houses the genetic information of the cell encoded in the sequence of DNA nucleotides. This information is fundamental for cellular homeostasis leading to the synthesis of proteins that determine structure and function of the cell. Usually the DNA is associated with proteins forming a fine network of threads called chromatin. For cell division, the

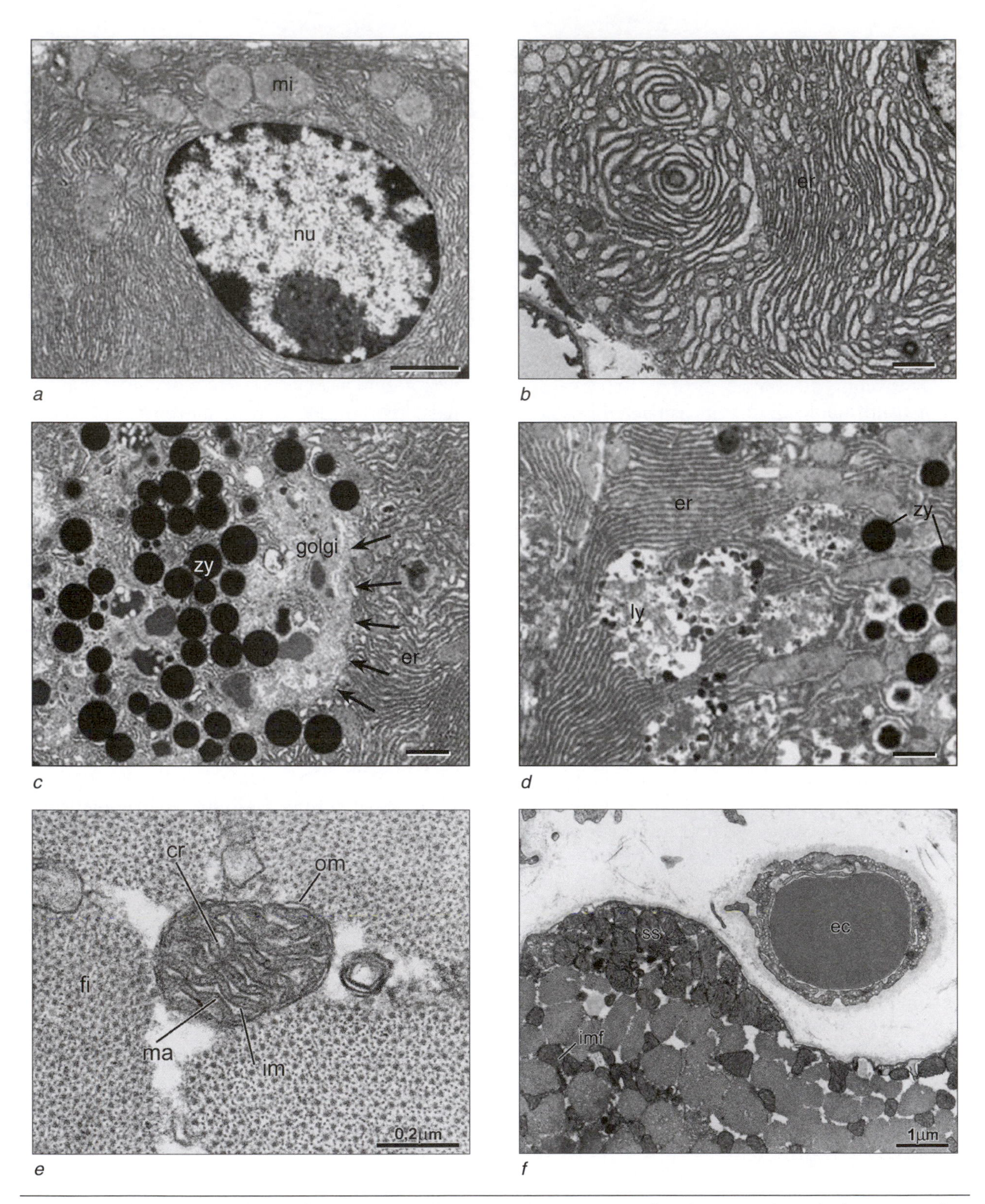

Figure 1.1 Electron micrographs of cellular organelles: *(a)* Nucleus (nu), mitochondria (mi); *(b)* endoplasmic reticulum (er); *(c)* Golgi apparatus (golgi); *(d)* lysosomes (ly), zymogen granula (zy); *(e)* mitochondrium in a cross-section of human skeletal muscle outer membrane (om), cristae (cr), intermembrane space (im), matrix (ma), and myofibrils (fi); *(f)* subsarcolemmal (ss) and intermyofibrillar (imf) mitochondria of a human muscle fiber. Micrographs *a-d* were taken from pancreatic acinar cells, which are an epithelial cell type. The bar in figures a-d indicated 1 μm.

Micrographs A-D by the courtesy of Dr. M.M. Lerch, Ernst-Moritz-Arndt Universität Greifswald; micrographs E-F from H. Hoppeler & M. Flück, Med. Sci. Sports Exerc. 35:95-104, 2003, with permission.

DNA protein aggregates will be condensed, forming the 46 chromosomes of a human cell.

During the interphase the genetic information is transcribed into RNA, which has to be translocated from the nucleus to the cytosol for translation into an amino acid sequence at the ribosomes. Additionally, RNA and proteins that build the ribosomes have to be transferred to the cytosol. This macromolecular exchange between the nucleus and the cytoplasm takes place through nuclear pores that result from the fusion of the two membranes of the nuclear envelope. These circular openings are approximately 800 angstroms in diameter and occur at regular intervals along the surface of the nuclear envelope. A supramolecular structure referred to as the nuclear core complex and of approximately 125 MDa is an important controlling element for the import and export of macromolecules (Weis 2003). Finally, the nucleolus can be easily localized within the nucleus. This filamentous region is responsible for decoding ribosomal RNA and assembling RNA and protein components of ribosomal subunits.

Ribosomes

Decoding of the genomic information and translating it into an amino acid sequence are performed at the ribosomes. Ribosomes are macromolecular complexes consisting of more than 50 different ribosomal proteins and a number of ribosomal RNA molecules. Typically each ribosome is composed of a large and a small subunit. The ribosomal subunits are assembled in the nucleolus region of the nucleus. Ribosomal proteins therefore have to be imported into the nucleus to associate with RNA. The complete subunits are then exported through the nuclear pores into the cytosol, where the subunits form the functional ribosome. The ribosomes either are located free in the cytosol or are attached to the endoplasmic reticulum to produce housekeeping or secretory proteins, respectively. For protein synthesis, however, a number of additional enzymes, like RNA polymerases, are required.

Endoplasmic Reticulum

The endoplasmic reticulum (ER) is a membrane-bound organelle that consists of a network of cisternae and tubules stretching throughout the cytosol (figure 1.1*b*). Morphologically a smooth and a rough endoplasmic reticulum can be distinguished, the latter being characterized by the attachment of ribosomes. This apparent structural simplicity should not obscure the fact that the ER handles a number of different functions such as

- post-translational modification of proteins,
- support of the vesicular transport machinery,
- assistance in generating tertiary protein structures,
- lipid biosynthesis,
- calcium storage, and
- detoxification.

One can speculate that the apparently homogeneous tubular system of the ER consists of several microdomains and subdomains characterized by specific biochemical markers. Recent investigations of the intracellular traffic between the ER and Golgi apparatus describe these organelles as highly dynamic in structure, making a precise definition and description difficult. Likewise, subdomains such as the transitional elements or the so-called ER-Golgi intermediate compartment (ERGIC) have been defined. It was found that any part of this secretory pathway has its unique set of resident or marker proteins that define its structural and functional properties. These resident proteins either do not enter transport vesicles or are retrieved from the Golgi apparatus, when they have escaped the ER. For this purpose resident proteins possess amino acid sequences like the dilysin motif targeting a protein to the ER (Teasdale & Jackson 1996).

The Golgi Apparatus

The Golgi apparatus consists of about six to eight membrane-bound flattened cisternae arranged in parallel order. It is a dynamic organelle that receives, post-translationally modifies, and sorts proteins for delivery to various destinations (figure 1.1*c*). Functionally, the two sides of the Golgi apparatus behave differently. At one side, the so-called cis-Golgi network (CGN), newly synthesized material from the ER is imported. At the opposite side, the trans-Golgi network (TGN), the material is exported after sorting and packaging. Between these steps the post-translational modification of proteins takes place, for example the linkage of carbohydrates to proteins to form glycoproteins.

Morphology and function of the Golgi apparatus are highly dependent on intact cytoskeletal structures. Depolarization of microtubules results in a fragmentation of the Golgi apparatus (Thyberg & Moskalewski 1999). The different organization of microtubule network in polarized epithelial cells and in fibroblasts therefore results in different localizations of the Golgi apparatus (McNiven & Marlowe 1999). In addition, maintaining the proper structure and function of the Golgi apparatus depends on a number of mechanochemical enzymes like dyneins, kinesins, and myosins, which are involved in formation and movement of Golgi-derived vesicles (Allan et al. 2002).

Mitochondria

Mitochondria are oval-shaped organelles measuring 1 to 2 μm in length and 0.1 to 0.5 μm in diameter. Mitochondria are separated from the cytosol by an outer membrane while their interior is further compartmentalized by a membrane system. The inner membrane divides the mitochondria's interior into the intermembrane space and the matrix (figure 1.1*e*). The inner membrane is folded several times, forming sheets or tubules that have been named christae mitochondrialis. Novel insights into the mitochondrial structure owing to recent technical advances revealed that mitochondria's shape and structure are variable and depend on source and conformational state. The christae structure can vary from tubular to complex lamellar structures. It has been shown that they are connected to the intermembrane space by tubular structures that might have important implications on metabolite diffusion and reaction rates (Frey & Mannella 2000). The two mitochondrial membranes are different with respect to their ion permeability. While the outer membrane is highly permeable for molecules up to 5,000 kDa, the inner membrane demonstrates low ion permeability comparable to that of the plasma membranes. Therefore the composition of the inner membrane space is comparable to that of cytosol. While the mitochondria matrix contains the enzymes of the Krebs cycle, the inner membrane carries the enzymes responsible for oxidative phosphorylation like cytochrome C.

Mitochondria are the only organelles that contain separate and semiautonomous genomic information on the mitochondrial DNA. This DNA codes for about 13 different proteins involved in oxidative phosphorylation and for ribosomal and transfer RNAs.

The role of mitochondria in cell homeostasis, however, does not seem to be limited to energy metabolism. There is evidence for a role of mitochondria in intracellular signaling. Mitochondria's role as a calcium storage and buffering organelle is well known (Rutter et al. 1998). Recent observations from several different cell types indicate a certain spatial distribution of mitochondria within the cytosol that results in the modulation of intracellular calcium signals (Tinel et al. 1992). Moreover, mitochondria are involved in the initiation of programmed cell death (apoptosis). Numerous close contacts between mitochondria, the ER, and the cytoskeleton were observed, emphasizing the dynamic interaction between these structures (Rizzuto et al. 1998). Finally, there is evidence for a spatial heterogeneity of the mitochondria population within a cell. In the skeletal muscle cell, mitochondria from the subsarcolemmal (SS) and the intermyofibrillar (IMF) region can be distinguished that differ with respect to their biochemical properties and their adaptational response to exercise (see later).

Vesicular Structures

Within the cytosol, at least three different vesicular structures can be defined: peroxisomes, lysosomes, and endosomes (figure 1.1*d*). These membrane-bound organelles account for about 3% of the total volume. Peroxisomes contain a number of different enzymes that are involved in lipid degradation and detoxification. Lysosomes are the final site of degradation and digestion of incorporated extracellular macromolecules and of obsolete intracellular material. However, there is increasing evidence that lysosomes and their precursor structures may have additional functions such as cytotoxicity and antigen processing/presentation. Lysosomes are loaded with a number of degrading enzymes like hydrolases and esterases and are characterized by an acidic pH value (Hunziker & Geuze 1996).

Endosomes represent a heterogeneous population of intracellular vesicular structures. Endosomal compartments can be defined according to various aspects such as function, morphology, or composition. The definition of the endosomal compartment is sometimes difficult because there are overlaps with other compartments, especially the lysosomal compartment. The term "early endosome" reflects an operational definition for this primary vesicle docking site either for endocytotic vesicles from the plasma membrane or as an acceptor compartment for vesicles from the trans-Golgi network. While early endosomes usually are located toward the cell periphery, late endosomes can be found closer to the nucleus (Gruenberg 2001). From the early endosomes, material can be directed back to the plasma membrane via the recycling endosome, as it occurs during receptor recycling. Other pathways are the transfer of material to the late endosomes, which may fuse with lysosomes for final degradation, and transcytosis across the cytosol from one plasma membrane domain to the opposite domain (Katzmann et al. 2002). Morphologically, early endosomes tend to be tubular while late endosomes are more spherical (Gruenberg 2001). It is becoming more and more apparent that every stage is defined by special markers or coat proteins. Vesicles coated with clathrin represent the best-defined internalization pathway, which is used, for example, for internalization of ligand-occupied receptors. Ligand–receptor coupling induces the exposure of signaling motifs that are recognized by a complex of adapter and clathrin molecules. The rate of clathrin-dependent endocytosis at the plasma membrane is also regulated by rab 5, a small GTPase. Rab 5 is also a marker of the early endosome or sorting endosome and is involved in promoting early endosome fusion. Rab 11 and rab 7 and 9 are other small GTPases and marker proteins for the recycling and the late endosome, respectively. Another marker of the late endosome is the (ci)-mannose 6-phosphate receptor (M6PR). It is found concentrated on late endosomes but is not present on lysosomal membranes (Clague 1998). Finally, it has been observed that the material during the traffic through the endosomal compartment is more and more acidified. This is due to the activity of a K^+-H^+-ATPase. The pH of the early endosomal compartment is about 6.0; late endosomes and lysosomes are characterized by pH values of about 5 and even below. This decline in pH seems to be important for dissociation of incorporated receptor complexes as well as for the shift of vesicular marker proteins (Mukherjee et al. 1997).

Cytosolic Traffic

Cytosolic traffic refers to the multiple exchange of vesicles between the intracellular organelles as well as with the plasma membrane (figure 1.2). There exist several routes in order to establish the communication between organelles and plasma membrane and to ensure that none of the organelles will be depleted of membrane components. In the synthetic route, proteins are first translocated into the ER where they undergo folding and maturation. ER quality control guarantees that only correctly folded proteins are released to the cell surface. Otherwise proteins are retro-translocated

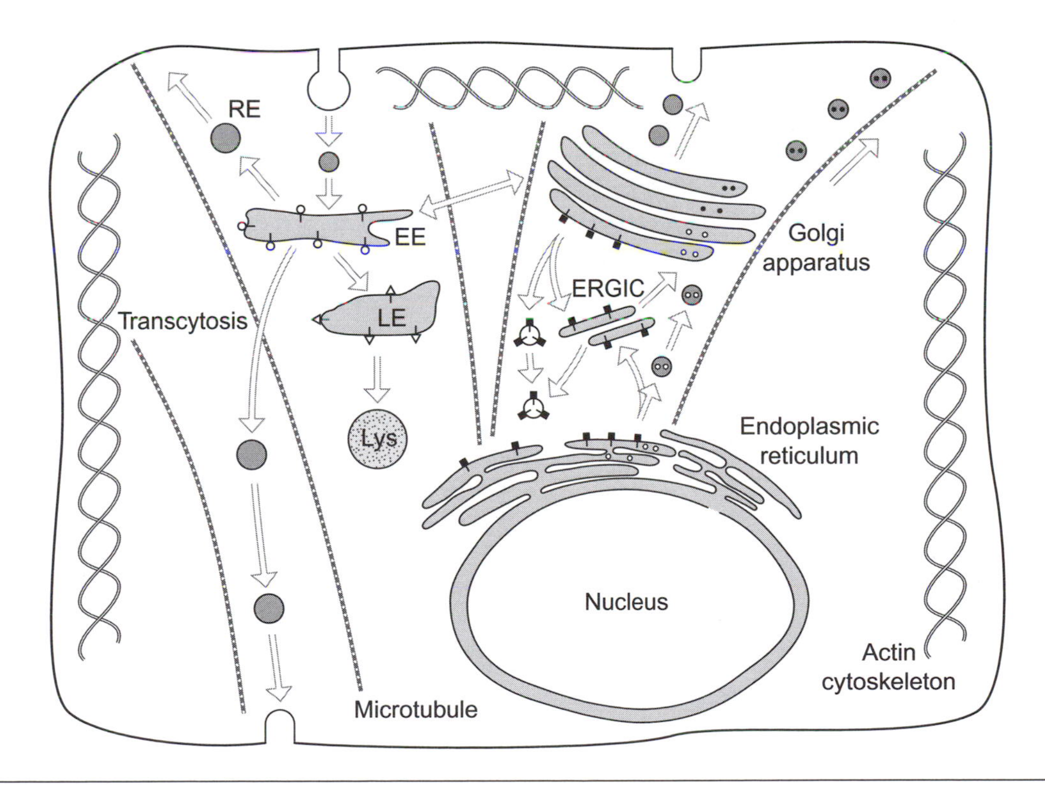

Figure 1.2 Schematic diagram of the cytosolic traffic organization (for details see text). ERGIC, endoplasmic reticulum-Golgi intermediate compartment; EE, early endosome; LE, late endosome; RE, recycling endosome; Lys, lysosome; ■ ○, organelle/vesicle targeting motifs; ⊚/⊚ cargo unmodified/modified.

to the cytosol for degradation by the proteasome (Bonifacino & Weissmann 1998). Cargo transport from the ER to the Golgi network is performed via the ER-Golgi intermediate compartment. During the cargo passage through the Golgi apparatus, post-translational processes are performed and cargo is packed again and transported for secretion to the plasma membrane. Another pathway directs digestive enzymes like hydrolases and esterases toward the lysosomal compartment via the endosomal compartment. To avoid membrane depletion of the ER and to retrieve lost ER proteins, a retrieval pathway exists from the Golgi apparatus to the ER as well as from the intermediate compartment to the ER.

In the endocytotic route, the imported material is directed first to the endosomal compartment, from which it can be transported via different routes. For example, internalized receptors can be recycled to the plasma membrane; material can be sorted for final digestion in the lysosomes or can be transferred to the trans-Golgi network; and finally material can be transported from one side of the plasma membrane to the opposite side.

Such vesicular transport is highly dependent on a number of factors. The similarity in membrane structure of the plasma membrane and the intracellular organelles is an important prerequisite for their structural communication via vesicles. Budding, movement, and fusion of vesicles are moreover dependent on an intact cytoskeletal network that arranges the different pathways and that, in combination with the cytoskeletal-associated proteins, is necessary for vesicle movement. To address vesicles to a final target, a number of code and marker proteins are used. Various vesicle coat proteins like clathrin or COP I and II have been described that have different functions in vesicular transport. Clathrin-coated vesicles, the first coated vesicles to be observed (in 1964), are involved in the internalization pathway (Roth & Porter 1964). However, some non-clathrin-dependent endocytotic events have been identified also. COP I and COP II vesicles seem to be involved in the transport to and from the ER (Allan et al. 2002; Teasdale & Jackson 1996). Moreover, a number of marker proteins have been characterized, such as small GTPases (rab 5,11, 7, 9) or special amino acid motifs like the d-lysine or d-arginine motifs that characterize the different organelles and vesicles and drive them to the target (Clague 1998). These marker sequences exist not only for membrane proteins, but also for soluble proteins of the organelle's matrix.

Finally, a complex protein machinery has been described that is responsible for membrane fusion. Studies from synaptic vesicle exocytosis in neurons have given insight into a protein complex known as the core complex, which seems operative also in other cell types. After docking, the first contact between the vesicular and target membrane, vesicle and target SNARE molecules interact with each other. Synaptobrevin is the relevant vesicular SNARE molecule, while the target SNARE consists of syntoxin 1 and SNAP-25. After the fusion event the core complex is released by the action of NSF (n-ethylmaleimide-sensitive factor), which is recruited to the membrane by adaptor proteins called SNAPs (no relation to SNAP-25). Despite the progress in understanding the mechanical and structural basis of membrane fusion, a number of questions remain open—for example, how specificity of membrane fusion arises (Rizo & Südhof 1998).

Exercise and the Cell

Acute and chronic exercise bouts are important physiological stimuli known to affect metabolism as well as organ structure. Acute exercise bouts, especially those that are maximal or supramaximal, often result in structural damage to tissues followed by a more or less long period of impaired function. Tissue repair leads to an adaptive response, usually making the tissue more resistant to subsequent stimuli. Regular and chronic exercise at a moderate workload is followed by an adaptive training response that results in an enhanced functional capacity of the organism, presenting clinically as an increased maximal oxygen capacity and enhanced muscular power and fatigue resistance. Both local damage and adaptive response have a cellular and molecular component, which is discussed in the following chapters. The present chapter is an overview of current knowledge about the effects of exercise on aspects of the cellular structure such as membrane composition, cytoskeletal network, and organelle function.

Exercise Effects on Plasma Membrane Integrity and Composition

Both acute and chronic exercise affect the integrity and the composition of the plasma membrane. Depending on the intensity and duration, acute exercise damages cellular membranes (Temiz et al. 2000; Overgaard et al. 2002). This effect is either delayed or blunted in trained individuals, suggesting an adaptational process during chronic exercise that can at least in part counteract the harmful effects associated with acute exercise (Senturk et al. 2001; Vincent et al. 1999; Yalcin et al. 2000; Venditti & Di Meo 1997).

In red blood cells, the exercise-induced plasma membrane damage is indicated by an enhanced osmotic fragility and an impaired deformability (Senturk et al. 2001; Yalcin et al. 2000). After an exercise test, structural alterations of plasma membrane of leucocytes have been described, which were associated with an increased cellular calcium content (Caimi et al. 1997). Azenabor and Hoffman-Goetz (2000) reported increased intracellular calcium levels and calcium influx in mouse thymocytes after exhaustive exercise. Recently, we could confirm these data in human lymphocytes immediately after an exhaustive exercise test at 80% $\dot{V}O_2max$. These alterations were time dependent, since these effects were not observed at later time points after the test. The increase in basal calcium was interpreted as an enhanced calcium influx across the plasma membrane due to exercise-associated structural damage, because the calcium sequestration was unaffected in these cells (Mooren et al. 2001).

A number of factors have been proposed to be responsible for plasma membrane structural changes observed during and after exercise. Oxidative stress is one factor that can induce structural and functional alterations. Since it is difficult to measure free radicals because of the short life span of these species, scientists focus on determining free radical tissue damage such as lipid peroxidation. Exercise-induced lipid peroxidation has been demonstrated in a number of studies (Dillard et al. 1978; Child et al. 1998; Cooper et al. 2002). Moreover, a correlation between lipid oxidation and exercise intensity was reported (Azenabor & Hoffman-Goetz 2000). Markers of free radicals could be diminished after application of antioxidants and allopurinol, an inhibitor of xanthine oxidase (see chapter 9) (Senturk et al. 2001; Vina et al. 2000).

Other factors that must be considered are metabolic and mechanical stress. Increased plasma lactate levels and, subsequently, an enhanced lactate influx into red blood cells may also affect cell rigidity (Szygula 1990). Red cell deformability decreases with increasing lactate levels under in vitro conditions. This effect might be mediated by an osmotic shrinkage of lactate-loaded erythrocytes that could be further aggravated by an exercise-associated fluid shift (Brun 2002).

Finally, mechanical stress affects plasma membranes. In endothelial cells, shear stress has been shown to be beneficial because it results in decreased adhesion of blood cells and therefore prevents atherogenesis (Cooke 2003). On the other hand, the mechanical stress in contractile tissue can result in a decreased cellular integrity. Structural damage is pronounced after eccentric exercise as indicated by the release of intracellular enzymes and macromolecules like creatine kinase and myosin heavy chains into the blood (Overgaard et al. 2002).

Exercise-induced structural changes in the plasma membrane can be delimited by training (Yalcin et al. 2000). One reason might be the up-regulation of defense mechanisms such as radical scavenging and depleting mechanisms (for details see chapter 9) (Vincent et al. 1999). Exercise-induced cellular hypertrophy is associated with an enlargement of the plasma membrane. Cell capacitance measurements indicate that the cell membrane surface area of cardiomyocytes increases under endurance training-induced cardiac growth (Mokelke et al. 1997). Furthermore it could be shown that regular exercise alters the lipid composition of both plasma and mitochondrial membrane (Dohm et al. 1975). In long distance runners, an increased ratio of phosphatidylcholine to phosphatidylethanolamine and a decreased ratio of cholesterol to total phospholipids were observed (Nakano et al. 2001). This finding was associated with an enhanced deformability of the athletes' erythrocytes. After four weeks of treadmill training, a change in the structural composition of rat skeletal muscle membrane was observed with a decrease in polyunsaturated fatty acids (Helge et al. 1999). These data were partially confirmed in humans after six-week exercise training of low intensity (Andersson et al. 1998). Furthermore, a muscle type-specific phospholipid fatty acid pattern was observed. The white quadriceps, a muscle dominated by fast-twitch glycolytic fibers, demonstrated a significantly higher degree of unsaturated fatty acids than soleus muscle, which consists mainly of slow oxidative fibers, or the red quadriceps muscle, which contains both slow oxidative and oxidative glycolytic fibers. This might be explained by the contribution of lipids that the fiber type utilizes for fueling oxidative phosphorylation, and emphasizes the impact of metabolism on membrane composition.

Mechanical Stress and Cytoskeleton

Mechanical stress is an important effector and modulator of the cytoskeleton's structure and organization. Cytoskeletal alterations depend on the intensity of the stimulus. Submaximal stimuli usually result in a reorganization of the structure, while supramaximal stimuli often disrupt the cytoskeletal network followed by regenerating and remodeling processes. Examples of both responses are given here while a distinction is made between muscle and non-muscle cells.

Exercise may damage fibers in the active muscles, especially when the exercise is relatively intense, is of long duration, and includes eccentric contractions. Clinically this presents as muscular discomfort and pain in the stressed muscles that peak 24 to 48 h after exercise, known as delayed-onset muscle soreness (DOMS; Friden & Lieber 2001). DOMS is directly related to the eccentric component of exercise, a form of exercise characterized by forcible lengthening of a contracting muscle. This type of exercise generates higher tension than any other form of contraction (Proske & Morgan 2001).

Which is the primary event of muscle injury after exercise is still controversial (figure 1.3). One hypothesis claims a primary intracellular event in which overstretched and disrupted sarcomeres represent the initial damage. Cytoskeletal disruption, especially of the desmin intermediate filament network, has been shown to occur early after eccentric exercise. Subsequently, morphological studies show sarcomeres out of register with one another and Z-disc streaming and broadening (figure 1.4; Friden & Lieber 2001). Whether these events are solely responses to the mechanical stimulus or the effect of activated intracellular proteases remains unclear. Several studies have demonstrated an increase in basal intracellular calcium concentrations in muscle fibers after exercise, which can activate calcium-dependent proteases such as calpain. Belcastro (1993) reported both an increased calcium affinity of calpain and an increased calpain activity in hindlimb muscles after exercise. Interestingly, calpain preferentially affects cytoskeletal proteins while it interferes less with the myofilaments. Another proposed pathway involves calcium-dependent activation of phospholipase A2, known to damage cell membranes by cracking membrane lipids (Armstrong 1990). While pharmacological inhibition of phospholipase A2 was effective in attenuating the stimulus-induced membrane damage in skeletal muscle, it was not responsible for myofibrillar disruption (Jackson et al. 1984; Duncan 1988).

The other hypothesis focuses on a primary sarcolemmal damage. This should result in both disintegration of the lipid bilayer and disconnection of the calcium release units (feet structure), which consists of membranous (voltage-dependent Ca^{2+} channel; dihydropyridine receptor, DHPR) and submembranous components (intracellular Ca^{2+} channel; ryanodine receptor, Ryr) together with cytoskeletal elements. A disturbance of ion homeostasis and of the excitation–contraction (EC) coupling follows (Warren et al. 2001). Supporting evidence comes from studies in mouse muscle. Exercise-induced tension deficit was reversible after application of caffeine, known to

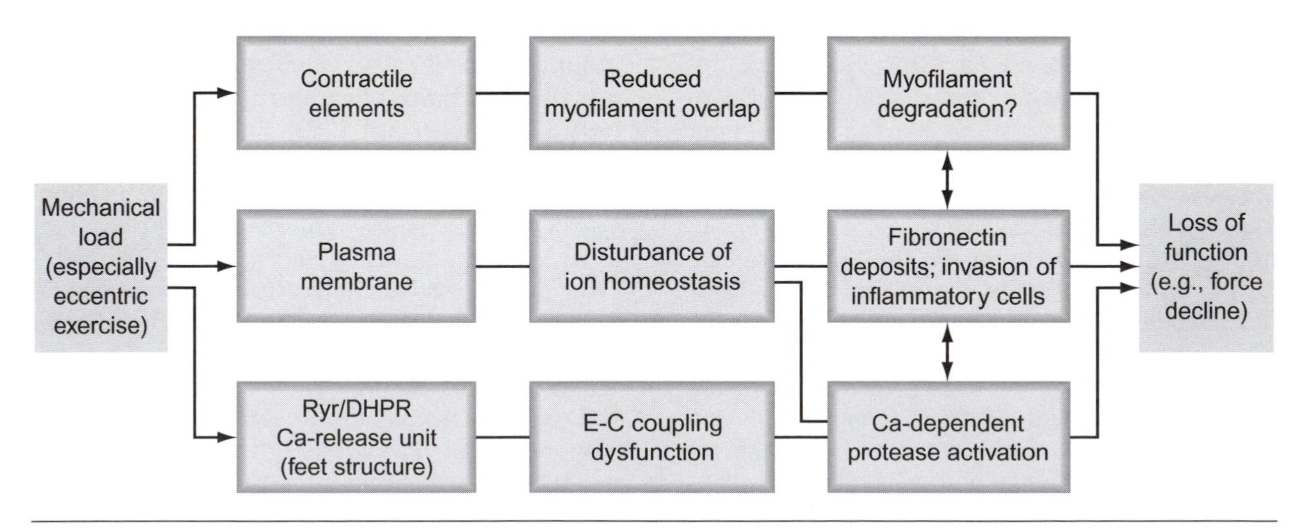

Figure 1.3 Postulated sequences of events leading from mechanical load to muscular loss of function.
Adapted from U. Proske & D.L. Morgan, 2001, J. Physiol. 537.2:333-345.

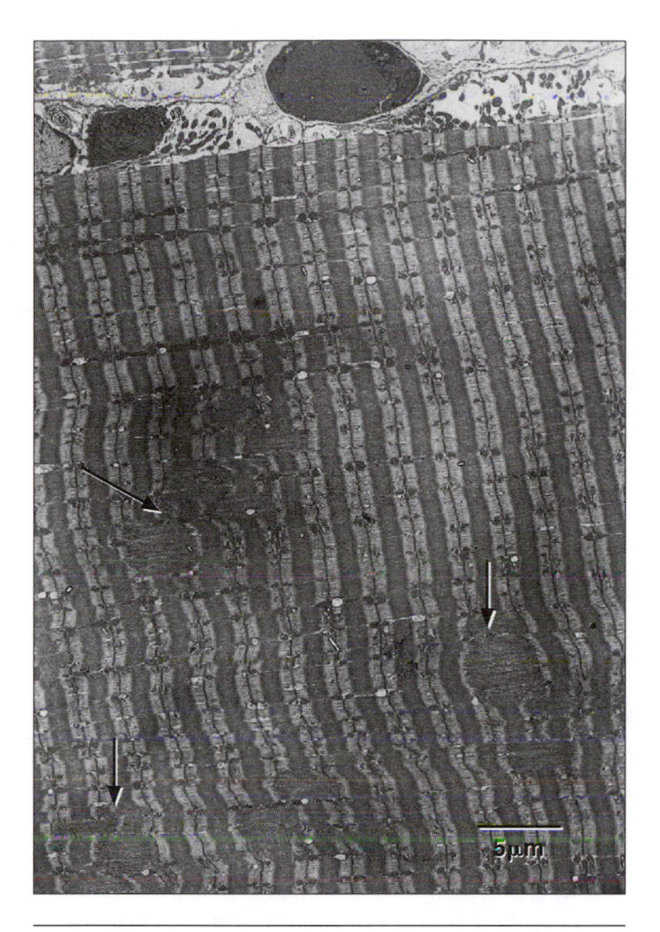

Figure 1.4 Electron micrograph of eccentrically damaged human skeletal muscle. Note the lack of registry of neighboring myofibrils, the Z-disc streaming and dissipation, and the sarcomeric disruption.

Micrograph courtesy of Dr. J. Friden, Department of Hand Surgery, Sahlgrenska University Hospital, SE-413 45, Göteborg, Sweden. Reprinted, by permission, from J. Friden.

release the sarcoplasmic calcium pool and thereby bypassing the voltage-dependent calcium-induced calcium release (Warren et al. 1993; Balnave & Allen 1995).

While the exact series of initial events leading to muscle damage remains open, there are obvious signs of impaired cellular integrity as indicated by the release of intracellular molecules such as creatine kinase, lactate dehydrogenase (LDH), or myosin heavy chains. Subsequently, an inflammatory response starts with deposits of fibronectin and the invasion of leucocytes, leading to tissue remodeling and adaptation of the muscle fibers. A second bout of eccentric exercise, repeated within days or weeks, produces much less damage. The structural basis of this adaptational process is proposed to be the formation of extra sarcomeres, which should improve the muscle's working range and should help the muscle to withstand further stress. Other hypotheses focus on an increased resistance against stress of the cytoskeleton and the cell membrane (Gibala et al. 1995).

Currently there is almost no information available about the effect of exercise on the cytoskeleton of non-muscle cells. However, one can find analogies in other areas such as hemodynamics and cardiovascular research. A number of interesting observations have been made recently about the effect of shear stress on the endothelial cell cytoskeleton. The blood flow constantly exposes endothelial cells in vivo to hemodynamic forces that change with the onset of exercise and depend on the type and intensity of exercise. The shape and the cytoskeletal organization of the endothelial cells depend on the intensity of the shear stress applied. In regions of the aortic tree exposed to low shear stress, the cells are polygonal and have only a few stress fibers. In contrast, in regions of high shear stress, the cells are more elongated and contain numerous stress fibers. These findings could be confirmed by in vitro experiments using flow chambers (Barakat 1999; Galbraith et al. 1998). Morphologically, enhanced shear stress results in more stress fibers and a rearrangement of the cytoskeleton. This includes thicker intercellular junctions, more apical microfilaments, and more microtubule organizing centers (Galbraith et al. 1998).

Several mechano-sensors have been identified in the cell membrane. On the luminal side of endothelial cells, vascular endothelial growth factor receptor (a receptor tyrosine kinase) can serve as a mechano-sensor. Similarly, integrins work on the abluminal side (Fisher et al. 2001; Shyy & Chien 2002). Dynamic interaction between mechano-sensitive integrins and extracellular matrix proteins results in the activation of many downstream signaling molecules. This includes activation of serine/threonine and tyrosine kinases located in the cell membrane and focal adhesions and induction of phosphorylation cascades of signaling molecules; examples are the Ras-MEKK-JNK signaling pathway and members of the Rho small GTPase family such as RhoA, which is functionally linked to MAPK signaling, cell migration, and the organization of the actin-based cytoskeleton (Shyy & Chien 2002). Shear stress-induced formation of stress fibers and cell elongation and alignment could be inhibited by RhoN19, a dominant negative mutant of RhoA (Li et al. 1996). Other endothelial cell shear stress mechano-sensors that have been proposed include G-proteins, intercellular junction proteins, membrane lipids, and ion channels (Ali & Schumacker 2002). Some ion channels have been identified that change their open probability upon membrane stretching, leading to a subsequent influx of calcium concomitant with an activation of calcium-dependent intracellular signaling pathways (Sachs & Sokabe 1990).

Mechanical Stimuli and Membrane Traffic

Mechanical stimuli including shear stress, osmotic pressures, and the like apply forces to the plasma membrane resulting in alterations of membrane tension, one intrinsic variable of plasma membranes. A rough approximation of plasma membrane tension is given by LaPlace's law. Other contributing factors are the attachment of cytoskeleton and hydrostatic pressure across the membrane (Dai & Sheetz 1999). Usually cells respond to increases in membrane tension with an enhanced exocytosis rate, leading to an enhanced cell surface as indicated by capacitance measurements. This mechanism is used as a trigger for secretory cells, for example the release of atrial natriuretic factor (ANF) and angiotensin II from stretched cardiac myocytes (Sadoshima & Izumo 1997). In contrast, a decrease in membrane tension is followed by an enhanced endocytosis rate resulting in a recovery of membrane tension. Several exceptions to these rules can be found, leading to the assumption that mechanical stimuli predominantly affect the rate of membrane turnover (Apodaca 2002). Exocytosis and endocytosis seem to be coupled processes leading to a continuous replacement of the cell surface. Membrane tension therefore can be adjusted depending on which process predominates.

Cellular mechano-sensors include integrins, ion channels, and the cytoskeleton, while downstream signaling pathways involve intracellular messengers like calcium, cyclic AMP, and the activation of phosphorylation/dephosphorylation cascades. For a more detailed description the reader is referred to some excellent recent reviews (Hamill & Martinac 2001; Davies 1995; Morris & Homann 2001).

Mitochondrial Biogenesis

Chronic stimulation of skeletal muscle results in structural and functional adaptations of mitochondria. With respect to exercise, endurance training is the most powerful stimulus for mitochondrial biogenesis. Moreover, a number of other physiological and pathophysiological conditions have been shown to enhance mitochondrial biogenesis, such as electrical stimulation, creatine depletion, hyperthyroidism, and mitochondrial respiratory deficiency.

Mitochondrial biogenesis is a complex process covering various steps such as signaling events leading to transcription of nuclear and mitochondrial genes, protein and lipid synthesis, protein import into mitochondria, and their assembly into the enzyme complexes (figure 1.5).

Initial Signaling Mechanisms

Muscle contraction is activated by and associated with numerous intracellular signaling events, one of which is increases in cytosolic calcium. Calcium is known to be an important second messenger. Voltage-induced calcium influx releases calcium from the sarcoplasmic reticulum. Besides facilitating actin/myosin cross-bridging, calcium serves as an activator of a number of kinases and phosphatases. The differential activation of transcription factors by distinct calcium signaling patterns seems to be responsible for myosin isoform expression, thereby modulating the muscle phenotype (Chin et al. 1998; Wu et al. 2001; Allen & Leinwand 2002; Chin et al. 2003) (see chapter 13). For a long time it was unclear whether increases in cytosolic calcium would be propagated into intracellular organelles. Using the recombinant calcium-sensitive photo-protein aequorin, modified by addition of a mitochondrial target in the sequence, the calcium concentration within the mitochondrial matrix of living cells could be determined (Rizzuto et al. 1992; Rutter et al. 1998). This approach revealed that upon physiological stimulation, the cytosolic calcium increase was accompanied by an increase of the mitochondrial matrix calcium concentration, which reached levels sufficiently high to activate matrix dehydrogenases. Another calcium-dependent pathway was demonstrated by Wu et al. (2002) using a transgenic mouse approach. The authors could show that a downstream effector kinase of the calcium signaling pathway, the calcium/calmodulin-dependent protein kinase, augmented mitochondrial DNA replication and mitochondrial biogenesis as well as up-regulation of mitochondrial enzymes. This effect is induced by expression of the peroxisome proliferator-activated receptor γ coactivator-1 (PGC-1).

There is evidence, however, that the calcium increase alone is not sufficient for induction

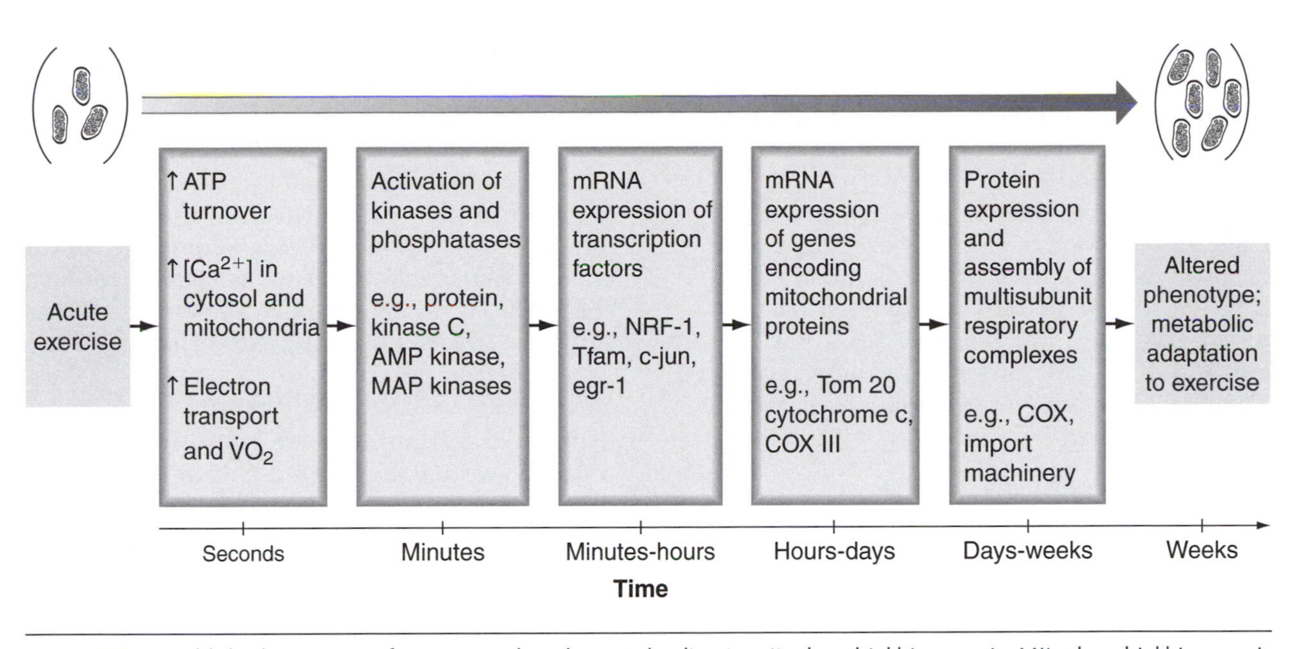

Figure 1.5 Established sequence of exercise-induced events leading to mitochondrial biogenesis. Mitochondrial biogenesis is a complex process depending on a sequence of signaling and anabolic events with different time constants (for details see text).

Reprinted, by permission, from D. Hood, 2001, "Plasticity in skeletal, cardiac, and smooth muscle: Invited review: Contractile activity-induced," *Journal of Applied Physiology* 90: 1137-1157.

of mitochondrial biogenesis. Cellular calcium increase subsequent to application of a calcium ionophore resulted in the expression of only a few nuclear genes encoding mitochondrial proteins, while others that would be critical to mitochondrial biogenesis were not affected. Therefore other co-stimuli seem to be necessary for mitochondrial biogenesis, one of which is the cellular pool of high-energy phosphate bonds. Interestingly, Jouaville et al. (1999) found that mitochondrial calcium levels correlate with an enhancement in mitochondrial ATP concentration.

ATP/AMP Ratio and ATP Turnover

Cellular energy status is reflected by the ATP/AMP ratio and phosphocreatine content, and appears as a central stimulator of mitochondrial biogenesis. In addition to the physiological stimuli such as endurance training, numerous recent experimental conditions leading to a disturbance in energy metabolism increased the mitochondrial content of skeletal muscles. Depletion of cellular phosphocreatine and ATP pools by chronic feeding of β-guanidinopropionic acid was followed by an increase of mitochondrial oxidative capacity of fast-twitch skeletal muscles due to an enhanced activity of mitochondrial enzymes, an up-regulation of cytochrome C mRNA, and activation of the nuclear respiratory factor-1 (NRF-1) transcription factor (see later) (Lai & Booth 1990; Freyssenet et al. 1994; Bergeron et al. 2001). An important downstream effector molecule that is activated by decreases in ATP and phosphocreatine is the AMP-activated protein kinase (AMPK). AMPK activity is increased during exercise in the skeletal muscle. However, recent observations indicate that this response depends on the muscle type. Endurance training caused an enhanced expression of a subunit of AMPK in rat red quadriceps muscle but not in soleus or white quadriceps (Durante et al. 2002). Activation of AMPK by administration of 5-amino-imidazole-4-carboxamide ribofuranoside (AICAR) increases expression of some mitochondrial enzymes like citrate synthase and cytochrome C (Ojuka et al. 2000; Winder et al. 2000). On the other hand, expression of fatty acid synthase and the pyruvate kinase gene is inhibited (Foretz et al. 1998). In addition to its role in mitochondrial biogenesis, enhanced AMPK activity was associated with increases in GLUT4 protein content.

Interrupting ATP production by pharmacological dissipation of the electrochemical gradient using mitochondrial uncoupling agents is another way to activate gene transcription important for mitochondrial biogenesis. However, mitochondrial biogenesis can be activated without substantial changes in cellular ATP level as indicated by the fact that exercise of moderate intensity can induce mitochondrial biogenesis as well. This could be an indication that in addition to the role of depleted energy pools, the turnover rate of energetic phosphates can induce mitochondrial biogenesis (Hood 2002).

The role of intracellular kinases and phosphatases for induction of mitochondrial biogenesis is far from clear. Although there is much evidence that contractile activity activates a number of different kinases, for example protein kinase C and ERK and MAP kinases (for details chapter 13), their roles in mitochondrial biogenesis have to be determined.

Transcription of Nuclear and Mitochondrial Genes

Although there are only a few mitochondrial gene products (two ribosomal proteins, 22 transfer RNAs, and 13 polypeptides), they are nonetheless important for establishing complete oxidative capacity. This is supported by the fact that the number of mitochondrial DNA copies changes in response to endurance training, suggesting a role at the pretranscriptional level (Murakami et al. 1994; Iwai et al. 2003). The presence of two genomes responsible for mitochondrial biosynthesis makes coordination between the transcription of both genomes necessary. In chronically stimulated rat anterior tibialis muscle, Hood et al. (1989) could demonstrate a similar time course in mRNA levels of the cytochrome oxidase subunit III, which is mitochondrially encoded, and IV, which is nuclearly encoded, in relation to cytochrome oxidase activity.

During the last decade a number of transcription factors have been discovered, such as nuclear respiratory factor 1 and 2 (NRF-1, NRF-2), that activate several nuclear genes encoding for mitochondrial enzymes. However, these transcription factors are also involved in the coordinated transcription of the nuclear and mitochondrial genomes, since NRF-1 sites exist in the promoter region of the mitochondrial transcription factor A (TFAM), which stimulates mitochondrial DNA transcription and replication. With the discovery of PGC-1 (PPARγ-coactivator-1), an important coactivator of mitochondrial biogenesis seems to have been identified (Wu et al. 1999). PGC-1 coactivates NRF-1 and is involved in the stimulation of GLUT4 expression. Recent studies suggest that single bouts of exercise induced the expression of these transcription factors (Pilegaard et al. 2003). After 3-h swimming, PGC-1 mRNA was detected in red muscles while PGC-1 protein was increased in the triceps muscle about 18 h after the exercise. Likewise, NRF-1 and NRF-2 were detected between 12 and 18 h after exercise (Baar et al. 2002). Additional transcription factors involved in mitochondrial biogenesis are activator protein-1 (AP-1) and peroxisome proliferator-activated receptor α and γ (PPARα/γ). The latter factor appears to be responsible for the transcriptional control of enzymes involved in mitochondrial beta oxidation (figure 1.6) (Hoppeler & Fluck 2003).

Regulation of Cellular mRNA Levels

Protein expression depends on the amount of mRNA available within a cell, which in turn depends on the one hand on the transcriptional activity and on the other hand on mRNA stability. For mitochondrial biosynthesis both processes seem to be important. Recent experiments by Freyssenet et al. (1999) indicated that the postexercise levels of mRNA increased initially due to an increase mRNA stability followed by an enhanced transcriptional activation. These results confirmed prior experiments that had shown an increase of cytochrome oxidase activity that was higher than the increase in its mRNA (Hood et al. 1989).

Mitochondrial Protein Synthesis

Besides amino acid sequence coding regions, regions coding for ribosomal RNA and tRNA can be found on the mitochondrial genome, suggesting an autonomous translational regulation of protein synthesis. This is supported by measurements in skeletal muscle that revealed different rates of protein synthesis and different adaptations to exercise between the two populations of mitochondria, IMF and SS mitochondria. Under nonadaptive steady state conditions, rates of protein synthesis were found to be higher in IMF than in SS mitochondria. Acute exercise resulted in a decrease of protein synthesis in both fractions, but with different time constants. In contrast, chronic exercise reduces protein synthesis only in the IMF subpopulations, but without any impairment in the chronic contractile activity-induced cytochrome C oxidase activity increase (Connor et al. 2000). There is evidence from a number of studies that the endurance training-induced increase in mitochondrial oxidase capacity is more pronounced in SS than in IMF mitochondria (Krieger et al. 1980; Hoppeler et al. 1973). Recent observations revealed a more differentiated regulation, since single enzyme activities in both subpopulations seem to respond independently to the training stimulus (Bizeau et al. 1998). Furthermore, high-intensity interval training

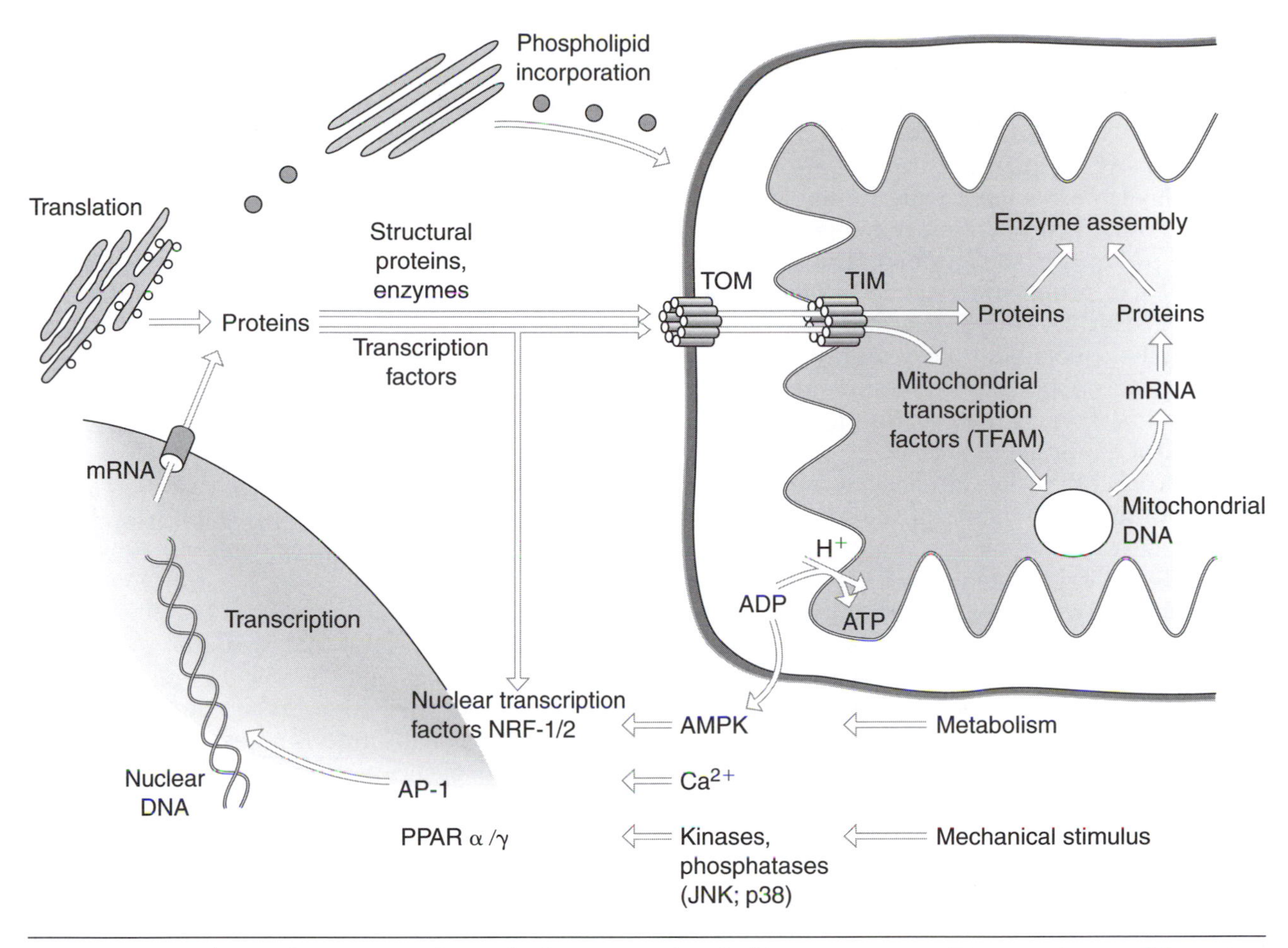

Figure 1.6 Signaling and transport pathways leading to mitochondrial biogenesis. TOM/TIM, translocase of the outer/inner membrane; JNK, c-Jun NH2-terminal kinase; AMPK, AMP-activated protein kinase; PPAR, peroxisome proliferator-activated kinase; p38, mitogen activated protein kinase; AP-1, activator protein 1; NRF, nuclear respiratory factor.

in comparison to continuous endurance training was shown to be more effective in stimulating fatty acid oxidation in IMF than in SS mitochondria, suggesting a differential effect of the training regimes on the biogenesis of mitochondrial subpopulations (Chilibeck et al. 1998, 2002).

Mitochondrial Protein Assembly

An optimal oxidative phosphorylation capacity requires the structural assembly and interaction of both nuclear- and mitochondrial-encoded proteins. Therefore biogenesis of mitochondria depends on an effective transfer system responsible for the import of precursor proteins from the cytosol as well as for the export of mitochondrially coded proteins. For this purpose, mitochondrial membranes contain several protein translocases along with a number of chaperones and processing enzymes for assistance in the translocation process. The first step for translocation of a protein into mitochondria is to attach a targeting sequence to the amino terminus. Next, the protein can bind to the translocation complex TOM (which stands for translocase in the outer mitochondrial membrane), which consists of several subunits responsible for recognition, binding, and translocation of the protein (figure 1.6; Lithgow 2000).

A number of cytosolic chaperones guide the precursor proteins to the translocases and prevent their misfolding and aggregation. Examples of these chaperones are cytosolic HSP70 and mitochondrial import stimulating factor (MSF).

After passing the TOM complex, proteins targeted for the mitochondrial inner membrane and for the matrix have to pass another translocase system named TIM (translocase of inner membrane). TIM consists of two different complexes, the TIM-23 complex responsible for transport of proteins into the matrix and the TIM-22 complex responsible for mediating protein insertion into the inner membrane (Koehler 2000). Again this translocase consists of several subunits closely assembled with mitochondrial HSP70 and the nucleotide exchange factor MGRP E, which function as an ATP-dependent translocation motor. TIM-22 is also associated with a number of proteins (TIM-18, TIM-54). With the help of a family of small proteins in the mitochondrial intermembrane space, proteins are guided to the TIM-22 complex. Insertion of the precursor into the inner membrane is dependent on membrane potential. Therefore dissipation of the ion gradient by uncoupling agents reduces the protein import into the inner mitochondrial membrane.

In chronically stimulated skeletal muscles, an adaptation of members of the protein import machinery (e.g., MSF, cytosolic and mitochondrial HSP70, TOM 20) has been shown (Takahashi et al. 1998). Twenty-week treadmill training increased the HSP60 expression in rat muscle (Samelman et al. 2000). A differential regulation of mitochondrial protein import has been shown for the mitochondrial membranes and for the mitochondrial subpopulations. The rate of protein import is higher in IMF mitochondria than in SS mitochondria (Takahashi & Hood 1996).

Mitochondrial Phospholipid Synthesis

As indicated in the preceding discussion, the structural basis of mitochondria is formed by a double membrane system. Therefore expansion of the mitochondrial reticulum during mitochondrial biogenesis requires the enhanced synthesis of various phospholipids as components of the membrane system. Lipid synthesis is performed in the ER, followed by transport and sorting of the lipids to the outer and inner mitochondrial membranes. In contrast, cardiolipin, which is involved in the translocation process of the TIM-23 complex, is synthesized at the inner mitochondrial membrane (Schlame & Haldar 1993).

Endurance training is followed by a significant increase of mitochondrial lipid synthesis. The ratio of protein to lipid plus protein content of mitochondria, however, remains stable during endurance training, suggesting a correlation between protein and lipid synthesis (Davies et al. 1981). Measurements of Hood and colleagues, however, indicated a different time constant of protein and lipid synthesis, the latter being faster (Hood et al. 1994; Takahashi & Hood 1993).

Conclusion

The development of novel techniques has enabled fascinating insights into the cell's structure and function. Its description as a steady state, a term generally used for biochemical reactions, seems applicable for the structural components of a cell as well. Under resting conditions, and even more under activated conditions, dynamic interactions and transitions between various cellular components exist and allow rapid and extensive adaptational responses. The following chapters give an overview of our current knowledge regarding how cells and subcellular structures face and respond to the exercise challenge.

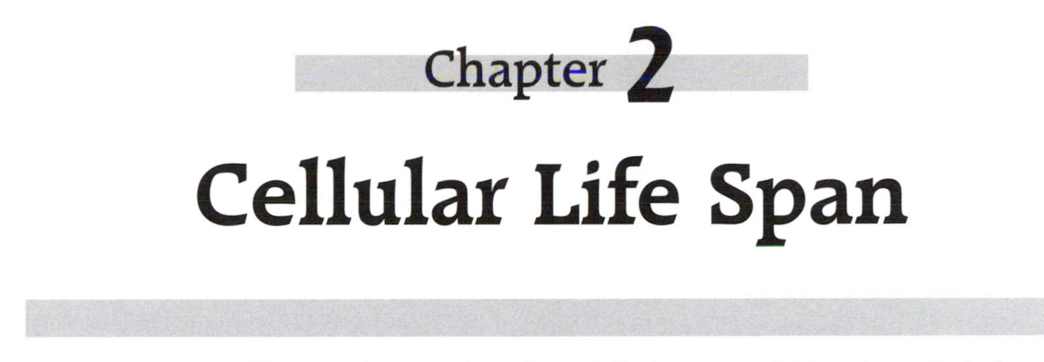

Chapter 2
Cellular Life Span

Laurie Hoffman-Goetz, Joe Quadrilatero, and Harshna Patel

Research supported by a grant from the
Natural Sciences and Engineering Research Council of Canada

Cell Cycle and Tissue Turnover

In order for proper growth to occur and life to be sustained, cell renewal and reproduction are required (Singh et al. 2000). Tissue turnover and homeostasis are dependent on the relationship between cellular proliferation, differentiation, and death and are essential for the generation and maintenance of complex tissue architecture. Alterations in the balance between the signals that lead to cell death and cell renewal can lead to undesired tissue atrophy or tissue growth (King & Cidlowski 1998; Pucci et al. 2000). The cell cycle is an integral factor controlling cellular proliferation and tissue turnover. The cell cycle is a highly ordered and tightly regulated process that assesses both the intracellular and extracellular environments, including factors such as extracellular growth signals, cell size, and DNA integrity, at multiple checkpoints. Progression through the cell cycle ultimately results in the replication and distribution of genetic material from one cell generation to the next (Ho & Dowdy 2002; Israels & Israels 2001; Schafer 1998).

Cell Proliferation

Extracellular signals (such as growth factors, hormones, and mitogens) play an important role in cell proliferation, development, differentiation, and survival (Talapatra & Thompson 2001). Growth factors accomplish this by interacting with specific receptors. Steroid hormones, such as testosterone, estrogen, and cortisol, influence cellular growth and differentiation by penetrating the plasma membrane and acting on nuclear receptors, thereby exerting direct effects on genomic regulatory mechanisms. Polypeptide growth factors, such as epithelial growth factor (EGF), insulin-like growth factor-1 (IGF-1), and interleukin-2 (IL-2), cannot penetrate the cell membrane and instead act on specific plasma membrane receptors. These cell surface receptors include enzyme-linked receptors, such as tyrosine kinase and serine/threonine kinase receptors, receptors associated with cytoplasmic kinases, and the seven transmembrane domain receptors (Martinez Arias & Stewart 2002).

Steroid hormone receptors are located in the cytoplasm of the cell or nucleus and include the receptors for androgens, estrogens, glucocorticoids and mineralocorticoids, thyroid hormone, retinoids, and vitamin D. Ligand–receptor binding results in an allosteric change that leads to heat shock protein dissociation, receptor dimerization, the binding of the receptor to specific DNA elements, and gene transcriptional activities (Kumar & Thompson 1999; Sheppard 2002; Weigel 1996). Tyrosine kinases are a family of enzymes that phosphorylate or transfer a phosphate group from ATP to tyrosine residues on specific cellular proteins, upon binding of ligand or growth factor. Tyrosine kinases can be divided into two categories, receptor (e.g., epithelial growth factor receptor, fibroblast growth factor receptor, insulin-like growth factor-1 receptor) and nonreceptor (e.g., *src, fgr, ctk, abl,* and *fak*) tyrosine kinases. The activity of tyrosine-specific protein kinases is known to be associated with various oncogenes that alter cellular growth (Cross & Dexter 1991; Goel et al. 2002; Haluska & Adjei 2001; Hubbard 1999; Hubbard & Till 2000).

Growth factors stimulate the cell to enter the cell cycle; however, absence or early withdrawal of growth factors will result in the cell's returning to a quiescent state (Schafer 1998). Growth factors prevent cell death through various signaling pathways, including the Jak/STAT, PI3K/Akt, and Ras/MAPK pathways (Talapatra & Thompson 2001). However, lack of growth factor availability results in the loss of mitochondrial function and leads to cell death (Talapatra & Thompson 2001).

Cell Cycle and Mitosis

The cell cycle is divided into the G_0, G_1, S, G_2, and M phases. The G_0 phase is a quiescent nonproliferating state. The G_1 and G_2 are gap phases during which required cellular components such as RNA, proteins, and enzymes are synthesized and accumulated for use in the subsequent stage. During the S or synthesis phase DNA replication occurs, while M phase or mitosis involves the division of nuclear and cellular components (Zafonte et al. 2000). The G_1, S, and G_2 phases can be grouped into the "interphase" whereas the M phase can be further delineated as six distinct subphases each with specific morphological characteristics. These include prophase, prometaphase, metaphase, anaphase, telophase, and cytokinesis. During prophase, nuclear chromatin begins to condense and centrosomes separate to form the mitotic spindles. During prometaphase, the nuclear envelope is broken down, which allows the chromosomes to associate with the spindle fibers. Metaphase is characterized by the lining up of the chromosomes at the equator of the cell and disappearance of the nuclear membrane. During anaphase, the centromeres split, allowing the sister chromatids to separate and move toward opposite poles of the cells. At telophase, the nucleolus reappears and there is a formation of a new nuclear membrane. The final result, known as cytokinesis, culminates with the formation of two genetically identical daughter cells (Bolsover et al. 1997; Singh et al. 2000; Van De Graaff & Fox 1995).

Upon appropriate growth stimulation, G_0 cells can enter the cell cycle at early G_1; similarly, G_1 cells deprived of external growth stimulation can exit into G_0. Although growth factor stimulation is needed for a cell to enter into G_1 and proliferate, this stimulus is essential only in the first two-thirds of the G_1 phase. This point between the early and late G_1 represents an irreversible commitment to undergo one cell division and is termed the restriction point (Ho & Dowdy 2002). This restriction point divides the cell cycle into a growth factor-dependent phase (early G_1) and growth factor-independent phase (late G_1 to M). Therefore, following initial stimulation, cells can complete the remainder of the growth cycle through mitosis in the absence of further exposure to growth factors or mitogens (Jones & Kazlauskas 2001; Lundberg & Weinberg 1999).

Normal cells do not progress from one phase to the next unless the events of the preceding phase have been correctly completed. Progression of cells through the cell cycle is controlled at specific stages, particularly at the G_1 to S and G_2 to M transitions. These transition stages are referred to as cell cycle checkpoints (Foster et al. 2001; Sherr 1996). Specifically, cyclins are the major regulators of the cell cycle because of their association with cyclin-dependent kinases (cdks). Cdks belong to a family of serine and threonine protein kinases and are dependent on the presence of cyclins for activation (Pucci et al. 2000). Normally, cdks are inactive, but once bound to their cyclin counterpart they form active cyclin-cdk complexes. These complexes phosphorylate critical serine and threonine sites on the target cells, which ultimately stimulate gene expression of transcription factors and critical cell cycle components (Foster et al. 2001; Lundberg & Weinberg 1999; Zafonte et al. 2000). A variety of cyclin-cdk complexes are formed during distinct phases of the cell cycle and are responsible for the phosphorylation of a particular set of target proteins (Lundberg & Weinberg 1999). Currently, 16 cyclins (cyclin A, B_1, B_2, C, D_1, D_2, D_3, E, F, G_1, G_2, H, I, K, T_1, and T_2) as well as 9 cdks (cdk 1-9) have been identified in mammalian cells. Table 2.1 summarizes selected mammalian cyclins, their associated cdks, cell cycle stage of involvement, and defined functions (Johnson & Walker 1999; Kong et al. 2000; Leclerc & Leopold 1996).

During the G_1 phase there is an increased expression of cyclin D (D_1, D_2, D_3) as well as cdk4 and cdk6. Progression through late G_1 is coupled with an increased expression of cyclin E and cdk2

Table 2.1 Mammalian Cyclins, Associated cdks, Cell Cycle Stage of Involvement, and Function During Cell Cycle Regulation

Cyclins	Associated cdk	Cycle phases	Function
A	cdk1(cdc2), cdk2	S	Participates as a negative feedback loop for E2F regulation
B_1, B_2	cdk1	G_2 to M	Regulation of cyclin degradation during mitotic exit
C	cdk8	G_0 to S	Phosphorylates RNA polymerase II
D_1, D_2, D_3	cdk4, cdk6	G_0 to S	Phosphorylates *Rb* and *Rb*-related proteins that regulate the activity of transcription factors (e.g., E2F family)
E	cdk2	G_1 to S	Maintains *Rb* in a hyperphosphorylated state therefore acting as a positive feedback loop for E2F
F	?	G_2 to M	Regulates cyclin B_1 localization to the nucleus

and is required for transition from G_1 to S phase. Binding of cyclin A with cdk2 allows DNA synthesis to occur during the S phase. Cyclin A associates with cdk1 later on in S phase, and increased expression of cyclin A and B with cdk1 propels the cell to exit G_2 and complete mitosis (Israels & Israels 2001). These cyclin-cdk complexes are regulated by the varying expression of cyclins and cdk inhibitors. Two families of cdk inhibitors are involved in cell cycle control. The Cip/Kip family includes *p21* and *p27,* which function at various points of the cell cycle. In particular, these cdk inhibitors target cdk2, -4, and -6. The second family includes the inhibitors of cyclin-dependent kinase 4 (INK4) genes, $p16^{INKa}$ and $p19^{ARF}$ (Israels & Israels 2001). Figure 2.1 shows a schematic representation of the cell cycle and various regulator components affecting progression through the various stages.

Telomeres

Telomeres are DNA-protein structures found at the ends of all linear eukaryotic chromosomes. This "cap" protects the chromosome ends from degradation that would normally occur during double DNA strand breaks. Telomeres also function to prevent illegitimate recombination of chromosome ends, which would lead to unstable chromosome forms. Therefore, telomeres are critical for the proper function, integrity, and stability of the chromosome and ultimately establish a stable linear chromosome end. In addition, it has been proposed that telomeres have a central role in determining the number of times a cell can divide (Blackburn 1991; Perrem & Reddel 2000).

Telomere DNA usually consists of double-stranded sequences, with a single G-rich strand that protrudes at the 3N end and forms a single-strand overhang (Price 1999). In mammalian cells, the length of the double-strand repeat (TTAGGG/CCCTAA)n ranges from a few to tens of kbp, whereas the length of the single-strand overhang (TTAGGG)n is usually no more than a few hundred nucleotides (Collins 2000). The number of telomeric repeats differs from one organism to another. Telomeric length also varies in different mammalian tissues as well as from one chromosome to another (Buchkovich 1996). Friedrich et al. (2000) found that telomere length was significantly shorter in leukocytes compared to skin (fibroblasts) and synovial tissue (fibrocytes) of the same patient and was consistent with the faster replicating properties of leukocytes compared to the other tissues.

During DNA replication, a gradual shortening of the telomere occurs. Because DNA polymerases work only in the 5N to 3N direction and require short oligonucleotides as primers, shortening of the telomeric DNA is expected during each cell cycle (Buchkovich 1996). Telomere terminal transferase or telomerase is a telomeric-specific ribonucleoprotein reverse transcriptase that is important in telomere maintenance. Telomerase is composed of a structural RNA and two proteins, telomeric repeat binding factor 1 (TRF1) and TRF2. Telomerase utilizes an integral RNA component as a template for de novo synthesis and adds single-stranded telomeric repeats to the chromosome's 3N end. Both TNF1 and -2 have been hypothesized to aid in the docking and holding of the enzyme to the telomeric DNA (Collins 2000; Counter 1996; Perrem & Reddel 2000).

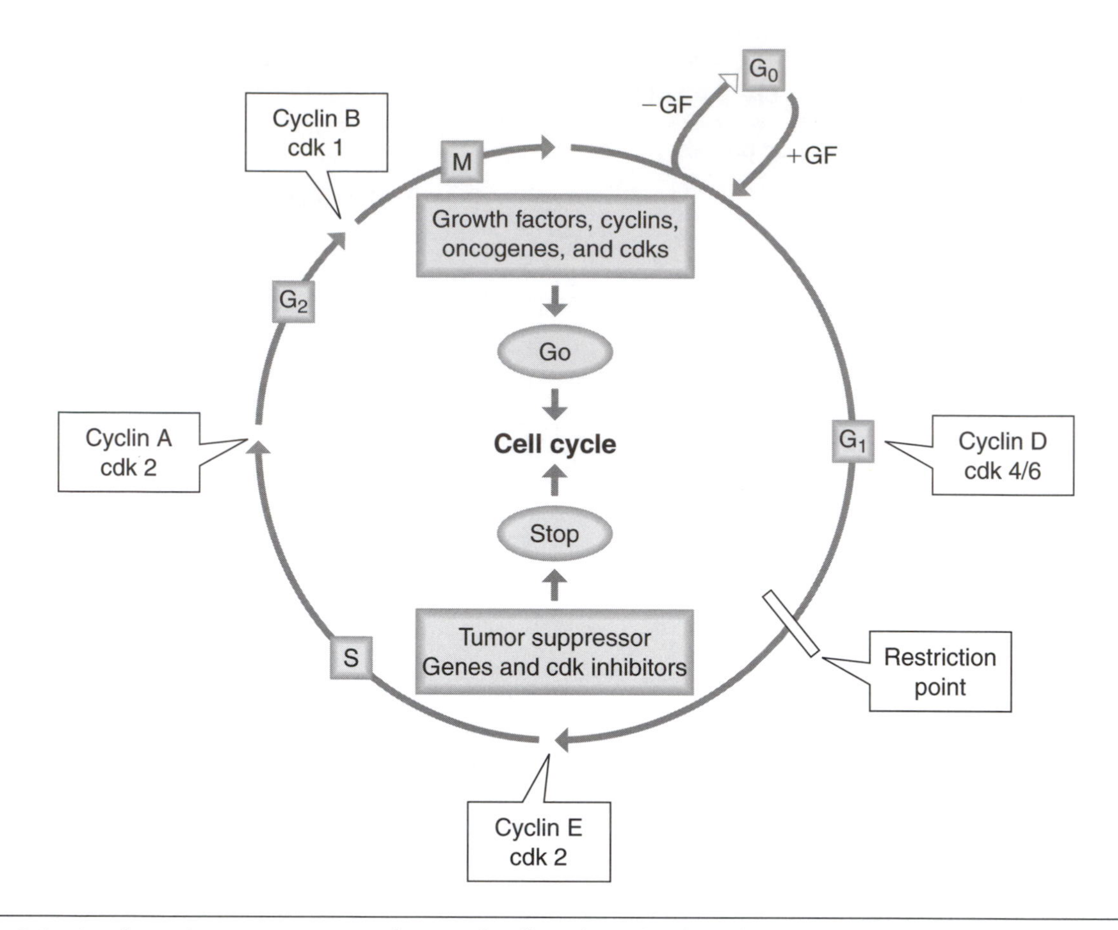

Figure 2.1 A schematic representation of normal cell cycle and selected regulatory elements involved in cellular progression. Following growth factor stimulation, normal cells progress through the stages of the cell cycle with the aid of various regulatory factors. Cycle progression is positively influenced by growth factor (e.g., insulin-like growth factor-1) availability (prior to the restriction point), cyclins (e.g., cyclin D), cyclin-dependent kinases (e.g., cdk4), and oncogenes (e.g., c-myc) and is inhibited by tumor suppressor genes (e.g., *p53*) as well as cdk inhibitors (e.g., *p21*).

Progressive shortening of chromosome ends is intimately coupled with cellular division of normal somatic cells and results in the so-called "end-replication" problem. This is attributable to the inability of the DNA polymerases to complete replication of the 3N end due to their requirement for an upstream RNA primer, ultimately resulting in the shrinking of the chromosome ends with each successive cell cycle. This affects cell cycle reproductive capacity (McEachern et al. 2000; Wynford-Thomas 1996). Cessation of cell proliferation is indicative of cellular aging and cellular senescence. Normal cells in culture possess a maximum number of cellular divisions called the "Hayflick limit" (Perrem & Reddel 2000). Telomeric length is predictive of replication capacity in normal cells in vitro (Allsopp et al. 1992; Harley et al. 1990), and telomerase activity is either absent or at very low levels in many adult somatic cells (Chiu et al. 1996; Hiyama et al. 1995). Moreover, if telomerase activity is low or absent in replicating cells, telomere lengths of the chromosome will gradually decrease (Goyns & Lavery 2000).

Immune response is dependent on the expression and differentiation of specific responsive cells following an antigenic challenge. Stimuli that result in the proliferation and expansion of normal lymphocytes should also result in the induction of telomerase activity. This would serve to maintain telomere length and reproductive potential. Hathcock et al. (1998) found that unstimulated mouse T-lymphocytes expressed low to virtually undetectable levels of telomerase activity but expressed significant telomerase activity after in vivo antigen stimulation. Weng et al. (1995) found that as donor age increased, from 24 to 75 years, terminal restriction fragment (TRF) length (an estimate of telomere size) decreased in both naive and memory $CD4^+$ lymphocytes. Furthermore, TRF lengths of naive $CD4^+$ lymphocytes were found to be 1.4 ± 0.1 kb longer than those of memory cells from the same donors and were independent of age. These findings suggest that cell division is ongoing throughout the life span and that memory cells undergo more extensive cell divisions than naive cells. A later study by Weng and colleagues (1997) showed that mean TRF lengths were significantly greater in germinal center B-lymphocytes compared to naive and memory cells. Differentiation from a naive to a germinal center lymphocyte results in significant lengthening of the telomere, while subsequent differentiation to a memory B-lymphocyte results in significant telomeric shortening that may be controlled by telomerase activity.

Angiogenesis

Angiogenesis, the formation of new capillary blood vessels from preexisting vasculature, involves complex interactions among endothelial cells, matrix proteins, and soluble factors, leading to endothelial cell proliferation, migration, and vessel formation. Normal tissue growth is characterized by dependence on new vessel formation for the supply of oxygen and nutrients as well as removal of waste products (Ziche & Morbidelli 2000). However, regardless of the cause, new vessels always grow out of preexisting vessels.

The blood vessels in the body function as a transport compartment for the blood and play a major role in maintaining homeostasis in various ways. The blood directly contacts the endothelial cells on the vessels composed of pericytes, smooth muscle cells, fibroblasts, basement membranes, and an extracellular matrix within the subendothelium (Griffioen & Molema 2000). The endothelial cells form a monolayer and are actively involved in several regulatory processes in the body including coagulation and metabolism. Tissue integrity and homeostasis are regulated by a balance between cell proliferation and cell death. Modulation of angiogenic function is achieved through regulation of the survival of endothelial cells during vessel repair and angiogenesis, whereby unnecessary, newly formed vessels undergo regression and apoptosis (Nor & Polverini 1999).

The adult vasculature is usually quiescent. In response to angiogenic stimuli, such as vascular endothelial growth factor (VEGF) and basic fibroblast growth factor (bFGF), endothelial cells undergo a series of tightly controlled events to form new capillary vessels. These steps include (1) initiation of the angiogenic response, (2) dissolution of the basement membrane by proteases, (3) endothelial cell migration and proliferation, and (4) maturation of the neovasculature.

Angiogenesis is initiated in response to hypoxic or ischemic conditions that result in the release of cytokines from various sources (Griffioen & Molema 2000). Relaxation of the blood vessel is a prerequisite for endothelial cells to enter the angiogenic cascade. This is mediated by nitric oxide (NO), an endothelium-derived relaxing factor (EDRF), and by up-regulation of fibroblast growth factor-2 (FGF-2) (Ziche & Morbidelli 2000). VEGF is then able to stimulate vasodilation and affect the permeability and vascular tone via endothelial NO production (Griffioen & Molema 2000; Ziche & Morbidelli 2000). The increase in microvascular permeability to proteins is a crucial step in angiogenesis. Leakage of plasma proteins and the formation of an extravascular fibrin gel are sufficient for endothelial cell growth (Folkman 1997). VEGF can also induce the expression of proteases and receptors that are crucial for tissue remodeling and prevention of cell death in endothelial cells (Farrara 2001).

For endothelial cells to enter into new areas of the body, they must detach from the basement membrane via proteolysis. Matrix metalloproteinases (MMP) are endopeptidases that are secreted, activated, and then degraded in the extracellular matrix. MMP facilitate the invasion of endothelial cells through the basement membrane and entrance into avascular tissue in response to angiogenic stimuli (Stelter Stevenson 1999). MMP are secreted by a number of cells including fibroblasts, inflammatory cells, epithelial cells, and endothelial cells.

As endothelial cells leave their original sites, they proliferate and can invade both vascular and avascular tissues. Angiogenic factors such as VEGF and bFGF can directly stimulate endothelial cell proliferation via cGMP-mediated activation of the mitogen-activated protein kinase (MAPK) family (Nie & Honn 2002). However, in order for endothelial cells to create a network, they must migrate toward and within the vascular tissue. As the cells proliferate, plasminogen is converted to plasmin by the plasminogen activators u-PA and t-PA. Plasmin degrades fibronectin, laminin, and the protein core of proteoglycans. Plasmin is considered the most important protease for mobilization of FGF-2 from the heparin sulfate proteoglycans in the extracellular matrix pool (Griffioen & Molema 2000). Aside from stimulating cell proliferation, VEGF and bFGF play a role in stimulating endothelial cell migration. However, both processes can be inhibited by angiogenesis inhibitors such as angiostatin and endostatin (Nie & Honn 2002).

Endothelial cells have the intrinsic ability to form tubelike structures after endothelial cell proliferation and migration through extracellular matrix. Endothelial cells elongate and align to form a sprout, and the lumen is formed by a curvature within each endothelial cell (Qian et al. 1997). Individual sprouts elongate and eventually join, forming loops through which blood begins to flow. However, in order to form a stable vasculature and prevent cell death by apoptosis, endothelial cells must interact with mural cells such as pericytes in small vessels and smooth muscle cells within large vessels (Griffieon & Molema 2000). This can be achieved through the production of platelet-derived growth factor (PDGF), which acts as a mitogen and chemoattractant. PDGF induces differentiation of the mural cells into pericytes or smooth muscle via cell-to-cell contact-dependent processes. Transforming growth factor-β (TGF-β) is found on both mural cells and endothelial cells, which interact to produce activated TGF-β that further induces changes in the pericytes and myofibroblasts. TGF-β is a homodymeric polypeptide that is strongly expressed in sites of tissue morphogenesis (Roberts et al. 1986).

Angiogenesis is controlled by a balance between angiogenic and angiostatic factors. Some angiogenic factors act directly by stimulating cells to migrate, proliferate, or form a vessel, whereas others can indirectly activate host cells (macrophages, mast cells, lymphocytes) to release endothelial growth factors. The direct-acting factors include acidic and basic fibroblastic growth factors (aFGF and bFGF), VEGF, and angiopoietin (Nie & Honn 2002). Indirect-acting factors include tumor necrosis factor-α (TNF-α), TGF-β, and platelet-derived endothelial cell growth factor/thymidine phosphorylase (PD-ECGF/TP). These factors are briefly described in table 2.2. However, angiogenesis inhibitors exist to counterbalance the effects of the angiogenic factors. These factors include thrombosponsdin, interferon, tissue inhibitors of MMP (TIMP), angiostatin, endostatin, and platelet factor-4 and are summarized in table 2.3.

Table 2.2 Angiogenesis Growth Factors

Factor	Function	Receptor
	DIRECT ACTING	
Basic fibroblast growth factor (bFGF/FGF-2)	Endothelial mitogen, survival factor, angiogenesis inducer; inducer of Flk-1 expression	FGFR-4
Acidic fibroblast growth factor (aFGF/FGF-1)	Endothelial mitogen and angiogenesis inducer	FGFR-4
Vascular endothelial growth factor/Vascular permeability factor (VEGF/VPF)	Endothelial mitogen, survival factor, and permeability inducer	Flk-1 Flt-1
Angiogenin	In vivo-acting angiogenesis inducer with RNase activity	Angiogenin receptor
	INDIRECT ACTING	
Epidermal growth factor	Weak endothelial mitogen, inducer of VEGF expression	EGFR
Tumor necrosis factor-α (TNF-α)	In vivo-acting angiogenesis inducer; endothelial mitogen (low concentrations) or inhibitor (high concentrations); inducer of VEGF expression	TNFR-55
Platelet-derived growth factor (PDGF)	Mitogen and motility factor for endothelial cells and fibroblasts; in vivo angiogenesis inducer	PDGFR
Thymidine phosphorylase (TP)/Platelet-derived endothelial cell growth factor (PD-ECGF)	In vivo-acting angiogenesis factor	Receptor unknown
Transforming growth factor-β (TGF-β)	In vivo-acting angiogenesis inducer or endothelial growth inhibitor; inducer of VEGF expression	TGF-BR I, II, III

Table 2.3 Inhibitors of Angiogenesis

Inhibitors	Function
Angiostatin	Plasminogen fragment; systemically acting angiogenesis inhibitor. Inhibits endothelial cell proliferation by unknown mechanism.
Endostatin	Proteolytic fragment of collagen XVIII. Systemically acting angiogenesis inhibitor. Inhibits endothelial cell proliferation and angiogenesis by unknown mechanism.
Thrombospondin-1 (TSP-1)	Binds to CD36 on endothelial cells and inhibits angiogenesis by unknown mechanism.
Tissue inhibitors of metalloproteinases (TIMPs)	Blocks breakdown of extracellular matrix and endothelial cell invasion by inhibiting MMPs.
Platelet factor 4 (PF4)	Inhibits angiogenesis by interfering with bFGFR.
Prolactin	Inhibits angiogenesis by bFGF and VEGF-induced proliferation to endothelial cell (S phase arrest).
EGF fragment	Inhibits EGF and laminin-dependent endothelial cell motility and angiogenesis.
Interferon-α (IFN-α)	Inhibits angiogenesis by antimigratory and antimitotic effect on endothelial cells; blocks bFGF production by parenchymal cells.
Interferon-β (IFN-β)	Inhibits endothelial cell proliferation and migration.
Interferon-γ (IFN-γ)	Inhibits endothelial cell proliferation and migration.
Interleukin-12 (IL-12)	Inhibits angiogenesis by stimulating IFN-γ.

Angiogenesis and Telomeres

The relationship between angiogenesis and telomere length/telomerase activity has been investigated. Franco et al. (2002) indicated that telomerase-deficient mice (Terc $^{-/-}$) have shorter telomere lengths and impaired angiogenic potential. Mean TRF lengths were found to be shorter in endothelial cells from the iliac arteries (a site of higher hemodynamic stress and cell turnover) compared to iliac veins. Furthermore, telomeric DNA loss was greater in the iliac artery compared to the internal thoracic artery (a site of low hemodynamic stress and normally free from atherosclerotic plaques), and was more dramatic with increasing donor age (Chang & Harley 1995). Of clinical importance is the finding that human telomerase reverse transcriptase (hTERT) mRNA is expressed by vascular endothelial cells of astrocytic tumors but is absent in normal brain (Pallini et al. 2001); hTERT mRNA was related to the histological grade of the tumor. Collectively, these studies suggest that telomere length and telomerase activity play a key role in angiogenic potential and may have implications for tumor therapy and age-associated vascular diseases, such as arteriosclerosis.

Cell Death

The processes of cell death, cell proliferation, and the cell cycle are inexorably linked (Evan et al. 1995). Two terms have been used to describe the process of cell death, *necrosis* and *apoptosis*. Necrosis, a term introduced by Virchow in the 19th century, refers to a passive process induced by catabolic, toxic, or pathological events (Grogan et al. 2002). Recently, it has been suggested that necrosis is not a mechanism of cell death at all but is rather the collection of morphological changes occurring in cells postmortem (Majno & Joris 1995). Necrotic cell death is associated with specific inflammatory changes in dying and adjacent cells. In contrast, apoptosis refers to the set of cell death events with particular morphological characteristics such as blebbing and nuclear fragmentation (Kerr 1971; Kerr et al. 1972). Wyllie and colleagues (1980) linked the observation of stereotyped patterns of induced DNA fragmentation in gel electrophoresis with the term apoptosis. Furthermore, apoptosis describes an active cell death process that occurs under both normal physiological and pathophysiological conditions. These processes are described in further detail in the sections that follow.

Apoptosis

Apoptosis is a highly regulated, evolutionarily conserved pathway that is essential for normal embryonic development, regulation of the immune system, hormone-dependent atrophy, chemically induced cell death, and tissue homeostasis (Cohen 1997; Phaneuf & Leewenburgh 2001; Sjöström & Bergh 2001). Apoptosis can also be induced by a number of deleterious stimuli or stressors including DNA damage, heat, hormonal triggering, anticancer drugs, and physical stress. Since apoptosis does not result in the release of intracellular material into the extracellular space, it usually does not lead to an inflammatory response. Apoptosis is an active process that requires participation of the dying cell and changes in cellular biochemistry and morphology, leading to DNA cleavage, nuclear condensation and fragmentation, and changes in membrane lipid distribution (Bidere & Senik 2001). Apoptotic cells are rapidly endocytosized by neighboring healthy cells or phagocytosized by professional phagocytes that recognize a number of signals on the surface of condemned cells, such as phosphatidylserine residues that pass through the plasma membrane. Apoptosis results in the irreversible, internucleosomal fragmentation of genomic DNA and the fragmentation of the cell into membrane-bound apoptotic bodies (Los et al. 2001; Phaneuf & Leewenburgh 2001). The complex sequential process of apoptosis involves an effector phase, a point of no return, a degradation phase, and a clearance phase. In the effector phase, endonucleases and cellular caspases are activated and mitochondrial dysregulation occurs (Salvesen & Dixit 1999). At the point of no return, cells are destined to die. Reorganization of the cytoskeleton and nuclear chromatin breakdown characterize the degradation phase. Finally, the removal of apoptotic cells by macrophages and other cells occurs as part of the clearance phase (Aigner 2002).

The effector phase of apoptosis involves the activation of cytosolic enzymes known as caspases. Caspases are cysteine proteases that cleave the carboxy terminal of aspartic acid (Asp) residues (e.g., cysteine aspartases). Caspases are present in an inactive form in the cytoplasm in virtually every cell (Newton & Strasser 2001; Villa et al. 1997). To date, 14 caspases have been identified and implicated in cell death. Caspases cleave precursors to produce mature cytokines (caspase 1, caspase 11), initiate the propagation of apoptotic death signals (caspase 8, caspase 9), and execute the apoptotic program through cleavage of several vital proteins (caspase 3, caspase 6, caspase 7). Three pathways result in caspase activation and subsequent apoptosis. Figure 2.2 is a simplified illustration of the pathways to apoptosis and necrosis.

In the first pathway, cell death is associated with mitochondria permeability changes. This is known as the death-by-neglect pathway because it involves deprivation of cells of necessary survival stimuli (e.g., growth factors or co-stimulators), which then leads to the rapid increase in the permeability of mitochondria membranes and the release of cytochrome C (Budd 2001). Various factors, including irradiation, glucocorticoids (GC), nitrogen monoxide, reactive oxygen species (ROS), and certain chemotherapeutic drugs, induce apoptosis by this biochemical mechanism. Cytochrome C, localized on the outside of the inner mitochondrial membrane and intermembrane space (Hirsch et al. 1997), functions as a cofactor with a protein known as apoptosis activating factor-1 (Apaf-1) to stimulate caspase 9 and initiate the apoptosis pathway. Cytochrome C is released into the cytosol, binds to Apaf-1, and forms a complex known as dATP. This complex then activates procaspase 9 to caspase 9 and eventually results in the activation of caspase 3 (van Cruchten & van den Broeck 2002). This pathway seems to be controlled by a group of proteins that regulate apoptosis. These proteins (e.g., Bcl-2, Bax, and Bcl-x) are coded for by the genes of the Bcl-2 (B-cell lymphoma) family and may function to regulate mitochondrial permeability via the mitochondrial permeability transition pore, with consequent release of cytochrome C into the cytosol.

The second pathway of apoptosis involves triggering by binding of ligands to death-inducing membrane receptors (Newton & Strasser 2001). This pathway is known as the receptor–ligand-mediated pathway of apoptosis. Two important ligands in this pathway are Fas ligand (FasL), which binds to the Fas receptor, and TNF-α (an inflammatory cytokine), which binds to the TNFR1 receptor. Fas and TNFR1 have corresponding cytoplasmic regions known as conserved "death domain" (DD) responsible for signaling transduction in apoptosis. Transduction occurs through the binding of the Fas-associated death domain with procaspase 8, the activation of caspase 8, and the eventual conversion of procaspase 3 into caspase 3.

Not surprisingly, lymphocytes express Fas. FasL is mainly expressed by T-lymphocytes after

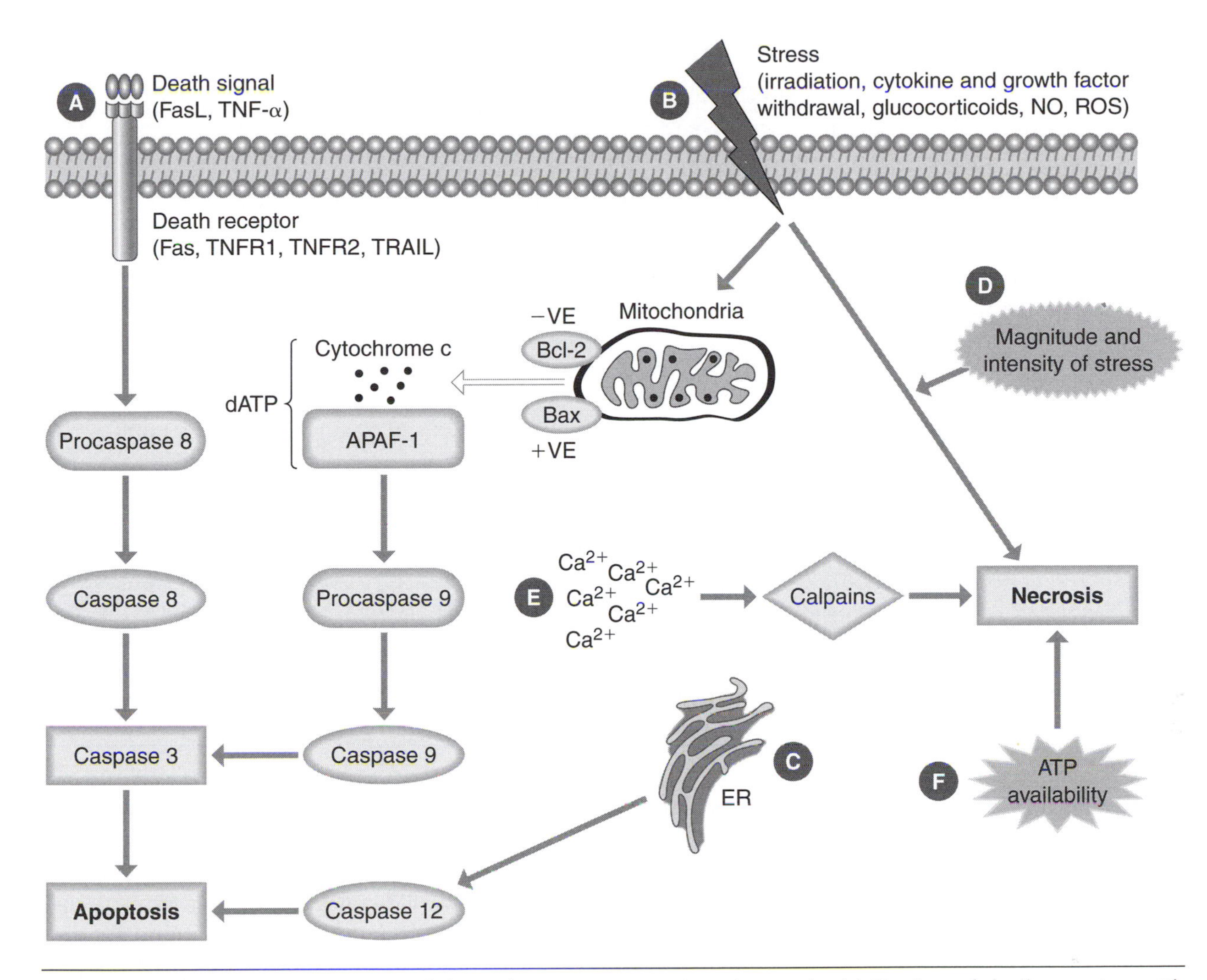

Figure 2.2 Schematic representation of the pathways to apoptosis and necrosis. *(a)* Ligation of death receptors with cytokines, such as FasL and TNF-α, leads to activation of death receptors (e.g., Fas, TNFR1, TNFR2, TRAIL), activation of the caspase cascade, and apoptosis. *(b)* Stress to the cell (e.g., ROS, GC, and growth factor withdrawal) leads to MPT formation and cytochrome C release into the cytosol, which is in part regulated by the expression of Bcl-2 and Bax. The formation of the dAPT complex initiates the activation of the caspase cascade and leads to apoptosis. *(c)* Stress to the endoplasmic reticulum (ER) may lead to cellular apoptosis via activation of caspase 12. *(d)* Greater magnitudes of stress will bypass the apoptotic pathway and lead to necrosis. *(e)* Sustained accumulations of intracellular calcium (Ca^{2+}) can activate calpains, leading to necrosis. *(f)* Decreased intracellular ATP availability can lead to cell death via necrosis.

activation by antigens and the cytokine IL-2. When mature T-lymphocytes are repeatedly stimulated by antigens, Fas and FasL are co-expressed. Cytotoxic ($CD8^+$) T-lymphocytes have a specialized form of Fas-Fas ligand-mediated apoptosis. In $CD8^+$ T-cells, binding of Fas-FasL leads to the release of perforins at the target cell membrane and the release of granzyme B into the target cytosol and induction of apoptosis (Martin & Green 1995; van Cruchten & van den Broeck 2002).

A third pathway for apoptosis has been termed the endoplasmic reticulum (ER)-mediated mechanism and has been identified by studies in caspase $12^{-/-}$ transgenic knockout (KO) mice (Nakagawa et al. 2000). The basis for this pathway is the finding that when mice are challenged with the ER-stressing agents, tunicamycin and thapsigargin, apoptosis occurs in fibroblasts from heterozygous (caspase $12^{+/-}$) mice but not in caspase $12^{-/-}$ animals. In contrast, when homozygous knockouts were given Fas-specific antibodies or synthetic GC (which activate the receptor ligand pathway and the mitochondrial pathway of apoptosis, respectively), no differences were found in apoptosis. Thus, it has been suggested that an ER-mediated mechanism of apoptosis must be independent of the receptor–ligand pathway and the mitochondrial pathway.

The commitment to cell death is based on a number of factors. For example, receptor clustering results in caspase activation to generate the first proteolytic signal. Commitment to other pathways is dependent on whether the extent of damage outweighs the ability of the cell to repair itself, which is in itself dependent on the ratio of proteins that initiate apoptosis (e.g., Fas, *p53*) to the proteins that inhibit or block apoptosis (e.g., Bcl-2) (Phaneuf & Leewenburgh 2001). Low levels of damage beyond the ability of the cell to repair itself result in postmitotic arrest and cell survival. More severe damage leads to the activation of apostat, induction of death signals, and the downstream activation of executioner caspases.

As noted earlier, several factors trigger apoptosis, including growth factor deprivation, cytokines, members of the TNF family, TGF-β, GC, retinoids, cytotoxic chemicals, and many more. Apoptosis can also be induced by cell injury resulting from toxicity, ischemia and reperfusion injury, oxidative stress, inflammation, and irradiation (Fellstrom & Zezina 2001). The apoptotic pathway in lymphocytes can be prevented by various activating stimuli, including specific antigen recognition, provision of essential growth factors (such as IL-2), and engagement of co-stimulators (such as CD28 on T-cells by B7 molecules on antigen-presenting cells). In response to apoptotic stimuli, inhibitory genes may be up-regulated, leading to the synthesis of apoptosis inhibitory proteins. One of the best characterized apoptosis inhibitory protein is Bcl-2, which blocks cytochrome C release from mitochondria, binds Apaf-1, and inhibits activation of caspase 9 (Newton & Strasser 2001). Bcl-2 inserts into the outer mitochondrial membrane and maintains the integrity of the mitochondria by allowing the export of H^+ ions through ion channels (Budd 2001). A family of proteins that contain the Bcl-2 homology domain 3 region (e.g., Bax, Bad, Bim, Bid) counteracts the actions of Bcl-2. Apoptosis signaling by death ligands is also regulated in a number of ways. In humans and a variety of animal species, secreted and membrane-anchored decoy receptors prevent apoptosis signaling by neutralizing FasL and TNF-related apoptosis-inducing ligand (TRAIL) (Newton & Strasser 2001). Moreover, in T-lymphocytes, delivery of FasL to the cell surface is under strict control. FasL is stored in lytic granules, and degranulation in response to target cell recognition leads to the rapid delivery of FasL to the cell surface. Silencer of death domain (SODD), another cytoplasmic protein, may be involved in limiting apoptosis signaling by death receptors since it also has a death domain and interacts with Fas (Musci et al. 1997).

Regardless of the pathway through which apoptosis occurs (i.e., through death by neglect or death by triggering of death receptors), certain morphological changes occur in the dying cell (Hacker 2000). These morphological consequences of apoptosis are (1) loss of attachment of the apoptotic cell to other cells and the extracellular matrix, (2) occurrence of protrusions from the plasma membrane ("blebs"), (3) condensation of the nucleus and fragmentation of genomic DNA, (4) dilation of the endoplasmic reticulum and release of ribosomes, and (5) disintegration of the cell into apoptotic bodies that are membrane-bound vesicles varying in size and composition (Hacker 2000). Apoptotic bodies are engulfed by neighboring cells, especially macrophages. Under some circumstances apoptotic cells that are not phagocytosized show features of necrosis without inflammation (van Cruchten & van den Broeck 2002).

Necrosis

The term necrosis means the irreversible and dramatic changes that are visible by microscopy after cell death (Majno & Joris 1995). Most pathology textbooks define necrosis based on whether the morphological changes are due to enzymatic digestion of cells (e.g., liquefactive necrosis) or due to denaturation of proteins (e.g., coagulative necrosis). Necrosis is characterized by osmotic swelling, rupture of the plasma membrane, and the release of cytosolic contents into the extracellular environment. Necrotic cells retain the shape of the nucleus, although chromatin condensation, late DNA degradation, and pyknosis (nuclear shrinkage and basophilia) can occur (Chautan et al. 1999; Schweichel & Merker 1973). Moreover, necrosis is characterized by the relatively slow disintegration of the cell in which the plasma membrane integrity breaks down and the cellular contents are enzymatically degraded and released (Hacker 2000).

The end result of necrosis is pathological inflammation. Whereas apoptosis is considered immunologically silent, a hallmark of necrotic death is the involvement of inflammatory cells. The involvement of these inflammatory cells (especially macrophages and neutrophils) and the release of chemoattractants to stimulate neutrophil emigration distinguish necrosis from apoptosis. However, this distinction may be more apparent than real. When apoptosis occurs on a large scale, for example during embryonic development, large numbers of phagocytic cells appear, although it is unclear whether these are "professional" phagocytes or simply neighboring parenchymal cells that take up the cellular debris (apoptotic bodies) (Majno & Joris 1995).

Unlike apoptosis, necrotic cell death does not involve activation of caspases. However, the absence of caspase activity does not eliminate the involvement of other enzyme cascades involved in proteolysis. Calpains (such as calpain μ) have been implicated in cell death (Wang 2000), since they inactivate caspases (Chua et al. 2000) and shift cells from an apoptotic phenotype to a necrotic one. Calpains can be activated physiologically by stimuli that induce sustained elevation of Ca^{2+} in the cytosol. The endogenous inhibitor, calpastatin, protects against calpain-mediated necrotic death. Nonetheless, the distinction between caspase-mediated apoptosis and calpain-mediated necrosis is far from absolute. For example, various calpain inhibitors have been shown to protect against apoptosis in lymphoid cells (Sarin et al. 1995; Squier et al. 1994; Wang 2000).

What factors trigger cells to respond with necrosis compared to apoptosis? Although many of the same stimuli trigger both cell death pathways, the more toxic the stressor, the more likely it is that necrosis will ensue. Additionally, which cell death pathway (apoptosis or necrosis) occurs following injury depends upon the intensity of the injury and the level of intracellular ATP available (Leist et al. 1997; Nicoterra et al. 1998). Other cellular factors, including short or prolonged opening of the mitochondrial permeability transition pore (Kowaltowski et al. 2001), the concentration of pro-apoptotic to antiapoptotic proteins (Denecker et al. 2001), and the magnitude of oxidative stress (Lemasters 1999), influence the decision point of death by apoptosis or death by necrosis (see figure 2.2). The "decision" to commit to apoptosis, necrosis, or cellular stasis is shown diagrammatically in figure 2.3.

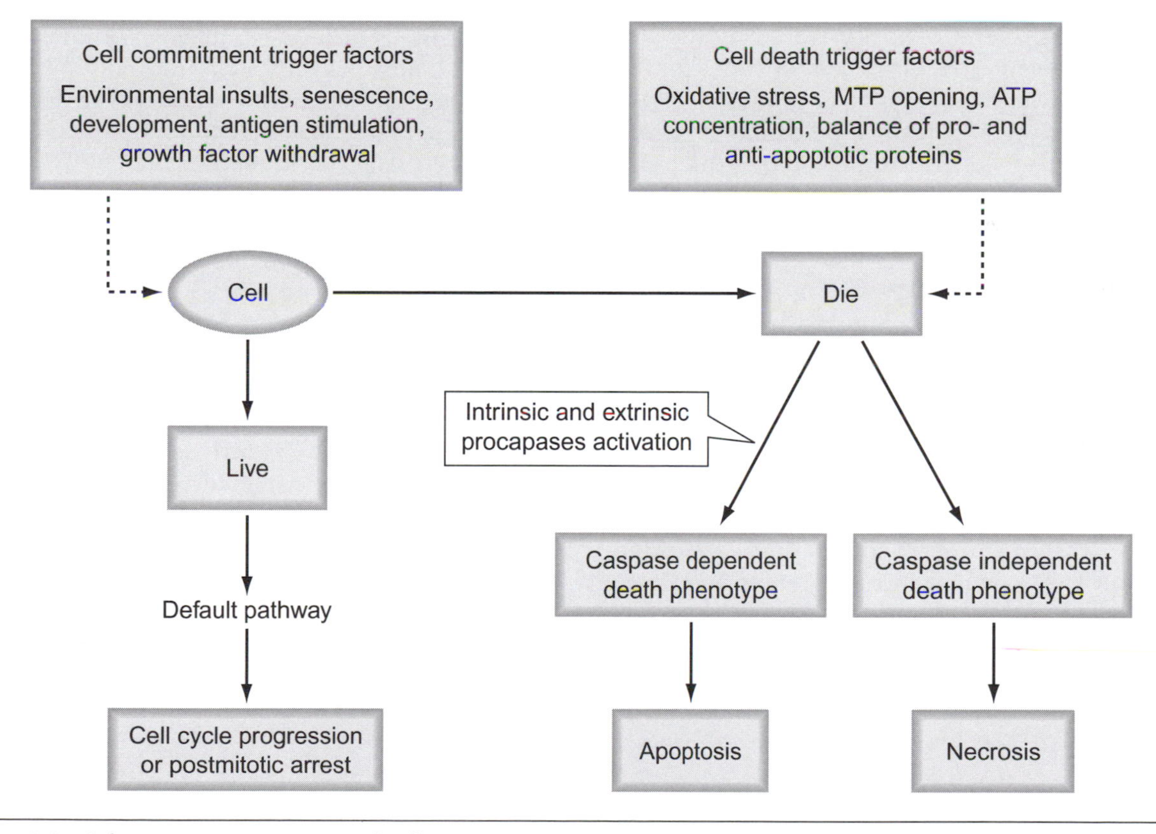

Figure 2.3 Schematic representation of cell progression to necrosis or apoptosis. Following triggering by extracellular (e.g., growth factors and antigens) and/or intracellular factors (e.g., pro- vs. antiapoptotic proteins), cells commit to cycle progression, to maintenance (stasis), or to a death pathway (involving necrosis or apoptosis). MTP = mitochondrial permeability transition pore; ATP = adenosine triphosphate.

It is apparent that although the mechanisms and the circumstances of cell death vary, the end result is always the same: the disappearance of cells. The impact of cell death for tissues, organs, systems, and organisms depends upon the biological context in which it occurs. In some cases, cell death is a normal physiological consequence of development and differentiation during embryogenesis. In other cases, cell death occurs as a pathophysiological response to injury or toxic insult.

Apoptosis and Cell Cycle

As described earlier, the cell cycle consists of the G_0 or rest phase, the G_1 or protein synthesis phase, the S or DNA replication phase, the G_2 phase extending from the end of DNA replication to eventual cell division, and the M or mitosis phase. The control of the cell cycle is through the action of cyclin proteins (e.g., A, B, D, E) and by cdks. Cyclin and cdks form complexes; and by changing the concentration and activity of these complexes, cell division can be initiated or stopped, or cells can be induced to undergo apoptosis. The relationship between apoptosis and the cell cycle is best illustrated by the *p53* tumor suppressor gene. This gene encodes for the *p53* protein, whose expression is increased following DNA damage (e.g., following oxidative stress or irradiation) (Mathieu et al. 1999). Expression of the *p53* protein induces the expression of the *p21* gene, which blocks cyclin-cdk complexes and ultimately leads to G_1 to S phase cell cycle arrest (Stewart & Pietenpol 2001). This interaction is further strengthened by the findings that mutations to *p16,* which are often seen in tumors, result in a loss of activity by the *p53* tumor suppressor gene. The loss in *p53* activity cripples the cell's ability to mobilize *p21* for inducing senescence and to mobilize pro-apoptotic proteins such as Bax to induce apoptosis (Lundberg & Weinberg 1999). Elevated levels of the *p53* protein induce the expression of Bax proteins, which induce greater apoptosis (Kamesaki 1998).

Other genes, including the retinoblastoma, *Rb* gene, regulate the cell cycle and influence apoptosis (Evan et al. 1995; van Cruchten & van den Broeck 2002). The *Rb* gene and its protein products *(pRb/p105, pRb2/p130* and *p107)* have growth-suppressive properties in that they inactivate transcription factors and negatively control the cell cycle between the G_1 and S phases (Tonini et al. 2002). *pRb* also functions as an antiapoptotic factor. For example, TGF-β_1 induces apoptosis in target cells through the mitochondrial mechanism (cytochrome C and Apaf-1) as well as through the receptor-mediated ligand mechanism, resulting in caspase activation (Cain & Freathy 2001). TGF-β_1 also suppresses *pRb* expression (Fan et al. 1996), and experiments with *pRb*-deficient mice indicate widespread apoptosis in a variety of tissues, including skeletal muscle (Zachsenhaus et al. 1996).

Detection of Apoptosis

There are many techniques for the detection of apoptosis in cells, each with advantages and limitations. The major approaches for the detection of apoptosis are via morphological changes by microscopy, Western blotting, gel electrophoresis to identify characteristic DNA patterns, histochemical techniques, ELISA, and flow cytometric analysis.

Morphological changes in condensed DNA of apoptotic cells can be detected by microscopy. For example, staining of cells with hematoxylin and eosin or with acridine orange or Hoeschst can be used to identify apoptotic cells but requires careful interpretation and technical experience, since apoptotic and necrotic cells may not be easily distinguished (van Cruchten & van den Broeck 2002). The molecular biology technique of Western blotting using antibodies against caspase substrates, apoptosis-related proteins, cyclins, and tumor suppressor genes can be used to identify apoptosis in cell populations. Limitations of this method have been reviewed by McCarthy and Evan (1998). Gel electrophoresis to demonstrate a characteristic DNA ladder pattern of internucleosomal DNA fragmentation detects apoptotic cells. DNA fragmentation as demonstrated in situ by the technique known as TUNEL (terminal deoxynucleotidyl transferase mediated dUTP nick end labeling) has been used as the "gold standard." The TUNEL method has been used to detect the early stages of cell death that cannot be visualized by routine histologic staining with hematoxylin-eosin. DNA fragmentation also occurs in necrotic cells; hence the specificity of this method is problematic (Saikumar et al. 1999; van Cruchten & van den Broeck 2002). Enzyme-linked immunosorbant assay (ELISA) methods can be used to measure caspase activity in apoptosis. Although caspase activity is indicative of apoptosis, limitations are similar to those of Western blotting: measurement of apoptosis in cell populations but not in individual cells. Flow cytometry techniques are a recent approach to identify apoptotic cells using flurochrome-labeled Annexin V, which binds to phosphatidylserine (PS) that is externalized on the membrane of damaged or apoptotic cells (Vermes et al. 1995). However, because Annexin V binds to PS residues within damaged cells, this technique may pick up both apoptotic and necrotic cells. To provide specificity in distinguishing apoptotic from necrotic cells, the use of propidium iodide (PI) is recommended since PI enters necrotic cells but is excluded from apoptotic ones. PI is a DNA stain and can therefore be used both for cell cycle analysis and to differentiate necrotic from apoptotic cells. Table 2.4 summarizes approaches used to measure apoptosis in cells.

Effect of Exercise on Cell Proliferation and Cell Death

Exercise has the potential to modulate cell proliferation and cell death through cytokines, hormones, growth factors, and metabolic pathways. Some of these factors have been well characterized in exercise models, such as VEGF and NO in vascular endothelial responses whereas others, such as expression of cyclins and telomerase in lymphocytes, have not been systematically investigated. The following section briefly describes exercise effects on angiogenesis, proliferation, and death in skeletal muscle and immune cells.

Angiogenesis

The formation of new capillaries is an essential adaptive response of skeletal muscle to repeated exercise and represents physiological angiogenesis within mature differentiated tissue. Exercise enhances O_2 transport conductance between the microcirculation and mitochondria, and increases the number and size of mitochondria in skeletal muscle and the activity of enzymes controlling oxidative metabolism. Skeletal muscle angiogenesis, determined by an increase in capillary density, capillary-to-fiber ratio, or both, reflects a vital adaptation to exercise and allows for improved aerobic capacity (O_2 transport, conductance, and extraction) of skeletal muscles (Amaral et al. 2001). These changes occur after aerobic endurance training characterized by low-resistance and high-repetition contractile activity (Amaral et al. 2001; Gustafsson et al. 1999). Angiogenesis is the result of prolonged imbalances between the metabolic requirements of the tissue and perfusion capabilities of the vasculature (Gavin et al. 2000a). Therefore, decreased oxygenation resulting from this imbalance causes the tissues to become hypoxic and produces a variety of metabolites implicated in vessel growth, including adenosine, ADP, lactic acid, nicotinamide derivatives, and prostaglandins of the E series. The mechanisms underlying capillary proliferation remain unknown, but it is likely that a number of factors are required to initiate physiological angiogenesis in response to exercise. However, it is suggested that these factors act in concert to increase vascularity to promote oxygen delivery to the tissue cells by increasing the capillary-to-fiber surface area interface and increasing maximal blood flow.

During exercise, local muscle oxygen tension falls; and exercise, under conditions of restricted blood flow, further reduces the oxygen tension. A

Table 2.4 Commonly Used Methods to Detect the Occurrence of Apoptosis in Cells

Method to detect apoptosis	Strengths	Limitations
Morphological—microscopy	Technique well validated; uses common histological dyes; most labs have access to microscopy equipment.	Difficulty in identifying apoptotic versus necrotic cells.
TUNEL–gel electrophoresis	Considered "gold standard" for verification of DNA laddering.	Difficulty in identifying apoptotic versus necrotic cells.
Caspases–ELISA	Caspase(s) activity definitive indication of apoptosis.	Application to cell populations and not single cells; requires microplate reader which may be expensive; will not detect caspase-independent apoptosis.
Phosphatidylserine (PS) externalization-flow cytometry	Useful for detecting early apoptotic cells.	May pick up PS in necrotic cells; requires expensive flow cytometry equipment.

possible mechanism that mediates angiogenesis is muscle hypoxia during exercise. Breen et al. (1996) showed that VEGF mRNA is increased fourfold after a single treadmill run in rat muscle and is increased further when exercise is performed under hypoxemia. It was also demonstrated that FGF-2 mRNA was increased with exercise under hypoxemia and that bFGF and TGF-β were increased after 1 h of exercise. Similar findings were also obtained in human muscle biopsies taken after 40 to 60 min of knee extensor exercise (Richardson et al. 1999). Chronic nerve stimulation has also been shown to increase VEGF mRNA levels in skeletal muscles after one and three days (Breen et al. 1996; Hang et al. 1995). This increase in VEGF gene expression is also correlated with increase in hypoxia-inducible factor (HIF) mRNA in working muscles after 30 to 45 min of knee extensions (Gustafson et al. 1999). However, change in mRNA is not necessarily followed by an increase in protein, but VEGF protein is increased almost threefold after three days of chronic electrical nerve stimulation (Annex et al. 1998). Asano and colleagues (1998) also showed that serum levels of VEGF were increased during high-altitude swim training in humans. VEGF has been considered a potent angiogenic factor since it is up-regulated after both exercise and electrical stimulation. Thus, increased VEGF protein levels should provide an important stimulus for angiogenesis, especially during the early phase of the training program. At present there is only one study that shows both increased VEGF and vessel density with training (Amaral et al. 2001). Lloyd et al. (2002) did not detect increased muscle capillarity during the early time period of training when the VEGF mRNA changes were most apparent. However, increase in capillarity was observed beginning at day 12 of training. Since angiogenesis has been observed as early as five days after the onset of chronic muscle stimulation, it is likely that the angiogenic process was already ongoing at earlier time points, during the dominant phase of the VEGF response. However, a sustained exercise stimulus, occurring over a number of days, may be necessary in order for the development of angiogenesis to be fully manifested.

Furthermore, inhibition of VEGF with a neutralizing antibody completely blocked angiogenesis induced by exercise, further supporting the relationship between exercise, VEGF, and angiogenesis. The angiogenesis-promoting factor action of VEGF is produced primarily through VEGF binding to its two receptors, Flk-1 and Flt-1 (Gavin et al. 2000a). Early gene expression of Flt-1 and not Flk-1 in response to exercise suggests that differences exist in their functions. Flk-1 is critical for embryonic endothelial cell differentiation and vasculogenesis. Flt-1 is important in the organization of the developing vasculature. However, both receptors may be needed in angiogenesis since Flt-1 and Flk-2 gene expression can be induced by hypoxia (Sandner et al. 1997). Unfortunately, the exact mechanisms of action of both receptors are not clear yet and need to be further elucidated.

In addition, constitutive VEGF protein expression was confirmed in muscle fibers, vascular smooth muscle (VSM) of larger vessels, interstitial fibroblasts, and other cell types (Brown et al. 2001). Expression of VEGF in VSM of arterial vessels in exercise-induced angiogenesis is hypothesized to be due to adenosine availability, which helps produce MMP from VSM. The MMP mobilize VEGF or encourage migration of mesenchymal cells participating in the arteriolization of capillaries. Endothelial cell-stimulating angiogenic factor (ESAF) is linked with capillary growth in stimulated muscles; this activates the pro-forms of metalloproteinase gelatinase A, collagenase, and streptolysin, which are implicated in capillary growth (Hansen-Smith et al. 1998).

Hypoxia-inducible factor 1 (HIF-1) has been shown to be required for the hypoxia-induced increase in VEGF (Forsythe et al. 1996). Both HIF-1α and HIF-1β mRNA levels increase in vivo along with increases in VEGF. However, a cause–effect relationship cannot be established, since a similar stimulus or mechanism affects the transcription or stabilization, or both, of VEGF and HIF-1 mRNA (Gavin et al. 2000b).

Nitric oxide (NO) also appears to be an important regulator of endothelial cell growth and angiogenesis. It is known to be important in blood flow regulation during exercise and is a cellular signal regulating mitochondrial respiration (Gavin et al. 2000b). As mentioned previously, hypoxic exercise produces a greater increase in VEGF mRNA levels than does exercise alone. However, NO has been shown to inhibit VEGF up-regulation through inhibition of HIF-1 in aortic smooth muscle cells (Liu et al. 1995). Therefore, since NO is important for vasodilation, it is expected that NO should increase VEGF levels rather than cause a decrease. Nonetheless, NO may regulate VEGF gene expression differentially depending on the specific tissue. NO increases VEGF mRNA levels via guanylate cyclase activity in human A-172 glioblastoma cells and human Hep G_2 hepatocellular carcinoma cells. NO stimulates soluble guanylate cyclase, which converts GTP to the intracellular second messenger cGMP (Pilz et al. 1995). Thus, VEGF does not function alone in exercise-induced angiogenesis in skeletal muscle. Capillary growth in stimulated muscles can be suppressed by treatment with inhibitors of either nitric oxide synthase (NOS) or prostaglandin production (Brown et al. 2001). Inhibition of NOS attenuates the exercise-induced increases in VEGF mRNA (Gavin et al. 2000b). However, the evidence is still preliminary and further investigation is needed to elucidate the effects of exercise on VEGF and VEGF receptor expression.

Although the presence of VEGF appears to be essential to initiate and facilitate angiogenesis, its actions are not sufficient as the sole agent to complete the process. The angiopoietins are essential for normal vascular remodeling. Both angiopoietins (Ang1, Ang2) bind to the Tie-2 receptor and thus compete with each other. Ang1 functions to ensure vessel stability while Ang2 has the opposite effect (Maisonpierre et al. 1997). Therefore, the Ang2:Ang1 ratio is thought to determine whether the net effect of the angiopoietins is to stabilize or destabilize the vasculature. An increase in the Ang2:Ang1 ratio (destabilization) is indicative of being proangiogenic. For example, Ang2 is up-regulated at the leading edge of new vessel growth (Maisonpierre et al. 1997). Preliminary evidence indicates that exercise, particularly ischemic exercise, leads to a shift in Ang2:Ang1 mRNA expression ratio that would support destabilization of the vasculature. Further, studies in transgenic mouse myocardium show that Ang2 has a synergistic effect with VEGF on capillary growth, while Ang1 antagonizes VEGF action (Visconti et al. 2002). In the model proposed by Lloyd and colleagues (2002), the Ang2:Ang1 mRNA ratio was elevated after a single day of exercise, thereby suggesting that destabilization of existing muscle vasculature is an early event in exercise-induced angiogenesis. In accordance with VEGF mRNA expression profiles, the up-regulation of angiopoietin receptor (Tie-2) mRNA was delayed relative to the up-regulation of ligand mRNA. However, the initial increase in the Ang2:Ang1 mRNA ratio coincided with the up-regulation of VEGF mRNA (Lloyd et al. 2002). The work by Lloyd and colleagues is the first to show that a coordinated activation of VEGF and the angiopoietin system occurs during physiological angiogenesis in skeletal muscle. Thus, a balance among these three factors appears to be necessary for normal vessel growth.

Although hypoxia is a known regulator of VEGF mRNA, its role in the exercise response has been questioned. Breen et al. (1996) showed that VEGF mRNA was unregulated in aerobic and hypoxic exercise conditions. This suggests that other signals associated with muscle contraction are sufficient to increase VEGF mRNA levels. These signals could be produced in response to flow-dependent stimuli, mechanical changes induced by load bearing within the muscle during exercise, or both. Increased shear stress during exercise may play an important role in modulating angiogenic factor gene expression. Shear stress increases NO levels, and Noris et al. (1995) found that NO activates HIF-1, a major regulator of VEGF expression, in the absence of hypoxia. In addition, increased stretch has been shown to increase VEGF mRNA expression in skeletal muscle (Rivilis et al. 2002). Although shear stress, hypoxia, and mechanical stimuli are all potential stimulators of VEGF gene expression in response to exercise, it is presently unknown which of these stimuli actually account for the higher concentrations of this important angiogenic growth factor.

Differential effects of angiogenesis factors on various tissue sites play an important role in exercise-induced angiogenesis. For example, the lungs respond to hypoxia with vasoconstriction of the vasculature whereas in skeletal muscle, hypoxia causes vasodilation of the vasculature. In addition, capillary growth following exercise takes several weeks to appear, making it difficult to ascertain the effects of angiogenic factors in the long term.

Proliferation in Skeletal Muscle

Satellite cells are undifferentiated mononuclear precursor muscle cells that are located between the sarcolemma and basal lamina of the fiber (Bodine-Fowler 1994). Evidence suggests that muscular activity can stimulate satellite cells to advance from their initial quiescent state (Yan 2000). This has led several authors to investigate the effect of various exercise protocols on muscle satellite cell activity. Smith et al. (2001) found that 30 min of decline running (–16° grade, 15 m/min) significantly increased soleus muscle satellite cell proliferation (1.0 ± 0.2% of fibers) compared to that in nonexercise control rats (0.4 ± 0.2% of fibers) following two but not one, four, or seven bouts of the same task. McCormick and Thomas (1992) found that progressive treadmill training for 10 weeks resulted in an increase in soleus satellite cell mitotic activity from 0.52 ± 0.13 nuclei/mm^2

to 1.28 ± 0.33 nuclei/mm^2 in exercised rats. Furthermore, Tamaki et al. (2000) found that a single bout of heavy hindlimb weightlifting significantly increased muscle mitotic activity in young (14-20 weeks) but not in old (>120 weeks) rats. It is currently unclear whether and how exercise affects cell cycle-specific components.

Proliferation in Immune Cells

Following mitogenic and antigenic stimulation, quiescent immune cells (G_0) progress through the cell cycle and are governed by the same molecules as other cell lineages. However, some cell cycle components interact with lymphocytes in a unique manner and therefore play a critical role in immune function. For example, T-lymphocytes do not express cyclin D_1. Therefore proliferation through the G_1 phase is dependent on cyclins D_2 and D_3 as well as cdk4 and cdk6. Furthermore, *p27* is constitutively present in quiescent T-lymphocytes and down-regulated in the G_1 phase following stimulation (Balomenos & Martinez-A 2000; Chitko-McKown & Modiano 1997).

Exercise has been shown to modulate many aspects of immune function, including cell proliferation (Pedersen & Hoffman-Goetz 2000). Several investigators have examined the role of physical activity on lymphocyte proliferation. Here we describe some representative findings rather than providing a comprehensive and exhaustive review of the area. Nieman et al. (1994) found that 45 min of high (80% $\dot{V}O_2max$)-intensity treadmill running was associated with a 21% decrease in concanavalin A-stimulated lymphocyte proliferation 1 and 2 h postexercise compared to baseline values following correction for changes in CD3$^+$ lymphocytes. However, moderate (50% $\dot{V}O_2max$)-intensity running had no effect on lymphocyte proliferation. Short-duration (20 min) submaximal cycling at 50% peak work capacity significantly increased lymphocyte proliferation in young (26 ± 3 years) but not elderly (69 ± 5 years) subjects (Mazzeo et al. 1998). A recent study (Green et al. 2002) showed that 60 min of running at 95% ventilatory threshold significantly decreased phytohemagglutinin-induced peripheral blood mononuclear cells (PBMC) proliferation but had no effect on NK-depleted PBMC or PBMC adjusted for CD3$^+$ percentage. Although studies that measure lymphocyte proliferation, as determined by ^{3}H-thymidine incorporation (or other radionucleotides), are suggestive of changes in the cell cycle with exercise, further research is needed to pinpoint specific regulatory components (e.g., cyclin-cdks).

Telomeres and Telomerase Activity

The proliferation of satellite cells following muscle injury plays a critical role in muscle recovery (Smith et al. 2001). In skeletal muscle degenerative conditions such as muscular dystrophy, satellite cells are forced to undergo repeated bouts of cellular proliferation to repair continuous skeletal muscle damage. This leads to satellite cell exhaustion due to repeated cycles of regeneration and telomeric loss. Indeed, muscular dystrophy is associated with a 14-fold greater rate (187 bp/year vs. 13 bp/year) of telomeric loss than that in healthy controls (Decary et al. 2000; Di Donna et al. 2000). Exercise also causes muscle to undergo repeated bouts of regeneration that could potentially affect telomeric length. Moreover, exercise affects lymphocyte and satellite cell proliferation, and these relationships have led some authors to suggest a possible link with alterations in telomere length and telomerase activity in these tissues with exercise.

Two recent studies have addressed the effect of exercise on telomere length or telomerase activity. Radak et al. (2001) found that 60 min of swimming five times per week for eight weeks (moderate exercise), or a gradual increase in swimming duration by 30 min/week (from 60 min during the first week to 270 min by the last week), did not alter skeletal muscle telomerase activity in Wistar rats. Furthermore, BDF1 mice implanted with the S-180 sarcoma and exposed to various exercise conditions did not have any differences in liver telomerase activity. Bruunsgaard et al. (1999) found that mean TRF length in blood mononuclear cell (BMNC), CD4$^+$, and CD8$^+$ lymphocytes was significantly shorter in elderly (median age 78) compared to younger (median age 23) subjects. Furthermore, TRF lengths were significantly decreased in BMNC and CD8$^+$ lymphocytes of younger subjects and CD4$^+$ lymphocytes of elderly subjects following maximal bicycle exercise lasting 17 to 20 min. Further research is warranted given the limited data on exercise-induced alterations in telomere length and telomerase activity.

Apoptosis in Heart and Skeletal Muscle

Cardiomyocytes undergo apoptosis in response to ischemia and reperfusion injury. Gottlieb et al. (1994) were among the first to demonstrate apoptosis in rabbit myocardium subjected to metabolic inhibition and recovery. Others have shown increased Fas mRNA undergoing hypoxia-induced myocyte death in rats (Tanaka et al. 1994), as well as ultrastructural changes (e.g., DNA fragmentation, membrane blebbing) indicative of apoptosis that were especially pronounced during reperfusion (restoration of oxygen, serum, and glucose following a designated period of deprivation) compared with the initial ischemia (deprivation of oxygen, serum, and glucose) (Umansky et al. 1995). Expression of Bcl-2, Bax, and Fas proteins in rat myocytes has been reported post-ischemia (Kajstura et al. 1996). As noted earlier, Fas and Bax are potent cell death triggers whereas Bcl-2 functions to increase cell survival. While it is clear that ischemia can lead to cardiac apoptosis and necrosis, what accounts for apoptotic injury to myocytes during reperfusion? Reperfusion of ischemic tissue is associated with the generation of oxygen free radicals and acute leukocyte activation (Kloner et al. 1989; Sussman & Bulkley 1990). ROS are generated through a variety of mechanisms, including the metabolism of arachidonic acid and respiratory burst activity from leukocytes (which generates superoxide anion, singlet oxygen, hypochlorous acid, and other reactive species). ROS can damage DNA by oxidation of purine and pyrimidine bases, most especially guanine (Loft & Poulsen 1996), leading to apoptosis of the injured cell.

Ca^{2+} also plays a role in cell death during reperfusion injury. Following reperfusion, a large increase in intracellular Ca^{2+} occurs with resultant mitochondria Ca^{2+} accumulation. These events lead to decreased ATP production and MTP formation, which can lead to cell death (Suleiman et al. 2001; Xu et al. 2001; Wang et al. 2002). Narayan et al. (2001) found that rat ventricular myocytes subjected to simulated ischemia and reperfusion showed increased apoptosis (Annexin V staining), but not when exposed to simulated ischemia alone. Following simulated reperfusion, free Ca^{2+} concentrations within the mitochondria increased from 111 ± 14 nM during baseline to 214 ± 22 nM and 382 ± 22 nM in Annexin-negative and Annexin-positive cells, respectively. Furthermore, calcium preconditioning of myocytes maintains mitochondria Ca^{2+} homeostasis, preventing MPT formation and apoptosis (Xu et al. 2001). However, ischemia–reperfusion injury is not the only trigger for non-necrotic cell death of cardiomyocytes. Overstretching results in apoptosis in rat myocytes and non-myocytes, together with increased expression of Fas (Cheng et al. 1995). However, cardiac adaptation to treadmill training in rats is not associated with increased apoptosis of myocytes (Jin et al. 2000).

Cell death occurs in immature skeletal muscle in response to a variety of injurious stimuli. Some have suggested that whether the response is apoptosis or necrosis depends upon the maturity of the muscle cell. Carraro and Franceschi (1997) hypothesized that the normal response to injury by immature muscle cells is apoptosis whereas the typical response in adult muscle fibers to the same inducing stimulus is necrosis. Evidence that age and development influence apoptosis or necrosis pathways comes from studies in neonatal rat skeletal muscle (Kaminska & Fidzianska 1996) and from observations of human embryos at different stages of development (Carraro & Franceschi 1997).

Apoptosis has been demonstrated under non-pathological conditions that stress muscle cells, such as eccentric exercise, or conditions of atrophy, such as prolonged traction and limb unweighting. As a rule, damage to skeletal muscle is greater with repeated contractions in which the muscle undergoes lengthening (eccentric exercise) than with activity involving shortening (concentric) or isometric contractions. Biral and colleagues (2000) used a model of sustained eccentric exercise, with repeated contractions while the muscle was extended. With use of the TUNEL method to detect apoptosis in rat soleus and extensor digitorum longus, the disappearance of the dystrophin-based membrane skeleton and an elevated expression of Bax and caspase 3 in individual muscle fibers were demonstrated. Thus, apoptosis occurs in normal, mature skeletal muscle fibers in response to extreme functional demands.

Is there any evidence for apoptosis of skeletal muscle fibers with less extreme workload demands? Although tentative, evidence of apoptosis in adult C57BL/6 mouse tibialis anterior muscle following 12 h of voluntary wheel running has been reported, with myofiber content of Bax, Fas, and ubiquitin correlated with the appearance of apoptotic myonuclei (Podhorska-Okolow et al. 1998). Apoptosis results in the elimination of myonuclei or satellite cells, or both, from atrophying rat muscle fibers with hindlimb unweighting (Allen et al. 1997).

Apoptosis of Immune Cells

An area of increasing research is the effect of exercise on apoptosis of immune cells. Strenuous exercise regulates several factors that alter lymphocyte apoptosis in a number of ways. For example, increased GC secretion, growth factor withdrawal, ROS generation, and increased plasma TNF-α are some of the signals that induce apoptosis in immune cells (Phaneuf & Leewenburgh 2001); and some of these factors (GC, TNF-α) are notably elevated after strenuous, prolonged, or muscle-damaging exercise. Apoptotic loss could contribute to the well-documented decrease in the number and function of lymphocytes postexercise (Hoffman-Goetz & Quadrilatero 2003; Pedersen & Hoffman-Goetz 2000).

The effect of high-intensity exercise on the induction of lymphocyte apoptosis has been repeatedly demonstrated. One of the earliest studies to address this was that of Concordet and Ferry (1993), who found DNA fragmentation (using the TUNEL assay) of thymus following two treadmill runs to exhaustion (separated by a 24-h rest) in rats. Thymocyte apoptosis could be blocked by administration of a GC receptor antagonist, RU-486 (mifepristone). In vitro studies indicate that corticosterone exposure (at levels observed after exercise) is associated with increased expression of Annexin V-positive thymocytes (Hoffman-Goetz & Zajchowski 1999). Increased oxygen consumption and hyperventilation with exhaustive exercise lead to the generation of ROS; ROS contribute to thymocyte cell death as measured by lipid peroxides and the influx of Ca^{2+} ion into thymocytes (Azenabor & Hoffman-Goetz 1999, 2000). Antioxidant supplementation inhibits exercise-induced leukocyte DNA damage (Hartmann et al. 1995) and thymocyte apoptosis (Lin et al. 1999). Increased caspase 3 activity was demonstrated in developing thymocytes (but not mature splenocytes) after a single intensive bout of exercise in mice (Patel & Hoffman-Goetz 2002), providing evidence that apoptosis occurs in GC-sensitive lymphoid tissues via caspase-dependent mechanisms. Lagranha et al. (2004) found that a single bout of intense treadmill running significantly increased Bax and Bcl-x_S (pro-apoptotic) gene expression while at the same time decreasing Bcl-x_L (anti-apoptotic) gene expression and mitochondrial membrane potential in neutrophils from mature rats.

What is the mechanism for the apoptosis of thymocytes in rats or mice after strenuous exercise? We hypothesize that intense exercise leads to apoptosis through both receptor–ligand signaling and mitochondrial pathways (shown diagrammatically in figure 2.2). Which pathway is activated depends on the whether the exercise triggers a hormonal response involving ACTH and cortisol/corticosterone release, a metabolic response involving ROS generation, or generation of inflammatory cytokines such as TNF-α. In an exercise bout of longer duration (e.g., 60-120 min) and moderate intensity (e.g., <75% of $\dot{V}O_2max$) associated with activation of the hypothalamic-pituitary-adrenal (HPA) axis (Ronsen et al. 2001), corticosterone release will trigger the mitochondrial pathway of apoptosis, activating procaspase 8 and, ultimately, caspase 3. In an exercise bout of high intensity (e.g., >75% of $\dot{V}O_2max$) and shorter duration (e.g., <30 min), respiratory oxidant stress will occur (Raidal et al. 2000), also leading to induction of mitochondrial damage and cytochrome C release and triggering of the caspase pathways. The type of exercise also modifies the apoptotic pathway that ensues once the minimum intensity and duration thresholds have been exceeded. For example, exercise that involves muscle damage (e.g., downhill running) leads to the release of inflammatory cytokines (e.g., TNF-α) and the induction of the receptor–ligand mechanism of cell death. In contrast, intense exercise that occurs without muscle damage could induce apoptosis in lymphocytes primarily through the mitochondrial pathway (see figure 2.4).

According to this model, short-duration submaximal treadmill exercise (Hoffman-Goetz et al. 1999) should not induce significant apoptosis in thymocytes or other susceptible lymphocytes. Similarly, high-volume exercise of very low intensity, which occurs with wheel running, would also not induce apoptosis in lymphocytes (Hoffman-Goetz & Fietsch 2002). Thus, variation in findings of apoptosis effects among exercise studies may be due to exercise intensity and duration continuum; exercise that fails to lead to either an elevation in GC or the generation of ROS or the release of inflammatory cytokines will not be associated with apoptotic events in lymphocytes. Moreover, differences in the sensitivity of lymphocytes to these apoptotic-inducing factors will also influence the extent of cell death. For example, the primary lymphoid compartments (thymus and bone marrow) contain many double positive ($CD4^+CD8^+$) and double negative ($CD4^-CD8^-$) lymphocytes compared with secondary lymphoid compartments (spleen, nodes, blood), which have primarily single positive cells ($CD4^+$ or $CD8^+$). Characterization of the sensitiv-

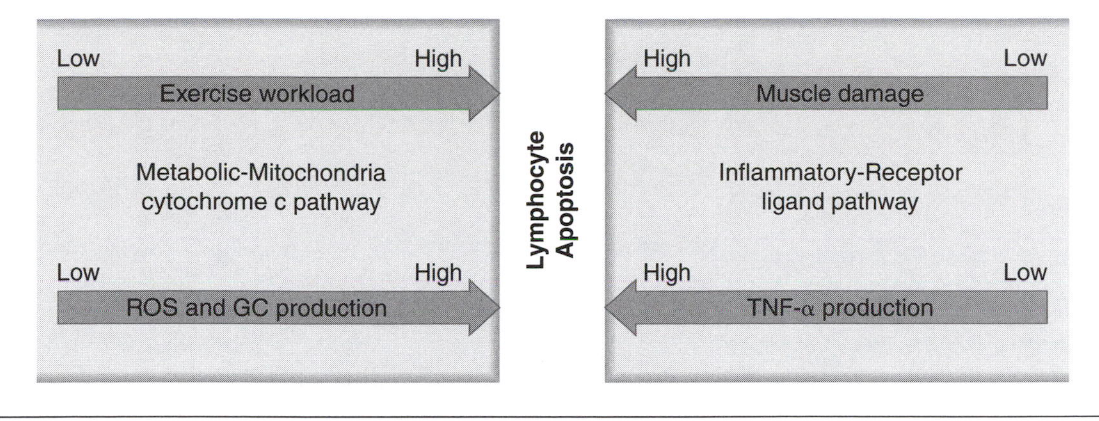

Figure 2.4 Model of exercise-induced lymphocyte apoptosis. Model predicts general exercise conditions under which apoptosis will occur. Reactive oxygen species (ROS) and glucocorticoids (GC), released in response to high-intensity or long-duration exercise or both, trigger the mitochondrial-cytochrome C pathway of apoptosis in lymphocytes. Other hormonal triggers (e.g., catecholamines) may be involved. Tumor necrosis factor-α (TNF-α), released during muscle damage or exercise associated with inflammatory events, triggers the receptor–ligand pathway of apoptosis in lymphocytes. Below some intensity and duration threshold, exercise-induced apoptosis of lymphocytes does not occur because of absence of physiological triggers.

ity of these different subpopulations to ROS, GC, and other apoptotic stimuli may help to explain the differential apoptosis in lymphocyte compartments after exercise.

There are limited data from human studies on the effects of exercise on the induction of apoptosis in lymphocytes. Treadmill exercise to exhaustion (Mars et al. 1998; Niess et al. 1996), treadmill exercise above the aerobic–anaerobic threshold (Hartmann et al. 1994), and treadmill exercise (80% of $\dot{V}O_2max$) to exhaustion (~30 min) (Mooren et al. 2002) were associated with induction of apoptosis in human lymphocytes. In exercise of lower intensity (60% of $\dot{V}O_2max$) for a short duration (~30 min), no apoptosis was observed in human lymphocytes. Steensberg et al. (2002) found an increase in the percentage (but not absolute numbers) of human blood lymphocytes expressing early markers of apoptosis after exercise at 75% of $\dot{V}O_2max$ for 2.5 h. Since an increase in both plasma cortisol and plasma isoprostanes (a measure of oxidative stress) was observed, it is likely that the mitochondrial pathway was activated, leading to the percentage increase in apoptotic lymphocytes. Further, Hsu et al. (2002) found decreased mitochondrial membrane potential and increased DNA fragmentation in peripheral blood leukocytes from subjects who performed consecutive bouts of aerobic exercise at 60% and 85% but not 35% $\dot{V}O_2max$.

Conclusion

Diverse factors (cytokines, growth factors, cell cycle components) regulate cell proliferation. An equally diverse and often bewildering group of intracellular and extracellular factors are involved in the signaling of cell apoptosis. How these factors converge to trigger the physical processes of cellular replication, cellular stasis, or cellular apoptosis has been the object of intense scientific research in the past 20 years. Better understanding of the relationship between level of expression of oncoproteins such as *c-myc* and *p53,* survival proteins such as the products of Bcl-2, and components of the cell cycle (cyclins-cdks) will be important for understanding how apoptotic resistance develops in cancer cells and how cellular senescence occurs. It is clear that with appropriate signals, cardiac and skeletal muscle cells, as well as lymphocytes, undergo apoptosis. The decision to commit to the cellular apoptosis program in response to an exercise challenge will ultimately reflect a combination of extracellular influences (e.g., glucocorticoids, generation of ROS, presence of inflammatory cytokines) and intracellular influences (e.g., ATP availability, short or prolonged opening of the mitochondrial permeability transition pore, concentration of pro-apoptotic to antiapoptotic proteins, activation of oncogenes) acting upon muscle and immune cells.

Chapter 3

Genes, Genetic Heterogeneity, and Exercise Phenotypes

Tuomo Rankinen and Claude Bouchard

This chapter focuses on the nuclear genome, the regulation of gene expression, and the effects of DNA sequence variation on functions. The first part of the chapter deals with molecular genetics concepts with an emphasis on those that have relevance for an understanding of the cellular adaptation to the demands of exercise. The second part reviews recent advances in understanding of the role of gene expression in response to exercise, as well as of the effects of genetic differences in the response to acute exercise and on trainability.

Genes and Genome

The blueprint of the human body is contained in the genetic code specified in the deoxyribonucleic acid (DNA) sequence of the 23 pairs of chromosomes found in every nucleated cell. The term human genome refers to the total genetic information in human cells. The human genome consists of 22 pairs of autosomes (non-sex-specific chromosomes) and two sex chromosomes. In addition, mitochondria carry a simple circular DNA sequence that contains 37 genes, 13 of them encoding peptides of the oxidative phosphorylation system, 2 encoding mitochondrial ribosomal ribonucleic acid (rRNA) molecules, and 22 producing mitochondrial transfer RNAs (tRNAs). Individuals inherit half of their genome from their father (22 autosomes and either X or Y sex chromosome) and half from their mother (22 autosomes and X chromosome). Mitochondrial DNA is derived from the mother.

A chromosome is constructed of two complementary strands of DNA. DNA molecules are large polypeptides in which the backbone of the molecule is composed of five-carbon sugar residues, deoxyribose. The adjacent deoxyribose molecules of a DNA strand are linked by covalent phosphodiester bonds. The genetic information in each chromosome is stored in a long string of the four DNA bases: adenine (A), cytosine (C), guanine (G), and thymine (T). The order and number of the bases determine the product of each gene. The DNA bases are attached to a deoxyribose molecule by a covalent bond with a carbon atom in the 1' position of the molecule. A unit formed by a deoxyribose sugar, an attached base, and a phosphate group forms a basic repeat unit of a DNA strand, a nucleotide. The complementary DNA strands are held together with relatively weak hydrogen bonds between the complementary nucleotides (C pairs with G, A with T). The linear structure of bases in the DNA strands is called the primary structure of the chromosome. The secondary structure of the chromosome arises when the two complementary strands of DNA twist to form a double helix. One turn of the helix is called a pitch and accommodates 10 nucleotides.

A typical gene consists of coding sequences (exons), noncoding regions (introns), and regulatory sequences located both before (5' end; promoter region) and after (3' untranslated region [UTR]) the coding regions of the gene. The number of exons is quite heterogeneous, with a range from one (e.g., intronless G-protein-coupled receptor genes) to several hundreds (e.g., titin gene with

363 exons). Introns were originally considered to be nonfunctional stretches of DNA between the coding regions, but it has been shown that introns may harbor several regulatory elements, such as alternative promoters, and splicing enhancers and suppressors.

Gene Expression

The process that allows the instructions contained in a given gene to be converted to a final gene product is simple on the surface: The DNA sequence is first converted to an RNA sequence, which is then translated to a polypeptide giving rise to the final protein. The first step of the process, transcription, takes place in the nucleus of the cell. The DNA sequence of a gene is used as a template to generate a messenger RNA (mRNA) molecule. The transcription process is accomplished through use of an RNA polymerase enzyme and ATP, CTP, GTP, and UTP molecules as RNA precursors. Three different RNA polymerase enzymes are required to synthesize different classes of RNA in eukaryotic cells. All genes that encode polypeptides, as well as most small nuclear RNA genes, are transcribed by RNA polymerase II. RNA polymerases I and III transcribe rRNA and tRNA genes.

Transcription

The initiation of transcription requires the presence of transcription factors that bind to specific DNA sequence elements located in the immediate vicinity of a gene. These sequence elements are usually clustered upstream of the coding sequence and form the promoter region of the gene. Once the necessary transcription factors are bound to the promoter, an RNA polymerase binds to the transcription factor complex and is activated to start the synthesis of RNA. The common promoter elements recognized by several transcription factors include a TATA box (usually TATAAA) located about 25 base pairs before (–25 bp) the transcriptional start site, a CAAT box (–80 bp), and a GC box. In addition to the common promoter elements, enhancers, silencers, and response elements form a group of regulatory sequences that can enhance or inhibit the transcriptional activity of specific genes. These sequence elements are usually located quite far from the transcription initiation site (several thousands of base pairs). They bind gene regulatory proteins and the DNA strand between the promoter and enhancer/silencer folds in a loop, allowing the regulatory proteins to interact with the transcription factors bound to the promoter.

Once the synthesis of an RNA molecule from the DNA template is finished, the primary RNA transcript (i.e., the full copy of the original template DNA) undergoes various post-transcriptional modifications. These include removal of the unwanted internal segments (i.e., intronic sequences) and rejoining of the remaining segments (exonic sequences), as well as addition of a 7-methylguanosine triphosphate (m7G) to the 5' end of the transcript and addition of several AMP residues to the 3' end of the mRNA molecule. The removal of the intronic RNA segments is called RNA splicing. The process is directed by specific nucleotide sequences at the exon/intron boundaries (splice junctions). Introns usually start with a GT dinucleotide and end with an AG sequence. In addition to splice junctions, a third conserved intronic sequence called branch site is required for RNA splicing. The splicing process starts with the cleavage of the 5' splice junctions, followed by the formation of a lariat-shaped structure through an interaction between the terminal G nucleotide of the splice donor site and the A nucleotide of the branch site. The process is completed by cleavage at the 3' splice junction and splicing of the exonic RNA segments. The splicing process is managed by a structure known as a spliceosome, which is a large molecular complex containing several small nuclear RNA (snRNA) molecules attached to specific proteins and protein splicing factors. The addition of the m7G molecule to the 5' end of the mRNA (capping), as well as polyadenylation of the 3' end, protects the transcript from degradation by exonucleases, facilitates transport of the mRNA molecules to the cytoplasm, and may facilitate interaction between the mRNA molecule and the ribosomal machinery of the translation process.

Translation

Translation, or the synthesis of polypeptides from the mRNA template, takes place in the cytoplasm, although translation in the nucleus has also been described. After mRNA molecules migrate from the nucleus to the cytoplasm, they bind with the ribosomes where translation takes place. The sequences in the beginning (5') and at the end (3') of the mRNA strand, which are copied from the 5' and 3' terminal exons, are not translated, but they contribute to the binding and stabilization of the mRNA on the ribosomes. Ribosomes are large RNA-protein complexes and consist of a large 60S subunit and a smaller 40S subunit; they provide a structural framework for polypeptide synthesis.

The order of the amino acids in a new polypeptide is determined by the triplet genetic code of the mRNA sequence. Each amino acid binds to a transfer RNA (tRNA) molecule. Each tRNA molecule has a specific sequence of three nucleotides (anticodon) located in the center of one arm of the molecule. The anticodon matches a specific complementary trinucleotide sequence (codon) of the mRNA and thereby allows translation of the genetic code of the mRNA into the correct amino acid sequence in the growing polypeptide chain. Translation process starts when the ribosome finds an initiation codon in the mRNA, usually AUG corresponding to methionine. The new amino acids are incorporated in the polypeptide chain through a peptide bond formed between the amino group of the incoming amino acid and the carboxyl group of the last amino acid of the polypeptide. Translation continues until the ribosome reaches a termination codon (UAA, UAG, or UGA in the nuclear-encoded mRNA). The primary polypeptides produced by translation usually undergo post-translational modifications, such as phosphorylation, methylation, acetylation, carboxylation, or glycosylation. The polypeptides may also be enzymatically cleaved to produce smaller functional products.

Regulation of Gene Expression

Every nucleated cell contains a complete copy of the human genome. Some genes are expressed in most or all cells because the gene product is essential for the function of the cell (housekeeping genes). However, the majority of the genes in the human genome are expressed only in specific organs, tissues, or cells; and some are expressed only at certain stages of a cell cycle or during specific periods of the development of an organism. Moreover, the expression of a gene may be enhanced or depressed in response to external stimuli, such as changes in the metabolic milieu of the cell or in extracellular concentrations of certain ions and nutrients. To accommodate all these specific demands, gene expression must be closely controlled and coordinated. The adaptability and coordination of the gene expression process are achieved through several regulatory mechanisms, such as transcription factors, alternative splicing, alternative promoters, genetic imprinting, and gene silencing.

Transcription Factors

The promoter region of a gene contains binding sites for various transcription factors, which are necessary for the initiation of gene transcription. Activation of RNA polymerases I and III, which transcribe housekeeping genes, requires a host of ubiquitous transcription factors. However, the transcription of polypeptide-encoding genes by RNA polymerase II utilizes complex sets of general and tissue-specific transcription factors. The complex of RNA polymerase II and general transcription factors are sufficient to initiate gene transcription at a minimum rate, which can be increased or turned off by additional positive or negative regulatory elements. However, the majority of the genes transcribed by RNA polymerase II show tissue-specific expression. The tissue specificity is achieved by interactions between special enhancer and silencer sequences and a variety of promoter sequence elements that are recognized only by tissue-specific transcription factors. The number of transcription factors, rather than the number of genes, has been suggested to be a key determinant of the biological complexity of an organism (Szathmary et al. 2001). For example, the human genome contains over 2,000 genes for transcription factors, whereas they number 500 and 700 in the worm and fruit fly genomes, respectively (Szathmary et al. 2001; Tupler et al. 2001).

Gene expression may also be acutely regulated in response to several external stimuli. These stimuli activate specific transcription factors, which bind to specific regulatory sequences in the promoter region (response elements) of the target genes leading to their activation. Some hydrophobic hormones (e.g., sex hormones, thyroxine) can enter the target cell directly and bind inducible transcription factors called intracellular receptors (hormone nuclear receptors). The activated nuclear receptor associates with a specific promoter response element and activates gene transcription. Hydrophilic molecules that cannot diffuse through the plasma membrane can induce gene transcription by binding to specific cell membrane receptors. The transmembrane receptor activated by a ligand molecule initiates an intracellular signal transduction pathway, which ultimately leads to activation of specific target transcription factors and initiation of gene transcription.

Alternative Splicing

One of the most striking surprises generated by the first draft of the human genome sequence was the low number of genes. The previous estimates based on the number of expressed sequence clusters ranged from 80,000 to 150,000, whereas the human genome draft revealed 32,000 annotated and predicted genes (Lander et al. 2001). This finding emphasized the fact that several genes must produce more than one transcript and that the phenomenon of one gene, several gene products, was more common than previously thought.

The main factor contributing to the disparity between the number of genes and gene transcripts is alternative splicing (Modrek & Lee 2002). Alternative splicing refers to a situation in which a single gene produces multiple mRNA isoforms through different combinations of exons (figure 3.1). It is estimated that 40% to 60% of the human genes have alternative splice forms, and that 70% to 88% of splice isoforms change the characteristics of the gene product (Modrek & Lee 2002). Alternative splicing may cause either inclusion or exclusion of one or several exons, or it may induce the use of alternative 5' or 3' sites. These changes may induce an in-frame addition or deletion of functional units in the gene product (alternative exons), change the amino terminus of the polypeptide (alternative initiation), or modify the size and carboxy terminus of the gene product due to frameshift or alternative termination (Modrek & Lee 2002). The regulation of alternative splicing is still poorly understood, but the available data suggest that the process involves a complex interaction between cis regulatory elements (e.g., exonic splice enhancers) and trans-acting factors, such as SR proteins (Lopez 1998; Liu et al. 2000; Smith & Valcarcel 2000; Modrek & Lee 2002).

Small RNAs

The latest breakthrough in understanding of the regulation of gene expression is the discovery of the role played by small RNAs. Micro RNAs (miRNA) and small interfering RNAs (siRNA) are short (~22 nt) RNA molecules that are cleaved from longer precursor mRNA sequences by a ribonuclease called Dicer. The precursors for miRNAs are about 70-nt-long hairpin-shaped RNAs transcribed from small noncoding genes, whereas the siRNA is produced from apparently aberrant double-stranded RNAs. The miRNAs can suppress gene expression either by binding to target mRNA and repressing translation, or by degrading the mRNA. The effects of siRNA on gene expression are mediated through its binding on the RNA-induced silencing complex, which gives rise to an endonuclease that cleaves target mRNAs. Recent studies also suggest that small RNAs can regulate gene expression by affecting the form of chromatin. Changes in the compactness of chromatin can alter which genes are expressed without affecting the coding sequence of the genes.

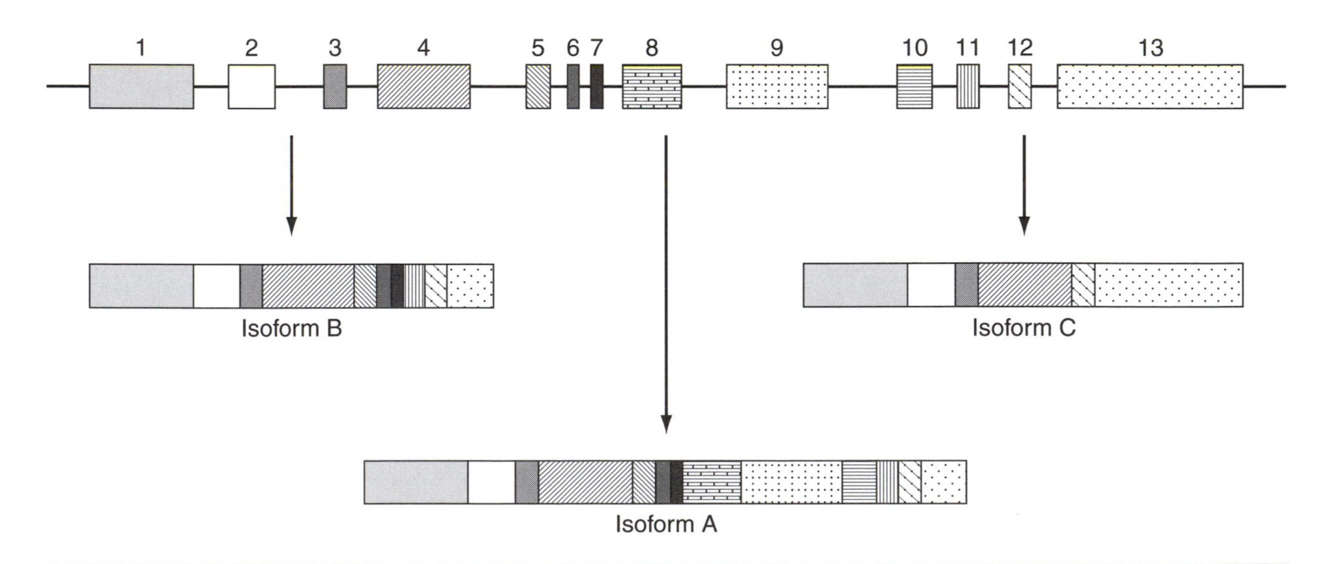

Figure 3.1 An example of alternative splicing. A gene consisting of 13 exons encodes three distinct mRNAs. Isoform A contains all 13 exons, whereas exons 8-10 and 5-11 are spliced off from isoforms B and C, respectively. Isoform C also utilizes an alternative stop codon.

Exercise and Gene Expression

Acute exercise and regular exercise training induce several physiological responses that necessitate increased protein synthesis. The latter is necessary to accommodate increased requirements for a specific gene product (e.g., enzymes of energy production pathways), to replenish proteins that are catabolized during exercise, or to support the adaptive changes associated with the improvement in the capacity to perform physical activity (e.g., enhanced blood flow in working muscles). It has been argued that the human genome evolved over a period when high levels of physical activity were fundamental for survival, which supports the view that there is a link between exercise and the regulation of gene expression (Booth et al. 2002).

An example of the effect of exercise on gene expression is muscle contraction-induced skeletal muscle hypertrophy. Muscle contraction has been shown to increase both transcription and translation of myofibrillar proteins, such as α-actin. Data from animal models suggest that a transcription factor called serum response factor (SRF) and its response element in the promoter of the α-actin gene are involved in muscle contraction-induced α-actin expression and consequently muscular hypertrophy. Muscle overload also increases expression levels of several growth factors, such as insulin-like growth factor I (IGF-I) and signaling molecules (e.g., calcineurin).

Several studies have shown that endurance exercise training improves both cardiac and peripheral vasodilation. This effect is mediated by decreased sympathetic activity and increased production of endothelial vasodilators, especially nitric oxide (Kingwell et al. 1997; Delp 1998; Higashi et al. 1999). In addition, increased laminar shear stress induced by greater blood flow is a major factor in exercise-induced vasodilation. Increased shear stress is known to regulate gene expression in endothelial and vascular smooth muscle cells (e.g., endothelial nitric oxide synthase). Recent studies have revealed that cytoplasmic protein-tyrosine kinase (PTK2; a.k.a. focal adhesion kinase) and c-src tyrosine kinase are key upstream molecules that regulate the signaling cascade activated by shear stress (Li et al. 1997; Davis et al. 2001).

These examples illustrate that the acute or chronic adaptation to exercise is dependent on the expression level of specific genes. However, the net effect of exercise-induced gene expression on exercise-related phenotypes is determined by the integrated action of multiple genes. Recent advances in microarray technology have opened new opportunities to investigate the expression levels of thousands of genes simultaneously in a single experiment. This allows the exploration of the effects of a specific stimulus, such as exercise, on the expression patterns of several gene families, such as transcription factors or genes involved in specific metabolic or physiologic pathways.

The first applications of microarray technology to exercise-related questions are starting to emerge. Roth et al. (2002) investigated the effects of a nine-week strength training program on the vastus lateralis gene expression profile in 20 sedentary subjects (5 young males, 5 young females, 5 older males, 5 older females). The strength training program consisted of unilateral knee extension exercises of the dominant leg. The subjects exercised three times per week, and each training session consisted of four sets of high-volume, heavy-resistance knee extensions. A total of 69 genes showed >1.7-fold difference in expression levels after the training period in the pooled data. Fourteen of these genes were identified in all age-by-sex subgroups (table 3.1), 12 of them showing decreased and 2 showing increased expression levels after the training program.

Bronikowski and coworkers (2003) investigated cardiac expression levels of 11,904 genes in middle-aged and old male mice derived from sedentary and spontaneously physically active selective breeding lines. In the sedentary animals, 137 genes showed significant aging-related changes in expression levels. These genes were involved in inflammatory and stress responses, signal transduction, and energy metabolism. However, in the physically active animals the number of genes showing expression changes with age was significantly lower ($n = 62$) than in the sedentary animals. Moreover, physically active animals showed smaller fold changes as compared to the sedentary animals in 32 of the 42 genes that were common to both groups (Bronikowski et al. 2003). These results suggest that regular exercise may retard several aging-related changes in cardiac gene expression.

The effect of unloading and low-intensity activity on the gene expression pattern of the soleus muscle was investigated in rats (Bey et al. 2003). A 12-h hindlimb unloading protocol induced significant changes in the expression patterns of 63 genes (~1% of the genes on the array). The expression of 21 genes decreased during the

Table 3.1 Summary of Genes Showing >1.7-Fold Change in Expression Levels in Response to a Nine-Week Strength Training Program

Gene name	Gene symbol	Fold change
DOWN-REGULATED GENES		
Four-and-a-half LIM domains 1	FHL1	0.248
Myosin, light polypeptide 2	MYL2	0.262
Cold shock domain protein A	CSDA	0.265
Glyceraldehyde-3-phosphate dehydrogenase	GAPD	0.297
Actin, alpha 2	ACTA2	0.405
Myosin, light polypeptide 3	MYL3	0.442
Dynactin	ACTB	0.446
Eukaryotic translation elongation factor 1 gamma	EEF1G	0.484
ATP synthase, mitochondrial F1 complex, beta polypeptide	ATP5B	0.505
Troponin I	TNNI1	0.508
Actin-related protein 1, centractin alpha	ACTR1A	0.513
Topoisomerase (DNA) I	TOP1	0.547
UP-REGULATED GENES		
Tetraspan 5	TM4SF9	2.131
TNF receptor-associated factor 6	N/A	1.852

Adapted from Roth et al. (2002).

unloading period, but only 4 of these remained down-regulated after the reloading phase. On the other hand, 27 of the 38 genes that were up-regulated by unloading remained significantly above the baseline control levels despite the reloading stimulus. These genes were mainly involved in protein synthesis and degradation, whereas the up-regulated genes that returned to the baseline level represented transcription factors and glucose metabolism proteins (Bey et al. 2003).

DNA Sequence Variation

The major source of genetic variation is the variety of heritable changes (mutations) in the nucleotide sequence of the DNA. These changes may vary from a substitution of a single base to a loss or gain of entire chromosomes or large chromosomal regions. Mutations that occur in the germline cells can be transmitted to future generations, whereas somatic cell mutations are restricted to a single individual and the population of cells derived from the mutated cell. Large-scale germline chromosomal abnormalities are relatively rare but usually show a clear phenotype, which is generally pathogenic and often lethal. Large-scale somatic mutations are more common and often occur in tumor cells.

Small-scale mutations can be grouped into three mutation classes: (1) base substitutions, (2) deletions, and (3) insertions (figure 3.2). Base substitutions usually involve the replacement of a single base. Synonymous or silent substitutions, which do not change an amino acid in the final gene product, are the most frequently observed in coding DNA. Nonsynonymous substitutions result in an altered codon that specifies either a different amino acid (a missense mutation) or a termination codon (a nonsense mutation). A missense mutation can induce either a conservative or a nonconservative amino acid substitution. A conservative substitution refers to a situation in which the new amino acid is chemically similar to the old amino acid, whereas the amino acid introduced by a nonconservative substitution has different chemical characteristics. Thus, nonconservative substitutions are more likely to change the properties of the gene product than conservative substitutions. Deletions and insertions refer to the removal of one or a few nucleotides from the DNA sequence or the addition of one or more nucleotides, respectively. These variations are relatively common in noncoding DNA. They are less frequent in exons, where they may introduce frameshifts, that is, alter the normal translational reading frame of the gene and thereby change the final gene product (figure 3.2).

Traditionally the small-scale mutations have been considered functionally significant only if they induce changes in the amino acid sequence. However, it is now acknowledged that silent substitutions in exons as well as mutations in the noncoding sequence may have strong effects on gene transcription and on the final gene product. Mutations in the 5' regulatory region may disrupt a transcription factor binding site, a response element, or an enhancer or silencer sequence and thereby affect the rate at which a gene is transcribed. Although the 3' untranslated region is not under as strict structural control as the coding and promoter regions, it harbors several sequence elements that affect nuclear transport, polyadenylation, subcellular targeting, and stability of mRNA (Conne et al. 2000). Mutations in these sequences could also potentially influence gene transcription and translation. Both synonymous and nonsynonymous substitutions in the coding sequence may alter splicing sites as well as splicing enhancers and silencers and thereby influence the properties of the mature polypeptide (Cartegni et al. 2002).

Genetics and Responsiveness to Exercise Training

The favorable effects of endurance training on physical performance and on several risk factors for chronic diseases have been well documented. However, it is also clear that there are marked interindividual differences in the adaptation to exercise training. For example, in the HERITAGE Family Study, 742 healthy but sedentary subjects

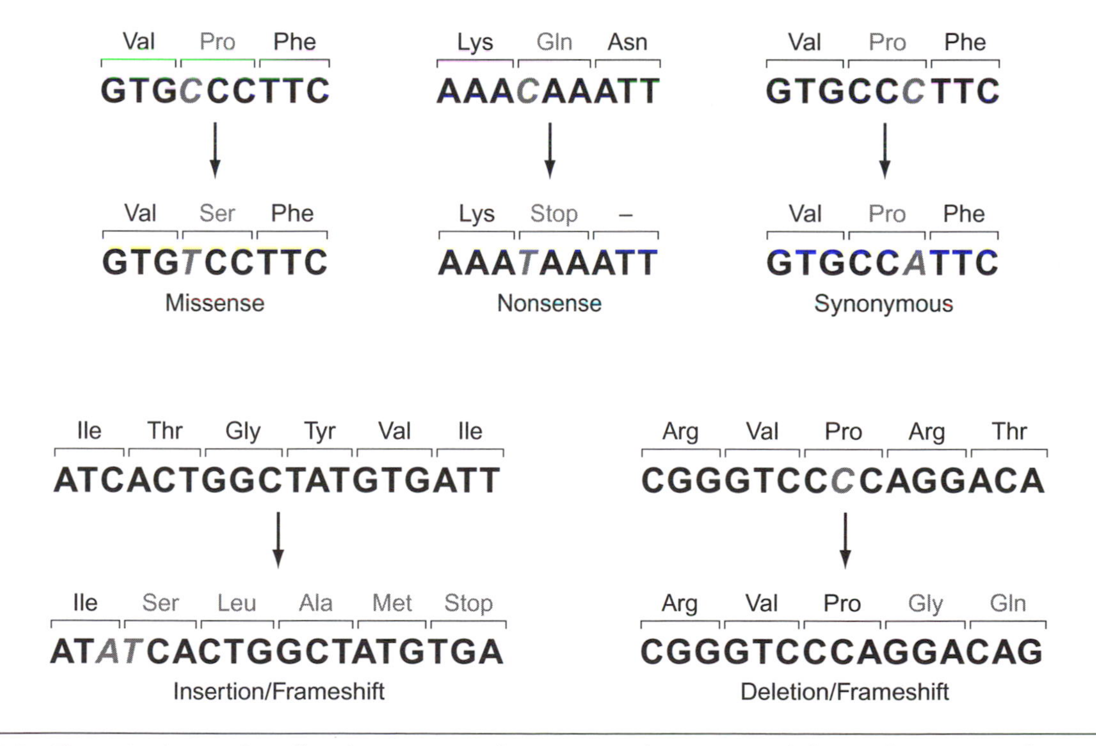

Figure 3.2 The main classes of small-scale mutations. The upper panel presents single base substitutions, where a change of a single nucleotide either induces a change in amino acid (missense) or a premature stop codon (nonsense) or has no effect on the gene product (synonymous/silent). The lower panel shows examples of small-scale insertion (left) and deletion (right) mutations that alter the translational reading frame of the gene (frameshift). Polymorphic nucleotides and resulting changes in amino acids are indicated with gray font.

followed an identical, well-controlled endurance training program for 20 weeks. Despite the identical training program, increases in maximal oxygen consumption ($\dot{V}O_2max$), a measure of cardiorespiratory fitness, varied from no change to increases of more than 1 L/min (Skinner et al. 2000; Bouchard & Rankinen 2001). A similar pattern of variation in training responses was observed for several other phenotypes, such as plasma lipid levels and submaximal exercise heart rate and blood pressure (Leon et al. 2000; Bouchard & Rankinen 2001; Wilmore et al. 2001). These data underline the notion that the effects of endurance training on cardiovascular traits should be evaluated not only in terms of mean changes but also in terms of response heterogeneity. Furthermore, high and low responsiveness to training in the HERITAGE Family Study was characterized by significant familial aggregation; that is, some families were mainly low responders whereas in some families all the members showed significant improvements. These observations support the notion that individual variability is a normal biological phenomenon, which may reflect genetic diversity (Bouchard 1995; Bouchard & Rankinen 2001).

In the remainder of this chapter we review the evidence for a role of genetic diversity in the response to exercise. First, the concepts of genetic epidemiology applied to exercise-related phenotypes are introduced. Then, we review the current status of the association and linkage studies addressing responsiveness to acute exercise and exercise training using molecular genetic markers. Finally, we summarize the monogenic disorders featuring exercise intolerance as a clinical symptom.

Genetic Epidemiology

The maximal heritability of a trait, that is, the combined effect of genes and shared environment on a phenotype, can be estimated using data from family and twin studies. The heritability estimates are based on comparisons of phenotypic similarities between pairs of relatives with different levels of biological relatedness. For example, biological siblings, who share about 50% of their genes identical by descent (IBD), should be phenotypically more similar than their parents (biologically unrelated individuals) if genetic factors contribute to the trait of interest. Likewise, a greater phenotypic resemblance between identical twins (100% of genes IBD) than between dizygotic twins (50% of genes IBD) indicates genetic effect on the phenotype.

The heritability of $\dot{V}O_2max$ in the sedentary state has been estimated from a few twin and family studies, the most comprehensive of these being the HERITAGE Family Study (Bouchard et al. 1998). An analysis of variance revealed a clear familial aggregation of $\dot{V}O_2max$ in the sedentary state. The variance in $\dot{V}O_2max$ (adjusted for age, sex, body mass, and body composition) was 2.7 times greater between families than within families, and about 40% of the variance in $\dot{V}O_2max$ was accounted for by family lines. Maximum likelihood estimation of familial correlations (spouse, four parent–offspring, and three sibling correlations) revealed a maximal heritability of 51% for $\dot{V}O_2max$. However, the significant spouse correlation indicated that the genetic heritability was less than 50% (Bouchard et al. 1998). Data from the twin studies have yielded very similar heritability estimates, ranging from 25% to 66% (Bouchard et al. 1986; Fagard et al. 1991; Sundet et al. 1994).

In pairs of monozygotic twins, the $\dot{V}O_2max$ response to standardized training programs showed six to nine times more variance between genotypes (between pairs of twins) than within genotypes (within pairs of twins) (Bouchard et al. 1992). Thus, gains in absolute $\dot{V}O_2max$ were much more heterogeneous between pairs of twins than within pairs of twins. In the HERITAGE Family Study, the increase in $\dot{V}O_2max$ in 481 individuals from 99 two-generation families of Caucasian descent showed 2.6 times more variance between families than within families, and the model-fitting analytical procedure yielded a maximal heritability estimate of 47% (Bouchard et al. 1999).

In addition to $\dot{V}O_2max$, the HERITAGE Family Study investigated the heritability of training-induced changes in several other phenotypes, such as submaximal aerobic performance (Perusse et al. 2001); resting and submaximal exercise blood pressure, heart rate, stroke volume, and cardiac output (An, Rice, Gagnon et al. 2000; An, Rice, Perusse et al. 2000; Rice, An et al. 2002; An et al. 2003); body composition and body fat distribution (Rice et al. 1999; Perusse et al. 2000); and plasma lipid, lipoprotein, and apolipoprotein levels (Rice, Despres et al. 2002). The maximal heritabilities for these traits ranged from 25% to 55%, further confirming the contribution of familial factors to the person-to-person variation in responsiveness to endurance training.

Molecular Genetics

The evidence from the genetic epidemiology studies suggests that there is a genetically determined component affecting exercise-related phenotypes. However, since these traits are complex and multifactorial in nature, the search for genes and mutations responsible for the genetic regulation must not only target several families of phenotypes, but also consider the phenotypes in the sedentary state and in response to exercise training. It is also obvious that the research on molecular genetics of exercise-related phenotypes is still in its infancy. We have begun the publication in *Medicine and Science in Sports and Exercise* of an annual review on the Human Gene Map for Performance and Health-Related Fitness phenotypes (Rankinen, Perusse et al. 2001). The 2002 update of the map included 90 gene entries and quantitative trait loci (QTL) on the autosomes and 2 on the X chromosome (Perusse et al. 2003) for physical performance (cardiorespiratory endurance, elite endurance athlete status, muscle strength, other muscle performance traits, exercise intolerance) and health-related fitness (hemodynamic traits, anthropometry, and body composition; insulin and glucose metabolism; blood lipids and lipoproteins and hemostatic factors) phenotypes. As a comparison, the latest version of a similar map for obesity-related phenotypes included more than 300 loci (Chagnon et al. 2003). These numbers demonstrate that relatively little has been accomplished to date. For instance, no gene contributing to human variation in endurance performance has even been identified as a result of studies based on model organisms. Now that we have entered the era in which large fractions of the human, mouse, and rat genomic sequences have became available, the field of exercise science and sports medicine will need to devote more attention to molecular and genetic research.

Candidate Gene Studies

The majority of the exercise-related molecular genetic studies published so far have utilized a candidate gene approach; that is, a gene has been targeted based on its potential physiological and metabolic relevance to the trait of interest. Statistical tests for an association are based on the comparison of allele and genotype frequencies of genetic markers between two groups of subjects, one having the phenotype of interest (e.g., high $\dot{V}O_2$max or endurance athletes, that is, the "cases"), the other one not (the "controls"). However, with continuous traits, the test is done via comparison of mean phenotypic values across genotype groups, or between carriers and noncarriers of a specific allele.

The 2002 update of the Human Gene Map for Performance and Health-Related Fitness (Perusse et al. 2003) included 16 autosomal genes from 29 studies with evidence of associations with cardiorespiratory endurance or muscular strength phenotypes (table 3.2). Only 2 of these genes (5 reports) were investigated in athletes, whereas 12 genes (19 reports) were tested in healthy nonathletes. Five genes were investigated on patients with chronic diseases. Most of these autosomal gene entries (14/16) were based on only one study. A clear favorite gene among exercise scientists has been the angiotensin-converting enzyme (ACE) gene, which was investigated in 13 reports.

A 287-bp insertion (I)/deletion (D) polymorphism in intron 16 of the ACE gene was first reported to be strongly associated with plasma ACE activity in 1990 (Rigat et al. 1990). The ACE activity was lowest in subjects with two copies of the I allele (I/I homozygotes), highest in the D/D homozygotes, and intermediate in the I/D heterozygotes. In subsequent studies, the D allele was reported to be associated with increased risks of coronary heart disease and other chronic disorders, although the evidence to date is still inconsistent.

In exercise-related studies, the I allele has been reported to be more frequent in Australian elite rowers and in Spanish endurance athletes than in sedentary controls (Gayagay et al. 1998; Alvarez et al. 2000). Likewise, British long distance runners tended to have a greater frequency of the I allele than sprinters (Myerson et al. 1999). In a group of postmenopausal women, who were selected on the basis of their physical activity levels (sedentary, recreationally active, endurance athletes), the I/I genotype was associated with greater $\dot{V}O_2$max and maximal arterial–venous oxygen difference as compared to the D/D homozygotes (Hagberg et al. 1998). The I allele was also associated with greater increase in muscular endurance and efficiency after 10 weeks of physical training in British Army recruits (Montgomery et al. 1998; Williams et al. 2000; Woods et al. 2002).

In contrast, the frequency of the D allele was found to be higher in elite swimmers than in sedentary controls (Woods et al. 2001). Moreover, in almost 300 sedentary but healthy white offspring from the HERITAGE Family Study, the D/D homozygotes showed the greatest improvements in

Table 3.2 Summary of Candidate Genes Reported to Be Associated With Cardiorespiratory Endurance and Muscular Strength Phenotypes in Humans

Gene name	Abbreviation	Location
Angiotensin I converting enzyme	*ACE*	17q23
Alpha-2A-adrenergic receptor	*ADRA2A*	10q24-q26
Beta-1-adrenergic receptor	*ADRB1*	10q24-q26
Beta-2-adrenergic receptor	*ADRB2*	5q31-q32
Apolipoprotein E	*APOE*	19q13.2
ATPase, Na^+/K^+ transporting, alpha-2 polypeptide	*ATP1A2*	1q21-q23
Cystic fibrosis transmembrane conductance regulator, ATP-binding cassette (subfamily C, member 7)	*CFTR*	7q31.2
Creatine kinase, muscle	*CKM*	19q13.2-q13.3
Ciliary neurotrophic factor	*CNTF*	11q12.2
Collagen, type I, alpha 1	*COL1A1*	17q21.3-q22.1
Growth differentiation factor 8 (myostatin)	*GDF8 (MSTN)*	2q32.2
Major histocompatibility complex, class I, A	*HLA-A*	6p21.3
Haptoglobin	*HP*	16q22.1
Insulin-like growth factor-2	*IGF2*	11p15.5
Uncoupling protein 2	*UCP2*	11q13
Vitamin D (1,25-dihydroxyvitamin D3) receptor	*VDR*	12q12-q14

For further details see Perusse et al. (2003).

$\dot{V}O_2$max and maximal power output following a controlled and supervised 20-week endurance training program (Rankinen, Perusse et al. 2000). Furthermore, some studies suggest that the D allele is associated with greater muscular strength gains in response to resistance training (Folland et al. 2000). It should also be noted that several reports show no associations between the ACE genotype and performance phenotypes (Taylor et al. 1999; Rankinen, Wolfarth et al. 2000; Sonna et al. 2001).

The observations on the ACE gene need to be put in perspective. It is clear that the statistical evidence for an association or against an association is not very strong in the published reports to date. As noted by Cardon and Palmer (2003), the most likely explanation for seemingly inconsistent results from association studies in different populations is typically the overinterpretation of marginal statistical evidence. Thus, the available data are still too sparse and fragmented to allow full evaluation of what role the ACE gene plays in the variation of human physical performance level.

The genetic data hold great promise to help us understand why some individuals respond favorably to exercise training in terms of reduction of chronic disease risk factor levels whereas others do not. The 2002 update of the Human Performance and Health-Related Fitness Gene Map included 16 genes from 22 studies that had been investigated in relation to exercise training-induced changes in hemodynamic (8 genes, 11 studies), body composition (8 genes, 8 studies), plasma lipid and lipoprotein (2 genes, 2 studies), and hemostatic (3 genes, 3 studies) phenotypes (table 3.3). The genes associated with body composition, plasma lipid, and hemostatic phenotype training responses were all based on a single study. However, with hemodynamic phenotypes, some candidate gene findings have been replicated in at least two studies. For example, an association between blood pressure training response and the angiotensinogen (AGT) M235T polymorphism has been reported in both the HERITAGE Family Study and the DNASCO study (Rankinen, Gagnon et al. 2000; Rauramaa et al. 2002). In white HERITAGE males, the AGT M235M homozygotes showed the greatest reduction in submaximal exercise DBP following a 20-week endurance training program (Rankinen, Gagnon et al. 2000), whereas in middle-aged eastern Finnish men, the M235M homozygotes had the most favorable changes in resting SBP and DBP during a six-year exercise intervention trial (Rauramaa et al. 2002).

Similarly, an association between the ACE I/D polymorphism and training-induced left ventricular (LV) growth has been reported in two studies

Table 3.3 Summary of Candidate Genes Reported to Be Associated With Health-Related Fitness Training Response Phenotypes in Humans

Gene name	Abbreviation	Location
HEMODYNAMICS		
Angiotensinogen	AGT	1q42-q43
Nitric oxide synthase 3 (endothelial cell)	*NOS3*	7q36
Lipoprotein lipase	LPL	8p22
Guanine nucleotide binding protein (G-protein), beta polypeptide 3	GNB3	12p13
Bradykinin receptor B2	BDKRB3	14q32.1-q32.2
Angiotensin I converting enzyme	*ACE*	17q23
Apolipoprotein E	*APOE*	19q13.2
Peroxisome proliferative activated receptor, alpha	PPARA	22q13.31
BODY COMPOSITION		
Peroxisome proliferative activated receptor, gamma	PPARG	3p25
Lipoprotein lipase	LPL	8p22
Beta-3-adrenergic receptor	ADRB3	8p12-p11.2
Uncoupling protein 3	UCP3	11q13
Guanine nucleotide binding protein (G-protein), beta polypeptide 3	GNB3	12p13
Vitamin D (1,25-dihydroxyvitamin D3) receptor	*VDR*	12q12-q14
Insulin-like growth factor-1	IGF1	12q22-q23
Angiotensin I converting enzyme	*ACE*	17q23
LIPIDS AND LIPOPROTEINS		
Lipoprotein lipase	LPL	8p22
Apolipoprotein E	*APOE*	19q13.2
HEMOSTATIC FACTORS		
Fibrinogen, A alpha polypeptide	FGA	4q28
Fibrinogen, B beta polypeptide	FGB	4q28
Plasminogen activator inhibitor 1	PAI1	7q21.3-q22

For further details see Perusse et al. (2003).

(figure 3.3) (Montgomery et al. 1997; Myerson et al. 2001). In 1997, Montgomery and coworkers reported that the ACE D allele was associated with greater increases in LV mass and in septal and posterior wall thickness after 10 weeks of physical training in British Army recruits (figure 3.3*a*) (Montgomery et al. 1997). In 2001, the same group reported a new study using a similar training paradigm in British Army recruits (Myerson et al. 2001). The cohort included 62 ACE I/I and 79 ACE D/D homozygotes, and the training-induced increase in LV mass was 2.7 times greater in the D/D genotype as compared to the I/I homozygotes (figure 3.3*b*). Interestingly, the association between the ACE genotype and LV mass response was not affected by angiotensin II type 1 receptor inhibitor (losartan) treatment (Myerson et al. 2001).

Linkage Studies

An alternative strategy to identify genes affecting performance-related phenotypes relies on linkage analysis. The basic idea of genetic linkage is to test if a genetic locus is transmitted from one generation to the next together with a trait (or another genetic locus) of interest. The process is fairly straightforward when the trait is influenced by only one (major) gene. In these cases, one can deduce the underlying genetic architecture by observing the transmission of the trait in affected families, which allows one to generate powerful models for linkage testing with genetic markers. However, multifactorial and oligogenic traits such as performance phenotypes rarely follow a specific inheritance model. In this case, it is not possible to use traditional parametric or model-based linkage methods. Instead, the linkage test for oligogenic traits is based on the idea that a pair of relatives (usually siblings) who are genetically similar should also be alike in terms of phenotypic values. The genetic similarity of the pair is determined by estimating how many common alleles at a given locus the individuals have inherited from the same ancestors (allele sharing identical by descent [IBD]).

The statistical testing of linkage is usually done using either regression-based methods, as originally outlined by Haseman and Elston in 1972, or variance components modeling, which has gained popularity since the early 1990s. Briefly, in the original Haseman-Elston regression method the phenotypic resemblance of a sibling pair, modeled as a squared sib-pair trait difference, is regressed with the genotypic resemblance (alleles shared IBD). Recently, Elston and coworkers (2000) published a revised version of the method, in which the phenotypic resemblance of the sibs is modeled as the mean-corrected cross-product of the sibs' trait values. In the variance component linkage methods, the total phenotypic variance is decomposed into additive effects of a trait locus, a residual familial background, and a residual nonfamilial component; and the phenotypic covariance

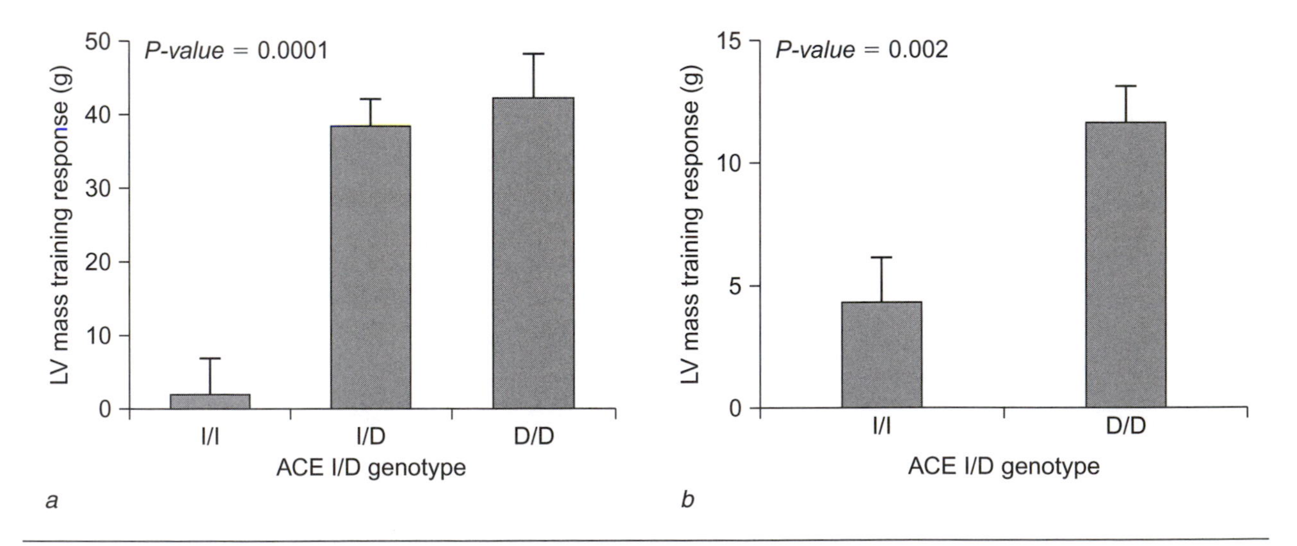

Figure 3.3 Associations between the angiotensin-converting enzyme (ACE) I/D polymorphism and exercise training-induced increases in left ventricular mass. *(a)* Data from 140 healthy army recruits who participated in a 10-week basic training program; *(b)* summary of data from a replication study (Myerson et al. 2001) using a similar training program with 141 healthy recruits.

Adapted from H.E. Montgomery et al., 1997, "Association of angiotensin-converting enzyme gene I/D polymorphism with change in left ventricular mass in response to physical training," *Circulation* 96: 741-747.

of the relative pair is modeled as a function of allele sharing IBD. The linkage testing is performed using likelihood ratio test contrasting a null hypothesis model of no linkage with an alternative hypothesis model in which the variance due to the trait locus is estimated. For more details regarding linkage methods, see Rao and Province 2001.

The major conceptual difference between linkage and association is that association targets a specific allele or a genotype at a given gene locus whereas linkage refers to a chromosomal region rather than a specific gene or mutation. Thus, the linkage analysis can be used to identify chromosomal regions that harbor gene(s) affecting the phenotype, even if there is no a priori knowledge of the existence of such genes. By definition, the linkage analysis always requires family data. Therefore, data collection is more challenging than in case-control and cohort studies with unrelated subjects. The usefulness of linkage analysis to study elite athletes is predictably quite limited. However, in studies on interindividual differences in responsiveness to acute exercise and exercise training, the family design can provide exploratory opportunities that are not available in association studies with independent subjects.

The HERITAGE Family Study is the first and thus far the only study employing a family design in exercise-related questions (Bouchard et al. 1995). The primary objective of HERITAGE is to investigate the factors contributing to the interindividual differences in responsiveness to endurance training. To date, six genome-wide linkage scans dealing with 16 endurance training response and 7 acute exercise response phenotypes have been published (Bouchard et al. 2000; Chagnon et al. 2001; Rankinen, An et al. 2001; Rankinen et al. 2002; Rice, Chagnon et al. 2002; Rice, Rankinen et al. 2002). The promising linkages, defined as a logarithm of odds (LOD) score equal to or greater than 1.75 or a p-value less than 0.0023, from these studies are summarized in table 3.4.

An example of a QTL for endurance training-induced changes in submaximal exercise stroke volume on chromosome 10 is given in figure 3.4. The QTL covers approximately 15 million base pairs in the short arm of chromosome 10 (10p11.2), and the maximum linkage (LOD = 1.96) was detected with a microsatellite marker, D10S1666 (Rankinen et al. 2002). Promising linkages were also found on chromosome 8q24.3 for submaximal exercise SBP training response and on 10q23-q24 for SBP measured during submaximal exercise at 80% of $\dot{V}O_2max$ (Rankinen, An et al. 2001). In addition, promising evidence of linkage was reported for training-induced changes in fat mass and percent body fat on chromosomes 1q31.1, 11q13.4-q21, and 18q21-q23 and for BMI and fat-free mass training responses on 5q21.1 and 12q23.2, respectively (Chagnon et al. 2001).

Table 3.4 Summary of Promising Linkages (LOD ≥ 1.75) for Acute Exercise and Endurance Training Response Phenotypes in the HERITAGE Family Study

Trait	Location	LOD score
HEMODYNAMICS		
ΔSBP at 50 W	8q24.3	2.36
ΔSV at 50 W	10p11.2	1.96
SBP at 80% $\dot{V}O_2max$	10q23-q24	1.84
BODY COMPOSITION		
Δ%fat, ΔFM	1q31.1	2.2
ΔBMI	5q21.1	2.4
Δ%fat	11q13.4-q21	2.2
ΔFFM	12q23.2	2.3
ΔFM, Δ%fat	18q21-q23	1.9

The first genome-wide linkage scan for $\dot{V}O_2max$ did not produce any promising QTLs; but several suggestive linkage signals were found, such as for the sedentary state $\dot{V}O_2max$ on chromosome 4q12 and for $\dot{V}O_2max$ training response on 4q26 and 6p21.33 (Bouchard et al. 2000). The lack of particularly strong QTLs most likely reflects the polygenic nature of these traits. The maximal oxygen consumption capacity is influenced by several intermediate phenotypes (cardiac output, oxygen transport capacity, oxidative capacity of the working muscles, etc.), and multiple genes contribute to the variation in each of these subphenotypes. Consequently, the detection of these gene effects requires denser microsatellite marker sets for linkage analyses, or a dense panel of single nucleotide polymorphisms for genome-wide association scans.

The identification of a QTL is only the first step in the gene discovery process. Since linkage analysis provides information about a genomic region, a typical QTL may span over several millions of base pairs. Such a vast region may contain dozens

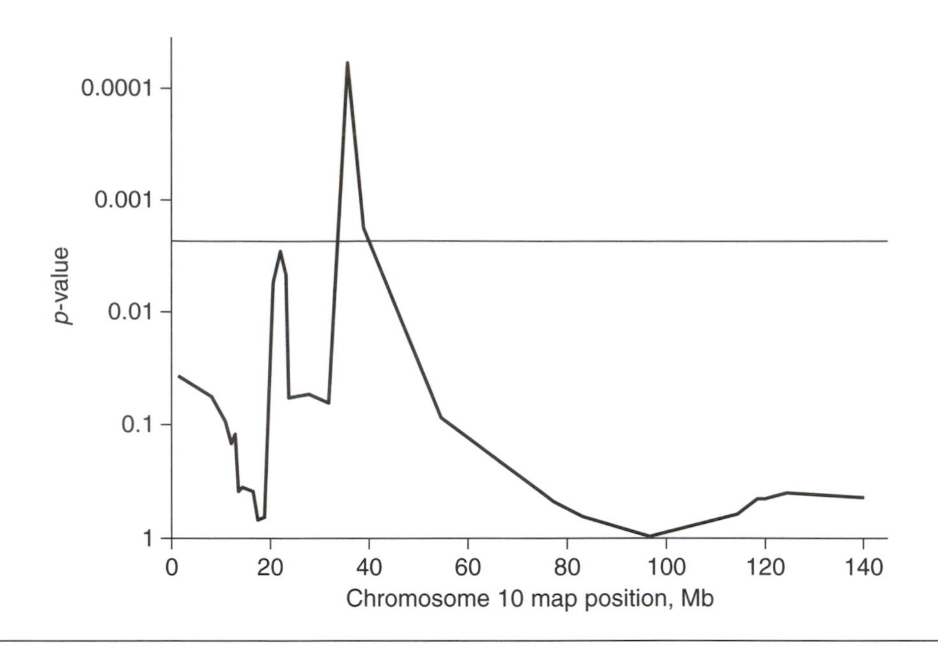

Figure 3.4 A quantitative trait locus for submaximal exercise (50 W) stroke volume training response on chromosome 10 in the HERITAGE Family Study. *P*-values from the regression-based multipoint linkage analysis are presented on the y-axis, and chromosomal location derived from the Location Database on the x-axis. The reference line indicates the criteria for promising linkage ($p < 0.0023$). For further details see Rankinen et al. 2002.

or even hundreds of genes. The procedure commonly used to identify the causal gene(s) within a QTL is called positional cloning, and the principal steps of the positional cloning strategy are summarized in figure 3.5. Briefly, the initial detection of the QTL is followed by the addition of more microsatellite markers within the QTL region with a view to narrow down the region as much as possible using linkage analysis. Once the resolution limit of the linkage approach is reached, the next step involves fine mapping using association analyses with single nucleotide polymorphisms (SNPs). The rationale of fine mapping is that the greater number of SNPs and the greater sensitivity of the association tests enable one to derive more detailed information about the region. Even if the SNPs used for fine mapping do not include the specific DNA sequence variant(s) affecting the trait, they likely provide useful leads regarding the underlying causal mutation(s). The closely located yet functionally neutral SNPs may cosegregate with the trait-influencing DNA variant and thereby be part of the haplotype that contains also the causal allele. This phenomenon is called allelic association or linkage disequilibrium, and it forms the theoretical basis for association mapping.

Once the results from the association analyses are deemed strong enough, the next step is to screen by sequencing the candidate genes for DNA sequence variation. The confirmation of the relevance of the detected mutations includes additional association studies in the original and other populations, as well as functional assays in vitro (expression studies in different cell lines) and in vivo (transgenic and knockout animal models). The positional cloning approach has been used successfully to identify genes causing several diseases, such as long QT syndrome 1 (Wang et al. 1996) and autosomal dominant familial polymorphic ventricular tachycardia (Swan et al. 1999; Laitinen et al. 2001). Both traits are characterized by increased incidence of cardiac events during exercise. The positional cloning efforts of the training response QTLs in the HERITAGE Family Study should yield a panel of new candidate genes that play a role in the adaptation to regular exercise.

Exercise Intolerance

Although exercise-related traits are mainly polygenic and multifactorial in nature, much can be learned from some monogenic disorders characterized by compromised exercise capacity or exercise intolerance. These disorders affect only a few individuals, but they provide interesting examples of genetic defects that have profound effects on the ability to perform physical activity, usually due to compromised energy metabolism. Although the

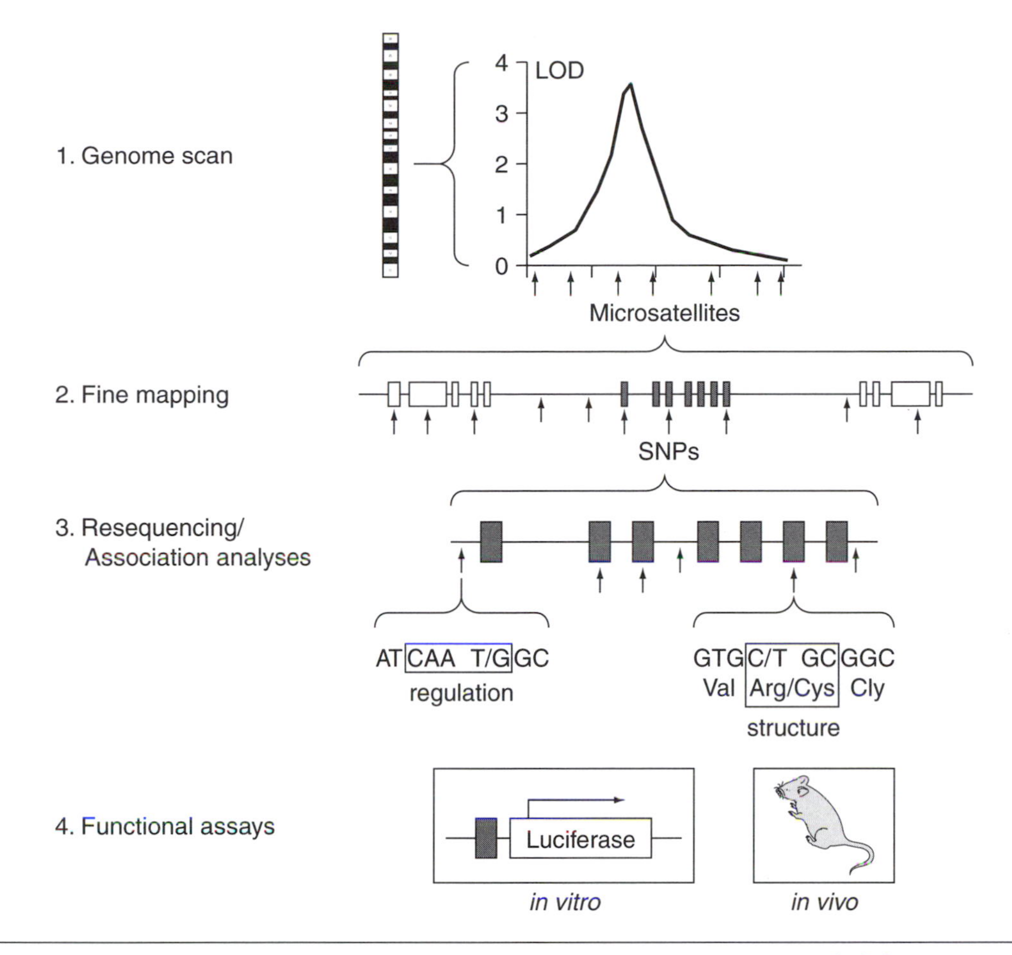

Figure 3.5 Schematic presentation of the positional cloning strategy. The initial genome-wide linkage scan is followed by fine mapping with additional microsatellite and single nucleotide polymorphism (SNPs) markers. Genes showing significant associations with the SNPs are then resequenced to identify all relevant DNA sequence variants in the study population. The relevance of the uncovered mutations is verified with additional association studies in other populations (replications) and using in vitro and in vivo functional studies.

Reprinted from E. Boerwinkle, J.E. Hixson, and C.L. Hanis, 2000, "Peeking under the peaks: Following-up genome-wide linkage analyses," *Circulation* 102: 1877-1878.

genetic defects described in this section compromise exercise capacity, it should be noted that there is no evidence that overexpression of these genes automatically leads to improved physical performance. However, it is important to understand the molecular mechanisms contributing to both ends of the distribution of physical performance and therefore there is justification for paying attention also to conditions characterized by exercise intolerance.

Muscle glycogen phosphorylase deficiency (McArdle disease) was the first documented carbohydrate metabolism disorder associated with exercise intolerance (McArdle 1951). The condition is characterized by partial or total absence of glycogen phosphorylase in skeletal muscles, which results in impaired glycogenolysis. Identification of three mutations in the muscle glycogen phosphorylase (PYGM) gene in 1993 established the molecular genetic basis for McArdle disease. These mutations were found in almost 90% of the investigated patients (Tsujino, Shanske & DiMauro 1993). Additional mutations have been reported in subsequent studies, and currently the Online Mendelian Inheritance in Man (OMIM, entry #232600) database lists 15 PYGM mutations related to McArdle disease. Phosphorylase kinase is a multimeric protein composed of four subunits that regulates the activity of glycogen phosphorylase. The gene encoding the muscle-specific isoform of the α subunit of phosphorylase kinase (PHKA1) is located on chromosome Xq12-q13, and two mutations have been identified in patients with phosphorylase kinase deficiency: a nonsense mutation

(G3334T) changing a glutamic acid residue to a stop codon (Wehner et al. 1994) and a guanine to cytosine splice-junction substitution at the 5' end of an intron, which causes skipping of the preceding exon (Bruno et al. 1998).

Muscle phosphofructokinase deficiency is a hereditary disorder of glycolysis featuring exercise intolerance as a clinical symptom. Phosphofructokinase catalyzes the conversion of fructose 6-phosphate to fructose 1,6-biphosphate and is a rate-limiting enzyme in the glycolytic energy production pathway. Several mutations in the gene encoding phosphofructokinase (PFKM) have been reported in patients with muscle phosphofructokinase deficiency (Sherman et al. 1994; Tsujino, Servidei et al. 1994; Vorgerd et al. 1996). Phosphoglycerate kinase 1 and phosphoglycerate mutase are also involved in glycolysis by catalyzing conversions of 1,3-diphosphoglycerate to 3-phosphoglycerate and 3-phosphoglycerate to 2-phosphoglycerate, respectively. Deficiencies of both enzymes have been shown to cause myopathy and exercise intolerance. At least four mutations in the muscle phosphoglycerate mutase (PGAM2) gene have been identified in the phosphoglycerate mutase deficiency patients, one causing a premature stop codon and three changing amino acids in the final gene product (Tsujino, Shanske, Sakoda et al. 1993; Toscano et al. 1996; Hadjigeorgiou et al. 1999). Several mutations have been identified in the phosphoglycerate kinase (PGK1) gene. However, only four of these have been found in the myopathic form of phosphoglycerate kinase deficiency (Tsujino, Tonin et al. 1994; Ookawara et al. 1996; Sugie et al. 1998; Hamano et al. 2000).

Comi and coworkers have reported two heterozygous missense mutations in the beta enolase (ENO3) gene in a patient with severe muscle enolase deficiency, exercise intolerance, and myalgias (Comi et al. 2001). Enolase is a glycolytic enzyme catalyzing the interconversion of 2-phosphoglycerate and phosphoenolpyruvate. Also muscle lactate dehydrogenase deficiency is characterized by exercise intolerance. Three mutations in the lactate dehydrogenase A (LDHA) gene have been identified, and they all induce a premature stop codon and result in a truncated gene product (Maekawa et al. 1990; Tsujino, Shanske et al. 1994).

Carnitine palmitoyltransferase II (CPT II) deficiency is the most common recessively inherited lipid metabolism disorder affecting skeletal muscle. Patients with CPT II deficiency have recurrent episodes of myoglobinuria triggered by exercise, fasting, or infection. Some patients also manifest muscle stiffness and soreness following exercise (Bonnefont et al. 1999). The most prevalent mutation in the carnitine palmitoyltransferase II gene in CPT II deficiency patients is the Ser113Leu missense mutation, which is found in about 50% of mutant alleles (Bonnefont et al. 1999). Another lipid metabolism-related gene pertaining to exercise intolerance is very long chain acyl-CoA dehydrogenase (ACADVL). A patient with a long history of exercise intolerance, myoglobinuria, low fasting ketogenesis, impaired palmitoylcarnitine oxidation by muscle mitochrondia, and lack of palmitoyl-CoA dehydrogenase enzyme in muscle and fibroblasts was found to be homozygous for a 3-bp deletion in exon 9 of the ACADVL gene. The mutation deleted lysine residue in position 238 of the mature gene product (Scholte et al. 1999).

Conclusion

The past decade has witnessed remarkable progress in human genetics. The availability of the almost complete DNA sequence of the human genome has changed our ability to study the genetic basis of complex multifactorial traits and to develop novel treatments for several chronic diseases. The recent advances in molecular genetics are starting to have an impact on exercise physiology. Although the research on molecular genetics of physical performance and health-related fitness is still in its infancy, we recognize that understanding the effects of DNA sequence variation on interindividual differences in responsiveness to acute exercise and regular exercise training holds great promise. Such data would help to improve both athletic training and the use of physical activity in the prevention and treatment of chronic diseases. Furthermore, the availability of powerful methods such as microarray technology to measure gene expression, targeting specific genes in knockout and transgenic animal models, will greatly add to our research capability in the investigation of basic and applied exercise physiology issues.

Chapter 4

Proteins and Exercise

Sataro Goto and Zsolt Radák

Life of an organism is dependent on proteins with a variety of functions: enzymatic, structural, and regulatory. Still other proteins have roles in oxygen transport (hemoglobin) and storage (myoglobin), provide protection against invading bacteria and viruses (antibodies), protect against denaturation of proteins and renaturation of denatured proteins (heat shock proteins), and so on.

Proteins are synthesized according to information in messenger RNA (mRNA) that is synthesized on DNA template by transcription and processed. The most recent estimation of the number of structural genes in humans is 25,000 to 35,000 based on sequencing of the whole genome (International Human Genome Sequencing Consortium 2001; Venter et al. 2001), a value much less than previous and expected estimates. The actual number of protein species should, however, be much greater than the number of genes, due to alternative splicing of mRNA precursors to generate multiple mature mRNA and post-translational modifications of proteins. Protein synthesis is a finely controlled process for meeting cellular requirements in different types of cells and during development and differentiation, as well as in response to internal and external stimuli.

Proteins once synthesized are subjected to alterations by internal factors such as unavoidable oxidation due to oxidative energy generation and other processes requiring oxygen, thermal perturbation at body temperature, and external stresses such as irradiation and smoking. Thus, there are ample opportunities for them to lose their biological activities or possibly even gain harmful functions, and cellular proteins therefore need to be continuously replaced to maintain cellular integrity. In addition, proteins required during only a limited time in the cell cycle or under specific situations such as hormonal stimulation, nutritional influences, and exercise may be degraded when they are no longer needed.

In this chapter we describe the basic molecular mechanisms of protein turnover (synthesis and degradation) and exercise-associated changes in protein metabolism with special reference to the skeletal muscles.

Protein Synthesis

Proteins are synthesized in accordance with codon sequences on mRNAs that are transcribed copies of genetic information encoded in DNA as described in detail in the previous chapter.

Immediate transcription products are mRNA precursors that contain sequences for both exons (nucleotide sequences that are retained in mature mRNA) and introns (nucleotide sequences that are removed from the precursors) of the DNA (figure 4.1). The mRNA precursors undergo processing (post-transcriptional modifications) by cap formation and poly (A) addition followed by splicing to give rise to functional mature mRNA that is transported out into the cytoplasm for translation (figure 4.1). Messenger RNA is eventually degraded in the cytoplasm and re-synthesized in the nucleus. Protein is also continuously replaced, mostly in the cytoplasm. Such dynamic states of the macromolecules appear to constitute a most fundamental feature of an organism, ensuring better maintenance of life against ever changing internal and external conditions.

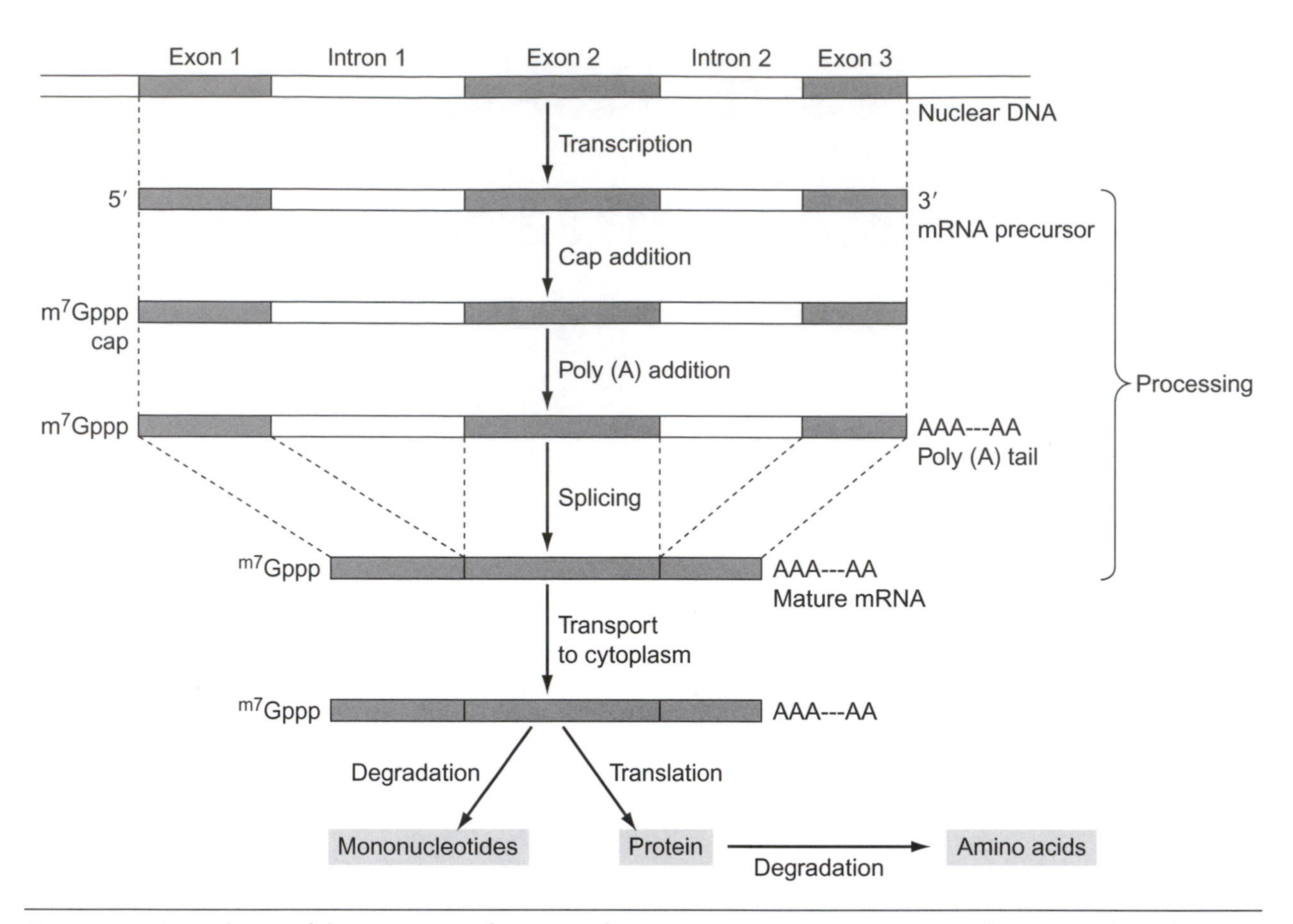

Figure 4.1 General view of the expression of genetic information: transcription, processing of mRNA, and translation.

Processing of mRNA Precursors: Capping, Adding a Tail, and Splicing

Messenger RNA precursor is "capped" by modification of the phosphate of the 5' terminal nucleotide with a 7-methyl guanosine at an early stage of transcription. The capping reaction consists of condensation of the triphosphate group of guanosine triphosphate (GTP) with a diphosphate at the 5' end of the initial transcript, resulting in the formation of unusual 5'–5' linkage of the nucleotides followed by methylation at position 7 of the guanine (m^7Gppp). The cap is important for the initiation of translation. At near-completion of transcription of the RNA, polyadenylate (poly (A) of about 200-nucleotide length) is added to the 3' end of the transcript by catalysis of poly (A) polymerase. This is achieved after removal of some length of the 3' terminal portion of the primary transcript. The poly (A) or poly (A) "tail" appears to have a role in stabilizing mRNA from degradation. Some eukaryotic mRNAs such as those for histones and prokaryotic mRNA are devoid of poly (A). These post-transcriptional modifications are followed by splicing of the "capped" and "tailed" (polyadenylated) mRNA precursors.

In eukaryotic cells, capped and tailed mRNA precursors are mostly spliced to form functional mature mRNA. The RNA splicing is a process in which transcribed intron sequences are excised from the capped and polyadenylated transcript and the remaining transcribed exon segments are rejoined to give rise to mature mRNA. The process is catalyzed by a protein complex called spliceosome. The consensus sequences at the 5' (donor) end and the 3' (acceptor) end of introns are recognized by the spliceosome consisting of proteins and small specific RNA molecules (U1, U2, U5, etc.) in addition to the mRNA precursor to be spliced. In some cases splicing is performed in different combinations of transcribed exon sequences (alternative splicing) to generate multiple mature mRNAs for synthesis of related species of proteins from a single mRNA precursor. For example, the primary transcript from a single smooth muscle myosin heavy chain gene generates one or more of four isoforms by alternative splicing in different muscle types and during muscle development. Troponin T and α-tropomyosin are other examples of muscle proteins for which mRNA precursors are alternatively spliced to generate multiple protein products that differ from one cell type to another (figure 4.2). The gene expression of the sarcoplasmic reticulum Ca^{2+}-ATPase is also regulated at alternative splicing. Thus, alternative splicing is a general means to generate multiple functionally similar proteins from a single gene.

Recent draft sequencing of the human genome suggests that 60% of the genes may have multiple splicing variants (Wolfsberg et al. 2001). In addition to alternative splicing, mRNA species may be diversified by alternative transcription that starts from different promoters of a single gene (see the previous chapter for details). Thus, the overall "complexity" of mRNA, and therefore that of protein, is manyfold higher than the number of genes in mammalian cells.

Messenger RNA Transport From the Nucleus to the Cytoplasm

Mature mRNA must be transported to the cytoplasm to be translated into protein. The mature mRNA in the form of ribonucleoprotein complex is exported out through the nuclear pore complexes only after the splicing process is completed and the spliceosome components are dissociated. A nuclear pore complex consisting of nearly 100 different proteins serves as a gate for export (RNAs and ribonucleoproteins) and import (nuclear proteins) of macromolecules. The transport of mRNA protein complex is an active process requiring specific recognition by proteins in the nuclear pore structure.

Translation

Translation is a process of converting information on an mRNA, present as a sequence of triplet codons consisting of trinucleotides, into the amino acid sequence of a protein. This process

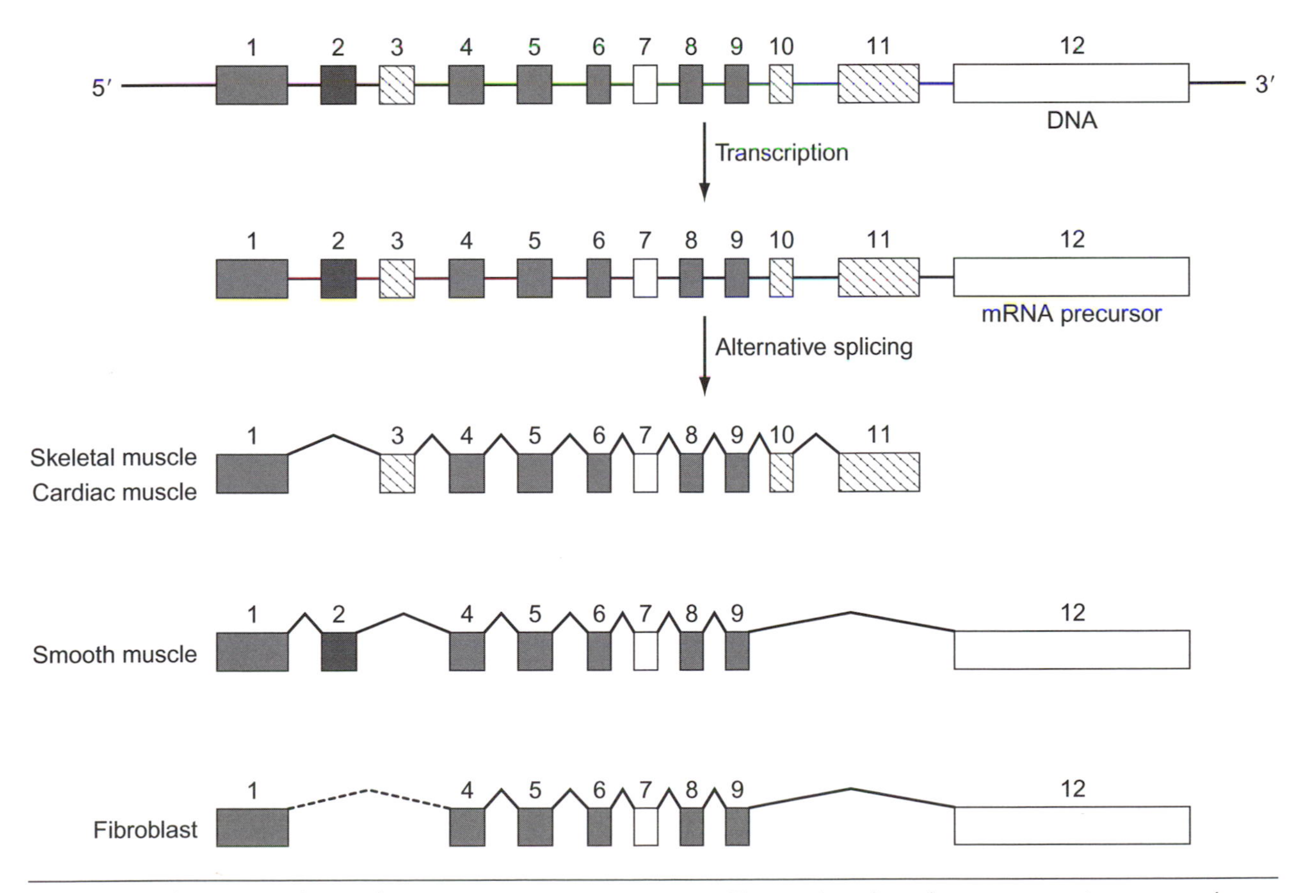

Figure 4.2 Alternative splicing of α-tropomyosin mRNA precursor. The number above boxes represents exon number.

is performed on the ribosome with the aid of specific adapter transfer RNA (tRNA) that makes a specific connection between a codon and the cognate amino acid. A peptide bond is formed by dehydration between a carboxyl group of one amino acid and an amino group of another amino acid, thereby extending the chain of peptide from the amino terminus to the carboxyl terminus of a polypeptide or a protein. The entire translation process consists of two steps: synthesis of aminoacyl tRNA and decoding of codons on mRNA.

Synthesis of Aminoacyl Transfer RNA

There is no characteristic chemical structure in a specific codon and the cognate (corresponding) amino acid by which they can recognize each other. The role of tRNA is to mediate the recognition between a codon and an amino acid. A codon makes hydrogen bonds with a complementary anti-codon trinucleotide in the cognate tRNA that is bound to the amino acid. Amino acid is first activated with energy generated by hydrolysis of ATP to form the aminoacyl AMP and pyrophosphate. The activated amino acid (aminoacyl AMP) then reacts with the cognate tRNA to form an ester bond between the 2' or 3' hydroxyl group of ribose of the 3' terminus of the tRNA and the carboxyl group at the alpha carbon of the amino acid. The two successive reactions are catalyzed by an aminoacyl tRNA synthetase or synthase. Each combination of an amino acid and a tRNA is catalyzed by different enzymes; for example, leucine is bound to $tRNA^{leu}$ catalyzed by the specific leucyl tRNA synthetase. The specificity of each enzyme is extremely high, making a highly specific combination between an amino acid and the tRNA; otherwise incorrect amino acids would be incorporated into protein (translational error), resulting in generation of less functional or nonfunctional molecules. Nevertheless, it is possible for structurally similar amino acids such as leucine and isoleucine to be misrecognized by a noncognate enzyme that might catalyze an incorrect combination of an amino acid and a tRNA. The enzymes are therefore equipped with a proofreading mechanism to correct the error by hydrolyzing an ester bond in incorrect combination, thereby reducing translational error frequency or ensuring translational fidelity.

Recognition of Codons by Transfer RNA

Codons on mRNA are recognized by tRNAs through complementary base-pairing with anticodons. There are 61 sense codons for 20 species of amino acids that may base-pair with a possible 61 anti-codons, and therefore 61 different tRNAs. Actually, however, one tRNA can recognize two or more degenerated codons because of a wobble base pair that is less stringent but specific enough for correct decoding. For example, inosine (I, a modified base in tRNA) in the first letter of an anti-codon can make a hydrogen bond with either cytosine (C), uracil (U), or adenine (A) of the third letter of a codon. As an example, out of four codons (GCC, GCU, GCA, and GCG) for alanine, three codons other than GCG can base-pair with one anti-codon CGI. Only two species of tRNA are therefore required for four different codons for this amino acid. Thus, in theory, only 46 species of tRNA with each different anti-codon are required for 61 sense codons.

Ribosome and Messenger RNA

Decoding of codons in mRNA is carried out on the ribosome. The 80S (Svedberg unit S) eukaryotic ribosome consists of 60S and 40S subunits. The two subunits are composed of 28S/5.8S/5S rRNAs and 18S rRNA, respectively, together with more than 30 different protein components each.

Messenger RNA consists of noncoding regions at the 5' and 3' termimal portions and a coding region in between. The noncoding and coding regions in mRNA are transcribed exon sequences. The 5' terminal noncoding region contains a specific nucleotide sequence that binds 18S rRNA in the 40S ribosome subunit for initiation of translation (see next section). The coding region starts with the initiation codon (AUG) for methionine and ends in the last sense codon just before one of three termination codons (UUA, UAG, UGA) that are also called nonsense codons.

Decoding of Codon

Decoding of information on mRNA starts at the initiation codon (AUG) and proceeds from the 5' side to the 3' side of mRNA. The translational process consists of three steps: initiation, elongation, and termination.

- Initiation. Prior to decoding of information on mRNA, the 40S ribosome subunit binds to methionyl $tRNA_i$ (initiator $tRNA_i$, which is different from other $tRNA^{Met}$ used for the methionine codon other than initiation) together with several initiation factors (eIF; e = eukaryote, IF = initiation factor). This 40S initiation complex then binds mRNA at the initiation codon AUG via specific codon–anticodon interaction. The 60S ribosome subunit then binds to the 40S subunit associated with mRNA together with other initiation factors. Formation of the initial complex is dependent on GTP that is hydrolyzed upon binding of the 60S ribosome subunit. The result is formation of the 80S initiation complex.

• Elongation. In the second step of decoding, a peptide bond is formed between the initiator methionine and an amino acid corresponding to the second codon. While the initiator methionyl tRNA is bound to the P site (peptidyl tRNA binding site) on the 60S subunit, tRNA carrying the second amino acid binds to the A site (aminoacyl tRNA binding site) next to the P site via codon–anticodon base-pairing. This step requires a protein factor called elongation factor 1 (EF1). Then, a peptide bond is formed between methionine and the second amino acid by catalysis of peptidyl transferase located on the 60S subunit using GTP, presumably as an allosteric effector for conformational change of the ribosome. The peptide bond formation is catalyzed by 28S rRNA, but not by protein in the 60S subunit. The resulting deacylated tRNA on the P site moves to the third tRNA binding site, the E site (exit site), and then is detached from the ribosome. Dipeptidyl tRNA on the A site is moved to the vacant P site catalyzed by translocase (also called elongation factor 2, EF2) while it is still bound to the second codon. Another molecule of GTP is hydrolyzed at this step. The third aminoacyl tRNA enters into the free A site by base-pairing with the third codon on mRNA that now faces to the A site after forward movement of the ribosome. Such elongation steps are repeated until the ribosome reaches the termination codon.

• Termination. At the end of elongation, a termination codon is recognized by termination factors while peptidyl tRNA is base-paired with the last sense codon. The final complex consisting of mRNA, tRNA covalently linked to the completed polypeptide, and 80S ribosome dissociates into each component with concomitant dissociation of the ribosome into 40S and 60S subunits, releasing the completed polypeptide by hydrolysis of the ester bond with the tRNA.

When cells are actively engaged in protein synthesis, multiple ribosomes are attached to a single chain of mRNA, forming polyribosome or polysome.

Translational Control

Translation efficiency of mRNA is controlled according to the demands of cells, thereby changing the concentration of particular proteins. Initiation is a common major step of translational control. Two key proteins are involved in the control of efficiency of the initiation: the translational repressor eukaryotic initiation factor 4E (eIF4E) binding protein (4E-BP) and the 70-kDa ribosomal protein S6 (a protein component in the 40S small subunit) kinase (p70S6K or S6K) (Volarevic & Thomas 2001). The hyperphosphorylation of 4E-BP results in enhanced availability of the mRNA cap binding protein eIF4E for binding eIF4G and forming the active eIF4F complex necessary for the formation of the 80S initiation complex. Phosphorylated S6K enhances the S6 phosphorylation that activates translational initiation (Anthony et al. 2001). These signaling events are mediated by the protein kinase mammalian target of rapamycin (mTOR). Rapamycin, a bacterial macrolide, is an immunosuppressant. Increase in the concentration of amino acids that inhibit the kinase leads to increase in the active dephosphorylated form. Branched amino acids, particularly leucine, stimulate mTOR signaling pathway, resulting in reduction of translation inhibition by 4E-BP and also leading to activation of S6K, thereby promoting translation (Proud 2002). The other means of translational control is seen in the phosphorylation (inactivation) and dephosphorylation (activation) of eEF2 by eEF2 kinase and protein phosphatases, respectively. Amino acids cause inhibition of the kinase, thus activating translation.

In the early stages of development of fertilized eggs, maternal mRNAs stored in oocytes are switched on to be translated to make proteins required for the rapid cell division. The dormant maternal mRNAs are in the form of ribonucleoprotein (a complex of mRNA and proteins) from which protein components are detached in fertilized eggs so that the mRNAs are then available for translation.

Another remarkable example of translational control is found in the synthesis of ferritin, an intracellular iron storage protein. The ferritin mRNA in an inactive ribonucleoprotein complex under the iron-deficient condition is converted to a translationally active form when iron is available in cells. The iron-response element in the 5' nontranslated region of the mRNA binds an iron-binding protein when the protein is not bound with iron, thereby suppressing translation in the absence of iron (figure 4.3*a*). When iron is present, however, the protein is dissociated from the mRNA that is then used for translation.

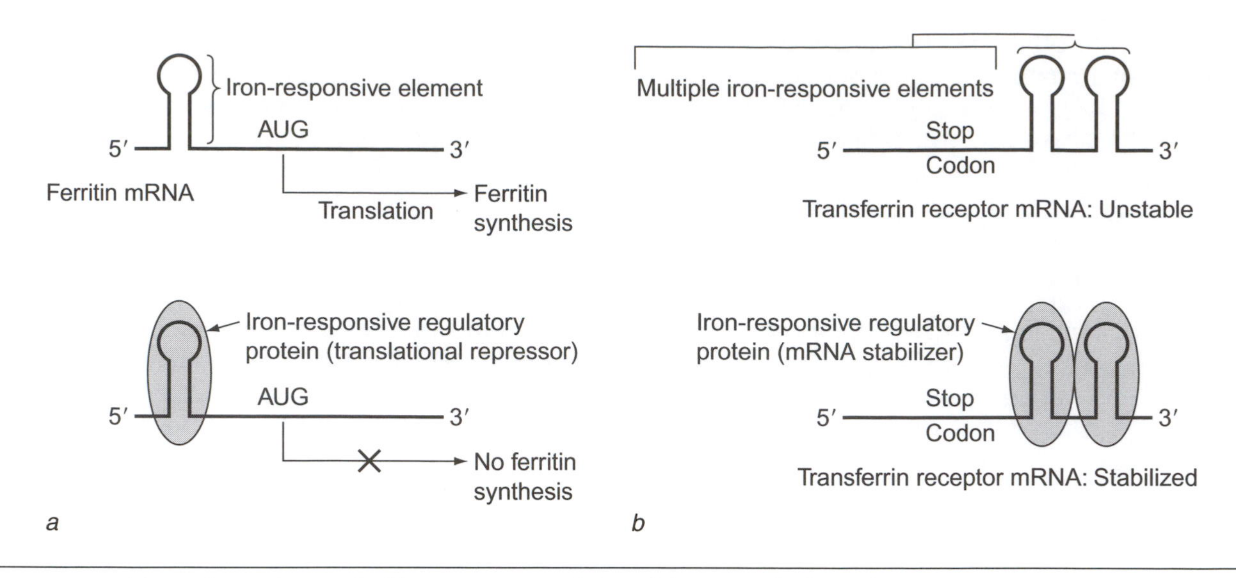

Figure 4.3 *(a)* Translational control of ferritin mRNA and *(b)* stability control of transferrin receptor mRNA. In both cases, the regulatory protein dissociates from the iron-responsive element of the mRNAs when iron concentration in cells is high (above), and the protein binds to the mRNAs when the concentration is low (below).

Post-Translational Physiological Modifications

The N-terminal amino acid of proteins is always methionine, corresponding to universal initiation codon AUG at the beginning of translation. The majority of proteins lack the N-terminal methionine residue, however, because it is removed during translation. In addition, proteins are often otherwise processed or modified after translation (post-translational modifications) to become functional, localize in a particular compartment of cells, or both. Some proteins are synthesized as proproteins or pre-proproteins as precursors of functional molecules. They are processed by cleavage at specific amino acid residues. For example, insulin is synthesized as pre-proinsulin on rough endoplasmic reticulum (ER) that is incorporated into the lumen of ER, where signal peptide at the N terminus is removed to generate proinsulin. Proinsulin is further cleaved at two sites, generating three fragments, the middle fragment being discarded as C-peptide; and disulfide bonds are formed between and within the remaining two (A and B) polypeptide chains, thus forming the functional insulin molecule.

Other modifications include addition of sugar moieties (glycosylation) in membrane and secretory proteins. Digestive enzymes and proteins secreted from the pancreas and other secretory organs are mostly glycosylated. These proteins include antibodies and interferons. Collagens, extracellular proteins secreted from fibroblasts, are highly glycosylated proteins.

4-Hydroxyproline and 5-hydroxylysine residues in collagens are formed by hydroxylation of proline and lysine residues post-translationally. Muscle contractile proteins contain 3-methylhistidine residues and other unusual methylated amino acids. 3-Methylhistidine found in urine is used to estimate degradation of muscle contractile proteins.

Reversible phosphorylations at serine, threonine, and tyrosine residues by a variety of protein kinases and phosphoprotein phosphatases are physiological modifications crucial to the control of activities of enzymes and other proteins in signal transduction and other biological processes (see chapter 7 for detail).

Stability of mRNA

One of the factors that control the amount of a protein synthesized is the concentration of its mRNA. The concentration of mRNA is determined by the rate of transcription/processing in the nucleus and the rate of degradation in the cytoplasm. This dynamic state of mRNA synthesis and degradation is called metabolic turnover or simply turnover. In spite of its obvious importance, for a number of years the degradation of mRNA has not attracted much attention compared with transcription and processing.

The stability of mRNA is usually measured as a half-life ($T_{1/2}$) because degradation normally follows first-order kinetics; that is, the time required to attain half the concentration of the original value is independent of the concentration. The half-life of mRNA varies greatly, ranging from several minutes to days in eukaryotic cells. Generally, mRNAs for proteins involved in control functions such as those in the cell cycle (e.g., cyclin) and transcription factors (e.g., c-fos and c-myc) are short-lived. The 3' noncoding region of these short-lived mRNAs contains a nucleotide sequence rich in A and U that is likely responsible for the instability. Messenger RNAs for proteins expressed in differentiated cells such as hemoglobin in erythroid cells and contractile proteins in muscle cells are long-lived.

Certain mRNAs are stabilized or destabilized in the presence or absence of hormones such as estrogen and insulin or other factors. As an example, the stability of transferrin receptor mRNA is controlled by the same regulatory protein described earlier that controls the translational efficiency of ferritin mRNA. Transferrin is a protein used for the transport of iron in the blood, and its receptor is required for uptake by cells. When the concentration of iron is high, the iron-bound regulatory protein dissociates from the mRNA, making it less stable, down-regulating its concentration, and thus reducing the rate of synthesis of the receptor to decrease the incorporation of iron into cells (figure 4.3*b*).

Protein Degradation

Degradation of a protein is the final stage of gene expression that starts in transcription of a gene and is followed by translation to produce a protein, abolishing the protein function. Mechanism and control of protein degradation is as important as transcription and translation, because this is one of crucial steps for a protein to function or loose function or even gain harmful function if damaged molecules are not removed.

Historical Background and an Overview

That cellular proteins are continuously replaced (metabolic turnover) was not recognized until 1935, when Rudorf Schönheimer discovered the degradation of proteins in a body by metabolically labeling them with a stable isotope nitrogen-15 that became available at that time. However, in contrast to synthesis, which has been extensively studied, the importance of protein degradation in cells has been largely ignored until recently. The reason may be that degradation was regarded as a process that can occur spontaneously with an increase in entropy without the use of energy. Now it is evident, however, that degradation of proteins requires ATP energy and is a highly controlled process. In recent years, impairment of protein degradation has been implicated in diseases, particularly neurodegenerative disorders, and in aging.

Intracellular protein degradation is divided into two major categories: lysosomal and nonlysosomal pathways (proteasome systems and calpains). Both lysosomal and proteasomal pathways that are involved in a large majority of cytoplasmic protein degradation are ATP dependent. All of these proteolytic systems are present in the muscle, and the activities are influenced by exercise, immobilization, hormonal and nervous stimuli, and nutritional status, damage, diseases (inflammation, dystrophy), and so on.

Proteasome System

Proteasomes are large proteinase complexes that have multiple catalytic sites (Coux et al. 1996). Proteasomes are responsible for degradation of most cytosolic proteins and also of specific regulatory proteins such as transcription factors or the associated proteins (e.g., c-jun, Ik-B, p53) and cell cycle regulators (e.g., cyclins). Proteasomes are also involved in the generation of oligopeptides for antigen presentation in immune cells (Groettrup et al. 2001). There are two forms of the enzymes. A 20S proteasome has a molecular weight of about 700 kDa and consists of seven different α subunits and seven different β subunits (figure 4.4). Each group of subunits forms a seven-member ring structure, and four rings (two α and two β rings) line up in the order α-β-β-α, thus adopting the form of a hollow cylinder. A 26S proteasome of about 2,000 kDa contains additional regulatory subunits (the 19S regulatory complex, also called the cap) attaching at both ends of the 20S proteasome cylinder, giving the whole structure a dumbbell shape (figure 4.4). There are at least five peptidase activities in the interior of the cylinder in the β subunits. The major activities include those for cleaving peptide bonds after basic, acidic, and hydrophobic amino acids, often called trypsin-like, peptidylglutamyl peptide-hydrolyzing, and chymotrypsin-like activities, respectively.

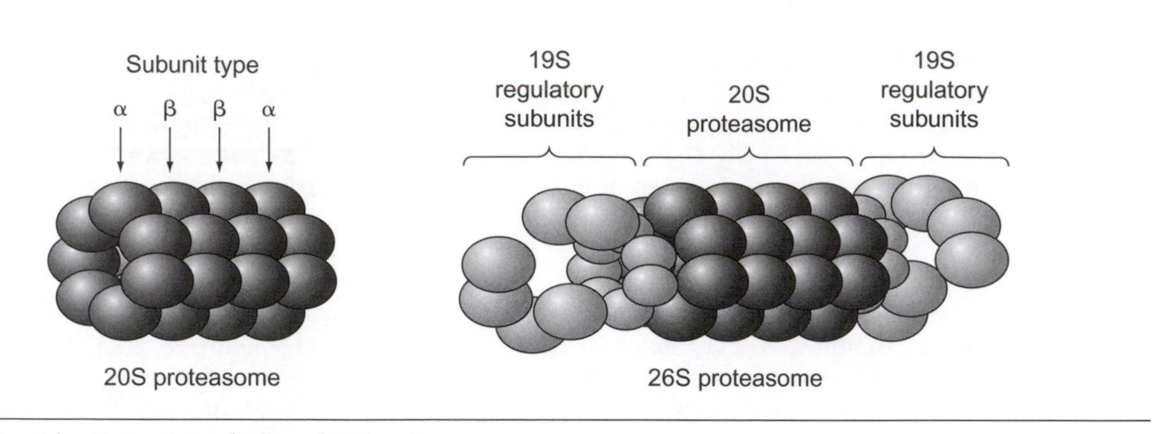

Figure 4.4 Subunit structure of 20S and 26S proteasomes.

Many, but not all, proteins to be degraded by proteasomes are marked by a covalent linkage with ubiquitin (Ub) (figure 4.5). Ub is a small protein of 76 amino acid residues that is found in any eukaryotic cell. It has multiple functions in addition to roles in proteolysis such as in signal transduction pathways (Ben-Neriah 2002). In proteolysis, it is conjugated to a target protein (substrate) through its C-terminal glycine residue to form isopeptide bonds with the ϵ-amino group in lysine residues of the substrate. The substrates are ubiquitinated at multiple available lysine residues forming polyubiquitin (poly Ub) chains in which additional Ub is covalently attached to the lysine residue at position 46 of preceding molecules. Polyubiquitination is a signal for degradation of substrate proteins by 26S proteasome. The ubiquitination consists of three steps (figure 4.5). Ub-activating enzyme E1 catalyzes conjugation of Ub to itself via a high-energy thioester bond between a cysteine residue in E1 and the C-terminal glycine residue of Ub that is accompanied by hydrolysis of ATP. The Ub is then transferred to Ub-conjugating enzyme E2. In most cases Ub ligase E3 mediates transfer of the Ub on E2 to ϵ-amino groups of lysine residues of a substrate protein through E3-Ub intermediate, or in some cases the Ub on E2 is directly transferred

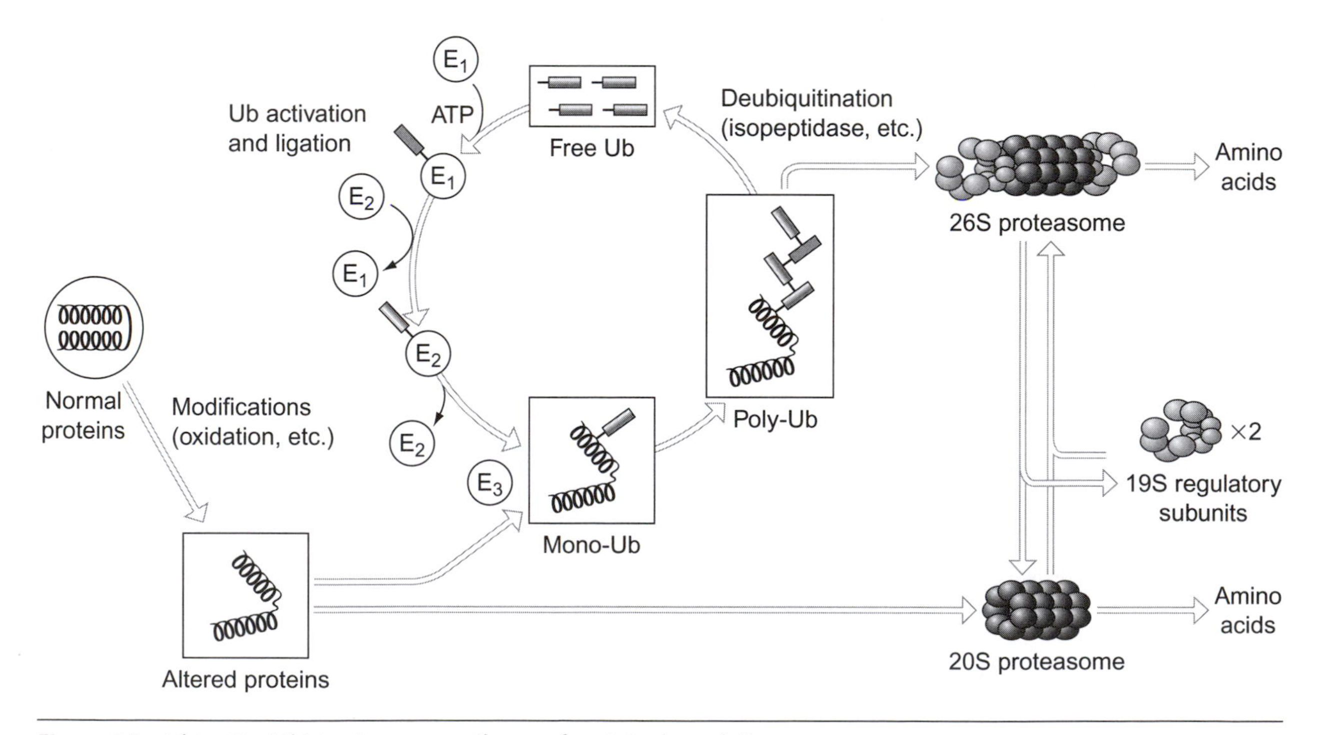

Figure 4.5 Ubiquitin (Ub)/proteasome pathway of protein degradation.

to the substrate. The polyubiquitin chain on the substrate is recognized by 19S caps of the 26S proteasome. The cap contains several ATPases that are presumably used to supply energy by hydrolysis of ATP when substrates are possibly unfolded to get into the cylinder to be degraded. The Ub monomers are released from the polyubiquitin by catalysis of isopeptidases (deubiquitination enzymes) and recycled. Some substrate proteins are not ubiquitinated and are degraded directly by the 20S or 26S proteasome. The degradation products are usually octa- to decapeptides. The resulting oligopeptides are likely further degraded by other cytoplasmic proteases except in immune T-cells, where the oligopeptides are processed for antigen presentation as epitopes in major histocompatibility complex I on the cell surface.

While the proteasomes are involved in the degradation of cytoplasmic proteins in general, they have an important role to eliminate altered proteins such as oxidatively damaged, misfolded, or unassembled proteins that are potentially harmful to cells (Grune et al. 1997). The quality control system has been extensively studied in the case of protein synthesis and processing in ER (Chevet et al. 2001). A considerable proportion of secretory or membrane proteins synthesized on rough ER and translocated into the lumen or inserted in the membrane are conceivably improperly post-translationally modified, folded, and/or assembled. Such potentially detrimental altered proteins undergo retrograde translocation back to the cytoplasm and are degraded by the proteasome systems. This system is called ER quality control.

Lysosome Pathways

Lysosomes are single-membrane organelles that contain a number of hydrolytic enzymes: nucleases, proteases, glycosidases, lipases, and others. In protein metabolism, the lyosomal pathway is mainly involved in the degradation of organellar proteins via autophagic processes and extracellular proteins that are incorporated into cells by endocytosis. Lysosomal proteolysis is responsible for the degradation of bulk proteins and is influenced by nutritional supply and hormones such as insulin, glucagon, and glucocorticoid. Generally, proteins with long half-lives (longer than about 45 h) are thought to be degraded via lysosomal pathways, albeit this is not always the case (Rogers & Rechsteiner 1988). Lysosomes contain several cysteine or aspartate endopeptidases (cathepsin B, D, L, H, etc.) and exopeptidases (carboxypeptidases and aminopeptidases) that are most active at acidic pH (pH 5-5.5) (Turk et al. 2001). The inside of the organelle is kept at pH 5 to 5.5 by a proton pump (H^+-ATPase) that utilizes ATP as energy. Substrate proteins are incorporated into lysosome via processes termed macroautophagy (wrapping of organelles by membranes) or microautophagy (incorporation of proteins into a multivesicular body).

Calpains

Calpains are cytoplasmic neutral cysteine-proteases that have a calmodulin-like Ca^{2+} binding domain. Two major calpains, μ- and m-calpains (also called calpain 1 and 2), are activated by micromolar and millimolar concentrations of Ca^{2+}, respectively, and expressed ubiquitously. Both μ- and m-calpains consist of a distinct large subunit (about 80 kDa) that has the catalytic and Ca^{2+} binding domains, and a common small regulatory subunit (about 30 kDa), forming a heterodimer. These so-called conventional calpains coexist with the specific endogenous inhibitor calpastatin. Substrates for calpains include cytoskeletal and plasma membrane proteins such as epidermal growth factor (EGF) receptor and platelet-derived growth factor (PDGF) receptor, as well as regulatory proteins such as transcription factors and protein kinase C. Of note here is an atypical calpain, p94 (also called calpain 3), which is a homolog of the catalytic large subunit of the conventional calpain (Sorimachi et al. 2000). This calpain is expressed predominantly in the skeletal muscle as a homodimer not complexed with the small subunit. It interacts with connectin/titin that spans between the Z- and M-lines in the muscle. Although in vivo substrates and the biological roles for calpain 3 have not been identified, it is suggested that calpain 3 influences the expression of conventional calpains such that a low level of calpain 3 results in the activation of conventional calpains, leading to the degradation of muscle proteins.

Exercise and Protein Metabolism

Exercise has a capability to significantly alter both protein synthesis and degradation. This part of this chapter, following the general section, will deal with the exercise-associated changes in protein turnover.

General View of Protein Synthesis and Degradation in Exercise

In order for an organism to survive, cells have to respond to changes in their environment in a way that maintains homeostasis. Physical exercise provokes different kinds of stimuli, including hormonal, metabolic, biochemical, and mechanical stimuli. These signals activate transcription factors, and as a result adaptive proteins are synthesized. However, the number of sensors and the signaling pathways are limited; the interactions or cross-talk between these pathways results in a superbly orchestrated network and finely regulated protein metabolism. Indeed, the up-regulation of synthesis of certain skeletal proteins is dependent on particular stimuli and is highly specific. Our genes have been selected for a physically active life—to survive chasing prey and escaping from predators—but not for a sedentary lifestyle, which is a result of our civilized environment. For modern-day life, the physical exercise-induced alteration in protein synthesis and degradation must be an important physiological adaptive process that aims to strengthen the homeostasis to cope with a variety of stressors.

Generally speaking, the activity and content of proteins with short half-lives, say minutes to a few hours, can be readily adapted in response to changes in situations. On the other hand, proteins (such as lens crystalline and histones) with longer half-lives, which can extend to days and sometimes weeks or months, are not easily replaced in a manner consistent with their biological roles. Therefore, the half-life of a protein is closely linked to function of the protein. Structural proteins such as myosin and actin in the muscle often have long half-lives, while inducible enzymes are generally very short-lived. Half-lives of proteins increase as a function of age, while caloric restriction, the only known method of retarding aging and increasing maximal life spans of various animals, results in faster protein turnover even at old ages (Goto et al. 2001). Accumulating evidence suggests that physical exercise, similar to caloric restriction, can shorten the half-life of proteins and thus likely increase the turnover rate of proteins. This cellular response might be an important part of the beneficial adaptive response by which exercise increases various physiological functions and overall improves the quality of life (Radak et al. 2001b).

Hormones

Physical exercise causes significant changes in the endocrine system, and protein metabolism is in turn greatly affected by hormones (see figure 4.6). Endocrine glands release hormones into the blood, and specific receptors transfer the hormonal message into target cells, resulting in effects on the expression of specific genes. At the cellular level, hormones can modify membrane properties and activate second messengers that lead to changes in transcription and translation. In the muscle, growth hormone (GH), testosterone, and insulin-like growth factor-1 (IGF-1) are anabolic hormones that stimulate protein synthesis; and glucocorticoids, such as cortisol, are catabolic hormones that are involved in up-regulation of protein degradation. Generally, exercise of high intensity increases the secretion of the anabolic hormones, while exercise of moderate intensity and long duration results in increased levels of the catabolic hormones.

A mounting body of evidence suggests that IGF-1 is heavily involved in proliferation and differentiation of myogenic cells. IGF-1 is a 70-amino acid peptide whose secretion is stimulated by GH. As in the case of other growth factors, it stimulates tyrosine kinase activity after receptor binding; and a series of phosphorylations of different cascades are involved, including MAP-kinase, c-fos, and c-jun. As a result of IGF-1-mediated cellular signaling, protein synthesis is increased. Jacob et al. (1996) demonstrated that IGF-1 injection into rats significantly induces protein synthesis in the skeletal muscle. The positive nitrogen balance is partly due to the suppressive effects of IGF-1 on proteasome complex. The loading-induced hypertrophy in the skeletal muscle is associated with an increased level of mRNA of IGF-1 (DeVol et al. 1990). Similarly, stretch-induced hypertrophy is accompanied by the up-regulation of mRNA of IGF-1. Muscle contraction, especially eccentric loading, can result in muscle damage. The regeneration damaged muscle and the compensatory hypertrophy are involved in the multiplication of satellite cells. IGF-1 is responsible for differentiation and fusion of satellite cells to regenerate functional muscle fibers. Thyroid hormone has been shown to stimulate mitochondrial biogenesis. The hormone increases synthesis of the nuclear DNA-coded proteins and import of the proteins into mitochondria.

Growth hormone is a polypeptide hormone with molecular weight of 22 kDa and a half-life of

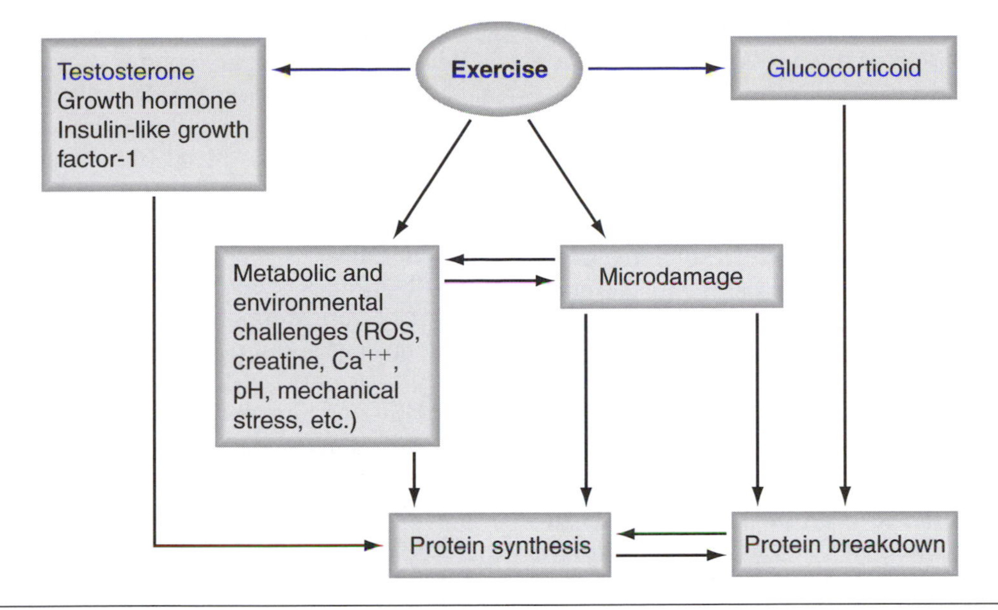

Figure 4.6 Relationship between exercise, hormones, macromolecular damage, protein synthesis, and degradation.

15 to 20 min in the circulation. Growth hormone receptors are present in all kinds of cells; and dimerization takes place after GH binding, which is essential for intracellular signaling. Growth hormone exerts multiple physiological functions in both the cytoplasm and nucleus. In the skeletal muscle, GH induces glucose and amino acid uptake and stimulates protein synthesis, possibly using the energy derived from its lipolytic action. The protein synthesis is stimulated by mobilization of amino acid transporters. This results in an increased intracellular amino acid level. The effect of GH is enhanced by the synergistic effects of IGF-1. Exercise-associated metabolic changes can up- or down-regulate GH secretion depending on the intensity and duration of the exercise. Administration of GH has been shown to increase protein synthesis; and the effect is especially significant in tendon proteins, including collagen.

Testosterone is secreted by the interstitial cells of the testis and is controlled by interstitial cell-stimulating hormone, produced in the anterior pituitary. The plasma testosterone level can be increased up to a 40% higher level by high-intensity exercise, which could mean increased demand for protein synthesis. Increases in total serum testosterone appear to occur immediately after exercise, affecting large muscle groups. The powerful effects of the anabolic hormone testosterone on the muscle can be seen by its widespread abuse as a doping drug to increase muscle mass. Indeed, administration of exogenous testosterone has been shown to significantly induce hypertrophy of exercising skeletal muscle. It should be mentioned here that uptake of this or analogous synthetic hormones has caused a variety of disorders including hypertension and ischemic heart disease.

Glucocorticoids are essentially catabolic hormones that primarily promote protein degradation and inhibit protein synthesis in the skeletal muscle. The acceleration of protein breakdown is mostly due to the ubiquitin/proteasome pathway involving the 26S proteasome complex (Hasselgren 1999). Muscle wasting associated with uremia is caused by increased proteolysis dependent on the ubiquitin/proteasome pathway that is stimulated by acidosis and a physiological level of glucocorticoids. Inhibition of amino acid transport in the muscle by pro-inflammatory cytokines such as tumor necrosis factor α and interleukin-1 is mediated by glucocorticoids. Glucocorticoids are potent and frequently used anti-inflammatory drugs. Their anti-inflammatory action is complex, but one of the major mechanisms is thought to be the suppression of prostaglandin synthase at the translational level. This translational control is likely due to proteins that are induced by glucocorticoids and bind to the 3' untranslated region in the prostaglandin synthase mRNA, thus interfering with the efficient translation. Glucocorticoids appear to be involved also in the inhibition of phospholipase A2 that releases arachidonic acid, the precursor of prostaglandins. The effects are exerted at transcriptional and translational levels as well as at the post-translational level through dephosphorylation of the active phosphorylated form of the enzyme.

Nutrition

It is well known that administration of exogenous amino acids and resistance exercise increase the net muscle protein synthesis and muscle mass (Rasmussen et al. 2000). The increased muscle mass is likely due to the combined effect of increase in protein synthesis and decrease in protein breakdown. Net muscle protein balance has been shown to increase dramatically in response to resistance exercise and amino acid ingestion. The increased net protein balance could be the result of increased activity of p70 S6 protein kinase that regulates the initiation of translation (see *Translational Control* in this chapter), since it has been reported that amino acid–carbohydrate administration and resistance training enhance the activation of this kinase (Baar & Esser 1999). Dietary creatine administration combined with resistance training results in marked muscle hypertrophy during recovery from immobilization (Hespel et al. 2001). At the same time the muscle glucose transporter GLUT4 gene expression is also increased.

Enzymes and Proteins in Mitochondrial Biogenesis

Endurance exercise or chronic use of skeletal muscle stimulates mitochondrial biogenesis, increasing the number and volume of the organelle as a result of an adaptive response. Sedentary lifestyle and aging reduce the activity of the process. Mitochondria are never formed de novo. The number or size of the organelle increases through the growth and division of existing mitochondria. In the skeletal muscle, mitochondria are located either under the sarcolemma (sarcolemmal mitochondria) or between myofibrils (intermyofibrillar mitochondria).

In 1967 John O. Holloszy published an influential seminal paper showing that endurance training increases the content of mitochondria in skeletal muscle. The initial factor, the stimulus for mitochondrial biogenesis, might be the Ca^{2+} ion that can activate protein kinase C and mitogen-activated protein kinase (MAPK) pathways. The calcium–calmodulin-dependent protein kinase signaling pathways also have been shown to induce mitochondrial biogenesis (Wu et al. 2002). Reactive oxygen species (ROS)-induced activation of the transcription factor, activated protein-1 (AP-1) pathway is also implicated as a triggering factor for mitochondrial biogenesis. Recently it has been suggested that the nuclear respiratory factors (NRF) play a role in the biogenesis of mitochondria, since they activate promoters of nuclear genes encoding components of the respiratory chain. Murakami et al. (1998) demonstrated that an acute bout of exercise increased the NRF-1 by 35% (in rats), indicating that exercise initiates the NRF signaling pathway. However, regular training or a single bout of exercise did not result in increased mRNA level of NRF-1 in human subjects (Pilegaard et al. 2003). It is well established that endurance training can significantly increase mitochondria biogenesis, which can occur within days (Hood et al. 2000). On the other hand, strength training fails to result in an increase in mitochondrial density. A large majority of more than 100 species of mitochondrial proteins are nuclear coded, only 13 being coded in mitochondrial DNA. Therefore, a very well-developed specific protein transport system is required to transfer nuclear DNA-encoded proteins into the mitochondria. The transported proteins must be unfolded in order to move into the organelle at special contact sites and refolded in the mitochondria. In fact, the increase in mitochondrial biogenesis must be associated with the enhancement of chaperones that mediate the unfolding and refolding. Mitochondrial heat shock protein 70 (HSP70) pulls cytoplasmically synthesized precursor proteins into the matrix by unfolding them, and HSP60 in the mitochondria refolds the proteins to mature forms after cleaving off the mitochondria-targeting signal sequences. Indeed, mitochondrial import stimulating factor and heat shock protein mRNA levels are increased significantly by contractile activity as a result of endurance training.

The biogenesis of skeletal muscle mitochondria is mainly dependent on metabolites, contractile activity, and Ca^{2+} and Fe^{2+} ions. Iron is essential for heme in cytochromes and iron-sulfur clusters in the electron transport chain. Recently, it was suggested that an intermittent rise in cellular calcium concentration increases mitochondrial biogenesis in muscle cells in culture (Ojuka et al. 2002). The morphological changes are accompanied by biochemical changes, namely with increased activity of certain mitochondrial enzymes. It is known that the selectivity of enzyme activity, as well as protein content of structural proteins, is dependent on the characteristics of physical activity, intensity, and duration. Aerobic exercise appears to have a more significant effect on the biogenesis of subsarcolemmal mitochondria, and this kind of mitochondria has a higher level of respiratory rate than intramyofibrillar mitochondria. Long-term endurance training results in increased activity and content of enzymes (Acetyl-CoA carboxylase, malonyl-CoA decarboxylase) involved in β-oxidation of fatty acids to save glycogen and decrease lactic acid production. AMP-activated protein kinase (AMPK) plays a major role in energy metabolism in the skeletal muscle, sensing the increase in AMP/ATP and decrease in phosphocreatine/creatine ratios. The down-regulation of Acetyl-CoA carboxylase and up-regulation of malonyl-CoA decarboxylase are mediated by AMPK pathway. Voluntary exercise of 30 min increased and decreased the activities of malonyl-CoA decarboxylase and Acetyl-CoA decarboxylase, respectively, through the AMPK pathway—not only in the skeletal muscles but also in liver and fat tissues (Park et al. 2002).

The activity and content of enzymes in the Krebs cycle are also up-regulated by endurance running (Holloszy 1967). A main function of mitochondria is to produce ATP (six ATP generated per mole of O_2), and to achieve this, electron and proton transport takes place on the mitochondrial inner membrane. Uncoupling proteins (UCP) in the membrane that mediate a decrease in the proton gradient are believed to be an active contributor to mitochondrial functions; a recent paper showed that endurance training decreases the mRNA content of uncoupling proteins (UCP2 and 3) in the rat (Boss et al. 1998), and similar results were obtained in a human study (Schrauwen et al. 2002). On the other hand, a recent study by Jones et al. (2003) suggests an increase in UCP3 content in the skeletal muscle as a result of endurance training, due to the increased biogenesis of whole mitochondria. It appears that there is a discrepancy between the increase in mRNA and content of UCP3 that are up-regulated by exercise. The increase in mRNA level is much greater, reaching 7- to 16-fold, while the gain in protein content is only up to 35% (Jones et al. 2003). There seems to be translational control. UCP3 is implicated in curbing the effect of ROS (Ricquier & Bouillaud 2000), as it can lower the mitochondrial membrane potential that reduces possible leakage of electrons from the respiratory chain and therefore potentially attenuates generation of superoxide anion radicals, hence working as an antioxidant.

Changes in the cellular redox milieu serve as signals to activate transcription factors including nuclear factor-κB (NF-κB) and AP-1. As a result of this activation a number of proteins can be regulated, including those involved in inflammation and apoptosis. ROS can regulate the activity and content of antioxidant proteins. Even a single bout of exercise can up-regulate the content of Mn-SOD in the skeletal muscle, liver, and kidney (Radak et al. 1996). Physical activity has been shown to regulate the activity of NF-κB. A single bout of exercise induces activation of NF-κB, while regular-term exercise tends to decrease it (Radák et al. 2004).

Muscle Structural Proteins

Mechanical stress, resistance training, and high tension, especially eccentric exercise, can cause up-regulation of the synthesis of certain cellular proteins in the skeletal muscle—mostly structural proteins—via a series of cross-talk between signal transduction pathways, thus resulting in structural changes of the tissue. The increase in protein content in the skeletal muscle could be due to activated protein synthesis, which is often observed in fast fibers, or could result from decreased protein breakdown. Hypertrophy of slow fibers is due to a decreased level of proteolysis (Goldspink 1991). In hypertrophy associated with protein synthesis, newly synthesized actin and myosin filaments are added to the preexisting myofibrils, consequently forming larger myofibrils and increasing the density of cross-bridge spacing. It is believed that this kind of hypertrophy is a result of longitudinal splitting of Z-discs, creating two or more daughter myofibrils. Therefore, some hypertrophy of the skeletal muscle can occur as a process associated with mechanical damage. Indeed, passive stretching is one of the most often used methods of inducing hypertrophy of myofibrils and myotubes. One study showed that the total RNA content and actin mRNA in stretch-induced hypertrophy of the skeletal muscle peaked after some days of stretch overload (Fink et al. 2000). Simultaneous treatment by chronic low-frequency electric stimulation and stretch load for three days resulted in a remarkable (more than fourfold) increase in protein synthesis in rat skeletal muscle (Goldspink 1991). Resistance training does not have consistent effects on the skeletal muscle except that α-actin, a component of the thin filament, increases without exception during hypertrophy.

Stretch-induced hypertrophy in anterior latissimus dorsi muscle resulted in a shift to the slow muscle phenotype, since the slow myosin-2 isoform increased significantly. In mice, the plantaris, which is predominantly composed of fast muscle, exhibited a slow α-myosin heavy chain (MHC) and light chain mRNA level after overloading, suggesting a fast-to-slow muscle shift (Gregory et al. 1987). Isometric resistance training in rodents caused increased mRNA of type IIx MHC. Hortobagyi and coworkers (2000) demonstrated that immobilization and retraining resulted in up-regulation of type IIx MHC mRNA level by 2.9 fold and 1.2-fold, respectively, using different types of muscle action. No change was reported in the level of type IIa MHC mRNA isoform. The type I MHC mRNA level was down-regulated by both immobilization and retraining. These data suggest that stretch-induced hypertrophy, resistance training, or immobilization is capable of shifting fiber phenotypes.

Animal studies suggest that the increase in the MHC mRNA content after resistance training occurs almost immediately in the skeletal muscle. MHC gene expression is regulated by myogenic regulatory transcription factors. Myo-D and myogenin are major members of this regulatory protein group and facilitate gene expression related to muscle development. Myo-D and myogenin levels are significantly higher in fast-type fibers and seem to be associated with the expression of MHC type IIx isoform. It appears that type IIx mRNA isoform responds most remarkably to heavy-resistance exercise compared with type I and IIa. Human subjects performed three sets of 8 to 10 repetitions with 75% to 80% of the maximum load on leg muscles (Willoughby & Nelson 2002). The single session of resistance exercise up-regulated the mRNA of all three MHC isoforms (type-I, IIa, IIx) at 6 h postexercise. The up-regulation of type IIx MHC mRNA was correlated with the protein expression of Myo-D, while that of type I and IIa MHC mRNAs was correlated with myogenin. Thus, the fast and effective increase in myofibrillar protein concentration after a single bout of resistance exercise is likely due to the increased expression of type I, IIa, and IIx MHC isoform genes.

Recovery from muscle damage is associated with increased protein synthesis. The most striking morphological feature of muscle damage is extensive sarcomeric disruption and Z-band streaming (Friden 1984). It has been shown that after eccentric exercise the disruption of cytoskeleton can occur very rapidly (as fast as in 5 min), most probably as a result of mechanical stress. The content of the structural protein desmin, which synchronizes the contraction of myofibrils by connecting adjacent myofibrils at the Z-band level, and Z-band to the sarcolemma, decreases immediately after the eccentric exercise; then the concentration rebounds to the control level 72 h after eccentric bouts. Then there is a huge increase in the desmin/actin mass ratio. It is well known that eccentric loading can precondition skeletal muscle against the damaging effect following eccentric exercise. It seems appropriate to suggest that the increased desmin concentration is a significant part of this protective adaptive response to eccentric muscle action, since desmin is anchored to actin and the Z-band. These structures are the front-line targets of the disruption induced by mechanical stress.

Effects of Exercise on Brain Proteins

Exercise with a wide range of effects induces protein synthesis not only in the skeletal and cardiac muscles, but also in the brain. The beneficial effects of physical exercise on cognitive functions are well known. The mechanism behind this is not fully understood, but it seems to be associated with an increased expression of certain proteins. Recent evidence indicates that physical exercise induces the expression of neurotrophic factors in selected regions of the brain. Hence, it seems reasonable to suggest that increased expression of neurotrophic factors is related to better memory and improved cognitive function. Brain-derived neutrotrophic factor (BDNF) can enhance the survival and differentiation of neurons. A low level of BDNF results in increased vulnerability of neurons to death. An increased level of BDNF protects against the neuronal degeneration that appears to occur in Alzheimer's disease and in ischemia–reperfusion injury. Voluntary exercise has been shown to increase BDNF content (Neeper et al. 1996). Six hours after voluntary running, the concentration of mRNA of BDNF was significantly increased. A positive correlation has been found between running distance and BDNF mRNA expression (Oliff et al. 1998). Our recent observation was that the BDNF content increased significantly in the brain of moderately trained rats; however, by contrast, overtraining did not affect the content of BDNF significantly.

Exercise and Protein Degradation

The turnover of proteins in the skeletal and cardiac muscles occurs in ordinary life without strenuous exercise or bed rest or immobilization, as is the case in other tissues (see *Protein Degradation* in this chapter). In resistance exercise, not only synthesis but also degradation increases for up to 48 h (Phillips et al. 1997). The increased turnover is an adaptive response to the stress that results in the increase of muscle mass. Overall increase in muscle mass is due to the excess synthesis over degradation that is necessary for remodeling of the muscle tissue.

Under microgravity conditions as seen in astronauts or animals under hindlimb or tail suspension, slow-twitch muscles such as soleus preferentially undergo atrophy in contrast to fast-twitch muscles that are resistant to such challenges. Ikemoto et al. (2001) studied the degradation of fast-type MHC of gastrocnemius muscle in young rats that stayed in a weightless condition in a space shuttle for 16 days. They found that the MHC was degraded in a specific way and that mRNAs for proteasome subunits were increased while no significant change was observed in cathepsin L and calpain mRNAs. The ubiquitination of muscle proteins was increased also. Such phenomena could be reproduced in tail-suspended rats. Since a potent inhibitor of cathepsin L, E-64, did not prevent the MHC degradation and since the levels of calpains and cathepsins were not changed, it was concluded that the proteasomal pathway—but not the lysosomal pathway or calpains—is responsible for increased proteolysis in disuse atrophy of the skeletal muscle. Exhaustive treadmill running was shown to stimulate decrease of [^{3}H]leucine from prelabeled proteins of skeletal muscle (Ji et al. 1988). This shows an increased proteolytic process as a result of a single bout of exercise. A similar finding was reported in calpain-like protease activity after 1 h of running (Raj et. al. 1998).

There is accumulating evidence that exercise training increases the activity of the proteasome enzyme complex. Nine weeks of swimming training increased the trypsin-like activity of proteasome in the brain of rats (Radak et al. 2001a). A similar increase was observed in the skeletal muscle of rats (Radak et al. 1999). In a related study, we administered hydrogen peroxide every other day for three weeks to trained and sedentary rats. Both hydrogen peroxide and physical exercise increased the activity of the proteasome complex in the myocardium of rats. Swimming training further up-regulated the effects of hydrogen peroxide administration on the activity of the proteasome complex (Radak et al. 2000). These findings suggest that a moderate increase in ROS production can up-regulate proteasomal activity that removes damaged proteins. Wakshlag and coworkers (2002) studied the effect of exercise (hunting activity) on the proteolytic capacity of skeletal muscle in dogs. The results indicated pronounced up-regulation of ubiquitinated conjugates and the p31 regulatory capping subunit during the peak hunting period compared with the preseason period. In contrast, the catalytic core of the proteasome (β subunits) showed no apparent up-regulation in response to increased physical activity. A human study, in which applied eccentric muscle action caused muscle damage, showed that eccentric exercise caused a decrease in calpain 3 mRNA immediately after the exercise, whereas the calpain 2 mRNA level was increased at day 1. Both mRNA levels returned to control values by day 14. By contrast, cathepsin B+L and proteasome enzyme activities were increased at day 14 (Feasson et al. 2002). These findings suggest that the proteolytic process occurs in a selective manner and plays an important role in muscle remodeling. It is suggested that exercise is capable of increasing the rate of protein turnover, hence the proteolytic activity, although not all the available results support this hypothesis (Kee et al. 2002). Further research is required in this important area.

Conclusion

Exercise may be viewed as a very complex stress to cope with and at the same time a beneficial stimulus to maintain homeostasis. As a result of exercise-induced adaptation, a variety of proteins change in quality and quantity, being produced and degraded in muscles and other tissues where selective alteration in the rate of protein turnover is a stimulus-dependent process. Proteins with short half-lives are readily capable of responding to exercise-induced alteration of the resting homeostatic state. Exercise of low intensity and relatively long duration can stimulate mitochondrial biogenesis, thereby elevating the activity and content of proteins that are involved in the efficient generation of ATP. Resistance exercise, on the other hand, can result in significant gain in the content of structural proteins, such as actin, myosin, collagen, and so on, to achieve a greater power output. It appears that regular exercise regulates the rate of protein turnover (not only in the skeletal and cardiac muscles but also in tissues such as the brain) that is an important part of the systemic adaptive process, thus removing potentially detrimental damaged proteins and replacing them with newly synthesized active molecules.

Chapter 5

Extracellular Matrix and Exercise

Vuokko M. Kovanen

In multicellular animals, the integration of tissues is made possible by their microenvironment, the extracellular matrix (ECM) with its sparse population of cells. ECM classically provides mechanical stability, physical strength, and elasticity to tissues like skin, tendon, cartilage, and bone. The framework of ECM consists of insoluble fibrils primarily of collagen, whose structure varies depending on the demands the tissue has to meet. An important determinant is the proportion of different collagen types. However, the presence of other ECM glycoproteins and proteoglycans strongly affects the shape, organization, and mechanical properties of the collagen fibrils.

The given mission of ECM, however, has been extended via its explored role as a dynamic action zone affecting, for example, the instruction of gene expression, cell cycle progress, cell shape, anchorage versus migration, and apoptosis. The information within ECM can be interpreted by cells via specific ECM motifs and cell surface receptors. The enzymes involved in ECM remodeling are pivotal in the decisions to release ECM-bond growth factors, expose new cryptic sites of ECM, and release bioactive ECM fragments. The magnitude of the guiding value of ECM has been explored during the last 5 to 10 years, and intense research focuses on the mechanisms involved. The significance of ECM has become apparent in various physiological phenomena like tissue development, differentiation, and adaptation as well as diseases and tissue repair.

Extracellular Matrices

The development of an ECM, a complex mesh of various proteins, was a crucial event in the emergence of multicellular organisms. Currently, the various ECMs can be divided into two major groups, the interstitial ECM and the basement membranes (BMs). BMs are thin sheets of highly specialized ECM that are found at the interfaces between mesenchyme and epithelia/endothelia and in some cases around individual cells. The non-basement membrane part of the ECM can collectively be defined as the interstitial ECM. In the following pages the general composition of interstitial ECM and that of BMs are presented separately.

Composition of the Interstitial Extracellular Matrix

The interstitial matrices are highly specialized in structure and composition to bear different types of mechanical stress: tension, compression, shear, and combinations of these. They are known as the classical "connective tissues" that can be histologically/anatomically classified into loose and dense connective tissues, cartilage, and bone in adults. The interstitial ECM consists of insoluble fibrils built up of collagen, elastin, or both. The structure of the fibrils varies depending on the demands the tissue has to meet. An important determinant is the proportion of different collagen types. However, the shape, organization, and mechanical properties of the fibrils also depend on the presence of other ECM glycoproteins and proteoglycans. The macromolecules of ECM are presented in figure 5.1. Additionally, interstitial ECMs can also affect the properties/behavior of adjacent cells, including cell shape, gene expression, proliferation, migration, and even apoptosis (see, e.g., Aumailley & Gayraud 1998 for review).

Collagens

Collagens constitute a highly specialized family of ECM proteins. A protein is defined as collagen if it is a structural matrix protein and has the following structural characteristics: (1) It is composed of three α-chains, either identical or homologous; (2) its primary structure contains the glycine-x-y repeat sequence where x often is proline and y hydroxyproline; and (3) based on the continuity of the (gly-x-y) repeat sequence the protein is assembled into either one continuous collagenous triple helix (COL domain) or several COL domains separated by noncollagenous (NC) domains. More than 20 collagen types containing in total at least 38 distinct α-chains have been identified. Their genes, with large number of exons and long intron areas, are dispersed among at least 15 chromosomes (Myllyharju & Kivirikko 2001; Sato et al. 2002; Hashimoto et al. 2002).

The collagens can be divided into two major classes, the fibrillar and nonfibrillar collagens (Myllyharju & Kivirikko 2001). Table 5.1 presents all the known collagen types classified into subgroups, their occurrence in tissues, and known or suggested functions. The interstitial collagens include all other collagens except type IV collagen and the transmembrane collagens. The major fibril-forming collagen types I, II, and III account for 80% to 90% of all collagens in the human body. They form highly ordered quarter-staggered structures when assembling into fibrils (see figure 5.2) and serve as scaffold for numerous associated proteins, proteoglycans, and cells in ECM. Type V and XI collagens, classified as fibrillar collagens on the basis of their homology with type I, II, and III collagens, are quantitatively minor ones. As core proteins, they contribute to the fibrillogenesis of the major collagens by limiting the increase of fiber diameter. Furthermore, the spatial organization of type I collagen fibrils is affected by the FACIT collagens XII and XIV via their free aminoterminal NC portions (see figure 5.2). Type I collagen

Collagens

- Fibril-forming collagens
  - Collagens I, II, III, V, and XI
- Nonfibrillar collagens
  - Collagens IV, VI-X, XII-XIX, XXV, XXVI

Glycoproteins

- Interstitial connective tissue
  - Fibronectins
  - Tenascins
  - Fibrillins
  - Elastin
  - Matrillins
  - Thrombospondins
- Basement membranes
  - Laminins
  - Nidogen/entactin
  - Fibulin

Proteoglycans

- Small leucine-rich proteoglycans, SLRPs
  - Decorin
  - Biglycan
  - Fibromodulin
  - Lumican
  - Epiphycan
- Modular proteoglycans: nonhyaluronan binding
  - Perlecan
  - Agrin
  - Testican
- Modular proteoglycans: hyaluronan- and lectin-binding
  - Aggrecan
  - Versican
  - Neurocan
  - Brevican

Figure 5.1 Macromolecules of the extracellular matrix.

Reprinted, by permission, from M. Aumailley and B. Gayraud, 1998, "Structure and biological activity of the extracellular matrix," *Journal of Molecular Medicine*, 76: 253-265. Copyright © 1998 Springer-Verlag.

Table 5.1 Collagen Types, Their Tissue Distribution, and Major Functions

Type	Occurrence	Major function
Fibril-forming collagens		
I	Most connective tissues; abundant in bone, tendon, ligaments, dermis, intramuscular ECM	Provides tensile strength for connective tissues
II	Cartilage, intervertebral disc, cornea, vitreous humor	Provides tensile strength for connective tissues
III	Extensible connective tissues, for example, skin, lung, vascular system; ubiquitous with type I except in bone and tendon	Provides tensile strength for connective tissues
V	Tissues containing collagen I, widespread in low quantities	Controls the fibril diameter
XI	Tissues containing collagen II, quantitatively minor component	Controls the fibril diameter
Non-fibril-forming collagens--* Network-forming collagens		
IV	All basement membranes	Forms the skeleton for basement membranes
VIII	Many tissues, especially in endothelium	???
X	Hypertrophic mineralizing cartilage	Calcium binding
*** FACITs, the fibril-associated collagens**		
IX	Tissues containing collagen II, quantitatively minor component	Maintains structural integrity of tissue
XII	Tissues containing collagen I or II, quantitatively minor component	Controls the fibril diameter
XIV	Tissues containing collagen I or II	Controls the fibril diameter
XVI	Many tissues	—
XIX[1]	Many tissues	—
XX	—	—
XIX	—	—
XXI[2]	Blood vessel wall	Contributes to blood vessel ECM formation
*** Beaded-filament-forming collagens**		
VI	Widespread, quantitatively minor component in many soft tissues, skin, cornea, tendon, ligaments	Bridging between cells and ECM
*** Anchoring fibril-forming collagens**		
VII	Skin, cornea, cervix, oral and esophageal mucosa	Strengthens dermal-epidermal junction
*** MACITs (membrane-associated collagens with interrupted triple helices)**		
XIII	Widespread in low quantities	Cell—CM anchorage
XVII	Hemodesmosomes in epithelium of cornea, lung, skin, for example	Cell adhesion
XXV[3]	Brain, neurons	Amyloidogenesis, neuronal degeneration in Alzheimer's disease

(continued)

Table 5.1 *(continued)*

Type	Occurrence	Major function
*** Multiplexins (multiple triple helix domains and interruptions)**		
XV	Many tissues in basement membrane zone, particularly in skeletal and heart muscle, kidney, and placenta	Anchorage of BM to ECM; angiogenesis inhibition by NC1 domain
XVIII	Many tissues in basement membrane zone, particularly in kidney, liver, and lung	As for type XV
XXVI[4]	Adult and neonate testis and ovary	Development of ovary and testis

* = Subgroups of nonfibril-forming collagens.

Adapted from Snellman 2000, Sumiyoshi et al., 2001, Chou and Li 2002, Hashimoto et al., 2002, and Sato et al., 2002.

characteristically forms strong parallel fibers and confers tensile strength and rigidity, whereas type III collagen forms a looser meshwork of fibers and confers compliance to the tissue.

The non-fibril-forming collagens all share the common property of having one or more imperfections in the COL sequence in the triple helical domains. Type VI collagen has shortest COL domains and type VII collagen the longest. As indicated by table 5.1, both fibril-forming and non-fibril-forming collagens show extremely wide tissue distribution as well as diverse functions within the body. More detailed information about the different collagen types can be found in the reviews by Vuorio and Crombrugghe (1990), Aumailley and Gayraud (1998), and Myllyharju and Kivirikko (2001), for example.

Proteoglycans

The classification of proteoglycans is presented in figure 5.1. They are a subset of large ECM proteins containing an exceptionally high amount of carbohydrates called glycosaminoglycans (GAG), which are covalently attached to a central core protein. Due to the interaction of the GAGs with water, the proteoglycans are well suited to resist compressive forces such as those that occur, for example, in cartilage. Additionally, via GAGs the proteoglycans interact with cells, collagens, and other ECM proteins as well as some growth factors. Many proteoglycans function as modulators of growth factors such as transforming growth factor-β (TGF-β), epidermal growth factor cellular receptor, and basic fibroblast growth factor (bFGF) (Moscatello et al. 1998; Yamaguchi et al. 1990) and thus affect intracellular signaling.

Decorin, one of the small leucine-rich proteoglycans (SLRP), specifically has a role in the fibrillogenesis of type I collagen (see figure 5.2) (see Neame et al. 2000). Interestingly, decorin is up-regulated in connection with experimentally induced postmyocardial infarction, which typically involves significant fibrosis during the recovery phase. Decorin expression was found to increase at an appropriate time to play a role in the organization of the collagen matrix (Zimmerman et al. 2001).

Some heparan sulfate proteoglycans, like syndecans, are cell surface molecules that penetrate the plasmalemma via their core protein, while their heparan sulfate chains bind interstitial ECM molecules like collagens I and III, fibronectin, tenascin, and thrombospondin. Syndecans, additionally, function as receptors, for example for bFGF, subsequently exposing it for the effective signaling receptor (e.g., Jalkanen et al. 1991; Zhang et al. 2003). For the wide area of proteoglycans the reader is referred to reviews such as Oldberg et al. 1990, Jalkanen et al. 1991, Heinegard and Pimental 1992, and Nakato and Kimata 2002.

Other Glycoproteins

In addition to collagens and proteoglycans, many NC glycoproteins are building blocks of ECM, either in the interstitial connective tissue or restricted to BMs (see figure 5.1). Among the various interstitial glycoproteins, only fibronectin and tenascin are discussed briefly here.

Fibronectin (FN), one of the best-studied interstitial ECM glycoproteins with wide tissue distribution, is present in the body as an insoluble tissue form as well as a soluble form in blood and in the interstitial space (Ruoslahti & Vaheri 1974). The alternatively spliced FN variants originating from a single gene diverge functionally from each other

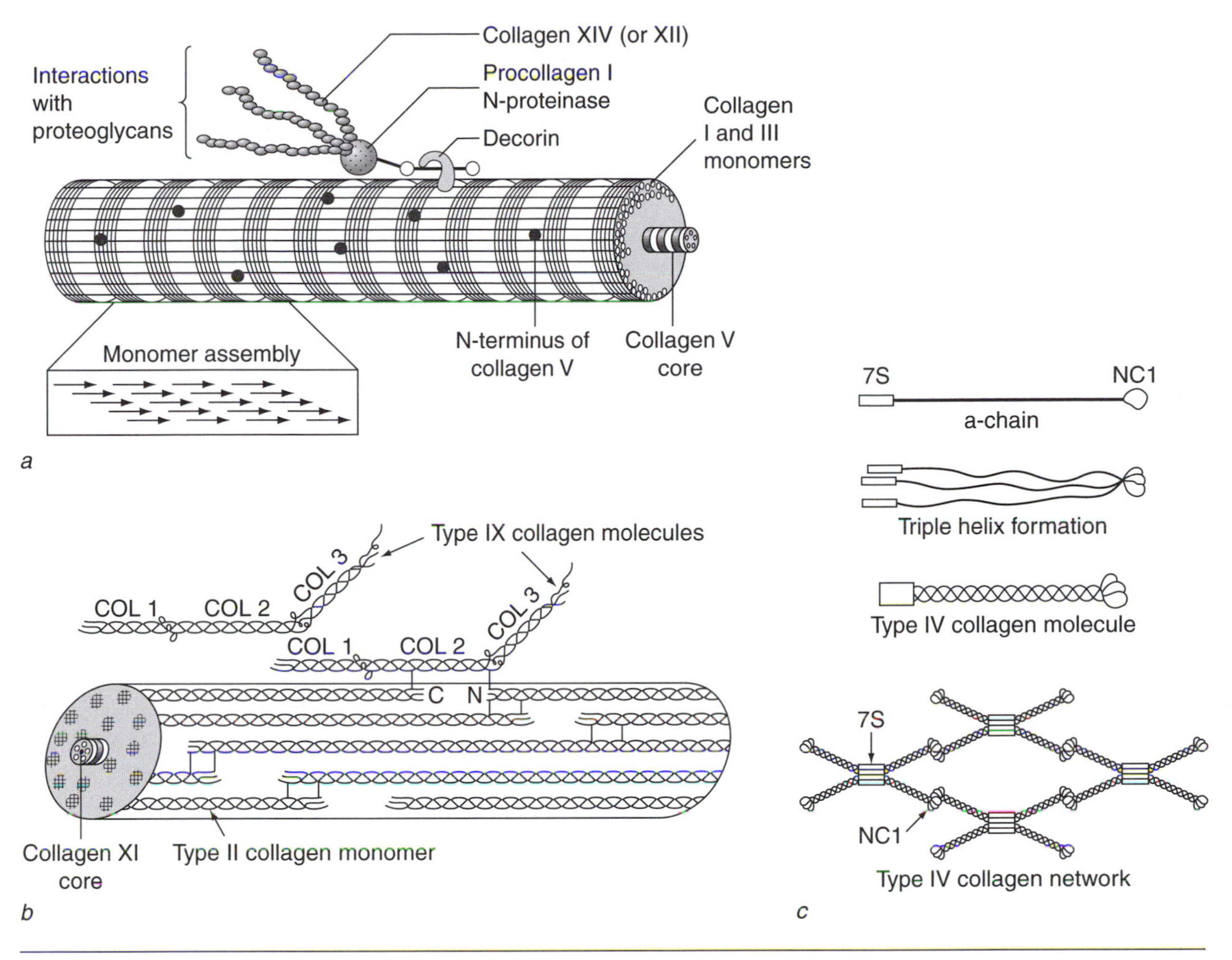

Figure 5.2 Structure of collagens I/III, II, and IV. *(a)* Type I and III collagen can exist as heterotypic fibrils with type V collagen copolymerized inside the fibril. Decorin binds at the surface of the fibril with collagens I, XII, and XIV and affects fibril diameter. *(b)* The structural constituents of type II collagen fibrils are type XI collagen as the fibril core and type IX collagen as the FACIT collagen. *(c)* The monomeric α-chains of collagen IV include several short interruptions in the long COL sequence, N-terminal 7S domain, and C-terminal NC1 domain. Four trimers interact through their 7S domains in a spider-shaped structure, and two trimers interact head to head through their NC1 domains. Several trimers can also lace together along their triple helical domain, thickening the sheetlike structure of type IV collagen.

Part *a* reprinted, by permission, from M. Aumailley and B. Gayraud, 1998, "Structure and biological activity of the extracellular matrix," *Journal of Molecular Medicine*, 76: 253-265. Copyright © 1998 Springer-Verlag. Part *b* reproduced with permission, from Diab et al., 1996, *Biochemical Journal* 314: 327-332. Copyright © the Biochemical Society. Part *c* reprinted, by permission, form Nathalie Ortega and Zena Werb, 2002, "New functional roles for non-collagenous domains of basement membrane collagens" *Journal of Cell Science* 115: 4201-4214.

with regard to the presence or absence of certain functional domains (e.g., Dickinson et al. 1994). FN specifically mediates the attachment of cells to the other ECM molecules. The ability of several cell types to migrate along surfaces covered by FN is fundamental, for example, for tissue repair-involved processes. In connection with muscle cell injury, leakage of soluble fibronectin from the interstitial space into muscle cells is found as an early sign of muscle cell membrane damage (Lieber et al. 1996).

Tenascins are their own gene family of interstitial ECM glycoproteins (e.g., Chiquet-Ehrismann et al. 1994) including tenascins-C, -R, -X, -Y, and -W, among which tenascin-C appears to be the most widely distributed and often coexpressed with fibronectin. Tenascins function as adhesive proteins for several cell types and attach ECM molecules like collagens, fibronectin, and proteoglycans. However, tenascin can also prevent cell attachment to fibronectin (see Orend & Chiquet-Ehrismann 2000). Tenascin induces MMP-2 (matrix metalloproteinase-2) to cleave fibronectin, after which an anti-adhesive site in fibronectin is exposed. For tissue homeostasis and tissue development, cellular de-adhesion and cell adhesion are equally important throughout the lifetime. Interestingly tenascin-C, often co-expressed with collagen XII in tissues subjected to high tensile stress, appears to be directly and reciprocally controlled by mechanical load (Flück et al. 2000). In connection with inflammatory diseases in several tissues, increased tenascin expression has been found; and in myocarditis, for example, tenascin-C expression appears as a useful marker for evaluation of disease activity (Imanaka-Yoshida et al. 2002).

Composition of Basement Membranes

BMs in the body are widespread, bordering the epithelial/endothelial cell layers. BMs can also delineate individual cells as in the case of muscle fibers, fat cells, and the Schwann cells in peripheral nerves. The fine structure and composition vary from tissue to tissue, as well as within the same tissue at different developmental stages and during repair. The synthesis of BMs is both spatially and temporally strictly feedback-regulated by the neighboring cells. The regulatory functions are mediated by cell matrix receptors, for example integrins (Streuli 1999; Heino 2000).

The outer electron-dense layer of BM, lamina densa, interacts with the surroundings via fibrillar structures in the lamina fibrillaris. Originally BMs were thought to have structural roles only, but recent discoveries of its complex molecular structure allow it significant regulatory functions affecting, for example, cell shape, gene expression, proliferation, and apoptosis (Aumailley & Gayraud 1998; Erickson & Couchman 2000; Ortega & Werb 2002).

All BMs contain laminins, nidogen-1/entactin-1, type IV collagen, and heparan sulfate proteoglycans, whereas collagens XV, XIII, and XVIII show tissue-specific distribution. The basic BM model involves two separate networks, one consisting of type IV collagen and the other of multiple laminins. The laminin network extends through both lamina rara and densa. Type IV collagen network associates with the laminin network within the lamina densa, which thus has an electron-dense appearance. Figure 5.6 presents some characteristics of muscle cell BM (for review see, e.g., Yurchenco & O'Rear 1994; Timpl & Brown 1996).

The heterotrimeric members of the laminin (LM) gene family are formed by combinations of different α-, β-, and γ-chains that form cross- or T-shaped structures. Currently 15 different heterotrimeric laminins have been detected due to the presence of five α-chains, three β-chains, and three γ-chains, all coded by their own genes (Colognato & Yurchenco 2000). Laminin molecules polymerize into the BM network via the interaction of the laminin short arms. Networking is assisted by the specific LM receptors, dystroglycan, and β1-integrins and involves intracellular actin reorganization and signaling events like tyrosine phosphorylation (Colognato et al. 1999; Colognato & Yurchenco 2000; Henry 2001). The fundamental significance of laminin for life is verified by the early embryonic lethality in γ1 (the most common laminin γ-chain) null mice (Smyth et al. 1999). Laminins affect, for example, myogenesis and synaptogenesis in skeletal muscles. Complete laminin α2 deficiency causes approximately half of human congenital muscular dystrophy (CMD) cases, while single missense mutations have been found in milder CMD with partial laminin α2 deficiency (e.g., Tezak et al. 2003). Interestingly, laminins guide motor nerve axonal behavior, laminin 11 being a critical organizer of synaptic development in developing and regenerating synapses in vivo (Patton et al. 1997).

The type IV collagen α-chains, α1 through α6, are coded by six different genes and have unique patterns of tissue distribution (for review see, e.g., Kühn 1995; Sado et al. 1998). The most common type IV collagen is composed of two α1(IV)-chains and one α2(IV)-chain (see figure 5.2). The specific and significant functions of the α3-α6-chains of type IV collagen are illustrated by their spatial and temporal regulation in physiological as well as in pathological processes. Mutations in their structure induce severe symptoms in humans, for example Alports syndrome, a heritable progressive hematuric nephritis with changes such as hearing loss and ocular changes (Kashtan 2000). The significance of the type IV collagens for growth, development, and cellular functions is also highlighted by the conserved expression of these proteins both in invertebrates (Zhang et al. 1994) and in humans (Zhou et al. 1993).

In addition to type IV collagen, collagens XV and XVIII are major components of various BMs (Rehn & Pihlajaniemi 1997; Tomono et al. 2002). Both collagens undergo intensive post-translational glycosylation. Consequently, type XV collagen is characterized as BM chondroitin sulfate (Li et al. 2000) and type XVIII collagen as BM heparan sulfate (Halfter et al. 1998) proteoglycan. The nonvascular BMs are shown to contain one of the two types, while subepithelial BMs contain in general collagen XVIII. Skeletal and cardiac muscle cell BMs appear to harbor mainly collagen XV. Continuous capillaries contain both collagen types, whereas specialized capillaries like those in glomeruli express only type XVIII. This distribution pattern suggests that the functions of BMs are finely adjusted by their collagen composition (Tomono et al. 2002).

Very recent research highlights the importance of all the BM collagens, IV, XV, and XVIII, via their C-terminal proteolytic NC1 fragments, tunstatin, endostatin, and nestin, respectively (see Ortega & Werb 2002; Tomono et al. 2002). They are released as a consequence of the degradation of BM structures during numerous physiological and pathological processes. For example, endostatin is released from collagen XVIII via MMPs. Once released, the fragments induce their regulatory signals, differing from those of the parent molecules, through the function of different integrin receptors. Specifically, the NC1 fragments function as angiogenesis inhibitors both in vitro and in vivo; but they can also stimulate migration, proliferation, apoptosis, or survival of different cell types and suppress various morphogenetic events (e.g., Sasaki et al. 2000; Marneros & Olsen 2001; Ortega & Werb 2002).

Type XIII collagen is a transmembrane protein that has several alternatively spliced forms. The protein is characterized by a short N-terminal cytosolic domain, a single transmembrane domain, and an extensive ectodomain with three COL and four NC domains (e.g., Hägg et al. 1998). Type XIII collagen has wide tissue distribution and is located at many cell matrix adhesion sites such as myotendinous junctions in skeletal muscle and at cell–cell adhesion sites like the intercalated discs of cardiac cells (Hägg et al. 2001). Its in-vitro binding with fibronectin, nidogen-2, perlecan, and heparin further supports its function in cell–matrix adhesion (Aumailley and Gayraud 1998) Mutant type XIII collagen mice express myopathic muscle phenotype, and mutations in this collagen type are suggested to result in certain inherited muscular myopathies whose cause has not yet been identified (Kvist et al. 2001).

Synthesis of Collagens

Like many secretory proteins, collagens undergo extensive enzymatic co- and post-translational modification within the rough endoplasmic reticulum (RER), where, however, their folding into the triple helix is unique. Once folded, collagen is secreted through Golgi as procollagen molecules and further enzymatically processed extracellularly to produce the collagen molecules. The assembly of individual collagen molecules in a quarter-staggered manner into fibrils occurs extracellularly within the crypts of the folded cell membrane, in interaction with cell membrane receptors and other ECM proteins such as fibronectin (e.g., Velling et al. 2002). Collagen biosynthesis-related matters are schematically presented in figure 5.3.

The unique folding of collagen pro-α-chains within the RER is initiated by the hydrophobic and electrostatic interactions among the globular C-terminal propeptides. This ensures that the three genetically distinct pro-α-chains combine in the correct composition to build up the specific collagen type. Before the nucleated growth of collagen triple helix propagates, collagen-specific RER resident enzymes catalyze several post-translational modifications such as prolyl and lysyl hydroxylation, hydroxylysyl glycosylation, and disulfide formation. For more details of collagen biosynthesis, readers are referred to reviews such as those by Kivirikko and Myllylä (1979, 1984) and Kivirikko et al. (1992).

A rate-limiting step in collagen biosynthesis is the hydroxylation of specific prolyl residues by the enzyme prolyl 4-hydroxylase (P4H). Consequently, P4H activity is commonly used to indicate enhanced collagen biosynthesis both in vitro and in vivo. The folding of the three polypeptides into the triple helix within the RER is delayed until the chains acquire the necessary content of hydroxyproline, about 100 residues per chain. This modification is required for the formation of stable and correctly folded triple helical procollagen. The active P4H enzyme is a tetramer composed of two pairs of nonidentical subunits, $\alpha 2\beta 2$. The α subunit contains the major portion of the catalytic site and incorporates into the tetramer directly after its synthesis, and its concentration limits the rate of formation of active P4H (see, e.g., Kivirikko & Myllylä 1984; Kivirikko et al. 1992).

Lysyl hydroxylase (LH), present in tissues as three isoenzymes (LH1, LH2, and LH3) (e.g., Ruotsalainen et al. 1999), hydroxylates specific lysyl residues, some of which are mono- or diglycosylated by specific glycosyl transferases, GT and GGT, respectively (e.g., Kivirikko & Myllylä 1979). Recently, LH3 has been recognized as a multifunctional enzyme having both LH and

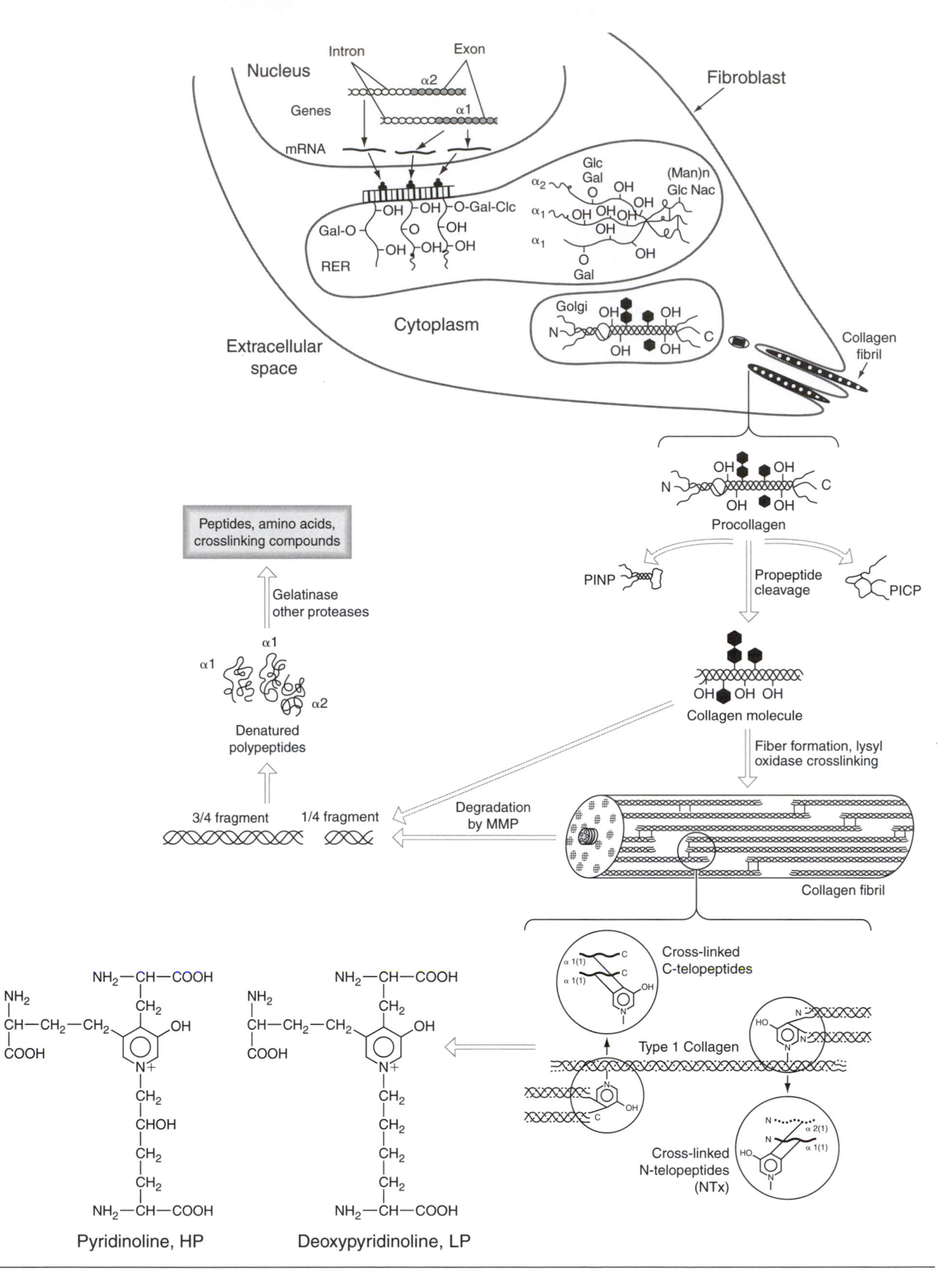

Figure 5.3 Collagen biosynthesis and degradation. Genes α1 and α2 of type I collagen are transcribed into respective mRNAs to become translated on the ribosomes at the rough endoplasmic reticulum (RER). During extracellular processing of the protein, the N- and C-terminal propeptides, PINP and PICP, respectively, are released and can be measured as indicators of, for example, type I collagen synthesis both in tissues and in blood. Single collagen molecules and collagen in fibrils are degraded by MMPs. The denatured fragments are susceptible to gelatinases and some other proteases. Consequently peptides, cross-linking compounds, and amino acids are released. Free and peptide-bound pyridinoline cross-links can be measured from serum and urine as indicators of degradation of mature collagen in fibrils.

Reprinted, by permission, from D. Eyre, 1995, "The specificity of collagen cross-links as markers of bone and connective tissue degradation," *Acta Orthopedica Scandinavica* (Suppl 266) 66: 166-170.

GGT activity (Wang et al. 2002). The amount of hydroxylysine in collagens varies from 17% in type III collagen to almost 90% in type IV collagen, further variations being found within the same collagen type and even in the same tissue under different physiological and pathological conditions. The extent of glycosylation, too, varies considerably among different collagen types and within the same collagen type from various sources (Kivirikko et al. 1992). The high level of hydroxylysyl-linked carbohydrate units associates with decreased fibril formation rate and a decrease in fibril diameter both in vitro and in vivo (Brinckmann et al. 1999; Notbohm et al. 1999), and thus significantly affects the mechanical properties of collagen fibers.

Folding, maturation, and efficient and accurate procollagen trafficking through Golgi and secretion from cells involve interaction with the RER resident collagen-specific chaperone, the heat shock protein 47 (HSP47) (e.g., Nagata 1996; Tasab et al. 2002). HSP47 expression is induced, for example, in tissue wounding, emphasizing its role in collagen synthesis (Keagle et al. 2001).

Extracellularly the assembly of collagen fibrils into larger fibers occurs via the lateral fusion of discrete ~4-nm subunits. The lateral fusion during fibrillogenesis appears important in generating resistance to deformation at low strain, while linear subunit fusion leading to longer fibrils appears important in the ultimate mechanical properties at high strain (Christiansen et al. 2000).

The molecular maturation of collagen molecules during fibrillogenesis includes the formation of intra- and intermolecular cross-links (see figure 5.3). This essential extracellular process affords tensile strength to the collagen fibrils and fibers. It is a precise, enzymatically controlled series of reactions. Initially, lysyl oxidase catalyzes the oxidative deamination of the ϵ-amino groups of specific lysyl and hydroxylysyl residues, producing reactive lysyl aldehydes. They are subsequently converted into intermediate divalent cross-links, which further mature to trivalent cross-links. Histidinohydroxymerodesmosine, pyrrole, and the hydroxypyridinium cross-links pyridinoline and deoxypyridinoline are the three best-known functional mature cross-links of collagens (see figure 5.3) (e.g., Last et al. 1990; Hanson & Eyre 1996). The nature of collagen cross-links depends primarily on the extent of lysyl hydroxylation of the N- and C-terminal telopeptide regions, those derived from hydroxylysines being more stable than those derived from lysine residues (Kivirikko & Myllylä 1984).

With advancing age, a second more adventitious non-enzymatic process produces glycation cross-links (Sell et al. 1996; Bailey 2001). The cross-linking of collagen alters its mechanical and biochemical properties, leading to increased tensile strength, decreased solubility, and enhanced resistance to some proteases. Especially the effects of glycation cross-links are regarded as deleterious to the tissue. As a consequence of the unique maturation process of collagens, there exist different pools of collagens in tissues with variable stabilities.

Degradation of Collagens

Degradation of collagen is essential in the turnover and remodeling of different connective tissues under both physiological and pathological conditions. Extracellular collagen concentration is controlled both intracellularly and extracellularly. Intracellular procollagen turnover may be influenced by altering synthesis rate—transcription, translation, or both—and the intracellular degradation rate of newly synthesized procollagen molecules. In skeletal muscle, as much as about 30% and in heart muscle about 60% of procollagen may be degraded intracellularly within minutes of synthesis (Laurent et al. 1985). In the extracellular space, the newly synthesized forms of collagen are degraded more quickly than the mature, cross-linked collagen fibrils and fibers. A generalized presentation of extracellular collagen degradation is included in figure 5.3.

A variety of proteolytic enzymes are involved in the degradation of collagens both intra- and extracellularly (see table 5.2). MMPs play a pivotal role in the extracellular degradation of collagens under conditions of rapid remodeling, as in connection with inflammation. Under steady state conditions, such as during normal turnover of soft connective tissues, collagen degradation is likely to take place particularly within the lysosomal apparatus after phagocytosis of the fibril (Creemers et al. 1998a,b).

MMPs represent a family of zinc- and calcium-dependent enzymes that are collectively responsible for remodeling of ECM (for review see, e.g., Birkedal-Hansen 1995; Borkakoti 2000; Ravanti & Kähäri 2000). Currently, at least 25 MMPs are identified that can be loosely subdivided on the basis of their substrate specificity into collagenases, gelatinases, stromelysins, matrilysin, macrophage elastase, and membrane-type MMPs (MT-MMPs), the last of these being localized to the cell surfaces as receptors (see table 5.3).

Table 5.2 Enzymatic Pathways Involved in Extracellular Matrix Degradation

Pathway	Tissue degraded	Effector enzyme	Cellular location
Plg-dependent pathway	Interstitial connective tissues, basement membranes	Pln	Peri/extracellular
MMP pathway	Interstitial connective tissues, basement membranes	MMPs	Peri/extracellular
PMN serine pathway	Interstitial connective tissues, basement membranes	PMN elastase, cathepsin G	Peri/extracellular
Phagocytic pathway	Interstitial connective tissues	Cathepsins	Intracellular
Osteoclastic pathway	Bone, cementum, dentin	Cathepsins	Peri/extracellular

Plg = plasminogen; Pln = plasmin; PMN = polymorphonuclear leucocytes.

From H. Birkedal-Hansen et al., 1993, "Matrix metalloproteinases: A review," *Critical Reviews in Oral Biology and Medicine* 197-250. Reprinted, by permission, from CRC Press, Inc.

Table 5.3 Collagenases, Gelatinases, and Membrane Type 1 Matrix Metalloproteinases (MMP) With Their Substrates

Collagenase type	Substrate
Collagenases	
Collagenase-1, MMP-1	Collagen I, II, III, VII, X, aggregan
Collagenase-2, MMP-8	Collagen I, II, III, aggregan
Collagenase-3, MMP-13	Collagen I, II, III, IV, IX, X, XIV, gelatin, laminin, fibronectin, aggregan, fibrillin, large tenascin C, osteonectin
Gelatinases	
Gelatinase A, MMP-2	Gelatin, collagen I, IV, V, VII, X, tenascin, fibronectin, fibrillin, osteonectin, α2-macroglobulin (α2M)
Gelatinase B, MMP-9	Gelatin, collagen IV, V, VII, XI, XIV, elastin, fibrillin, osteonectin, α2M
Membrane type MMPs	
MT1-MMP, MMP-14	Collagen I, II, III, gelatin, laminin, aggregan, vitronectin, tenascin, nidogen, perlecan, α2M, fibrin

Adapted, by permission, from Ravanti, L. & V-M. Kähäri 2000. Matrix metalloproteinases in wound repair (Review). *International Journal of Molecular Medicine* 6: 391-407.

One of the three collagenases, primarily, initiates the extracellular degradation of triple helical collagen fibrils specifically into 3/4- and 1/4-length fragments (Netzel-Arnett et al. 1991). Additionally, the MT1-MMP and MT2-MMP can degrade fibrillar collagens. The degradation of the denatured collagen fragments is continued by gelatinases, MMP-2, and MMP-9, which also degrade type IV collagen (Parks & Mecham 1998). Recently, a novel biphasic collagenolytic mechanism of action has been suggested for MMP-2 (Patterson et al. 2001). However, the function of MMP-2 as well as that of MT-MMPs in the degradation of fibrillar collagens might be locally limited due to their presence in association with cell surfaces.

MMPs have an important role in tissue remodeling during, for example, development, growth, and tissue repair as well as in different disease states. Their expression is low in healthy tissues in steady state conditions, but the expression can be transcriptionally induced in response to cytokines, growth factors, ECM components, and hormones, for example. Additionally, MMPs, which are secreted as inactive proenzymes, can be post-translationally activated through a plasmin or MT-MMP-dependent pathway; the latter is schematically presented in figure 5.4 (e.g., Ellerbroek et al. 2001; Valtanen et al. 2000; Itoh et al. 1998).

TIMP -1, -2, -3, and -4, the tissue inhibitors of MMPs, are also transcriptionally induced in response to cytokines and growth factors, for example. The pericellular TIMP-2 plays a dual role. It inhibits either MMP-2 or MT1-MMP catalytic activity via formation of a 1:1 noncovalent inactive enzyme-inhibitor complex. Somewhat paradoxically, TIMP-2 is also critical in the ternary complex formation for proMMP-2 activation, where MT1-MMP is the third player (see figure 5.4). TIMPs are important regulators in the ECM turnover via the regulation of MMP activity, but multiple other biological functions have been suggested for them as well. TIMPs have been shown to inhibit cellular invasion, tumorigenesis, metastasis, and angiogenesis and affect growth of several cell types as well as involving apoptosis, independent of their MMP-inhibitory activity. The mechanisms underlying the cellular effects of TIMPs are poorly understood but are under intensive research. Currently it can be stated that these relatively small proteins illustrate well the multiple and complex roles of proteins associated with the metabolism and structure of the ECM (for review and references see, e.g., Brew et al. 2000; Nagase & Woessner 1999).

ECM of Skeletal Muscle

A specialized setup of matrix molecules as presented in the previous pages builds up the intramuscular ECM. Individual muscle fibers are enclosed by ECM structure called endomysium, while bundles of muscle fibers are surrounded by perimysium. Loose networks of collagen fibrils appear to connect perimysium and endomysium. Collagenous reticular layer of endomysium separates adjacent muscle fibers from each other (see figure 5.5). Epimysium is the ECM layer covering

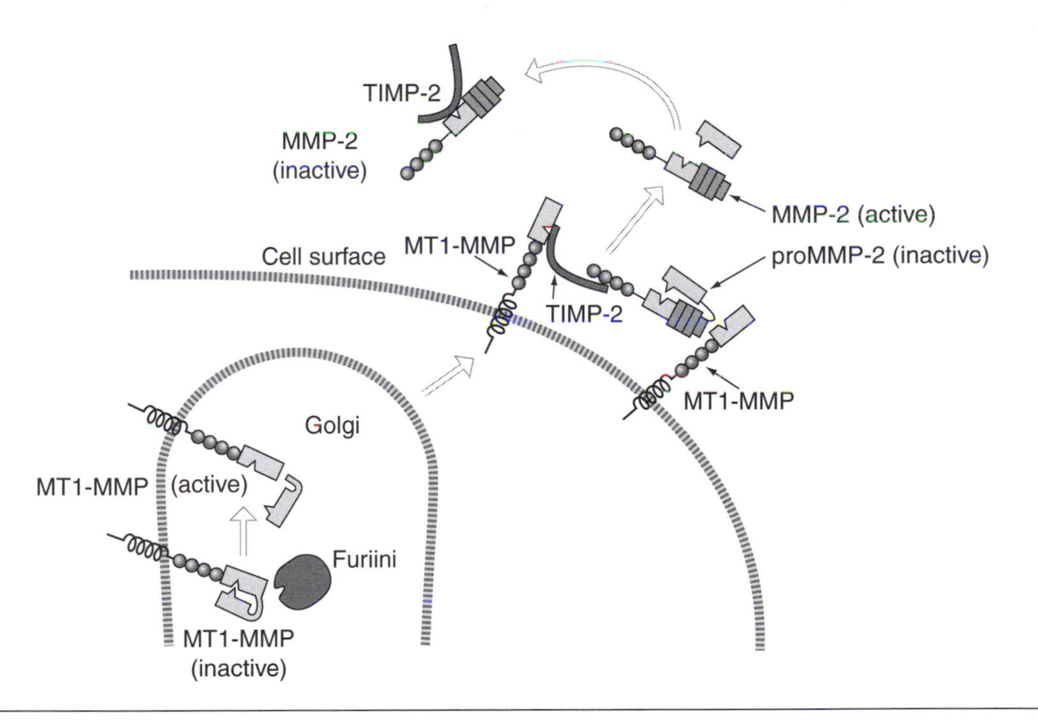

Figure 5.4 Schematic presentation of the activation of matrix MMP-2 in the TIMP-2- and MT1-MMP-dependent pathway. MT1-MMP is activated by the furin enzyme while still within the Golgi. At cell surface MT1-MMP binds the inactive proMMP-2 with the aid of TIMP-2. Paradoxically, TIMP, which has been determined to inhibit MMPs, here is necessary in the process, activating proMMP-2. Another MT1-MMP is needed to activate the proMMP-2 by cleaving off the prodomain. Consequently, activated MMP-2 is released and degrades its substrate proteins; but TIMP-2, among other inhibitors, limits the amount of the active MMP-2 enzyme. Accordingly, the local concentrations of TIMP-2 and MT1-MMP are critical for MMP-2 activation.

Adapted, by permission, from J. Heino and M. Vuento, 2002, *Solubiologia* (Helsinki/WSOY), 204.

the whole muscle. The general structure, morphology, and composition of these entities are reviewed by, for example, Mayne and Sanderson (1985), Kovanen (1989), Purslow and Duance (1990), Kovanen (2002), and Purslow (2002).

The molecular complexity of the intramuscular ECM as a dynamic action zone around muscle cells has been explored in line with the overall ECM research. Several different ECM molecules have been shown to be necessary components in the development of healthy muscles during embryogenesis (see, e.g., Patton et al. 1997) and during regeneration of muscle cells (Grounds et al. 1998). Morphologically, the beauty of the intramuscular collagen network (figure 5.5) was introduced via SEM (scanning electron microscope) pictures published by Nishimura et al. (1994) and Purslow and Trotter (1994). Via those images it became easier to figure out how collagenous structures provide an extensive three-dimensional set of organized and connected tunnels within which muscle fibers as well as fibroblasts operate.

Force transmission in muscles occurs not only at myotendinous junctions (MTJ); also lateral force transmission via ECM between neighboring fibers and fascicles has been verified (see Huijing 1999; Purslow 2002; Trotter 2002). Mechanical forces are coordinated and passed between adjacent muscle cells via cell matrix interactions and via the endomysial ECM that links the cells together. An emerging concept is that division of a muscle into fascicles by the perimysial ECM is related to the need to accommodate shear strains as muscles change shape during contraction and extension. Muscle tissue can be viewed as an example of a biologically "smart material" in which the endo-

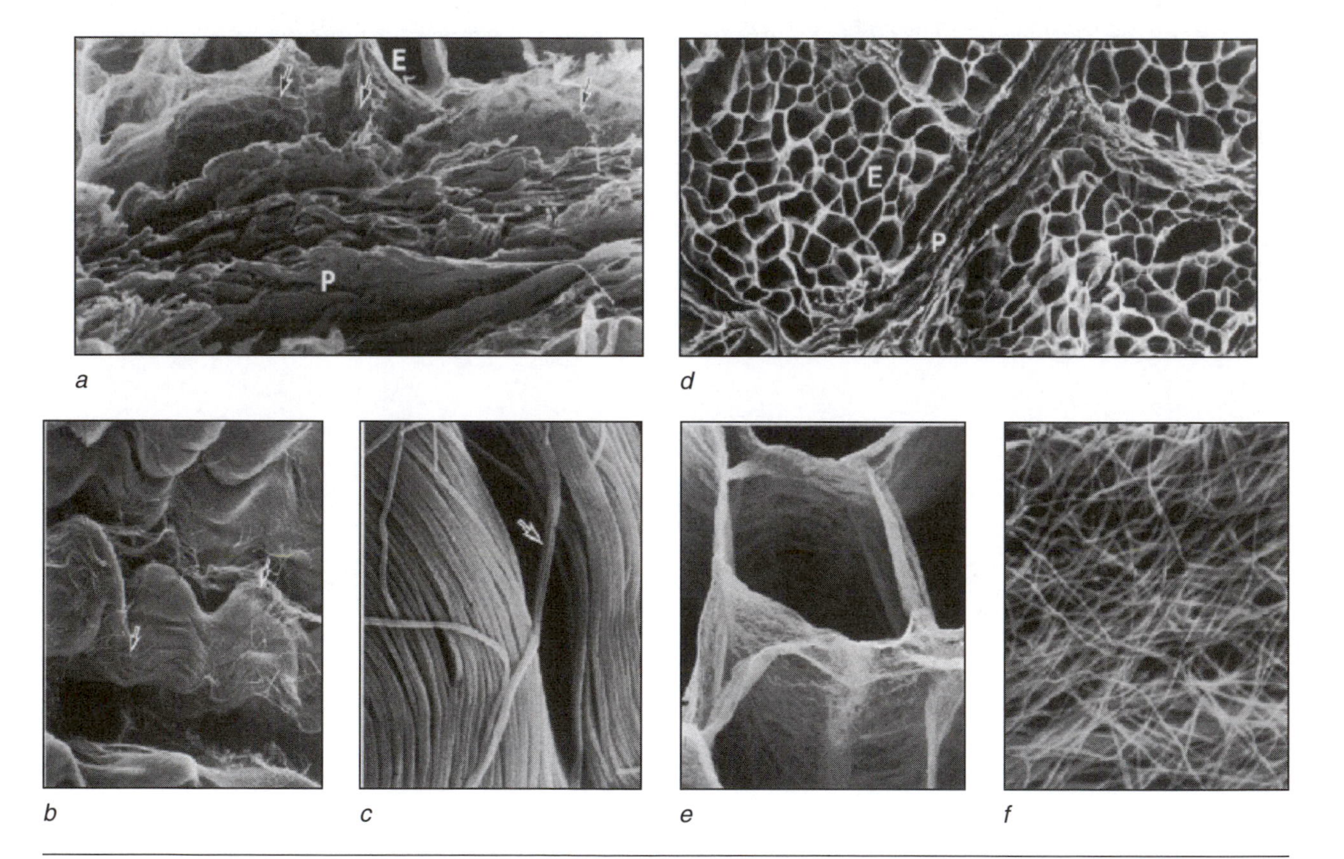

Figure 5.5 (SEM) micrographs of intramuscular arrangement of collagenous structures in bovine semitendinous muscle. Muscle cell contents, BMs, and proteoglycans have been removed chemically *(a and b).* Perimysium (P) is composed of several sheets of collagen fibers *(a)* that have an extremely wavy pattern covered with loose networks (arrow) of collagen fibrils *(b)*. *(c)* Wavy perimysium with higher magnification shows the arrangement of collagen fibrils, some branching off the fiber and fusing with the adjacent collagen fiber. *(d)* Endomysium looks like cylindrical sheets housing individual muscle fibers as honeycomb-like structures. *(e and f)* The membranous endomysial sheets are composed of fine fibrous meshwork of collagen fibers making up the reticular layer of the endomysium.

Part *a* from Nishimura, et al., 1994, "Ultrastructure of the intramuscular connective tissue in bovine skeletal muscle: A demonstration using the cell-maceration/scanning electron microscope method" *Acta Anatomica* 151: 250-257. Reprinted by permission of S. Karger AG, Medical and Scientific Publishers.

mysial and perimysial structures significantly reduce stress inequalities in the fiber and at the fiber matrix interface, increasing the overall efficiency of force transmission (Trotter et al. 1995).

Composition of Muscle ECM

The total amount of collagen in muscle is about 1% to 9% of the fat-free dry mass (Lawrie 1979). Perimysial ECM contains about 95% collagen, while endomysium contains about 40% collagen (Light & Champion 1984). The vast majority of collagen is composed of the interstitial fibrillar collagens, which represent as much as 95% to 97% of the total muscular collagen (Bailey & Nicholas 1989). Type I, III, and V collagens are the fibril-forming collagens in each anatomic entity in skeletal muscle. Type I collagen predominates in epi- and perimysium, whereas types I and III are more equal in endomysium (Foidart et al. 1981; Light & Champion 1984). It is not known whether type I and III collagens are present in muscle as "pure" fibers, as I/III heterotypic fibers, or both.

The proteoglycans of skeletal muscle contain both chondroitin sulfate and heparan sulfate. About 70% of the proteoglycans are of the chondroitin sulfate type and are associated with collagen fibrils in both the perimysium and the endomysium (Purslow & Duance 1990). Heparan sulfate proteoglycans (HSPG) represent about 30% of total muscle proteoglycan content, are associated with endomysial BMs, and appear to be five times concentrated at the neuromuscular junction (e.g., Bayne et al. 1984; Nishimura et al. 1996). Decorin also interacts with the HSPGs syndecan and glypican and significantly contributes to muscle development and differentiation, for example via modulating the cell's responses to growth factors such as TGF-β and bFGF (Larrain et al. 1998; Velleman 1998; Riquelme et al. 2001; Osses & Brandan 2001). Mice deficient in SLRPs (small leucine-rich proteoglycans) have been shown to develop muscular dystrophy, among other diseases, suggesting that these molecules are part of a yet undiagnosed predisposing genetic factor (see Ameye & Young 2002). Despite their obvious significance in skeletal muscle ECM, hardly any studies have been published about proteoglycans in different loading conditions in skeletal muscle—in contrast to the situation in articular cartilage, for example.

Fibronectin (FN) is present throughout the different ECM layers in muscle. In endomysium as a component of BM, FN extends to the reticular layer and thus interconnects the cell surface and the intercellular matrix (Hantai et al. 1983). In connection with regeneration of muscle following trauma, FN appears to play an important role (Lehto 1983).

Collagens IV, VI, XIII, XV, XVIII, and XIX, at least, are found at the BM zone in skeletal muscles. Type IV collagen as the major BM collagen forms the skeleton. The interrupted triple helical structure of type IV collagen molecules and the networking of the molecules provide BMs with flexibility, strength, and rigidity (see, e.g., Kühn 1995). Regional differences exist within muscle cell BMs in the α-chain composition (α1-α6) of type IV collagen (Sanes et al. 1990; Miner & Sanes 1994). In human muscle, α3(IV)- and α4(IV)-chains are restricted to the neuromuscular junction, whereas α1(IV) and α2(IV) are more abundant extrasynaptically. Collagen α5 (IV) is also concentrated in human muscle synaptic BM. As indicated in figure 5.6, type IV collagen network is connected via nidogen to the laminin (laminin 2 or 4).

Collagen VI is one of the most abundant, widespread, and complex collagens. The microfilamentous collagen VI network is abundant close to cells, and the microfibrils can interact with, for example, collagens I and IV, fibronectin, biglycan, and decorin—suggesting that collagen VI involves anchorage of BM to the underlying endomysium. Additionally, the three-dimensional arrangement of fibronectin in ECM is dependent on secreted collagen VI (see Sabatelli et al. 2001). Type VI collagen deficiency results in early onset of Bethlem-type myopathy in mice (Bonaldo et al. 1998). Bethlem myopathy is an autosomal dominant myopathy with contractures and implies a significant role for collagen VI in stabilizing contractile muscle cells through the ECM-sarcolemma-cytoskeletal interface.

Collagen XV is expressed widely in tissues, but the highest levels in the mouse have been found in the heart and skeletal muscle. Immunoelectron microscopy of skeletal muscle and peripheral nerve has localized collagen XV to the outermost layer of lamina densa and to the endomysial collagen fibers near the BMs (see Eklund et al. 2001). Collagen XV null mice show progressive histological changes that are characteristic for mild muscular diseases after three months of age, and the muscle cells are more vulnerable than in controls to acute running exercise-induced muscle injury. These mice also have abnormalities in the morphology of heart microvasculature, which most probably cause marked ischemic-like damage after increased loading (Eklund et al. 2001).

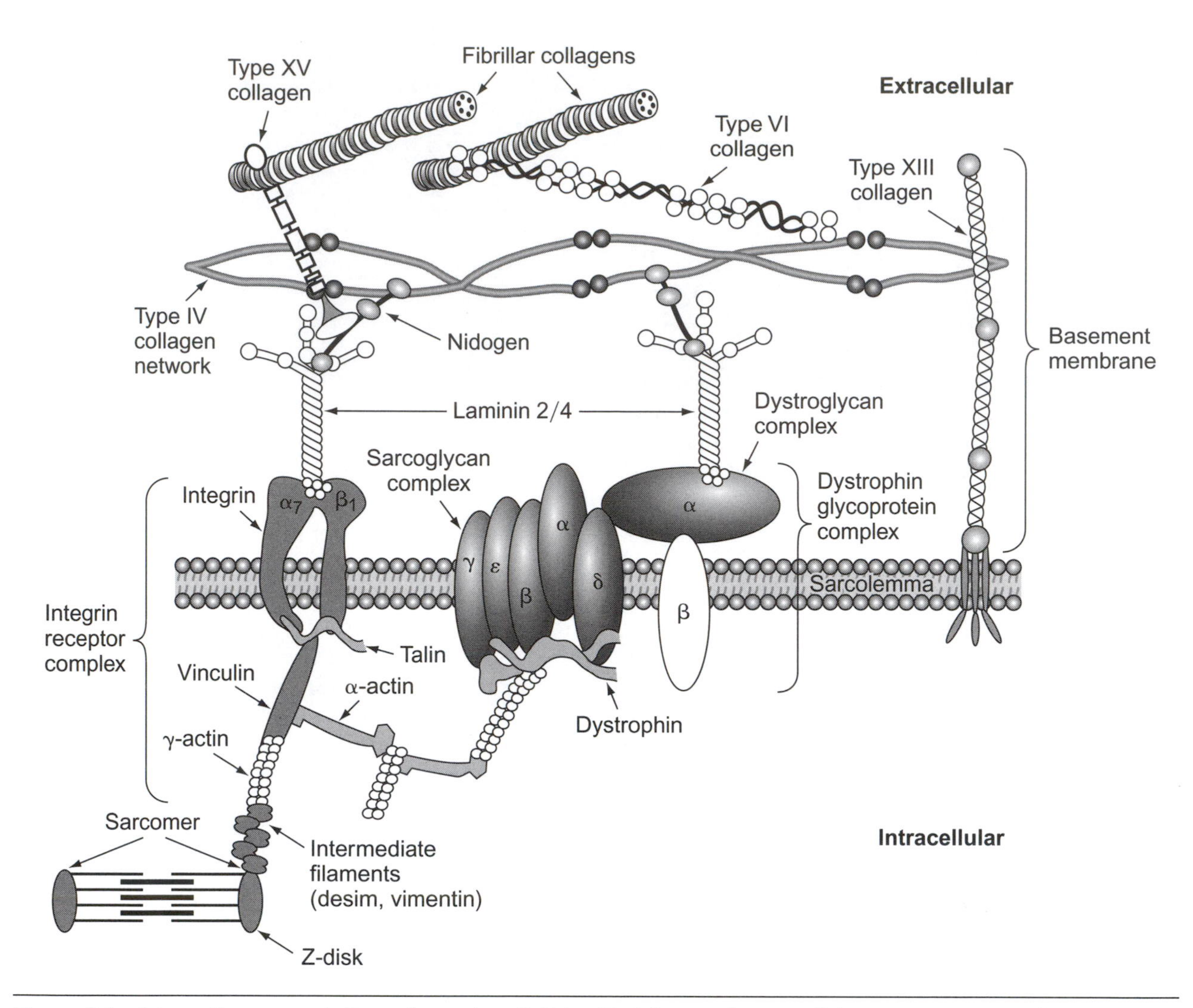

Figure 5.6 Schematic representation of muscle cell BM with intra- and extracellular associations. At the costamere, two laminin receptors, a dystrophin/glycoprotein complex and an integrin receptor complex, are associated with the sarcolemma. They link the contractile apparatus of muscle fibers with the surrounding BM. Intracellularly, both dystrophin and the integrin-associated cytoskeletal proteins (talin, vinculin, α-actin) connect via γ-actin and the intermediate-filament proteins desmin and vimentin to the Z-disc of muscle cell sarcomeres. Type VI and XV collagens are suggested to link BM to endomysium via interacting with type IV collagen and nidogen, respectively.

Reprinted, by permission, from M. Aumailley and B. Gayraud, 1998, "Structure and biological activity of the extracellular matrix," *Journal of Molecular Medicine*, 76: 253-265. Copyright © 1998 Springer-Verlag.

The transmembrane type XIII collagen is one of the minor collagens within the BM zone. Specifically, type XIII collagen participates in the linkage between muscle fiber and the BM (see figure 5.6). Its extensive extracellular domain could potentially penetrate deep into the matrix and strengthen the cell matrix association. Type XIII collagen is found in a range of integrin-mediated adherent junctions, including the myotendinous junctions and costameres of skeletal muscle (e.g., Hägg et al. 2001). Transgenic mice lacking the N-terminal cytosolic and transmembrane domains of type XIII collagen develop a progressive muscle disorder phenotype (Kvist et al. 2001). The affected skeletal muscles show abnormal myofibers with disorganized myofilaments, as well as streaming of z-lines characteristic for myopathy; and collagen XIII molecules are found outside the cell in the matrix near the vicinity of the plasmalemma but not embedded within it. A single bout of excessive physical exercise induced more prominent muscle damage in the mutant than in the wild-type mice. This increased sensitivity to exercise-induced damage was present already at 8 to 10 weeks of age; and in view of the progressive nature of the muscular disorder, older animals would likely show even more severe changes (Kvist et al. 2001).

Laminin molecules within the BM interact with each other via their short arms forming a second network structure in addition to that formed by type IV collagen molecules. Similar to the situation with type IV collagen, regional variability exists along the muscle cell BM in the expression of the different laminins. In adult muscles the extrasynaptic muscle cell BM is rich in α2β1γ1 (laminin 2) synaptic BM containing α2β2γ1 (laminin 4), α4β2γ1 (laminin 9), and α5β2γ1 (laminin 11) (see Patton et al. 1997). Myotendinous junction, as well, is enriched with laminin 4, which interacts with muscle cells via the specific integrin receptors α7Aβ1D and α7Bβ1D (Belkin et al. 1997; Patton et al. 1997). Similarly, at the neuromuscular junction and along the lateral muscle cell membrane, the laminins attach to the muscle cells via specific spliced variants of α7β1 integrin (Belkin et al. 1997). Interestingly, α7β1 integrin spliced variants show temporal and spatial differences in their expression during muscle regeneration following transsection injury. Furthermore, α7 integrin expression, but not that of laminin 2, is affected by mobilization during muscle cell regeneration following the transsection injury (Kääriäinen et al. 2001).

ECM-Synthesizing and -Degrading Cells in Muscle

To understand how muscle ECM functions, it is important to know which cells make the various ECM-related proteins. Both myogenic and fibrogenic cells have been shown to be able to synthesize interstitial collagens I, III, and V and fibronectin (Sasse et al. 1981), as well as the BM components collagen IV and laminins, during myogenesis (Kühl et al. 1982; Patton et al. 1997). When mononuclear myoblasts fuse into myotubes, the synthesis of ECM molecules drastically decreases (Sasse et al. 1981; Kühl et al. 1982); and, consequently, muscle fibroblasts remain the main source at least for the interstitial ECM molecules in mature muscle. Although fibroblasts do not synthesize BM, muscle fibroblasts are able to synthesize type IV collagen that can be incorporated into the BM surrounding the myotubes (Kühl et al. 1984). This further supports the statement that the mononucleated myogenic and fibrogenic cells rather than muscle cells themselves would be the source of muscle cell BM type IV collagen (Lipton et al. 1977; Kühl et al. 1984). In agreement with this, really low levels of type IV collagen mRNA with the aid of RT-PCR are found in adult rat muscle fibers (Väliaho et al. 2000). On the other hand, myotubes have been shown not only to synthesize multiple laminins, but also to differentially distribute them into the specialized BM regions of muscle fibers (Patton et al. 1997).

A recent study by Balcerzak et al. (2001) with bovine skeletal muscle suggests that muscle cells themselves, intramuscular fibroblasts, and the myogenic cell cultures derived from satellite cells express several MMPs and related proteins, but their expression levels are very low. The expressed molecules include MMP-1, -2, -9; MT1-MMP and MT3-MMP; TIMP-1, -2, -3; and plasminogen activator and its receptor.

An important source of mononucleated myogenic cells in mature muscles is the satellite cells between the muscle cell plasmalemma and BM. These could contribute to the synthesis of ECM-related molecules following their activation and proliferation, which can occur, for example, subsequent to more or less severe muscle damage. Accordingly, different types of physical activities can induce satellite cells to proliferate and contribute, for example, to ECM synthesis (see, e.g., Kadi & Thornell 2000; Vierck et al. 2000; Cameron-Smith 2002).

Muscle ECM and Physical Activity

The discussion in this section is limited, in the first place, to the effects of physical activity on the ECM of skeletal muscle and is based mainly on in vivo studies. Most of these studies have been carried out with experimental animals, while human studies on muscle ECM are very scant. The ECM responses appear to be loading type dependent and muscle specific (e.g., Kovanen 1989; Palokangas et al. 1992; Zimmerman et al. 1993; Koskinen et al. 2001; Ahtikoski et al. 2001; Flück et al. 2002; Ahtikoski et al. 2003). Consequently, the specific ECM responses are discussed here in a loading type-dependent context. The main focus is on the metabolism and expression of collagens, but some data on tenascin expression are also presented. To the best of my knowledge, hardly any studies exist that deal with the major proteoglycans and other ECM glycoproteins of skeletal muscles in connection with changed physical activity.

The significance of mechanical loading on cell behavior has been explored over the past 10 years.

Specifically, the development of devices for in vitro studies has allowed isolated cells to be subjected to a variety of mechanical loads (shear stress, mechanical strain, tension, or compression). Accordingly, evidence accumulating at the cellular and molecular level indicates that the manner and magnitude of mechanical stress acting on ECM-synthesizing cells affect the quality and quantity of different ECM molecules. On the other hand, the ECM molecules significantly influence gene expression and the behavior of adjacent cells. The way in which cells respond to mechanical stimulus appears to involve the cytoskeleton and ECM via integrin receptors at focal adhesion complexes (e.g., Gordon et al. 2001; Flück et al. 2002) and is currently described as the tensegrity model (e.g., Ingber et al. 2000).

In vivo, cell regulation is complex due to the involvement of humoral, neuronal, and nutritional factors in addition to the mechanical load under which the cell is functioning. Nevertheless, specifically changed physical loading can be used to investigate the regulations and functions of different proteins.

Endurance-Type Training

It is well known that endurance training characterized by "high-repetitive low load" induces metabolic adaptation in skeletal muscle cells as well as phenotypic changes in the contractile proteins. On the other hand, intramuscular ECM responses to endurance-type training have received surprisingly little attention. The animal studies available so far suggest that at least collagenous structures of muscle ECM respond to endurance training and that consequently the mechanical properties of muscles are affected (figure 5.7) (Kovanen et al. 1984; Kovanen & Suominen 1988; Gosselin et al. 1998).

In human muscles, the enhancement of collagen biosynthesis due to endurance training was first reported by Suominen and Heikkinen (1975) among middle-aged and elderly male endurance-trained athletes. In that study, higher P4H activities were found in the vastus lateralis muscle of the trained athletes as compared to sedentary controls. Additionally, increased P4H activity was found in vastus lateralis muscle after two months of training in 69-year-old women (Suominen et al. 1977). Since then there has been a lack of human studies dealing with the adaptation of skeletal muscle ECM in connection with any kind of physical activity.

Recently, microdialysis technique has been used in human tendon studies to investigate tissue concentrations and release rates of compounds involved in type I collagen synthesis (PICP, C-terminal propeptide) and degradation (ICTP, C-terminal propeptide) (Langberg et al. 1999, 2001). These studies of the human peritendinous Achilles tendon region have given a surprisingly dynamic picture of tendinous type I collagen metabolism in connection with physical loading. Twelve weeks of intensive physical training (daily 2-4 h, marching, running, and combat training) resulted in an increased turnover of type I collagen (Langberg et al. 2001). Within 4 weeks of the training, both synthesis and degradation were elevated, whereas after 12 weeks the anabolic processes were dominating, causing a net synthesis of type I collagen in the tendon-related tissue. Together with the plasma data, the studies by Langberg et al. (1999, 2001) support the view that exercise induces a positive protein balance in the region around the Achilles tendon. Interestingly, Simonsen et al. (1995) showed that swim training, but not strength training, compensated for the age-related failure in the tensile properties of rat Achilles tendon. It can be speculated that the low-threshold, frequent muscle contractions expose the nearby tendon fibroblasts under mechanical loading, which induces the expression of COL1A1 and COL1A2 genes, thus enhancing type I collagen synthesis to strengthen the tendon structure.

Animal studies further support the activation of skeletal muscle collagen metabolism in response to endurance training at the protein level. Accordingly, enhanced collagen synthesis is suggested on the basis of increased P4H activity (Kovanen et al. 1980; Takala et al. 1983; Myllylä et al. 1986; Kovanen 1989) or decreased hydroxypyridinium cross-link concentration in muscle collagen (Zimmerman et al. 1993; Gosselin et al. 1998). Short-term endurance training, that is, 1-h daily running up to 10 weeks, seems not, however, to affect collagen concentration of skeletal muscles of young adult mice, or rats irrespective of age (Kovanen et al. 1980; Takala et al. 1983; Myllylä et al. 1986; Zimmerman et al. 1993; Gosselin et al. 1998).

Endurance training carried out throughout the life of rats resulted in higher P4H and GGT activities in the soleus muscle of two-year-old rats when compared to values in untrained rats of similar age, while no changes were found in rectus femoris muscle (Kovanen 1989). Additionally, total collagen and type IV collagen concentrations increased in the soleus muscle in concert with the increased

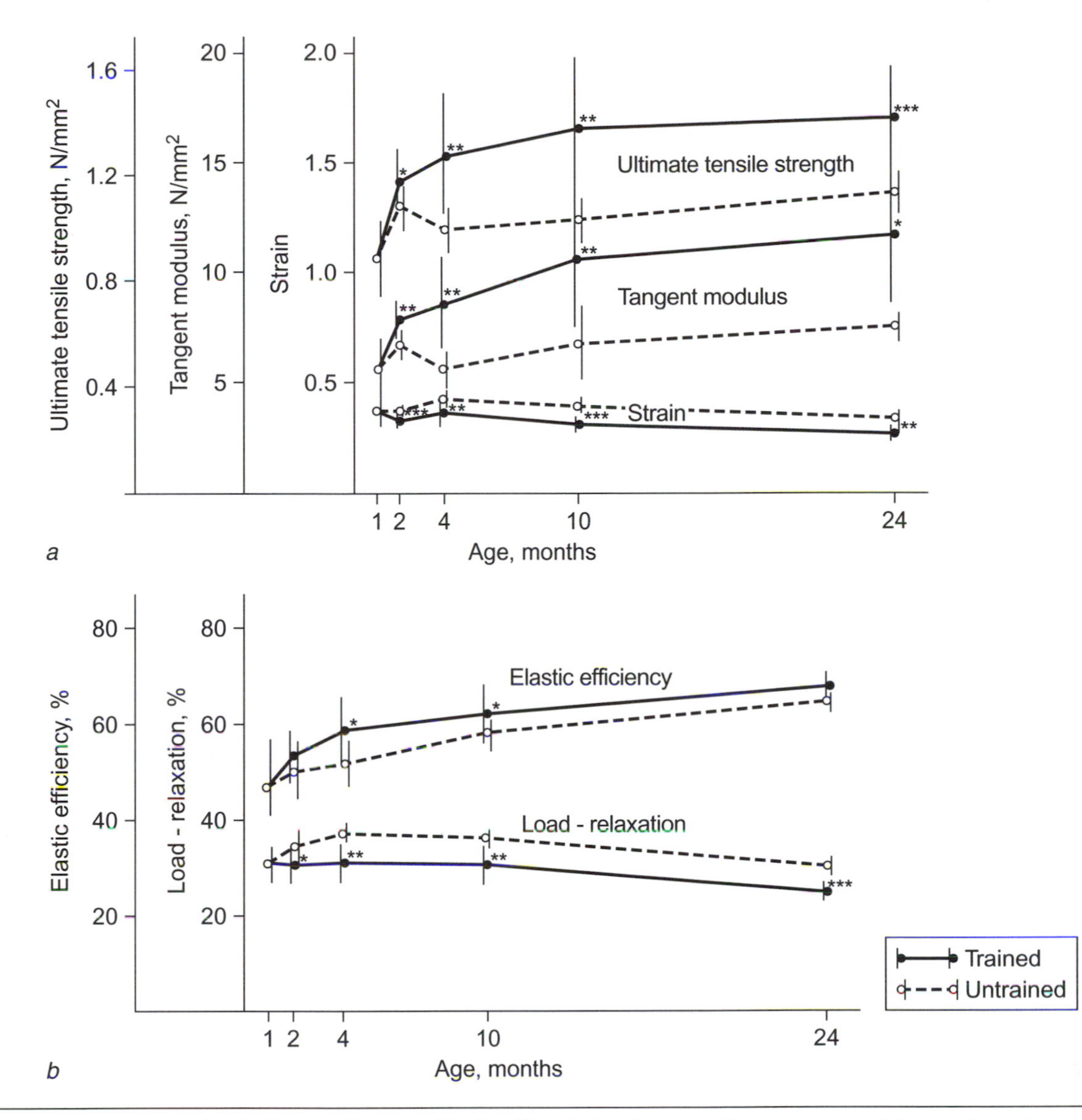

Figure 5.7 Passive mechanical properties, specifically ultimate tensile strength, tangent modulus, and maximum strain of soleus muscles in trained (two years, endurance-type treadmill running) rats, show that animals that have been kept active throughout their lives have stronger and stiffer soleus muscles than their age-matched untrained counterparts. The elastic efficiency (not shown here) increased concomitantly with the increase of tangent modulus, suggesting improved recovery of elastic energy.

From V. Kovanen and H. Suominen, 1988, "Effects of age and life-long endurance training on the passive mechanical properties of rat skeletal muscle," *Comprehensive Gerontology* A, 2: 18-23. Reprinted, by permission, from Blackwell Publishing.

mechanical stiffness of that particular muscle (see figure 5.7) (Kovanen 1989). One can question whether a less endurance-biased training program might have been more beneficial for the functional demands of that muscle. Interestingly, in a recent study by Gosselin et al. (1998), 10-week endurance training changed the stiffness properties of soleus muscles of old rats, resulting in muscles with stiffness characteristics not different from those of young untrained rats. Furthermore, the hydroxypyridinium cross-link concentration was reduced in response to training in the old animals, suggesting enhanced collagen metabolism and its contribution to the passive muscle viscoelastic properties (Gosselin et al. 1998).

The increased total collagen and type IV collagen concentration due to lifelong endurance training (Kovanen 1989) in the low-threshold slow-twitch rat soleus muscle, but not in the more phasic rectus femoris muscle, could be due to the more constant recruitment of soleus in this kind of training. Accordingly, its increased/maintained contractile activity exposed muscle fibroblasts to frequent stretch, which might have enhanced collagen biosynthesis in the fibroblasts via the TGF-β-dependent mechanism (see Lindahl et al. 2002). So far, no data are available on MMP activities in connection with endurance training. In cell culture conditions, MMP expression is not affected by stretch, whereas relaxed conditions induce MMP expression (e.g., Trächslin et al. 1999).

Mechanical stretch as a regulator of collagen synthesis, also in vivo, has been accepted (for review and references see Bishop & Lindahl 1999), but the mechanisms have been unknown. A very recent cell culture study with cardiac fibroblasts elegantly begins to open the black box. Lindahl et al. (2002) managed to show that cyclic mechanical strain induces transcriptional activation of the COL1A1 gene and involves at least two regions in the proximal promoter. The complex regulation is dependent on the release of active TGF-β and involves increased binding of the ubiquitous transcription factor CBF (also known as NF-Y) at the proximal inverted CCAAT-box element. On the basis of these findings, and the fact that functional CBF-binding sites in both COL1A1 and COL1A2 genes are highly conserved between species, Lindahl et al. (2002) propose that CBF may play an important role in the up-regulation of type I collagen expression in response to cellular stresses mediated by TGF-β stimulation. TGF-β, the most potent activator of collagen gene expression identified to date, is known to be activated by mechanical forces. There thus is a strong possibility that the induction of collagen transcription by mechanical load is mediated through the autocrine stimulation by TGF-β, emphasizing a synergistic model of action (see Lindahl et al. 2002).

Intensive Physical Training

Strength or resistance training characterized by "low-repetitive high-load" exercises is known to induce, in addition to phenotypic changes, hypertrophy of muscle cells, hyperplasia, or both, in humans. In humans, however, there are no studies thus far about possible ECM changes in connection with intensive physical training or development of muscle hypertrophy.

In animal studies, several models have been used to investigate ECM responses in connection with heavy physical loading. These include work-induced compensatory hypertrophy (tenotomy of the synergistic muscle) of rat soleus muscle (e.g., Jablecki et al. 1973; Turto et al. 1974), stretch- or overload-induced hypertrophy of chicken wing anterior latissimus dorsi (ADL) muscle (e.g., Laurent et al. 1985; Flück et al. 2000), and forced lengthening contractions of rat tibialis anterior muscle (Koskinen et al. 2002). A single bout of acute excessive treadmill running in untrained animals is a more physiological exercise model and involves changes in ECM protein expressions as well (Myllylä et al. 1986; Komulainen & Vihko 1998; Han et al. 1999; Koskinen et al. 2001). At the onset of loading, each of these models induces muscle cell damage involving, for example, clear inflammatory reaction. Contrary to the other models mentioned, acute exhaustive treadmill running has not been shown to induce muscle hypertrophy.

Response to physical loading in the ECM of skeletal muscle can occur as an immediate or a delayed reaction. The time course of the response can give some clue as to whether direct action of tensile stress or a secondary response to loading, such as inflammation and regeneration, is involved in the regulation of a specific protein.

Flück et al. (2000) studied the expressions of tenascin-C, tenascin-Y, and collagen XII in the stretch-overloaded chicken ADL muscles and found different spatial and temporal expressions for these proteins. In basic conditions, tenascin-C and collagen XII are predominantly expressed at myotendinous junctions, while tenascin-Y is found in both the endo- and perimysium. Among the striking findings by Flück et al. (2000) was the ectopic induction of tenascin-C at both the mRNA and protein level in the endomysial fibroblasts of the ADL muscle with concomitant down-regulation of tenascin-Y mRNA in the endomysium. Furthermore, the changes in the tenascin-C and tenascin-Y expression occurred rapidly, within only 4 h of loading. It is known that in this study model, the immigration of immune cells like lymphocytes and macrophages occurs after approximately 24 h; thus they have not contributed via their cytokines and growth factors to the increased tenascin-C gene expression. The significance of the rapid, locally changed tenascin-C and tenascin-Y expression is an interesting, open question awaiting further innovative studies.

The rapid change in expression patterns of the two tenascins after mechanical stimuli, as just described, is compatible with a direct action of tensile stress on their expression (Flück et al. 2000). The result of this in vivo study is in line with the in vitro study by Trächslin et al. (1999), which showed that tensile stress enhances tenascin-C and collagen XII expression in fibroblasts, at both the mRNA and the protein level, the enhancement being a rapid and reversible response to the mechanical stimuli per se. In fact, "stretch-responsive" enhancer regions in the promoter regions of tenascin-C and collagen XII genes have been identified (see Chiquet et al. 1998; Chiquet 1999). The cell culture studies by Trächslin et al. (1999) failed to show changes in fibronectin and MMP-2 mRNA expression due to tensile stress.

In the context of metabolism of collagens, enhancement of total collagen synthesis via increased P4H activity was found following work-induced hypertrophy in skeletal and cardiac muscles of rats already in the early studies by Turto et al. (1974) and Lindy et al. (1972), respectively. This was interpreted to indicate a connection between the synthesis of the components of muscle tissue itself and the intramuscular connective tissue in hypertrophying muscle. Additionally, Jablecki et al. (1973) showed that in compensatory muscle hypertrophy, especially the new RNA synthesis was highly activated in mononuclear cells in the interstitial space. The autoradiographic method used fails to separate fibroblasts from proliferating satellite cells, which are currently known to become activated following muscle damage. Accordingly, activation of ECM-producing cells, both myogenic and fibrogenic, can be suggested to be an important prerequisite to work-induced muscular growth.

In stretch-overloaded chicken ADL muscles, Laurent et al. (1978) showed enhanced collagen synthesis via increased uptake of radiolabeled proline into collagen, and consequently increased total collagen deposition in the hypertrophied ADL muscle. Furthermore, collagen degradation together with collagen synthesis appeared to play an important role in the regulation of collagen mass (Laurent et al. 1985). During the ADL hypertrophy, an increase of about fivefold was found in the collagen synthesis rate; there was a 60% decrease in the rate of degradation of newly synthesized collagen and an increase of about fourfold in the amount of degradation of mature collagen. Interestingly, the study by Laurent et al. (1985) shows that enhanced degradation of mature collagen is required for muscle growth, and suggests a physiological role for the pathway whereby in normal muscle a large proportion of newly produced collagen is rapidly degraded.

Acute excessive treadmill running exercise has been shown to enhance collagen biosynthesis in mouse muscle in line with the extent of muscle damage (Myllylä et al. 1986), which is more marked in red than in white parts of the quadriceps femoris muscle (MQF) (see Komulainen & Vihko 1998). The prolonged, 9-h running of untrained mice temporarily increased P4H activity 2, 5, and 10 days after exercise, more prominently in the red than in the white part of MQF. Twenty days after exercise, P4H activity was back to control level. Collagen concentration of the red muscle showed a mild increase 10 days after the exercise and that of white muscle 20 days afterward. These responses were regarded as reflecting the muscle regenerative process and were attenuated by previous training of the animals (Myllylä et al. 1986).

More recent studies have shown that following acute running exercise, collagen synthesis- and degradation-related proteins can change at both the mRNA and protein level in rat skeletal muscles (Han et al. 1999; Koskinen et al. 2001). The timing, level, and extent of the changes depend on which muscle, which collagen type (I, III, IV), and which synthesis- and degradation (P4H, MMP-2, TIMP-1, TIMP-2)-related proteins are concerned. Muscle type-dependent differences in the responses appear, in most cases, to depend on the extent of muscle cell injury (see figure 5.8).

The gene expressions for the fibrillar collagens I and III were found to be up-regulated, at the mRNA level, both in soleus (no or only slight damage) and in the red part of MQF (severely damaged) within one day following the exercise, peaking at two days in soleus and at four days in MQF. Furthermore, the increase of collagen I mRNA was about three times higher, and that of type III collagen about two times higher, in MQF compared to the respective increases in soleus (Han et al. 1999; Koskinen et al. 2001). The white part of MQF with minor damage also showed increased mRNA expressions for collagens I and III, the timing being similar to that for red MQF, while the extent of the increases was between values for soleus and red MQF (Koskinen et al. 2001). Consequently, the muscle injury involving the inflammatory reaction can be regarded as a significant contributor to increased mRNA expressions for fibrillar collagens I and III.

These activated gene expressions of type I and III collagens did not manifest at the protein level as increased total collagen or changed type I/III collagen ratio, at least up to the end of the follow-up period, which was 14 days (Han et al. 1999). The newly synthesized so-called soluble collagen forms were not measured in this study. Thus no statement can be made as to whether the fraction of newly synthetized collagen, which in the basic

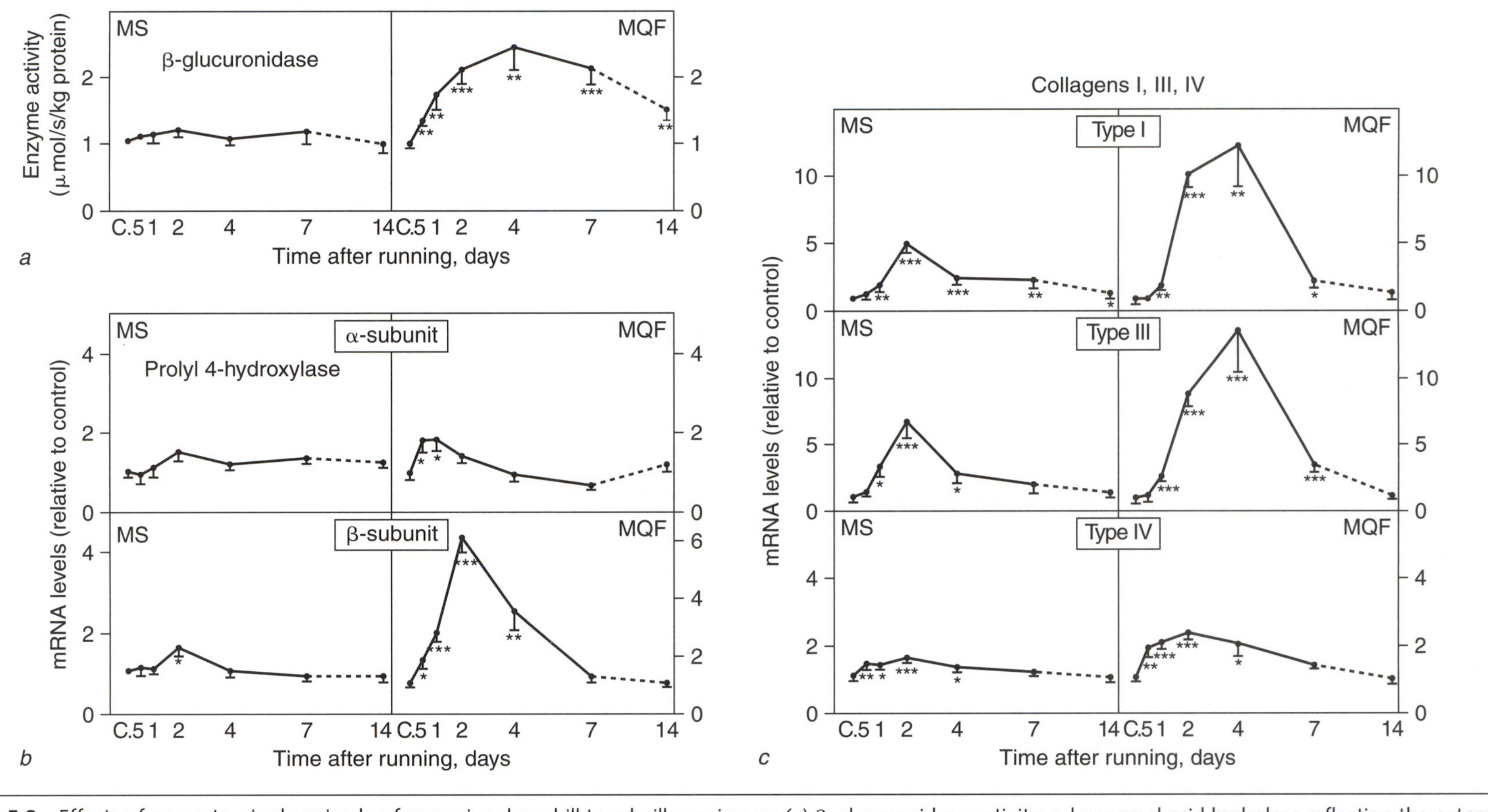

Figure 5.8 Effects of an acute, single episode of excessive downhill treadmill running on *(a)* β-glucuronidase activity, a lysosomal acid hydrolase reflecting the extent of muscle damage, *(b)* the steady state mRNA levels of prolyl 4-hydroxylase α and β subunits, as well as *(c)* those of type I, III, and IV collagens in rat soleus (MS) and the red part of quadriceps femoris muscle (MQF). The variables were measured at 12 h and 1, 2, 4, 7, and 14 days after the exertion.

Reprinted, by permission, from Han et al., 1999, "Increased mRNAs for procollagens and key regulating enzymes in rat skeletal muscle following downhill running," *European Journal of Physiology* 857-864.

state is rapidly degraded, had changed. One can recall that Laurent et al. (1985) found in a heavy hypertrophy model a 60% decrease in the rate of degradation of newly synthesized collagen with concomitant increase in collagen deposition in the chicken ADL muscle. It may also be worth keeping in mind that thick collagen fibrils, newly secreted soluble collagen molecules, and some degradation fragments of collagen molecules are known to differ as ligands in collagen and MMP gene regulation via integrin receptors (see Carragher et al. 1999).

The activity of P4H increased about twofold in the red MQF 2 and 4 days after the running exercise, while an increase of about 50% to 70% was found in the soleus muscle at days 2, 4, and 7 (Han et al. 1999). The enzyme activities recovered to control levels 7 days and 14 days after running in MQF and soleus, respectively. Interestingly, in the MQF muscle, significant increases at the mRNA level were found for both the α and β subunits of P4H. The increase for the β subunit was higher than that for the α subunit and was followed 1 to 2 days later by the increased P4H activity (see figure 5.8).

The temporally coordinated change in P4H activity and in the mRNA levels of P4H subunits is interesting, since according to earlier studies, α subunit synthesis limits the formation of α2β2-tetramer, the active form of the enzyme (Kivirikko et al. 1989). While signs of muscle damage were found in this study only in MQF muscle, the results suggest that P4H was up-regulated at the pretranslational level only in the damaged muscle (Han et al. 1999). The fractional synthesis rate of P4H is not known in rat skeletal muscle, but it is approximately 20% per day in chicken tendon cells (Berg et al. 1980). If the turnover of P4H occurs roughly at the same rate in rat skeletal muscle, the observed 80% to 85% increase in the level of P4H α subunit mRNA in MQF should lead to a net increase in the enzyme pool by 15% per day, assuming that no further adjustment occurs during translational or post-translational processes. However, the P4H activity of MQF doubled in two days, and a similar increase in the enzyme activity was found in the undamaged soleus muscle despite a nonsignificant increase in its α subunit mRNA level. The results thus suggest that in this study model, P4H expression and activity were also regulated at the translational level, the post-translational level, or both. Because P4H activity is known to limit total collagen synthesis rate, the observed increases in the enzyme activity strongly suggest that acute exercise could enhance collagen synthesis with or without biochemical or histological signs of muscle damage (Han et al. 1999). This statement agrees with earlier studies showing that enhanced P4H activities can also be found following non-damaging exercise like endurance-type training (Suominen & Heikkinen 1975; Kovanen et al. 1980; Kovanen 1989).

The gene expression of type IV collagen was affected at both the mRNA and protein level following the acute running exercise; but in comparison to the fibrillar collagens, the responses at the mRNA level were observed clearly earlier, at 12 (Han et al. 1999) or 6 h (Koskinen et al. 2001) after performance of the exercise (figure 5.8). Additionally, the increased mRNA levels were found in all studied muscles, while the extent of the increase appeared to be dependent on the severity of the muscle damage—red MQF showing the greatest increases. In this study model characterized by exercise-induced delayed muscle injury, both the neutrophil response and the invasion of monocytes into the sites of muscle injury usually occur within 24 h after the strenuous exercise (see Komulainen & Vihko 1998). Consequently, neutrophils, monocytes, and macrophages with their cytokines and growth factors appear to be unlikely contributors to the increased type IV collagen gene expression. Instead, mechanical load-induced gene expression can be suggested. In vitro studies would be needed to more exactly investigate type IV collagen gene regulation by mechanical load. Indirect support for mechanical load-induced gene expression can be found via the significantly increased collagen IV mRNA in the soleus muscle without microscopic signs of injury (Han et al. 1999).

At the protein level, type IV collagen was affected with a delayed time course in a muscle-specific way, differing according to the severity of damage, and in line with the degradation and remodeling of the BM (Koskinen et al. 2001). In soleus muscle, no changes in collagen IV concentration were found regardless of the increased mRNA expression. The white MQF with mild damage showed elevated collagen IV concentration seven days after the exercise, in line with the preceding enhancement of gene expression at mRNA level. Unexpectedly, the most severely damaged red MQF showed decreased collagen IV concentrations one and two days after the exercise despite the highly increased mRNA expression. Subsequently, the protein levels attained the control values within seven days. This behavior might be related to elevated degradation of BMs shortly after the severe damage and to the restoration of this structure with the regeneration of muscle cells.

Concerning the capacity for collagen IV degradation, MMP-2 expression increases at both the mRNA and protein levels in connection with postexercise injury, the increase being related to the extent of muscle cell damage (Koskinen et al. 2001). Elevated mRNA levels were found two days postexercise in the red MQF, values peaking at days 4 and 7. A minor transient increase in MMP-2 mRNA was found in soleus at day 2, while in white MQF a minor increase appeared in a more delayed manner. At the protein level, protein expression activities for proMMP-2 was found in all muscles studied; the increase was highest in the red MQF. Soleus muscle, with no microscopic signs of damage but a mild increase in β-glucuronidase activity, showed a smaller increase in proMMP-2 expression but one that was temporally similar to that for red MQF. However, the proteolytically activated form of MMP-2 effective in ECM proteolysis was restricted solely to the severely damaged red MQF. Altogether, the results suggest that in vivo MMP-2 expression is pretranslationally regulated in connection with acute physical loading, while post-translational regulation is induced in severely damaged muscle. Interestingly, a very recent study with α1 integrin knockout mice (Väliaho, Heino, & Kovanen unpublished data) shows that proMMP-2 expression in connection with postexercise injury is further enhanced in the absence of α1 integrin. This suggests a significant role for α1 integrin in the regulation of MMP-2 expression in damaged muscle.

At the post-translational level, the capacity of MMP-2 for ECM proteolysis is significantly affected by TIMP-2, which is known to have a dual role

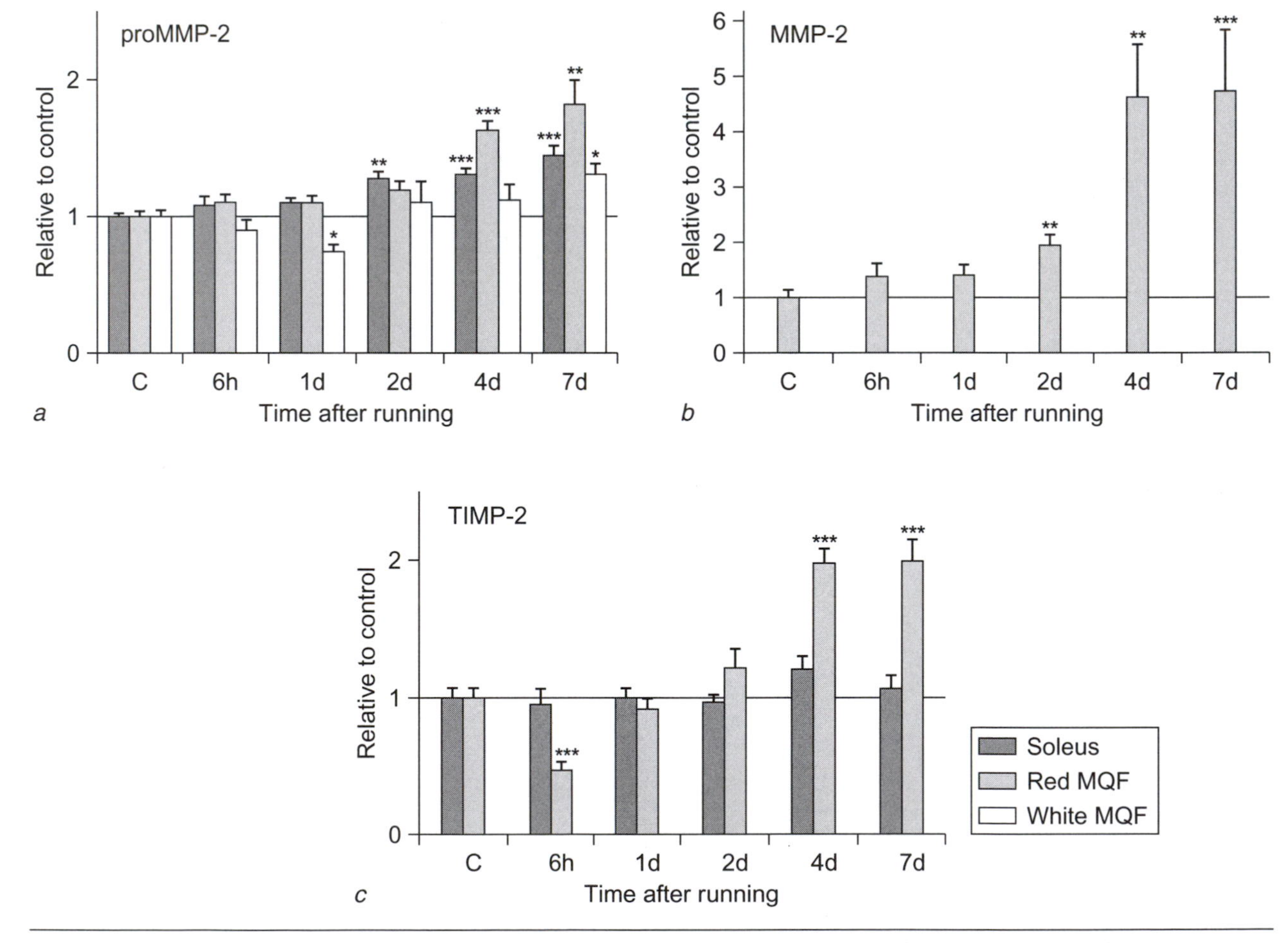

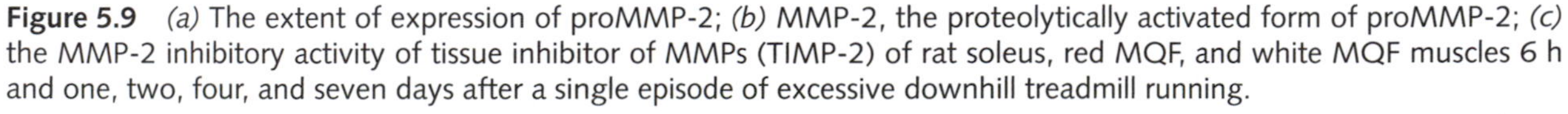

Figure 5.9 *(a)* The extent of expression of proMMP-2; *(b)* MMP-2, the proteolytically activated form of proMMP-2; *(c)* the MMP-2 inhibitory activity of tissue inhibitor of MMPs (TIMP-2) of rat soleus, red MQF, and white MQF muscles 6 h and one, two, four, and seven days after a single episode of excessive downhill treadmill running.

Reprinted, by permission, from Koskinen et al., 2001, "Acute exercise induced changes in rat skeletal muscle mRNAs and proteins regulating type IV collagen content," *American Journal of Physiology-Regulatory, Integrative and Comparative Physiology* 280: R1292-R1300.

in MMP regulation (Nagase & Woessner 1999; Brew et al. 2000). TIMP-2 binds noncovalently to the active form of MMP-2, thus inhibiting its activity. Paradoxically, however, TIMP-2 is also needed for the efficient activation of proMMP-2 in the MT1-MMP-dependent process. Recently, novel in vivo evidence for the importance of TIMP-2 in the efficient activation of proMMP-2 in skeletal muscle has been provided (Koskinen et al. 2001). In connection with acute running exercise, enhanced TIMP-2 expression was found at both the mRNA and protein level, temporarily in a temporally well-coordinated manner with MMP-2 expression. The extent of the enhancement was strongly associated with the severity of muscle damage, the TIMP-2 mRNA and protein levels being highly elevated only in the severely damaged red MQF muscle. The response at the mRNA level turned to a decrease after day 4, while protein levels remained highly elevated up to the end of the seven-day follow-up period (figure 5.9). Accordingly, the pretranslational induction of TIMP-2 expression seems to be connected to the elevated TIMP-2 protein levels in connection with postexercise muscle injury involving inflammatory reaction. Furthermore, the temporally coordinated expressions of MMP-2 and TIMP-2 would properly support degradation and restoration of damaged BMs.

It is worth recalling that MMP-2 utilizes, in addition to type IV collagen, several other ECM molecules as its substrate (see table 5.3). Accordingly, the action of MMP-2 could affect the BM zone widely. The tightly and appropriately regulated MMPs could, at their best, produce subtle changes within the ECM via specific enzymatic cuts. Consequently, remodeled ECM with exposed new cryptic sites and released bioactive fragments, as well as released, ECM-stored growth factors, could modify gene expression and behavior (see, e.g., Bishop & Lindahl 1999; Marneros & Olsen 2001) of proliferating satellite cells, fibroblasts, endothelial cells, and muscle cells themselves in favor of proper muscle regeneration (see figure 5.10).

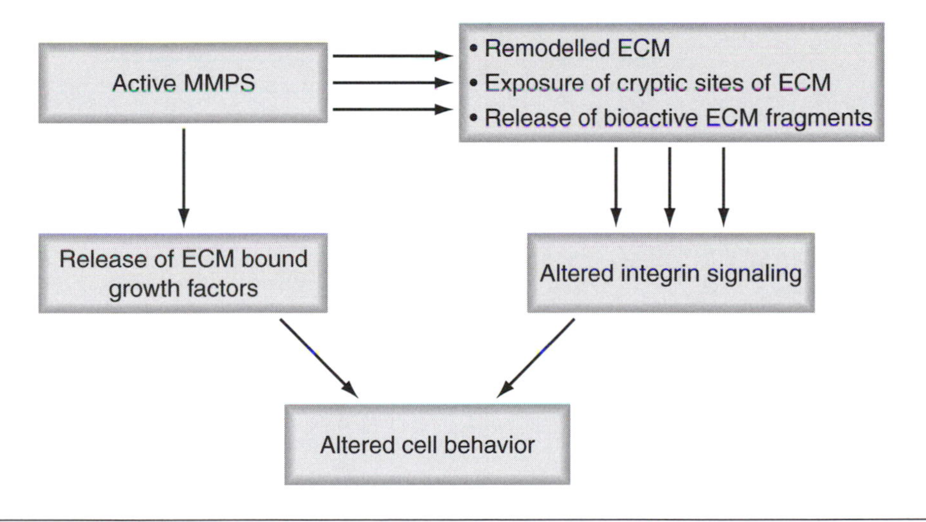

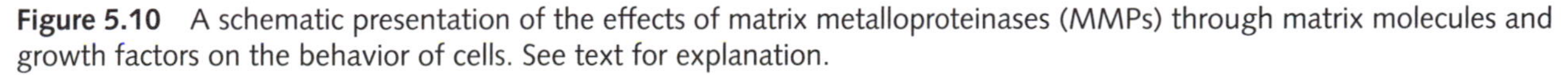

Figure 5.10 A schematic presentation of the effects of matrix metalloproteinases (MMPs) through matrix molecules and growth factors on the behavior of cells. See text for explanation.

Reprinted, by permission, from V. Kovanen, 2002, "Intramuscular extracellular matrix: Complex environment of muscle cells," *Exercise and Sport Sciences Reviews* 30(1): 20-25.

Chapter 6

Regulation of Intracellular Ion Composition and pH

Frank Mooren

There is a characteristic distribution of ions across the plasma membrane of eukaryotic cells, established by active ion transport mechanisms and selective membrane ion permeability. This leads to a potential difference across the membrane of about −70 mV during resting conditions, which is near the equilibrium potential for potassium. Therefore the gradient for sodium and calcium is directed inward while the potassium gradient is directed outward. Ionized magnesium is nearly equally distributed across the plasma membrane. However, due to the cell's negative resting potential, it is far from its electrochemical equilibrium. The physiological ranges of these ions in the extracellular and the intracellular compartment are given in table 6.1.

An optimalized ion homeostasis is fundamental for metabolism, cell growth, and function. Effective mechanisms regulate the ion concentrations within their narrow ranges. For this reason eukaryotic cells contain various membrane ion channels and transporters. Moreover, these mechanisms are closely related to intracellular pH and volume regulation.

Table 6.1 Distribution of Ions Across the Plasma Membrane of Erythrocytes

Ion	Intracellular concentration	Extracellular concentration
Potassium	99.1 ± 5.3	4.05 ± 0.34
Sodium	7.35 ± 1.26	140.0 ± 2.4
Calcium	19 ± 9 μmol/L	2.44 ± 0.10
Magnesium	1.95 ± 0.19	0.78 ± 0.04
Chloride	67.9 ± 4.9	103.7 ± 1.9

In mmol/L if not otherwise indicated.

According reference Wissenschaftliche Tabellen Geigy, 1991.

Potassium

Potassium is the most abundant cation in the body, amounting to about 3,500 to 4,000 mmol in total. The total quantity is related to the fat-free body mass and ranges from 50-70 mmol/kg. Of the total body potassium, 98% is localized intracellularly, while the rest is found in the various extracellular compartments (interstitium, 40 mmol; transcellular fluids, 35 mmol; plasma, 40 mmol). This potassium gradient is of fundamental importance for cell volume regulation as well as nerve and muscle excitability.

Potassium-Regulating Membrane Proteins

The unequal potassium distribution is predominantly the result of active transport processes, namely the sodium-potassium-ATPase (Na^+-K^+-ATPase). Moreover, its distribution depends on the permeability of the plasma membrane for potassium, which is affected by various potassium transport pathways. Therefore, intracellular potassium homeostasis is finely tuned by multiple potassium-carrying structures such as ion pumps, potassium channels, and potassium transporters. The most important ones with respect to exercise are discussed in the following paragraphs.

Potassium Channels

Potassium channels represent the largest and most diverse subgroup of ion channels. As an overview, the two dominating potassium channel superfamilies are the voltage-sensitive and inward rectifier potassium channels. Both channel types consist of four peptide chains or subunits that associate as homo- or heterotetramers to form the ion permeation pathway across the membrane. The peptide chains of the inwardly rectifying potassium channels consist of two transmembrane helices with a short amino acid segment between them. This has been called the P-loop because it dips into the membrane without fully crossing it. Such a structure is characteristic of all potassium channel types (Choe 2002).

Voltage-Sensitive Potassium Channels

In voltage-sensitive potassium channels (Kv), four additional peptide chains precede the two transmembrane helices containing the positively charged voltage-sensing unit. The voltage-sensitive channels show a great diversity in their voltage sensitivities and activation kinetics. At least 22 different genes encoding for voltage-sensitive potassium channels in mammals have been identified, while further varieties are produced by alternative splicing and heterodimerization. One example is the delayed rectifier potassium channel that has been described in nearly all excitable membranes. This channel opens upon membrane depolarization after a short delay, thereby restoring membrane potential back to resting levels. Another group of voltage-gated potassium channels is represented by the calcium-activated potassium channels (K_{Ca}). These types of channels are activated by both intracellular calcium increases and depolarization. Calcium-activated potassium channels are divided into at least three subtypes by their conductance: large (200-300 ps), intermediate (20-60 ps), and small (10-14 ps). The channel open probability is regulated by a wide range of intracellular calcium concentrations spanning from about 0.5 to 50 mmol, which suggests multiple regulatory calcium binding sites. In particular, the low-affinity calcium binding sites suggest the role of these channels only under conditions of increased intracellular calcium concentration such as during sustained muscle activity or in exhausted fibers.

Inward Rectifier Potassium Channels

Members of the inward rectifier potassium channels (Kir) share as a common property the regulation by various intracellular modulators or messengers such as nucleotides, phospholipids, kinases, polyamines, pH, and G-proteins. Inward rectifier potassium channels finely tune the membrane excitability. Channel opening stabilizes the membrane potential near the potassium equilibrium potential, making the cell less excitable. Another member of the Kir group is the ATP-sensitive potassium channel (K_{ATP}). This channel type is an example of how metabolism and excitability may be linked. In metabolically exhausted muscle fibers, a decrease in cytosolic ATP concentration as well as in pH results in a great increase of membrane potassium conductance due to opening of ATP-sensitive potassium channels. It is supposed that in conjunction with an inadequate potassium uptake via the Na^+-K^+-ATPase, this leads to a reduced capacity of cell membrane to generate action potentials concomitant with a reduced force development (Renaud 2002).

Na^+-K^+-ATPase

The unequal distribution of monovalent cations across the plasma membrane is generated and maintained by the Na^+-K^+-ATPase. This ATP-driven pump is electrogenic, since during each cycle only two potassium ions are imported in exchange for three exported sodium ions. The sodium-potassium pump consists of two subunits, an α-catalytic subunit and a 50-kDa β-glycoprotein subunit. Multiple isoforms of the Na^+-K^+-ATPase pumps exist. Presently four α-isoforms and three β-glycoprotein isoforms have been identified. In muscle tissue the expression of Na^+-K^+-ATPase isoforms is dependent on fiber type. In rats, the predominant isoforms of the red oxidative muscles are α1,β1 and α2,β1, while in white glycolytic muscles only α1,β2 and α2,β2 isoforms can be found. In mixed fiber muscles, both β1 and β2 subunits can be found (Thompson & McDonough 1996; Hundal et al. 1992).

During resting conditions the sodium-potassium pump activity equals only about 5% of the maximal pump capacity in isolated muscle fibers. With the onset of contractile activity, the sodium-potassium pump rate increases in order to maintain ion gradients. The sodium-potassium pump capacity can be enhanced principally by two different mechanisms, either by up-regulation of the number of active sodium-potassium pumps in the plasma membrane or by enhancement of the ion pump activity. Indeed, translocation of the sodium-potassium pump from intracellular compartments toward the plasma membrane has been demonstrated recently (Hundal et al. 1992; Lavoie et al. 1996; Juel et al. 2000). Furthermore, sodium pump activity is controlled by multiple mechanisms such as intracellular sodium concentration, cyclic AMP levels via the protein kinase A pathway, ATP availability, pH, membrane potential, and osmotic stress (Clausen 1998). Whether the contraction-associated drop in ATP levels is relevant to stimulating the sodium-potassium pump is unclear because the $K_{0.5}$ (i.e., the ATP concentration that stimulates the sodium-potassium pump to 50% of its maximum pump rate) is about 150 µmol/L, a range usually not reached even during exhaustive exercise.

Na^+-K^+-Cl^- Cotransporter

The Na^+-K^+-Cl^- cotransporter (NKCC) belongs to the family of cation-coupled chloride cotransporters (CCC). Another member of this family is the potassium chloride cotransporter, which is discussed further on. Two isoforms of the NKCC have been identified with molecular masses between 122 and 130 kDa. The protein consists of a central hydrophobic region with 12 transmembrane-spanning domains flanked by an amino and a carboxy terminal region. While so far the NKCC2 isoform has been identified only in the kidney, the NKCC1 isoform has been found in various epithelial and nonepithelial cells like skeletal muscle. The NKCC can be classified as a secondary active transport driven by the combined chemical gradient of the transported ions. Nevertheless, there is plenty of evidence that its function depends also on the availability of ATP (Russell 2000). However, the mode of ATP activation differs from the E1-E2-type ATPases, favoring the view that ATP affects NKCC activity via a protein phosphorylation/dephosphorylation mechanism. The NKCC transports ions electrically silent into the cell with a stochiometry of 1 Na^+: 1 K^+: 2 Cl^-. However, at least in squid axon, there is evidence for an unusual stochiometry of 2 Na^+: 1 K^+: 3 Cl^-. Loop diuretics such as bumetanide are important inhibitors of the NKCC. Furthermore, increases in intracellular chloride concentration and acidic intracellular pH inhibit the NKCC. Functionally the NKCC is involved in a variety of epithelial absorptive and secretory processes. Additionally there is evidence for an important role of NKCC in cell volume regulation.

K^+-Cl^- Cotransporter

Four different genes have been identified that encode for the four different isoforms of the K^+-Cl^- cotransporter, among which the KCC1 isoform is ubiquitously distributed in different tissues. Like the Na^+-K^+-Cl^- cotransporter, the K^+-Cl^- cotransporter is a secondary active ion transporter driven by the gradient of the translocated ions. Most properties of the cotransporter have been characterized in erythrocytes, namely the electroneutrality of transport, ion affinity, kinetics, and pharmacological inhibition (Lauf & Adragna 2000). The increase in intracellular chloride concentration stimulates cotransporter activity. Moreover, a variety of oxidants can stimulate the K^+-Cl^- cotransporter, such as H_2O_2 and diamide (Bize & Dunham 1995; Adragna & Lauf 1997). Similar to the situation with the NKCC, activation of the potassium chloride cotransporter is modulated by ATP and pH. Functionally, the K^+-Cl^- cotransporter is involved in cellular volume regulation and intracellular ion homeostasis. Recent investigations also propose a role of this cotransporter in the control of peripheral arterial resistance (Adragna et al. 2000).

Intracellular Potassium Homeostasis

The exact distribution of potassium between the intracellular and the extracellular compartment determines resting membrane potential and cellular excitability. Extracellular potassium concentration, which amounts only to about 2% of total body potassium content, is regulated within narrow ranges between approximately 3.5 and 5 mmol/L. Two pathways dominate in the fine adjustment of potassium concentration in the extracellular fluid:

- Potassium secretion by the kidney and to a minor extent the gastrointestinal tract. Potassium secretion via sweat plays a significant role only under conditions of heavy and ongoing exercise.
- Potassium shifts between the extracellular and the intracellular compartment. For further details about the regulation of extracellular potassium, the reader is referred to textbooks of physiology and recent excellent reviews (Giebisch 1999; Halperin & Kamel 1998).

Since nearly 98% of body potassium content is found in the intracellular compartment, the role of intracellular potassium regulation is emphasized

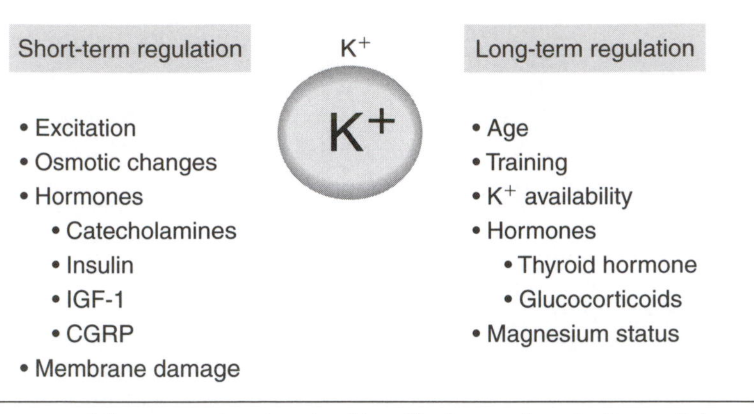

Figure 6.1 Schematic diagram of the parameters involved in affecting and controlling cellular potassium homeostasis. A distinction was made according to their kinetics in short- and long-term regulatory factors. IGF-1, insulin-like growth factor; CGRP, calcitonin gene-related peptide.

Reproduced, with permission, from T. Clausen, 1998, "Clinical and therapeutic significance of the Na+, K+ pump," *Clinical Science*, 95(1), 3-17. © the Biochemical Society.

in this section. Cellular potassium concentration is subject to short-term and long-term regulation as shown in figure 6.1. While the effects of acute and chronic exercise are discussed later, other regulatory factors such as osmotic stress and hormones also merit brief discussion. Cells facing osmotic stress respond with an immediate regulation of cell volume, that is, a regulatory volume decrease (RVD) after hyposmotic shock and a regulatory volume increase (RVI) after hyperosmotic shock. In these volume-regulatory processes, a number of different ion channels and transporters can be involved; therefore the specific response is highly dependent on the cell type investigated. During RVD, many cells release potassium chloride via ion channels or via the K^+-Cl^- cotransporter. In contrast, volume expansion during RVI depends often on the activity of the Na^+-K^+-Cl^- cotransporter (Lauf & Adragna 2000).

Insulin and catecholamines are the two major hormones shifting potassium into the cells (Halperin 1998). There is evidence that insulin promotes potassium movements by two different mechanisms. Insulin activates sodium-potassium pumps from submembranous pools that are inserted into the plasma membrane. Furthermore, insulin seems to enhance the affinity of the sodium-potassium pump for sodium. The latter effect might be mediated by activation of protein kinase C (Clausen 1998). Catecholamines act via β_2-adrenoceptors and a subsequent adenylate-cyclase activation followed by an increase of cyclic AMP. Cyclic AMP in turn activates protein kinase A, which is assumed to enhance the sodium affinity of the sodium-potassium ATPase (Clausen & Flatman 1977).

In contrast, hormones such as glucocorticoids and thyroid hormones exert more long-term effects on potassium regulation, predominantly via an enhanced expression of the Na^+-K^+-ATPase. Using tritium-labeled ouabain, which is a well-known effective Na^+-K^+-ATPase inhibitor, an increase in the concentration of Na^+-K^+-pumps in hyperthyroid subjects could be demonstrated followed by a return to control levels after standard treatment of the thyroid disorder (Kjeldsen et al. 1984). Alterations in the concentration of sodium-potassium pumps due to thyroid disorders have been found also in other cell types like platelets and leucocytes (Kahn & Baron 1987; Chan et al. 1991). Finally, glucocorticoid treatment was found to induce an up-regulation of the Na^+-K^+-ATPase in skeletal muscle cells (Dorup & Clausen 1997).

An important modulator of the cellular potassium concentration is the divalent cation magnesium, since magnesium has been shown to regulate a number of different ion transport systems (Mooren et al. 2001; Bara et al. 1993; Ryan 1993). Therefore magnesium deficiency, which may arise during prolonged and intensive physical activity (see later), leads to a decrease of intracellular potassium concentration most likely via a decreased activity of the Na^+-K^+-pump and an enhanced activity of membrane potassium channels.

Potassium Homeostasis During Exercise

As outlined previously in this chapter, cellular potassium concentration is under the control of several different membrane proteins whose activity and expression can be modulated by contractile activity as well as by various hormones related to exercise. The following paragraphs address a comprehensive view of the effects of exercise on potassium homeostasis and the interaction of the various effectors.

Effects of Acute Exercise

During contractile activity, potassium continuously leaves the muscle cell. Changes in intracellular potassium concentration occur depending on the intensity, duration, and type of exercise. Several pathways mediate potassium efflux.

The most important role is that of the delayed inward rectifier potassium channel, which is responsible for membrane repolarization during the action potential. The channel mediates potassium efflux and restores membrane excitability. Interestingly, the release of potassium per action potential seems to be higher in fast-twitch than in slow-twitch muscles (Sreter 1963). It remains to be shown that this is a compensating mechanism for an increased sodium influx during the action potential. Moreover, there is evidence that the increase in cellular sodium concentration during electrical stimulation is higher in fast-twitch fibers than in slow-twitch fibers (Nielsen & Clausen 1996). Ruff (1996) demonstrated, furthermore, that the number of sodium channels was larger and the sodium current density much higher in fast- than in slow-twitch muscles. Accordingly, the intracellular potassium concentration has been found to be higher and the resting membrane potential more negative in fast-twitch muscles than in slow-twitch muscles (Cairns et al. 1997).

Moreover, there is some evidence that during exercise, muscle cells release potassium also via the ATP-sensitive potassium channel and the calcium-activated potassium channel. Although a decrease in intracellular ATP concentrations to a level near the K_i value (half-maximum inhibition) of the channel is necessary for channel opening, this level is usually not reached during physiological contractions; instead, several additional factors have been found that modulate the activity of the ATP-sensitive potassium channels. Important modulators are the concentrations of intracellular magnesium and hydrogen ions, which shift the dose-response curve for inhibition by ATP toward higher concentrations. Indeed, the regulation of ATP-sensitive potassium channels was shown to take place in the presence of normal ATP levels if the intracellular pH was decreased (Standen et al. 1992). A similar effect has been found for lactate. Another modulator of K_{ATP} is the purine nucleoside adenosine. Adenosine increases during hypoxia and activates ATP-sensitive potassium channels via the A1 receptor in both cardiac and skeletal muscle cells (Costa et al. 2000).

In addition, an activation of calcium-activated potassium channels cannot be excluded, especially during exhaustive exercise. It has been found that the resting level of cytoplasmic calcium increases during electrical stimulation under in vivo and in vitro conditions. A final modulator of channel function common to both channel types is oxidative stress. Studies predominantly performed in cardiac myocytes and in vascular smooth muscles revealed that the various reactive oxygen species are to able to alter channel activity in both directions (Liu & Gutterman 2002).

The function of these two potassium channels, K_{ATP} and K_{Ca}, during exercise is not entirely clear. It has been proposed that the dissipation of cellular potassium gradient and the concomitant decrease in cellular excitability contribute to a safety mechanism during muscle contraction. ATP and calcium-sensitive potassium channels may act as a link between muscle metabolism and cellular excitability and consequently the cell's ability to develop force.

The potassium loss of the muscle cell occurring with the onset of contractile activity is counteracted after a short delay by the potassium uptake due to the activity of the Na^+-K^+-pump. Interestingly, enhancement of Na^+-K^+-pump activity is more pronounced in slow-twitch than in fast-twitch muscles (Everts et al. 1988). In rats, exercise has been shown to affect also transcriptional regulation of Na^+-K^+-ATPase subunits as indicated by enhanced mRNA levels for the soleus α_1 subunit and the β_2 subunit in white-type muscle after treadmill running for 1 h (Tsakiridis et al. 1996). During contractile activity, the Na^+-K^+-pump capacity in the contracting muscles is enhanced by different mechanisms (Nielsen & Harrison 1998). First, the increase in intracellular sodium concentration during contraction increases the pump activity. However, changes in intracellular sodium most likely are not the only factor responsible for the increase in Na^+-K^+-pump activity. Other sodium-independent regulatory mechanisms of the Na^+-K^+-ATPase involve hormones, intracellular substrates, and the cytoskeleton. Catecholamines are well known to activate the sodium-potassium pump via β_2 adrenoceptors and an activation of adenylate cyclase. In contrast, nonselective beta-blocker treatment was followed by an enhanced exercise-induced hyperkalemia most likely because of an increased pump lag of the Na^+-K^+-ATPase (Gullestad et al. 1995). Similar effects might at least in part account for the muscle fatigue associated with nonselective beta-blocker treatment. The stimulating effect of catecholamines on the Na^+-K^+-ATPase seems to depend on the activity state of the muscle. Catecholamines given to the practicing muscle have either a small or no effect on the sodium-potassium pump activity (Everts et al. 1988). Therefore electrical activation of muscles and catecholamines seem to activate the Na^+-K^+-pump most likely via a common mechanism. How the contractile activity-induced stimulation of Na^+-K^+-pump activity is mediated is unclear. One possibility might be activation by another important regulator of pump activity, calcitonin gene-related peptide (CGRP), since it is released at the neuromuscular junction in addition to acetylcholine and binds to postsynaptic CGRP receptors.

Finally, a role of the cytoskeleton on sodium-potassium pump activity is proposed. Na^+-K^+-ATPase is linked to the actin cytoskeleton by actin cytoskeleton-associated molecules such as ankyrin and spectrin. Moreover, in in vitro experiments, the increase in Na^+-K^+-pump activity has been shown to occur upon application of actin. Recently, in aortic smooth muscle cells, Songu-Mize et al. (2001) demonstrated that the stimulating effect of cyclic stretch on the activity of the Na^+-K^+-ATPase requires an intact actin cytoskeleton.

A completely different way to enhance Na^+-K^+-pump capacity is by insertion of additional Na^+-K^+-pump subunits into the plasma membrane. Such a redistribution of α_2- and β_1-isoforms of the Na^+-K^+-pump has recently been shown upon insulin stimulation (Hundal et al. 1992; Lavoie et al. 1996). Recently several groups could demonstrate that such a mechanism is of functional importance also during exercise in both rat and human muscles. At least in rats, this increase in Na^+-K^+-pump subunits could be demonstrated for both fiber types and after different exercise protocols (Tsakiridis et al. 1996; Juel et al. 2000, 2001). Moreover, the Na^+-K^+-pump redistribution was reversible upon termination of exercise (Juel et al. 2001).

Potassium Redistribution

The narrow range which the extracellular potassium concentration is usually regulated within challenges drastic alterations during exercise. Depending on the type and intensity of exercise, marked increases in plasma potassium concentration occur. During high-intensity exhaustive exercise, levels of about 10 mmol/L potassium have been described. In contrast, during repetitive bouts of intensive exercise, potassium concentrations below 3 mmol/L have been found in the exercise-free intervals. Such variations in extracellular potassium concentration may be harmful to cardiac function, especially in subjects suffering from heart diseases. The increase in plasma potassium concentration depends predominantly on the release of potassium from the exercising muscle. However, additional mechanisms such as a decrease in plasma volume and a possible release of potassium from damaged cells have to be considered. This raises the question whether a redistribution of potassium into noncontracting tissues occurs that may help to attenuate the exercise-associated amplitudes in extracellular potassium concentration. There is evidence that a redistribution of extracellular potassium into resting muscle cells takes place and that this is mediated by the same mechanisms as in the active muscle. The role of red blood cells as a sink for the enhanced extracellular potassium levels remains controversial. The conditions during exercise—lactate production and concomitant red blood cell shrinkage, increase in catecholamine levels—favor an ion uptake via the combined activities of the Na^+-K^+-ATPase and the Na^+-H^+ exchanger as well as the Na^+-K^+-Cl^- cotransporter. Such an increase in transporter activity has recently been shown by Lindinger et al. (1999); however, other investigators failed to show any effect (Maassen et al. 1998; Juel et al. 1999). Therefore the role of red blood cells in buffering alterations in extracellular potassium concentration remains speculative.

Training Effects on Cellular Potassium Regulation

Regular exercise training has differential effects on the potassium handling of skeletal muscle cells. Only minor effects of training on the resting potassium concentration have been described (Sejersted & Sjogaard 2000). However, there is evidence that training attenuates the potassium loss from the muscle during exercise as indicated by measurements of plasma potassium concentration. At the same absolute exercise workload, the increases of extracellular potassium were significantly lower in trained than in untrained subjects. However, it seems that at the same relative workload there is no difference in potassium loss between trained and untrained subjects (Harmer et al. 2000; McKenna et al. 1993; McKenna 1995).

The overwhelming majority of studies prove a significant up-regulation of the Na^+-K^+-pump after training in both animals and human subjects. In contrast, decreased muscle activity is associated with a down-regulation of Na^+-K^+-ATPase. The increase in Na^+-K^+-pump concentration was induced by different training regimes, sprint, endurance, and strength training (McKenna et al. 1993; Evertsen et al. 1997; McKenna 1995). Medbo et al. (2001) found a slight dependence of change in the Na^+-K^+-ATPase concentration on the training frequency. Doubling of training frequency resulted in a significant up-regulation of sodium pump concentration; further increases of training frequency had no additional effects (Medbo et al. 2001).

The kinetics of Na^+-K^+-ATPase up-regulation show that even short-term training, for example cycling 2 h/day at 65% $\dot{V}O_2max$ for six days, resulted in a significant increase of Na^+-K^+-ATPase (Green et al. 1993; McCutcheon et al. 1999). Interestingly, training-induced increases in Na^+-K^+-pump were converted into a down-regulation of the ion pump if training was performed in normobaric hypoxia (Green et al. 1999). This result has recently been confirmed in mountaineers after a high mountain expedition for 21 days (Green et al. 2000). After the expedition, Na^+-K^+-ATPase was down-regulated by about 14% while no other changes in mitochondrial enzyme content and muscle histochemical properties were reported. It is a likely speculation that this mechanism might help the cell to spare energy costs.

Only a few reports exist about the effect of exercise on potassium transport in other cell types. Chen et al. (2001) recently presented evidence that exercise training modulates potassium channels of vascular smooth muscle cells. In red ventricular cardiocytes, endurance training altered the repolarizing potassium currents and the anoxia-induced current through ATP-sensitive potassium channels (Jew et al. 2001; Jew & Moore 2002). However, it seems too early to draw further conclusions about underlying principles and mechanisms by which exercise affects potassium movements in these cell types.

Intracellular pH Regulation

Regulation of intracellular pH is a hallmark of cell homeostasis, since most biochemical reactions and cell functions rely on enzymes and mechanisms that are characterized by a pronounced pH dependency. Therefore maintenance of intracellular hydrogen concentration is indispensable for cell metabolism and is necessary to avoid cellular dysfunction. The normal intracellular pH is between 6.9 and 7.2, depending on cell type, and continuously faces acid loads from various sources including metabolic processes and passive hydrogen ion fluxes driven by the electrochemical gradient (Roos & Boron 1981). The acid load increases with an enhanced metabolic rate, such as during exercise, and may reach excessive levels during high-intensity anaerobic exercise (Lindinger 1995; Juel 1998a).

pH-Regulating Systems

Intracellular pH is balanced by two distinct mechanisms. One is through the intrinsic buffer capacity, which comprises two components, a CO_2-dependent and a non-CO_2-dependent buffering power. The other utilizes various acid-equivalent ion transporters, which can be characterized as either acid exporting (= alkaline loaders) or acid importing (= acid loaders) systems. Principally their mode of action depends on the electrochemical gradient. However, under physiological conditions, the group of alkaline loaders includes Na^+-H^+ exchanger (NHE), hydrogen ATPases, and proton-conductive pathways, while the Cl^-/HCO_3^--anion exchanger is among the acid loaders. The mode of other transporters, such as the Na^+/HCO_3^- cotransporter and the $lactate^+/H^+$ cotransporter (MCT), varies with cell type and rate of metabolism, respectively. Since most of the data available on pH regulation during exercise focus on NHE and MCT, the next two sections deal with these two ion transport systems in more detail.

Mammalian bicarbonate transporters are encoded by at least two different gene families, SLC4 and SLC26. The latter family consists of approximately 10 members that exchange Cl^- for several different anions like sulfate, bicarbonate, hydroxyl, and iodide. The SLC4 gene family includes the Na^+-independent Cl^-/HCO_3^- exchangers (AE), Na^+-HCO_3^- cotransporters (NBC), and Na^+-dependent anion exchangers (Alper et al. 2001).

The AEs promote the electroneutral exchange of a chloride anion for a bicarbonate anion. Four isoforms (AE1-4) have been identified, of which AE1 is expressed at high levels in the red cell membrane (band 3 protein) and at lower levels in heart, distal colon, and other tissues. AE2 can be found predominantly in the basolateral membrane of nearly all epithelial cells, while AE3 is highly expressed in excitable tissues but also found in kidney and gastrointestinal tract. An important inhibitor of the anion exchanger family is the stilbene derivative DIDS (4,4'-diisothiocyano-2,2'-stilbene disulfonate) (Alper et al. 2001, 2002).

Less extensively studied than the AEs is the Na^+-dependent Cl^-/HCO_3^- exchanger, which is expressed in the renal proximal tubule. It mediates the electroneutral transport of Na^+ and HCO_3^- into the cell in exchange for outward Cl^- movement.

Finally, it is important to mention the subfamily of NBC, which consists of at least three isoforms that have been found to be widely expressed. The stoichiometry of the NBCs and their mode of action vary among the different cell types. In the kidney proximal tubule, three base equivalents per Na^+ are exported, leading thereby to an intracellular acidification. In contrast, in heart or liver cells, NBC promotes intracellular alkalinization. A stoichiometry of two bases per Na^+ or even an electroneutral transport has been found. While usually silent under resting conditions, in the heart NBC has been shown to be active at resting pH. NBCs are inhibited in an alkaline milieu, while activity is enhanced by decreasing pH. NBCs are under hormonal control and are modulated by intracellular signaling pathways including cyclic AMP, protein kinase C, and calmodulin. Acute and chronic alterations in acid–base status were found to regulate NBC activity (Romero 2001; Romero & Boron 1999).

Some cell types such as leucocytes possess proton-conductive pathways, which are activated by decreasing cytosolic pH values and cellular depolarization. The H^+-ATPase becomes of particular importance under conditions of extracellular acidification. Driven by ATP, the pump is able to extrude protons from the cytosol against the proton gradient, thereby maintaining cellular functions like superoxide production also in the presence of an unfavorable environment (Hackam et al. 1996; Torigoe et al. 2002).

The Sodium Proton Exchanger (NHE)

Presently the NHE family of ion exchangers contains six isoforms (NHE1-NHE6). While the NHE1 is ubiquitously expressed in mammalian cell types, the other isoforms have a more limited distribution. NHE2 through NHE4 can be predominantly found in the kidneys and the gastrointestinal tract, while NHE5 is predominantly expressed in neuronal cells. NHE6, which shares only about 20% sequence identity with the other isoforms, is restricted to intracellular localization (Putney et al. 2002). It has been identified within mitochondria especially in heart and skeletal muscle cells. Interestingly, investigations of the NHE1 distribution in various rat muscle types revealed the highest expression in glycolytic fibers, whereas in oxidative fibers, NHE was less expressed (Juel 2000).

This section focuses on structure and regulation of the NHE1 exchanger since it is the most extensively studied isoform and is the most relevant isoform with respect to exercise. The NHE1 of mammalian cells has a molecular weight of about 91 kDa and consists of approximately 815 amino acids. The sequence involves an N-terminus of 12 transmembrane α-helices responsible for ion exchange and a C-terminal hydrophilic cytoplasmic regulatory domain that has been shown to modulate the activity of the ion exchange domain (Putney et al. 2002).

The sodium-hydrogen exchanger promotes the electroneutral exchange of intracellular hydrogen and extracellular sodium down the sodium gradient. NHE1 is activated by intracellular acidification. The Hill coefficient of about 2 for the hydrogen effect on NHE1 activity suggests an allosteric activation by hydrogen ions. Moreover, a number of NHE1 interacting proteins have been identified that regulate the NHE1 activity through modifications of the C-terminal cytoplasmic regulatory domain. This includes phosphorylation and the binding of regulatory proteins such as calmodulin, which is supposed to change hydrogen affinity by inducing conformational changes. Furthermore, NHE1 activity is under hormonal control. Several hormones such as growth factors and catecholamines have been shown to regulate NHE activity via different classes of cell surface receptors like receptor tyrosine kinases, G-protein-coupled receptors, and integrin receptors (Hall et al. 1998). Several intracellular signaling pathways such as Ras-Raf-MAPK, phosphatidylinositol-3 kinase (PI3K), protein kinase C, and phospholipase C-IP3-calcium-calmodulin, as well as RhoA-ROCK, can be involved that exert their effect on the regulatory domain via the NHE1 interacting proteins (Fliegel 2001).

Beside its role in cellular pH regulation, the NHE1 is involved in regulation of cell volume, cell proliferation, and cell death and cytoskeletal-dependent functions such as cell migration and adhesion. In pathophysiological events the NHE has been shown to play an important role in the development of myocardial damage during ischemia and reperfusion. Moreover, in hypertensive patients, NHE1 activity has been found to be enhanced (Fröhlich & Karmazyn 1997; Park et al. 1999).

Lactate-H$^+$ Cotransporter

It is vital to cells that lactic acid is rapidly moving across their membranes. In lactic acid-producing cells, its accumulation and the concomitant pH decrease, slow down, or even inhibit multiple biochemical pathways. In contrast, lactic acid-consuming cells such as in brain or heart use it as a major respiratory fuel under certain conditions. However, at physiological pH, lactic acid is almost entirely dissociated, making free diffusion across the plasma membrane difficult. Lactate transport therefore is mediated via a carrier system, the proton-linked monocarboxylate transporter (MCT), which promotes the facilitated diffusion of both lactate and proton. The MCT is stereoselective for L-(+)-lactate and is temperature and pH sensitive.

Several isoforms of the MCTs (MCT1-9) have been identified; these are not specific solely for lactate. Instead, MCTs are responsible also for the transport of other important monocarboxylates such as pyruvate and the ketone bodies acetoacetate, hydroxy-butyrate, and acetate (Halestrap & Price 1999). In skeletal muscle cells, MCT1 and MCT4 are the major isoforms. While MCT1 and type I fibers are positively correlated, no such relation seems to exist for MCT4 (Pilegaard et al. 1999b). For a detailed description of MCTs, their expression, and distribution within various tissues the reader is referred to chapter 8. Here only a few details about the general structure of MCTs and their role in pH regulation are given.

Hydropathy plots predicted 12 transmembrane-spanning helices for the MCT1-3 with intracellular C- and N-termini and a large intracellular loop between the transmembrane segments 6 and 7. Lactate-H$^+$ cotransport is performed electroneutrally in a 1:1 ratio (Poole et al. 1996). Kinetic measurements revealed an initial proton binding to the transporter followed by the lactate anion. After translocation, both ions are released in a sequential order again. The transport cycle is completed after return of the free carrier, which is the rate-limiting step in these sequences (De Bruijne et al. 1985).

The K_m values of MCTs for lactate show a significant variability between 0.5 and 40 mM depending on isoform, cell type, and analysis method. The ranges of K_m values for lactate are about 5 to 10 mM and 10 to 20 mM for MCT1 and MCT4, respectively. This suggests that these transporters play a significant role for pH regulation only in the presence of large lactate gradients as in intense exercise (Juel & Halestrap 1999). MCTs are activated either by a decrease in pH, if lactate is available on the same side, or by a pH increase on the opposite side, which enhances the rate of return of the free carrier. Several inhibitors of MCTs have been found including aromatic monocarboxylates, DIDS, and phloretin. However, none of these substances has been proven to be specific for MCTs (Halestrap & Price 1999).

Resting pH Regulation

As already mentioned, cellular pH is the result of an equilibrium between hydrogen-increasing and -decreasing mechanisms. During resting state a stable pH is present, indicating a balance between the two mechanisms. Whether this balance requires the activity of acid-equivalent membrane transporters can be determined using specific inhibitors. If inhibition of a transporter, for example, the NHE by application of amiloride, is followed by a decrease in pH, this indicates its involvement during resting conditions. The importance of the NHE for maintaining resting pH differs between cell types and varies with environmental conditions and with pathophysiological transition. While the NHE anti-porter is virtually quiescent under resting conditions in several cells such as lymphocytes (Grinstein et al. 1985), kidney cell lines (Montrose & Murer 1986), and resistant arteries, it seems to be active in skeletal muscle cells and parietal cells (Paradieso et al. 1989). In vascular smooth muscle cells, inhibition of NHE lowered resting pH if cells were incubated in bicarbonate-free buffer. Adding bicarbonate to the buffer resulted in a slight increase (0.08 units) in resting intracellular pH. In addition, inhibition of NHE was no longer effective in alteration of resting pH (Kahn et al. 1990). Finally, NHE activity is known to be enhanced under pathophysiological conditions such as hypertension. Using amiloride, Izzard and Heagerty (1989) demonstrated that in hypertensive animals NHE was much more important in maintaining resting intracellular pH of resistance arteries than in control animals, while the anion exchanger DIDS was without effect in both cases.

pH Regulation During Acute Exercise

During intense exercise, changes in a number of physicochemical factors—carbon dioxide (CO_2) partial pressure, strong ion difference, and total concentration of weak acids and bases—contribute to an intracellular acid load. A decrease in strong ion difference due to both intracellular potassium decrease and lactate increase is the predominant factor, since it contributes about 60% to the pH decrease (Lindinger 1995). Phosphocreatine breakdown into inorganic phosphate and creatine, which enhances the concentration of weak acids/bases, is responsible for a hydrogen increase of about 20%, while changes in P_{CO2} play only a minor role (figure 6.2).

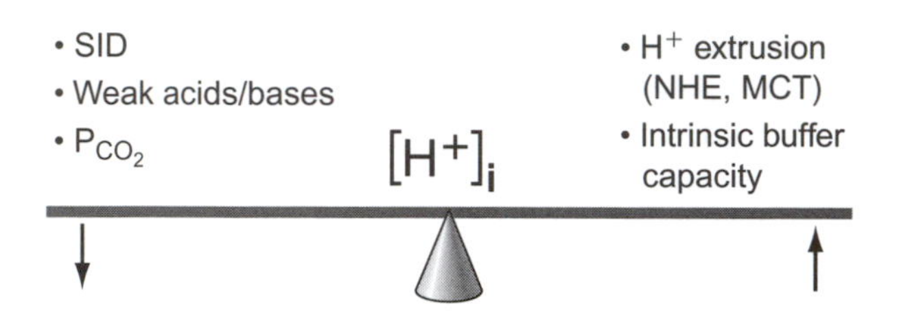

Figure 6.2 pH equilibrium during exercise. With the onset of exercise, hydrogen-generating mechanisms (left side) induce an intracellular acidification (pH decrease), which is counteracted by pH-buffering mechanisms (right side). For details, see text. SID, strong ion difference; P_{CO2}, CO_2 partial pressure; NHE, sodium-hydrogen exchanger; MCT, monocarboxylate transporter.

The extent to which exercise-induced acid loads influence intracellular pH changes depends on both the capacity of intracellular hydrogen-buffering molecules and the capacity of hydrogen-extruding pathways. The former is called the intrinsic buffer capacity or in vitro buffer capacity since it is determined by acid titration on homogenized tissues or cells. The latter are part of the functional buffer capacity or in vivo buffer capacity, which includes intrinsic buffer capacity, the HCO_3^-/CO_2 buffer system, and the functional operation of the various ion transport mechanisms.

During exercise, short-term changes in the intrinsic buffer capacity are unlikely to occur; therefore, the acute acid load should be responded to primarily by an enhanced activity of the hydrogen-extruding transporters. The role of NHE1 is suggested by measurements in electrically stimulated isolated mouse and rat muscle. Recovery of intracellular pH after exercise was delayed after application of amiloride (Juel 1988). Interestingly, the expression of NHE1 isoform was highest in glycolytic fibers, which have the largest capacity for production of lactic acid (Juel 2000). Furthermore, we found that after an acute bout of exhausting exercise there was an increase in the amiloride-inhibitable NHE activity in isolated human lymphocytes (Mooren manuscript in preparation).

Expression of MCT1 and MCT4 was enhanced after a single bout of exercise (60% $\dot{V}O_2$max for 5-6 h) for up to six days postexercise (Green et al. 2000). Whether lactate/proton transport capacity is specifically up-regulated after exercise is still controversial. Dubouchaud et al. (1999) presented evidence for a loss of saturation kinetics of the MCT. They suggested an enhanced passive diffusion of lactate resulting most probably from exercise-induced membrane damage (Eydoux et al. 2000). Interestingly, eccentric contractions were shown to decrease sarcolemmal lactate/H^+ transport capacity (Pilegaard & Asp 1998).

The role of MCTs in pH regulation of muscle cells depends on exercise intensity. During intense exercise, almost two-thirds of hydrogen export is mediated by MCTs (Juel 2001). In contrast, during moderate-intensity exercise the non-lactate-dependent hydrogen export becomes overweighted (Bangsbo et al. 1997). These data fit well with the high K_m values for lactate of the MCTs expressed in muscle cells, suggesting their significant role in pH regulation only in the presence of large lactate gradients (Juel & Halestrap 1999).

Training-Associated Adaptations in pH Regulation

Recent studies on training effects on the intrinsic buffer capacity of skeletal muscle cells showed contradictory results. Bell and Wenger (1988) reported an increase in intrinsic buffer capacity in the trained leg after seven-week one-legged sprint training, while others found no change in muscle buffer capacity after eight-week sprint training (Nevill et al. 1989). Both studies indicated an increase in exercise performance after training accompanied by an increase in achieved lactate values.

Measurements of the functional buffer capacity revealed more homogeneous results. Comparison of sedentary controls and subjects active in ball games revealed a higher buffer capacity for the trained subjects (Sahlin & Henriksson 1984). Eight weeks of sprint training induced an increase in functional buffer capacity of about 35%. Interestingly, the pretraining value was not significantly different from values obtained from endurance-trained athletes, suggesting that endurance training does not lead to adaptations in functional buffer capacity (Sharp et al. 1986). These data suggest that training modulates functional buffer capacity most likely by affecting the hydrogen extrusion capacity. MCT expression has been shown to be enhanced with exercise training, while K_m values of MCTs for lactate were probably not affected. Surprisingly, high-intensity as well as endurance training and chronic low-frequency electrical stimulation induced an MCT up-regulation (McCullagh et al. 1996; Pilegaard et al. 1999a). This makes acid load as a single stimulus for an enhanced MCT capacity unlikely and fits with data showing that lactate and hydrogen transport capacity is higher in oxidative fibers than in glycolytic fibers (Juel & Pilegaard 1998). Whether the training-induced increase in MCT content is transferred into an enhanced lactate/H^+ transport capacity remains controversial (Eydoux et al. 2000). Although some studies demonstrated a different up-regulation of the muscle MCT isoforms 1 and 4 with exercise, the underlying regulatory mechanisms remain unclear (for details see chapter 8). In contrast, denervation and inactivity resulted in a down-regulation of MCT transport capacity.

On the other hand, NHE exchange capacity was increased in rat skeletal muscle after six weeks of high-intensity treadmill training (Juel 1998b). Similar results were obtained in our previously mentioned study using isolated human lymphocytes. After five days of high anaerobic training, the amiloride-inhibitable NHE exchange capacity was significantly improved compared to pretraining levels (figure 6.3). In both studies it remained unclear whether the enhanced NHE capacity was due to an increased number of anti-porters or an enhanced activity of existing transporters. However, in a recent study, Juel (2000) demonstrated that three weeks of treadmill training induced an increased expression of NHE1 in rat skeletal muscles. Furthermore, the up-regulation was similar in both muscle types, oxidative and glycolytic fibers.

Magnesium

Magnesium (Mg^{2+}) is both a ubiquitous and an essential biological element that is found in abundant quantities in cells. It plays an important role in the regulation of cell function. Despite its wide range of important functions and distribution in the body, study of Mg^{2+} has been neglected in comparison to the second messenger, divalent cation, calcium. Reasons may be that in the past, accurate biological techniques were not available to measure intracellular free magnesium concentration and that in most instances, calcium seems to interfere with the techniques (Hurley et al. 1992). However, over the past 10 to 15 years, persistence in magnesium biology research alternative strategies has led to the study of magnesium homeostasis in different cell types. This has enabled scientists to characterize the transport of magnesium across the cell membrane, as well as its intracellular homeostasis especially with respect to its role in cellular signal transduction.

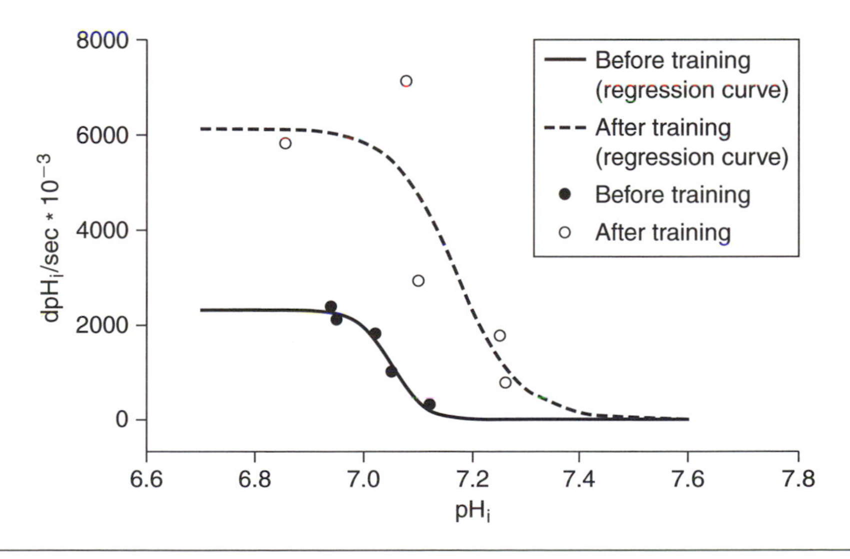

Figure 6.3 Adaptation of Na^+-H^+ exchanger (NHE) exchanger capacity in human peripheral lymphocytes to five-day anaerobic training (four 400-m runs per day). NHE exchanger capacity was measured as the initial rate of pH recovery after an intracellular acid load (dpH_i/sec) and was related to the pH_i at which the recovery originated. Symbols represent the measured points, which were fit by a sigmoidal function (regression curve) as reported by Reusch et al. (1993).

Mg^{2+} Chemistry

Magnesium is named after the Greek city Magnesia, where large deposits of magnesium carbonate were found. The element magnesium does not naturally occur in the free state but in its principal mineral forms as oxides, carbonates, chlorides, silicates, fluorides, sulfate, and phosphate. In water, Mg^{2+} occurs as a chloride at a concentration of 5.0 to 5.5 g/L.

Magnesium is extremely electropositive and readily loses two electrons to yield a divalent cation (Mg^{2+}), in common with other alkaline earth metals. The alkaline earth metals decrease in electronegativity with increased atomic weight. Usually, the alkaline earth metals occur as salts of commonly found anions. For example, magnesium, calcium, strontium, and barium occur in nature as sulfates and carbonates. Magnesium forms oxides on heating in air, and its nitrates have very low thermostability. Many Mg^{2+} salts are insoluble in water while others are deliquescent. Isotopes of Mg^{2+} and their abundance are $^{24}Mg^{2+}$ (78.60%), $^{25}Mg^{2+}$ (10.11%), and $^{26}Mg^{2+}$ (11.29%) (Williams 1993). Magnesium ions (Mg^{2+}) have an ionic radius approximately two-thirds that of calcium ions (Ca^{2+}) and sodium ions (Na^{+}) and half that of potassium ions (K^{+}). Therefore, due to its small size and relatively large charge, Mg^{2+} is the strongest metal ion in abundance in biological systems. Cations usually form the most stable complexes with anions or ligands of similar hardness, and hard ligands contain highly electronegative donor atoms. Therefore Mg^{2+} can form stable complexes with phosphate or carboxylate ions and with nitrogen's lone pair of electrons. Mg^{2+} has a longer hydration energy than Ca^{2+} because of its greater polarizing potential. This means that magnesium salts of large organic acids are more soluble than analogous calcium compounds (Williams 1993).

Biological Roles of Mg^{2+}

Mg^{2+} exists as an abundant divalent cation and is involved in several physiological and biochemical processes during cellular homeostasis through binding to organic substances, such as proteins, nucleic acids, and nucleotides. In general, Mg^{2+} is an important regulator/modulator of three main complexes: (1) enzyme activation, (2) membrane function, and (3) calcium signaling.

Mg^{2+} is an important cofactor for several hundreds of enzymes. It is involved in the synthesis and replication of RNA and DNA, as well as the secretion of enzymes and hormones (Henrotte 1988; Wisdom et al. 1996). As a component of the ATP-Mg complex, Mg^{2+} plays an important role in multiple metabolic pathways and ATP-generating and -consuming reactions, such as oxidative phosphorylation and muscle contraction.

Moreover, Mg^{2+} affects membrane function leading to a stabilization of membrane structure and electrical potentials. It modulates the transmembrane movements of ions by altering the activity of several ion transporters (e.g., Na^{+}-K^{+}-ATPase, Ca^{2+}-ATPase, K^{+}-Na^{+}-Cl^{-} and Na^{+}-Cl^{-} transporter and Ca^{2+}-Na^{+} and HCO^{-}_{3}-Cl^{-} exchangers) and ion channels, for example Na^{+}, K^{+}, Cl^{-}, and Ca^{2+} channels (Flatman 1991; Agus et al. 1989; Gunther & Vormann 1994).

Finally, Mg^{2+} is thought to regulate cellular signaling pathways. Whether Mg^{2+} itself may act as a second messenger seems unlikely. However, there is overwhelming evidence that Mg^{2+} acts as a natural calcium antagonist, thereby finely tuning cellular calcium signals.

Mg^{2+} Homeostasis

Mg^{2+} is the fourth most abundant cation in the human body and the second most abundant intracellular cation after K^{+}. It is predominantly located in the cellular and extracellular matrix of bones (approximately 52%); the remainder is located in muscle cells (28%), soft tissue (19%), serum (0.3%), and red blood corpuscles (0.5%) (Elin 1987). About 30% of Mg^{2+} in the bone is believed to represent an exchangeable pool of Mg^{2+}. Mobilization of Mg^{2+} from this pool is more rapid in children compared to adults (Mudge & Weiner 1992).

The concentration of total plasma Mg^{2+} is approximately 0.75 to 1.1 mmol/L (Mudge & Weiner 1992). It is now believed that serum Mg^{2+} is "in transit" between bone stores and actively metabolizing tissues (Vormann & Günther 1993). In serum about 40% of Mg^{2+} is bound to proteins, about 50% is ionized, and the remaining 10% is complexed to ions including phosphate and citrate (Speich et al. 1981). The total intracellular Mg^{2+} concentration has been estimated in similar ranges between 2 and 20 mmol/L. Again, an equilibrium between ionized and bound Mg^{2+} is established. However, about 90% to 95% of the cellular magnesium is forming more or less stable complexes, of which a high percentage are compartmentalized in cellular organelles, especially in mitochondria and endoplasmic reticulum. Mg^{2+} can be mobilized from these intracellular pools with different kinetics. Only the ionized Mg^{2+} is available to react in physiological and biochemical processes. Since only about 5% to 10% of the cellular Mg^{2+} is present in the ionized form, variations in the concentrations of Mg^{2+} buffers have significant effects on the concentration of free ionized Mg^{2+}. Thus our knowledge of both free Mg^{2+} concentrations and the regulation of cellular Mg^{2+} buffer capacity is essential for an understanding of the physiological role of this important divalent cation in physiological and biochemical processes. Recently, novel techniques have become available, such as magnesium-selective electrodes and magnesium-sensitive fluorescent dyes, which are used for the determination of free ionized magnesium in the extracellular and intracellular compartment, respectively.

Regulation of Intracellular Mg^{2+}

As already indicated, there are only minor concentration gradients between extra- and intracellular ionized Mg^{2+}. However, due to the inside negative resting membrane potential, an inward-directed electrochemical force for magnesium exists. Since the membrane permeability for Mg^{2+} is low, only small leak currents occur. In some cell types such as cardiac ventricular cells, the magnesium membrane permeability seems to be enhanced as indicated by alterations of basal intracellular Mg^{2+} upon variations of extracellular Mg^{2+}. Whether Mg^{2+} thereby crosses the membrane via specific Mg^{2+} influx pathways seems unlikely. The inhibitory effect of calcium channel antagonists on the magnesium movement suggests that Mg^{2+} penetrates via calcium channels (Quamme & Rabkin 1990).

Intracellular Mg^{2+} is furthermore under the control of a secondary active transport system, the Na^+-Mg^{2+} exchanger. Although to date no Mg^{2+} transporter on the plasma membrane has been purified or cloned, there is much evidence that a Na^+-Mg^{2+} exchanger is operative in a number of different cell types (Flatman 1991).

There is also evidence for hormonal control of intracellular Mg^{2+}. Hormone application resulted in changes of intracellular total and free Mg^{2+} in both directions (Mooren et al. 2001; Romani & Scarpa 1990; Dai et al. 1998; Touyz & Schiffrin 1996; Okada et al. 1992; Romani et al. 1993). This includes Mg^{2+} shifts across the plasma membrane as well as an intracellular Mg^{2+} shift between the cytosol and intracellular organelles, especially the mitochondria and the endoplasmic reticulum. Recent studies suggest that Mg^{2+} shifts between cytosol and mitochondria have a functional significance such as is known for Ca^{2+}. Removal of Mg^{2+} from the mitochondrial matrix stimulated severalfold levels of succinate and glutamate dehydrogenases. This indicates that matrix Mg^{2+} is able to control respiratory and metabolic pathways (Panov & Scarpa 1996). However, a definition of the role of Mg in mitochondrial biogenesis needs further investigation. The regulation of intracellular Mg^{2+} is summarized in figure 6.4. Within the three compartments—extracellular space, cytosol, intracellular stores—free Mg^{2+} concentration is determined by its affinity to various Mg^{2+} binding proteins and chelators and by the rate of exchange of free Mg^{2+} between the different compartments.

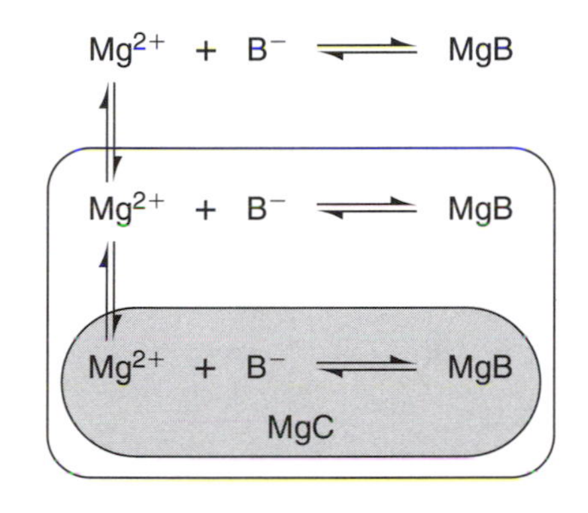

Figure 6.4 Magnesium homeostasis. The role of Mg^{2+} is dependent on the ratio of free to bound Mg^{2+} that is established in each of the different compartments—extracellular, cytosolic, and intracellular store. Mg^{2+} crosses the membranes primarily in the ionized form. The model emphasizes the important role of Mg^{2+} buffers, Mg^{2+} binding proteins, and chelators for regulating magnesium homeostasis. While some Mg^{2+} ligand complexes can release Mg^{2+} (MgB), others do not (MgC). Since only about 5% to 10% of the cellular Mg^{2+} is present in the ionized form, variations in the concentrations of Mg^{2+} buffers have significant effects on the concentration of free ionized Mg^{2+}.

Mg^{2+} and Exercise

Although the regulation of Mg^{2+} during exercise has attracted interest from scientists for many years, our view is still far from complete, for two primary reasons. Many studies have focused on the extracellular magnesium concentration only. However, as stated earlier, extracellular magnesium is only the tip of the iceberg and is not representative of the body's magnesium status. Secondly, most of the currently available data on the regulation of Mg^{2+} during exercise are unfortunately based only on measurements of total magnesium, using atomic absorbance spectroscopy. Moreover, these studies often determined Mg^{2+} only in a single compartment and were unable to measure ionized Mg^{2+} levels. Nevertheless, data available so far clearly demonstrate that exercise results in alterations in Mg^{2+} homeostasis that seem to depend on type, duration, and intensity of exercise.

After short-term, high-intensity exercise the majority of studies indicated an increase of extracellular Mg^{2+}, which seemed to be caused predominantly by an exercise-induced decrease in plasma volume (Joborn et al. 1985; Sjogaard 1983; Rayssiguier et al. 1990). Unfortunately many studies did not correct for volume shifts. Another parameter seems to be lactate acidosis. It has been proposed that acidosis might trigger a release of Mg^{2+} from the intracellular compartment (intracellular Mg^{2+} = $[Mg^{2+}]i$). However, to date no clear evidence for this hypothesis has been presented. In most human studies, no change in erythrocyte magnesium was measured after high-intensity exercise (Hespel et al. 1986; Deuster et al. 1987). The results from animal studies on this aspect are contradictory, although it must be admitted that different cell types were investigated. While Navas et al. (1997) reported a decrease in rat erythrocyte $[Mg^{2+}]i$ after an exhaustive swimming test, Cordova et al. (1992) found an increase of $[Mg^{2+}]i$ in muscle and liver tissue.

Exercise-induced changes in ionized extracellular magnesium have been measured so far only in one study. After anaerobic exercise, Porta et al. (1997) found an increase in ionized Mg^{2+}, while we found a decrease in our own study. Moreover, after correction for volume changes, the decrease in ionized extracellular Mg^{2+} was even pronounced (Mooren manuscript in preparation). In addition, we determined for the first time the ionized intracellular Mg^{2+} levels of blood cells after exercise. Interestingly, we found an increase in ionized $[Mg^{2+}]i$ in thrombocytes and erythrocytes after a progressive treadmill test, while total $[Mg^{2+}]i$ was unchanged. This suggests either an altered intracellular buffering for Mg^{2+} by intracellular ligands or a release of Mg^{2+} from intracellular pools. In vitro data suggest that the intracellular acidification by lactate might be a reason for the increase in ionized $[Mg^{2+}]i$. Similar results were reported for free ionized $[Mg^{2+}]i$ levels in the skeletal muscle during exercise through use of magnetic resonance spectroscopy (Iotti et al. 2000).

After prolonged submaximal exercise, most studies reported a hypomagnesemia (Hoffman and Böhmer 1988; Beller et al. 1975; Stendig-Lindberg et al. 1987; Lijnen et al. 1988). It seems unlikely that sweat Mg^{2+} losses, enhanced renal Mg^{2+} excretion, or a combination of the two accounts for this decrease in plasma Mg^{2+}. Indeed, during moderate exercise, Nishimuta et al. (1997) reported a slightly decreased urine excretion of Mg^{2+}. Some authors suggested that a shift of Mg^{2+} into the cellular compartment during prolonged exercise occurs because they measured increases in intracellular Mg^{2+} concomitant with the Mg^{2+} decrease in plasma levels. This may be caused by exercise-associated hormonal alterations. Likewise β-adrenergic stimulation has been shown to evoke a Mg^{2+} decrease in plasma in sedentary humans (Whyte et al. 1987; Joborn et al. 1985). In contrast, other investigators reported no changes in cellular Mg^{2+} content or reported a decrease (Laires et al. 1993; Lijnen et al. 1988). In vitro data revealed that hormonal stimulation leads to both Mg^{2+} extrusion from and Mg^{2+} accumulation in cells depending on hormone and cell type. Stimulation with isoproterenol, epinephrine, or norepinephrine resulted in a marked extrusion of Mg^{2+} from cardiac and liver cells (Romani & Scarpa 1990). In contrast, vasopressin and angiotensin-II induced an Mg^{2+} accumulation in hepatocytes, renal epithelial cells, and smooth muscle cells (Dai et al. 1998; Touyz & Schiffrin 1996; Okada et al. 1992; Romani et al. 1993). Therefore at the moment the nature of exercise-induced Mg^{2+} shifts remains elusive.

Longitudinal and cross-sectional studies suggest that intensive training may be followed by Mg^{2+} depletion and that athletes are prone to Mg^{2+} deficiency. Dolev et al. (1991-92) reported a decrease in mononuclear cell Mg^{2+} after 12 weeks of strenuous physical training of military recruits. Similar results were found in an animal training study (Navas et al. 1997). Recently a novel method, the Mg^{2+} loading test, was used to assess athletes' Mg^{2+} status. This method should be superior to the conventional measurements of Mg^{2+} content in erythrocytes or leukocytes (Gueux et al. 2001; Basso et al. 2000). A comparison between endurance-trained subjects and controls revealed an enhanced retention of the intravenously applied Mg^{2+} in the athlete group, indicating a depleted Mg^{2+} status. Interestingly, at the same time, measurements of plasma and erythrocytic Mg^{2+} levels showed no difference, emphasizing the unreliability of these measurements in reflecting the Mg^{2+} status (Saur et al. 2002).

Chapter 7

Inter- and Intracellular Signaling

Frank Mooren, Klaus Völker, Bente Klarlund Pedersen, Axel Schulz, and Hansjörg Teschemacher

Life is characterized by the maintenance of a balanced disequilibrium. This is indicated by multiple gradients between different compartments or even within the same compartment. For example, ion gradients across the cell membrane are fundamental for excitability, while pressure gradients drive the continuous blood flow. Thermodynamically such situations are defined by a high degree of order that requires a constant energy input to be maintained. Such situations can also be described as a steady state. Homeostasis circumscribes the perfect matching of multiple regulatory and integrative responses to maintain this situation and to adapt to external influences. Exercise could be such an exogenous factor affecting this steady state. Transition from a resting to an active state requires adaptational changes in circulation, respiration, and metabolism until a new steady state is reached. If no steady state can be established, exercise has to be terminated soon. To manage these challenges, homeostasis control of a multicellular organism depends on an effective communication system, including sensing and effector mechanisms, in order to regulate and orchestrate the functions of the different tissues and cell types. On the cellular level this means an information transfer between cells and within cells, often termed intercellular and intracellular signaling, respectively.

Principally, intercellular communication is carried out in three different ways:

- Gap junctions, which form direct and continuous connections between cells
- Contacts via membrane-anchored surface proteins
- Communication by chemical messengers released from cells into the extracellular fluid

The first part of this chapter addresses the latter of these forms of communication, including discussion of various hormones and their alterations during exercise. In the second part of the chapter we discuss the numerous intracellular signal transduction pathways and their role during exercise.

Hormones and Receptors*

A number of different intercellular chemical signals can be distinguished dependent of their way of action. Autocrine and paracrine messengers are locally secreted with a locally restricted effect. While autocrine signals influence the cell type they have been secreted from, paracrine signals also affect neighboring cells of other tissues. Neurotransmitters are produced within neurons and secreted into the synaptic cleft providing the communication between nerve cells or nerve and muscle cells. Hormones can be defined as chemical messengers that are produced in small amounts

* The text on this page beginning with the heading *Hormones and Receptors* and ending with the text under the heading *Catecholamine Response to Exercise* on page 114 is contributed by Frank Mooren.

by specialized cells (endocrine glands or tissues) transported via blood over some distance to their target cells, where they induce specific cellular responses and modulate cell activities. Whatever way of intercellular signaling is operative, the principle mechanisms of intercellular signal-target cell interaction are similar and will be discussed in the following paragraphs, considering as an example the hormone-receptor interaction.

Mechanisms of Hormone–Receptor Interaction

To fulfill their messenger function, hormones have to interact with sensing units, receptors, on their target cells. Hormone and receptor are noncovalently bound by one of the following mechanisms: (1) hydrogen bonds, (2) ionic bonds between oppositely charged ionic or polarized groups, (3) van der Waals forces, or (4) hydrophobic interactions. Affinity describes the strength with which the hormone binds to its receptor. It is determined thermodynamically by the resulting changes in energy and entropy levels. Another important attribute of the receptor is its specificity. This refers to the ability of only one ligand or a limited number of structurally related molecules to bind to the receptor and provides the basis for an organized information transfer.

Receptor ligands can be classified, according to their biological activity on the receptor, as agonists, antagonists, and partial agonists. Usually an agonist with the highest binding affinity induces the biological activity at the lowest concentration. Similarly, the antagonist with the highest affinity prohibits a biological response at the lowest concentration in competition with other present agonists. A partial antagonist binds specifically to the receptor but induces only a partial biological response. A quantitative description of the hormone–receptor interaction follows the law of mass action:

$$H + R \rightleftarrows HR$$

The biological effect is proportional to the concentration of the hormone–receptor complex. Analogous to the Michaelis-Menton equation, the relationship between effect and the concentration of the free hormone ([H]) can be determined as

$$\text{effect} = \text{maximal effect} \times [H] / (K_D + [H])$$

For $H = 0$, no effect is obtained; for $H = K_D$ the effect is half-maximal; and for H much higher than K_D the effect reaches asymptotically the maximal level. The concentration of drugs or inhibitors that induce a half-maximal effect is also called EC50 or I50, respectively. Plotting the effect versus hormone concentration results in the typical dose-response curve. This can be performed after various mathematical operations leading to, for example, the Scatchard or Eadie-Hofstee type of plot, which helps to interpret the hormone receptor binding behavior.

Receptor Types

A target cell expresses in principle two types of receptors—a membrane-bound surface receptor and a soluble intracellular receptor. The first receptor type is accessible to water-soluble molecules, which bind to the extracellular domain of the receptor. In general, receptor occupation is transmitted via the transmembrane/intracellular domain to the cytosolic site of the plasma membrane and initiates further signaling cascades. Ligands for intracellular receptors usually are lipophilic hormones, such as cortisol and thyroid hormones, that are able to penetrate the plasma membrane. The intracellular receptor is located either in the cytosol or in the nucleus. Upon hormone binding the hormone–receptor complex enters the nucleus. After uncovering DNA-binding proteins, the hormone–receptor complex binds to certain hormone response elements on the DNA in order to initiate transcription.

Nuclear Receptors

Ligand-activated nuclear receptors of the steroid/thyroid hormone receptor superfamily show many similarities in terms of their structure and their organization into different domains. Within the amino acid sequence, five regions with different functions have been identified. The NH2-terminal region A/B contains a transcriptional activation function. Downstream of A/B is the well-conserved C region, which contains two zinc finger motifs. These are involved in DNA binding and receptor dimerization. The D domain is responsible for nuclear localization. Finally, the C-terminal E domain binds the ligand and the heat shock proteins and seems also to be involved in dimerization and hormone-dependent transcriptional activation (Beato et al. 1996; Tenbaum & Baniahmad 1997).

The steroid receptor is associated with several proteins such as HSP90, HSP56, HSP70, and p23 that facilitate ligand–receptor binding. After ligand binding this hetero-oligomeric structure dissociates, and the ligand–receptor complex translocates into the nucleus (Hutchison et al. 1994; Rexin et al. 1992).

For tight binding of the complex to the response element of the target genes, a dimerization of the steroid receptors is necessary. The functional diversity of this signaling system is further enhanced by the formation of heterodimers containing both mineralocorticoid and glucocorticoid receptors (Trapp & Holsboer 1996). Together with coactivators and transcription factors like GRIP170 or OTF1, a stable preinitiation complex is formed that allows transcription initiation by RNA polymerase II (Beato et al. 1996; Bourguet et al. 2000; Eggert et al. 1995). However, nuclear hormone receptors may also have an inhibitory effect on transcriptional activity. This effect is mediated either by hormone receptor binding to so-called composite HREs or by direct protein–protein interactions between the nuclear hormone receptor and other transcription factors like AP-1 or NFκB (Jonat et al. 1990; Pearce & Yamamoto 1993; Yang-Yen et al. 1990). Thus, glucocorticoids mediate their anti-inflammatory and antiproliferative effects via both mechanisms—stimulation and inhibition of transcriptional activity.

Surface Receptors

Plasma membrane-bound surface receptors recognize either locally or distantly released soluble ligands (which are not membrane permeant), cell-attached ligands, or components of the extracellular matrix. This receptor type consists of three domains—an extracellular domain that is connected to an intracellular domain via a transmembranous domain. Receptor activation by ligand binding is transduced to the intracellular domain and initiates intracellular signaling processes. Cell surface receptors are classified according to the initial signal transduction pathways they activate:

- Receptors that couple and activate G-proteins
- Receptors with an intrinsic tyrosine kinase activity
- Receptors that themselves function as ion channels
- Receptors that bind cytoplasmic kinases with or without additional adaptor proteins

However, this does not exclude additional signal transduction pathways addressed at later stages. This is exemplified in the following, more detailed description of two different receptor types that are present on immune cells. First, the N-formylmethionine-leucyl-phenylalanine (fMLP) chemokine receptor, found on the surface of neutrophil granulocytes, is an example of a G-protein-coupled receptor. Second, the T-cell receptor functions via the association of intracellular adaptor proteins or kinases. Details about the different signaling pathways are given later.

The fMLP Receptor

N-formylated peptides such as N-formylmethionine-leucyl-phenylalanine, which are believed to be structural analogs of bacterial metabolites, are potent chemoattractants for neutrophils. As indicated by binding studies of intact cells, approximately 55,000 binding sites are present on granulocytes with an average Kd of 20 nM (Lew 1990; Remes et al. 1993). Their corresponding fMLP receptor belongs to the family of seven transmembrane domain receptors. Other members of this family are the receptors for adrenaline and noradrenaline as well as for histamine, serotonin, substance P, and bradykinin. These receptors mediate their activation via intracellular coupling to heterotrimeric G-proteins. These proteins consist of three subunits, α, β, and γ of 39-46 kDa, 37 kDa, and 8 kDa, respectively, which show a remarkable diversity. The GTPase activity is part of the α subunit, which usually determines the specificity. The different subunits, Gs/Gi/Gα and G12, can be distinguished by the inhibiting potential of various toxins such as pertussis and cholera toxin (Neer 1995). Depending on the α subunit, different effector molecules are stimulated or inhibited. For the fMLP receptor, its occupation results in a pertussis toxin-sensitive G-protein (Gs)-dependent activation of phospholipase Cβ, which generates diacylglycerol (DAG) and inositol 1,4,5-trisphosphate (IP3) (Krause et al. 1985). Moreover, it has been shown that the fMLP receptor is coupled to tyrosine kinases and adenylate cyclase (Mitsuyama et al. 1995; Nahas et al. 1996; Zu et al. 1996).

T-Cell Receptor

The T-cell antigen receptor (TCR) is of fundamental importance for activation and clonal expansion of T-lymphocytes. This receptor consists of the polymorphic α and β chains whose peptide structure determines the antigen specificity. It is associated with the CD3 complex consisting of γ, δ, ϵ, and ζ chains in a stochiometry of 1:1:

2:2, which are responsible for the transduction of signals across the T-cell membrane (Ashwell & Klusner 1990). Unlike the growth factor receptors, the TCR/CD3 complex does not have an intrinsic tyrosine kinase activity. Instead, after antigen binding, at least three phosphotyrosine kinases associate with the complex: p59fyn, p56lck, and ZAP-70 (Baniyash et al. 1988; Isakov et al. 1994). Phospholipase Cγ has been identified as a major target of the tyrosine kinases. After phosphorylation and recruitment of the enzyme to the plasma membrane, inositol phospholipids are cleaved into the two second messengers inositol 1,4,5-trisphosphate (IP_3) and 1,2-diacylglycerol (DAG) (June et al. 1990). While DAG is a potent activator of protein kinase C, IP_3 induces an increase in intracellular calcium by the release of calcium from intracellular stores. Another target of the TCR/CD3-associated tyrosine kinases is phosphatidylinositol 3'-hydroxy kinase (PI-3K), which generates a number of polyphosphoinositides. PI-3K is also involved in the CD28 signaling pathway, which has been shown to be necessary for optimal T-cell activation (Azuma & Lanier 1995; Ward et al. 1995). TCR/CD3 and CD28 co-stimulation is required for activation of CD45 protein phosphatase (Leitenberg et al. 1999), which is able to control the generation of specific phosphorylation patterns during cell activation (Hegedus et al. 1999). A further signaling pathway addressed by the TCR/CD3 complex is regulation of the activity of the guanine nucleotide binding protein p21Ras. This is performed most likely by tyrosine kinase-dependent phosphorylation of p21Ras-regulating proteins like p120-GAP and C3G (Baldari et al. 1994; Zhang & Samelson 2000). The different signaling pathways induce the activation of several transcription factors responsible for clonal proliferation and cytokine expression.

Target Cell Adaptation

The continuous presence of ligands confronts the cells with the problem of avoiding overshooting responses and preserving further excitability to repetitive stimuli. Therefore cells are endowed with several adaptational mechanisms that regulate surface receptor activity, density, or both, concomitant with an attenuation of both intracellular signals and cellular responses to agonists. These mechanisms can be divided according to their attenuation kinetics into those that are short-term acting and those that are long-term acting (figure 7.1).

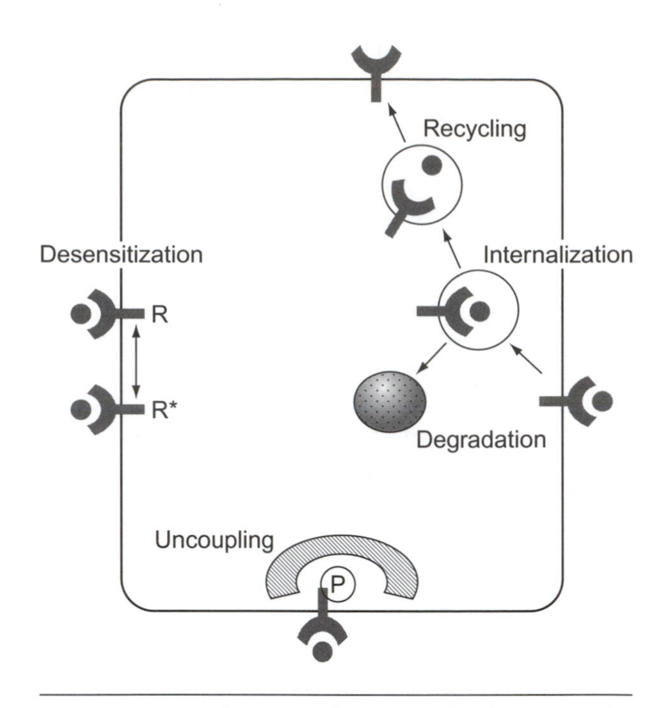

Figure 7.1 Mechanisms of target cell adaptation (for details see text). P = phosphate group.

Short-term mechanisms of signal attenuation (seconds to minutes) include ligand removal from the extracellular fluid, receptor desensitization, and internalization (Grady et al. 1997). The first process is very common for synaptic signal transmission. Cell surface and luminal peptidases degrade neuropeptides in the extracellular fluid, thereby terminating their biological activities. Receptor desensitization refers to the functional impairment of receptors after prolonged or repetitive stimulation concomitant with an attenuation or loss of biological response. The causal mechanism for this process shows a relation to the receptor types. In the case of neuronal nicotinic receptor, which is one of the ligand-gated ion channels, the model assumes for the receptor a number of discrete conformations: R (resting or activatable), D (desensitized), and I (intermediate desensitized). Their relative distribution changes upon ligand exposure. Since the desensitized state has a relatively higher affinity for the agonist, all receptors finally end up in this state. The time course of desensitization thereby depends on the rate constants for the transition between the different states (Quick & Lester 2002).

For G-protein-coupled receptor, desensitization means the uncoupling of the receptor from the G-protein or other related signaling pathways (Bunemann et al. 1999). Several studies provide evidence that in either case, receptor phosphorylation plays an important role in the regulation of desensitization. Protein kinase A and protein kinase C have been shown to phosphorylate the nicotinic receptor (Paradiso & Brehm 1998). Upon activation, G-protein-coupled receptors are phosphorylated by GRKs (G-protein receptor kinases), which promote receptor association with β-arrestins (Miller & Lefkowitz 2001). Recruitment of β-arrestins to the G-protein-coupled receptor blocks coupling to G-proteins, thereby terminating or at least attenuating signal processing. β-arrestins are a gene family of at least four members and are among the adaptor or scaffolding proteins. Scaffold proteins can be defined as proteins that promote or block the interaction of two or more partners (Hall & Lefkowitz 2002).

Besides their role in regulation of receptor desensitization, β-arrestins have been shown to be involved in receptor internalization. Among the nonreceptor proteins β-arrestins interact with are endocytotic proteins such as clathrin, AP-2, and NSF, which facilitate agonist-promoted receptor internalization into clathrin-coated pits (Goodman et al. 1996). Such sequestration of receptor–hormone complexes to submembranous vesicles decreases the receptor density on the cell surface but does not change the total numbers of receptors present in cells or tissues. Ubiquitination serves as an important signal for the entry of the hormone–receptor complex into the endocytotic pathway and for its final degradation. Interestingly, internalization of certain tyrosine kinase receptors does not lead to receptor inactivation, but instead is an integral part of their signal processing. The function of this endosomal or compartment-restricted signaling is not entirely clear. However, there is growing evidence that it is an important mechanism for regulating the specificity of receptor tyrosine kinase-mediated responses. If a receptor can be activated by several ligands, such as the EGF receptor with its ligands EGF, TGF-α, amphiregulin, and others, such a mechanism would help to discriminate between them. Moreover, the intracellular ligand–receptor trafficking could be helpful for bridging greater distances as in the case of axonal neurotrophic signal transmission (Wiley & Burke 2001).

The short-term signal-attenuating mechanisms are often fully reversible upon removal of agonist from the extracellular fluid. After dephosphorylation, either receptor affinity is reestablished or the receptor regulates back to the plasma membrane (resensitization) (Tsao & Zastrow 2001).

In contrast, long-term signal-attenuating mechanisms occur after a prolonged stimulation (hours to days) of receptors by agonist ligands and involve a process called receptor down-regulation, which is characterized by a net decrease in the total numbers of receptors in cells or tissues (Tsao et al. 2001). Furthermore, this process shows a slow or incomplete reversibility following removal of agonist (Tsao & Zastrow 2001).

The number of receptors per cell depends on the equilibrium between rates of synthesis and degradation. Down-regulation of receptors occurs on both sides. In case of the β2-adrenoreceptor, regulation of both gene transcription and mRNA stability has been observed (Collins et al. 1989; Tholanikummel & Malbon 1997). On the other hand, degradation of receptor depends on lysosomal and non-lysosomal mechanisms (Tsao & Zastrow 2001). Whether internalizations during short- and long-term signal-attenuating mechanisms share common endosomal pathways or not remains to be shown. There is evidence that at least for certain receptor tyrosine kinases, surface receptors are degraded also by a non-lysosomal proteasome-dependent pathway (Levkowitz et al. 1999). In both pathways, the ubiquitination machinery can be involved, which targets receptors for the degradation processes (Hicke 1997, 1999; Tsao et al. 2001).

Cardiovascular Hormones

The cardiovascular system is the central transport system of the body composed of the heart, the vasculature, and the blood. It maintains a constant supply of oxygen and nutrients to different parts of the body, and it is responsible for dissipation of carbon dioxide, metabolites, and heat. The resting levels of blood pressure and flow face enormous challenges during exercise. In order to meet the increasing demands for oxygen and nutrients during exercise, cardiovascular functions have to be improved by the combined control of autonomic nervous system and the cardiovascular hormones, such as catecholamines, natriuretic peptides, and endothelin.

Catecholamines

The common structure shared by naturally occurring catecholamines and other important sympatho-mimetic drugs consists of a benzen-benzene ring and an ethyl-amine side chain. The naturally occurring catecholamines dopamine, norepinephrine, and epinephrine share hydroxyl groups in the three and four positions of the benzen ring. They differ regarding the substitutions on the ethyl-amine side chain.

Catecholamines belong to the group of biogenic amines representing important neurotransmitters in the nervous system. Norepinephrine and dopamine are major transmitters of the central nervous system. In addition, norepinephrine is a transmitter of most sympathetic postganglionic fibers. In contrast, epinephrine is not a common neurotransmitter in the central nervous system but is a major hormone of the adrenal medulla. Common precursor of the naturally occurring catecholamines is tyrosine, which is converted into L-dopa by the action of tyrosine hydroxylase, regarded as the rate-limiting step in the catecholamine synthesis pathway. Tyrosine hydroxylase activity is enhanced after phosphorylation through cyclic AMP and calcium–calmodulin-dependent protein kinases. L-dopa is further converted into dopamine, which is converted into epinephrine via norepinephrine. Catecholamines in both adrenergic nerve endings and adreno-medullary tissue are stored in electron-dense vesicles or granules and are released upon cellular stimulation.

The peripheral action of catecholamines is critically dependent on the distribution and mechanism of action of the various adrenergic receptor subtypes. Norepinephrine and epinephrine bind to two major classes of receptors: α-adrenergic and β-adrenergic receptors. These receptor types are subdivided into α1/α2 and β1/β2 receptors, respectively. Adrenergic receptors belong to the group of seven transmembrane-spanning domain receptors and mediate the signal via various G-proteins. Further downstream, signaling pathways that are addressed involve the cyclic AMP-PKA, IP_3-calcium, and DAG-PKC pathways. For further details the reader is referred to textbooks of physiology or recently published reviews.

Catecholamine Response to Exercise

Plasma concentrations of epinephrine and norepinephrine originate primarily from the adrenal gland and from terminal nerve endings of the sympathetic nervous system, respectively. Usually the ratio between resting epinephrine and norepinephrine concentrations is about 1 to 4 since the release of epinephrine from the adrenal gland is rather low. During exercise, both catecholamines increase exponentially in relation to increasing workload (Mazzeo 1991). Usually during short-term exercise, the ratio between the catecholamines is not affected. However, during prolonged exercise at submaximal workloads the adrenal response becomes overweighted as indicated by an increasing ratio of epinephrine to norepinephrine. At exercise intensities above the anaerobic threshold, plasma concentrations of epinephrine and norepinephrine increase markedly. After termination of exercise, both hormones return rapidly to pre-exercise levels due to the short half-life of catecholamines.

Exercise training has been shown to evoke adaptations in the catecholamine response to acute exercise. At a given submaximal absolute exercise intensity, the catecholamine concentrations are attenuated in endurance-trained subjects compared with sedentary controls. On the other hand, the adrenal epinephrine secretion capacity was reported to be enhanced in endurance-trained athletes compared to sedentary subjects if the individuals exercised at identical relative workloads (Kjaer 1998). These data have been confirmed by experiments in which the adrenal medulla was stimulated by other stress factors like hypoglycemia or hypoxia (Kjaer & Galbo 1988). In contrast, the clearance rate for catecholamines seemed to be similar in trained and untrained subjects (Kjaer et al. 1985). The greater adrenal secretion capacity in trained versus untrained subjects seems to be the result of two factors. First, in trained rats, it has been shown that the adrenal tyrosine hydroxylase activity is significantly enhanced. Moreover, some studies showed a training-induced hypertrophy of the adrenal medulla. Interestingly, there seems to be a sex difference in adreno-medullary adaptation to training. Stallknecht et al. (1990) found a slightly pronounced increase in adrenal medulla volumes in trained male compared to trained female rats. This finding is supported by a recent published study that did not indicate any influence of training status on the catecholamine response to a supramaximal exercise test in women (Jacob et al. 2002).

Finally, the control functions that catecholamines exert regarding various physiological and metabolic processes are not limited to the neurotransmitter's concentrations in plasma or tissues. They are furthermore affected by the density and the distribution pattern of adrenergic receptors and the downstream signaling pathways that they address. The effects of exercise on this side are described on page 130.

Atrial and Brain Natriuretic Peptides*

The cardiac natriuretic system is composed of atrial and brain natriuretic peptides (ANP and BNP). Both play a major role in blood pressure and fluid homeostasis, protecting the organism from pressure and volume overload (Geny et al. 2001).

Atrial natriuretic peptide (ANP) is composed of 28 amino acid residues. It was first isolated from human atria in 1984 (Kangawa & Matsuo 1984). The major portion of plasma ANP is synthesized in and secreted from atria. This cardiac hormone has a wide range of biological effects, among them diuresis, natriuresis, and vasodilatation. Therefore it plays an important role in body fluid and cardiovascular homeostasis (Kangawa & Matsuo 1984; de Bold 1985; Sugawara et al. 1985).

Brain natriuretic peptide (BNP) is composed of 32 amino acid residues (Kambayashi et al. 1990). It was first isolated from porcine brain in 1988 (Sudoh et al. 1988). Because it could be demonstrated that plasma BNP was predominantly synthesized in and secreted from the heart, mostly from the ventricles (Saito et al. 1989; Ogawa et al. 1990; Mukoyama et al. 1991), it is regarded as the second cardiac hormone.

Available evidence about the release of ANP by atrial myocytes into the circulation indicates that atrial stretch is the main mechanism. This mechanism is present in some physiological maneuvers such as physical exercise, postural change, as well as during modification of sodium intake (Brenner et al. 1990).

The mechanism regulating the release of BNP by ventricular myocytes in healthy subjects is an ongoing field of research. There are only a few data available on the effect of physiological maneuvers, such as differences in posture (Solomon et al. 1986; La Villa et al. 1993) and physical exercise, regarding which a small increase was observed (Nicholson et al. 1993; Marumoto et al. 1995).

Bicycle ergometric exercise increases the release of ANP and isometric exercise leads to an increase of lesser extent (Barletta 1998). For dynamic exercise, the findings are consistent with the literature (Bouissou et al. 1989; Brenner et al. 1990; Rogers et al. 1991). Short-term exercise (Tanaka et al. 1986; Saito et al. 1987; Thamsborg et al. 1987) seems to increase plasma ANP level, as well as endurance exercise (Lijnen et al. 1987; Passene 1996). The mechanisms regarded as responsible for the release are the increase of plasma catecholamines, heart rate, cardiac output, left atrial diastolic pressure, and left ventricular filling pressure (Tomiyama et al. 1995). Furthermore, the release is influenced by the intensity of exercise and the age of the examinee (Lijnen et al. 1987; Ohashi et al. 1987). One study showed a significant relationship between heart rate and plasma ANP for both dynamic and isometric exercise (Barletta et al. 1998).

Only a few data are available on the behavior of BNP in physiological conditions. Dynamic exercise induced a small increase in plasma BNP. There was a significant difference from the controls during the exercise, but no significant correlation with the workload was found (Barletta et al. 1998). In the literature there are consistent findings (Nicholson et al. 1993), but also other data showing no changes from pre-exercise levels (Marumoto et al. 1995). During isometric exercise also, a small significant increase in plasma BNP level was reported (Barletta et al. 1998). The small increase in plasma BNP induced by physiological maneuvers is consistent with previous data showing only small elevations of BNP—much smaller than the simultaneously measured larger increase of plasma ANP level (Lang et al. 1992). There was no correlation between changes in plasma ANP and plasma BNP (Barletta et al. 1998). For short-term exercise, some studies showed a nonsignificant tendency toward an increase in BNP level (Tanaka et al. 1995; Tanabe et al. 1999). However, even a small change in plasma BNP might be of physiological relevance. BNP infusions in healthy humans that caused moderate increases in plasma levels of about 50% were sufficient to increase sodium excretion and to reduce plasma aldosterone levels (La Villa et al. 1995).

Regarding the ANP and BNP response to extreme prolonged strenuous exercise, a 100-km ultramarathon, there was a 5.6-fold increase in the mean postrace BNP level, which was much higher than the simultaneously measured 2-fold increase in ANP. Stimuli or influences such as increase of plasma catecholamines, blood lactate, or age previously reported to be responsible for the increase of ANP and in part BNP were not significantly correlated with the postrace levels of ANP and BNP. Because subclinical myocardial cell necrosis was observed with such exercise, this mechanism was considered to be one of the factors influencing the significant increase in plasma ANP and especially in plasma BNP (Obba 2001).

* The text on this page beginning with the heading *Atrial and Brain Natriuretic Peptides* and ending with the text under the heading *Training* on page 117 is contribured by Klaus Völker.

Endothelin

Endothelin-1 (ET-1) (Cosenzi et al. 1996) is a 21-amino acid peptide that is synthesized by vascular endothelium cells (Yanagisawa et al. 1988) and by cardiac myocytes (Suzuki et al. 1993). It predominantly exerts a strong, prolonged vasoconstrictor activity. An ET-A receptor linked either to the phosphatidylinositol metabolism or to voltage-operated calcium channels (or both) is involved in increasing cytosolic calcium concentration (Masaki et al. 1991) and leading to activation of vascular smooth muscle.

For vascular endothelium cells it has been reported that the production of ET-1 is increased by some humoral factors like angiotensin II, arginine vasopressin, and catecholamines, as well as by mechanical factors like shear stress, endothelial stretching (Yoshizumi et al. 1989; Sumpio & Widmann 1990), and hypovolemia (Matzen et al. 1992). In cultured ventricular myocytes it could also be demonstrated that the expression of preproET-1 mRNA was increased by angiotensin II (Ito et al. 1993) and mechanical stretch (Yamazaki et al. 1996). It was hypothesized that the production of ET-1 in the heart in vivo may also be regulated by comparable neurohumoral and mechanical factors (Maeda et al. 1998).

Physiological Role in Peripheral Vessels

Many studies have demonstrated that endothelin may affect systemic and regional hemodynamics (Wright & Fozard 1989; Neubauer et al. 1990). ET-1 leads to sustained and powerful vasoconstriction, thus reducing regional blood flow in peripheral vessels as well as in organs like the kidneys (Yokokawa et al. 1989) and enhancing arterial blood pressure (Mortensen et al. 1990). Endogenously generated ET-1 is seen to contribute to basal vascular tone, because systemic administration of the endothelin receptor antagonist TAK-044 significantly decreases systemic blood pressure and peripheral vascular resistance (Haynes et al. 1996). Low concentrations of ET-1, which did not produce vasoconstriction, potentiated contractions to norepinephrine (Yang et al. 1990). Endogenous ET-1 seems to contribute to the regulation of vascular tonus through its direct vasoconstrictive effect and through an indirect effect by modulating the effect of norepinephrine (Maeda et al. 1997). The observation that the production of ET-1 especially in nonworking muscle is enhanced demonstrates that ET-1 contributes to the exercise-induced redistribution of blood flow. ET-1 decreases blood flow in nonworking muscle by its direct and indirect effects (Maeda et al. 1997).

Physiological Role in the Heart

ET-1 produced by cardiac myoctes, in addition to its potent vasocontractile effect, has a potent positive inotropic and chronotropic effect on isolated heart muscle (Ishikawa 1988a,b) and induces myocardial cell hypertrophy (Shubeita et al. 1990). ET-1 contributes to the regulation of coronary vascular tone at low myocardial metabolic demand, but its influence decreases during increased metabolic rates (Merkus et al. 2002). The expression of mRNA of ET-1 in the heart is increased by aging and is further increased by exercise training-induced cardiac hypertrophy and improvement of cardiac function (Iemitsu et al. 2002).

Single Bout of Exercise

Various stimuli such as shear stress, humoral factors, or hormones enhanced during exercise modulate the release of ET-1 (Ando & Kamiya 1993). Cultured endothelium cells exposed to shear stress increase expression of preproendothelin (Yoshizumi et al. 1989). Moreover, exposure to epinephrine, which is enhanced during exercise, modulates in vitro release of endothelin (Koeler 1987). Prolonged exercise of 45 min caused an increase in ET-1 production in the heart in rats. The expression of preproET-1 and the peptide level of ET-1 were markedly higher in the exercised rats. Expression of endothelin type A- and type B-receptor mRNA, as well as the endothelin-converting enzyme mRNA in the heart in rats, did not differ from values in untrained rats (Maeda et al. 1998).

For humans, changes in plasma volume or distribution, which may be caused by prolonged physical exercise, influence the release of endothelin (Cosenzi et al. 1996). From in vivo studies focusing on physical exercise there are conflicting data. For cross-country ski racers performing 5-, 10-, or 161-km runs, plasma endothelin concentration was increased (Appenzeller & Wood 1992). High-intensity exercise of 30 min significantly increased circulating plasma concentration of ET-1 and enhanced the production of ET-1 in nonworking muscle and in the kidneys (Maeda et al. 2001). After 15-min aerobic exercise (Predel et al. 1990) and after symptom-limited maximal treadmill exercise (McMurray et al. 1992), endothelin level was unchanged. A biphasic response during prolonged exercise was observed. A decrease after 30 min was followed by an increase, reaching initial concentration after 60 min (Richter et al. 1994). Unchanged plasma levels do not exclude the possibility of an increased release during exercise. Enhanced blood flow may also lead to greater uptake and inactivation, as has been documented for the lung and the skeletal muscle (de Nucci et al. 1988).

Training

ET-1 mRNA expression in the aorta of swim-trained aged rats was significantly higher than in sedentary aged rats. These findings indicate that the expression of ET-1 mRNA expression is decreased by aging but can be increased by training (Maeda et al. 2002). For the aged human heart it was demonstrated that exercise training induced an improvement in cardiac function, a development of cardiac hypertrophy, and a further increase of mRNA expression of ET-1 (Iemitsu 2002). The effect of exercise training of eight weeks and of detraining on plasma level of ET-1 was examined in healthy young humans. The plasma concentration of ET-1 significantly decreased, whereas the plasma concentration of NOx significantly increased. The was a significant negative correlation between plasma NOx and plasma ET-1. The effect of exercise training persisted for four weeks after cessation of training. ET-1 and NOx returned to basal levels in the eighth week after cessation (Maeda et al. 2001).

Opioid Peptide Precursor Derivatives*

The human organism is known to be exposed to a variety of psychic or physical stressors every day, ranging from simple fear to serious tissue injury. The consequences are moderate to serious disturbances of its homeostasis, which again provoke a variety of reactions—essentially targeting the previous homeostatic state. This is achieved through protection of the individual, for example against the attack causing the fear, or by repair of the tissue damage that has occurred. Physical exercise is a stressor leading to a number of local and systemic deviations from the organism's homeostatic state; however, the target of the subsequent stress response is not reduction of the imbalanced systems to the original state, but their conversion to an altered state of homeostasis in terms of a functional and morphological adaptation to physical stress.

The response to physical stress includes the release of a variety of neuronal or humoral signals from nervous, endocrine, or immune systems, ending up with morphological or functional alterations, for example of muscular tissue or of the cardiovascular system. In humans, analysis of neuronal signals or morphological alterations in response to exercise is limited, whereas there is more or less free access to metabolic, neuroendocrine, or immune cell-related signals mediated via the cardiovascular compartment. Thus, although proopiomelanocortin (POMC) and proenkephalin (PENK) systems are scattered all over the human organism, their reactions to physical exercise in practical terms have as yet been studied only at the endocrine level, that is, through screening of blood samples for POMC or PENK derivatives.

POMC Derivatives

Among POMC derivatives, adrenocorticotropic hormone (ACTH) and β-endorphin are the compounds most often studied under physical stress, but this is an arbitrary selection; further POMC fragments are also released with physical exercise, for example β-lipotropin (β-LPH).

Expression of POMC in the Human Organism

In humans, there is one POMC gene per haploid genome located on chromosome 2 (p23). It is 7,665 base pairs long and consists of three exons separated by two introns. In the human organism, transcription of the POMC gene into mRNA occurs in the pituitary, in the brain, and in a number of peripheral tissues such as thyroid gland, adrenal medulla, gonads, placenta, pancreas, kidneys, spleen, liver, gastrointestinal wall, skin, monocytes/macrophages, and T-cells. The mRNA in the pituitary is 1,072 bases long, whereas the transcript in the brain is longer and most of the transcripts in the periphery are shorter, that is, about 800 bases long. The POMC mRNA in the pituitary and in the brain are translated into the amino acid sequence of "pre-proopiomelanocortin," "pre-POMC." After cleavage of the N-terminally located signal sequence from pre-POMC, POMC is left; this is a protein of 241 amino acid residues, from which POMC fragments such as ACTH or β-endorphin are released (Höllt 1993; Bertagna 1994). Most information is available on the pituitary POMC system; there is much less about brain and skin and almost nothing about most of the other

* The text on this page beginning with the heading *Opioid Peptide Precursor Derivatives* and ending with the text under the heading *Summary* on page 124 is contributed by Axel Schulz and Hansjoerg Teschemacher.

peripheral POMC systems. This review concentrates on the pituitary POMC system in view of availability of data, as well as in connection with the aspect of release of POMC fragments from the pituitary, although release of POMC fragments from the adrenal medulla under physical exercise cannot be excluded (Evans et al. 1983).

In lower species such as the rat, POMC is expressed in two pituitary cell types, in the corticotroph cells of the anterior and in the melanotroph cells of the intermediate lobe. In the corticotrophs, POMC is glycosylated and phosphorylated and is subsequently enzymatically cleaved into three large fragments, a so-called 16 K fragment, ACTH, and β-LPH; the latter POMC fragment consists of the sequences of γ-LPH and β-endorphin, which are released in small amounts also (figure 7.2). Enzymes involved are "prohormone convertase 1" (PC 1) and possibly to a lesser extent "prohormone convertase 2" (PC 2) or "POMC converting enzyme" (PCE). In the melanotrophs, the same POMC fragments are released from POMC as in the corticotrophs. However, they are further processed to release a series of smaller fragments, which are in part N-terminally acetylated or C-terminally amidated; α-melanocyte-stimulating hormone (α-MSH) and acetylated β-endorphin fragments are typical examples for this type of POMC derivative (Loh 1992; Young et al. 1993; Castro & Morrison 1997).

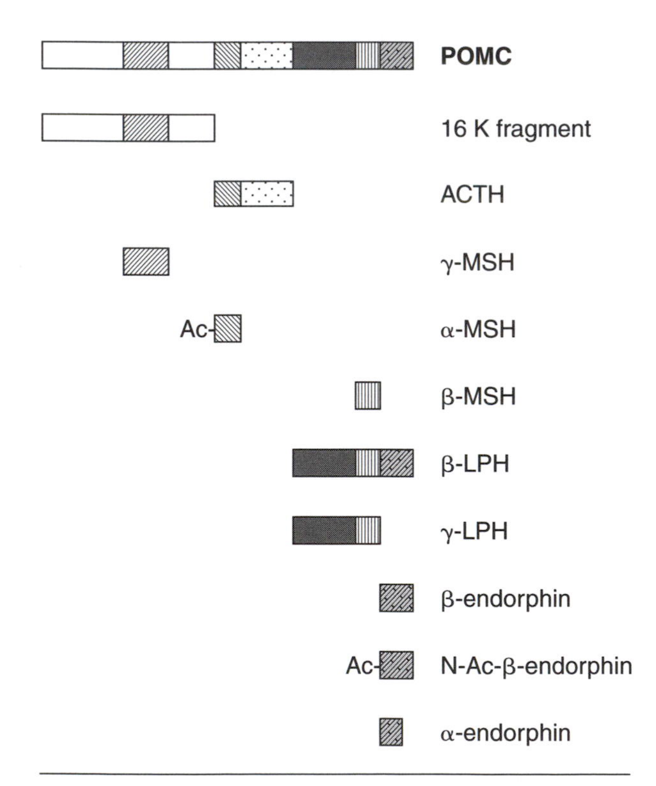

Figure 7.2 Various proopiomelanocortin (POMC) derivatives and their positions in the POMC precursor (Young et al. 1993; Bertagna 1994; Castro and Morrison 1997). ACTH, adrenocorticotropic hormone; α-MSH, α-melanocyte-stimulating hormone; β-MSH, β-melanocyte-stimulating hormone; γ-MSH, γ-melanocyte-stimulating hormone; β-LPH, β-lipotropin; N-Ac-β-endorphin, N-acetyl-β-endorphin.

The human pituitary lacks a well-defined intermediate lobe containing melanotroph cells. However, the POMC-expressing cells in the anterior lobe are able to process POMC according to either type of enzymatic cleavage pattern, the melanotroph as well as the corticotroph one. These enzymatic systems, however, apparently are still regulated separately (Evans et al. 1994).

Functional Significance of Pituitary POMC Derivatives

The pituitary gland is an endocrine organ that sends hormonal messages to tissues of the whole organism via the cardiovascular compartment receiving orders for this kind of message transfer mainly from the central nervous system, but also, to a considerably lesser extent, over the cardiovascular compartment itself. Apparently the pituitary contains several message transfer systems for specific transfer tasks such as the gonadotropic or the thyrotropic system. For each of these specific message transfer systems, specific orders based on specific stimuli can be postulated—as well as specific targets in the periphery to be met by the hormonal messengers released from the pituitary into the blood.

For the pituitary POMC system, in fact a large number of stimuli triggering its activation have been demonstrated, all of which can be classified under the term "stress." Psychic as well as physical stressors, ranging from pain or fear to metabolic shifts or serious tissue injury due to parturition, accidents, or operations have been shown to induce the release of POMC fragments into the cardiovascular compartment. However, it should be mentioned that psychic and physical stressors are processed in the organism in different ways. Psychic stressors or the psychic components of physical stressors are thought to be received and processed by supraspinal structures such as the limbic system in a fairly complicated way before the resulting information is transferred via hypothalamic structures to the pituitary for activation of the POMC system (Herman & Cullinan 1997). In contrast, the information launched upon peripheral tissue injury, in principle carrying the character of life threat, is immediately signaled via the spinal cord to the brain and is directly linked to the pituitary via hypothalamic structures. Thus, volatile anesthetics causing loss of consciousness are able to block pain sensation but are not able to block most of the hormonal response to peripheral injury, which, however, can be achieved by blockade of spinal cord information transfer using spinal anesthesia (Kehlet 1988).

ACTH was studied as the main representative of the POMC system for many years under the aspect of stress-induced activation of the "hypothalamic-pituitary-adrenal axis" (Ganong et al. 1987; Tache & Rivier 1993); later β-endorphin and β-LPH were recognized to be released from the pituitary under identical or similar stress conditions (Owens & Smith 1987; McLoughlin et al. 1993). Thus, it is mainly the "corticotroph" and not the "melanotroph part" of the pituitary POMC system that is activated under stress conditions; the "melanotroph part" occasionally seems to participate in this activation process, but the stimuli triggering its activation in humans are not clear as yet.

Thus whereas there is overwhelming evidence for a prompt response of the pituitary POMC system to all kinds of stressors, there is as yet no clear-cut information about the functional significance of the POMC fragments released under stress into the cardiovascular compartment. Although there are well-known targets such as the adrenal gland for ACTH or a variety of immune cells for β-endorphin (Sibinga & Goldstein 1988), the functional significance of the effects elicited by the POMC fragments at peripheral targets is not clear at all; effects on the immune system elicited by β-endorphin (Teschemacher et al. 1990) or by ACTH via corticosteroids, for example in response to fear, are still a matter of speculation. Thus, the pituitary POMC system clearly plays the role of a "stress responder," but as yet it cannot be classified as a "stress adapter," since it is unclear how it might contribute to the reduction of a disturbed function to a homeostatic state (Teschemacher 2003).

Release of Pituitary POMC Derivatives in Response to Exercise

Exercise can be regarded as one of many stressors provoking metabolic, cardiovascular, or immunological reactions of the organism and therefore activating the pituitary POMC system. However, in contrast to most of the other stressors, there is no doubt that the organism tries to adapt to this stressor not by reduction of morphology and functions to the original state before exposure to exercise, but by conversion of morphology and functions to an altered, that is, a stressor-adapted state, thus counteracting future stress situations of the same type.

Exercise-Specific Stimuli

Numerous studies (Steinberg & Sykes 1985; Cumming & Wheeler 1987; Sforzo 1988; Schwartz & Kindermann 1992; Hoffmann et al. 1996; Goldfarb & Jamurtas 1997) have shown that activation of the pituitary POMC system is reliably achieved by a certain degree of metabolic demand, which is characterized by blood lactate levels beyond the anaerobic threshold; these again are reached with either incremental or short-term anaerobic exercise. However, after aerobic endurance exercise performed over 1 h or more, β-endorphin levels have been reported to increase as well, although lactate levels did not increase (Schwartz & Kindermann 1992; Goldfarb & Jamurtas 1997). Thus, a metabolically imbalanced situation leads to the release of corticotropin-releasing hormone (CRH) and [Arg]vasopressin (AVP) (Inder et al. 1998), known to be released from the hypothalamic paraventricular nucleus (PVN) or from the posterior pituitary, respectively, for activation of the POMC system also under nonexercise stress conditions. As a consequence, POMC fragments such as β-endorphin are subsequently released into the cardiovascular compartment.

Thus, although the prerequisites or the consequences of anaerobic metabolism are apparently closely related to the activation of the POMC system, lactate by itself does not seem to directly stimulate the hypothalamic structures responsible for pituitary POMC system activation (Petrides et al. 1999). It is rather the acidotic state resulting from anaerobic exercise that is among the candidates assumed to be responsible for activation of the pituitary POMC system (Taylor et al. 1994). Other candidates are stimuli resulting from metabolic regulation involving pancreatic functions or lipolytic effects of POMC derivatives on fat tissue (Goldfarb & Jamurtas 1997).

Released POMC Derivatives

As shown for other stressors, physical exercise does not reliably lead to the activation of the "melanotroph part" of the pituitary POMC system. In contrast, under the conditions of physical exercise outlined earlier, the "corticotroph part" of the pituitary POMC system is activated, resulting in the release of ACTH and β-endorphin immunoreactive material as reported by a number of groups.

For the most part, ACTH and β-endorphin immunoreactive material concentrations in the plasma reached about the same levels under exercise conditions (Rahkila et al. 1988; Schwartz & Kindermann 1990; Heitkamp et al. 1993; Schulz et al. 2000). However, as shown in female marathon runners, basal levels of ACTH and β-endorphin immunoreactive material can be different, and the increase of ACTH levels under certain conditions, for example in running to exhaustion, can exceed the increase of β-endorphin immunoreactive material by a factor of five (Heitkamp et al. 1996).

Some methodological difficulties concerning the determination of β-endorphin should be mentioned in this context. Whereas the determination of ACTH always revealed clear-cut results, β-endorphin immunoreactive material as determined in many studies, in addition to β-endorphin, might have contained 10 or more β-endorphin derivatives. In fact, recent findings indicated that β-endorphin immunoreactive material was identical with β-LPH and not with β-endorphin. In studies wherein β-LPH was determined in comparison with ACTH or β-endorphin immunoreactive material using β-LPH-specific determination methods (Petraglia et al. 1988; Oleshansky et al. 1990), β-LPH reached about the same plasma concentrations as β-endorphin immunoreactive material or ACTH. In contrast, further studies with selective or even highly specific assays for authentic β-endorphin showed that under exercise conditions, the plasma concentrations of authentic β-endorphin were low (Farrell et al. 1987; Engfred et al. 1994) to minimal (Harbach et al. 2000; Schulz et al. 2000) in comparison with those of ACTH or β-LPH.

Besides ACTH and β-LPH, further POMC derivatives, which apparently represent small-sized β-endorphin fragments (Wiedemann & Teschemacher 1983), may be released with physical exercise under certain conditions; the proportions of the plasma concentrations of defined β-endorphin fragments may vary dependent on the state of fitness of the participants (Viru & Tendzegolskis 1995).

In comparisons of ACTH or β-endorphin plasma levels of female and male athletes or volunteers under basal or exercise conditions (Kraemer et al. 1989; Goldfarb et al. 1998), no major differences have been demonstrated as yet with the exception of a slightly elevated basal β-endorphin immunoreactive material concentration in males as compared to females (Kraemer et al. 1989). Even between plasma concentrations of β-endorphin immunoreactive material during the luteal phase and the follicular phase of the menstrual cycle, no significant differences were observed (Goldfarb et al. 1998).

Influence of Training

The influence of training on β-endorphin plasma levels under basal or under exercise conditions has been reviewed (Goldfarb & Jamurtas 1997); a number of studies differing in the composition of volunteer groups or using different protocols have revealed controversial results. Similar findings have been obtained for other POMC derivatives. In several studies comparing sedentary with trained subjects, no training-induced adaptation of POMC derivative release was observed under various exercise conditions (Kraemer et al. 1989; Goldfarb et al. 1991). A difference between trained and untrained volunteers, that is, a somewhat more pronounced response in trained subjects, was seen only under extreme exercise conditions (Farrell et al. 1987). In a further study on sedentary and trained subjects, basal ACTH levels were increased in highly trained as compared to sedentary subjects, whereas absolute ACTH levels observed with exercise did not differ (Luger et al. 1987). When absolute workload was considered, however, exercise-related ACTH levels in trained subjects proved to be decreased. In a further study of a similar type (Duclos et al. 1997), ACTH levels after running exercise proved to be elevated in marathon-trained athletes in comparison with untrained volunteers; however, since cortisol levels did not reflect this difference, the authors assumed a decreased sensitivity of the ACTH targets, that is, the adrenals, in the athletes.

A more clear-cut experimental design was presented by another group that did not simply compare trained and untrained volunteers, but looked at the same subjects before and after training that was conducted and controlled within their study framework (Engfred et al. 1994). ACTH and β-endorphin response to exercise workload proved to be dramatically reduced after a five-week training protocol. In a similarly designed study, untrained females underwent eight-week endurance training (Heitkamp et al. 1998), and the effects of running 30 min three times a week on ACTH and β-endorphin levels were measured. Basal β-endorphin levels were not altered by the training program, but basal ACTH levels increased. After exhaustive treadmill running, both maximum ACTH and maximum β-endorphin levels were lower in trained than in untrained subjects; however, ACTH levels proved to be even less elevated than β-endorphin levels. Thus, training appeared to provoke a sort of POMC fragment-specific response that was difficult to recognize upon acute exercise challenge of the POMC system. Results obtained with untrained and trained volunteers also showed differences between the two groups in ACTH and β-endorphin response, which perhaps can be interpreted in the same way (De Diego Acosta et al. 2001). Training also appears to influence β-endorphin metabolism in the resting state and during exercise depending on training previously performed (Viru & Tendzegolskis 1995).

Functional Significance

Although a couple of well-known exercise effects have been linked to POMC fragments, in particular to opioid peptides (Cumming & Wheeler 1987), the functional significance of the POMC fragments released under exercise conditions is still a matter of speculation. Some of the functions discussed for POMC derivatives released into the cardiovascular compartment appear to be unlikely. The reason is that the respective effects are not thought to be elicited in peripheral tissues, but in the central nervous system—for example a reduction in depressive state, reduced anxiety, improved self-esteem, improved well-being (all in all, an improvement in mood occasionally becoming manifest as "runner's high"). The same holds for an analgesic significance ascribed to the release of β-endorphin. Although, in addition to the well-known targets in the central nervous system, targets in peripheral tissues have been shown for narcotic analgesics, conclusive evidence recently presented allows rejection of the hypothesis of an analgesic function of β-endorphin released under stress conditions (Matejec et al. 2003). Further candidates for POMC fragment functions in the periphery are certainly influences on food uptake or reproduction, since fat tissue or gonads represent peripheral targets. However, central POMC systems in hypothalamic areas, in either case, appear to be more important candidates for influence on food intake (Schwartz et al. 2000) or on reproductive dysfunction (Rivier & Rivest 1991). A further exercise effect to be elicited in the central nervous system rather than in the periphery is addiction to exercise. However, in a study focusing on this question, scores on an exercise-dependence survey were not correlated with β-endorphin plasma levels (Pierce et al. 1993).

Efforts to clarify the question of β-endorphin involvement in exercise effects by blockade of opioid receptors during exercise (Strassman et al. 1989; Angelopoulos 2001) have revealed controversial effects on β-endorphin plasma levels and have not provided further insight into the mechanisms in question. In contrast, an attractive candidate for consideration would be a POMC fragment influence on the immune system. Interactions of β-endorphin (Sibinga & Goldstein 1988) or ACTH (via corticosteroid release from the adrenals; see textbooks) with cells of the immune system are well testified to, as are effects on the immune system by stress in general (Fricchione & Stefano 1994) and by physical exercise in particular (Jonsdottir et al. 1997). These effects could contribute to both activation and suppression of separate functions of the immune system responsible for defense against infections or necessary for morphological alterations in terms of elevated capability to cope with future exercise stress. Last, but not least, the metabolic response of the organism to exercise stress, including fat, carbohydrate, and protein metabolism (Weissman 1990), should be envisaged as a possible target of POMC fragment release. In view of the effects of ACTH, β-endorphin, and α- and β-MSH on glucagon, insulin, or glucose plasma levels observed in humans or animals (Knudtzon 1986), a functional target of POMC fragment release might well be counterbalancing the metabolic derailment for which exercise is responsible due to increased energy requirement, for example by increase of glucose uptake into the skeletal muscle as recently suggested (Evans et al. 1997).

PENK Derivatives

As compared to the POMC systems, much less is known about the PENK systems of the human organism and, in particular, the response of PENK systems to stress (e.g., to physical exercise), although either system might play an equally important role.

Expression of PENK in the Human Organism

The human PENK gene is located on chromosome 8 (q23-q24) and spans approximately 5,300 bp of DNA. It consists of four exons separated by one short and two large introns. Messenger RNA observed in humans or animals and post-translational processing products found in human tissues indicate that PENK is expressed in the human central nervous system, the pituitary, the adrenal medulla, male and female gonads, placenta, gastrointestinal tract, skin, pancreas, lung tissue, and monocytes/macrophages. The human mRNA is 1,248 bases long, but there is indication for longer as well as shorter DNA transcripts in brain or in peripheral tissues, respectively. Translation into the amino acid sequence leads to "pre-proenkephalin" (pre-PENK); and after release of the signal sequence, PENK is left, a protein of 243 amino acid residues (Comb et al. 1982; Noda et al. 1982; Legon et al. 1982; Litt et al. 1988; Höllt 1993; Rossier 1993).

PENK contains seven enkephalin (ENK, Tyr-Gly-Gly-Phe) copies: four sequences of (Met^5) ENK, one of (Leu^5) ENK, one of (Met^5, Arg^6, Phe^7) ENK, and one of (Met^5, Arg^6, Gly^7, Leu^8) ENK (figure 7.3). From PENK a series of relatively short peptides, each containing one or two of these ENK sequences, may be released. These are named BAM (bovine adrenal medulla) 12P, 20P, or 22P, peptide B, peptide E, peptide I, or peptide F; there is also an amidated fragment of peptide F, amidorphin. In addition to several large PENK fragments with lengths of 8.6, 12.6, 18.2, and 23.3 kDa (which also contain one or several ENK copies), the N-terminal fragment of PENK containing no ENK copy at all, synenkephalin, has also been isolated (Höllt 1993; Rossier 1993).

PENK
M-ENK
L-ENK
M-ENK-R-G-L
M-ENK-R-F
SYN-ENK
Peptide F
Peptide E
Peptide B
BAM 22 P
BAM 12 P

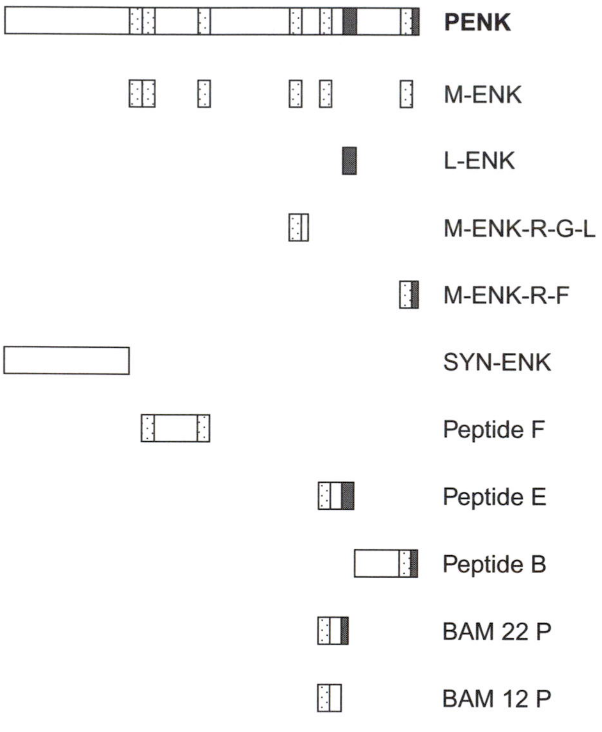

Figure 7.3 Various proenkephalin (PENK) derivatives and their positions in the PENK precursor (Höllt 1993; Rossier 1993). M-ENK, (Met^5) enkephalin (Tyr-Gly-Gly-Phe-Met); L-ENK, (Leu^5) enkephalin (Tyr-Gly-Gly-Phe-Leu); M-ENK -R-G-L, (Met^5, Arg^6, Gly^7, Leu^8) enkephalin (1-8); M-ENK -R-F, (Met^5, Arg^6, Phe^7) enkephalin (1-7); SYN-ENK, synenkephalin; BAM 22 P, bovine adrenal medulla 22 peptide; BAM 12 P, bovine adrenal medulla 12 peptide (BAM sequences identical for bovine and human species).

Functional Significance of Adrenal PENK Derivatives

Information available about a functional significance of specific PENK fragments in the human organism is less than meager. In addition to the fact that less research has been conducted in this area as compared to the POMC field, problems similar to those relating to certain POMC fragments may have played a role. First, conclusions concerning a functional significance have been drawn in various studies from the presence of PENK fragment immunoreactive materials determined in tissues or in plasma under certain conditions. However, it is clear that a "(Met^5) ENK immunoreactive material" might be identical with any PENK derivative containing the (Met^5) ENK sequence recognized by the antibody used to determine the immunoreactive material. Thus the PENK-derived active principle has not been clarified in such investigations, which means that the information is unclear. Frequently, the reports do not even refer to an "immunoreactive material," but instead the immunoreactive material is just called "(Met^5) ENK"; thus, this means that the information is not only unclear but that it is even wrong. In a further, frequently reported approach, information about a functional significance was derived from in vitro or in vivo effects of certain PENK derivatives—mostly the penta-, hepta-, or octapeptides of the ENK series. The problem here is that (1) longer peptides could have done the same job equally well and (2) effects that can be elicited only by non-opioid amino acid sequences additionally contained in the long PENK fragments have not been tested as yet. Nevertheless, some reliable statements on a functional significance (Khachaturian et al. 1993) can be made if one refers to "PENK derivatives" in general. Apparently they play an important role as "opioid growth factors" in growth or development of several tissues such as brain or skin (Zagon et al. 1996, 1999; Kilpatrick 1993). In addition, evidence has been presented for a functional significance of PENK derivatives in the male and the female reproductive tract, in the immune system, and in the heart (Sibinga & Goldstein 1988; Kilpatrick 1993). Findings in animal experiments favor a functional role in the central nervous system (Solbrig et al. 2002), in particular, under the aspect of stress (Przewlocki 1993).

The most attractive candidate for a functional significance of PENK or PENK derivatives in the framework of this review, however, appears to be a role relating to an information transfer system in the adrenal medulla, in analogy to the role played by the POMC system in the pituitary. In fact, the adrenal medulla has been extensively studied as a site of PENK expression, also under functional aspects (Höllt 1993). As an example, (Leu^5) enkephalin and, in particular, (Met^5) enkephalin, have been shown to induce the release of corticosterone or of aldosterone from rat adrenal cortex—the latter in range similar to that observed with angiotensin II (Hinson et al. 1994a,b); contradictory findings were reported as well (Delitala et al. 1991). In either case, PENK derivatives would play a paracrine rather than an endocrine role like POMC derivatives. Since, however, PENK derivatives were found to be co-stored with catecholamines in high concentrations in the adrenal medulla, it also was tempting to speculate about an endocrine role, that is, co-release of PENK derivatives together with catecholamines into the cardiovascular compartment. There is in fact strong evidence that PENK derivatives found in the plasma originate in the adrenal medulla. However, as a second candidate contributing to the release of PENK derivatives—in particular in the case of failure of adrenal medullary function—the autonomic nervous system has to be considered (McLoughlin et al. 1993). In humans, the PENK derivatives demonstrated in the plasma were (Met^5) enkephalin as well as a 3- to 5-kDa, an 8-kDa, a 13-kDa, and an 18-kDa PENK fragment, as well as PENK itself. PENK derivatives have been found, although to a moderate extent in comparison to POMC derivatives, in the cardiovascular compartment under "stress," for example in hypoglycemia, hypotension, hyperventilation, or thermal stress (McLoughlin et al. 1993).

Release of Adrenal PENK Derivatives in Response to Exercise

Whereas much information has been collected on the release of POMC fragments into the cardiovascular compartment under physical exercise conditions, the number of studies dealing with the release of PENK derivatives is very limited. Determination of PENK derivatives has concentrated on (Met^5) enkephalin and peptide F immunoreactive materials. However, it should be emphasized that either immunoreactive material might have contained one or several PENK derivatives. Usually, in view of the adrenal origin of PENK derivatives in the cardiovascular compartment and in view of co-localization of catecholamines and PENK derivatives in the adrenal medullary cells, in most studies epinephrine was determined in addition to the PENK derivatives.

With the exception of very few studies (Mougin et al. 1992), a moderate to dramatic increase in the concentration of (Met^5) enkephalin immunoreactive material (Howlett et al. 1984; DeSommers et al. 1989, 1990; Boone et al. 1992) or of peptide F immunoreactive material (Kraemer et al. 1985, 1988, 1990, 1991; Triplett-McBride et al. 1998; Bush et al. 1999) in plasma was regularly observed with physical exercise such as treadmill or heavy resistance exercise, marathon, or the like. However, again with very few exceptions (Kraemer et al. 1990), above a certain concentration of epinephrine in plasma the levels of (Met^5) enkephalin or peptide F immunoreactive materials declined again. Thus, depending on the level of epinephrine, an inverse behavior of PENK derivative and epinephrine concentrations became obvious in most of the studies insofar as they allowed a comparison between these two parameters (Kraemer et al. 1985, 1991; Boone et al. 1992; Triplett-McBride et al. 1998). Apparently, however, the specific parameter inversely correlated with the PENK derivative concentration was not the epinephrine, but the lactate concentration in plasma as could be derived from studies wherein either parameter had been determined (Kraemer et al. 1988; Bush et al. 1999). Thus, besides further parameters, anaerobic conditions might be responsible for switching over from epinephrine/PENK derivative co-release to sole epinephrine release. The mechanisms responsible for the release of PENK derivatives into the cardiovascular compartment are unclear as yet in any case. There is an opioid-controlled mechanism responsible for inhibition of PENK derivative release from the adrenal under exercise conditions (Marchant et al. 1994) and, even more important, there apparently exist release mechanisms replacing that in the adrenal after adrenalectomy (McLoughlin et al. 1993). Dramatically increased PENK derivative levels as observed in patients with ischemic heart disease indicate the functional importance of this PENK derivative system (Korkushko et al. 1988, 1990).

In most studies, the concentrations of (Met^5) enkephalin or peptide F immunoreactive materials increased with physical exercise such as treadmill running, as previously outlined. However, in response to a training program conducted and controlled within the framework of the study (Howlett et al. 1984), plasma levels of (Met^5) enkephalin immunoreactive material after treadmill running were shown to decrease almost to basal levels. In further studies comparing untrained volunteers with those who had already achieved an advanced state of fitness before starting the study, this finding was supported by (Met^5) enkephalin immunoreactive material plasma levels of untrained subjects that exceeded those of trained subjects after exercise (Boone et al. 1992). In contrast, in a further investigation of this type, postexercise plasma levels of peptide F immunoreactive material were shown to be higher in trained than in untrained subjects (Triplett-McBride et al. 1998). Moreover, in yet other studies, basal as well as postexercise plasma levels of both (Met^5) enkephalin (DeSommers et al. 1989) and peptide F immunoreactive material (Kraemer et al. 1985) proved to be higher in trained than in untrained subjects. Thus, further research is required to elucidate the influence of training on the release of (Met^5) enkephalin and peptide F immunoreactive materials into the cardiovascular compartment.

Summary

Humans confront a variety of stressors every day, from simple fear to life-threatening injury. The human organism is able to activate a variety of endogenous mechanisms to maintain its homeostatic state or to restore it in case of imbalance; two of these at the neuroendocrine level are the POMC system of the pituitary and the PENK system of the adrenal medulla. One of the most frequently experienced stressors is physical exercise, and this stressor is well known to activate either system. There is much information about exercise-induced activation of the pituitary POMC system, but information about the adrenal PENK system is marginal. Activation of the pituitary POMC system with physical exercise elicits release of several POMC fragments such as ACTH or β-LPH from the pituitary; and activation of the adrenal PENK system apparently leads to the release of two or more PENK derivatives, for example methionine-enkephalin and peptide F, into the cardiovascular compartment. Apparently the adrenal PENK system is readily activated upon aerobic exercise, whereas the pituitary POMC system is activated only after a longer time under aerobic conditions. Under anaerobic conditions, however, the pituitary POMC system is further activated dramatically, whereas the adrenal PENK system appears to be switched back. As yet, the most feasible candidate for the stimulus directly responsible for activation of the pituitary POMC system is the metabolically driven pH shift in acidic direction; the direct stimulus activating the adrenal PENK system is not known. Targets of pituitary POMC fragment release are possibly effects compensating for the metabolic derailment under anaerobic exercise by enhanced energy supply, as well as an intervention in the immune system perhaps relevant for defense of infections or the reorganization of muscular tissue. The functional significance of the activation of the adrenal PENK system is not known. In summary, the significance of the pituitary POMC and the adrenal PENK systems under physical exercise as "stress responder systems" has been clearly proven; however, the significance of these systems as "stress adapter systems" remains to be elucidated.

Cytokines*

We turn now to the cytokine response to exercise (Pedersen et al. 1998, 2001; Pedersen & Nieman 1998; Febbraio & Pedersen 2002). Cytokines are polypeptides, originally discovered within the immune system. However, it appears that many cell types produce cytokines and that the biological roles of cytokines go beyond immune regulation. Recent data suggest that several cytokines have important metabolic functions and that they exert their effects locally or work in a hormone-like fashion. Other cytokines, the chomokines, have a chemoattractant effect on blood leukocyte subpopulations. Most studies on cytokines come from sepsis research. In sepsis models, the cytokine cascade consists of increased plasma levels (named in order) of tumor necrosis factor (TNF)-α, interleukin (IL)-1-β, IL-6, IL-1 receptor antagonist (IL-1ra), soluble TNF receptors (sTNF-R), and IL-10 and the chemokines IL-8 (chemoattractant for neutrophils) and macrophage inflammatory protein-1 (MIP-1)-α and -β (chemoattractant for lymphocytes) (figure 7.4).

Until now, pro- and anti-inflammatory cytokines have been considered part of the acute (local or systemic) phase response to an infection or tissue injury. Thus, cytokines are released at the site of inflammation (caused by an infectious pathogen or traumatic injury) and facilitate an influx of lymphocytes, neutrophils, monocytes, and other cells that participate in the clearance of the antigen and healing. The local inflammatory response is accompanied by a systemic response known as the acute phase response. This response includes the production of a large number of hepatocyte-

* The text on this page beginning with the heading *Cytokines* and ending with the text under the heading *T-Cell Cytokine Production* on page 129 is contributed by Bente Klarlund Pedersen.

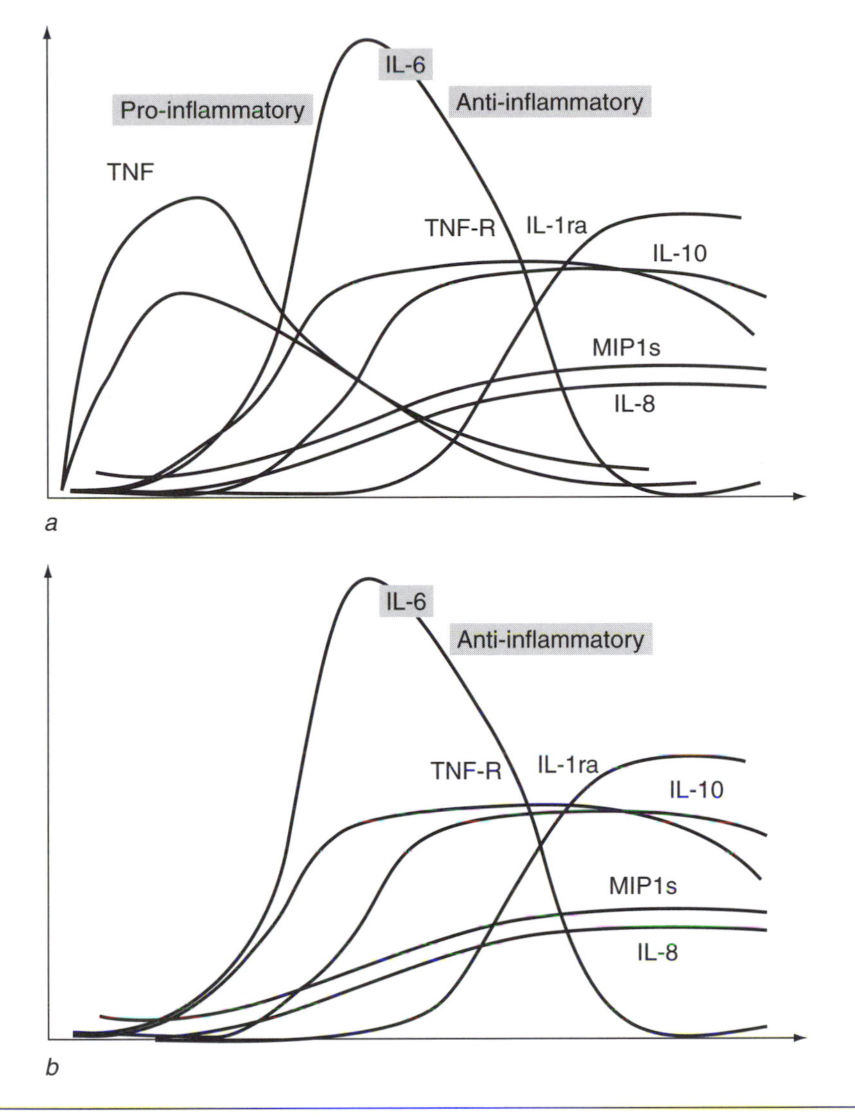

Figure 7.4 The cytokine response in *(a)* sepsis and *(b)* exercise: tumor necrosis factor (TNF)-α, interleukin (IL)-1-β, IL-6, IL-1 receptor antagonist (IL-1ra), TNF receptors (TNF-R), IL-10, IL-8, and macrophage inflammatory proteins (MIPs).

derived acute phase proteins, for example C-reactive protein (CRP). Injection of TNF-α or IL-1 into laboratory animals or humans produces most if not all aspects of the acute phase response (Dinarello 1992). These cytokines are therefore usually referred to as inflammatory or pro-inflammatory cytokines. IL-6 has been classified as both a pro- and an anti-inflammatory cytokine, but the recent view is that IL-6 has primarily anti-inflammatory effects (Tilg, Dinarello, & Mier 1997). Infusion of IL-6 into humans results in fever but does not cause shock or capillary leakage-like syndrome as observed with the prototypical pro-inflammatory cytokines, IL-1 and TNF-α (Mastorakos et al. 1993). Unlike IL-1 and TNF-α, IL-6 does not up-regulate major inflammatory mediators such as nitric oxide or matrix metalloproteinases (Barton 1997). Rather, IL-6 appears to be the primary inducer of the hepatocyte-derived acute phase proteins, many of which have anti-inflammatory properties (Barton 1997). IL-6 directly inhibits the expression of TNF-α and IL-1; and furthermore, IL-6 is a potent inducer of the interleukin-1 receptor antagonist (IL-1ra), which exerts anti-inflammatory activity by blocking IL-1 receptors and thereby prevents signal transduction of the pro-inflammatory IL-1 (Steensberg et al. 2003). There are a number of biologic inhibitors of the inflammatory cytokines, including IL-1ra, sTNF-R, and IL-10 (Richards et al. 1995). Initially, it was thought that the cytokine response to exercise represented a reaction to exercise-induced muscle injury or transposition of lipopolysaccharide (LPS) to the gut. However, recent data suggest that the massive response of IL-6 to exercise may primarily mediate important exercise-related metabolic changes.

The Cytokine Response to Exercise

Several cytokines can be detected in plasma during and after strenuous exercise (Ostrowski et al. 1998a,b, 1999; Suzuki et al. 2002). Most studies indicate that exercise does not induce an increase in plasma levels of TNF-α, although a few demonstrate that strenuous, prolonged exercise, such as marathon running, results in a small increase in the plasma concentration of TNF-α (Pedersen et al. 1998; Starkie et al. 2001b; Toft et al. 2000; Suzuki et al. 2000). Furthermore, most studies demonstrate that IL-1-α and IL-1-β do not increase in the blood, although a few have demonstrated minor changes (Suzuki et al. 2002). The fact that the classical pro-inflammatory cytokines, TNF and IL-1, do not increase, or increase to only a minor degree, is important and distinguishes the cytokine cascade induced by exercise from the cytokine cascade in response to infections (figure 7.4). The first cytokine present in the circulation following exercise is IL-6, which may increase 100-fold. The increase in IL-6 is followed by a marked increase in the concentration of IL-1ra (Ostrowski et al. 1998b, 1999). The cytokine inhibitors (IL-1ra and sTNF-R), as well as the anti-inflammatory cytokine IL-10 (Ostrowski et al. 1999), furthermore increase; and also the concentrations of the chemokines, IL-8, MIP-1-α, and MIP-1-β, are elevated after a marathon race (Ostrowski et al. 2001; Niess et al. 2000). When blood mononuclear cells are sampled during or following strenuous exercise and stimulated in vitro to produce cytokines, it has been reported that the in vitro cytokine production is either impaired (Drenth et al. 1995; Weinstock et al. 1997), not changed, or enhanced (Haahr et al. 1991). These findings are likely explained by exercise-induced altered blood mononuclear cell composition.

It is important to stress the fact that in relation to exercise, IL-6 is produced in larger amounts than any other cytokine examined. Furthermore, it has been demonstrated that contracting skeletal muscles produce IL-6. Most cytokine changes observed in response to exercise are a direct consequence of the exercise-induced IL-6 production. Therefore, this section of the chapter deals primarily with mechanisms underlying muscle-derived IL-6 and its possible biological roles.

The IL-6 Response to Exercise

Interleukin-6 is produced in larger amounts than any other cytokine in relation to exercise. The finding of increased levels of IL-6 after exercise is remarkably consistent (Ostrowski et al. 1998a,b, 1999, 2001; Starkie et al. 2001b; Toft et al. 2000; Suzuki et al. 2000; Drenth et al. 1995; Northoff & Berg 1991; Gadient & Patterson 1999; Nehlsen-Canarella et al. 1997; Castell et al. 1997; Rohde et al. 1997; Hellsten et al. 1997; Nieman et al. 1998b; Starkie et al. 2001a; Steensberg et al. 2000, 2001a; Steensberg, Toft et al. 2000, 2002; Nielsen et al. 1996; Ostrowski et al. 2000; Northoff et al. 1994). The finding of markedly increased levels of IL-6 after strenuous exercise has also been obtained consistently in many studies (Ostrowski et al. 1998a,b, 1999; Toft et al. 2000; Drenth et al. 1995; Northoff & Berg 1991; Castell et al. 1997; Rohde et al. 1997; Hellsten et al. 1997; Sprenger et al. 1992; Ullum et al. 1994; Nehlsen-Cannarella et al. 1997) (figure 7.5). A twofold increase in plasma IL-6

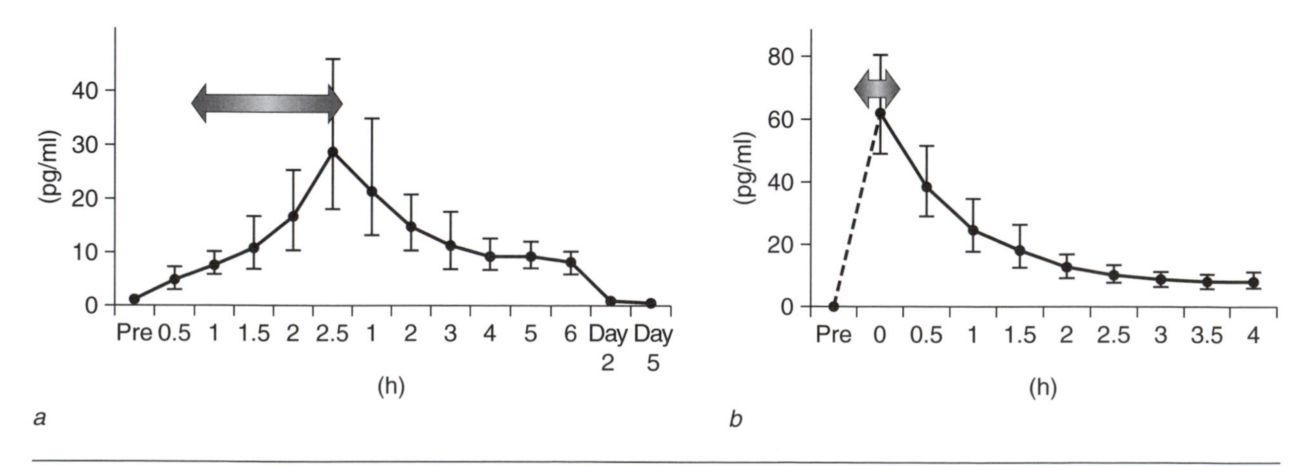

Figure 7.5 Plasma interleukin-6 (IL-6) concentrations *(a)* during and after 2.5 h of treadmill running at 75% of $\dot{V}O_2$max (Ostrowski et al. 1998a) and *(b)* following a marathon race (Ostrowski et al. 1999).

was demonstrated after 6-min intense exercise (Nielsen et al. 1996). In treadmill running, the IL-6 level in blood was significantly enhanced 30 min after the start of running, with the IL-6 peaking at the end of 2.5 h of running (Ostrowski et al. 1998a). In other studies, in which IL-6 was not measured during the running but at several time points afterward, maximal IL-6 levels were found immediately after the exercise, followed by a rapid decline. Thus, after a marathon run, maximal IL-6 levels (100-fold increase) were measured immediately after the 3- to 3.5-h race (Ostrowski et al.1998b, 1999).

Data from the Copenhagen Marathon (1996, 1997, and 1998, n = 56) suggest that there is a correlation between the intensity of exercise and the increase in plasma IL-6 (Ostrowski et al. 2001). A previous study suggested that the appearance of IL-6 in the circulation was related to muscle damage (Bruunsgaard et al. 1997); however, more recent studies clearly demonstrate that muscle contractions without any muscle damage induce a marked elevation of plasma IL-6 (Ostrowski et al. 1998a, 1999; Croisier et al. 1999; Toft et al. 2002). Apart from exercise, intensity duration, and mode, it has also been suggested that the exercise-induced increase in plasma IL-6 is related to the sympatho-adrenal response (Nehlsen-Canarella et al. 1997; Nieman et al. 1998b; Rhind et al. 2002). A study in animals suggested that the increase in epinephrine during stress was responsible for the increase in IL-6 (DeRijk et al. 1994). However, recent data show that when adrenaline was infused to volunteers to closely mimic the increase in plasma adrenaline during 2.5 h of running exercise, plasma IL-6 increased only fourfold during the infusion, but 30-fold during the exercise (Steensberg, Toft et al. 2001c). Thus, it seems that epinephrine plays only a minor role in the exercise-induced increase in plasma IL-6. It was previously demonstrated that peak plasma IL-6 during exercise correlated with plasma lactate (Ostrowski et al. 1998a).

Exercise-Induced IL-6—Where Is It Produced?

A number of studies (Ullum et al. 1994; Moldoveanu et al. 2000) have demonstrated that IL-6 mRNA in monocytes, the blood mononuclear cells responsible for the increase in plasma IL-6 during sepsis (Pedersen & Hoffman-Goetz 2000), did not increase as a result of exercise. Through determination of intracellular cytokine production it was demonstrated that the number, percentage, and mean fluorescence intensity of monocytes staining positive for IL-6 either does not change during cycling exercise (Starkie et al. 2000) or, in fact, decreases during prolonged running (Starkie et al. 2001b).

To test the possibility that working muscle produces IL-6, muscle biopsies were collected before and after exercise (Ostrowski et al. 1998b; Starkie et al. 2001a; Steensberg et al. 2001a). IL-6 mRNA is present in small amounts in resting skeletal muscle but is enhanced up to 100-fold in contracting muscle, thus indicating that exercise is responsible for the IL-6 gene induction (figure 7.6).

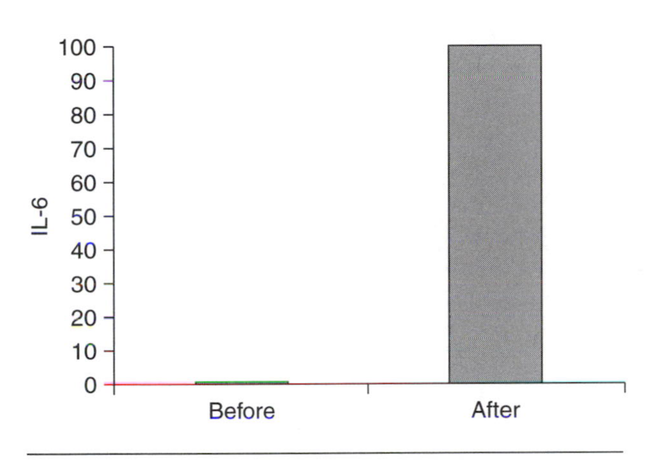

Figure 7.6 Schematic presentation of IL-6 mRNA in muscle biopsies obtained at rest and after exercise.

Recently, it was shown not only that the IL-6 gene is activated in working muscle, but also that the IL-6 protein is released in large amounts from a contracting limb, markedly contributing to the exercise-induced increase in arterial plasma concentrations (Steensberg et al. 2000).

Carbohydrate Loading and IL-6 Production

Several studies have shown that carbohydrate ingestion attenuates elevations in plasma IL-6 during both running and cycling (Nehlsen-Canarella et al. 1997; Nieman et al. 1998a). Recently, it was reported that carbohydrate ingestion did attenuate the increase in plasma IL-6 in response to both cycling and running (Starkie et al. 2001a). However, the IL-6 gene expression in the contracting muscles was not affected by carbohydrate ingestion. Thus, during exercise, carbohydrate ingestion exerts its effect at the post-transcriptional level of IL-6, although it is yet not known whether this effect occurs through translation, the release of IL-6, or both.

Biological Roles of Muscle-Derived IL-6

IL-6 was initially identified as a cytokine with immunoregulatory effects. Research during recent years, however, identifies IL-6 as an important player in metabolism.

Glucose Uptake As discussed earlier, the increase in IL-6 mRNA (Steensberg et al. 2001a; Keller et al. 2001), its nuclear transcriptional activity (Keller et al. 2001), and protein release from skeletal muscle (Steensberg et al. 2001a) are augmented when muscle glycogen availability is reduced. Furthermore, the increased expression of IL-6 was associated with increased glucose uptake during exercise (Steensberg et al. 2001a). This suggests that IL-6 may be involved, at least in part, in mediating glucose uptake during exercise.

Fat Metabolism Besides the possible glucoregulatory effect of IL-6, emerging evidence suggests that this cytokine may be involved in other metabolic pathways. Concomitantly with the increase in liver glucose output during IL-6 infusion, Stouthard et al. (1996) found an increase in circulating free fatty acids (FFA) with recombinant human (rh)IL-6 infusion. In a recent study (van Hall et al. in press), IL-6 was infused in normal, healthy volunteers at a dose that did not influence catecholamine and glucagon levels. In this study, physiological concentrations of rhIL-6 induced pronounced lipolysis, indicating that IL-6 should be classified as a novel lipolytic hormone. The suggestion that IL-6 is strongly involved in fat metabolism is supported by a very important study (Wallenius et al. 2002) showing that IL-6-deficient mice develop mature-onset obesity compared with wild-type control mice. In addition, treating the mice with IL-6 for 18 days elicited a significant decrease in body weight in transgenic, but not wild-type, mice. Thus it is evident that IL-6 is a powerful lipolytic factor, and it is suggested that during exercise the increase in arterial free fatty acid concentration is mediated at least in part by IL-6 released from the muscle. Hence, we propose that muscle-derived IL-6 acts in a neuroendocrine hormone-like manner.

Anti-Inflammatory Effects of Exercise

TNF-α and IL-6 are tightly linked; thus TNF-α stimulates IL-6 production. On the other hand, both in vitro (Fiers 1991) and animal (Mizuhara et al. 1994) studies have suggested that IL-6 may inhibit TNF-α production. Recently, it was demonstrated that both physical exercise and rhIL-6 infusion at physiological concentrations inhibit the production of TNF-α elicited by low-level endotoxemia in humans (Starkie, Ostrowski, Jauffred, Febbraio, & Pedersen unpublished data). The findings that exercise induces a pronounced increase in the production of IL-6, and that IL-6 infusion also inhibits TNF-α production, suggest that IL-6 may be involved in mediating the exercise effect on endotoxin-induced TNF-α production. In addition, IL-6 infusion enhances plasma levels of IL-1ra and IL-10 and thereby markedly contributes to mediate an anti-inflammatory response.

Immunoregulatory Effect of Exercise-Induced Cytokines

The lymphocyte concentration in blood increases during exercise and falls below pre-exercise values following intense long-duration exercise (McCarthy & Dale 1988). It appears that the acute exercise effect on lymphocytes is mediated by catecholamines, in particular epinephrine (Steensberg, Toft et al. 2001c). However, the postexercise decline of lymphocytes is mediated by both epinephrine and cortisol. The latter hormone is of particular interest if the exercise is of long duration. The increase in cortisol is mediated by IL-6 (Steensberg et al. 2001b). Linking exercise-induced lymphocyte changes to an effect of IL-6 on cortisol production is further supported by several studies demonstrating that carbohydrate loading during exercise attenuates the exercise effect on lymphocyte number and function (Nehlsen-Canarella et al. 1997; Nieman et al. 1997a,b). Interleukin-8 is produced by various cells upon stimulation and influences a variety of functions of leukocytes in particular neutrophils. Systemic administration of IL-8 induces a rapid neutropenia associated with sequestration of neutrophils in the lung that is followed by a neutrophilia characterized by the rapid release of neutrophils from the bone marrow. These cells are released predominantly from the bone marrow venous sinusoids (van Eeden & Terashima 2000). Other chemokines, for example MIP-1-α and -β, are involved in directing cell movements necessary for the initiation of T-cell immune responses (Luther & Cyster 2001).

Increased levels of plasma chemokines in response to exercise have been described only in studies in which the exercise has been extremely exhaustive as in a marathon (Ostrowski et al. 2001), whereas even 2.5 h of treadmill running did not cause an increase in plasma chemokines

(IL-8, MIP-1-α and -β) (Ostrowski et al. 1998a). Thus, it is not known if chemokines are involved in postexercise neutrophilia.

T-Cell Cytokine Production

T-cells can be divided into type 1 and type 2 cells according to their cytokine profile. Type 1 T-cells produce interferon (IFN)-γ and interleukin (IL)-2, whereas type 2 T-cells produce IL-4, IL-5, IL-6, and IL-10. Type 1 T-cell responses are stimulated by IL-12 and have been shown to protect against intracellular pathogens such as several viruses. IL-6 has been shown to induce Th2 polarization by stimulating the initial production of IL-4. Type 1 T-cells mediate protection against intracellular microorganisms such as virus, whereas type 2 T-cells are important in the defense against extracellular parasites such as several helminths. Two recent studies (Steensberg et al. 2001b; Ibfelt et al. 2002) demonstrate that the postexercise decrease in T-lymphocyte number is accompanied by a more pronounced decrease in type 1 T-cells, which may be linked to high plasma epinephrine. The relatively more pronounced decrease in type 1 compared with type 2 T-cells in the recovery period may explain the increased sensitivity to infections following strenuous exercise, as these infections are often caused by viruses.

In conclusion, exercise induces increase in a number of plasma cytokines. The first cytokine that appears in the circulation during exercise is IL-6; this is followed by an increase in a number of anti-inflammatory cytokines and chemokines. Most of the circulating IL-6 during exercise is due to local production of IL-6, which is further enhanced when the muscle glycogen content is low. Muscle-derived IL-6 is involved in metabolism and inhibits low-grade TNF-α production. By stimulating cortisol production, IL-6 together with chemokines appears to play an important role in exercise-induced immune regulation (figure 7.7). The production of anti-inflammatory cytokines during exercise may contribute to mediation of the beneficial health effects of exercise. On the other hand, the finding that exercise also suppresses production of the T-cell-derived cytokine IFN-γ may explain the increased risk of obtaining an infection following strenuous exercise.

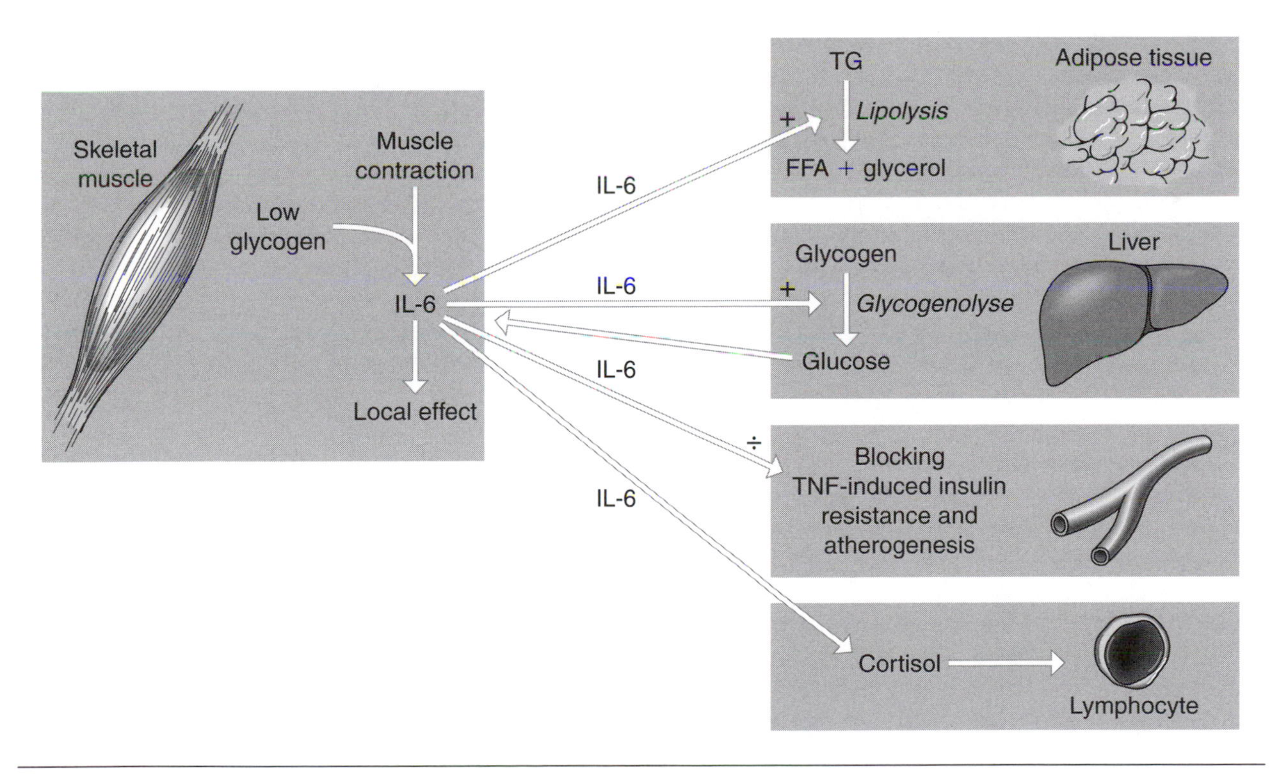

Figure 7.7 The biological roles of muscle-derived interleukin-6 (IL-6) are still not clear. However, accumulating evidence suggests that the role of IL-6 is to induce lipolysis, whereas it is less clear whether IL-6 also contributes to maintenance of glucose homeostasis during exercise. In addition, IL-6 may inhibit tumor necrosis factor (TNF)-induced insulin resistance. Furthermore, IL-6 stimulates cortisol production. Thereby IL-6 is an important player in exercise-induced immune impairment.

Intracellular Signal Transduction*

Intracellular signaling refers to the processes initiated from the occupation of a cellular receptor leading toward the cellular response. Thereby the intracellular signaling cascades consist of various proteins acting principally in two different ways (figure 7.8):

- As enzymes, thereby producing intracellular diffusible signaling molecules (second messenger)
- As providers or de-providers of activating phosphate groups (Guse 1998; Sun & Tonks 1994).

The reversible phosphorylation of proteins represents the major device by which ligand–receptor binding is converted into a cellular response. Intracellular signaling enables a modification of the incoming signal in terms of amplification, sensitivity, and divergence. Only a few ligand–receptor complexes can release a wave of intracellular messengers or initiate a cascade of phosphorylation steps. In addition, ligand–receptor binding is seldom connected to a single signaling pathway. Intracellular signals can diverge and address different pathways that furthermore may interact with each other (cross-talk).

Intracellular signals are coupled to cellular effector functions with different kinetics. Immediate or fast responses are the opening of ion channels or the exocytosis of cellular secretion products triggered by binding of diffusible messengers or the transfer of phosphate groups. Slower or delayed responses, such as proliferation or expression of receptors, often involve transcriptional processes. In such a case, both free second messengers and phosphorylation events can independently activate transcription factors that control the expression of specific genes. The aim is the long-term adaptation of cellular function or phenotype (Berridge 1997).

In the following section we discuss some of the major intracellular signals and pathways and their multiple interactions. Then we describe the effects of exercise on the signaling pathways and present some possible mechanisms that may be involved in the exercise-dependent modulation of intracellular cellular signaling pathways.

Intracellular Second Messengers

One event that may follow the ligand (first messenger) binding to its receptor is the formation/release of diffusible molecules termed second messengers. This mechanism serves as an amplifica-

* The text on this page beginning with the heading *Intracellular Signal Transduction* and ending with the text under the heading *Redox Status* on page 144 is contributed by Frank Mooren.

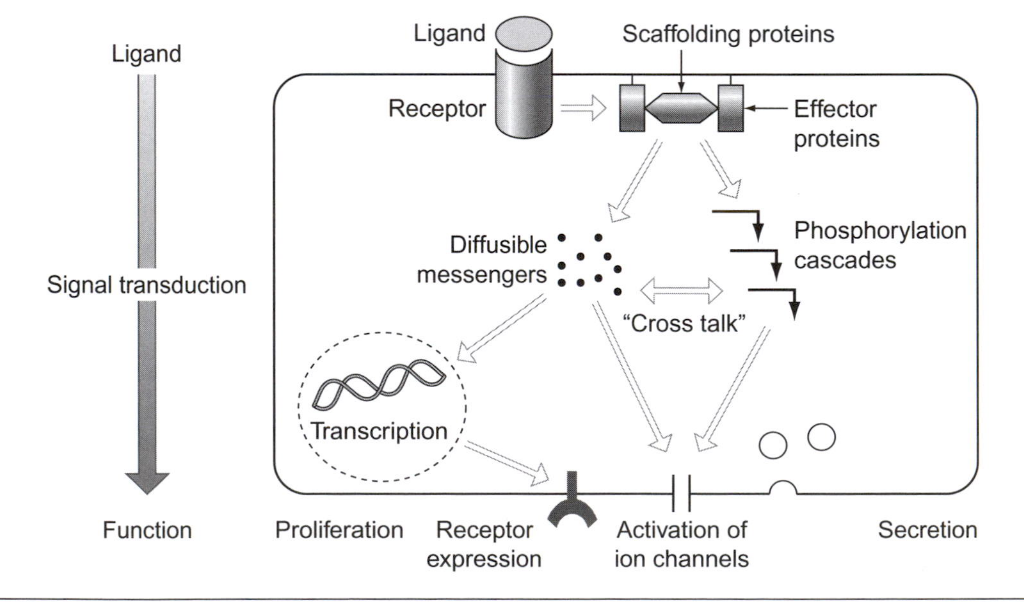

Figure 7.8 Intracellular signal transduction includes all processes that link the activated receptor to a cellular response such as proliferation or secretion. As schematically shown, this involves either the release of diffusible messengers or the activation of phosphorylation cascades or both.

tion of the first messenger, which often is in the nanomolar range, since the formation of second messengers is catalytic rather than stochiometric. The range of action is dependent on their mobility, determined by the diffusion coefficient and spatial distribution of release/formation and uptake/degradation sites. This section deals with three different messengers: the cyclic nucleotide compounds cyclic AMP and cyclic GMP, inositol-1,4,5-trisphosphate, and calcium.

Cyclic AMP

The archetypal second messenger is cyclic AMP, which was identified in the late 1960s. Initially the classical cyclic AMP signaling pathway was thought to be simple; today it is considered much more complex. Other cyclic nucleotide compounds, such as cyclic GMP, were found. Today cyclic AMP and cyclic GMP represent central diffusible intracellular second messengers.

The synthesis of cyclic AMP and cyclic GMP from ATP and GTP is catalyzed by adenylate cyclases and guanylate cyclases, respectively (Hurley 1999). To date at least nine closely related isoforms of adenylate cyclases (AC1-9) have been cloned and characterized. Distribution of the isoform AC6 is widespread; isoforms AC2,3,4,5, and 8 have been found in heart and AC9 in skeletal muscle (Hanoune & Defer 2001). ACs are membrane-integrated proteins consisting of two hydrophobic domains (each with six transmembrane spans) and two cytoplasmic domains (Hurley 1999).

Besides activation of ACs by the α subunit of the Gs protein and by forskolin, a number of stimulants and inhibitors such as protein kinase C, calmodulin kinase, calcium, and other G-protein subunits have been described (Simonds 1999). It has been proposed that the capacitative entry of calcium plays a major regulating role for ACs either positively (AC1 or AC8) or negatively (AC5 + 6). Moreover, AC5 and 6 have been shown to be inhibited by phosphorylation through protein kinase A (PKA), which functions therefore as a negative feedback loop (Fagan et al. 1996; Yu et al. 1993).

In analogy to the ACs there exist several isoforms of guanylate cyclases, which catalyze the formation of cyclic GMP from GTP. GCs are either bound to the plasma membrane or located in the cytoplasm, the latter form being regulated by nitric oxide (Garbers 1992; Lucas et al. 2000).

Cyclic AMP is an activator of PKA, which catalyzes the phosphorylation of a number of proteins like mitogen-activated protein kinases (MAP-kinases) involved in the control of cell growth and differentiation (Walsh & Van Patten 1994).

PKA is a tetrameric enzyme consisting of two catalyzing and two regulating domains. In this configuration the enzyme is inactive. Upon binding of cyclic AMP to the regulating domain, catalyzing and regulating domains dissociate, releasing the active catalyzing subunit (Beebe 1994). A family of PKA-anchoring proteins (AKAPs) plays an important role in the compartmentalization of PKA, bringing it in close proximity to its substrates. Cyclic GMP-dependent protein kinases show similar characteristics to PKA in terms of their regulation, but regulatory and catalyzing domains are located on the same molecule (Francis & Corbin 1994).

Finally, cyclic AMP is inactivated by cyclic nucleotide phosphodiesterases (PDE), thereby "turning off" this signaling pathway. In thyroid cells, for example, cyclic AMP has been shown to inhibit its own pathway by activation of PDE (Oki et al. 2000) To date the PDE superfamily has been subgrouped into 10 different PDE families, which can be distinguished by their enzymatic characteristics and their sensitivity toward pharmacological inhibitors (Soderling & Beavo 2000). In T-cells, levels of PDE7 are increased by co-stimulation of TCR/CD3 complex and CD28 receptor, and this seems to be involved in T-cell proliferation as indicated by inhibition of PDE7 through antisense oligonucleotides (Soderling & Beavo 2000).

Inositol 1,4,5-Trisphosphate (IP_3)

Inositol 1,4,5-trisphosphate (IP_3), which mediates the release of intracellular calcium (see next section), is generated from hydrolysis of phosphatidylinositol 4,5-biphosphate by phosphoinositide-specific phospholipase C (PLC; Streb et al. 1983). During this biochemical reaction another second messenger, diacylglycerol (DAG), is produced that is a direct activator of protein kinase C (PKC). About 10 PLC isozymes have been identified that are classified into three groups (β, γ, δ). While PLC-β-type isozymes are recruited by heterotrimeric G-proteins, PLC-γ isozymes are activated via receptor tyrosine kinases or nonreceptor tyrosine kinases. Therefore IP_3 signaling is involved in the response to various extracellular stimuli such as hormones, growth factors, and neurotransmitters. The target for the IP_3 molecule is an intracellular calcium release channel, the IP_3 receptor, a tetrameric intracellular calcium channel of about 260-kDa molecular weight, of which at present three isoforms are known (Mikoshiba et al. 1994).

For example, in Jurkat T-cells the TCR/CD3-mediated Ca^{2+} mobilization is mediated by the type I isoform of the IP_3 receptor while in B-lymphocytes the type III IP_3 receptor is present (Jayaraman & Marks 1997; Khan et al. 1996). The IP_3 receptor is predominantly located on the ER surface and can be regulated by a nonreceptor tyrosine kinase (Marks 1997). There is an IP_3 binding site on every IP_3 receptor subunit. IP_3-induced calcium release is modulated by calcium concentration following a bell-shaped curve (maximum open probability at 0.2 μM) with positive feedback at lower Ca^{2+} concentrations and negative feedback at higher Ca^{2+} concentrations (above 0.5 μM). The biphasic effect of Ca^{2+} on the IP_3 receptor provides the key to understanding certain calcium signaling patterns, such as calcium oscillations (see next section). However, recent investigations have also provided evidence for an oscillatory IP_3 formation (Hirose et al. 1999). In addition to its physiological role, there is growing evidence that altered IP_3 signaling is a part of the pathophysiological processes that underlie reperfusion arrhythmias (Woodcock et al. 2000).

Calcium

The role of calcium (Ca^{2+}) as an intracellular second messenger is based on its regulation on a nanomolar level. While the extracellular calcium concentration is about 1.2 to 2 mmol/L, intracellular calcium is usually about 100 to 200 nmol/L. In addition, in most cells, huge calcium gradients exist across the intracellular membranes of calcium storage organelles such as the endoplasmic/sarcoplasmic reticulum. Intracellular calcium stores do not have a homogeneous structure and cannot be attributed to a single cellular organelle. Likewise, mitochondria and intracellular secretory vesicles demonstrate calcium storage capacity (Petersen et al. 1999; Pettit & Hallett 1997). They contain large amounts of calcium binding proteins such as calsequestrin and calreticulin.

These different calcium gradients are established by active transport mechanisms. The calcium pumps, which belong to the P-type class of ion-motive ATPases, are located either in the plasma membrane (as plasma membrane Ca^{2+}-ATPases, PMCAs) or in the membrane of intracellular calcium stores (as sarco/endoplasmic reticulum Ca^{2+}-ATPases, SERCAs) in order to extrude calcium to the extracellular side or to sequester Ca^{2+} into the stores, respectively. For both types of Ca^{2+} pumps, various isoforms have been described that differ in their activation and regulation properties and, at least in polar cells, in their spatial distribution (Carafoli & Brini 2000). In a few cell types like cardiac myocytes, the Na^+-Ca^{2+} exchanger is also operating. This secondary active transporter uses the inward-directed sodium gradient, which is established by the Na^+-K^+-ATPase for the export of calcium. However, this exchanger may also operate in a "reversed mode" leading to a calcium influx.

Upon stimulation, cellular calcium increases because of a release of Ca^{2+} from intracellular stores, or through a Ca^{2+} entry across the plasma membrane, or both. The sequence of these processes depends on the cell type and type of stimulation.

Pharmacological stimulation of epithelial or endothelial cell surface receptors usually triggers the release of calcium from intracellular stores followed by an influx of calcium across the plasma membrane (figure 7.9*a*). Various calcium-releasing factors can address different calcium pools (Clementi et al. 1994). IP_3 was the first Ca^{2+}-releasing factor to be described (Berridge & Irvine 1989). Recently two other Ca^{2+}-releasing factors were identified by Lee and coworkers: cyclic ADP-ribose (cADPR) and nicotinic acid adenine dinucleotide (NAADP) (Lee 1997). Cyclic ADPR is the endogenous ligand of the ryanodine receptor (RYR), the second type of intracellular calcium channel, which in T-lymphocytes is found in a non-endoplasmic reticulum, caffeine-sensitive Ca^{2+} pool (Guse 2000; Guse et al. 1997). Upon CD3 stimulation, formation of cADPR was demonstrated probably via the involvement of the ecto-enzyme CD38, a type II transmembrane protein (Lee 2000). In granulocytes, for example, two different Ca^{2+} storage sites have been identified by their response to different stimuli. Stimulation with fMLP addressed a storage site in the juxtanuclear space, whereas clusters of CD11b/CD18 integrin induced Ca^{2+} release from a pool located peripherally under the plasma membrane (Pettit & Hallett 1997). So far it has remained unclear which calcium-releasing factors are involved in this case.

The amount of Ca^{2+} inside the stores is limited. Therefore Ca^{2+} release from stores is only a transient signal. However, for cellular activation, often a sustained change in intracellular calcium concentration is needed, making a calcium entry via plasma membrane Ca^{2+} channels necessary. Different types of Ca^{2+} channels have been described with regard to their structure and their mode of activation: voltage dependent, stretch activated, receptor operated, and store operated (Krause et al. 1993).

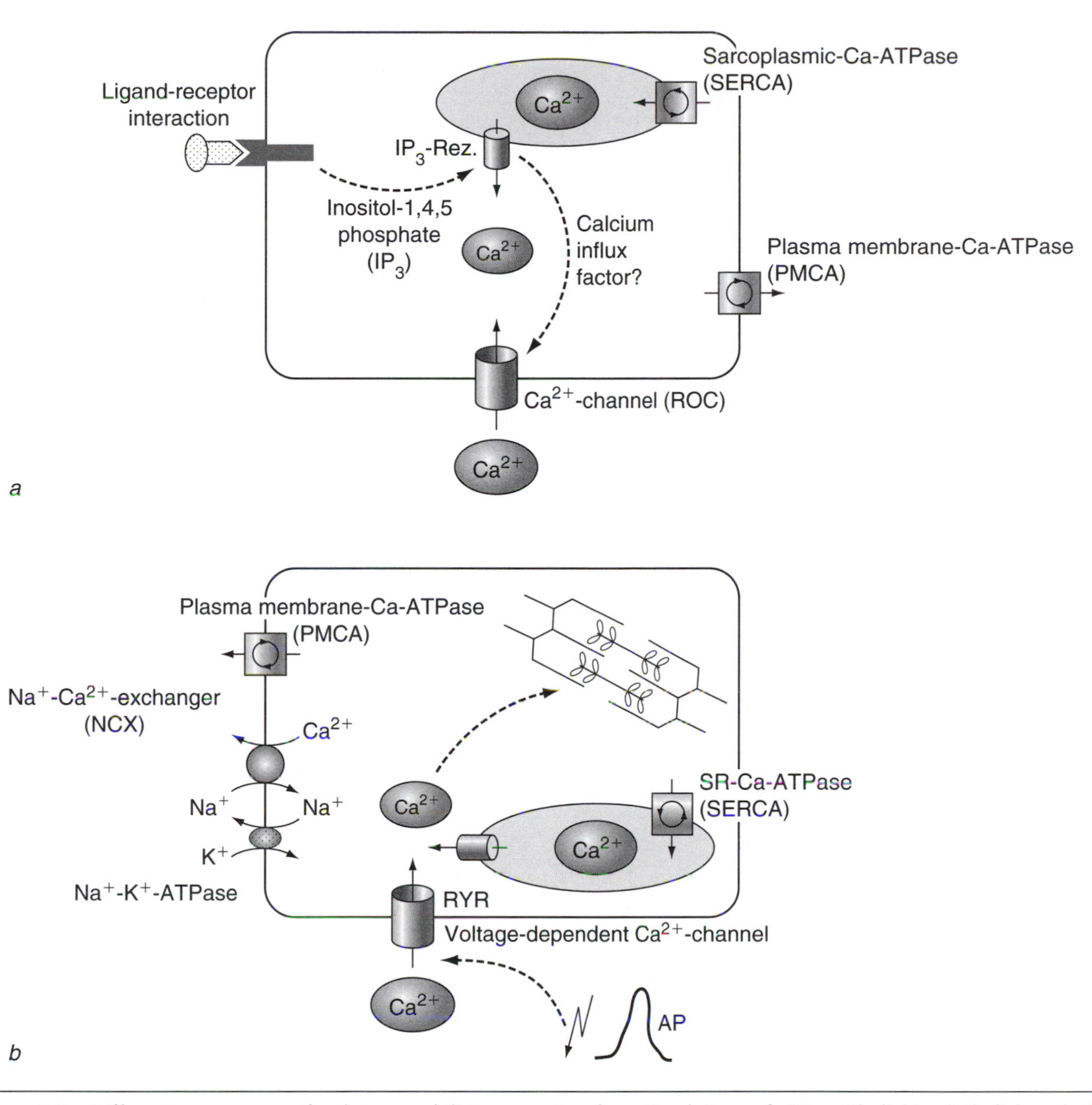

Figure 7.9 Different sequences of calcium-mobilizing events after stimulation of *(a)* epithelial/endothelial and *(b)* contractile cells.

In electrically nonexcitable cells like leukocytes, the mechanism of Ca^{2+} entry remains unclear. There is much evidence that the Ca^{2+} entry pathway is activated by Ca^{2+} release from stores (CRAC, Ca^{2+} release-activated Ca^{2+} channels); or, in other words, Ca^{2+} influx is controlled by the filling state of the Ca^{2+} stores (SOCE, store-operated Ca^{2+} entry) (Zweifach & Lewis 1993). The coupling mechanism between the two events is unclear. A diffusible calcium influx factor (CIF) released by the depleted stores in order to open the Ca^{2+} channel has been reported. However, there is also evidence for a sterical protein–protein interaction between the intracellular IP3 channels and the Ca^{2+} entry pathway (Putney & Ribeiro 2000).

Electrical stimulation of cells, for example cardiac myocytes, induces an alternative sequence of events (figure 7.9*b*). Calcium influx across the plasma membrane triggers calcium release from intracellular pools. Changes in membrane potential trigger the opening of voltage-activated calcium channels (VOCs), leading to calcium influx, which in turn activates the sarcoplasmic ryanodine channels (calcium-induced calcium release). This sequence is facilitated by the spatial arrangement of both channels, resulting in a tight structural coupling between the two components.

The nature and pattern of the intracellular calcium transients, however, depend not only on the balance between calcium-increasing and -decreasing mechanisms. Another important role is also played by the intracellular buffer capacity for free calcium ions. Cellular calcium is buffered by at least two different mechanisms:

- Molecular buffers
- Organelle buffers

The former can be subdivided into inorganic, small organic, and macromolecular Ca^{2+} ligands. The formation of Ca^{2+} salts usually occurs under pathophysiological conditions. L-glutamate, citrate, and ATP are examples of small organic ligands, while macromolecular ligands include molecules such as calmodulin and calsequestrin. Changing both the affinity and the capacity of the calcium buffers can have significant effects on intracellular calcium transients.

The complex nature of cellular calcium regulation and signaling is responsible for temporally and spatially distinct calcium signals such as Ca^{2+} transients and for plateaus as well as Ca^{2+} oscillations and Ca^{2+} waves (Petersen et al. 1999; Pettit & Hallett 1997). There is now growing evidence that these different Ca^{2+} signaling patterns encode for different cellular effector functions; for example, in B-lymphocytes, the nature of the amplitude and duration of the cellular calcium signal determines which transcription factor is activated (Dolmetsch et al. 1997). Changes in intracellular calcium concentrations can evoke a number of cellular responses including proliferation, motility, oxidative burst, secretion of cytokines, and gene expression and are involved in the induction and regulation of apoptosis (Davies et al. 1994; Elferink & Deierkauf 1985; Lew et al. 1984; Negulescu et al. 1994). Within these responses, important downstream targets of Ca^{2+} are the calcium binding protein calmodulin and the serine/threonine phosphatase calcineurin (see later), which in turn activates the transcription factor NFAT (nuclear factor of activated T-cells). Besides its physiological role as an intracellular second messenger, alterations in the regulation of Ca^{2+} are involved in the pathogenesis of several diseases such as inflammatory diseases, neurologic disorders, and cardiovascular diseases (Davies et al. 1991, 1994; Mooren & Kinne 1998).

Transfer of Phosphate Groups

The phosphorylation of an amino acid side chain is a mechanism of fundamental importance in regulating enzyme activities and controlling cellular signal transduction pathways. Furthermore, one has to realize that phosphorylation is a dynamic and reversible equilibrium process (Sun & Tonks 1994). The phosphorylation level is balanced by the activity of certain phosphatases that are responsible for dephosphorylation of either serine/threonine or tyrosine residues. In order to avoid a defective signal transduction, an integrative and coordinative action of both kinases and phosphatases is indispensable.

Depending on the amino acid to be phosphorylated, namely, serin/threonin (Ser/Thr), tyrosine (Tyr), histidine (His), and aspartate/glutamate (Asp/Glu), four classes of protein kinases could be differentiated (Hunter 1991). Since only limited information is available about the latter two, this section deals with Ser/Thr- and Tyr-specific kinases.

Protein Kinase C

PKC is an element of cellular signaling pathways involved in cell growth and differentiation. Originally it was described in 1984 by Nishizuka and coworkers as a Ca^{2+}- and phospholipid-dependent protein that is part of the Ser/Thr kinase superfamily (Keenan et al. 1997; Nishizuka 1984).

The PKC family consists of at least 12 different subtypes, which have been classified into three groups according to their biochemical characteristics. The first group, the conventional PKCs α, $\beta 1$, $\beta 2$, and γ, is dependent on Ca^{2+} and phospholipids like the originally described kinase activity. The second group, the novel PKCs δ, ϵ, η, and θ, is dependent on phospholipids but not on Ca^{2+}. A third group, termed the atypical PKCs ξ and $\tau\lambda$, does not require either Ca^{2+} or phospholipids. In general, within the primary structure of the 63- to 84-kDa PKC molecules, a C-terminal catalytic domain and an N-terminal regulatory domain can be discriminated that consist of both variable and conserved sequences (Nishizuka 1992).

The major intracellular activator of PKC is DAG, which is generated via different enzymatic reactions with different time constants. Initially DAG is recruited transiently from hydrolysis of phosphatidylinositol-4,5-bisphosphate, followed by a more sustained increase due to hydrolysis of phosphatidylcholine (Besterman et al. 1986).

Upon activation, PKC is redistributed from cytosol to the membrane in a calcium-dependent manner. The N-terminal regulatory domain seems to be necessary for this enzyme translocation, which is facilitated by an interaction with anchoring proteins like RACK1 (receptor for activated PKC) (Mochly-Rosen et al. 2000). The activation-induced membrane association is responsible for some of the specific functions of PKC such as down-regulation of receptors, modulation of ion channel activity, and rearrangement of cytoskeletal structures during exo- and endocytosis (Keenan & Kelleher 1998). Additional substrates of PKC are the Raf-kinase within the Ras-MAP-kinase pathway and IκB, which is an inhibitor of the transcription factor NFκB. Both links indicate the regulatory effects of PKC on the mitogenic pathway, transcriptional processes, and the cell cycle (Genot et al. 1995; Siegel et al. 1990). Moreover, recent investigations of the expression pattern of the PKC isoforms in lymphocytes suggest an individual regulation and a differentiated function of the isoforms within the single cell, as well as lineage- and differentiation stage-specific patterns of PKC isozyme expression (Keenan et al. 1997; Mochly-Rosen & Kauvar 2000; Wilkinson & Nixon 1998). TCR-mediated activation of PKC isoforms showed different kinetics. While PKC α and θ peaked after a few minutes, activation of PKC β, δ, and ϵ started about 2 h later. Application of isoenzyme-specific antibodies revealed that PKC α and θ were responsible for interleukin-2 receptor expression, whereas inhibition of the other isoenzymes suppressed interleukin-2 synthesis (Szamel et al. 1998). The novel PKC θ seems to be restricted to T-lymphocytes (Keenan et al. 1997; Keenan & Kelleher 1998). Its translocation and close association to the TCR/CD3 complex is essential for TCR-mediated T-cell activation, NFκB activation, and cytokine production of mature lymphocytes; but it seems to be dispensable in thymocytes (Sun et al. 2000). On the other hand, PKC $\beta1$ and δ seem to be major targets of B-cell receptor signaling (Kawakami et al. 2000). Finally, there are many, sometimes controversial, reports about the involvement of the PKC isoforms in apoptosis. The working hypothesis so far is that the classical and atypical PKC isoforms are associated with cell survival, while the novel PKC isoenzymes play a pro-apoptotic role (Cross et al. 2000).

Protein Ser/Thr Phosphatases

Protein phosphatases (PP) that dephosphorylate phosphoserine and phosphothreonine residues can be classified into at least four groups: PP1, 2A, 2B, and 2C. Protein Ser/Thr phosphatases consist of a highly conserved catalytic domain and an associated regulatory or targeting subunit of considerable structural and functional diversity (Barford 1996). As in other metalloenzymes, two metal ions are located at the center of the catalytic domain (Mn^{2+} and Fe^{2+}) and are involved in the catalyzing process (Chu et al. 1996; Egloff et al. 1995). As the catalyzing subunits are able to catalyze a broad range of substrates in vitro, the specificity of the PP in vivo is dependent on the associated subunit. Therefore this subunit functions as a regulatory subunit, either by modulating the specificity through protein–protein interactions or by subcellular localization of the enzyme near or close to the substrate.

An important member of the protein Ser/Thr phophatase family is the PP2B or calcineurin, which plays a crucial role during cardiac and skeletal muscle hypertrophy and during activation of T-cells. After PP2B is stimulated by Ca^{2+} binding to its regulatory B subunit/calmodulin complex, it dephosphorylates the transcription factor NFATp. The dephosphorylated factor enters the nucleus and induces expression of the IL-2 gene (Masuda et al. 1998). Calcineurin is the target of cyclosporin and FK506, which in complex with their appropriate binding proteins inhibit the phosphatase and thereby suppress the T-cell activity (Dumont 2000; Matsuda & Koyasu 2000).

Protein Tyrosine Kinases

Proteins with tyrosine kinase activity are grouped into receptor tyrosine kinases and nonreceptor tyrosine kinases (Horowitz & Klein 2000). Growth factor receptors such as receptors for platelet-derived growth factor (PDGF), epidermal growth factor (EGF), and transforming growth factor (TGF) contain an intrinsic tyrosine kinase activity and therefore belong to the first group. They consist of an extracellular domain for ligand binding, a transmembrane domain, a juxtamembrane domain for receptor transmodulation, the cytoplasmic protein tyrosine kinase domain, and the carboxy terminal tail carrying several autophosphorylation sites (Ullrich & Schlessinger 1990). Upon ligand binding and receptor dimerization, a conformational transition enables autophosphorylation and activation of the intrinsic tyrosine kinase activity and creates a substrate docking place. Examples of target molecules for receptor tyrosine kinases are the phosphoinositide-3-kinase, phospholipase C-γ MAP-kinases, and multiple ion channels (Kazlauskas 1994; Lepple-Wienhues et al. 2000). Nonreceptor tyrosine kinases are grouped into different families, such as the Src or Syk family, due to their sequence homologies (Neet & Hunter 1996). They are permanently or transiently associated with different membrane-related subcellular structures, which are determined by the SH2 and SH3 domains within the kinases (Li & She 2000).

Protein Tyrosine Phosphatases

Similar to the situation for protein Ser/Thr kinase/phosphatase interaction, specific protein tyrosine phosphatases exist that are responsible for controlling and regulating cellular phosphorylation of tyrosine residues (Hunter 1995). Enzymes able to hydrolyze protein phosphotyrosines share an active-site motif consisting of an arginine and a cysteine residue and are inhibited by vanadate (Mourey & Dixon 1994; Stone & Dixon 1994). The tyrosine phosphatase superfamily is grouped into four families: the tyrosine-specific phosphatases, the VH1-like dual-specificity phosphatases, cdc25, and low molecular weight phosphatases. The tyrosine-specific phosphatases are the best-studied group and are subdivided into receptor-like and nonreceptor-like phosphatases as the corresponding tyrosine kinases (Fauman & Saper 1996). CD45, known to be essential for B- and T-cell activation, is an example of the first group (Altin & Sloan 1997). Nonreceptor-like phosphatases contain catalytic domains and extra-catalytic domains responsible for their subcellular localization (Dixon 1996).

Small GTPases/Ras Proteins

Ras proteins belong to the group of small GTPases that act as important molecular switches involved in controlling such fundamental cellular processes as proliferation, differentiation, and apoptosis. In mammals, four different Ras proteins (H-Ras, K-Ras 4A, K-Ras 4B, and N-Ras) are coded by at least three genes. Initially Ras genes were identified through their mutated forms in certain human tumors; for example, K-Ras can be found in adenocarcinomas of the colon and the pancreas in about 50% and 90%, respectively, of cases.

Ras proteins cycle between two conformations, being inactive when bound to GDP (Ras-GDP) and active when bound to GTP (Ras-GTP). GTPase-activating proteins (GAPs) promote the hydrolysis of GTP leading to inactive Ras-GDP (off-reaction). The opposing reaction, GDP dissociation and GTP association, is catalyzed by guanine nucleotide exchange factors (GEFs) (on-reaction). The activation of Ras is induced by various signals predominantly after ligand binding to receptors with intrinsic or associated tyrosine kinase activity, such as growth factor receptors. Alternative Ras-stimulatory signals involve increases in calcium and DAG. A requisite for Ras function is localization to the plasma membrane. This is achieved by post-translational modifications of Ras—farnesylation and carboxymethylation—by the enzyme farnesyltransferase (FTase). The farnesyl moiety anchors the Ras proteins at the inner leaflet of the plasma membrane, thus allowing a close spatial relationship between intracellular receptor domains, GEFs, and adaptor proteins.

Ras activation addresses a number of different effector proteins; among these, the three component kinase cascades consisting of MAP-kinase kinase kinases (e.g., RAF proteins), MAP-kinase kinases (e.g., MEKs), and MAP-kinases (e.g., ERKs) are the best-characterized pathways. Other downstream effectors include the phosphoinositol-3 kinase (PI-3K), phospholipase C-γ, and RAL-GEF, which leads to activation of RAL, another small GTPase.

Mitogen-Activated Protein Kinases

Mitogen-activated protein kinases (MAP-kinases) are ubiquitous proteins that link the perception of extracellular signals or environmental alterations to a variety of cellular responses. Today more than 20 different MAP-kinases have been identified, which are grouped into at least four different families:

- Extracellular signal-related kinases (ERK) 1 and 2
- Stress-activated protein kinases 1-C-JUN NH2-terminal kinases (SAPK1-JNK)
- Stress-activated protein kinases 2-p38 (SAPK2-p38)
- Extracellular signal-regulated kinases 5 or BIGMAP-kinase-1 (BMK)

All MAP-kinases share a common activation modus, namely the concomitant phosphorylation of a threonine and a tyrosine residue within a conserved THR-X-TYR motif in the activation loop of the kinase domain. The X stands for different amino acids among the different MAP-kinases, for example glutamine in the case of ERK1 and 2. MAPKs are the final elements of parallel phosphorylation cascades, each demonstrating a conserved architecture consisting of the three kinase modules. They are phosphorylated by a MAP-kinase kinase (MAPK-ERK-kinase = MEK), which in turn is phosphorylated by a MAP-kinase kinase kinase (MEK-kinase = MEKK). MAPK pathways are activated by the recruitment of the

MEKKs to the plasma membrane where they are phosphorylated through small GTPases such as Ras, Rac, and Rho. Various input signals regulate the MAPK pathways, among which the ERK pathway is the best characterized. The ERK pathway is activated through many different receptor types including receptor tyrosine kinases (e.g., insulin, EGF, PDGF, ETC), G-protein-coupled seven transmembrane-spanning domain receptors (angiotensin II, ET-1), or serin-threonine kinase receptors (TGF-related polypeptides), which all address the Ras protein. Ras in turn recruits and activates MEKK of the RAF family, which activate the two MEKs MEK1 and MEK2. Finally, the MEKs activate ERK1 and 2.

Environmental stress such as hypoxia, osmotic changes, heat shock, or oxidant stress and inflammatory cytokines are usually the stimulatory signals for the other MAPK pathways. However, some exceptions exist. While JNK and p38 pathways are activated by inflammatory cytokines (TNF family), ERK5 is not; p38 is activated during both ischemia and reperfusion, whereas JNK pathway is activated only during reperfusion.

The target substrates of the MAPK pathways include both transcription factors and other protein kinases. While some substrates are selectively recruited (ERK1-2 MAPK-activated protein kinase 1, MAPKAP1; p38 MAPKAP2-3), others like mitogen- and stress-activated protein kinases 1 and 2 (MSK1-2) are commonly addressed. Similarly the AP1 transcription factor is a downstream target of both stress and mitogenic signaling pathways. Other transcription factors activated by MAPK include cyclic AMP response element binding protein (CREB), ELK-1, activating transcription factor (ATF) 1 and 2, and myocyte enhancer factor 2 (MEF2). In contrast, there is evidence that JNKs phosphorylate NFAT (nuclear factor of activated T-cells), therefore preventing its stimulus-induced nuclear translocation.

Activation of transcription factors usually requires the translocation of activated MAPK from the cytosol to the nucleus, emphasizing the need for their prolonged activation. Magnitude and duration of MAPK activation are therefore critical determinants of the biological effect, and reflect the balance between upstream activating kinases and protein phosphatases. Since MAPK activation requires phosphorylation on both tyrosine and threonine residues, deactivation can be performed by both types of phosphatases. Moreover, dual-specificity protein phosphatases (threonine-tyrosine) have been identified.

Exercise and Intracellular Signaling

Research over the last century has provided much evidence that exercise is able to change the cellular phenotype and to affect various cell functions. However, many questions remain, especially regarding the underlying mechanisms that lead to these changes. From pathophysiological studies it is well known that during the pathogenesis of many diseases, alterations of signaling pathways occur often initiated by exogenic factors. Likewise, intracellular signaling pathways seem to be important targets that are addressed by exercise (i.e., by exercise-associated stress factors such as changes in the redox status). Therefore, this part of the chapter aims to summarize what is currently known about alterations of signaling pathways by exercise.

This section focuses on the effects of physical exercise on intracellular signaling pathways. Most studies so far have concentrated on the effects of exercise on one of several signaling branches—intracellular calcium, PKC, cyclic AMP, and MAP-kinases; less information is available on IP_3 and Ras proteins.

Exercise and Second Messengers

This section focuses on the effects of physical exercise on intracellular second messenger pathways. Most studies so far have concentrated on intracellular calcium and cyclic AMP, while less information is available about IP_3. This is most likely the result of different analytical methods used for the messenger detection and which are most difficult for IP_3.

Cyclic AMP

It is well documented that acute exercise induces an up-regulation of β-adrenergic receptors on platelets, lymphocytes, and monocytes, while for granulocytes the results are inconsistent (Brodde et al. 1984; Butler et al. 1983; Fujii et al. 1996; Lehmann et al. 1983; Maki 1989; Stock et al. 1995). Increase in β-2-AR (adrenoreceptor) number has been shown after both dynamic and isometric exercise (Graafsma et al. 1989) and was abolished after treatment with a β-2-AR-selective antagonist (Brodde et al. 1988). Interestingly, β-2-AR density and cyclic AMP synthesis (see later) were unaffected by exercise in patients with primary hypertension. After hypertension treatment, exercise-induced changes similar to those in healthy controls were observed (Middeke et al. 1994. Data on acute exercise-induced changes of β-2-AR on cardiomyocytes are rare and less clear. From comparative approaches there is evidence that β-AR densities of lymphocytes and cardiomyocytes are correlated (Sbirrazzuoli & Lapalus 1989). Likewise, Izawa et al. (1989) reported an increase in β-2-AR number in sarcolemmal membrane fractions of rat heart after an acute treadmill run. In contrast, after a 2-h swim test, a decrease in sarcolemmal β-2-AR number was reported (Werle et al. 1990). Consequently, there is no agreement about the underlying mechanisms of receptor regulation. Fujii et al. (1997) reported that exercise-induced receptor up-regulation in lymphocytes was accompanied by increases in β-2-AR mRNA levels, which suggests an enhanced transcriptional activity. On the other hand, in the study by Izawa et al. (1989), β-2-AR numbers in both sarcolemmal membranes and light vesicles were analyzed, and the authors found a transfer between the two fractions suggesting a receptor translocation from intracellular sites to the plasma membrane. Most likely, both mechanisms contribute to the exercise-induced increase in β-2-AR number (Fujii et al. 1996).

Several studies indicate that isoprenaline-stimulated cyclic AMP stimulation is enhanced after exercise, concomitant with the increase in β-2-AR number (Butler et al. 1983; Fujii et al. 1996; Mäki et al. 1989). Whether adenylate cyclase activity increases too remains unclear. Indeed, Graafsma et al. (1990) found that after exercise there was an increase in cyclic AMP synthesis following application of isoproterenol but not of forskolin. In cardiac myocytes a positive correlation between cyclic AMP synthesis and acute exercise intensity and duration was found (Dunbar & Kalinski 1994). However, after long-lasting exercise (3 h), a decrease in isoproterenol-stimulated cyclic AMP was found, which might be attributable to a depletion of catecholamines in the adrenals and cardiac muscle (Mäki 1989). In platelets, the enhanced cyclic AMP responsiveness was correlated to the increase in agonist-induced aggregation, which is reported in most studies (Naesh et al. 1990; Winther & Trap-Jensen 1988). The enhanced cyclic AMP production may also attributed to changes in subpopulation composition. The greater the proportion of natural killer cells, the higher the cyclic AMP production. Likewise, Stock et al. (1995) found no effect of exercise on cyclic AMP formation in CD4+ cells.

Interestingly, cyclic AMP production per receptor seemed to be reduced after acute exercise, suggesting a desensitization of the adrenoreceptor, which was confirmed by an in vitro study of Davies (1988) using membrane preparations containing β-2-adrenergic receptor and plasma obtained before and after exercise (Graafsma et al. 1990; Fujii et al. 1993).

The influence of chronic exercise on the β-2-AR number is less well documented in lymphocytes than in cardiomyocytes. While some studies showed no change in lymphocyte β-2-AR number or affinity to catecholamines with training, others showed a down-regulation (Fell et al. 1985). Most studies agree that endurance training reduces the β-adrenoreceptor density on both platelets and cardiac myocytes (Sylvestre-Gervais et al. 1982; Lehmann et al. 1986; Ohman et al. 1987; Werle et al. 1990), although again, others report no changes (Bohm et al. 1993; Moore et al. 1982). β-adrenoreceptor number seems to be affected in a similar way (Lehmann et al. 1986). A tissue- and cell type-specific regulation of β-2-AR number seems to exist. Ten-week treadmill training of rats resulted in a desensitization of the heart β-adrenergic system as indicated by a decrease of receptor number and coupling to downstream pathways, while the adipose β-adrenergic system was more sensitized (Nieto et al. 1996a,b). The β-2-AR density distribution varies among muscle fiber types, with the highest expression in type I fibers and lowest in type IIb. Furthermore, the training-associated up-regulation seems to be more pronounced in type I and type IIa fibers (Plourde et al. 1993). Training also affects β-2-AR responsiveness after acute exercise bouts as indicated by a report from Schaller et al. (1999). After endurance training, the increase in the adrenoreceptor density on human lymphocytes induced by an acute exercise test was blunted.

After endurance training, the exercise-induced decrease of the cyclic AMP production per β-AR turned into an increase, indicating an enhanced receptor sensitivity (Schaller et al. 1999). Likewise Nieto et al. (1996a) found an increase of adenylyl cyclase activity in rat liver cells after endurance training. Interestingly, they could demonstrate that this effect was related to an enhanced content of G-proteins (G50, Giα, Gβ) in liver cell membranes (Nieto et al. 1996a). Similar results have been obtained for the adenylyl cyclase system in adipose tissue and rat testis (Nieto et al. 1996b; Lu et al. 1997). In addition, it was suggested recently that the underlying mechanism for an improved lipolytic response after exercise was an enhanced expression of the AKAP150 protein facilitating the interaction of PKA and lipase (Nomura et al. 2002). Contrary results were reported for myocardial cells. Nieto et al. (1996b) found in the study just mentioned that densensitization was evident in the heart after endurance training, suggesting tissue- and cell-specific regulatory mechanisms. Others found an enhanced AC activity after training. In aged rats the increase in AC activity could be attributed to a reduced expression of the G-protein Giα (Bohm et al. 1993). A sensitization of the cyclic GMP system in platelets was evoked by two months of bicycle training, which was reversible after a detraining period (Wang et al. 1997).

Taken together, these data suggest a desensitization of AR receptor coupling after acute exercise while the opposite is true for chronic exercise. However, due to the different modes, intensities, and durations of exercise and various training regimes and different species, the picture of the exercise effects on cyclic AMP pathway signaling is still blurred. Moreover, a differential regulation within the different cell and tissue types seems to occur.

Inositol 1,4,5-Trisphosphate (IP_3)

To our knowledge, only one report describes a modulation of cellular IP_3 generation by exercise. Following strenuous exercise, an increase in IP_3 was found in erythrocytes, which was reversible within 2 h after termination of exercise (Piacentini et al. 1996). Furthermore, case reports indicate that inhaled heparin, an antagonist of IP_3 action at the IP_3 receptor, might prevent exercise-induced asthma, suggesting a role for this second messenger in the pathophysiological steps within the pro-inflammatory cells, for example during allergen-induced mast cell degranulation (Jerzynska et al. 2000).

Intracellular Calcium

Acute and chronic exercise affect the properties of intracellular calcium signaling. The maintenance of basal intracellular calcium concentration ($[Ca^{2+}]_i$) after acute exercise seems to be affected dependent on exercise intensity and duration. In platelets Haller et al. (1996) and Barr et al. (1988) found no change in basal $[Ca^{2+}]_i$. However, in the study of Barr et al., the workload was 120 W for 30 min, while in the study of Haller et al. subjects exercised at a heart rate of around 150/min for 6 min. In other studies using exhaustive exercise protocols, a slight increase of resting $[Ca^{2+}]_i$ was found in platelets and in thymocytes (Azenabor & Hoffman-Goetz 2000; Wang & Cheng 1999). In our own study we found a time-dependent change of resting $[Ca^{2+}]_i$ in lymphocytes (Mooren et al. 2001a). Immediately after the exhaustive exercise test, a significant increase in $[Ca^{2+}]_i$ was found, while later no alterations were detected. The rise in basal $[Ca^{2+}]_i$ is partly caused by exercise-induced alterations of membrane integrity resulting in an enhanced inward-directed calcium leak (Caimi et al. 1997). However, Ca^{2+} entry proceeds also via specific pathways. In skeletal muscle cells, increased basal $[Ca^{2+}]_i$ was induced after prolonged electrical stimulation (120 min). Gissel and Claussen (2000) found an association of the calcium increase with the electrical stimulation-induced cellular sodium increase. $[Ca^{2+}]_i$ increase could be avoided by application of tetrodoxin, a specific inhibitor of sodium channels. Corresponding to the higher expression of sodium channels in fast-twitch muscles, the basal $[Ca^{2+}]_i$ increase was more pronounced in fast-twitch than in slow-twitch muscles. Some studies support the view that exercise-induced increases in basal $[Ca^{2+}]_i$ are also the result of defective or impaired Ca^{2+} extrusion mechanisms. Both PMCA and SERCA proteins can be found oxidized after exercise, leading to pump dysfunction. Marked changes in basal $[Ca^{2+}]_i$ with a delay of up to 48 h were observed after eccentric contractions that resulted in a significant functional impairment such in as power output (Lynch et al. 1997). Besides functional dysfunction, increases in cellular Ca^{2+} levels may result even in cell death, as Ca^{2+} is an important apoptosis trigger. Important mechanisms involve the activation of proteolytic enzymes such as calpains as indicated in rat muscle cells after running (Belcastro 1993). Another requisite is sufficiently high Ca^{2+} concentrations, which may occur as localized Ca^{2+} changes, for example in submembranous regions or within cellular organelles. Likewise, changes in mitochondrial calcium concentration have been found after prolonged downhill walking (Duan et al. 1990).

Acute exercise is also a modulator of the agonist-induced calcium transients, affecting the transients in both directions dependent on the agonist itself and its coupling to the calcium signaling pathways. Thrombin and adrenaline-induced Ca^{2+} responses in platelets were decreased after exercise, while angiotensin II-induced transients were unchanged and ADP-induced Ca^{2+} responses were enhanced (Haller et al. 1996; Wang & Cheng 1999). This suggests that exercise differentially affects the signaling branches coupled to the various receptors.

In endothelial cells, the acetylcholine-induced calcium transients were enhanced after exercise, most probably due to enhanced calcium influx across the plasma membrane. The effect was further dependent on the synthase, since its inhibition caused an attenuation of the calcium influx (Jen et al. 2002).

In two studies on leukocytes, we found exercise-induced effects on cellular calcium regulation. Exhaustive exercise evoked enhanced cellular Ca^{2+} responses to fMLP and PAF in granulocytes, which were, however, not transduced into an enhanced functional response. Furthermore, in lymphocytes we found a time dependency of the exercise-modulated Ca^{2+} responses. Immediately after the test, the Ca^{2+} responses induced by anti-CD3 antibody (OKT3) and phytohemagglutinin (PHA) were reduced while at 1 and 24 h the Ca^{2+} signals were enhanced. Moreover, stimulation with the inhibitor of the sarco/endoplasmic reticulum Ca^{2+}-ATPase thapsigargin revealed unchanged calcium transients at any time before and after the test, suggesting two points (figure 7.10): (1) exercise does not change the Ca^{2+} load of intracellular stores; and (2) exercise affects upstream parts of the signaling pathway—for example, exercise might affect formation of IP3 by modulating the activity of phospholipase C or tyrosine kinases.

Another aspect was added by the study of Wang et al. (1997). The investigators found that the alterations in calcium signaling by acute exercise depend on fitness level. In platelets, training diminished the increasing effects of acute strenuous exercise on both basal and agonist-induced Ca^{2+} levels, an observation that we could confirm in a comparative study between long distance runners and sedentary subjects in lymphocytes (Wang et al. 1997; Mooren unpublished observation).

The effects of chronic exercise on basal Ca^{2+} levels seem to negligible. Most investigations have focused on the effects of training on stimulus-induced Ca^{2+} transients. However, the direction of changes seems to be quite different, suggesting a cell- and tissue type-dependent regulation. This is best illustrated by recent investigations of training effects on vasocontraction and -relaxation. In vascular endothelial cells, chronic exercise enhanced the acetylcholine- and ATP-induced Ca^{2+} transients. This effect was mediated most probably by facilitating the Ca^{2+} influx. The enhanced intracellular calcium levels are supposed to enhance the release of endothelial-derived relaxing factors, namely NO (Chu et al. 2000). On the other hand, in vascular smooth muscle cells of endurance-trained animals, the endothelin-evoked Ca^{2+} transients were attenuated after training. This is all the more surprising since whole-cell voltage clamp experiments demonstrated an enhanced voltage-gated Ca^{2+} channel density in smooth muscle cells from coronary arteries of exercise-trained pigs (Bowles et al. 1998). The enhanced Ca^{2+} influx, however, did not increase the free $[Ca^{2+}]_i$, suggesting either an enhanced recirculation of Ca^{2+} across the plasma membrane or an enhanced sequestration into cellular Ca^{2+} stores. Another possibility would be a difference in intracellular Ca^{2+} binding capacity, for example by an increased expression of calcium binding proteins or an enhanced uptake into organelles like mitochondria. Whatever the reason, the data clearly demonstrate the differ-

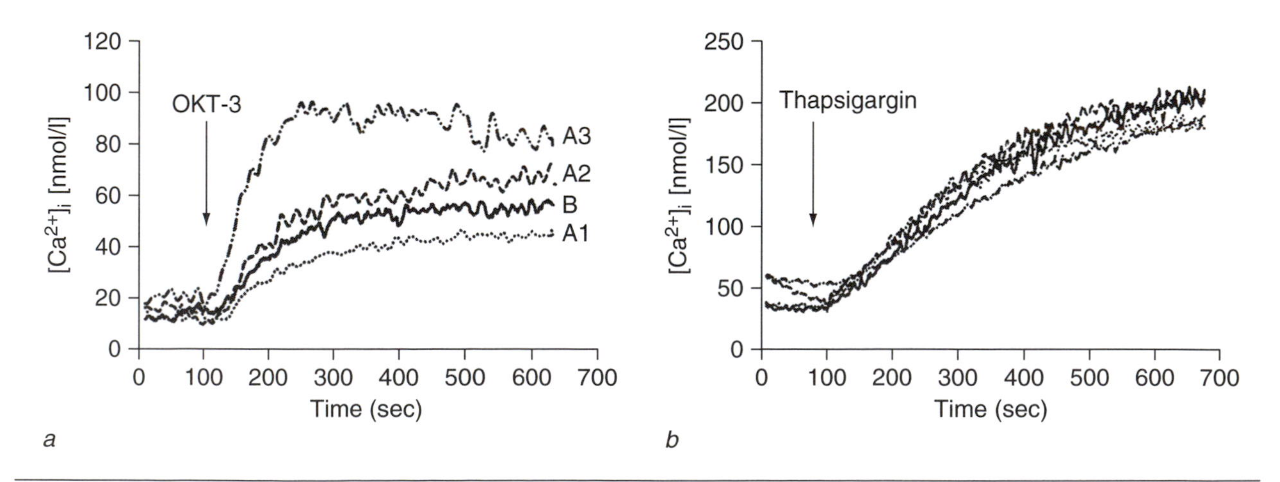

Figure 7.10 *(a)* Effect of acute exhaustive exercise on calcium signaling in human lymphocytes. Initially after exercise the calcium transients evoked by CD3 stimulation (OKT3) were reduced, followed by a reversal 1 and 24 h after the test (B, before; A1, immediately after; A2, 1 h after; A3, 24 h after test). *(b)* In contrast, thapsigargin-induced transients were unaffected by exercise, suggesting that exercise does not alter the loading of calcium stores.

ential effects of chronic exercise on cells within the same tissue, however, with a similar functional result. Enhanced Ca^{2+} transients in the endothelial cell enhance the release of NO by activating NO synthase, thereby leading to vasorelaxation. Similarly, the vasotonus is reduced as the endothelin-induced Ca^{2+} transients in the smooth muscle cells are attenuated (figure 7.11).

Other cell types whose calcium signaling was affected by training include platelets, adipocytes, and cardiac myocytes. After a two-month training period, basal and agonist-induced calcium transients were reduced in platelets. The effect was fully reversible after a three-month sedentary period (Lehmann et al. 1986; Wang et al. 1997).

In addition to the exercise effects on the cyclic AMP system of adipocytes as discussed previously, Izawa et al. (1989) demonstrated alterations in calcium signaling. Endurance training resulted in an enhanced basal $[Ca^{2+}]_i$ while phenylephrine- and ACTH-induced calcium transients were inhibited. Furthermore, lipolysis in adipocytes from trained animals was even more impaired after application of a calmodulin antagonist, suggesting an enhanced sensitivity of lipolytic processes toward the Ca^{2+}-calmodulin complex.

Finally, the cardiac myocyte response to endurance training should be discussed. Structural and functional adaptations of the heart to chronic exercise have been well known for more than 100 years. During the past 5 to 10 years it has become more and more evident that these changes are associated, perhaps even induced, by alterations in Ca^{2+} signaling. Results on the electrical-induced calcium transients in cardiac myocytes of trained animals are inconsistent. However, overwhelming evidence indicates an enhanced loading of sarcoplasmic Ca^{2+} stores after training concomitant with an enhanced expression of SERCAs. Likewise, there is evidence that chronic exercise can reverse age-related and heart failure-associated down-regulation of SERCA, which is involved in diastolic dysfunction (Cain et al. 1998; Tate et al. 1994). Furthermore, in rat hearts after myocardial infarction, exercise training restored Ca^{2+} transients by phospholamban-dependent modulation of SERCA activity (Zhang et al. 2000). For more details about exercise effects on Ca^{2+}

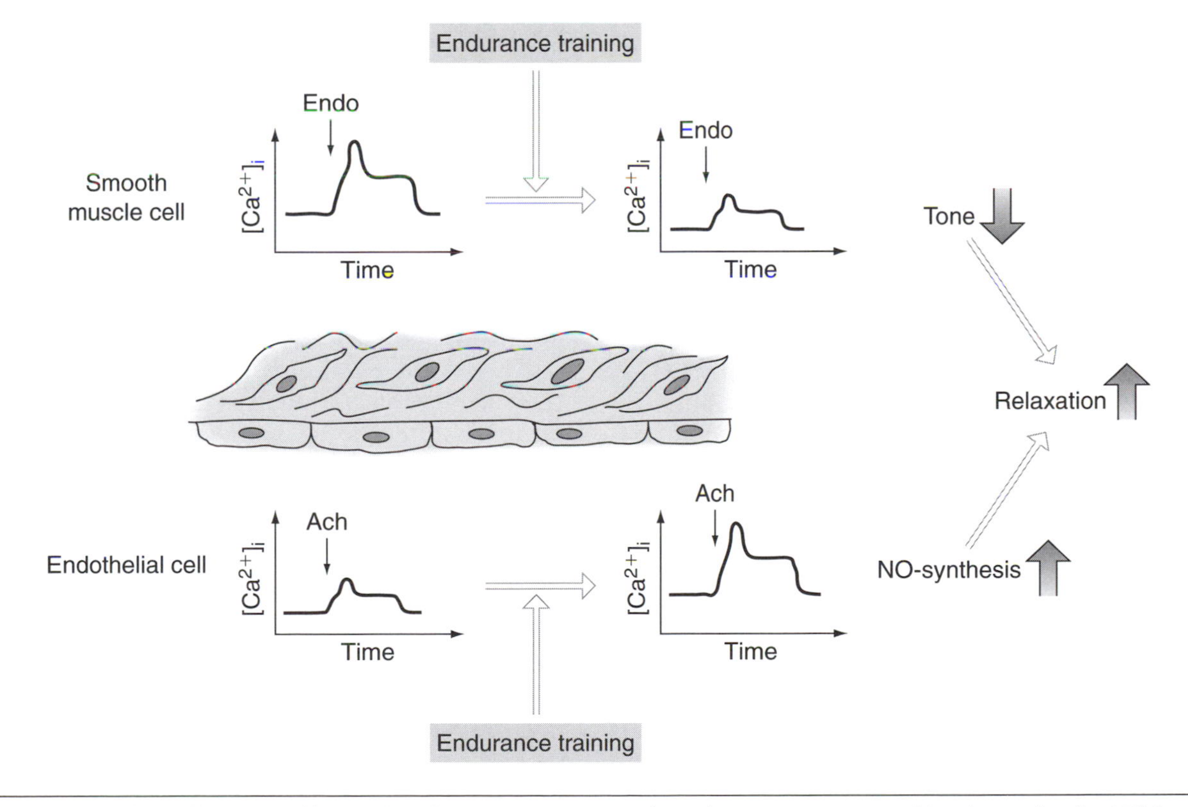

Figure 7.11 The differential effects of endurance training on the calcium transients in blood vessel cells. While the hormone-induced calcium response in vascular smooth muscle cells is reduced after training, the opposite is true for endothelial cells. However, both signals act synergistically, inducing blood vessel relaxation (Endo, endothelin; Ach, acetylcholine) (for details see text).

regulation and related downstream processes in cardiac myocytes, the reader is referred to chapter 11.

Exercise and Phosphorylation Cascades

Only a few reports are available about exercise effects on PKC activity. Lymphocyte and platelet PKC activity have been shown to be decreased in older human subjects (Wang et al. 2000). Likewise, the agonist-induced PKC translocation was impaired. Regular exercise training prevented these age-related declines in PKC function (Wang et al. 2000). Similar results were obtained with a more indirect approach. Mild hyperthermia as occurs during exercise or fever induced the activation of PKC and the expression of HSP70 in lymphocytes. The latter was dependent on PKC activity as indicated by incubation with the PKC inhibitor calphostin (Di et al. 1997).

Finally the enhanced activity of PKC after exercise could be confirmed in a study of Yamashita and coworkers. In addition, they demonstrated that this mechanism mediated the protection against the ischemia–reperfusion injury (Joyeux et al. 1997; Yamashita et al. 2001). In summary, the available data suggest that acute exercise enhances total PKC activity, while nothing is known about the regulation of the different PKC isoforms.

Exercise and Small GTPases/Ras Proteins

The knowledge about the relationship of exercise and Ras proteins is very limited, too. Goalstone et al. (1999) demonstrated in vitro that insulin increases amounts of cellular farnesylated Ras, for example in adipocytes and vascular smooth muscle cells, by activation of farnesyltransferase (FTase). Likewise in tissues of hyperinsulinemic animals, the amounts of farnesylated Ras was increased (Goalstone & Draznin 1999). Regular exercise resulted in a decrease of hyperinsulinemia followed by a decrease in the amount of farnesylated Ras, indicating that the amount of Ras recruitable for hormonal stimulation had been restricted. Obviously the exercise effect on Ras protein is only an indirect one, and this study does not indicate anything about Ras activity itself and its modulation by exercise.

Exercise and MAP-Kinases

Exercise results in cellular stress via different mechanisms, for example alterations in redox status, body temperature, and O_2 partial pressure. MAPK pathways can be regarded as an important link between the exercise stimulus and cellular adaptations. So far most studies have focused on skeletal muscle cells and indicated in both humans and animals an enhanced MAPK phosphorylation after exercise of different types and intensity. ERK1-2 phosphorylation increased within minutes after exercise in an exercise intensity-dependent manner. In contrast, acute exercise-induced ERK1-2 phosphorylation was unrelated to training status and muscle type (Wretman et al. 2000). Interestingly, both endurance and resistance exercise lead to increase in ERK1-2 phosphorylation. Furthermore there seems to be an age dependency of MAPK signaling. In muscles of old subjects (mean age 79 ± 3 years), resting level of ERK1-2, p38 MAPK, and so on were enhanced; and, contrary to findings in young controls, MAPK phosphorylation decreased after exercise (Williamson et al. 2003). While there is evidence that in the signaling cascade upstream of ERK1-2 exercise seems to recruit RAF1 and MEK1, activation of other steps remains elusive.

Whether acute exercise increases ERK1-2 protein expression seems controversial. In humans, two groups reported no changes in ERK1-2 expression in vastus lateralis muscle biopsies after a marathon run (Yu et al. 2001; Boppart et al. 2000). In contrast, Lee et al. (2002) found in trained rats that after a single bout of exercise, ERK1-2 protein content was enhanced in skeletal muscle. In the same study, ERK1-2 was found to be largely enhanced after the training regime independently of training intensity (Lee et al. 2002). Finally, ERK1-2 expression seems to be related to muscle fiber type, with a higher content in slow-twitch muscle (Wretman et al. 2000).

Besides the ERK1-2 pathway there is evidence that exercise also addresses the stress-activated protein kinases, both p38 and JNK pathway (Aronson et al. 1997; Boppart et al. 2000). However, the effects of exercise on p38 MAPK phosphorylation seem to be less robust than for ERK1-2 phosphorylation (Yu et al. 2001; Widegren et al. 1998). Furthermore, the different p38 isoforms (α, β, δ, γ) seem to respond differentially to exercise (Boppart et al. 2000). The activation of SAPK seems to be transient in nature, since 24 h after a marathon run, phosphorylation levels of JNK and p38 attain pre-exercise levels (Boppart et al. 2000). After one-leg cycle ergometry at about 70% $\dot{V}O_2max$ for 1 h, activation of MAPK attains resting levels within 1 h. Interestingly, in this study it could be demonstrated that p38 was also activated in the nonexercising leg while ERK1-2 was not. This suggests that besides local factors, systemic factors are also responsible for recruiting p38 MAPK. Another difference in the activation pattern between mitogen- and stress-activated protein kinases was recently reported by Lee et al. (2002). Training selectively enhanced ERK1-2 phosphorylation while p38 phosphorylation was significantly decreased, which may represent a mechanism to induce specific training adaptations. The exercise-induced alteration of MAPK results in further activation of downstream targets such as p90RSK and MAPKAPK2. For further details, the reader is referred to an excellent recent review (Widegren et al. 2001).

Exercise Effect on Intracellular Signaling Pathways—How Is It Mediated?

Without doubt, intracellular signaling pathways are modulated by exercise and thereby many exercise-induced adaptations of cellular function and phenotype can be explained. The question that remains concerns the mechanisms and processes through which exercise affects the signaling pathways. Clarification of this point might help either in the treatment of negative side effects of exercise or in establishment of specific training regimes.

Physical exercise is associated with remarkable alterations in metabolism as well as hormonal and biochemical functions of the body (Coyle 2000). Intracellular signaling pathways may represent important targets for these exercise-associated stressors. The aim of this section is to highlight what is currently known about the effects of these "stress factors" on cellular signaling. Since the role of hormones has been emphasized in the preceding sections of this chapter, here we discuss the influence of metabolism and redox status on cellular signaling. The reader, however, should realize that most of these data have been collected from in vitro experiments and that therefore their validity under exercise conditions remains to be proven.

Metabolism

The two major substrates for cellular ATP generation are fatty acids and carbohydrates. Besides their role in energy metabolism, recent investigations indicate that both fatty acids and glucose affect cellular signaling pathways.

During exercise, catecholamine-induced lipolysis releases non-esterified fatty acids or free fatty acids (FFA) into the blood, which are then transported to the utilization sites in liver and skeletal and heart muscle (Horowitz & Klein 2000; Turcotte 1999). Common fatty acids are stearic acid, oleic acid, and linoleic acid. In the circulation, FFA are bound to albumin, thereby preventing increases in the concentration above 2 mmol/L (Jeukendrup et al. 1998a, b). Availability of FFA depends on the rate of lipolysis, fatty acid uptake, and oxidation. Fat oxidation is lower in high-intensity exercise than in moderate-intensity exercise, probably because of an impaired release of FFA due to changes in adipose tissue perfusion. After termination of high-intensity exercise, however, a marked increase of FFA in blood has been shown without changes in lipolysis (Hodgetts et al. 1991; Jeukendrup et al. 1998a, b; Ranallo & Rhodes 1998).

FFA have been shown to affect leukocyte function in vitro and in vivo (Horowitz & Klein 2000; Richieri & Kleinfeld 1989). Lymphocyte adhesion, known to be important in lymphocyte recirculation and homing, is changed in either direction depending on the type of fatty acid and the extracellular matrix protein investigated (Stephen et al. 1997). In elderly men, Rasmussen and coworkers found a negative correlation between polyunsaturated fatty acids and NK cell activity (Rasmussen et al. 1994). Production of Il-1β, IL-2, IFN-γ, and TNF-α by human peripheral lymphocytes was even more affected by saturated than by unsaturated fatty acids (Karsten et al. 1994). DNA synthesis was enhanced by low and inhibited by high FFA concentration (Calder 1998). There is evidence that the impairment of leukocyte function is caused by the action of FFA on intracellular signaling pathways (Chakrabarti et al. 1997).

Recently, Stulnig and coworkers could demonstrate that FFA inhibited the Ca^{2+} response to stimulation through the T-cell/CD3 receptor. This effect was achieved predominantly by cis-unsaturated fatty acids, suggesting that the steric conformation of these types of FFA is important for the effect (Stulnig et al. 2000). Furthermore, these effects were achieved with FFA concentrations lower than expected during endurance exercise. Unsaturated FFA have been shown to inhibit the capacitative Ca^{2+} entry pathway and to stimulate Ca^{2+} extrusion (Breittmayer et al. 1993; Gamberucci et al. 1997; Nordstrom et al. 1991). Both mechanisms result in a decreased Ca^{2+} signal. In Jurkat T-cells, IP_3-independent mobilization of Ca^{2+} from both thapsigargin-sensitive and -insensitive Ca^{2+} pools by oleic and docosahexanoic acid has been described (Bonin & Khan 2000; Gamberucci et al. 1997). The direction of FFA-induced Ca^{2+} changes seems to be dependent on the type of FFA and on the cell type (James et al. 1990). Sperling et al. reported that polyunsaturated fatty acids were able to attenuate the formation of IP3 by phospholipase C (Sperling et al. 1993). Moreover, there is evidence that saturated FFA are able to stimulate de novo synthesis of DAG followed by activation of PKC and MAP-kinase (Yu et al. 2001). The described effects of FFA on leukocyte signaling and function are believed to be mediated by an alteration in the lipid composition of the plasma membrane. It can be assumed therefore that even more signaling processes may be affected, since any membrane-integrated and membrane-associated structure represents a potential target for FFAs.

Only recently has it been recognized that the amount of intracellularly available glucose has significant influences on cellular signaling pathways. Sustained physical activity causes the depletion of muscle glucose and glycogen stores, which has been shown to induce the expression of glucose-related proteins (GRP) and inflammatory cytokines, such as interleukin-6 (see previous section) (Locke 1997). Whether changes in intracellular calcium are involved awaits further investigations. However, in other cell types, glucose has been shown to modulate calcium transients. In lymphocytes, glucose starvation enhanced the capacitative Ca^{2+} entry (Marriott & Mason 1995; Wu et al. 1997). Moreover, intracellular application of glucose via a micropipette has been shown to affect intracellular Ca^{2+}-mobilizing messengers and to alter the frequency of intracellular calcium oscillations in pancreatic acinar cells (Cancela et al. 1998).

Redox Status

Exercise is known to enhance the formation of free radicals, especially reactive oxygen species (ROS). A major radical-generating mechanism is the mitochondrial electron transport chain because of the enhanced oxygen consumption. However, there is also evidence that exercise-associated tissue hypoxia can elevate free radicals via xanthine oxidase (Alessio 1993; Giuliani & Cestaro 1997). Because free radicals are short-lived, only few studies have directly measured free radical production; most studies to date have determined their indirect effects on subcellular structures like lipid peroxidation or DNA damage (Hawkins & Davies 2001; Leaf et al. 1997; Sjödin et al. 1990). Enhanced exercise-induced radical formation is balanced by increased expression of cellular defense systems (Oh-ishi et al. 1997).

Our final focus is on the signaling functions of radicals in combination with cellular gluthathione levels and on radical-sensitive signaling pathways. There is evidence that ROS may act as intracellular second messengers. PDGF stimulation of smooth muscle cells induced a transient intracellular elevation of H_2O_2 (Sundaresan et al. 1995). H_2O_2 application to T-lymphocytes increased both tyrosine phosphorylation and tyrosine phosphatase activity, modulated the activity of PLC-γ, and activated the MAP-kinase pathway (Goldstone & Hunt 1997; Monteiro & Stern 1996). In T- and B-cells, H_2O_2 activated both tyrosine kinases of the Syk and Src family (Lowe et al. 1998). H_2O_2 has been shown to inhibit mitogen-induced cell proliferation by interference with cellular calcium signaling. Mobilization of calcium from intracellular stores and influx of extracellular calcium were affected most likely by oxidant-sensitive PTK enzymes via an impairment of the PLC-IP3 pathway (Duncan & Lawrence 1989; Qin et al. 1996; Schieven et al. 1993).

In addition, H_2O_2 has been shown to affect activities of the three transcription factors NFκB, AP-1, and NFAT. However, different responses of NFκB activities to oxidative stress have been reported from T-cell lines and peripheral T-cells. While oxidizing steps seem to be helpful in the initial steps of NFκB activation such as IκB phosphorylation and degradation, reducing conditions are necessary for binding of NFκB to DNA (Anderson et al. 1994; Flescher et al. 1998; Schoonbroodt & Piette 2000).

A key role in the relation of redox status to intracellular signaling pathways is played by the tripeptide glutathione. The balance of the levels of reduced (GSH) to oxidized (GSSG) glutathione is particularly important for lymphocyte function (Ginn-Pease & Whisler 1998). Low GSH levels result in decreased Il-2 production and cytotoxic activity. Several diseases, as well as acute exercise and overtraining syndrome, are characterized by reduced levels of GSH, which seems to act at least in T-cells as a molecular switch between different modes of cellular responses (Dröge et al. 1998; Hack et al. 1997; Roederer et al. 1991; Staal et al. 1994).

Chapter 8

Energy Turnover and Substrate Utilization

Juleen R. Zierath, John A. Hawley, David J. Dyck, and Arend Bonen

Exercise and physical activity induce a substantial increase of the body's energy needs. This results predominantly from the enhanced contractile activity of the skeletal muscles but also includes the altered functions of other organs and tissues of the body (e.g., the heart, liver, and so on). Energy demands are fulfilled by two major sources: carbohydrates and fatty acids. These are stored in muscle cells as macromolecules and in liver and adipose tissues as glycogen and triacylglycerols. The total amounts of energy, energy flow, and energy sources vary widely depending on the type and intensity of exercise. Therefore an exact matching of glucose and free fatty acid uptake and metabolism, as well as the export of metabolites such as lactate, is indispensable for a graded and finely tuned adaptation to and coping of the various physical challenges. This chapter highlights selected aspects of fuel metabolism during exercise. The first part focuses on the molecular mechanisms governing CHO utilization, and the second part focuses on the uptake mechanisms of free fatty acids. The final part addresses the routes of lactate transport in skeletal muscle and their regulation by exercise.

Skeletal Muscle Carbohydrate Metabolism During Exercise*

Physical exercise can increase whole-body energy metabolism 20-fold above basal levels. Such a metabolic perturbation results in a significant depletion of endogenous carbohydrate (CHO) stores, with muscle glycogen and blood glucose utilization a function of the relative intensity and duration of exercise and the initial (pre-exercise) glycogen concentration. The importance of CHO availability for sustaining muscle contraction is demonstrated by the observation that fatigue during prolonged exercise is often associated with depletion of muscle glycogen stores, hypoglycemia, or both. As skeletal muscle represents 40% to 50% of body mass and accounts for up to 80% of glucose disposal under insulin-stimulated conditions, it is not surprising that intense research interest has focused on elucidating the mechanisms that regulate glucose transport into this tissue. While the acute effects of exercise on substrate utilization has been a topic of great interest in the last decades, the signaling pathways by which muscle contraction promotes glucose uptake and fatty acid (FA) metabolism have not been completely described. One pathway by which muscle contraction may directly regulate CHO and fat metabolism is the 5' AMP-activated protein kinase (AMPK) signaling cascade. In addition to the acute metabolic actions of exercise, multiple bouts of exercise/contraction (i.e., training) lead to changes in protein expression

---

* The text on this page beginning with the heading *Skeletal Muscle Carbohydrate Metabolism During Exercise* and ending with the text under the heading *Summary* on page 155 is contributed by Juleen R. Zierath and John A. Hawley. The work in the authors' laboratories is supported by the Swedish Medical Research Council, the Swedish Diabetes Association, the Foundation for Scientific Studies of Diabetology, and the Swedish National Centre for Research in Sports (JRZ) and Glaxo SmithKline (U.K.), EFFEM Foods (Australia Ltd), the Australian Research Council, and RMIT Faculty Research Grants (JAH).

of genes believed to be important for the regulation of glucose homeostasis. Thus, exercise training may enhance insulin sensitivity through mechanisms largely dependent on a coordinated change in gene expression. Candidates for a role in exercise-induced changes in gene expression include the AMPK and the mitogen-activated protein kinase (MAPK) pathways that, collectively, govern cellular proliferation, growth, and differentiation in many cell types. It will be important for the exercise physiologist to provide evidence supporting a physiological role for these kinases in mediating exercise-induced responses on gene expression in skeletal muscle. With information from the Human Genome Project, and through efforts in comparative genomics, the physiological role(s) for components of known and novel signaling pathways that regulate glucose and fat metabolism, as well as gene expression, will be of utmost interest to biological scientists involved in both whole-body physiology and molecular/cellular biology. This chapter highlights selected aspects of fuel metabolism during exercise, with a particular emphasis on molecular mechanisms governing CHO utilization.

Carbohydrate-Based Fuels for Exercise

The CHO stores of the body are principally located in skeletal muscle and liver, with small amounts present in the systemic circulation in the form of blood glucose and lactate. The glycogen concentration of skeletal muscle from sedentary individuals consuming a mixed diet is ~350 mmol/kg muscle dry wet weight (d.w.) (reviewed in Hawley et al. 1997). However, in well-trained endurance athletes, resting glycogen concentrations are ~500 mmol/kg d.w. If we assume an active muscle mass of ~12 to 14 kg, then stored (muscle) glycogen in individuals who undertake prolonged, regular endurance training would amount to 400 to 500 g (figure 8.1). After several days of a high-CHO diet (i.e., >10 g/kg body mass), muscle glycogen content in trained subjects can be elevated to values >800 mmol/kg muscle d.w. and after prolonged, exhaustive exercise can be depleted to <100 mmol/kg muscle d.w. Resting glycogen concentrations in diabetic subjects are ~280 to 300 mmol/kg muscle d.w. (Maehlum et al. 1977).

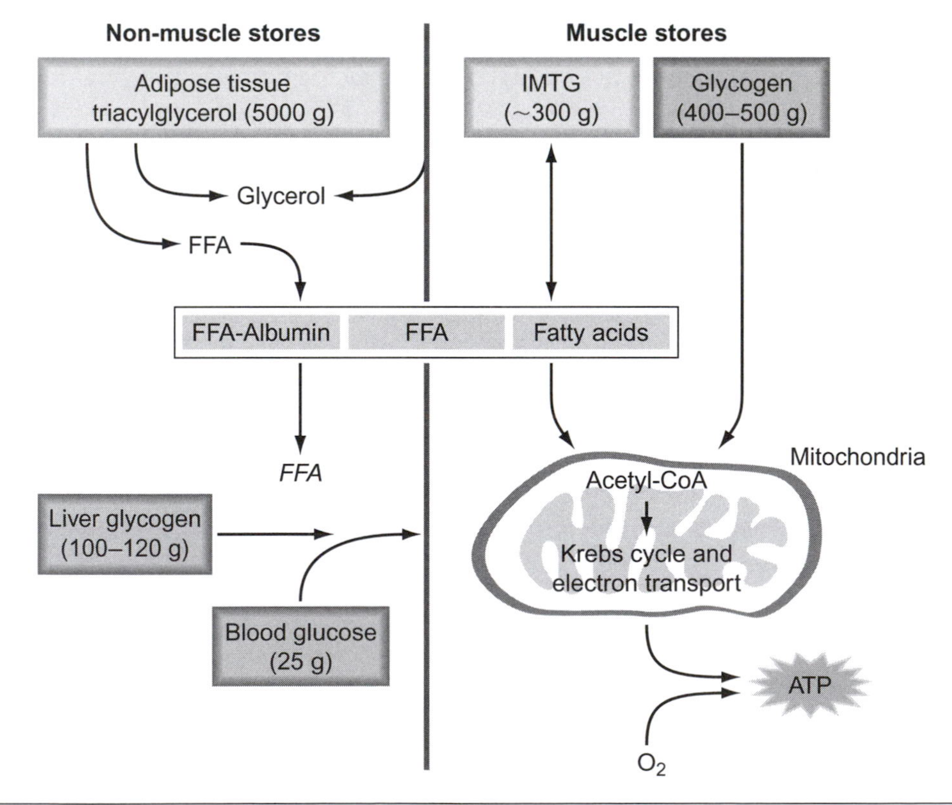

Figure 8.1 A schema of the major muscle and non-muscle storage sites of fat and carbohydrate in a well-nourished, endurance-trained subject. IMTG, intramuscular triglyceride; FFA, free fatty acids.

Adapted from E.F. Coyle, 1977, "Fuels for sports performance," *Perspectives in: Ex. Sci. Sports Med.* 10: 95-129.

The major source of the CHO located outside the muscle is found in the liver, which contains 100 to 120 g of glycogen (~360 g/kg d.w.) in the well-nourished endurance-trained individual (Nilsson et al. 1973). Liver cells can dephosphorylate glucose-6-phosphate via the action of G-6-phosphatase: The release of glucose into the bloodstream is an essential mechanism for maintaining euglycemia (i.e., 5 mM) during exercise and between meals. Though skeletal muscle lacks the enzyme G-6-P, it can release CHO either as free glucose (produced by cleavage at the glycogen branch points) or as a glucose precursor, lactate. Traditionally, lactate has not been considered a readily available fuel source. However, it is an important metabolic intermediate, serving as both a gluconeogenic precursor for the liver (Miller et al. 2002; Trimmer et al. 2002; Wasserman & Cherrington 1991) and an oxidative substrate for contracting heart as well as skeletal muscle (Brooks 2002; MacRae et al. 1992; Rauch et al. 1995). Indeed, lactate production can make a significant contribution to the overall energy production of muscle (Rauch et al. 1995). The energy derived from lactate formation varies between muscle fiber types as a function of the concentration of enzymes for the Embden-Meyerhof (glycolytic) pathway: The maximal rate of lactate production by type II fibers is about double that of type I fibers (Terjung et al. 1974). As exercise intensity increases, there is a shift from type I oxidative fibers to type II glycolytic fibers, with a concomitant increase in lactate production such that during brief, high-intensity exercise, maximal lactate production can cover nearly half the energy requirements for maximal contractile activity.

Effects of Exercise Intensity on Carbohydrate Metabolism

Both the absolute and relative (i.e., percentage of maximal aerobic power [$\dot{V}O_2$max]) intensity of exercise play important roles in the regulation of fuel metabolism. The absolute work rate or energy flux determines the total quantity of fuel required by the working muscles, while the relative exercise intensity dictates the proportions of CHO- and fat-based fuels for oxidative phosphorylation. During exercise at low to moderate intensities (i.e., 50-60% of maximal oxygen uptake [$\dot{V}O_2$max]), the proportions of total energy derived from the oxidation of CHO and fat are similar. However, as exercise intensity increases, energy demand rises as a power function of work rate/speed. Accordingly, there is a "crossover" to CHO-based fuels (Brooks & Mercier 1994), with muscle glycogen and glucose utilization scaling exponentially to the relative work rate (figure 8.2). The shift to CHO at high power outputs is due to increased sympathetic nervous system drive, the relatively greater abundance of glycolytic as opposed to lipolytic enzymes in skeletal muscle, and a change in

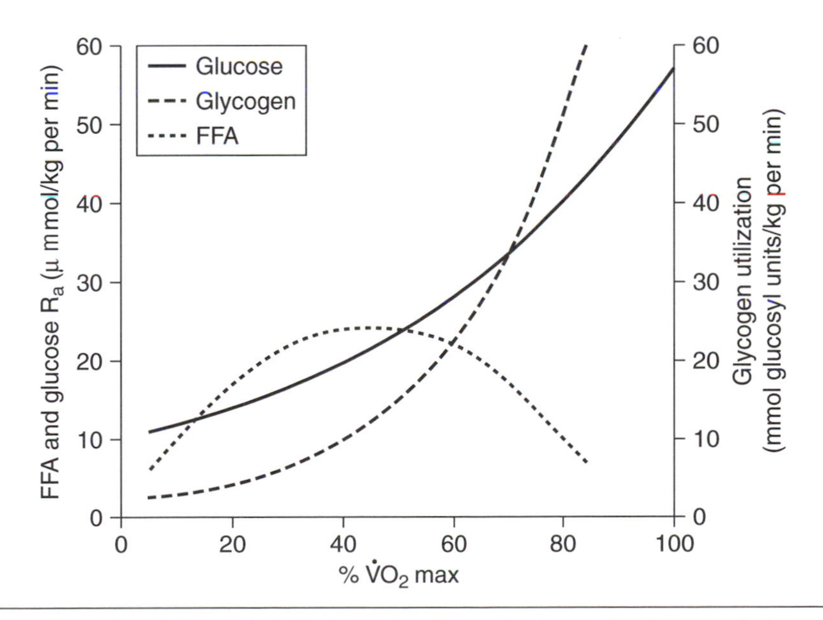

Figure 8.2 Blood glucose and free fatty acid (FFA) flux rates (R_a) and net muscle glycogenolysis as a function of relative exercise intensity. $\dot{V}O_2$max, maximal oxygen uptake.

Reprinted, by permission, from G.A. Brooks and J.K. Trimmer, 1996, "Glucose kinetics during high-intensity exercise and the crossover concept," *Journal of Applied Physiology* 80(3): 1073-1075.

the pattern of fiber recruitment to involve type II (fast-twitch) glycolytic motor units (Brooks & Mercier 1994). The increased metabolism in type II fibers does not involve a proportional increase in fat oxidation, but rather increased glycogenolysis and glycolysis leading to net lactate production (Connett et al. 1990).

Glucose taken up from the bloodstream or derived from glycogen stored inside the muscle cell is metabolized in the glycolytic pathway with the production of pyruvate. One of the major fates of pyruvate is to be transported into the mitochondria where the non-equilibrium enzyme pyruvate dehydrogenase (PDH) regulates the conversion of pyruvate to Acetyl-CoA. This reaction regulates the entry of CHO into the tricarboxylic acid (TCA) cycle and is the first irreversible step in the oxidation of CHO-derived carbon (reviewed in Spriet & Hiegenhauser 2002). Whereas the rate of CHO oxidation during exercise is closely matched to the energy demands of the working muscles (discussed subsequently), there are no mechanisms for matching the availability and metabolism of FA to the prevailing energy expenditure (Holloszy et al. 1998).

Carbohydrate Availability and Exercise Capacity

The availability of CHO as a substrate for ATP resynthesis within contracting skeletal muscle and central nervous system function is critical for the performance of prolonged (>90 min) submaximal and intermittent high-intensity exercise, and plays a permissive role in the performance of brief high-intensity work (Burke & Hawley 1999). The importance of CHO availability is demonstrated by the observation that fatigue during prolonged exercise is often associated with depletion of muscle glycogen stores, hypoglycemia, or both (reviewed in Coggan & Coyle 1991). A lack of CHO results in a reduction in pyruvate levels (Sahlin et al. 1990). Pyruvate is a substrate for Acetyl-CoA formation and for reactions that supply TCA cycle intermediates and is necessary for the continued oxidation of free fatty acids (FFA) and amino acids. Accordingly, one strategy for delaying the onset of CHO depletion is to ingest CHO before and during exercise (Coggan & Coyle 1991; Hawley et al. 1995). Such a practice rapidly increases blood glucose concentration and preserves the rate of CHO oxidation late in exercise (Coyle et al. 1986). Although CHO ingestion during exercise reduces liver glucose output (Bosch et al. 1994; McConell et al. 1994), it fails to attenuate the rate of muscle glycogen utilization during either submaximal running or cycling (Arkinstall et al. 2001). Despite one early study reporting that CHO ingestion before exercise resulted in marked hypoglycemia and a reduction in exercise capacity (Foster et al. 1979), that result is the exception, rather than the rule, and many subsequent investigations have shown the benefits to performance of pre-exercise CHO feedings (Burke & Hawley 1999).

Regulation of Skeletal Muscle Glucose Uptake During Exercise

As skeletal muscle represents 40% to 50% of body weight and accounts for up to 80% of glucose disposal under insulin-stimulated conditions (DeFronzo et al. 1981), it is not surprising that intense research interest has focused on elucidating the mechanisms that regulate glucose transport into this tissue. Glucose transport across the plasma membrane is, under most physiological conditions, considered the major rate-limiting step in glucose utilization and an important regulatory point in muscle CHO metabolism (Furler et al. 1991; Holloszy 2003). However, glucose uptake into skeletal muscle is the final outcome of several distinct regulatory steps such as the delivery of glucose from the systemic circulation to the interstitial space, transmembrane transport from the interstitial space to the inside of the (muscle) cell, and intracellular metabolism of glucose. Each step (supply, transport, and/or metabolism) may be rate limiting under specific circumstances and has the potential to limit glucose uptake into contracting muscle. For the maximal enhancement of glucose utilization to occur, all three processes need to be increased simultaneously. However, it is generally assumed that the rate-limiting step in glucose utilization during exercise is transport across the surface membrane (Furler et al. 1991; Holloszy 2003).

Glucose transport into muscle occurs by means of a passive transport mechanism that does not require ATP. This process is saturable and in skeletal muscle is mediated by two isoforms of the glucose transporter proteins (GLUT), GLUT1 and GLUT4 (Douen et al. 1990). Providing that the signaling pathways that activate glucose transport are intact, skeletal muscle GLUT4 concentration determines the capacity for stimulated transport. Indeed, there is a close correlation between GLUT4 content and maximally stimulated glucose transport (Henriksen et al. 1990; Kern et al. 1990).

Exercise is a potent stimulus for skeletal muscle glucose uptake (reviewed in Holloszy 2003). Both exercise- and insulin-stimulated glucose transport are mediated by translocation of glucose transporters from intracellular storage sites to the surface membrane (reviewed in Holloszy 2003; Richter et al. 2001). The contraction-mediated increase in glucose transport is independent of insulin action, with the maximal effects of insulin and muscle contraction on glucose transport being additive. This suggests that these stimuli activate glucose transport via different pathways.

At rest, skeletal muscle accounts for ~20% of total peripheral glucose uptake. During submaximal (55-60% of $\dot{V}O_2$max) cycling, leg muscle glucose uptake can account for as much as 80% to 90% of total body glucose utilization (Wahren et al. 1971) and may be even greater at higher intensities (Katz et al. 1986). An enhanced glucose uptake by skeletal muscle with an increase in exercise intensity arises from an increase in glucose delivery (i.e., blood flow) and an increase in extraction (i.e., arteriovenous [a-v] glucose difference). The increase in blood flow is by far the most important contributor to the exercise-induced increase in glucose uptake: Blood flow may increase 20-fold from rest during intense exercise (Ahlborg & Jensen-Urstad 1991; Katz et al. 1986), whereas the a-v glucose difference increases only two- to fourfold at a constant arterial glucose concentration (Katz et al. 1986; Wahren et al. 1971).

During exercise at a fixed, submaximal power output, blood flow to the working muscles increases rapidly to reach a plateau during the first 5 min and thereafter remains remarkably constant for the remainder of exercise (Ahlborg et al. 1974; Wahren et al. 1971). Accordingly, for any increase in muscle glucose uptake to occur, there must be an increase in glucose supply. Although glucose may actually be released from contracting muscle at the onset of exercise (Jorfeldt & Wahren 1970), such a contribution to overall CHO metabolism is trivial. Therefore, glucose supply (i.e., plasma glucose concentration) during exercise is a function of hepatic glucose production (i.e., liver glycogenolysis), exogenous CHO supplementation, or both. During prolonged continuous, submaximal exercise in the absence of an exogenous CHO supply, glucose delivery is probably limiting for sustaining muscle contraction: Arterial blood glucose concentration declines as hepatic glycogen stores become depleted (Ahlborg et al. 1974; Wasserman et al. 1991). In contrast, when CHO is ingested (Angus et al. 2002) or infused (Hawley et al. 1994), glucose uptake increases progressively during prolonged exercise and does not appear to limit exercise capacity.

Regulation of Skeletal Muscle Glycogenolysis During Exercise

Muscle glycogen utilization is a function of the intensity and the duration of exercise (figure 8.2) and the initial (pre-exercise) glycogen concentration (Blomstrand & Saltin 1999; Hargreaves et al. 1995; van Hall et al. 1995; Weltan et al. 1998). Glycogenolysis is most rapid during the early stages of exercise, with the rate of utilization being exponentially related to the relative exercise intensity. The first step in the breakdown of glycogen is the cleavage and phosphorylation of one glucose residue from glycogen to form glucose-1-phosphate. This reaction is controlled by the rate-limiting enzyme, glycogen phosphorylase (PHOS). This enzyme exists as a less active *b* form and a more active *a* form. PHOS *b* is phosphorylated to the *a* form by phosphorylase kinase and returned to the *b* form by phosphorylase phosphatase. Phosphorylase in skeletal muscle is bound to a glycogen-enzyme-sarcoplasmic reticulum complex that also contains phosphorylase kinase (Entman et al. 1980; Haschke et al. 1970; Heilmeyer et al. 1970; Meyer et al. 1970).

At the onset of intense exercise there is a rapid burst of glycogenolysis resulting from activation of PHOS *a* by the increase in cytosolic Ca^{2+} released during excitation–contraction coupling (Heilmeyer et al. 1970). However, the Ca^{2+}-mediated effect on glycogenolysis shuts off after several minutes once a large supply of pyruvate becomes available to the mitochondria, and the proportion of PHOS in the *a* form declines to close to resting levels (~10%) with a corresponding increase in the *b* form (Chasiotis 1983; Conlee et al. 1979; Richter et al. 1982). In addition to the activation of PHOS by Ca^{2+} and an initial rise in pH, there are many other factors that control the breakdown of glycogen to pyruvate, including increases in adenosine monophosphate (AMP), inorganic phosphate, ammonium, and fructose-1,6-biphosphate, all of which modulate the activity of PFK. The precise mechanisms by which phosphorylase activation reverses during prolonged, continuous exercise have not been fully elucidated. However, the reversal is, in part, due to release of phosphorylase from the glycogen particle as glycogen breaks down, thus uncoupling phosphorylase kinase and the Ca^{2+}-activating mechanism from phosphorylase (Constable et al.

1986), and also inhibition of phosphorylase kinase by glucose-1-phosphate.

In rodents, the level of glycogen content exerts a regulatory effect on glucose uptake during contraction (Hespel & Richter 1990; Richter and Galbo 1986), with the effect of glycogen occurring at the glucose transport step (Hespel & Richter 1990). However, muscle glycogen levels also influence contraction-induced GLUT4 translocation to the surface membrane (Derave et al. 1999, 2000). In humans, evidence for regulation of muscle glucose uptake during exercise by glycogen availability is equivocal (Richter et al. 2001), possibly due to different exercise/diet protocols employed to manipulate pre-exercise glycogen levels and their subsequent effects on hormonal and substrate availability, as well as experimental methodology (i.e., a-v balance techniques vs. isotopic tracers). In this regard, pre-exercise glycogen content only influences glucose uptake during subsequent exercise when the delivery of substrates and hormones remains constant (Steensberg et al. 2002).

Effect of Endurance Training on Carbohydrate Metabolism

Regularly performed endurance training (i.e., more than three days per week) induces striking adaptations in skeletal muscle metabolism that modify the rate at which various fuels are used during exercise. Specifically, during exercise performed at the same absolute intensity, endurance training leads to a greater utilization of lipid-based fuels and a concomitant "sparing" of CHO-based fuels (Coggan et al. 1990; Hurley et al. 1986; Kiens et al. 1993; Klein et al. 1994; Phillips et al. 1996b). The decrease in CHO utilization is due to both a decrease in muscle glycogenolysis (Azevedo et al. 1998; Green et al. 1995; Kiens et al. 1993) and a decrease in muscle glucose utilization (Coggan et al. 1990; Mendenhall et al. 1994; Phillips et al. 1996c; Richter et al. 1998). Training also significantly reduces the rate of gluconeogenesis; this effect is mostly due to a slower rate of hepatic glycogenolysis (Coggan et al. 1995). Although the "sparing" of muscle glycogen accounts for most of the training-induced reduction in whole-body CHO utilization, this is mostly because muscle glycogen is the major CHO-based fuel for muscle metabolism both before and after training (Coggan et al. 1990, 1992; Mendenhall et al. 1994; Turcotte et al. 1992). Indeed, if submaximal exercise is continued for >30 min, the training-induced reduction in muscle glucose uptake becomes quantitatively as important as the reduction in muscle glycogenolysis (Coggan et al. 1990, 1992; Mendenhall et al. 1994; Turcotte et al. 1992). In spite of a training-induced increase in total GLUT4 concentration, the reduction in glucose uptake is partly due to a blunted exercise-induced translocation of GLUT4 protein to the sarcolemma, which in turn leads to a diminished exercise-induced sarcolemmal glucose transport capacity (Richter et al. 1998). Regardless of the precise mechanisms, the slower utilization of CHO-based fuels (particularly muscle glycogen) during submaximal exercise in trained individuals is an important means by which training enhances endurance capacity.

Mechanisms by Which Exercise Regulates Carbohydrate Metabolism

The molecular mechanisms by which exercise directly increases glucose uptake and metabolism are elusive. Diverse efforts are currently being made to elucidate the cellular and molecular mechanisms by which exercise, and specifically muscle contraction, regulate CHO metabolism. Many of the current concepts regarding the regulation of glucose uptake in skeletal muscle have been derived from efforts in the insulin-signaling field. Mammalian homologues of fuel sensors from yeast genetics, such as AMP-activated protein kinase AMPK, are also considered.

The Insulin Signaling Cascade

The insulin signaling pathways have provided an entry point into studies dealing with the regulation of glucose uptake in skeletal muscle, and efforts have been made to determine if any of the components of the insulin signaling cascade mediate exercise-induced responses on glucose uptake (figure 8.3). While mechanisms governing insulin signaling have been reviewed in detail (Saltiel & Kahn 2002), a brief overview is warranted.

The insulin receptor is a heterotetrameric membrane glycoprotein composed of two α and two β subunits, linked together by disulfide bonds. Insulin binds to the extracellular α subunits and this leads to activation of the transmembrane β subunits and autophosphorylation of the receptor. The predominant mechanism by which the insulin receptor transmits further intracellular signaling to metabolic and mitogenic events involves the recruitment of insulin receptor substrate (IRS) docking proteins. IRS isoforms (IRS-1 to -4), Gab-1, and Cbl are critical mediators of the insulin signal and link the initial event of this signaling

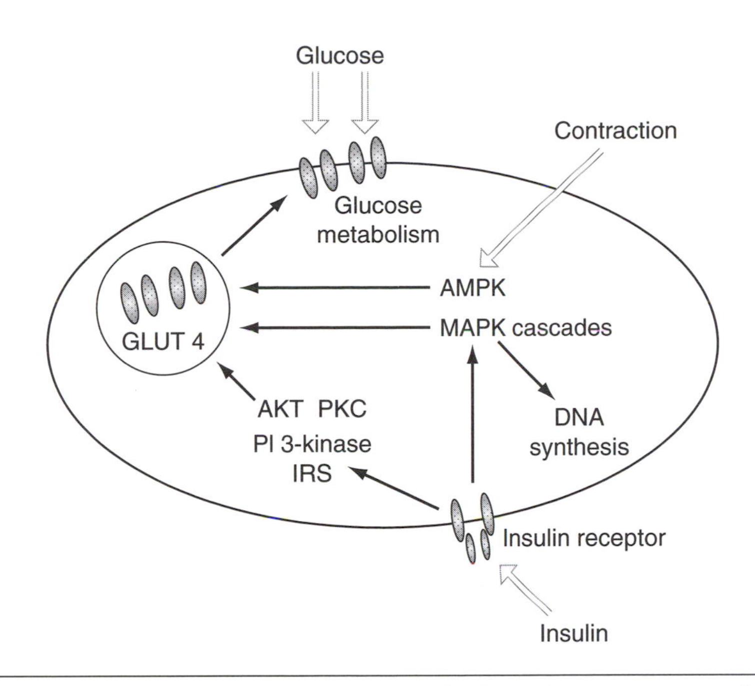

Figure 8.3 Schematic description of cellular signaling events in skeletal muscle in response to either insulin or contraction. Insulin-stimulated glucose transport is mediated though the insulin receptor IRS-1/2 and PI 3-kinase pathway. Acute effects of exercise on glucose uptake are insulin independent and are partly mediated through the AMP-activated protein kinase (AMPK) pathway.

cascade to downstream events. IRS proteins contain multiple tyrosine phosphorylation sites that become phosphorylated after insulin stimulation. Tyrosine-phosphorylated IRS-1 associates with phosphatidylinositol 3-kinase (PI 3-kinase), which plays an important role in insulin-stimulated glucose transport and GLUT4 translocation in skeletal muscle. PI 3-kinase catalyzes the formation of phosphatidylinositol-3,4,5-trisphosphate, which serves as an allosteric regulator of phosphoinositide-dependent kinase, leading to the activation of downstream signaling proteins including protein kinase B (PKB/AKT) and the atypical protein kinase C (PKC) isoforms PKCζ and PKCλ. While there is still a limited understanding of the nature of the signaling molecules downstream of AKT and PKC that link insulin signaling to glucose transport, clearly multiple signaling networks converge on glucose transport and contribute to the regulation of glucose uptake in insulin-sensitive tissues (Zierath et al. 2000). The MAPK cascade constitutes a parallel pathway of the insulin signaling cascade and mediates cell growth and mitogenesis.

Exercise Effects on Glucose Transport

Glucose transport in skeletal muscle is increased immediately after an acute bout of exercise. The exercise-mediated glucose uptake occurs through an unknown insulin-independent translocation of GLUT4 to the cell surface (Lund et al. 1995). Thus, immediate effects of acute exercise on glucose homeostasis occur primarily at the level of an elusive signaling cascade that mediates GLUT4 traffic, rather than through the established insulin signaling pathway mediating signal transduction at the level of insulin receptor IRS-1, IRS-2, or PI 3-kinase (Lund et al. 1995; Wojtaszewski et al. 1997). Revealing the nature of the insulin-independent, exercise signaling pathway to glucose transport has been both a challenge and an opportunity for biological scientists. The identification of the components of this pathway will reveal a novel regulatory mechanism of GLUT4 translocation in skeletal muscle.

AMPK

The molecular mechanisms by which exercise directly increases glucose uptake and metabolism are elusive. AMPK is one candidate that may mediate part of the exercise effect on glucose uptake and metabolism. AMPK also plays a role in the regulation of fatty acid metabolism.

AMP-Activated Protein Kinase

AMPK has been identified as one of the key signaling proteins in the contraction-mediated pathway to glucose transport. AMPK is a heterotrimeric protein, composed of one catalytic (α) and two noncatalytic (β and γ) subunits (Winder 2001). In mammalian cells, two isoforms of the α and β subunits and three isoforms of the γ subunit have been identified. To achieve full activation of the kinase, the β and γ subunits are required to form a complex with the α subunits (Woods et al. 1996). Regulation of AMPK activity involves several mechanisms. AMPK activity can increase in response to an increase in the AMP/ATP ratio and a decrease in the PCr/Cr ratio and in response to phosphorylation of the AMPK kinases (Ruderman et al. 1999). When AMP binds to residues on the α and β unit of AMPK, these subunits are cross-linked together and expose an AMPK kinase phosphorylation site, leading to the activation of AMPK. AMPK activity does not appear to be increased in response to insulin. AMPK has multiple roles in mediating metabolic and mitogenic events (figure 8.4).

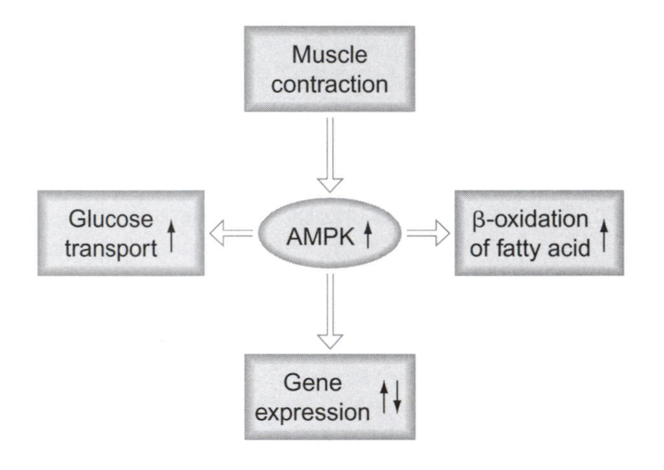

Figure 8.4 Schematic description of AMP-activated protein kinase (AMPK) signaling events in response to muscle contraction. The AMPK pathways may constitute a signaling network in the regulation of glucose and fatty acid metabolism, gene transcription, and protein synthesis in response to exercise.

AMPK in the Regulation of Glucose Metabolism

The most compelling evidence to support a role for AMPK in the regulation of glucose transport comes from correlative studies. An increase in AMPK activity in response to muscle contraction or exercise has been correlated with GLUT4 translocation and glucose transport in skeletal muscle (Bergeron et al. 1999, 2001; Hayashi et al. 1998, 1999; Kurth-Kraczek et al. 1999; Merrill et al. 1997). However, results from studies of transgenic mice overexpressing a negative dominant inhibitory AMPK challenge the hypothesis that AMPK mediates contraction-induced glucose uptake. Transgenic overexpression of a dominant inhibitory mutant of AMPK in skeletal muscle completely blocks the ability of hypoxia to activate glucose uptake, while only partially reducing contraction-stimulated glucose uptake (Mu et al. 2001). This highlights that hypoxia and 5-aminoimidazole-4-carboxamide ribonucleoside (AICAR; an adenosine analog that can be taken up into intact cells and phosphorylated to form 5-aminoimidazole-4-carboxamide ribonucleotide, the monophosphorylated derivative that mimics the effects of AMP on AMPK) increase glucose transport via an AMPK-dependent pathway, while exercise/muscle contraction recruits an AMPK-independent pathway. This is especially intriguing since exercise, hypoxia, and AICAR all increase glucose uptake via an insulin-independent pathway and none of these stimuli leads to an additive effect on glucose uptake. Thus in addition to insulin-mediated pathways, AMPK-dependent and AMPK-independent pathways contribute to the regulation of glucose uptake in skeletal muscle.

AMPK in the Regulation of Fatty Acid Oxidation

Increased AMPK activity has also been correlated with increased FA oxidation in skeletal muscle (Bergeron et al. 1999), decreased lipogenesis and lipolysis in adipocytes (Sullivan et al. 1994), and decreased FFA and cholesterol synthesis in hepatocytes (Henin et al. 1995). Activation of AMPK during muscle contraction is associated with phosphorylation of the β-isoform of coenzyme A carboxylase (ACC-β), which leads to the inhibition of ACC activity and a reduction in the malonyl-CoA content, thereby de-repressing carnitine palmitoyltransferase-1 activity and increasing FA oxidation (Yamauchi et al. 2002). Thus, AMPK may function in the regulation of glucose uptake and FA oxidation through increased GLUT4 translocation and inhibition of Acetyl-CoA carboxylase, respectively.

Physiological Aspects of AMPK

In response to exercise, there is an isoform-specific and intensity-dependent change in AMPK activity in skeletal muscle (Fujii et al. 2000; Wojtaszewski et al. 2000; Yu et al. 2002). Low- to moderate-intensity exercise induces an isoform-specific and intensity-dependent increase in AMPK α_2 but not AMPK α_1 activity in moderately trained subjects (Fujii et al. 2000; Wojtaszewski et al. 2000). However, both AMPK α_1 and α_2 activity are increased in response to sprint exercise (Chen et al. 2000). As AMPK α_1 activity remains at resting levels after prolonged, continuous, low-intensity (44% of $\dot{V}O_2max$) cycling leading to exhaustion and glycogen depletion (Wojtaszewski et al. 2000), this strongly suggests that activation of the AMPK α_1 isoform is related to the *rate* of fuel utilization rather than the *magnitude* of substrate depletion. These exercise intensity differences may also be related to the finding that AMPK complexes containing the α_2 rather than the α_1 isoform have a greater dependence on AMP (Chen et al. 2000; Salt et al. 1998). One interesting feature of AMPK is its role in mitochondrial biogenesis in skeletal muscle (Zong et al. 2002), and this may provide a mechanism by which exercise training increases mitochondrial capacity. Therefore, AMPK may be the initial signaling step linking the acute metabolic responses to exercise (i.e., increased glucose transport and FA oxidation) to the chronic adaptive responses of increased mitochondrial biogenesis.

Candidates for Exercise-Mediated Changes in Protein Expression

The master regulators of exercise responses on gene expression have not been completely defined. Changes in gene expression in response to exercise training may be mediated via AMPK and MAPK signaling to downstream substrates. AMPK and MAPK pathways provide a putative molecular mechanism for exercise-induced transcriptional regulation in skeletal muscle.

Effects of Chronic Exercise on Protein Expression

Even though the acute effects of exercise on glucose transport are independent, several hours after an exercise bout there is a persistent increase in insulin sensitivity to glucose transport into skeletal muscle. Even more striking are the effects of exercise on gene expression. Exercise training increases insulin-mediated whole-body glucose disposal (Dela et al. 1995; Houmard et al. 1991; Hughes et al. 1993), primarily through increased cellular glucose transport in skeletal muscle (Chibalin et al. 2000; Ren et al. 1994). This effect is correlated with increased protein expression of GLUT4 (Chibalin et al. 2000; Dela et al. 1995; Houmard et al. 1991; Hughes et al. 1993; Ren et al. 1994; Yu et al. 2001b), as well as with adaptive responses in expression and function of key insulin signaling molecules including IRS-1, IRS-2, and PI 3-kinase (Chibalin et al. 2000; Houmard et al. 1999; Kirwan et al. 2000; Yu et al. 2001a).

AMPK in the Regulation of Gene Expression

Clearly, multiple mechanisms contribute to the regulation of insulin action and protein expression. In addition to the metabolic actions of AMPK, this kinase is also important in the regulation of gene expression. AMPK complexes containing the $\alpha2$ isoform can be directly targeted to the nucleus (Salt et al. 1998), and this may initiate a series of events leading to changes in gene expression. AMPK participates in transcriptional regulation by repressing genes involved in the glucose signaling system in hepatocytes (Salt et al. 1998; Woods et al. 2000) and up-regulating genes involved in glucose uptake and substrate metabolism in skeletal muscle (Holmes et al. 1999; Ojuka et al. 2000; Winder et al. 2000). As previously noted, activation of AMPK mimics several classic exercise-mediated responses on gene expression including increases in GLUT4 mRNA and protein content, hexokinase II mRNA and activity, uncoupling protein 3 mRNA, mitochondrial enzymes, and glycogen content in skeletal muscle (Holmes et al. 1999; Ojuka et al. 2000; Song et al. 2002; Winder et al. 2000; Zhou et al. 2000; Zong et al. 2002). In skeletal muscle from insulin-resistant diabetic rodents, activation of AMPK by repeated daily AICAR injections for seven days is associated with increased hexokinase II and GLUT4 protein expression, as well as in vitro myocyte enhancer factor 2 (MEF2) sequence-specific binding activity (Song et al. 2002). A similar increase in MEF2 sequence-specific binding activity has also been observed in human skeletal muscle after marathon running (Yu et al. 2001b). Thus, increased MEF2 sequence-specific binding activity may confer exercise-specific changes in gene expression. Consistent with this hypothesis, the MEF2 site appears to be essential for GLUT4 expression, since deletions or point mutations within the MEF2 consensus binding sequence of the human GLUT4 promoter completely prevent tissue-specific and hormonal/metabolic regulation of GLUT4 (Thai et al. 1998). Another target of the AMPK pathway is peroxisome proliferator-activated receptor-γ coactivator 1 (PGC-1α) (Zong et al. 2002). PGC-1α stimulates the expression of the nuclear respiratory factors and mitochondrial transcriptional factor A, and these responses are linked to activation of nuclear and mitochondrial genes encoding mitochondrial proteins (Wu et al. 1999). Thus, in addition to the metabolic effects,

AMPK may be a central player in the regulation of gene expression in response to exercise training.

MAPK Cascades in the Regulation of Gene Expression

MAPK activation is an important mechanism governing cellular proliferation and differentiation in many cell types, and there is evidence for activation of MAPK signaling cascades in response to muscle contraction and exercise (Widegren et al. 2001). Members of the MAPK family form at least three parallel signaling cascades that include the extracellular regulated kinase (ERK1/2 or p42 and p44 MAPK), p38 MAPK, and c-jun kinase. MAPK signaling cascades are directly activated in human skeletal muscle in response to acute, short-term exercise (Aronson et al. 1997; Krook et al. 2000; Widegren et al. 1998, 2000; Yu et al. 2002) or endurance running (Boppart et al. 2000; Yu et al. 2001b). Downstream substrates of ERK and p38 MAPK signaling cascades, including MAPK-activated protein kinase (MAPKAP-K-) 1 and 2 and the mitogen- and stress-activated kinase (MSK) 1 and 2, are increased immediately after acute sprint (Krook et al. 2000) or endurance exercise (Yu et al. 2001b). Elucidation of substrate specificity along the ERK or p38 pathways in human skeletal muscle is challenging, since these signaling cascades function as networks rather than linear pathways, with a high degree of cross-talk between the upstream activator and downstream substrates in response to cellular stress. Substrate specificity for MAPK signaling cascades has been determined in response to contraction in isolated rat epitrochlearis muscle, combined with the use of chemical inhibitors of ERK and p38 MAPK (Ryder et al. 2000). Thus, contraction-induced induction of MAPKAP-K1 and MAPKAP-K2 occurs via separate pathways, reflecting ERK and p38 MAPK stimulation, respectively. In contrast, induction of MSK1 and MSK2 requires simultaneous activation of ERK and p38 MAPK (Ryder et al. 2000). Even more challenging is establishing a direct link between MAPK activation and changes in gene expression in skeletal muscle after exercise (figure 8.5).

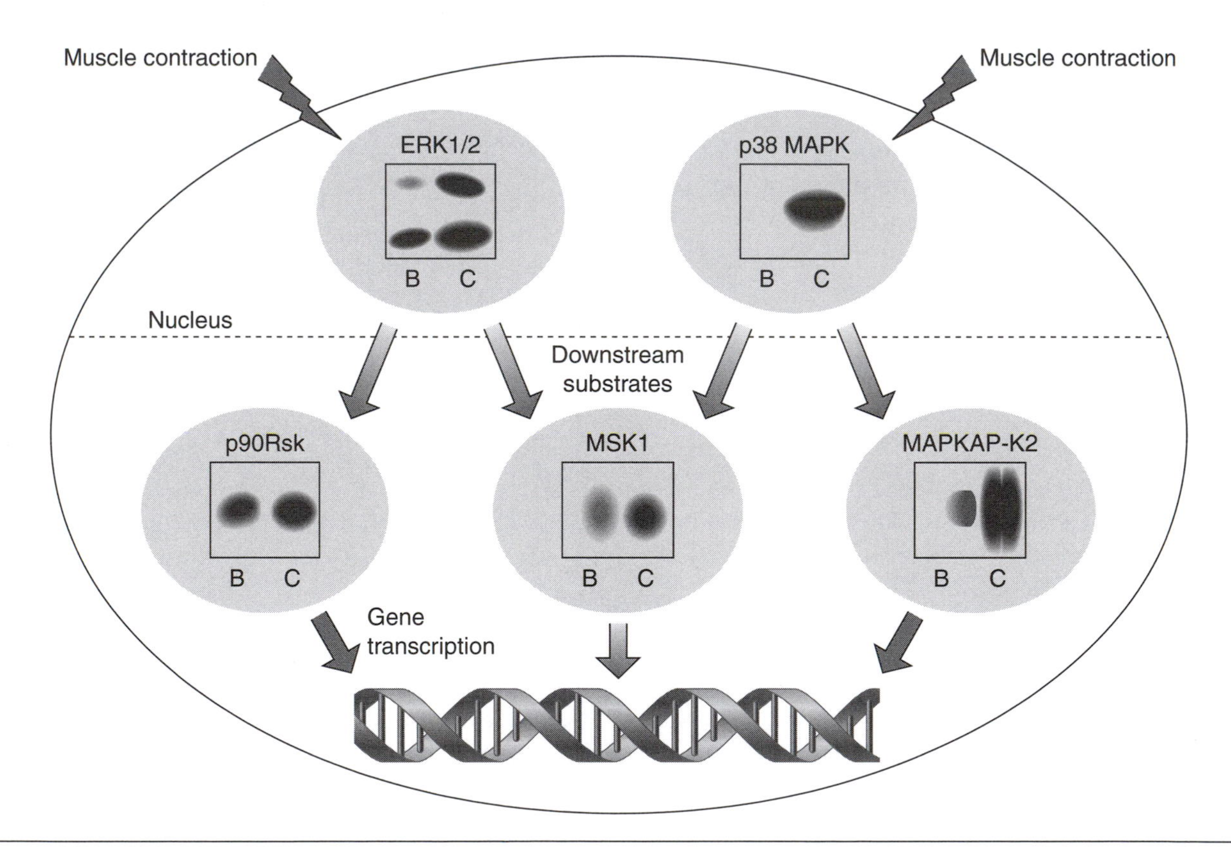

Figure 8.5 Exercise increases phosphorylation or activity of ERK MAPK and p38 MAPK and downstream substrates. Filled arrows indicate the direct substrates of ERK or p38, respectively, in response to contraction as determined using pathway-specific MAPK inhibitors. Shaded arrows indicate a putative relationship between MAPK substrates and transcriptional machinery. The MAPK pathways may constitute a signaling network in the regulation of gene transcription and protein synthesis in response to exercise.

Adapted from Krook et al. 2000; Ryder et al. 2000; Widegren et al. 2000; Yu et al. 2001.

Summary

The signaling pathway(s) that lead to changes in glucose transport and gene expression in response to either insulin or exercise have not been fully elucidated. Accordingly, a major aim of future research in this area will be to clarify the components of, and the interrelationships between, these signaling networks. Comparative analysis of such networks identified through inhibitor studies, use of genetically modified model organisms, and translational studies in humans can be used to assign the physiological role(s) for components of known and novel pathways that regulate CHO and fat metabolism, as well as gene expression, in response to exercise. Since many of the approaches used thus far have taken advantage of cell-based studies, or genetically manipulated animal models in which selective candidate genes have been either overexpressed or ablated, approaches using applied physiology will be essential to validate the identified components of the proposed signaling pathways to elucidate the role of these genes in the regulation of metabolism and gene expression. These efforts can be facilitated through technology platforms developed by both the academic and biotechnology sector. A greater understanding of the physiological response to exercise will be important, as there has been a drastic worldwide increase in the incidence of metabolic disease such as obesity and non-insulin-dependent diabetes mellitus.

Recent Advances in Long Chain Fatty Acid Transport and Metabolism*

In the past 25 years our understanding of the complexities of skeletal muscle carbohydrate metabolism and the molecular basis of glucose uptake at rest, during exercise, and in disease states has greatly increased. In contrast, our understanding of the regulation of skeletal muscle long chain fatty acid (LCFA) uptake and lipid metabolism has been much slower to develop, although in very recent years there has been a substantially better understanding of these processes. The purpose of this section of the chapter is to review basic mechanisms of LCFA transport into the muscle and its subsequent metabolism, including oxidation, incorporation into the triacylglycerol (IMTG) depots, and hydrolysis of these depots (figure 8.6). In addition, the effects of exercise, training, and obesity/diabetes on these key metabolic steps are commented upon as appropriate. The intention is not to review in detail all aspects of FA metabolism. Discussion of other topics relevant to lipid metabolism, such as gender differences, fasting, and high-fat diets, is beyond the scope of this discussion. For these topics, the reader is directed to several excellent reviews (Van der Vusse & Reneman 1996; Tate & Holtz 1998; Turcotte 2000; Watt et al. 2002b; Jeukendrup et al. 1998a,b,c).

LCFA Transport Into the Muscle Cell

For many years it was thought that LCFAs entered muscle cells via passive diffusion (Hamilton & Kamp 1999), and that therefore the uptake into the tissue was determined by the delivery (concentration × blood flow) of LCFAs to the tissue (Hagenfeldt 1979). While a portion of the LCFAs undoubtedly enter the cell via passive diffusion, in the past decade considerable evidence has also accumulated showing that LCFAs enter cells via a protein-mediated mechanism involving a number of FA binding proteins (Bonen et al. 2002; Schaffer 2002). This is not dissimilar to the regulated entry of substrates such as lactate and glucose, which can also diffuse into the muscle cell but for which well-defined transport proteins have now been identified. The protein-mediated uptake of LCFAs does not likely occur in quite the same manner as for glucose and lactate, whose transporters form pores in the sarcolemma, through which these substrates enter the cell. Instead, it is believed that protein-mediated LCFA uptake occurs through a linked sequence of binding proteins, although direct experimental evidence for such a process is currently lacking. Nevertheless, this binding system exhibits characteristics that are similar to those of the classic substrate transport systems (see later). This observation enables us, for convenience, to also term the protein-mediated LCFA uptake LCFA transport.

Characterization of a Model for Examining LCFA Transport in Skeletal Muscle

Examination of LCFA transport had been hampered by lack of an appropriate model system. Studies of LCFA transport in adipocytes, hepatocytes, and cardiac myocytes were confounded by

* The text on this page beginning with the heading *Recent Advances in Long Chain Fatty Acid Transport and Metabolism* and ending with the text under the heading *Summary* on page 168 is contributed by David J. Dyck and Arend Bonen.

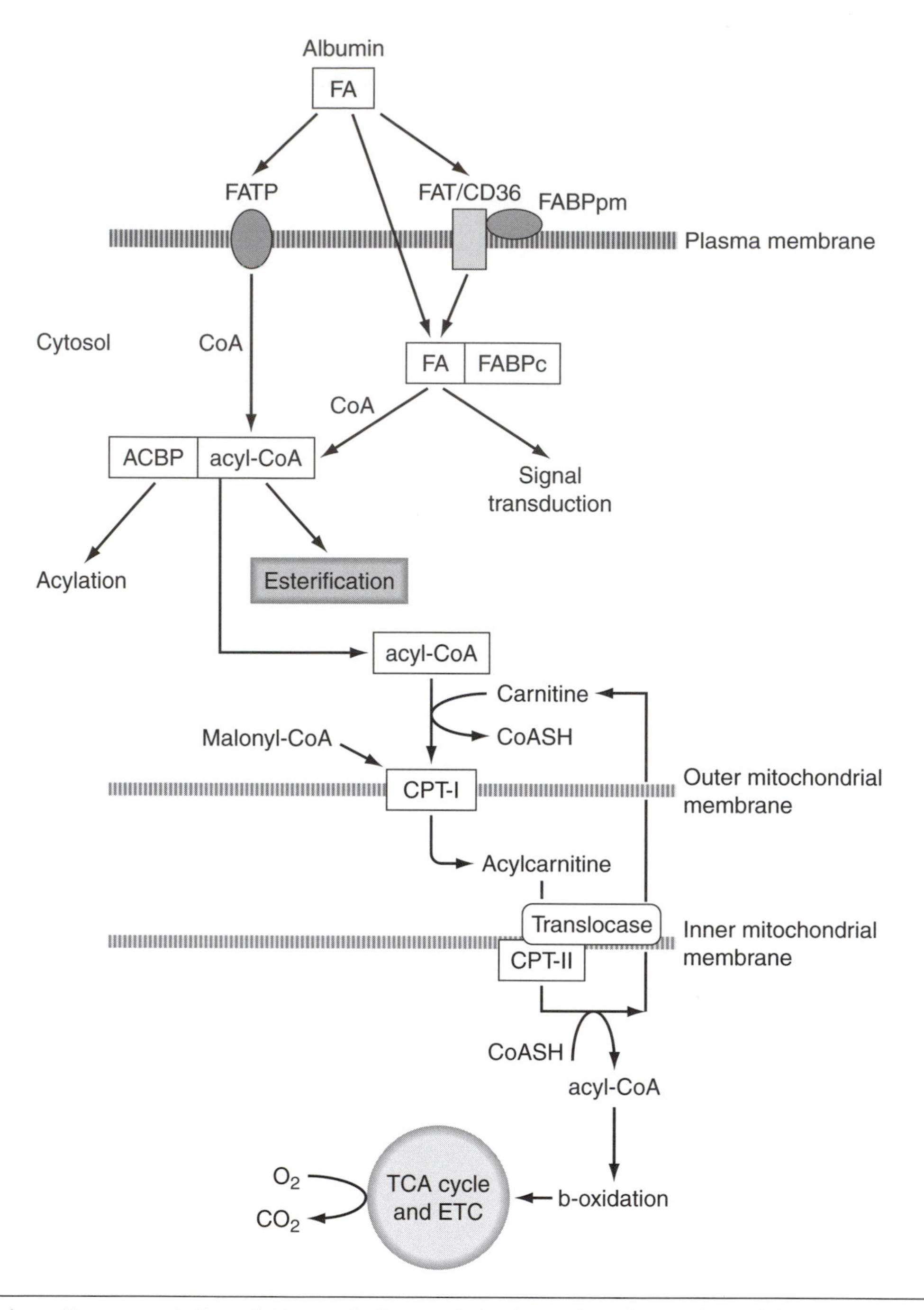

Figure 8.6 Schematic representation of FA metabolism in skeletal muscle. Albumin-bound fatty acids (FA) are released and taken into the muscle across the plasma membrane via one or more FA transporters (FTP, FAT/CD36, FABPpm) and by diffusion. Once in the cytosol, FA are bound to cytosolic FA binding protein (FABPc), are complexed to FA-CoA, and are bound to FA acyl-CoA binding protein (ACBP), leading to acylation, esterification, and oxidation. The latter route requires entry into the mitochondria by forming FA-acylcarnitine and being carried across the outer mitochondrial membrane via carnitine palmitoyltransferase-1 (CPT-1); at the inner mitochondrial membrane, the process is reversed with carnitine being shuttled back into the cytosol and FA-acyl-CoA being committed to β-oxidation forming CO_2.

the concurrent LCFA metabolism (i.e., oxidation and esterification). To resolve this difficulty, giant sarcolemmal vesicles obtained from muscle were used (Bonen et al. 1998a). Characterization studies of these vesicles have shown that they are oriented 100% right side out, have a large diameter (10-15 μm), and contain an intact sarcolemma with many proteins, including transport proteins (Bonen et al. 1998a). Importantly, for functional studies of LCFA transport, these giant vesicles also contain excess quantities of the 16-kDa cytosolic FA binding protein (H-FABPc), which acts as a cytosolic FA sink once FA have crossed the plasma membrane (Bonen et al. 1998a).

In the giant sarcolemmal vesicles there was no evidence of any FA metabolism, either oxidation or esterification; and all of the FA taken up into the giant vesicles was recovered as FA in the lumen of the vesicle, as none was associated with the plasma membrane (Bonen et al. 1998a). It was also demonstrated that vesicular palmitate transport was inhibitable by protein-modifying agents (phloretin, trypsin), by inhibitors of FA transport proteins (FABPpm antisera, sulfo-*N*-succinimidyl oleate [SSO]), by reduced assay temperatures, and by excess long chain fatty acid (oleate) (Bonen et al. 1998a). As expected, palmitate transport was not altered by glucose (Bonen et al. 1998a). Finally, the rates of FA transport in heart and in red and white skeletal muscles (heart >> red > white) reflect the known differences in the rates of FA metabolism in these tissues (Bonen et al. 1998a).

Collectively, these studies have shown that giant sarcolemmal vesicles obtained from rodent skeletal muscle provide a suitable preparation for the study of FA uptake in the absence of FA metabolism. Moreover, these studies (Bonen et al. 1998a) have also shown that in skeletal muscle, FA are taken across the plasma membrane largely via a protein-mediated mechanism. Additional evidence for a regulatable LCFA transport system has also been obtained in more recent studies (see later).

Identification of Fatty Acid Transporters

A number of FA binding proteins appear to carry LCFAs across the plasma membrane. Among these are the 40-kDa peripheral membrane fatty acid binding protein (FABPpm) (Isola et al. 1995), the family of the 63-kDa FA transport proteins, FATP1-5 (Schaffer & Lodish 1994; Hirsch et al. 1998), and the 88-kDa heavily glycosylated FA translocase (FAT/CD36, the rat analog of the human CD36) (Abumrad et al. 1993). FATP1, FAT/CD36, and FABPpm mRNAs are each expressed in rodent (Luiken et al. 2001) and human muscle (Bonen et al. 1999b). Importantly, for each of these transporters there is evidence that they can independently increase FA transport when overexpressed in heterologous cells (Abumrad et al. 1993; Schaffer & Lodish 1994; Isola et al. 1995). However, their precise mechanism of action in the transmembrane transport process and their exact roles in biological tissues are not yet fully understood. In recent years, most of the knowledge concerning LCFA transport has been developed for FAT/CD36 and FABPpm, proteins for which suitable antibodies are available.

Linking Fatty Acid Transport to Fatty Acid Transporters

The rates of FA transport have been compared with the content of FA transporters at the plasma membrane, the site where transporters exert their physiologic role. Among muscle tissues, the maximal FA transport rates, as assessed in giant vesicles derived from these tissues, differed substantially (heart >> red > white muscle) (Bonen et al. 1998a; Luiken et al. 1999). Importantly, these maximal rates of transport in these muscle types paralleled the well-known differences in the rates of FA metabolism (esterification and oxidation) among these tissues (Lopaschuk & Saddik 1992; Dyck et al. 1997). There was also a good relationship between either plasma membrane FAT/CD36 or plasma membrane FABPpm and FA transport (Luiken et al. 1999). Interestingly, there was an inverse relationship between FATP1 and FA transport (Luiken et al. 1999). However, one cannot exclude the possibility that these observations for FATP1 may have been due in part to the problems encountered. In this study the FATP1 antibody did not work well in biologic tissues, although it was adequate in some cell lines (unpublished data). Others have questioned the transport function of FATP1, believing it to be a very long chain fatty acyl-CoA synthetase (Watkins et al. 1998; Coe et al. 1999). However, very recent studies, in which the FATP1 structure has been altered using site-directed mutagenesis, have shown that FATP1 has both a transportlike function and a very long chain fatty acyl-CoA synthetase function in *Saccharomyces cerevisiae* (Zou et al. 2002). Based on our studies in rodents, it seems that in heart and muscle, both FAT/CD36 and FABPpm are physiologically important transporters in moving FA across the plasma membrane. It has been proposed, based on some indirect evidence, that FAT/CD36 and FABPpm may cooperate to transport FA across the plasma membrane (see Luiken et al. 1999), but more direct experimental evidence is currently lacking.

Genetic alterations of LCFA transport proteins (FAT/CD36 and FABPpm) confirm the key role of these proteins in regulating LCFA transport into the muscle cell. In transgenic mice that overexpress FAT/CD36 in muscle, circulating FA levels were reduced and FA oxidation was increased, although only during muscle contraction (Ibrahimi et al. 1999), indicating that LCFA uptake was increased only when the muscle's energetic demands were increased. Conversely, in FAT/CD36 null mice, circulating FA levels are increased (Febraio et al. 1999) while the tissue uptake of the iodinated FA analog β-methyliodophenylpentadecanoic acid (BMIPP), an indirect measure of FA transport, is reduced in heart, muscle, and adipose tissue (Coburn et al. 2000). In recent studies in which FABPpm was overexpressed in muscle, LCFA transport was also increased (Clarke & Bonen unpublished data). Thus, collectively, these studies indicate that altering the expression of the FA transport protein FAT/CD36 or FABPpm results in parallel changes in skeletal muscle FA uptake rates.

Regulation of Fatty Acid Transport by Fatty Acid Transporters

Once it had been demonstrated that LCFA transport across the sarcolemma was protein mediated, it was important to ascertain whether LCFA transport rates could be altered acutely (i.e., within minutes) or chronically (e.g., adaptive responses to training and in metabolic diseases). FA transport may be regulated by transport proteins in a number of ways; these include (a) altering the expression of the FA transporters, (b) altering their subcellular distribution, or (c) possibly altering their activity at the plasma membrane. Since protein expression levels are likely not changed within a very short period of time (<30 min), the acute regulation of FA transport (i.e., within minutes) presumably occurs through one or both of the latter two mechanisms, while chronic adaptive responses taking place after a few hours or days most likely occur through the first or second mechanism, or both.

Acute Regulation of LCFA Transport by Muscle Contraction and by Insulin

To examine whether LCFA transport was increased during muscle contraction, the sciatic nerve was electrically stimulated in order to cause all rat hindlimb skeletal muscle fibers to contract, without inducing fatigue (Bonen et al. 2000a). Subsequently, giant vesicles were prepared and transport studies were performed. This work showed that with muscle contraction, FA transport into giant vesicles was increased rapidly, in direct relation to the muscle contraction intensity (Bonen et al. 2000a). Kinetic studies of FA uptake showed that Vmax was increased while there was no change in Km. From fractionation studies of muscle tissue it was shown that FAT/CD36 is present both in an intracellular pool and in the plasma membrane (figure 8.7). Importantly, with muscle contraction, a portion of this intracellular pool was translocated to the plasma membrane (figure 8.7), thereby increasing FA uptake acutely. During the postcontraction recovery period, FA transport returned to control levels while concomitantly FAT/CD36 was internalized from the plasma membrane (Bonen et al. 2000a).

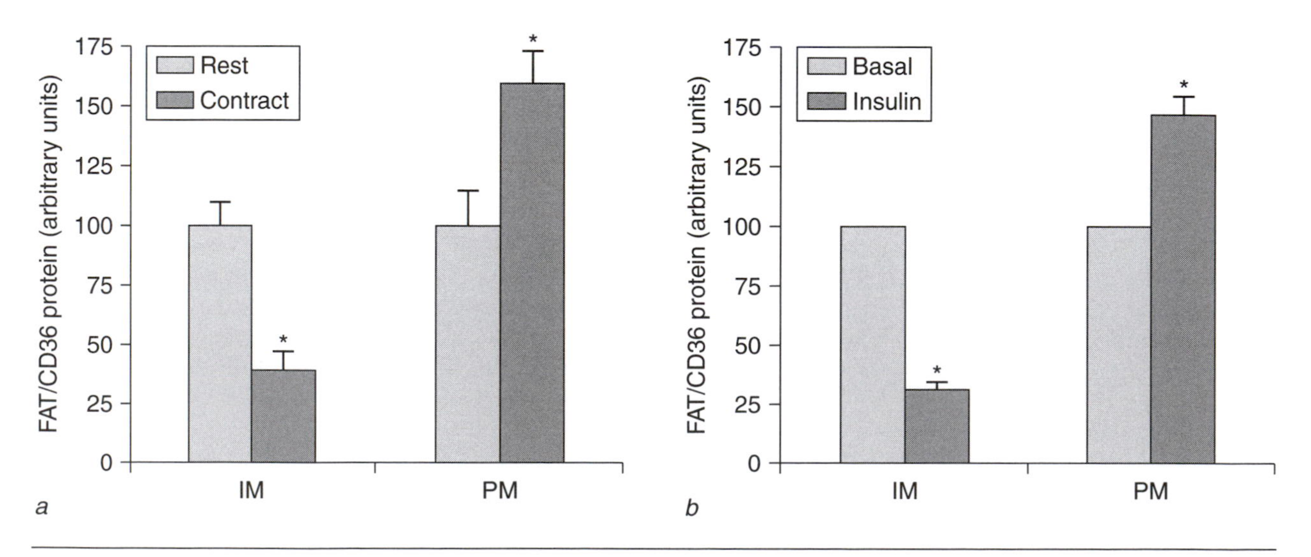

Figure 8.7 Effects of *(a)* muscle contraction (30 min) and *(b)* insulin (60 min) on the translocation of FAT/CD36 in rat muscle. PM, plasma membrane; IM, intracellular membrane. *$P < 0.05$, contraction versus rest, or insulin versus basal.
Reprinted, by permission, from Bonen et al., 2000, and Luiken, Dyck et al. 2002.

In more recent work, rat hindlimb muscles were perfused with insulin. This markedly increased the total quantity of LCFAs that entered the muscle due to a markedly increased rate of esterification, despite a concomitant reduction in LCFA oxidation (Luiken et al. 2002b). This augmented LCFA uptake was attributable to an insulin-stimulated translocation of FAT/CD36 from an intracellular pool to the plasma membrane (figure 8.7). This appears to entail the postreceptor signaling pathway involving PI 3-kinase, since blocking PI 3-kinase activity prevented the insulin-induced translocation of FAT/CD36. In other studies it has been shown that the effects of insulin and muscle contraction on LCFA uptake into intact muscle are additive (Dyck et al. 2001); this may suggest that there are separate intracellular pools of FAT/CD36 that are sensitive to insulin- or contraction-activated signals. Such insulin- and contraction-sensitive intracellular transport protein pools have already been identified for GLUT4 (Goodyear & Kahn 1998).

Importantly, these studies (Bonen et al. 2000a; Luiken et al. 2002b) have revealed an entirely unsuspected mechanism, namely that FA transport rates into skeletal muscle are regulated acutely, within minutes, by muscle contraction (Bonen et al. 2000a) and by insulin (Luiken et al. 2002b). The regulation mechanism involves the cellular redistribution of FAT/CD36 from an intracellular depot(s) to the plasma membrane. However, whether FABPpm is redistributed by these physiologic stimuli is not known, although it has been observed that this protein also is present in an intracellular pool (Bonen et al. unpublished data).

Regulation of LCFA Transport With Chronically Altered Muscle Activity

It is well known that chronically increased muscle activity such as exercise training increases the capacity for LCFA oxidation. In addition, in metabolic diseases such as obesity and diabetes (type 1 and type 2) there is also a pronounced increase in LCFA metabolism, largely reflected by an increased storage of triacylglycerols, since LCFA oxidation is reduced in these individuals. This raised the question whether LCFA transport and transport protein expression were also coordinately up-regulated when muscle activity was altered.

To examine whether LCFA transport was increased with chronically increased muscle activity, rat muscles were stimulated with seven days of low-frequency electrical stimulation (Bonen et al. 1999a). This resulted in an increase in FAT/CD36 mRNA and protein (Bonen et al. 1999a) as well as in FABPpm mRNA and protein (Benton & Bonen unpublished data). Consequently, there was also an increase in plasma membrane FAT/CD36 and in LCFA transport, as assessed in giant sarcolemmal vesicles. Exercise training has also been shown to increase the FA transporters FABPpm (Kiens et al. 1997) and FAT/CD36 in humans (Tunstall et al. 2002), although changes in FA transport were not determined in these studies.

When muscle activity was eliminated by denervation (seven days), LCFA transport across the plasma membrane was reduced. This, however, was not attributable to a repression of the total pool of LCFA transporters. Instead, the reduction in LCFA transport was associated with a reduction in the LCFA transporters located at the sarcolemma (Koonen & Bonen unpublished data), suggesting that these proteins were resequestered into their intracellular depots when muscle activity was eliminated. This then indicated that just as for GLUT4, the cycling of LCFA transporters between the plasma membrane and the intracellular depots might be altered without changing the expression of the LCFA transporters. Specifically, in this inactivity model it appears that the rate of FAT/CD36 exocytosis is lower than the rate of endocytosis, resulting in the segregation of the LCFA transporters at the plasma membrane. In this manner LCFA transport rates may be altered without necessarily altering LCFA transporter protein levels.

Thus, it appears that when muscle activity is chronically altered there are corresponding changes in LCFA transport. Increased and decreased rates of LCFA transport are attributable to concomitant changes in the LCFA transporters at the plasma membrane. Importantly, these changes in sarcolemmal LCFA transporters may (chronically increased muscle activity) or may not (denervation) be associated with altered levels of LCFA transport expression in the muscle, indicating that chronically altered muscle activity can also alter the cycling of these LCFA transporters between the plasma membrane and their intracellular compartments. This also indicates that simply measuring the expression of LCFA transporter proteins or their mRNAs may not necessarily provide any information about the rates of LCFA transport.

Fatty Acid Metabolism in Skeletal Muscle

Beyond moving LCFAs across the plasma membrane, the metabolism of LCFAs is regulated by a variety of factors, including hormones, substrates, and muscle activity. The key metabolic pathways in muscle include the storage or incorporation of LCFA to intramuscular triacylglycerols (IMTG), as well as the β-oxidation of IMTGs and LCFA. Other aspects of lipid metabolism (e.g., peroxisomal lipid metabolism and membrane biosynthesis) are beyond the scope of this discussion.

Glucose-Fatty Acid Cycle

For many years, the glucose-fatty acid (G-FA) cycle (figure 8.8*a*) has been used to explain the apparent reciprocal relationship between lipid and CHO metabolism in skeletal muscle. In the 1960s, Randle et al. hypothesized that the observed impairments in glucose uptake and oxidation in obesity and diabetes were due to an elevation of plasma LCFA and its ensuing oxidation (Randle et al. 1963). Subsequent to increased LCFA availability and oxidation, there are elevations in the intracellular citrate and Acetyl-CoA/CoA ratio, which are potent in vitro inhibitors of phosphofructokinase (PFK) and pyruvate dehydrogenase (PDH), respectively. The consequence of such inhibitions would be a buildup of glucose-6-phosphate, which in turn would inhibit hexokinase. Thus, the underlying premise of the G-FA cycle was the notion that LCFA oxidation was primarily regulated by its own delivery to muscle and that the subsequent, specific loci of metabolic regulation existed on the CHO side. Although the observations leading to the formulation of the G-FA cycle theory originated from resting diaphragm and contracting heart muscles, Rennie and Holloszy (1977) also provided evidence supporting the existence of this cycle in perfused contracting rodent muscle. However, other studies failed to support the existence of the G-FA in contracting muscle in rodents (Dyck & Spriet 1994; Dyck et al. 1996a). Evidence for the existence of the G-FA cycle in humans is also controversial. The increased provision of circulating lipids generally results in reduced CHO utilization both at rest (Boden et al. 1994) and during exercise of moderate intensity, that is, 65% and 85% $\dot{V}O_2$max (Costill et al. 1977; Dyck et al. 1993; Vukovich et al. 1993; Dyck et al. 1996b; Odland et al. 1998b). However, the accumulation of muscle citrate and acetyl-CoA contents is not responsible for the observed decrease in glycogenolysis during exercise when lipid provision is enhanced (Dyck et al. 1993; Dyck et al. 1996b). Rather, reduced accumulation of free inorganic phosphate and AMP (Dyck et al. 1993; Dyck et al. 1996b) may have resulted in reduced post-transformational (i.e., allosteric) activation of glycogen phosphorylase, accounting for the observed glycogen sparing. Although it is not clear how the increased lipid provision resulted in a reduced accumulation of Pi and AMP, it should be noted that these metabolic changes are also observed during exercise following endurance training, which also results in reduced glycogen utilization at a given absolute power output (Phillips et al. 1996b).

Reverse Glucose-Fatty Acid Cycle

While there is little doubt that under conditions of rest and exercise, the utilization of CHO can be affected by the availability of circulating LCFA, there is also evidence that the fuel selection by muscle is influenced by CHO availability ("reverse" G-FA cycle) (figure 8.8*b*). Malonyl-CoA is a potent inhibitor of CPT-1, the enzyme regulating the flux of LCFA into the mitochondrial matrix. Malonyl-CoA is regulated by the relative activities of Acetyl-CoA carboxylase (ACC, which synthesizes malonyl-CoA from Acetyl-CoA) and malonyl-CoA decarboxylase (MCD, which catalyzes the decarboxylation of malonyl-CoA).

In rodent muscle, malonyl-CoA is increased by enhancing CHO availability (i.e., glucose and insulin) (Elayan & Winder 1991; Duan & Winder 1993) and is decreased in contracting muscle (Winder et al. 1989, 1990), presumably resulting in altered rates of FA oxidation. However, FA oxidation was not actually determined in these studies. In resting humans, the elevation of glucose and insulin also results in a reduction in FA oxidation, which corresponds to a decrease in muscle malonyl-CoA content (Bavenholm et al. 2000; Rasmussen et al. 2002a). Of interest is the observation that insulin fails to increase muscle malonyl-CoA and diminish FA oxidation when glucose uptake is maintained at a constant rate (Yee & Turcotte 2002), indicating that it is the actual CHO availability, and not insulin, that regulates FA metabolism.

In humans, the total oxidation of FA is reduced during intense (i.e., 85% $\dot{V}O_2$max) exercise (Romijn et al. 1993), which is only partially restored by the artificial elevation of plasma FA, indicating that there must also be a direct impairment of oxidation, presumably at the level of the mitochondrion. It was suggested that an increase in muscle malonyl-CoA may have been responsible for the decrease in FA oxidation. However, this was not actually measured; in fact, there are no reports that muscle malonyl-CoA content increases during exercise (i.e., corresponding to a decrease in FA oxidation). Spriet and colleagues were unable to

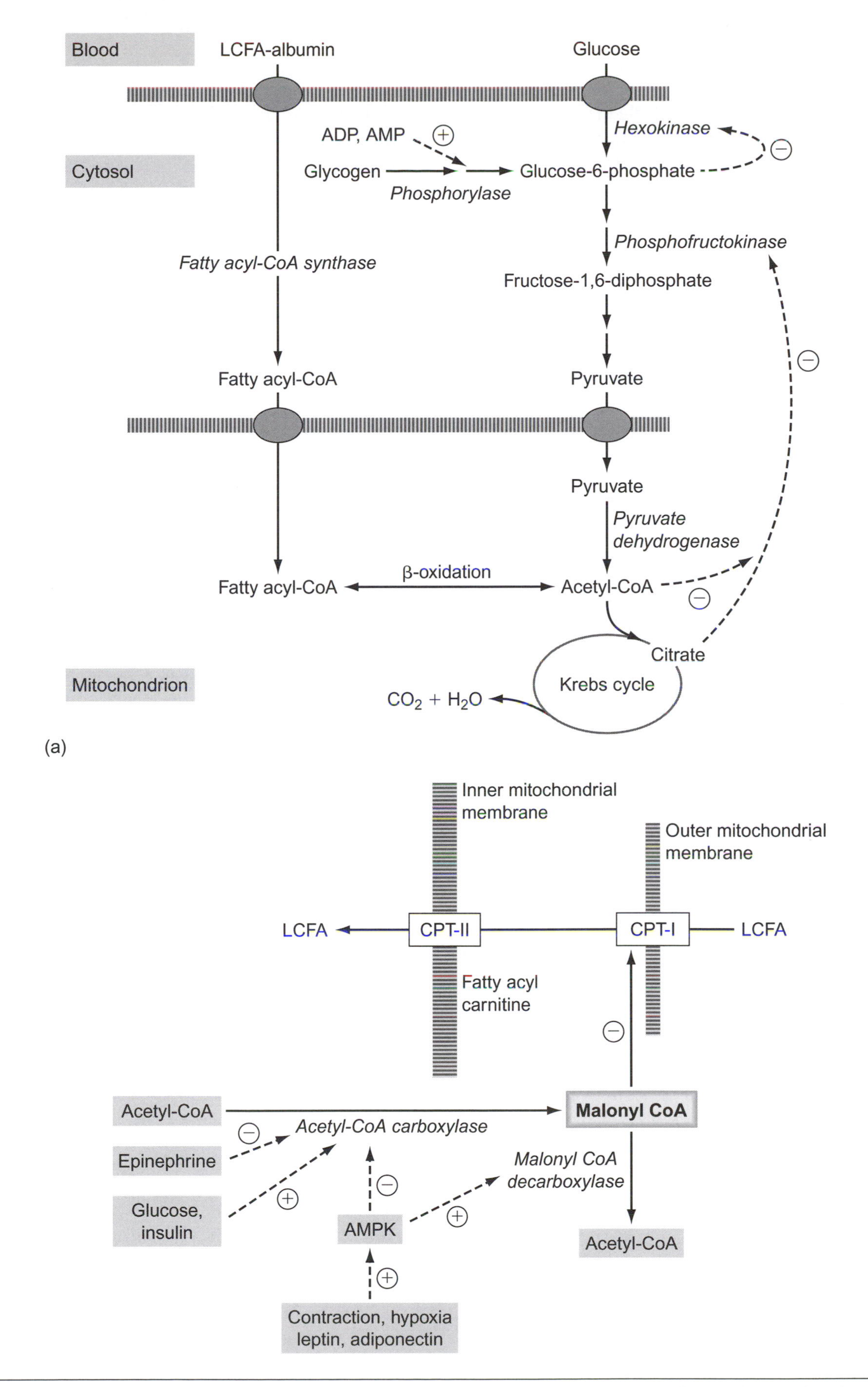

Figure 8.8 Schematic representation of *(a)* the glucose-fatty acid cycle, and *(b)* the reverse glucose-fatty acid cycle. LCFA, long chain fatty acid; CPT, carnitine palmitoyltransferase; AMPK, AMP-activated protein kinase.

demonstrate changes in muscle malonyl-CoA content over a range of exercise intensities up to 85% $\dot{V}O_2$max, in spite of the changes in the proportions of fat and CHO used at these intensities (Odland et al. 1996, 1998b). Dean et al. (2000) demonstrated a small but significant decrease in malonyl-CoA content during knee extensor exercise, which was not predictive of the shift toward CHO oxidation. Surprisingly, CPT-1 from trained muscle is more sensitive to changes in malonyl-CoA than from untrained muscle (Starritt et al. 2000). Taken together, these data suggest that the regulation of CPT-1 is more complex than presently understood, and in exercising humans likely involves factors other than malonyl-CoA.

AMP-Activated Protein Kinase

Recently, considerable research has focused on the role of AMP-activated protein kinase (AMPK) as an acute regulator of substrate metabolism in skeletal muscle. AMPK is primarily responsive to a decrease in the energy charge of the cell (i.e., increases in the AMP/ATP and Cr/PCr ratios) (Hardie & Carling 1997; Hardie et al. 1998), and has been shown to stimulate both glucose uptake and LCFA oxidation in muscle (Merrill et al. 1997; Alam & Saggerson 1998; Bergeron et al. 1999; Muoio et al. 1999b). Acute activation of AMPK using the cell-permeable compound, 5-amino-4-imidazolecarboxamide ribofuranoside (AICAR), has been shown to repartition FA toward oxidation and away from storage (Muoio et al. 1999b). Stimulation of AMPK inhibits ACC while simultaneously activating malonyl-CoA decarboxylase (MCD) (Saha et al. 2000; Kaushik et al. 2001; Park et al. 2002) in oxidative muscle, resulting in a decrease in malonyl-CoA content. Furthermore, AMPK-induced partitioning of FA toward oxidation and away from esterification is achieved in part by the inhibition of glycerol-3-phosphate acyltransferase (GPAT) in rodent muscle (Muoio et al. 1999b). This is in agreement with observations in rodent and human muscle during contraction, when the relative proportion (%) of FA directed toward incorporation into IMTG is greatly reduced despite the fact that there is actually a net increase in IMTG esterification per se (+30%) (Dyck & Bonen 1998; Lau et al. 2001; Sacchetti et al. 2002). The role of the AMPK-ACC-malonyl-CoA axis as a regulator of FA oxidation in resting muscle is likely to be significant, as shown by the previously cited studies documenting an increase in FA oxidation in resting muscle when stimulated by AICAR (and thus, AMPK). In addition, mice that do not express ACC2 (ACCβ), the predominant isoform in cardiac and skeletal muscle, demonstrate up to 30-fold lower contents of malonyl-CoA in cardiac and skeletal muscle and 30% greater FA oxidation rates in soleus muscle (Abu-Elheiga et al. 2001).

During exercise in humans, the significance of this axis is less clear. As previously noted, there has yet to be a study documenting changes in muscle malonyl-CoA content that are consistent with the changes in substrate utilization. Initial studies of the role of AMPK in exercising human muscle were able to demonstrate that AMPK α_2 activity was increased during exercise at 70% to 90% $\dot{V}O_2$max (Fujii et al. 2000; Wojtaszewski et al. 2000), but not at lower intensities such as 50% $\dot{V}O_2$max, an intensity at which fat oxidation would be expected to be a significant contributor to energy provision. In a more recent study (Stephens et al. 2002), the activity of AMPK α_2 progressively increased during exercise at 60% $\dot{V}O_2$max; this was related to an increase in whole-body fat oxidation, suggesting that AMPK activation may be an important regulator of fat oxidation at this power output. However, taken together these studies suggest a threshold for AMPK activation, that is, 60% $\dot{V}O_2$max, which is difficult to reconcile with numerous observations that fat oxidation is an important contributor at work intensities below 60% $\dot{V}O_2$max, particularly as the duration of exercise increases. The effect of endurance training on the expression and activity of AMPK and its various isoforms has not been examined in humans. In trained rodents, the activation of AMPK α_2 is blunted in contracting red quadriceps muscles (Durante et al. 2002), likely due to a reduced metabolic stress and blunted changes in the AMP/ATP and Cr/PCr ratios. Thus, the overall contribution of AMPK as a regulator of fat oxidation in exercising human muscle remains controversial.

Hormonal Regulation of LCFA Metabolism

Several hormones have been demonstrated to alter muscle LCFA metabolism. Insulin, leptin, and adiponectin are three such hormones and are discussed in the following paragraphs.

Insulin

Recently, it has been shown that CHO availability may regulate skeletal muscle FA metabolism (i.e., the "reverse" G-FA cycle). When glucose availability is enhanced, FA oxidation in muscle is suppressed (Sidossis et al. 1996; Sidossis & Wolfe 1996), an effect that may be mediated by an increase in malonyl-CoA content and subsequent inhibition of CPT-1 (Elayan & Winder 1991). Hyperinsulinemia, whether during eu- (Kelley et al. 1990) or hyperglycemic conditions (Sidossis et al. 1996; Sidossis & Wolfe 1996), also inhibits whole-body FA oxidation. Insulin-induced inhibition of FA oxidation has also been observed in isolated rodent muscle both at rest (Muoio et al. 1997; Alam & Saggerson 1998; Muoio et al. 1999a,b) and during contraction (Dyck et al. 2001; figure 8.9). Insulin also increases FA esterification in resting rodent skeletal muscle (Muoio et al. 1997; Muoio et al. 1999a; Luiken et al. 2002b). Insulin's effects on lipid partitioning (i.e., enhanced esterification, decreased oxidation) are eliminated in the presence of AICAR, suggesting that insulin may alter muscle lipid metabolism by decreasing AMPK activity (Muoio et al. 1999b; Winder & Holmes 2000). Since insulin also stimulates the activity of acylglycerol-3-phosphate acyltransferase (GPAT), one of the rate-limiting steps in skeletal muscle TG esterification (Coleman et al. 2000), this may also contribute to insulin's repartitioning effects on LCFA metabolism. Interestingly, there is recent evidence that the overexpression of human GPAT in muscle cell lines enhances TG storage (Ruan & Pownall 2001), suggesting that the increased expression or activation of GPAT, or both, may be a contributing factor to muscle TG accumulation, and hence insulin resistance. It is tempting to suggest that the hyperinsulinemic condition prevalent in obese, insulin-resistant individuals is responsible for the repartitioning of FA away from oxidation and toward storage. Interestingly, the infusion of insulin to supraphysiological concentrations suppressed whole-body lipid oxidation in lean, but not in obese, humans (Kelley et al. 1999). Along similar lines, in perfused muscles of obese Zucker rats, insulin failed to stimulate further the already augmented rates of LCFA esterification, while in lean rats insulin increased LCFA esterification to rates observed in the obese Zucker rats (Han & Bonen unpublished data). These studies

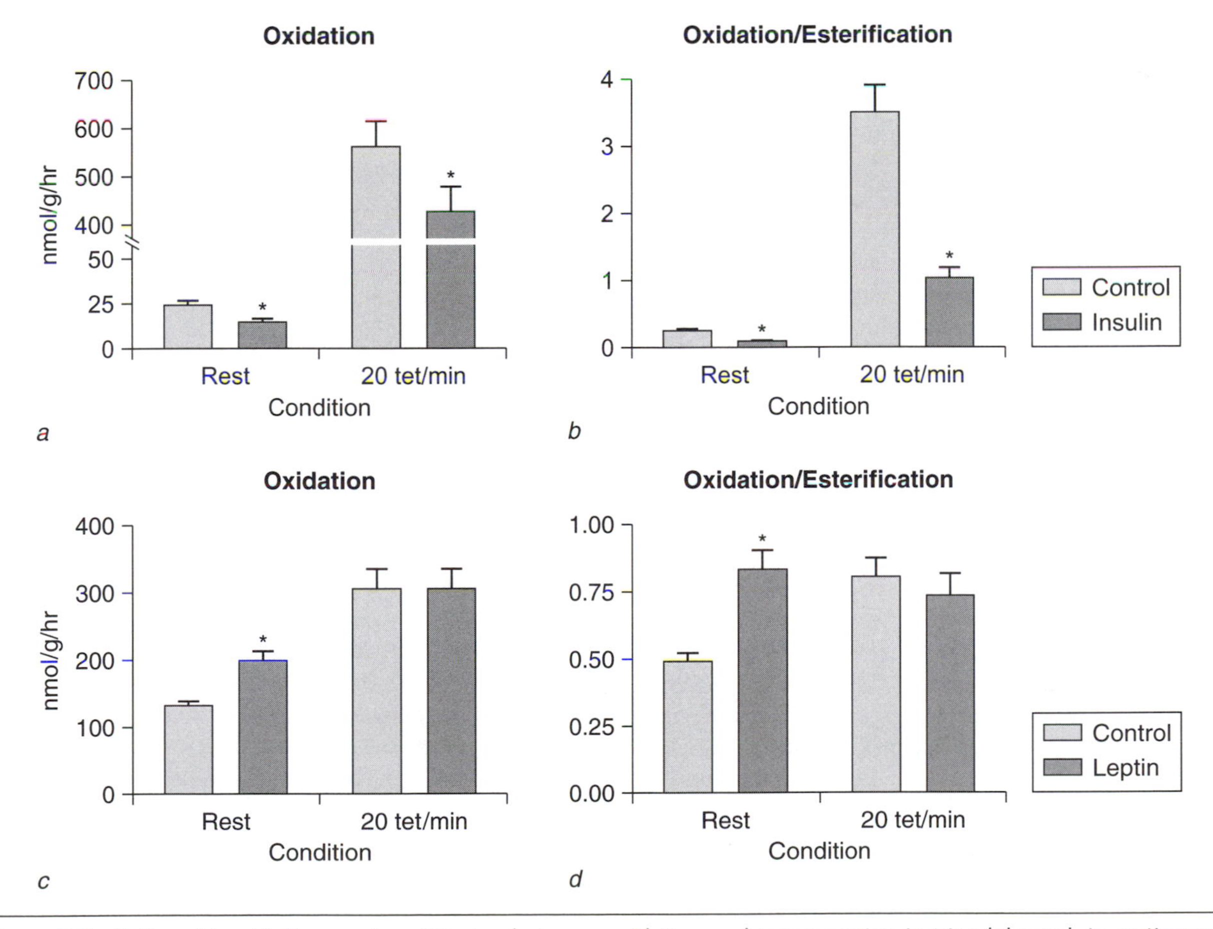

Figure 8.9 Fatty acid oxidation, and partitioning between oxidation and incorporation in triacylglycerol, in resting and contracting rodent soleus with and without insulin and with and without leptin. * Significantly different from control.

Adapted from Bonen et al. 2000 and Luiken, Dyck et al. 2002.

point to a "metabolic inflexibility" in the insulin-resistant conditions.

Leptin

In addition to its central effects on appetite control and energy expenditure (Zhou et al. 1997), leptin is able to interact with numerous peripheral tissues expressing receptors, including skeletal muscle (Ghilardi et al. 1996; Kellerer et al. 1997). Acutely (60-90 min), leptin's effects on FA metabolism completely oppose that of insulin; that is, they partition FA toward oxidation and away from esterification and enhance TG hydrolysis (Muoio et al. 1997; Muoio et al. 1999a). Chronically, leptin has similar metabolic effects (Steinberg et al. 2002a), which ultimately results in a decrease in intramuscular TG content (Shimabukuro et al. 1997). This reduction in stored intramuscular lipids may in part be responsible for the improvement in insulin sensitivity observed following chronic leptin treatment (Yaspelkis et al. 1999). The mechanisms underlying leptin's acute and chronic effects on FA metabolism have not been clearly elucidated, but appear to involve the stimulation of AMPK (Minokoshi et al. 2002).

Contrary to expectation, high levels of circulating leptin characterize most cases of human obesity (Lonnqvist et al. 1997), suggesting the development of central, peripheral, or both central and peripheral resistance to leptin. However, until recently, direct evidence for the development of leptin resistance in skeletal muscle did not exist. Studies from our laboratory have demonstrated that following four weeks of a high-fat diet (60% kcal safflower oil) in rodents, the stimulatory effects of leptin on muscle lipid oxidation and hydrolysis are completely eliminated (Steinberg & Dyck 2000). Furthermore, the loss of leptin's stimulatory effect on FA oxidation has also been demonstrated in isolated muscle from obese females (Steinberg et al. 2002b). During exercise, leptin does not increase. Furthermore, it has been shown, in isolated skeletal muscles, that leptin does not appear to exert any significant regulatory control over FA oxidation (figure 9.4), as leptin's effects are not additive to that of contraction (Lau et al. 2001).

Adiponectin

Like leptin, adiponectin (also known as AdipoQ, Acrp30, apM-1) is released from adipocytes. Collectively, hormones released from adipocytes are often referred to as adipocytokines and also include resistin, tumor necrosis factor, adipsin, and others. Several of these cytokines are proposed to provide a link between the accumulation of fat mass and the development of insulin resistance. However, only leptin and adiponectin have been demonstrated to have a direct effect on skeletal muscle FA metabolism. But, unlike what occurs with leptin and the other cytokines, adiponectin levels are reduced in human obesity (Hotta et al. 2001; Lindsay et al. 2002; Matsubara et al. 2002) and parallel a decrease in insulin sensitivity (Hotta et al. 2001). Furthermore, the administration of adiponectin to obese mice restores insulin sensitivity and corrects the hyperglycemia/hyperinsulinemia (Yamauchi et al. 2001). Recently, adiponectin has also been shown to stimulate FA oxidation in rodent skeletal muscle through the activation of AMPK (Yamauchi et al. 2002). It remains to be determined whether adiponectin, like leptin, also decreases FA esterification and increases IMTG hydrolysis.

Hormone-Sensitive Lipase and Intramuscular Triacylglycerol Utilization

The hydrolysis of TG, in both adipose tissue and muscle, is regulated by the activity of hormone-sensitive lipase (HSL). The expression of HSL has recently been confirmed in rodent and human skeletal muscle, and its activity is regulated both by epinephrine and by contraction (Langfort et al. 1999; Kjaer et al. 2000). In adipocytes, recent attention has focused on the role of perilipins, a family of proteins that coat the lipid droplets and prevent the docking of activated HSL with the TG molecule (Clifford et al. 2000). Thus, in vivo IMTG lipolysis may not simply be a matter of HSL activation, but also of its physical interaction with the TG droplet. However, to date, perilipins have not been identified in skeletal muscle.

Role of Contraction and Hormones

Increases in IMTG lipolysis have been demonstrated in isolated contracting rodent muscle in the absence of any hormones (Hopp & Palmer 1990; Dyck & Bonen 1998), indicating that the intracellular changes associated with contraction (e.g., energy charge, AMPK, calcium, other metabolites) are important in the activation of HSL.

The hormonal regulation of HSL/lipolysis in skeletal muscle has not yet been well defined. HLS activity from rodent and human muscle is increased by epinephrine (Langfort et al. 1999), as is the rate of IMTG hydrolysis in oxidative rodent muscle (Peters et al. 1998). The regulation of muscle HSL by other hormones such as leptin and insulin has not been determined, although each of these hormones regulates IMTG hydrolysis in rodent muscle. Leptin has been shown to be a potent stimulator of IMTG hydrolysis both acutely (Steinberg & Dyck 2000) and following chronic leptin exposure (Steinberg et al. 2002a) in rodents. But in humans the effects of chronic leptin treatment on skeletal muscle FA transport and metabolisms have not yet been investigated. Insulin inhibits IMTG hydrolysis both at rest and during contraction in isolated rodent muscle (Dyck et al. 2001). In humans, depressed glycerol release from muscle, an index of reduced intramuscular lipolysis, has been demonstrated during a hyperinsulinemic clamp (Maggs et al. 1995; Enoksson et al. 1998; Jacob et al. 1999), although a direct effect of insulin on HSL activity or IMTG hydrolysis has not been examined.

IMTG As an Energy Substrate

The specific contribution of the IMTG pool as a metabolic substrate during prolonged submaximal exercise in humans has been debated. The majority of evidence from isotopic tracer and ^{1}H-magnetic resonance spectroscopy (MRS) studies demonstrates a net utilization of IMTG during 90 to 120 min of exercise, while the findings of studies measuring IMTG content directly from skeletal muscle biopsies have been controversial (for review, see Watt et al. 2002b). The controversial findings are likely due primarily the high variability caused by contamination with extracellular TG, as demonstrated by a relatively high coefficient of variation (~23%) when IMTG content is compared between duplicate biopsies from the muscle of untrained individuals (Wendling et al. 1996). This variability is dramatically improved in trained individuals (~12%), allowing for the detection of net IMTG utilization during prolonged exercise (Watt et al. 2002a). There are numerous pros and cons to each of the three mentioned techniques available for the assessment of IMTG use. Although such a discussion is beyond the scope of this review, one disadvantage common to all of these techniques is that only net IMTG utilization is measured. In contrast, studies from our laboratory have utilized the dual-label pulse chase technique in isolated rodent muscle to simultaneously assess both the incorporation and loss of labeled FA from endogenous lipids (Dyck et al. 1997; Dyck & Bonen 1998; Dyck et al. 2000, 2001). This technique was also recently adopted to assess IMTG turnover and oxidation during prolonged, moderately intense exercise in humans (Guo et al. 2000). Results from this study confirmed our previous findings in isolated rodent muscle that (1) both exogenous FA and IMTG significantly contribute to lipid oxidation during exercise and (2) incorporation of FA into IMTG occurs during muscle contraction, although in an amount that is small relative to the loss of FA from this lipid pool.

Role of Uncoupling Proteins

The recent identification of novel uncoupling proteins (UCP2 and UCP3) has stimulated considerable interest in, as well as controversy surrounding, their potential role as regulators of energy expenditure in skeletal muscle. Although UCP2 is ubiquitously expressed, it is UCP3 that is expressed predominantly in skeletal muscle. The ability of UCP1 (the originally discovered uncoupling protein) to uncouple oxidative phosphorylation and the electron transport chain is well substantiated; however, the physiological role of UCP3 remains controversial. In humans, situations such as fasting (Millet et al. 1997) and acute exercise (Giacobino 1999), in which FA oxidation is increased, result in rapid increases in the mRNA expression of UCP3. Stresses to the muscle, such as hypoxia, as well as activation of AMPK, also rapidly stimulate the expression of UCP3 protein (Zhou et al. 2000). It seems paradoxical that in situations such as fasting in which energy expenditure is decreased, UCP3 expression is increased, while with endurance training in humans, which results in increased energy expenditure, there is a decrease in UCP3 mRNA (Schrauwen et al. 1999). Despite this, several investigators have suggested that UCP3 expression is increased in response to elevated FA oxidation. Specifically, it has been suggested that UCP3 may not act as an uncoupler but rather as an anti-porter of FA from the mitochondrion, thus preventing their accumulation when transport into the mitochondrion exceeds the ability to be oxidized (Millet et al. 1997; Schrauwen et al. 2001b). This is supported by recent evidence from Schrauwen et al. (2002) that the increase in mRNA and protein expression of UCP3 following exercise was blocked by the ingestion of glucose before, during, and after the exercise period, which prevented the normal rise in plasma FA. Nevertheless, the role of UCP3 as an uncoupler of mitochondrial respiration remains controversial. In a recent study, it was observed that UCP3 knockout (UCP3KO) mice had a fourfold greater rate of ATP synthesis in skeletal muscle but, strangely, did not have greater whole-body energy expenditure (Cline et al. 2001). Furthermore, the fact that UCP3 protein expression is not related to body mass (Schrauwen et al. 2001a) or altered by weight loss in humans (Schrauwen et al. 2000) creates further controversy regarding its role as a regulator of whole-body energy expenditure.

Fatty Acid Transport and Metabolism: Implications for Disease

Contrary to earlier beliefs, basal whole-body FA oxidation rates are lower in obese, insulin-resistant individuals relative to lean controls (Mandarino et al. 1996). In muscle, the ratio of aerobic (citrate synthase, β-hydroxyacyl-CoA dehydrogenase) to glycolytic (PFK) enzyme activity is depressed in obese individuals (Simoneau & Kelley 1997), and the fractional uptake of FA across the leg is reduced in obese individuals (Colberg et al. 1995) and those with non-insulin-dependent diabetes mellitus (Kelley & Simoneau 1994). A recent study has also demonstrated that the oxidation of long and short chain FA is impaired in muscle homogenates from obese individuals (Kim et al. 2000). There is a strong association between skeletal muscle insulin resistance and (a) plasma FA concentration (Boden et al. 1991, 1996; Santomauro et al. 1999) and (b) intramuscular TG accumulation (Phillips et al. 1996a; Pan et al. 1997; Perseghin et al. 1999; Manco et al. 2000). How these lipids might contribute to insulin resistance remains unknown. Studies in our laboratories have begun to examine some of the possible changes associated with changes in LCFA metabolism in obesity and diabetes.

At the level of LCFA transport it has been shown in obese Zucker rats that the rates of FA transport in skeletal muscle are markedly increased (Luiken et al. 2001). However, this was not attributable to changes in the expression of FA transporters, at either the mRNA or protein level. Rather, the increased FA transport was attributable to an increase in FAT/CD36, but not FABPpm, at the plasma membrane (Luiken et al. 2001). In a model of rodent type 1 diabetes (streptozotocin-induced diabetes) it was found that LCFA transport was also increased; but this was attributable to the increased expression of the LCFA transporters in muscle, and their consequent increase at the plasma membrane (Luiken et al. 2002a). Thus, as discussed previously, LCFA transport in muscle can be increased in rodent models of obesity and diabetes through an increase in sarcolemmal LCFA transporters that arise through several mechanisms: (a) an altered cycling between the endosomal pool and the sarcolemma (e.g., obesity), (b) an increased expression of the LCFA transporters (e.g., type 1 diabetes model), or (c) both these mechanisms.

It has now also become possible to extend studies of FA transport in animals to obese humans and those with type 2 diabetes. In recent studies by Steinberg et al. (2002b) using isolated human muscle strips, it was shown that in obese individuals, total LCFA uptake (esterification + oxidation) was increased, coinciding with a greater rate of LCFA esterification (figure 8.10*a*). Steinberg et al. (2002b) also presented the first evidence of the development of leptin resistance in muscle from obese humans, which may account for the commonly observed reduction in LCFA oxidation in vivo.

The increase in total LCFA uptake into isolated muscles strips obtained from obese individuals appeared to be due to an increased rate of LCFA transport across the plasma membrane (figure 8.10*b*). This increase was associated with an increase in plasma membrane FAT/CD36, but not FABPpm, while the expression of FAT/CD36 and FABPpm was not altered (Bonen et al. unpublished data). In type 2 diabetics, LCFA transport rates were also increased due to the increased expression of FAT/CD36 and its consequent increase at the plasma membrane (Bonen et al. unpublished data) (figure 8.10*c*). Importantly, there was a good relationship between rates of FAT/CD36, LCFA transport, and intramuscular triacylglycerol accumulation in all groups (figure 8.10*c*) (Bonen et al. unpublished data).

Summary

In the past several years, considerable evidence has accumulated to indicate that LCFA uptake into rodent and human muscle is regulated by an LCFA transport system. While the exact mechanisms of facilitating LCFA transport across the sarcolemma are unclear, there is good evidence that at least two proteins, FAT/CD36 and FABPpm, are involved, and that they may collaborate to facilitate LCFA transport into the muscle cell. FATP1 may also be involved, but there are no data on this transporter in skeletal muscle. Quite unexpectedly, it is now also clear that LCFA uptake is regulated acutely (within minutes) by muscle contraction and by insulin, involving the translocation of FAT/CD36 from an intracellular pool to the plasma membrane. Chronic (i.e., days to weeks) adaptation of

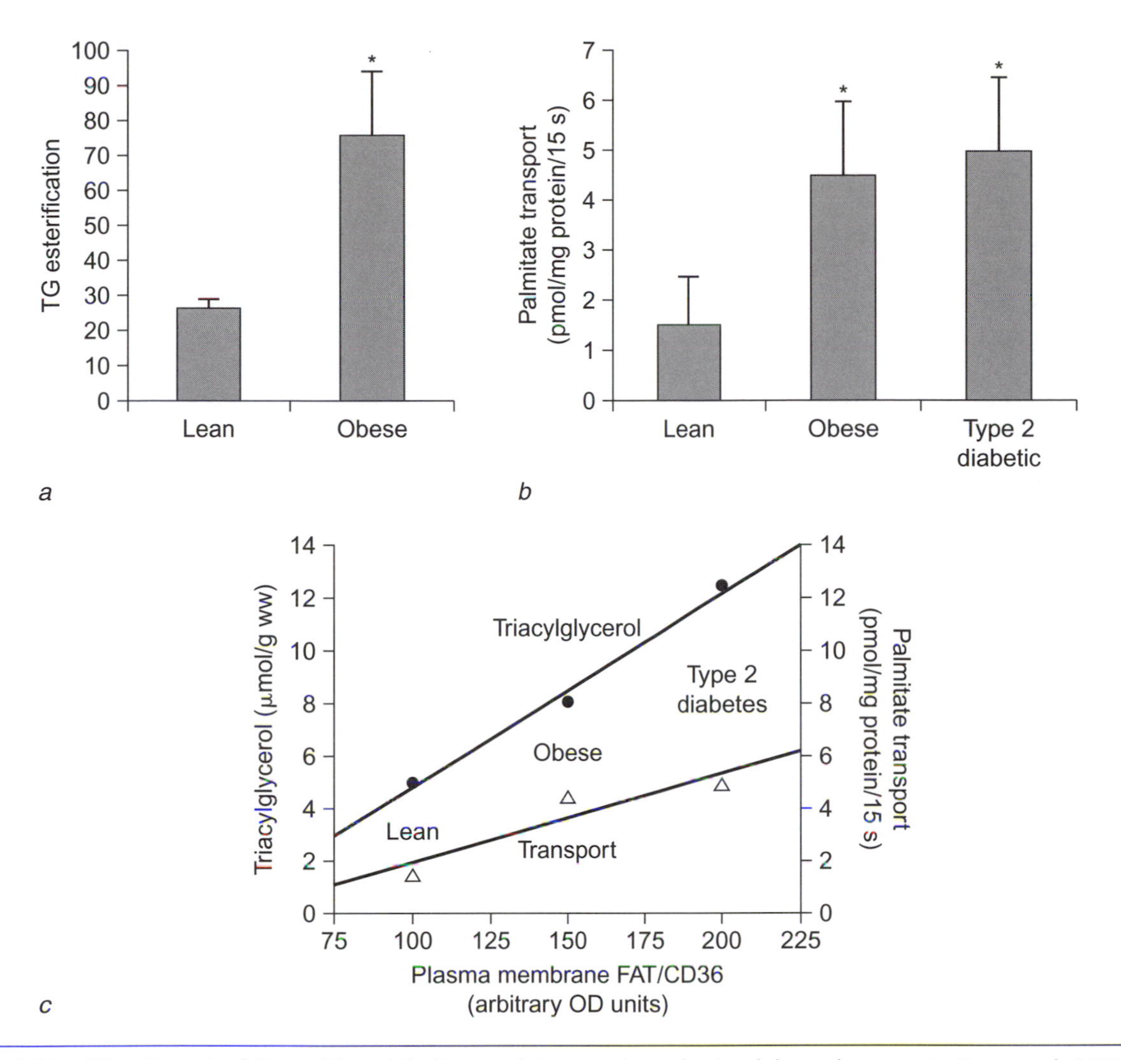

Figure 8.10 Alterations in fatty acid metabolism and transport, and triacylglycerol concentrations and FAT/CD36 expression, in insulin-resistant human skeletal muscle. *(a)* Rates of palmitate incorporation into triacylglycerol in isolated human muscle strips from lean and obese subjects (Dyck et al. unpublished data). *(b)* Rates of palmitate transport into giant sarcolemmal vesicles from human muscle (Bonen et al. unpublished data). *(c)* Relationship between FAT/CD36 and intramuscular triacylglycerol concentrations and palmitate transport (Bonen et al. unpublished data). * Significantly different from lean.

LCFA transport, either an increase or decrease, can occur via altered expression, a subcellular relocation of FAT/CD36 and FABPpm, or both. These mechanisms are observed when muscle activity is altered chronically (i.e., either increased or decreased), or in diseases such as obesity and diabetes, in both rodents and humans. Thus, it is important to realize that changes in LCFA transport can occur without any changes in their expression levels. This is completely analogous to what occurs with the glucose transporter GLUT4, whose expression level is not altered in insulin-resistant states such as obesity, type 2 diabetes, or gestational diabetes but whose signaling machinery is impaired, making it unresponsive to insulin. How alterations in LCFA transport impact on LCFA metabolism is not yet clear, but as with glucose transport, regulation of substrate entry into the muscle cell may be expected to have important consequences.

For many years, the glucose-fatty acid (G-FA) cycle has been used to explain the apparent reciprocal relationship between lipid and CHO metabolism in skeletal muscle. However, research over the past several years has demonstrated that the regulation of LCFA utilization/metabolism is a much more complex process. While there is little doubt that the utilization of CHO can be affected by the availability of circulating LCFA, there is also evidence that the fuel selection by muscle is influenced by CHO availability ("reverse" G-FA cycle). In rodent muscle, malonyl-CoA is increased by increased CHO availability (i.e., glucose and insulin) and decreased in contracting muscle, leading to alterations in LCFA oxidation. AMP-activated protein kinase, which is primarily responsive to a decrease in the energy charge of the cell, has been shown to be an activator of LCFA oxidation in skeletal muscle. However, the effect of AMPK on other aspects of LCFA metabolism, including IMTG hydrolysis, has not been well studied. Furthermore, there does not appear to be any activation of AMPK in human muscle at power outputs below 60% $\dot{V}O_2$max, when the contribution of LCFA oxidation to overall energy production would be substantial. A decrease in malonyl-CoA in exercising human muscle has also not been demonstrated to date. Thus, the significance of the AMPK-ACC-malonyl-CoA axis as a regulator of LCFA metabolism in contracting human muscle remains to be confirmed. At rest, muscle LCFA metabolism (esterification and oxidation) is under the control of numerous hormones, including epinephrine, insulin, leptin, and adiponectin. However, it is debatable whether muscle LCFA metabolism is under hormonal regulation during contraction, when cellular factors such as the increase in calcium and decrease in energy charge are likely to be primary regulators. To date, the regulation of HSL in contracting human muscle remains largely unstudied. Finally, the possible role of uncoupling proteins as a regulator of muscle LCFA metabolism is unknown, but may be a potential cause in the energy imbalances observed in obesity.

Recent work in humans has shown that FA metabolism is perturbed in insulin-resistant conditions such as obesity and type 2 diabetes. Obese muscle demonstrated increased rates of FA incorporation into intramuscular triacylglycerol depots, as well as leptin resistance, which may be related to reduced capacities for FA oxidation in vivo. At the same time there is also a marked up-regulation of FA transport into obese muscle and type 2 diabetic muscle. These observations have led us to the view that in obesity and type 2 diabetes, alterations in FA metabolism occur at the level of the plasma membrane and in the intramuscular trafficking of FA within the cell (see figure 8.11). Finally, the possible role of uncoupling proteins as a regulator of muscle LCFA metabolism is unknown but may be a potential cause in the energy imbalances observed in obesity.

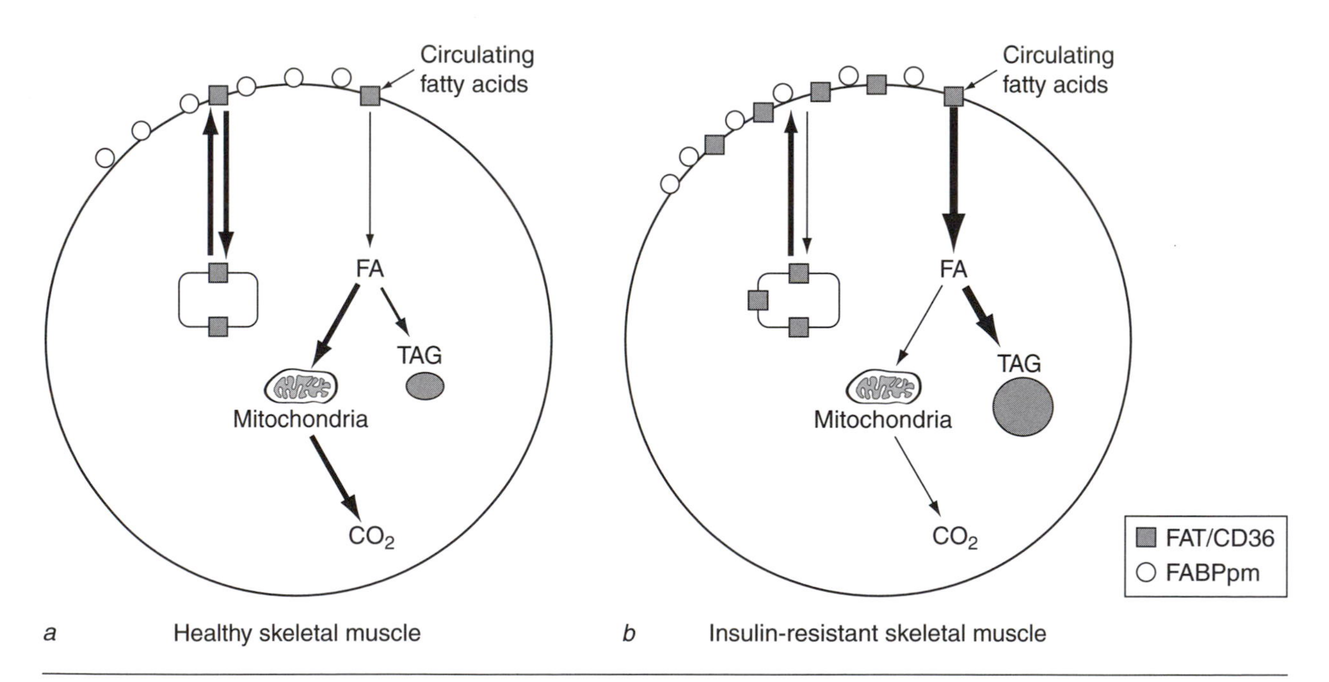

Figure 8.11 *(a)* Healthy skeletal muscle; *(b)* insulin-resistant skeletal muscle. Schematic representation of altered fatty acid metabolism in insulin-resistant human skeletal muscle. Arrows show changes in insulin-resistant muscles in (1) rates of FAT/CD36 cycling to and from the plasma membrane, (2) rates of fatty acid transport across the plasma membrane due an increase in sarcolemmal FAT/CD36, and (3) reduction in fatty acid oxidation. The effects of the increase in fatty acid transport and the reduction in fatty acid oxidation result in the increased concentrations of intramuscular triacylglycerol, which have been linked to insulin resistance. FA, fatty acyl-CoA; TAG, triacylglycerol.

Molecular Basis of Lactate Transport in Skeletal Muscle*

It has been known for over 150 years that lactic acid is produced in large quantities by mammalian skeletal muscle during muscular exercise. But, due to its low pKa (3.86), lactic acid rapidly dissociates (99%) into La^- and H^+ at physiologic pH. Because of the deleterious effects of La^- and H^+ on muscle performance (Fitts & Holloszy 1976; Metzger & Fitts 1987), it is desirable to remove these products from the muscle cell. Therefore, during muscular exercise, blood lactate is increased as the muscle attempts to rid itself of La^- and H^+. The La^- carried in the bloodstream is delivered to the liver, where La^- is reconverted to glucose (Cori cycle), and to the heart, where La^- is oxidized (Chatham & Forder 1996; Chatham et al. 1999). However, La^- released from a muscle cell can also be taken up and metabolized by other muscle cells, either by delivery of La^- via the bloodstream or movement of La^- through the interstitium to neighboring muscle cells. This latter removal route constitutes a pool of La^- that never appears in the blood. The La^- that is taken up by other muscle cells, independent of how the La^- is delivered to these cells, can be converted to glycogen (minor pathway [Bonen & Homonko 1994; Bonen et al. 1990; McLane & Holloszy 1979; Pilegaard et al. 1995]), passively stored in resting muscle (Kelley et al. 2001), and oxidized in contracting muscle (Bonen et al. 1979; Kelley et al. 2001; McGrail et al. 1978). Thus, muscle cells exhibit the capacity for not only extruding La^- but also taking up La^-, either from the circulation or from neighboring muscle cells. Despite the fact that the "lactate shuttle" (i.e., the movement of La^- among tissues such as muscle, liver, and heart) has been popularized as a new concept in recent years (Brooks 1986), the fact remains that La^- released from muscle and its utilization by other muscles and tissues have been recognized for at least 70 years (Owles 1930; and cf. McDermott & Bonen 1992).

The "shuttling" of lactate into muscle cells or other tissues does not occur via simple diffusion. Because the La^- ion diffuses much more slowly than lactic acid, a transport system, to aid the efficient movement (i.e., influx and efflux) of La^- across the plasma membrane, has evolved (Price et al. 1998). The purpose of this part of the chapter is to outline briefly the differences between a simple diffusion and a transport system and the importance of the transport system. In addition, we review the responses to exercise and training of this plasma membrane La^- transport system and the underlying molecular mechanisms involved. Finally, we discuss a major controversy that has recently arisen, regarding whether La^- can be directly metabolized within the mitochondria and whether by extension mitochondria form part of the lactate shuttle mechanism.

Comparison of La^- Diffusion with La^- Transport

The uptake of a substrate across the plasma membrane increases when its extracellular concentration is increased. In a purely diffusional system, the rate of transport across the plasma membrane increases linearly with increasing substrate (La^-) concentrations. Thus, the sole factor governing the rate of uptake is the external La^- concentration. However, crossing the lipid bilayer of the plasma membrane by the La^- anion can occur more rapidly if there is a transport system (i.e., a transport protein). Just as with diffusion, a facilitated La^- transport system, involving a transport protein, also imports La^- into the cell in relation to the increasing external La^- concentration. However, at some point no further increase in La^- uptake occurs, despite increasing concentrations of La^- (i.e., saturation of transport).

While at first glance a La^- transport system may appear to be more limiting than a diffusional system, this is not true. In the case of La^-, a facilitated transport system will move La^- across the plasma membrane much more rapidly than can occur via simple diffusion. The concentration of the transport protein(s) at the plasma membrane determines the rate of lactate transport across the plasma membrane at a given external concentration of La^-. The same general argument applies whether La^- is transported into or out of the cell. However, most studies have typically examined rates of La^- influx into the cell, as this is experimentally simpler. But, since La^- moves efficiently into and out of muscle cells, it may well be important to determine the separate vectorial rates of La^- transport—particularly since muscle tissue, for unknown reasons, expresses transport proteins with differing affinities for La^- (see further on).

* The text on this page beginning with the heading *Molecular Basis of Lactate Transport in Skeletal Muscle* and ending with the text under the heading *Summary* on page 178 is contributed by Arend Bonen.

Skeletal Muscle Lactate Transport

Despite many years of examining lactate responses to various types of exercise, only modest understandings had developed as to how this substrate moved into and out of muscle and other tissues. In recent years studies in isolated tissues and vesicle preparations have provided a basis for understanding lactate movement into and out of parenchymal cells.

Evidence for a La⁻ Transport System in Skeletal Muscle

In the early 1970s, several groups proposed that lactate efflux out of contracting dog muscles was limited by a transport protein or set of proteins in the plasma membrane, indicating that lactate traversed the sarcolemma via a facilitative transport system (Hirche et al. 1970, 1975; Karlsson et al. 1972). Similarly, lactate efflux out of human skeletal muscle during exercise was shown to be limited, exhibiting the characteristics of a saturable transport process (Jorfeldt et al. 1978).

In the early 1990s, studies in sarcolemmal vesicle preparations confirmed that indeed La^- movement across the sarcolemma does occur via a facilitated transport system, since (a) rates of La uptake reached a plateau despite increasing extracellular lactate concentrations and (b) the rates of La^- transport were inhibitable by protein modifying agents (McCullagh & Bonen 1995a; McCullagh et al. 1996a; McDermott & Bonen 1993a,b, 1994; Dubouchaud et al. 1996, 1999; Juel 1991a,b, 1994, 1995; Juel & Wibrand 1989; Roth 1991; Roth & Brooks 1990a,b). In addition, it was shown that (1) it is the La^- anion that is transported (McDermott & Bonen 1993b, 1994); (2) the system functions as a proton symport, in which La^- and H^+ are cotransported in an electroneutral manner (Juel 1997); (3) the system is stereoselective (e.g., transports L-lactate and not D-lactate) (Roth & Brooks 1990a,b; Juel 1997); and (4) transacceleration of La transport is also observed (Brown & Brooks 1994). These traits are associated with a facilitative transport system. Thus, in skeletal muscles there is a plasma membrane La^- transport system that is regulated by a protein or set of proteins.

A number of excellent detailed reviews are available on understandings of lactate transport in the era just prior to the discovery of monocarboxylate transporters (i.e., before the late 1990s: Gladden 1996; Halestrap et al. 1997; Juel 1997; Poole & Halestrap 1993).

Lactate Transport Into Different Muscle Fibers

Since fast-twitch glycolytic muscles exhibit high rates of lactate production, it had been expected that such muscle fibers would also exhibit the highest rates of La^- transport. But this was not the case. In the early 1990s, a number of studies demonstrated that, unexpectedly, rates of La^- transport were greater in muscles rich in fast-twitch oxidative glycolytic (FOG) fibers than in muscles rich in fast-twitch glycolytic (FG) fibers (Juel 1996b; McCullagh et al. 1996b; Pagliassotti & Donovan 1990). This discrepancy may be related to the differences in the co-expression of several La^- transporters that are more abundant in FOG muscle fibers than in FG muscle fibers (see further on).

Muscle Activity-Induced Changes in Lactate Transport

For exercise physiologists, an important question has been whether the La^- transport system is regulatable. This has been addressed by examining La^- transport after a single bout of exercise (acute exercise) and after a period of training or chronic muscle stimulation (chronically increased muscle activity).

Acute Exercise

It is well known that glucose transport is increased during exercise and that this lasts for some time after exercise (see Goodyear & Kahn 1998 for review). But whether lactate transport is also altered during or immediately after exercise remains controversial. When muscles had been electrically stimulated for 30 min there was no increase in lactate transport despite a marked increase in glucose transport (McDermott & Bonen 1994). In contrast, after treadmill exercise there was a small increase in lactate uptake (+10%) (Bonen & McCullagh 1994). Others have reported that the uptake of low concentrations of lactate (1 mM) is reduced after submaximal and exhaustive treadmill exercise (Dubouchaud et al. 1999; Eydoux et al. 2000a), whereas after the same exercise bouts La^- uptake is not altered at external lactate concentrations ranging from 5 to 75 mM (Dubouchaud et al. 1999; Eydoux et al. 2000a). But, unexpectedly, after exhaustive exercise, lactate uptake was increased at a very high lactate concentration (100 mM) (Dubouchaud et al. 1999; Eydoux et al. 2000b). Unfortunately, no molecular mechanisms were identified to account for these unusual observations.

Thus, at present it remains uncertain whether rates of lactate transport are altered with an acute bout of exercise. Important methodological considerations may be at the root of this uncertainty. For example, it may well be important to control for the La^- transport proteins in a given muscle that have known differences in the affinity for La^- (see later discussion). Moreover, in all of the published studies, the lactate transport measurements were delayed for some time after muscle contractions had ceased because of the time needed to prepare the muscles or vesicles for the transport measurements. Thus, whether lactate transport is altered *during* exercise remains to be determined. This will require studies with a non-metabolizable La^- analog either in contracting isolated or in contracting, perfused muscles.

Chronically Increased Muscle Activity

A number of studies have demonstrated that rates of La^- transport are increased or decreased when muscle activity is chronically altered (i.e., days to weeks) (figure 8.12). For example, La^- transport was increased either with chronically increased muscle activity (electrical stimulation, 10 Hz, 24 h/day for seven days) in rats (McCullagh et al. 1997), or with exercise training in rodents (McDermott & Bonen 1993a) and humans (Pilegaard et al. 1993) (figure 8.12). When muscle activity was reduced, either by denervation (McCullagh & Bonen 1995b; Pilegaard & Juel 1995) or by hindlimb suspension (Dubouchaud et al. 1996), rates of La^- transport were reduced. Presumably these changes in rates of La^- transport reflect concomitant changes in either the transport protein concentrations, their activity, or both. With the discovery of a family of monocarboxylate proteins, it has become possible to begin to identify some of the molecular mechanisms involved in the activity-induced changes in skeletal muscle lactate transport (see later).

Summary

In the 25-year period from about the early 1970s to the mid-1990s, it was firmly established that La^- is transported into and out of muscle. But it remains unclear whether rates of lactate transport are altered during or immediately after a single bout of exercise. There is, however, no doubt that when muscle activity is chronically altered for days or weeks, rates of lactate transport are also altered, increasing when muscle activity is increased (e.g., training, chronic muscle stimulation) and decreasing when muscle activity is reduced (e.g., denervation, hindlimb suspension). The molecular bases of these changes awaited the discovery of moncarboxylate transporters within the past decade.

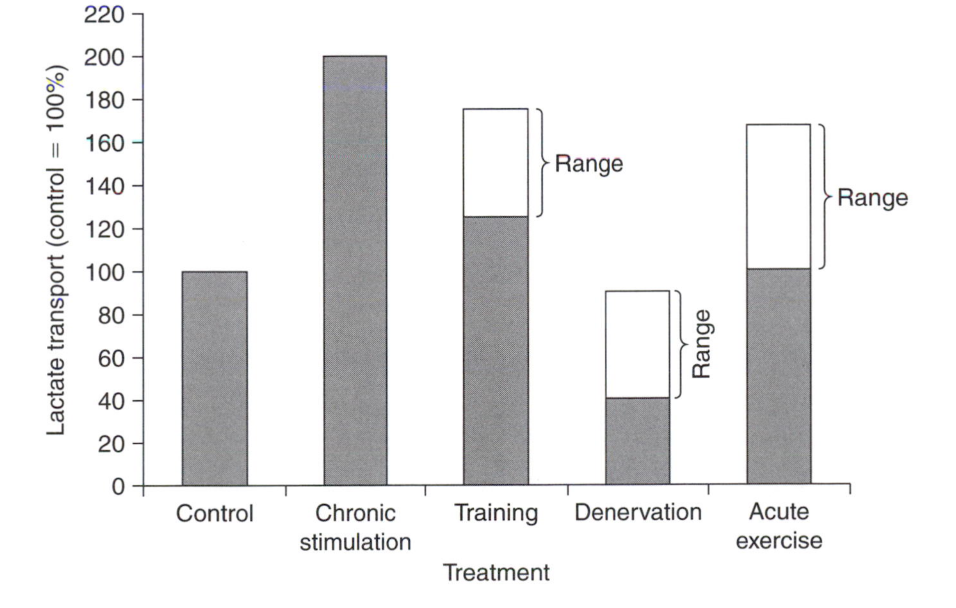

Figure 8.12 Effects of chronically altered muscle activity and a single bout of exercise (acute exercise) on La^- transport into muscle. Control rates of uptake were set to 100% and experimental perturbations are expressed relative to controls.

Data from McCullagh et al., 1997; McDermott and Bonen, 1993; McCullagh and Bonen, 1995; Pilegaard and Juel, 1995; McDermott and Bonen, 1994; Bonen and McCullagh, 1994; Tonouchi et al., 2002.

Molecular Basis of Lactate Transport

Because there was such overwhelming evidence of a lactate transport system, or more precisely a monocarboxylate transporter (MCT) system (i.e., this includes monocarboxylates such as lactate, pyruvate, α-ketoglutarate, β-hydroxybutyrate, acetoacetate) (Bolli et al. 1989; Roth & Brooks 1990a,b), a number of groups became interested in identifying the transport proteins involved. Based on work with glucose transporters, an expectation was that there might well be a family of such transport proteins.

Discovery of a Family of Monocarboxylate Transporters

In 1994 Garcia et al. (1994a,b) discovered, serendipitously, the first monocarboxylate transporter (MCT1), followed shortly thereafter by the discovery of MCT2 (Garcia et al. 1995). MCT3 was cloned in 1997 (Yoon et al. 1997). Shortly thereafter Price and Halestrap (1998) identified a family of MCTs (1-8). MCT9 was reported in 1999 (Halestrap & Price 1999). It has also been argued that the cell surface glycoprotein CD147 facilitates the proper expression of MCT1 and MCT4 at the cell surface (Kirk et al. 2000). The cloning of these MCT transporters provides novel avenues for further investigations of the control of lactate movement into and out of the muscle cell. Yet only a few of the MCT transporters have been functionally characterized.

Transport Kinetics of MCT1, MCT2, and MCT4

Knowledge concerning the transport capacities of different MCTs (table 8.1) can provide some indication as to their physiological role. Transport kinetics have been characterized for three transporters, namely MCT1, 2, and 4. Initial studies purporting to examine the transport kinetics of MCT1 and MCT4 appeared to indicate that the transport kinetics of lactate of MCT1 and MCT4 were almost identical (i.e., Km for MCT1 ~6.4 mM; Km for MCT4 ~10.1 mM) (Wilson et al. 1998). But, in more recent studies in which these MCTs were expressed in *Xenopus* oocytes, it was shown that the affinities for lactate by MCT1, MCT2, and MCT4 differ (table 8.1). The Km for MCT1 (table 8.2) suggests that it may be designed to deal with low levels of La^-, such as at rest and moderate exercise, while with intense exercise, when La^- production is plentiful, MCT4 may be more important. In contrast, MCT2 has a very high affinity for pyruvate (Km = 25-80 μM) when compared to lactate (740 μM) (Lin et al. 1998; Broer et al. 1999).

Table 8.2 Km Values of MCT1, MCT2, and MCT4 for Pyruvate and Lactate

Transporter	Pyruvate	Lactate
MCT1 Km (mM)	1-2	4-6
MCT2 Km (mM)	0.025-0.08	0.74
MCT4 Km (mM)	20-153	28-34

Data derived from studies by Manning-Fox et al. (2000), Dimmer et al. (2000), Lin et al. (1998), and Broer et al. (1998, 1999) in which individual MCTs were expressed in *Xenopus laevis* oocytes.

Tissue-Specific Expression of Monocarboxylate Transporters

It has become clear that just as for glucose transporters, MCT isoforms are expressed in a tissue-specific manner with kinetic transport characteristics that are presumably congruent

Table 8.1 Expression of MCT Proteins in Selected Rat Tissues

Tissue	MCT 1	MCT 2	MCT 3	MCT 4	MCT 5	MCT 6	MCT 7	MCT 8
Heart	+	+		----*	----	+	----	+
RG	+	+		+	+	+	----	----
WG	+	+		+	+	+	----	----
SOL	+	+		+	----	+	----	----

RG, red gastrocnemius; WG, white gastrocnemius; SOL, soleus; + protein signal detected using Western blotting; ---- no protein signal detected using Western blotting; ----* protein signal detected in 10-day-old rat hearts only (see Hatta et al. 2001).

Data from Bonen (unpublished data).

with the metabolic environment, demands of specific tissues, or both. MCT3 provides an excellent example of this, since uniquely among MCTs, MCT3 may be phosphorylated and has been found to be confined exclusively to the retinal pigment epithelium (Yoon et al. 1997). This suggests that the function of MCT3 is finely tuned to the physiologic and metabolic needs of the retina in a highly specialized organ such as the eye.

Based on dbEST, Northern blotting, or protein expression, MCTs are now known to be widely expressed among many tissues (table 8.1). MCT1 is ubiquitously expressed in many tissues (figure 8.13), whereas MCT4 is primarily expressed in skeletal muscles and vas deferens (figure 8.13). Recently MCT4 mRNA has also been detected in neonatal heart (Hatta et al. 2001), testis, small intestine, parotid gland, lung, and brain, but not liver (Dimmer et al. 2000). But, except for observing the expression of MCT4 protein in the neonatal heart (Hatta et al. 2001), we have not been able to confirm that the MCT4 protein is expressed in many of the tissues identified by Dimmer et al. (2000) (Bonen unpublished data). In rodents, MCT2 is expressed in brain, liver, testis, stomach, and kidney (Jackson et al. 1997; Koehler-Stec et al. 1998), as well as in heart and muscles of hamsters. Although MCT2 appeared initially not to be expressed in rat muscle (Jackson et al. 1997), we have recently found that MCT2 is present in rat muscle and in human muscle (Bonen unpublished data). We have also found that MCT5 and 6 are expressed in muscle and MCT6 and 7 in the heart (table 8.1). But nothing is known about their functional roles in these tissues or whether they transport La^- or some other monocarboxylate. The focus of the remainder of this section is on MCT1 and MCT4, which are the proteins that transport La^- in skeletal muscle and for which experimental data are available.

Expression of Monocarboxylate Transporters in Skeletal Muscle

The MCT distribution among different tissues indicates that lactate flux in the body is not regulated solely by skeletal muscles. But, given that muscle (a) is the largest organ (~40% of body mass) and (b) has a highly variable metabolic rate, this tissue is obviously a key site regulating whole-body lactate flux and metabolism. The metabolic heterogeneity of different types of skeletal muscle (figure 8.13) provides an opportunity to examine differential levels of MCT expression and thereby offer on a comparative basis some indication of the function of specific MCTs. In particular this may provide some insight as to why MCT1 and MCT4 are co-expressed in skeletal muscle, the most important site of La^- production and one of the key sites for La^- oxidation, along with the heart.

MCT1 and MCT4 in Rat Muscles

Comparisons of MCT1 and MCT4 in muscle tissues with vastly different oxidative capacities showed that MCT1 expression was highly correlated with the aerobic capacities of these different muscles (figure 8.14) (Bonen et al. 2000b; McCullagh et al.1996a). For example, much more MCT1 protein is present in the heart than in skeletal muscle (Baker et al. 1998; Bonen et al. 2000b), and in rat skeletal muscle MCT1 protein content varies threefold among fast-twitch muscles with varying oxidative capacities (figure 8.14). Moreover, MCT1 is highly correlated with the oxidative capacity of different types of rat muscles. For example, there are correlations between MCT1 and (a) oxidative fiber composition ($r = 0.98$), (b) H-LDH content ($r = 0.83$), and (c) citrate synthase activity ($r = 0.82$) (Bonen et al. 2000b; McCullagh et al. 1996a).

In contrast, MCT4 is present in high quantities in both FG and FOG muscle fibers, while in

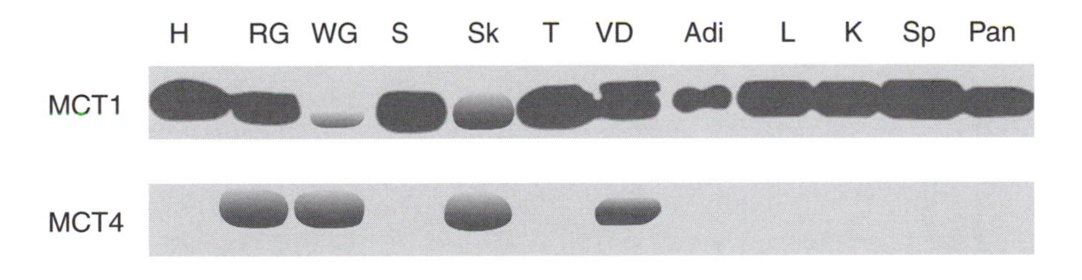

Figure 8.13 Tissue distribution of monocarboxylate transporter (MCT)1 and MCT4 in the rat. Heart (H), red (RG) and white gastrocnemius (WG), soleus (S), skin (Sk), testis (T), vas deferens (VD), adipose tissue (Adi), liver (L), kidney, (K), spleen (Sp), pancreas (Pan).

Bonen unpublished data.

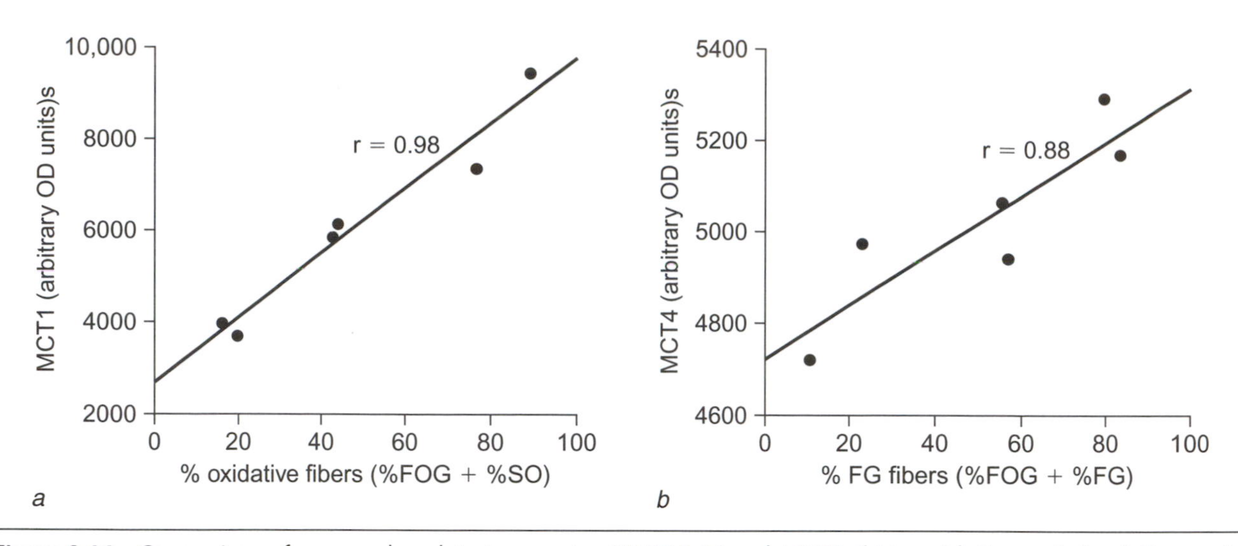

Figure 8.14 Comparison of monocarboxylate transporter (MCT)1 *(a)* and MCT4 *(b)* in oxidative and glycolytic muscle fibers, respectively. Note the correlation between oxidative capacity and MCT1 expression *(a)* and between glycolytic capacity and MCT4 expression *(b)*. (Note that the range of MCT4 expression is very small [~13%] over a wide range of %fast-twitch fibers.)

Adapted, by permission, from K.J. A. McCullagh et al., 1997, "Chronic electrical stimulation increases MCTI and lactate uptake in red and white skeletal muscle," *American Journal of Physiology* 273: E239-E246.

the highly oxidative adult rat heart MCT4 is not expressed and in the adult soleus muscle is very low (i.e., MCT4 is likely present in only the few fast-twitch fibers of soleus muscle; Bonen et al. 2000b). We have found a high positive correlation between MCT4 and %FT fibers (r = 0.88) in the various rat skeletal muscles (figure 8.14), although it must be noted that this correlation occurred over only a very narrow (13%) range of MCT4 differences among rat muscles over a large FT fiber range (10-80%) (Bonen et al. 2000b). However, in the immediate postnatal period when both heart and soleus muscle exhibit high glycolytic capacities, MCT4 is detected in both tissues. But MCT4 declined rapidly as the aerobic capacities of these tissues increased (Hatta et al. 2001).

It is unclear why both MCT1 and MCT4 are co-expressed in muscle. One suggestion is that because of their differing affinity for La^-, these transporters function optimally at low (MCT1) and high (MCT4) La^- concentrations. Evidence in our laboratory has also shown a high correlation between MCT1 expression and the rates of lactate uptake among perfused rat hindlimb muscles (figure 8.15*a*). This relationship between La^- uptake and MCT1 was preserved when only MCT1 was up-regulated (Bonen et al. 2000c). Notwithstanding the very narrow range of MCT4 expression among rat fast-twitch muscles as already discussed, there appeared to be a negative relationship between MCT4 and lactate uptake (figure 8.15*b*). On the basis of these observations and others (Hatta et al. 2001), we have speculated that the function of MCT1 is to facilitate the entry of La^- into the muscle while that of MCT4 is to export La^- out of the muscle.

Standing against this suggestion that MCT1 favors La^- import and MCT4 favors La^- export is the fact that transporters can transport substrates into or out of a cell. But it is possible that their vectorial transport capacities may differ (i.e., rates of influx may not necessarily equal rates of efflux). Older studies, performed before the discovery of the family of MCTs, showed that there was no asymmetry in lactate transport (i.e., rates of influx and efflux were similar) (Juel 1996b). However, somewhat different methods were used to measure rates of La^- influx and efflux, and importantly, the MCT1 and MCT4 content of the muscles was not controlled. We now know that these preparations contained varied proportions of MCT1 and MCT4. Nevertheless, it is necessary to confirm empirically, in studies with vesicle preparations, that the vectorial transport capacities of MCT1 and MCT4 do in fact differ. Such evidence is needed to validate our hypotheses for the roles of MCT1 and MCT4. These studies are currently in progress.

Although MCT1 mRNA is highly correlated with MCT1 protein expression in rat muscles (r = 0.94), no relationship is observed between MCT4 mRNA and MCT4 protein (Bonen et al. 2000a). Changes in MCT1 expression in chronically stimulated muscle

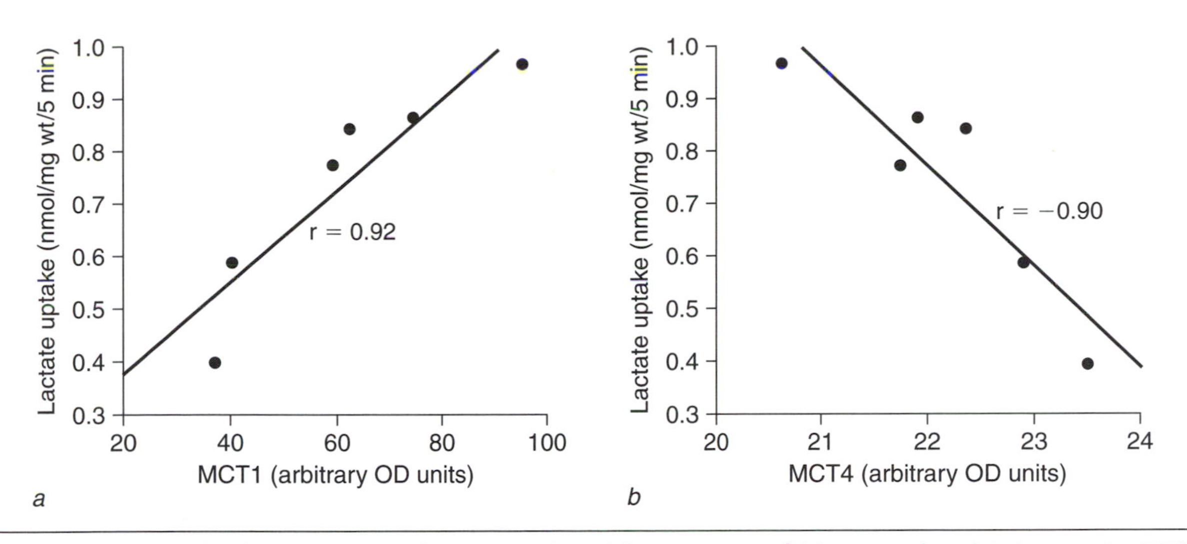

Figure 8.15 Relationship between La⁻ uptake into muscle and the expression of *(a)* monocarboxylate transporter (MCT)1 and *(b)* MCT4. (Note that the range of MCT4 expression is very small [~13%] over a wide range of %fast-twitch fibers.)

Adapted, by permission, from K.J. A. McCullagh et al., 1997, "Chronic electrical stimulation increases MCTI and lactate uptake in red and white skeletal muscle," *American Journal of Physiology* 273: E239-E246.

(seven days) appear to occur via post-transcriptional mechanisms, as MCT1 mRNA is minimally altered despite a two- to threefold increase in MCT1 (Bonen et al. 2000b). Others have also suggested that MCT protein expression is apparently controlled by post-transcriptional mechanisms (Halestrap & Price 1999).

MCT1 and MCT4 in Human Skeletal Muscle

Human muscle, just as rat skeletal muscles, co-express MCT1 and MCT4 (Baker et al. 2001; Dubouchaud et al. 2000; Pilegaard et al. 1999b). MCT1 was found in both type I and II human muscle fibers, while MCT4 is present only in type II human muscle fibers (immunofluorescent studies) (Pilegaard et al. 1999b). There is a good correlation between MCT1 and the percentage of type I fibers, or oxidative muscle fibers, in humans ($r = 0.66$), a negative correlation between MCT1 and IIX muscle fiber composition ($r = -0.73$), and no discernible correlation between MCT1 and the type IIa fibers (Pilegaard et al. 1999b). There was no correlation between MCT4 and human muscle fiber composition (Pilegaard et al. 1999b).

A reduced capacity for lactate transport (measured in erythrocytes, not in muscle) has been observed in patients who reported decreased exercise capacity and severe cramping after exercise. Some of these individuals exhibited a mutation in skeletal muscle MCT1 cDNA (Merezhinskaya et al. 2000), suggesting that perhaps an aberrant MCT1 expression in muscle, as well, may have been at the root of their reduced lactate transport capacity. But, MCT4 mRNA was not determined. In contrast, a patient with a mitochondrial myopathy had much higher circulating lactate levels (3.7 mM) than control subjects (<1 mM). In this patient the muscle MCT4 was 86% greater than in healthy controls, while MCT1 was increased somewhat less (+47%) (Baker et al. 2001). These observations suggest that high levels of lactate may up-regulate the expression of MCT4 or MCT1, or both, in muscle. This requires further investigation.

Subcellular Distribution of MCT1 and MCT4 in Muscle

From studies of the glucose transport system (Goodyear & Kahn 1998) as well as the FA transport system (Bonen et al. 2000b), it has become clear that transport proteins can exist both at the cell membrane and in one or more intracellular depots. The advantage of the latter is that such intracellular stores provide a reservoir to be mobilized during specific situations. It is known that muscle contraction and insulin can each translocate GLUT4 (Goodyear & Kahn 1998) or the FA transporter FAT/CD36 (Bonen et al. 2000b; Bonen, Luiken et al. 2002a) to the sarcolemma. Thus, it was of considerable interest to ascertain whether MCTs also were present in various subcellular compartments and whether some MCTs may be translocated in response to a stress such as muscle contraction. This would potentially reveal further insights into the physiologic roles of MCTs.

Both MCT1 and MCT4 are present in the plasma membranes and T-tubules, while the relative abundances of MCT1 and MCT4 in these subcellular locations differ (figure 8.16). Significant quantities of MCT4, but not MCT1, are also located in an intracellular compartment (figure 8.16). One group (Brooks et al. 1999a) has also found MCT1 at the mitochondrial membrane. However, this MCT1 localization appears to have been contaminated by the plasma membrane in these studies. In highly purified mitochondrial preparations, MCT1 and MCT4 are apparently observed only in the subsarcolemmal (SS) mitochondria, not in the intermyofibrillar (IMF) mitochondria (figure 8.17) (Bonen unpublished data).

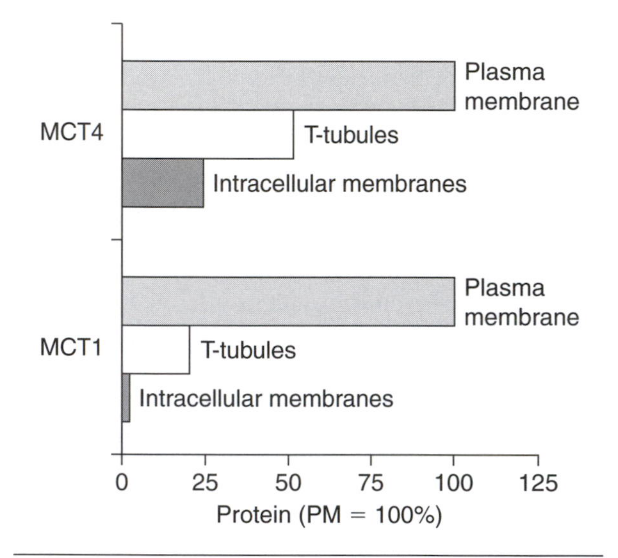

Figure 8.16 Subcellular distribution of monocarboxylate transporter (MCT)1 and MCT4 in rat muscle. The plasma membrane MCT content was set to 100% while the content of other subcellular fractions was set relative to this level. Quantitative comparisons between subcellular compartments could not be made as the data have not been corrected for protein recoveries. However, the figure does identify the presence of MCT1 and MCT4 at the cell surface (plasma membrane and T-tubules) and in an intracellular depot (MCT4 only, not MCT1).

Data from Bonen et al. 2000a.

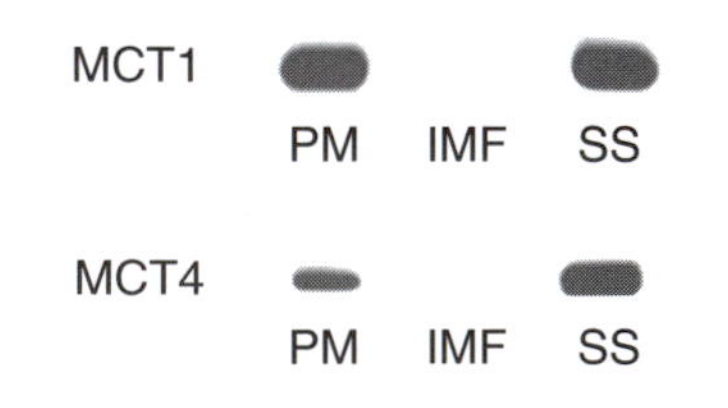

Figure 8.17 Monocarboxylate transporter (MCT)1 and MCT4 in plasma membrane (PM) and in subsarcolemmal (SS) and intermyofibrillar (IMF) mitochondria. Note the absence of MCT1 and MCT4 in IMF.

Effects of Muscle Activity on La⁻ Transport and MCTs

With the identification of MCTs, a number of studies have examined how changes in MCT1 and/or MCT4 affect rates of La^- transport. These studies have been performed commonly by altering muscle activity either acutely or chronically.

Acute Exercise

Since there is no intracellular MCT1 pool (figure 8.16), MCT1 cannot be translocated to the cell surface by a physiologic stimulus, such as a bout of exercise, to increase La^- transport. Therefore, the functional capacity for lactate-proton cotransport by MCT1 is solely dependent on the number of the MCT1 transporters at the plasma membrane, assuming that the activity of MCT1 cannot be altered. But the situation is less clear for MCT4, since there is an intracellular MCT4 pool. This may be a reserve pool of MCT4 that can perhaps be translocated to the cell surface to alter lactate flux. However, studies to address this matter have yielded unexpected findings. With intense, electrically induced muscle contraction (2×5 min, with 1 min rest), plasma membrane MCT4 was reduced (~25%) and MCT1 was unaltered, while lactate transport was increased (figure 8.18) (Tonouchi et al. 2002). This may suggest that the activity of the surface MCTs is altered by muscle contraction, or alternatively, that other MCTs are involved in up-regulating La^- transport. Whether MCT4 can be translocated to the plasma membrane under different physiologic circumstances (e.g., hypoxia, less intense exercise) still needs to be examined.

Chronically Altered Muscle Activity

In the past 40 years it has been shown that skeletal muscle is remarkably adaptable to exercise training. Therefore it has been of obvious interest to determine whether MCTs can also be regulated by exercise training. In initial studies, we used chronic low-frequency muscle stimulation (24 h/day for seven days at 10 Hz), in which all muscle fibers are contracting. This mimics the innervation patterns in slow-twitch muscles. Hence, this type of contractile activity can be seen as an exclusively aerobic stimulus. With this chronic contraction model we observed that only MCT1 expression was up-regulated, not MCT4 (figure 8.19*a*) (Bonen et al. 2000c), due to post-transcriptional mechanisms, as MCT1 mRNA abundance was not altered. With the up-regulation of MCT1 but not MCT4, the rates of lactate influx are increased (figure 8.19*b*) (Bonen

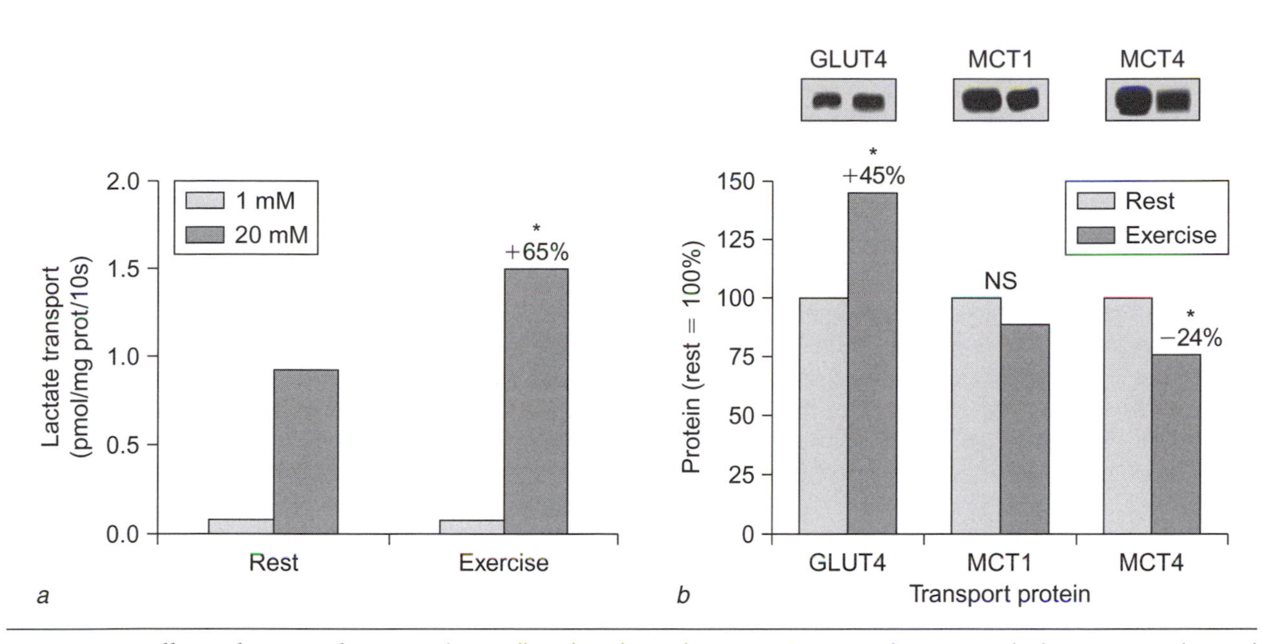

Figure 8.18 Effects of 10 min of intense, electrically induced muscle contraction on *(a)* lactate uptake by giant sarcolemmal vesicles and *(b)* plasma membrane content of glucose transporter protein (GLUT)4, monocarboxylate transporter (MCT)1, and MCT4. Note that lactate transport was measured at external lactate concentrations of 1 and 20 mM at rest and after 10 min of contraction, and that the increase in plasma membrane GLUT4 was a positive control in these experiments. Note also the unexpected decline in plasma membrane MCT4.

Adapted, by permission, from M. Tonouchi, H. Hatta, and A. Bonen, 2002, "Muscle contraction increases lactate transport while reducing sarcolemmal MCT4, but not MCT1," *American Journal of Physiology Endocrinology and Metabolism* 282: E1062-E1069.

et al. 2000c; McCullagh et al. 1997). The increased rate of influx is directly proportional to the change in MCT1 transporters (figure 8.19*b*) (McCullagh et al. 1997). Thus, these studies (Bonen et al. 2000c; McCullagh et al. 1997) establish clearly that merely through increasing only MCT1, lactate flux into muscle is increased. Similar studies in which only MCT4 is increased have not yet been performed. However, it has been shown that eliminating muscle activity (denervation) reduces rates of

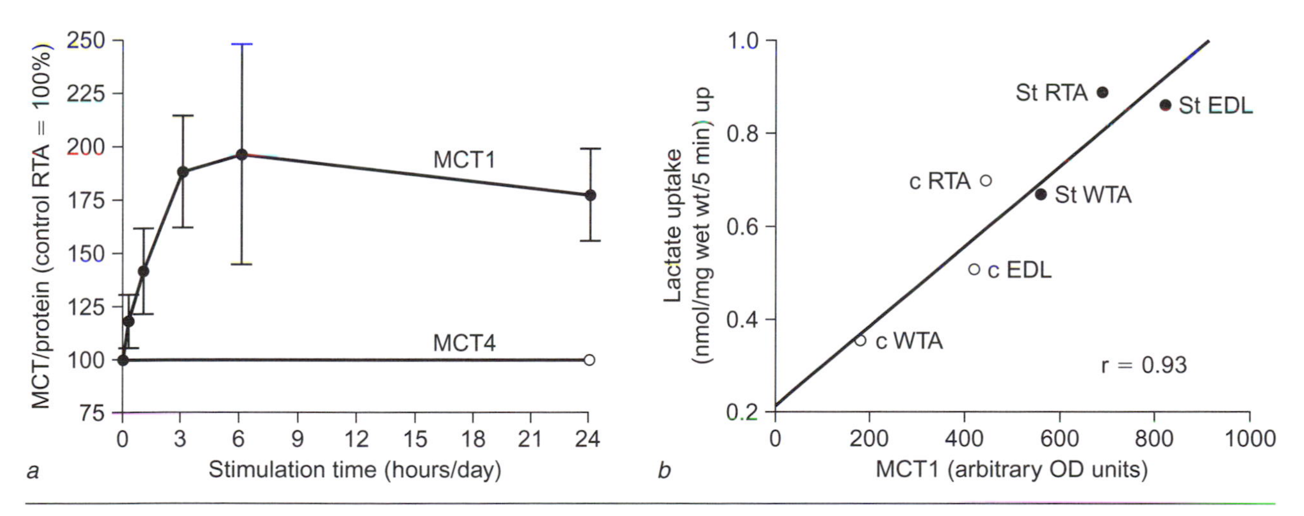

Figure 8.19 *(a)* Effects of chronic low-frequency muscle stimulation on monocarboxylate transporter (MCT)1 and MCT4 expression in red tibialis anterior muscle stimulated for varying periods of time (0, 0.25, 1, 3, 6, and 24 h/day) for seven days, and *(b)* the effects of seven days of chronic stimulation (24 h/day) on the relationship between MCT1 and La^- transport into muscle. RTA, red tibialis anterior; WTA, white tibialis anterior; EDL, extensor digitorum longus; c, control muscle; St, chronically stimulated muscle.

Part *a-b* data from McCullagh et al. 1997 and Bonen et al. 2000.

lactate transport (McCullagh et al. 1996a) due to reductions in both MCT1 and MCT4 (Wilson et al. 1998).

Exercise training can increase both MCT1 and 4 in human muscle (Bonen et al. 1998b; Dubouchaud et al. 2000; Pilegaard et al. 1999a) and in rat muscle and heart (Baker et al. 1998). Animal studies have shown that the training stimulus must be sufficiently intense to demonstrate an increase in MCT1. Low-intensity exercise failed to increase MCT1 in muscle, although in the same animals, heart MCT1 was increased with low-intensity exercise training (Baker et al. 1998). This suggests that the increases in MCT1 in the heart occur more easily than in muscle. Training-induced increases in rat muscle occurred primarily in oxidative muscles (i.e., soleus and RG, but not EDL and WG) (Baker et al. 1998), which may reflect that these muscle types were primarily recruited by the exercise training stimulus. Interestingly, training responses in MCT1 were already observed within the first week of training, and in more recent studies we have shown that a single bout of exercise increases both MCT1 and MCT4 (Coles & Bonen unpublished data). In human studies, training-induced increases of up to 76% to 90% have been reported for MCT1 and of 34% to 47% for MCT4 after eight to nine weeks of high-intensity training (Dubouchaud et al. 2000; Pilegaard et al. 1999a). Collectively, the animal and human studies indicate that the expression of MCT1 and 4 are independently regulated. Thus, while muscle contractions that stress only aerobic metabolism result in an increase only in MCT1, not in MCT4 (figure 8.19*a*) (Bonen et al. 2000c), more intense muscle contractions that may stress both the oxidative and glycolytic capacities of muscle result in an increased expression of both MCT1 and MCT4 (Dubouchaud et al. 2000; Pilegaard et al. 1999a).

Is La⁻ Metabolized Directly by Mitochondria?

While it has long been known that heart and skeletal muscle can oxidize La⁻, it was also long believed that an obligatory step was the conversion of La⁻ to pyruvate in the cytosol, after which pyruvate was transported into the mitochondrion to be metabolized to CO_2. However, provocative initial studies by Szczesna-Kaczmarek (1990) and a later study by Brooks et al. (1999b) have indicated that lactate may also be oxidized in mitochondria. Brooks et al. (1999b) have taken the apparent direct oxidation of La⁻ by mitochondria to suggest that mitochondria are a part of the lactate shuttle system. But whether or not La⁻ is directly metabolized in mitochondria is not a necessary, central tenet of the "lactate shuttle" concept.

Brooks' work (1999b) has been severely challenged recently, on both theoretical and empirical grounds. Neither Sahlin et al. (2002) nor Rasmussen et al. (2002b) have been able to replicate the finding that isolated mitochondria can oxidize lactate directly; and they find that mitochondrial LDH represents <0.7% of the total available LDH in the muscle cell, indicating that the first requisite enzymatic step in metabolizing lactate is lacking in mitochondria. Moreover, it also seems that direct oxidation of lactate in mitochondria is thermodynamically unlikely from a theoretical perspective due to the highly reduced state of the mitochondrial NAD^+-NADH redox couple (Sahlin et al. 2002). Thus, there is considerable disagreement as to whether La⁻ can be metabolized directly in mitochondria. The MCTs at the mitochondrial membrane may simply indicate that at this location these transporters are required to import pyruvate into the mitochondrion.

Summary

The molecular basis of La⁻ transport has only begun to be understood within the last five years. There is a family of nine MCTs. MCT1, 2, 4, 5, and 6 proteins are co-expressed in rat muscle, while in the adult rat heart MCT1, MCT2, MCT6, and MCT8, but not MCT4, are expressed. The reason for MCT co-expression in muscle is not entirely clear. The kinetic activities and fiber type distribution of MCT1 and MCT4 are quite dissimilar, suggesting that they may have different roles in moving lactate into and out of the muscle tissue. There is also evidence that MCT1 and MCT4 may have a different subcellular distribution pattern and that their expression may be regulated independently by different stressors. There is no doubt that exercise training increases MCT1 and MCT4 in muscle. Responses of MCT2, 5, and 6 to training have not been examined to date. There are, however, key questions that remain to be answered in the next few years. Among these are (a) whether acute exercise increases La⁻ transport and if so, which MCT protein(s) is involved; (b) whether MCTs transport La⁻ into the mitochondrion to be oxidized in this organelle; (c) what factors regulate the expression of MCTs; (d) what signaling proteins may be involved in the possible recruitment of MCTs to the cell surface; and (e) whether changes in MCT expression have a functional impact on skeletal muscle performance capacities.

Generation and Disposal of Reactive Oxygen and Nitrogen Species

Andreas M. Niess

I am grateful to Mrs. Lisa Neumann for her assistance in the preparation of the manuscript.

In the last 25 years, the role of reactive oxygen and nitrogen species (RONS) in exercise physiology has received considerable attention. Acute physical exertion has been shown to induce an augmented generation of RONS via various mechanisms. In this context, actions resulting from RONS appear to affect important mechanisms in the field of exercise physiology. Evidence exists that RONS formation in response to vigorous physical exertion can result in oxidative stress. However, the functional significance of exercise-induced oxidative stress is still open to discussion. More recent research has revealed the important role of RONS as signaling molecules. In this context, RONS modulate a broad array of physiological functions. RONS modulate contractile function in unfatigued and fatigued skeletal muscle. Furthermore, involvement of RONS in modulation of gene expression via redox-sensitive transcription factors represents an important regulatory mechanism, which has been suggested to be involved in the process of training adaptation. Adaptation of endogenous antioxidative systems in response to regular training may lead to a limitation of exercise-induced oxidative stress and reflects a potential mechanism responsible for augmented tolerance to exercise. Effects of training also seem to include alterations in RONS generation, which may exert beneficial effects in the therapy and prevention of chronic disease processes. At present, many of the available data regarding the effects of exercise on RONS-associated mechanisms at the cellular level in humans are derived from studies in peripheral immunocompetent cells. Consequently, immunological aspects are one dominant part of this chapter.

The first section of this chapter provides basic information about RONS by focusing on their generation properties, mechanisms of action, involvement in physiological functions, and the components of the antioxidant network. The second part of the chapter presents current knowledge about the specific impact of acute and chronic exercise on the formation, effects, and regulatory properties of RONS, as well as the response of the antioxidant systems.

Reactive Oxygen and Nitrogen Species in Living Organisms

By definition, free radicals are atoms or molecules that have one or more unpaired electrons in their orbitals and present very pronounced chemical reactivity as a result (Halliwell 1998).

Conventionally, a dot (·) is used to symbolize free radical species. Additional oxidating derivates without unpaired electrons are classified as so-called non-radicals. Such non-radicals also exert oxidizing effects and have similar reactivities and regulatory effects when compared to free radicals (Dröge 2002).

Figure 9.1 summarizes major types of RONS for whose generation enzymatically controlled, non-enzymatic, and iron- or copper-catalyzed mechanisms have been shown to be responsible. Chemical reactivity and resulting toxicity to cellular targets vary between the different types of RONS. By far the most reactive oxygen-containing species is the hydroxyl radical ($\cdot OH^-$) (Halliwell 1998). Superoxide ($\cdot O_2^-$) is the most thoroughly investigated among the biologically relevant ROS. Under basal conditions, total body generation of $\cdot O_2^-$ has been calculated to be close to 2 kg/year (Halliwell 1998). As quantified by measurements of nitrite and nitrate, nitric oxide (·NO) has been shown to be produced in larger amounts, reaching values of 9 kg/year (Halliwell 1998).

Mechanisms of RONS Generation

A terminological distinction is made between oxygen-centered *reactive oxygen species* (ROS) and nitrogen-derived *reactive nitrogen species* (RNS), whereas the term RONS unifies both species. Several mechanisms are known to be responsible for generation of ROS and RNS.

Oxygen-Centered Reactive Species and their Derivates

Production of superoxide anions via NADPH oxidase isoforms represents a mechanism that

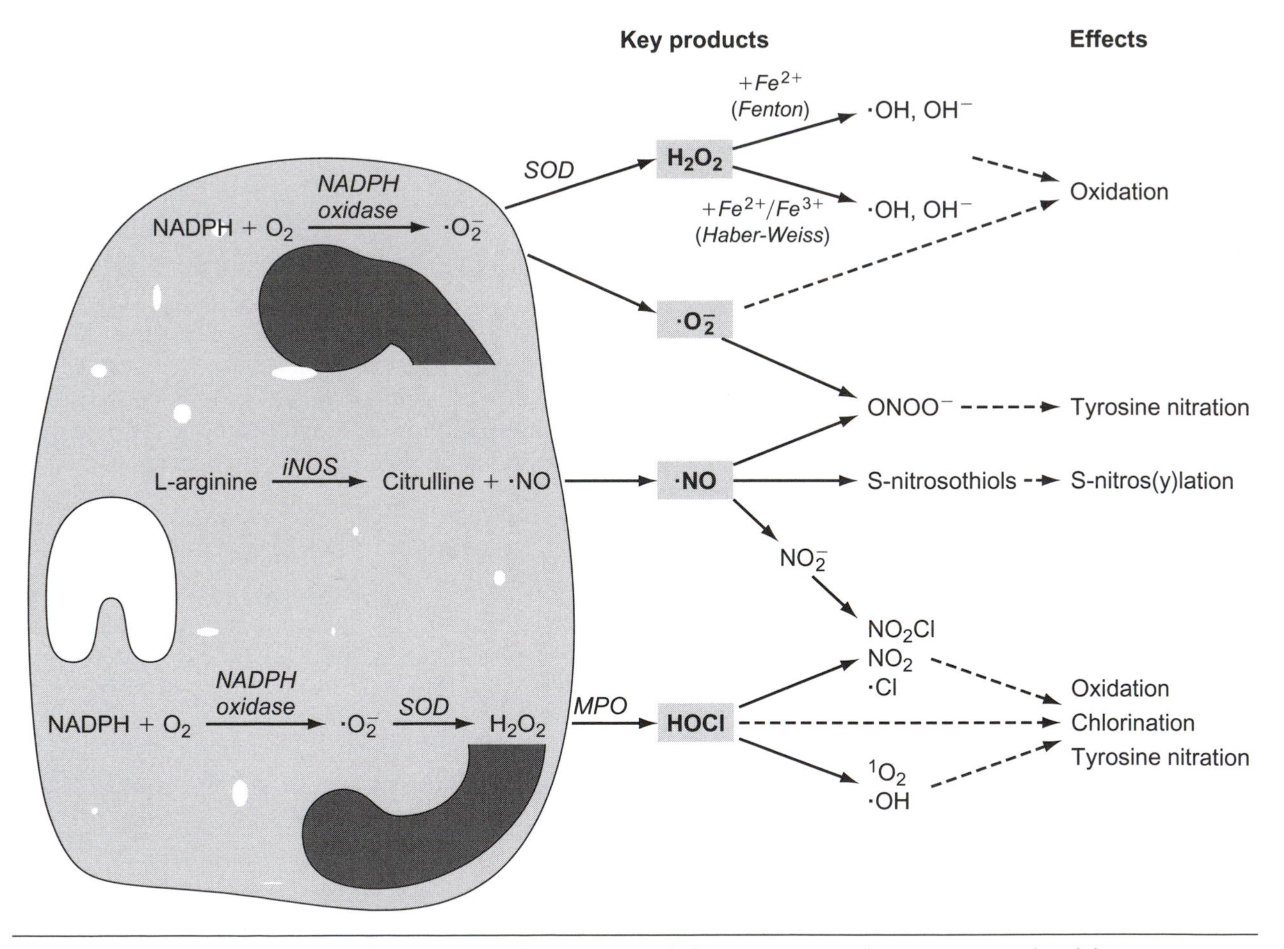

Figure 9.1 Pathways for RONS formation in phagocytes and their interaction during antimicrobicidal actions. MPO, myeloperoxidase; iNOS, inducible nitric oxide synthase; SOD, superoxide dismutase; $\cdot O_2^-$, superoxide radical; H_2O_2, hydrogen peroxide; HOCl, hypochlorous acid; ·OH, hydroxyl radical; OH^-, hydroxyl anions; $ONOO^-$, peroxynitrite; NO_2^-, nitrite; NO_2Cl, nitryl chloride; $\cdot NO_2$, nitrogen dioxide.

Adapted from, *Current Opinion in Immunology,* Vol. 12, C. Bogdan, M. Rollinghoff, and A. Diefenbach, Reactive oxygen and reactive nitrogen intermediates in innate and specific immunity, pgs. 64-76. Copyright 2000, with permission from Elsevier.

is present in phagocytes and a variety of other cell types such as vascular smooth muscle cells, endothelial cells, cardiac and skeletal myocytes, and fibroblasts (Griendling et al. 1994; Javesghani et al. 2002; Jones et al. 1996). The extent of ROS formation by nonphagocytic NADPH oxidase is lower, and plays an important role in the modulation of redox-sensitive signaling pathways. Activation of cardiovascular isoforms of the NADPH oxidase has been shown to occur in response to metabolic, hemodynamic, and hormonal changes (Dröge 2002). Platelet-derived growth factor, thrombin, and tumor necrosis factor-α (TNF-α) stimulate NADPH oxidase in vascular smooth muscle cells, while endothelial cells respond to mechanical forces. Recently, it was shown in rats that a nonmitochondrial NADPH oxidase enzyme complex is present inside skeletal muscle fibers. It has characteristics similar to those of nonphagocytic NADPH oxidases and contributes to basal ROS formation in skeletal muscle (Javesghani et al. 2002).

Superoxide radicals ($\cdot O_2^-$) are continuously formed by the redox-reactive semi-ubiquinone compound of the electron transport chain in mitochondria. In the process of mitochondrial oxidative phosphorylation, molecular oxygen undergoes four-electron reduction. This reaction is catalyzed by cytochrome oxidase and applies to about 90% to 98% of oxygen consumed. The remaining oxygen undergoes univalent reduction with the formation of $\cdot O_2^-$ (Halliwell 1998), a process that occurs in all mitochondria-containing cells and seems to be further enhanced with age (Desai et al. 1996). In a further step, generated $\cdot O_2^-$ can be converted to ·OH by an iron-catalyzed reaction. Estimations concerning the extent to which total electron flux in the mitochondria shows this leakage and leads to the formation of $\cdot O_2^-$ vary in the literature between 1% and 5% (Halliwell 1998). Therefore, it can be assumed that the mitochondrial electron transport chain is the quantitatively most important mechanism for ROS formation.

Xanthine oxidase (XO) is predominantly localized in the vascular endothelium and catalyzes the conversion of hypoxanthine to xanthine and from xanthine to uric acid. Xanthine oxidase normally exists in vivo as a dehydrogenase form (XD), which utilizes nicotinamide adenine dinucleotide (NAD) as an electron acceptor and is incapable of ROS formation (Roy & McCord 1983). Sulfhydryl oxidation or proteolysis via activation of calcium-dependent proteases can cause conversion from XD to XO, the latter of which uses molecular oxygen and forms superoxide radicals during catalytic action (Parks & Granger 1986). This ROS-generating process has been examined under metabolic stress and plays a special role during cellular hypoxia–reoxygenation injury (Granger 1988). Furthermore, hypoxia or ischemia is associated with an accumulation of adenosine monophosphate (AMP) as a result of impaired adenosine triphosphate (ATP) resynthesis. Further conversion of AMP to inosine monophosphate (IMP) leads to increased levels of hypoxanthine, which is the major substrate of XO. Free iron has been shown to catalyze the formation of ROS. A well-known mechanism of iron-catalyzed generation of ROS is the Fenton reaction, which converts hydrogen peroxide (H_2O_2) to the much more reactive hydroxyl radical (Fenton 1894). By reducing Fe^{3+} to Fe^{2+}, $\cdot O_2^-$ can increase the availability of iron ions for the Fenton reaction (Halliwell 1998). Copper also has the potential to react with H_2O_2 to form $\cdot OH^-$ (Haber & Weiss 1943). In addition, electron transfer to peroxides and hydroperoxides can be served by both iron and copper ions and leads to the formation of ROS.

Arachidonic acid plays an important role as a precursor of the prostaglandin and leukotriene pathway. Prostaglandin formation via cyclooxygenases and resulting further steps in prostaglandin biosynthesis go along with the generation of ROS. Similarly, conversion of arachidonic acid to leukotrienes by lipoxygenases has been identified as a source of ROS (Los et al. 1995). Nevertheless, the potential role of arachidonic metabolism in ROS formation is not yet clear.

Nitrogen-Centered Reactive Species and Their Derivates

Nitric oxide (·NO) is synthesized in the conversion of L-arginine to L-citrulline by three different ·NO-synthases (NOS). NOS have been designated as neuronal (nNOS or NOS1), endothelial (eNOS or NOS3), and inducible NOS (iNOS or NOS2), the latter of which mainly operates in the immune system. However, many tissues express more than one of these isoforms. Although the three NOS isoforms catalyze the same reaction, they differ in their regulation, site of preferred expression, and the amount and duration of ·NO generation (table 9.1) (Lincoln et al. 1997). Neuronal NOS and eNOS show a constitutive expression, and preformed proteins gain activity upon calcium influx and binding of calmodulin. Vascular formation of ·NO via eNOS is directly facilitated by increased flow and shear stress (Cooke et al. 1991). Expression of nNOS is increased by crush injury or muscle and mechanical activity, whereas it decreases following denervation (Stamler & Meissner 2001).

Table 9.1 Characteristics of ·NO-Synthase Isoforms

	NOS ISOFORM		
Characteristic	**Type I NOS (nNOS)**	**Type II NOS (iNOS)**	**Type III NOS (eNOS)**
Molecular weight	160 kDa	130 kDa	133 kDa
Cellular localization	Cytosol	Cytosol	Membrane bound
Exemplary inductors	Muscle activity Nerve injury	Cytokines Endotoxins Hypoxia	Shear stress
Ca^{2+}dependence	+	–	+
Amounts of NO generated	pmol	nmol	pmol
Exemplary functions	Neuronal messenger	Immunocytotoxicity	Vascular regulation

Adapted from Lincoln et al., 1997.

Activation of nNOS and eNOS results in low levels of ·NO, not exceeding the range of picomoles. In contrast, iNOS activity does not depend on intracellular calcium and is mainly regulated on the transcriptional level by signaling pathways that involve nuclear factor-κB (NF-κB), mitogen-activated protein kinases (MAPKs), protein kinase C, and hypoxia-inducible factor-1 (HIF-1) (Bogdan 2001; Wenger 2000). Once expressed, iNOS produces ·NO in high amounts, a process that can be sustained for several hours or even days.

Immunocompetent Cells

During stimulation, neutrophils and macrophages can generate large amounts of superoxide and its reactive derivates through activation of a phagocytic isoform of NADPH oxidase (Bogdan et al. 2000). Furthermore, NADPH oxidase is also present in eosinophils and can be induced for ROS formation (Lindsay & Giembycz 1997).

Formation of ROS in immunocompetent cells underlies a cascade of enzymatically controlled and non-enzymatic reactions (figure 9.1). In neutrophils, molecular oxygen is initially reduced to $\cdot O_2^-$ by NADPH oxidase. During the following step, H_2O_2 is produced by dismutation of superoxide via superoxide dismutase (SOD). Hydrogen peroxide is further converted to hypochlorous acid (HOCl), which is catalyzed by the enzyme myeloperoxidase (MPO). Hypochlorous acid is the most bactericidal oxidant and the essential active end product in the enzyme cascade of the oxidative burst (Weiss 1989).

Stimulation of neutrophils and monocytes/macrophages and resulting formation of ROS are subject to complex regulation mechanisms. Important activators and modulators of phagocytic NADPH oxidase are lipopolysaccharide (LPS) and several cytokines or chemokines. Interleukin-8 plays a special role, since it not only induces formation of ROS but also has many other biological properties, such as induction of transendothelial migration and expression of adhesion molecules (Baggiolini et al. 1995). Evidence exists that in both neutrophils and monocytes, intracellular signaling of the oxidative burst depends in part on an increase in free calcium (Ca^{2+}).

Expression of iNOS can be induced in several immunocompetent cells in response to inflammatory stimuli such as cytokines and LPS. Non-inflammatory iNOS inducers are hypoxia, hyperthermia, and low concentrations of ·NO itself. In contrast, cortisol and higher levels of ·NO exert inhibitory effects. In this context, the latter mechanism seems to reflect an important feedback mechanism to prevent ·NO overproduction. It is important to note that under certain conditions several non-immune cells such as hepatocytes, epithelial cells, and cardiac and skeletal myocytes are also capable of expressing iNOS (Lincoln et al. 1997; Stamler & Meissner 2001).

Reactive Oxygen and Nitrogen Species—Mechanisms of Action

Direct chemical action of most RONS is confined to the close vicinity of their site of generation and is determined by their reactivity and the availability of suitable reaction partners. The characteristics of RONS give rise to a potent destructive effect, which is the basis for their damaging effects on lipids, proteins, nucleic acids, and the extracellular matrix (Halliwell 1998). On the other hand, RONS, as generated in moderate levels under tightly controlled conditions, take on the role of regulatory mediators in a variety of signaling processes (Dröge 2002), and fulfill important physiological functions.

Oxidative Stress in the Biological System

Maintenance of the cellular redox homeostasis requires a balance between the generation rate of ROS and the capacity of the antioxidant system. The current paradigm proposes that cellular redox homeostasis is mainly regulated by redox-sensitive signaling mechanisms, which respond to an augmented formation of ROS by the induction of scavenger systems. However, if the generation of ROS is excessive or rapid, the system may not react sufficiently. In turn, the cellular redox state shifts toward a more pro-oxidant state with augmented oxidative modifications of lipids, proteins, and nucleic acids as a result (Halliwell 1998).

Lipid Peroxidation

The initial step of non-enzymatic lipid peroxidation processes is reflected by the abstraction of hydrogen from a polyunsaturated fatty acid (PUFA) side chain by ROS. In turn, resulting lipid radicals and oxygen are necessary for the intermediate and propagation steps in this radical chain reaction. Additional availability of metal ions decomposes lipid peroxides to peroxyl and alkoxyl radicals, which in turn abstract hydrogen and start new peroxidation cycles (Halliwell 1998).

During the process of lipid peroxidation, accumulating lipid peroxides exert destabilizing effects on cell membranes. This scenario gives rise to disturbances in cell integrity and results in further ROS-mediated reactions. Peroxyl radicals are capable of removing hydrogen not only from PUFA but also from nucleic acids and amino acids, explaining the occurrence of oxidative modifications of DNA and membrane proteins during the process of lipid peroxidation.

Protein Oxidation

Actions of ROS include a variety of oxidative modifications in amino acid residues such as arginine, methionine, cysteine, tryptophan, and lysine (Stadtman 2001). A major pathway of oxidative protein modification involves initial $\cdot OH^-$-induced abstraction of hydrogen from amino acid residues forming carbon-centered free radicals. Further reaction steps lead to the formation of alkyl, peroxyl, and alkoxyl radicals, which in turn may also abstract hydrogen from amino acid residues. The most frequently studied markers of oxidative protein damage are reactive carbonyl derivates (RCD).

Modification of proteins can impair their physiological function and accelerate proteolytic degradation, which can be demonstrated using glutamine synthetase as an example (Levine 1983). Inhibition of enzymes like glycerinaldehyde phosphate-dehydrogenase or mitochondrial ATP synthetase is elicited via oxidation of sulfhydryl groups by ROS, such as H_2O_2 (Cochrane 1992). Based on markers such as RCD, it is well established that accumulation of oxidized proteins occurs progressively during the process of aging and correlates with the severity of a number of diseases (Stadtman 2001). Interestingly, oxidatively damaged proteins exert scavenging effects, which have been shown to be larger than those observed in intact proteins (Dröge 2002).

Oxidative DNA Damage

DNA has been assumed to be the most biologically significant target of oxidative damage. Estimates regarding the rate of DNA damage suggest an average of at least a few hundred oxidative DNA lesions occurring in each human cell per day (Loft & Poulsen 1999). There have been further reports of a higher steady state amount of oxidative damage to mitochondrial DNA (mtDNA) than in nuclear DNA (nDNA). Oxidative damage to nDNA is considered a potential pathophysiological factor in the development of cancer. Oxidative modifications of mtDNA lead to an accumulating rate of mtDNA mutations, which in turn result in deficient mitochondrial respiratory function and disturbances in cellular energy supply (Johns 1995). For rapid removal of DNA lesions, cells are employed by a complex system of DNA repair enzymes (Croteau & Bohr 1997).

Oxidative damage to DNA involves DNA base modifications, sugar lesions, single or double strand breaks, abasic sites, and DNA–protein cross-links (figure 9.2). Oxidation of guanosine is one of the most abundant DNA modifications and leads to the formation of 8-hydroxy-2'-deoxyguanosine (8-OHdG). Due to its mispairing properties, 8-OHdG is directly involved in the process of carcinogenesis, and its detection has been adopted in many laboratories as a biomarker of oxidative DNA damage (Loft & Poulsen 1999). However, the carcinogenic effects of ROS also seem to be based

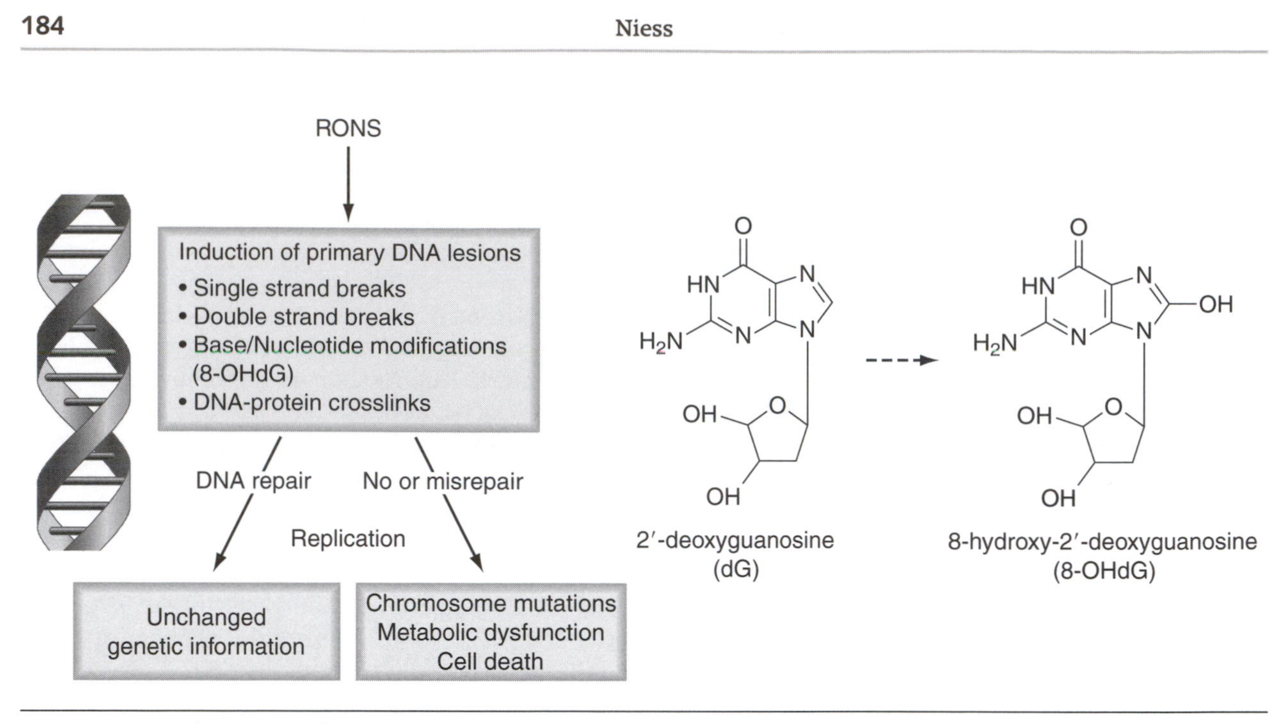

Figure 9.2 Oxidative DNA damage.

Adapted from *Handbook of oxidants and antioxidants in exercise*, edited by C.K. Sen, L. Packer, and O. Hanninen, DNA damage in exercise, pg. 196, Copyright 2000, with permission from Elsevier.

on oncogene activation or suppressor gene inactivation, and through formation of more long-lived, secondary chromosome breaking agents called clastogenic factors (Emerit 1994).

Reactive Nitrogen Species—Mechanisms of Actions

Nitric oxide rapidly diffuses and, owing to its high reactivity, has a number of cellular targets. Toxicity of ·NO is considerably enhanced by the reaction with $\cdot O_2^-$, which yields the highly reactive anion peroxynitrite ($ONOO^-$). The addition of $ONOO^-$ to cells leads to oxidation and nitration of lipids, proteins, and DNA, often resulting in cell death. In addition, formation of nitrite (NO_2) can involve intermediate generation of $\cdot NO_2$ radicals, which in turn can induce nitration of amino acid residues such as tyrosine (Tirosh & Reznick 2000). These effects of ·NO on cellular molecules or redox metal-containing proteins (Stadtman 2001) provide the basis for a broad array of signaling functions and cytotoxic actions (Lincoln et al. 1997). Furthermore, interaction of ·NO and ROS, in particular $\cdot O_2^-$, reflects an important scavenger effect. Interestingly, ·NO scavenges $\cdot O_2^-$ three times more rapidly than SOD. Thus, and likewise for ROS, actions of RNS can be characterized as a doubled-edged sword.

When produced under controlled conditions, ·NO fulfills various useful functions. It plays an important role in cell signaling pathways, primarily through activation of guanylate cyclase and modulation of ionic channels (Wink et al. 1993). By increasing intracellular levels of cyclic guanosine monophosphate (cGMP), ·NO causes vascular relaxation (for detail, see chapter 12). Furthermore, it acts as a neurotransmitter, antithrombotic or antiapoptotic molecule, or inhibitor of leukocyte adhesion to endothelium, and augments basal glucose uptake by skeletal muscle among many other physiological roles (Anggard 1994; Moncada & Higgs 1993; Stefanelli et al. 1999). In addition, ·NO is involved in host defense, but is also produced in tissues undergoing inflammatory responses (Bogdan 2001). In energy metabolism, the complex effects of ·NO include the inhibition of NADH-ubiquinon and succinate-ubiquinon-oxidoreductase in the mitochondrial respiration chain (Stuehr & Nathan 1989) and of the glycerinaldehyde-3-phosphate-dehydrogenase in glycolysis (Mohr et al. 1994), which may lead to cellular energy depletion.

Reactive Oxygen and Nitrogen Species and the Immune System

Destruction of invading microorganisms, tumor cells, or degenerated cells by RONS and related products represents an important mechanism of innate immunity (Bogdan et al. 2000). The importance of the oxidative burst in host defense is best exemplified in chronic granulomatous disease, where a deficiency in neutrophil NADPH oxidase goes along with an impaired or completely absent formation of ROS. As a result, affected individuals experience serious infections. To what extent phagocytes other than neutrophils utilize ROS for the destruction of microorganisms is still not completely clear (Bogdan et al. 2000). It has been assumed that the primary physiological role of the oxidative burst in macrophages seems to be in redox signaling rather than microbicidal activity (Forman & Torres 2001). It is important to note that ROS act in concert with RNS within the immune system, which may enhance the capacity of the host defense machinery. However, compared to the situation with ROS, the exact role of ·NO within the human immune system is less clear. In rodents, macrophage ability to kill tumor cells and microorganisms depends in part on an increased production of ·NO via iNOS. The significant role of iNOS in host defense is confirmed by the observation that iNOS knockout mice are more susceptible to infections than the wild type. Furthermore, macrophages from mice lacking iNOS failed to prevent proliferation of lymphoma cells (Lincoln et al. 1997).

Immunocompetent cells play an important role in chronic inflammatory reactions and induce damage via infiltration of tissue and subsequent release of ROS and other substances, like hydrolytic enzymes and antimicrobial polypeptides (Weiss 1989). Migrating granulocytes and monocytes are co-determinant factors in the process of ischemia–reperfusion injury (Becker 1993). Reperfusion injury is a good example of the interaction of various ROS-generating systems in causing tissue damage. Ischemia- and cytokine-induced activation of XO causes the adhesion and activation of neutrophils via generation of $\cdot O_2^-$, with the subsequent release of protease and other ROS (Bulkley 1994).

Reactive Oxygen and Nitrogen Species in Diseases

In recent years, evidence has increased that oxidative as well as nitrosative stress plays an important causative role in the pathogenesis of various acute and chronic diseases, such as cancer, arteriosclerosis, diabetes, and inflammatory and neurodegenerative diseases (Halliwell 1994b; Dröge 2002). In addition, many other diseases are associated with the deleterious action of RONS (Bogdan 2001; Halliwell 1998; Lincoln et al. 1997). Pathophysiological actions of RONS include oxidative damage to mitochondrial DNA and enzymes (Johns 1995), leading to energy disturbances of the cell. RONS also exert damaging effects on nuclear DNA, which may result in cell death or cancer. Nitric oxide produced in large amounts via iNOS has been proposed as a prime candidate for causing hypotension associated with sepsis (Lincoln et al. 1997). Furthermore, accumulation of oxidatively modified proteins, nucleic acids, and lipids increases as a function of age, which supports an important role of ROS in the process of aging (Beckman & Ames 2000). However, in a large number of disease processes, it remains to be clarified whether oxidative stress is a decisive causal factor, or more likely an epiphenomenon.

Redox-Sensitive Targets in Signaling Cascades

By activating redox-sensitive transcription factors and cellular signaling cascades, RONS take on the role of intracellular messengers. Among the key cellular components exquisitely sensitive to redox changes are NF-κB and activator protein-1 (AP-1) (Baeuerle & Henkel 1994). Further regulatory effects of RONS include activation of MAPKs, heat shock transcriptional factor-1 (HSF-1), and insulin receptor kinase by inhibition of protein tyrosine phosphatases (Dröge 2002; Martindale & Holbrook 2002) (table 9.2), (for details, see chapter 7). Inactivation of tyrosine phosphatases by ROS may represent a mechanism by which changes in the intracellular redox state can exert modulating effects on insulin receptor activation even under physiological conditions (Dröge 2002).

Activation of NF-κB requires the release of the inhibitory subunit I-κB from the NF-κB complex, which allows translocation of the p65/p50 unit to the nucleus and its subsequent binding to regulatory DNA sequences of related target genes (Baeuerle & Henkel 1994). At least two major mechanisms exist by which ROS can modulate gene expression via NF-κB. First, exposure of cells to extracellular stimuli such as inflammatory cytokines, LPS, viruses, mitogens, UV radiation, or oxidative stress increases intracellular formation of ROS, which results in rapid phosphorylation, ubiquination, and proteolytic degradation of I-κB. Several antioxidants were found to inhibit activation of NF-κB by blocking the degradation process of I-κB (Baeuerle & Henkel 1994). Thus, an enhanced pro-oxidant state seems to be crucial for initial activation of NF-κB. Second, in contrast, subsequent binding of activated NF-κB to its cognate DNA site requires more reducing conditions and can be suppressed by physiological levels of oxidizing agents (Galter et al. 1994). These site-

Table 9.2 Exemplary Transcription Factor Pathways That Respond to Changes in the Cellular Redox State

Example	Exemplary gene products/ Intracellular targets	Exemplary cellular effects
Nuclear factor κB (NF-κB)	Superoxide dismutase (SOD) Pro-inflammatory cytokines, iNOS, VCAM-1 Apoptosis genes	Antioxidant function Pro-inflammatory effects Pro-apoptotic effects
Heat shock transcriptional factor-1 (HSF-1)	Heat shock proteins (HSP)	Cellular protection, antioxidant function Chaperone function, antigen presentation
Mitogen-activated protein kinases (MAPKs)	Intracellular signaling molecules Transcription factors	Cell proliferation, myocyte differentiation Pro- and antiapoptotic effects
Hypoxia-inducible factor-1 (HIF-1)	Vascular endothelial growth factor (VEGF) Heme oxygenase-1 (HO-1) Inducible ·NO-synthase (iNOS) Erythropoietin Glucose transporter-1 (GLUT1) Aldolase, phosphofructokinase (PFK)	Angiogenesis Antioxidant function Vascular regulation Erythropoiesis Glucose transport and homeostasis Glucose metabolism

VCAM-1, vascular cell adhesion molecule-1; iNOS, inducible nitric oxide synthase.

specific requirements for NF-κB-associated gene expression may prevent uncontrolled transcriptional activation, for example during inflammatory processes (Dröge 2002).

Genes in which regulation NF-κB participates as an inducible transcriptional activator involve cell adhesion molecules, inducible nitric oxide synthase, acute phase proteins, cytokines, and hematopoietic growth factors (Baeuerle & Henkel 1994). In this context, sustained activation of NF-κB by oxidative stress has been suggested to play a central role in inflammatory processes and sarcopenia (Reid & Durham 2002). On the other hand, changes in gene expression through regulatory transcription factors are crucial components of the machinery that determines cellular protective responses to oxidative perturbations. Several proteins of the antioxidant network contain NF-κB and AP-1 binding sites in their gene promoter (Ji 2002). In this context, superoxide dismutase (SOD) and γ-glumatylcysteine synthetase (GCS) are potential targets for NF-κB-signaled gene expression. Interestingly, the gene of the antioxidative stress protein heme oxygenase-1 (HO-1) contains binding sites for several redox-sensitive transcription factors and responds to activation of NF-κB, AP-1, MAPKs, and HIF-1 (Morse & Choi 2002). However, gene expression is not only sensitive to oxygen-centered reactive species, but also reactive to ·NO and related nitrogen species. ·NO-related signal transduction involves activation of ERK, p38, and JNK subgroups of MAPKs (Sen 2000).

The Antioxidant Network

The effects of ROS are opposed by a complex network of antioxidant systems and endogenous, as well as alimentary, antioxidants (Powers & Sen 2000). Beyond their considerable role in ameliorating the harmful effects of ROS, antioxidant mechanisms also modulate a variety of redox-sensitive signaling processes. While the function of alimentary antioxidants depends in part on their ingestion with food, endogenous antioxidative systems are subject to complex regulation processes. Enzymatic antioxidative systems include superoxide dismutase (SOD), catalase (CAT), glutathione peroxidase (GPX), and thioredoxin (TRX). In addition, stress proteins such as heat shock proteins (HSP) and heme oxygenase-1 (HO-1) contribute to cellular protection. Importantly, antioxidant components often exist at specific sites and exert their functions in different compartments within and outside the cells.

Dietary and Endogenous Antioxidants

The protective role of dietary antioxidants in the biological system is well established. α-tocopherol is considered the most effective natural antioxidant (Meydani 1995), and due to its lipid solubility it contributes to membrane stability and fluidity by preventing lipid peroxidation. During its antioxidant action, α-tocopherol itself is oxidized and will partly be regenerated via reactions with ascorbic acid and reduced glutathione (GSH) (Powers & Sen 2000) (figure 9.3). Thus, the capacity of α-tocopherol to serve as a radical scavenger is closely related to the availability of other antioxidants. Hydrophilic ascorbic acid is the predominant dietary antioxidant in plasma and interstitial fluids and scavenges many ROS and RNS (Buettner & Jurkiewicz 1993). Similar to what occurs with α-tocopherol, scavenging of ROS by ascorbic acid results in the formation of its corresponding radical, which in turn can be reduced back by cellular thiols, such as dihydrolipoic acid and glutathione (GSH) (Powers & Sen 2000). Primarily membrane-bound carotenoids such as β-carotene are effective in scavenging several ROS including $\cdot O_2^-$ and peroxyl radicals (Rice-Evans et al. 1997), and carotenoid-related plant pigments exert similar effects. Flavonoids and other plant phenols inhibit lipid peroxidation and lipooxygenase enzymes, but may also prevent tyrosine nitration by scavenging RNS (Halliwell 1998).

Endogenous antioxidants include, but are not limited to, GSH, uric acid, bilirubin, ubiquinones, α-lipoic acid, ferritin, and lactoferrin. Similar to glutathione, some other endogenous antioxidants are closely related to the enzymatic antioxidant systems. Other endogenous antioxidants, such as uric acid or bilirubin, are metabolic by-products and are produced continuously within the organism.

Enzymatic Antioxidant Systems

Living organisms express a number of antioxidant enzymatic proteins, which remove excess ROS and modulate the cellular redox state. Important representatives of antioxidant enzymes include the glutathione peroxidase, superoxide dismutase, and catalase. In addition there are other antioxidant systems, the significance of which is partly unresolved.

The Glutathione System

Glutathione represents a redox active nonprotein thiol that is present in most mammalian cells at millimolar concentrations. Liver and skeletal muscle are responsible in large part for synthesis of GSH in the organism. Availability of cysteine, a major precursor of GSH biosynthesis, is a critical determinant of cellular levels of GSH. Agents such as N-acetyl-L-cysteine or α-lipoic acid can facilitate intracellular availability of cysteine and enhance cellular GSH content (Sen 2000). GSH fulfills various biological functions, which include

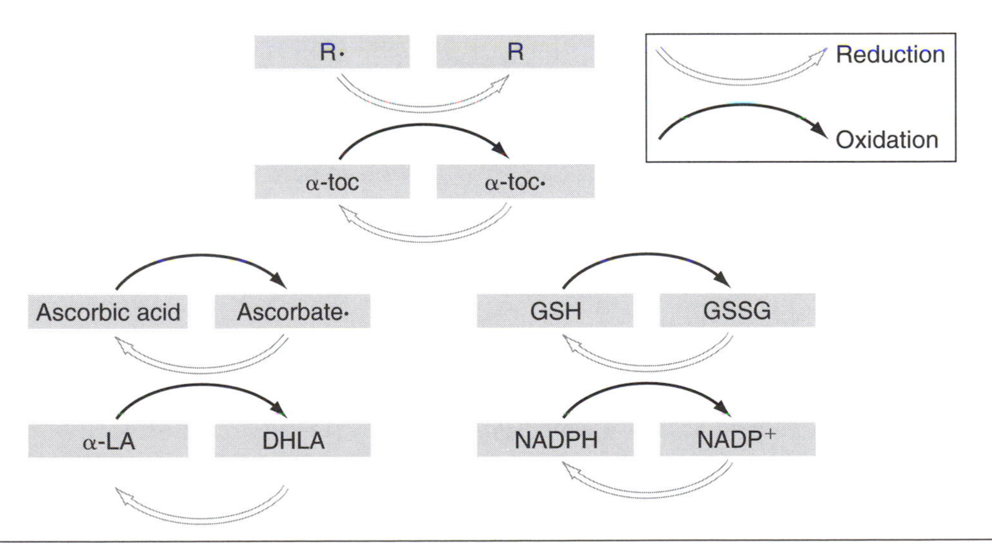

Figure 9.3 Interactions between α-tocopherol (α-toc), ascorbic acid, α-lipoid acid (α-LA), and the glutathione system. α-toc·, tocopheryl radical; ascorbate·, ascorbate radical; GSH, glutathione; GSSG, oxidized glutathione; GPX, glutathione peroxidase; GR, glutathione reductase; DHLA, dihydrolipoic acid.

Adapted from *Handbook of oxidants and antioxidants in exercise,* edited by C.K. Sen, L. Packer, and O. Hanninen, DNA damage in exercise, pg. 196, Copyright 2000, with permission from Elsevier.

detoxification of electrophilic xenobiotics, storage and transport of cysteine, modulation of redox-sensitive signaling processes, reduction of ascorbate or tocopheroxyl radicals, and other antioxidant effects (Sen 1998). GSH exerts its antioxidant function either via direct scavenging of ROS or by catalytic action of glutathione peroxidase (GPX), leading to the formation of oxidized GSH (GSSG) (figure 9.3).

The selenoprotein GPX is located in both the cytosol and mitochondrial matrix in close vicinity to cellular sources of H_2O_2 formation. GPX catalyzes the reduction of H_2O_2 and a wide range of complex organic hydroperoxides to H_2O and alcohol, respectively, using GSH as the electron donor. Importantly, GSSG can be quickly recycled to GSH by GSSG reductase in the presence of NADPH. As a result, in most tissues, levels of GSSG are very low and the ratio of GSH to GSSG is normally kept at higher values (Powers & Sen 2000). Enzymatic activity of GPX depends on the availability of selenium, and selenium deficiency can dramatically suppress the function of GPX in all tissues in a dose-dependent fashion (Ji 1998).

Regulation of several enzymes and cellular signaling molecules depends on the thiol-disulfide exchange between protein thiols and low molecular weight disulfides. For example, all protein tyrosine phosphatases exhibit a cysteine residue in the active center, which can be converted into a mixed disulfide by GSSG and results in the loss of catalytic activity (Dröge 2002). Similarly, physiological levels of GSSG can suppress binding of activated NF-κB to its cognate DNA site, underlining the important role of the cellular thiol redox state for the regulation of gene expression (Galter et al. 1994). Furthermore, a decrease of the GSH/GSSG ratio has been shown to deplete internal Ca^{2+} stores (Henschke & Elliott 1995) and represents a mechanism by which changes in the cellular thiol redox state inhibit signal transduction.

The Thioredoxin (TRX) System

The main components of the thioredoxin system are thioredoxin (TRX), TRX reductase, and NADPH. The TRX system plays a central role in keeping proteins in their reduced state (Arner & Holmgren 2000). TRX reductase from mammalian cells is a dimeric flavin enzyme and comprises a glutathione reductase-like equivalent. Antioxidant properties of the TRX system include removal of H_2O_2, free radical scavenging, and protection from oxidative stress. In addition, the TRX system contributes to ascorbate regeneration and may thus be a part of the complex antioxidant network (Sen 2000). Besides its important role within the antioxidant system, the TRX system has been shown to exert a key regulating function in cellular signaling and gene transcription.

Superoxide Dismutase (SOD) and Catalase (CAT)

SOD, a metalloprotein, catalyzes the reaction of $\cdot O_2^-$ to H_2O_2 and O_2. The three existing isoenzymes of mammalian SOD can be characterized by metal ions and their cellular locations. Cu- and Zn-SOD are cytosolic enzymes, whereas Mn-SOD is found in mitochondria (Suzuki, Ohno 2000). It has been estimated that up to 80% of the formed $\cdot O_2^-$ in the mitochondria are reduced by SOD.

Catalase (CAT) is mainly present in the peroxisomes of most mammalian cells, but mitochondria and other intracellular organelles such as endoplasmic reticulum also contain this enzyme. CAT catalyzes the decomposition of H_2O_2 by converting it to H_2O and O_2. Similar to the situation with SOD, the activity of CAT is highest in liver and rather low in skeletal muscle (Ji 1998).

Stress Proteins With Antioxidant Properties

Heat shock proteins (HSP) can be grouped into six families with molecular weights ranging between 18 and 110 kDa (for details, see chapter 10). Typical representatives of HSP are HSP27 or HSP70, whose expression is induced by heat shock and several other stressors including oxidative stress. Oxidation and depletion of nonprotein thiols result in HSF-1 activation, HSP gene expression, and HSP synthesis (Paroo et al. 2002). Up-regulation of HSP72 has been shown to provide protection from ischemia–reperfusion-induced lipid peroxidation in rat myocardium (Demirel et al. 1998). Furthermore, overexpression of mouse HSP25 and human HSP27 exerts augmenting effects on cellular levels of GSH by increasing the activity of glutathione reductase and glucose-6-phosphatase dehydrogenase (Preville et al. 1999) (figure 9.4). A specific antioxidant stress protein is represented by heme oxygenase, which exists in two isoforms, the inducible HO-1 (32 kDa) and the constitutive HO-2 (36 kDa) (Morse & Choi 2002). An elevated expression of HO-1 provides protection of the cell against oxidative stress by reducing the intracellular pool of free iron via induction of ferritin synthesis. Moreover, HO-1 catalyzes the initial step in the degradation of heme to bilirubin, a potent water-soluble antioxidant. HO-1 has been shown to be strongly

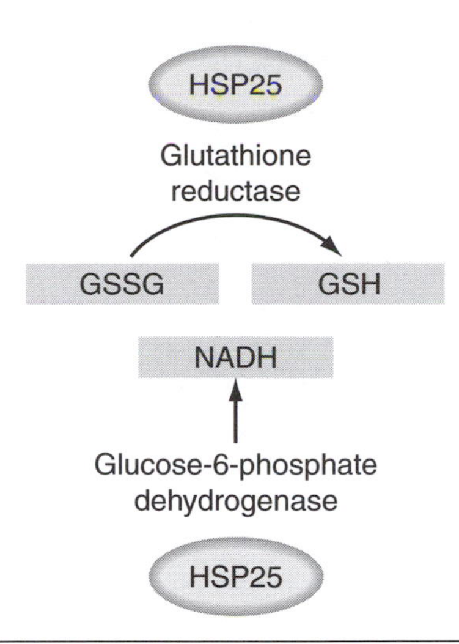

Figure 9.4 Effects of heat shock protein 25 (HSP25) on the glutathione system. HSP25 increase reduced glutathione (GSH) by inducing effects on glutathione reductase (GR) and glucose-6-phosphatase dehydrogenase (G6PD) activities. G6PD provides reducing power (NADH) for GR.

Reprinted, by permission, from T.J. Koh, 2002, "Do small HSP protect skeletal muscle from injury?" *Exercise and Sport Science Reviews* 30(3): 120.

induced by RONS, hypoxia, UVA irradiation, and various other inducers of oxidative stress. Thus, HO-1 seems to have potential as a useful marker of cellular oxidative stress (Dröge 2002). Regulation of HO-1 expression at the RNA level seems to involve transcription factors such as NF-κB, AP-1, MAPKs, and HIF-1 (Morse & Choi 2002).

Reactive Oxygen and Nitrogen Species in Exercise

Research over the last 25 years has revealed a complex link between exercise and RONS. In this context, RONS are implicated in cellular damage in terms of oxidative stress arising from strenuous exercise. Furthermore, RONS modulate various cellular functions, including contractile force and gene expression. Adaptation of antioxidative systems in response to regular training may lead to a limitation of exercise-induced oxidative stress and reflects a mechanism responsible for an augmented tolerance to exercise.

Effects of Acute Exercise on RONS Generation

Acute exercise has been shown to induce a complex stress response, which involves reactions on the cardiocirculatory, metabolic, hormonal, and immunological level. More recent research has shown that the exercise-induced stress-response is also paralleled by the formation of RONS.

Exercise-Induced Generation of Oxygen-Centered Reactive Species

While indirect methods initially hinted at an exercise-induced formation of free radicals (Alessio et al. 1988; Dillard et al. 1978; Lovlin et al. 1987; Sumida et al. 1988), direct evidence of increased ROS formation using electron spin resonance spectroscopy (ESR) was found in rat or cat muscle (Davies et al. 1982; Jackson et al. 1985; O´Neill et al. 1996) and more recently by Asthon et al. (1998) in human serum after exhausting exercise. The augmenting effect of exercise on ROS production is further supported by findings of changes in the thiol/disulfide redox state, as shown by decreases in skeletal muscle as well as serum GSH/GSSG ratios in response to strenuous exercise (Sen, Atalay et al. 1994; Sen, Rankinen et al. 1994).

Principally, different mechanisms seem to be responsible for an augmented formation of ROS in response to exercise. The metabolic rate in working skeletal muscle increases to up to 100 times higher than basal levels, leading to a marked increase of oxygen consumption. This increase of consumed oxygen gives rise to the electron flux through the mitochondrial electron transport chain, which results in an augmented formation of $\cdot O_2^-$ (Benzi 1993). Some authors assume that during intensive exercise, formation of $\cdot O_2^-$ may be further enhanced by a loss of cytochrome oxidase activity (Sjödin et al. 1990). However, some evidence exists that generation of ROS in response to increased muscle activity is not solely derived from the electron transport chain but also influenced by other mechanisms.

Under basal conditions, XO does not seem to play a major role in ROS formation. However, exercise has been shown to be associated with changes in purine metabolism (Sutton et al. 1980). The so-called ATP breakdown occurring during high-intensity exercise leads to an augmented formation of AMP via adenylate kinase, which is further degraded to IMP, inosine, and hypoxanthine. A loss of cellular calcium homeostasis due to vigorous exercise allows conversion of XD to its oxidase form and therefore enhances the amount of generated superoxide radicals (Jackson 1998). Increased levels of hypoxanthine and uric acid (Hellsten-Westing et al. 1991) indirectly indicate that XO is active in response to exercise, which has been confirmed recently by direct measurements of XO activity in the plasma of rats (Vina et al. 2000). In addition, treatment with allopurinol, an inhibitor of XO, prevents exercise-induced oxidation of GSH, providing evidence that XO contributes to ROS formation in response to exercise.

An augmented release of iron-containing heme proteins, such as hemo- or myoglobin, enhances the availability of free iron, which can stimulate formation of ROS (Halliwell 1998). Jenkins et al. (1993) found an increased concentration of loosely bound iron in muscle tissue of rats exercised to exhaustion, whereby evidence of increased lipid peroxidation was observed in parallel. This effect could be explained by an increased release of free iron from hemo- and myoglobin or other heme products as a result of exercise-induced damage to muscle tissue.

Whether contractile activity contributes to ROS generation by this NADPH oxidase in working muscle remains unclear at present. From a hypothetical point of view, hormonal, hemodynamic, and metabolic changes may be involved in the exercise-induced activation of nonphagocytic NADPH oxidases (Dröge 2002).

Exercise-Induced Generation of Nitrogen-Centered Reactive Species

The first evidence of an augmented formation of ·NO in response to exercise was derived from studies that assessed exhaled ·NO (Bauer et al. 1994; Phillips et al. 1996) or its major stable metabolites nitrate and nitrite in plasma or urine (Brouwer et al. 1997; Jungersten et al. 1997; Leaf et al. 1990). After endurance exercise, increases in plasma nitrate ranged between 20% and 100% (Brouwer et al. 1997; Jungersten et al. 1997; Suzuki, Yamada 2000). Although the precise mechanisms for augmented ·NO formation during exercise have not been completely elucidated, several indices exist to suggest that both eNOS, activated as a result of shear stress, and iNOS, as induced by the induction of inflammatory mechanisms, are involved (Jungersten et al. 1997; Shen et al. 1995).

More recent research has evoked skeletal muscle as one source of ·NO formation during exercise (Reid 2001). Under physiological conditions, constitutive expression of eNOS and nNOS in muscles appears to provide the basis for contraction-induced ·NO formation via mechanical stimulation. However, elevated levels of ·NO in response to eccentric muscle damage (Radák, Pucsok et al. 1999) may also result from up-regulation of iNOS (Sakurai et al. 2001), which could be induced by local inflammatory mechanisms.

Immuncompetent Cells

Few data are available on the contribution of immunocompetent cells to exercise-induced production of ROS. Cannon et al. (1990) described an increased generation of ROS in nonstimulated neutrophils one day after 45-min running exercise at 70% of $\dot{V}O_2$max. In another investigation, Guarnieri et al. (1992) detected a significant formation of $\cdot O_2^-$ in patients with stable angina directly after maximal physical exercise. This finding was also apparent to a lesser degree in a control group of healthy subjects.

Evidence exists that neutrophils become activated in response to vigorous endurance exercise, more intensive workouts, or eccentric muscle stress as induced by downhill running (Camus et al. 1992; Niess, Passek et al. 1999). An augmented release of neutrophil granule constituents, such as elastase and MPO, in response to exercise reflects direct neutrophil activation in vivo. Along with a parallel release of IL-8 and TNF-α (Niess, Passek et al. 1999; Northoff & Berg 1991), this may be the basis for an augmented production of ROS in neutrophils due to exercise.

In contrast to the spontaneous formation of ROS in response to acute exercise, a large number of studies have examined the exercise-related alterations of the respiratory burst activity as determined by in vitro stimulation of neutrophils. While moderate exercise has been shown to prime human neutrophils for ROS production, investigations that assessed the effect of intensive or prolonged exercise yielded considerable variation in results. Several studies have revealed decreases, whereas others have shown an increase or no change of the respiratory burst activity in response to such more vigorous exercise protocols (Peake 2002).

With regard to RNS, induction of iNOS mRNA and protein expression has been shown to occur in human leukocytes after vigorous endurance exercise and intensive exhaustive exercise, respectively (Niess, Sommer, Schlotz et al. 2000). This may contribute to an augmented endogenous formation of ·NO and reflects a systemic inflammatory response to heavy exertion.

Exercise-Induced Formation of RONS—Consequences and Functional Aspects

Exercise-induced formation of RONS can result in oxidative stress and exert modulating effects on cellular functions; it may also be involved in modulation of gene expression via redox-sensitive transcription pathways.

Exercise-Induced Oxidative Stress

Current literature shows that enhanced formation of ROS in response to exercise can lead to oxidative modification of lipids, proteins, nucleic acid, and other cellular compounds (Alessio 1993; Hartmann & Niess 2000; Niess, Dickhuth et al. 1999; Sjödin et al. 1990; Tiidus 1998). Exercise-induced oxidative stress seems to occur if the generation of RONS overwhelms tissue antioxidative defenses. The functional consequences of exercise-induced oxidative stress are only partly understood. It is important to note that features of oxidative stress in response to exercise are usually transient. Furthermore, exercise-induced oxidative damage exhibits a magnitude comparable to that observed under several pathophysiological conditions. Nevertheless, further research is necessary to elucidate the extent to which oxidative stress may negatively affect exercise performance or regeneration or affects health over the long term.

Exercise-Induced Lipid Peroxidation

One of the first studies concerning the occurrence of lipid peroxidation in response to exercise was performed by Dillard et al. (1978). They were able to detect an increase in expired pentane after 20-min cycling ergometry at 75% $\dot{V}O_2$max. In numerous further studies, exercise-induced lipid peroxidation was assessed by measuring thiobarbituric acid reactive substances (TBARS), malondialdehyde (MDA), lipid hydroperoxides, or F2-isoprostane in blood, urine, various tissues, or expired air (Alessio et al. 1988; Alessio 1993; Davies et al. 1982; Ji 1992; Kayatekin et al. 2002; Khanna et al. 1999; Maughan et al. 1989; Saxton et al. 1994; Venkatraman et al. 2001; Steensberg et al. 2002). Although the majority of the studies could show that exercise exerts an inducing effect on lipid peroxidation in animal models as well as in humans, the results vary considerably. This seems to be due to differences in factors such as training status, exercise protocols, analyzed medium or tissue, time of sampling, and assay techniques used. It appears that measurements performed in plasma or serum more often yield equivocal findings if compared to detection in cells such as myocytes or erythrocytes. Furthermore, vigorous endurance exercise seems to have rather inducing effects when compared to short-time but more intensive protocols. In addition, lipids are more likely to undergo peroxidation during exhaustive isometric exercise when compared to short-term treadmill exercise (Alessio et al. 2000).

Exercise-Induced Protein Oxidation

Initial results from Reznik et al. (1992) demonstrated that rats subjected to exhaustive exercise accumulated reactive carbonyl derivates (RCD) in muscle. Similar results were observed in the hind legs of rats following a two-month period of strenuous running exercise (Witt et al. 1992), indicating an increased rate of oxidative damage to proteins. Further studies confirmed these findings in muscle (Sen et al. 1997; Smolka et al. 2000) and lung tissue (Radák et al. 1998) of rats and in human plasma (Alessio et al. 2000; Saxton et al. 1994), whereas in another investigation accumulation of RCD occurred only if exercise was performed during hypoxia (Radák et al. 1997). Interestingly, immobilization similarly results in elevated protein oxidation (Tirosh & Reznick 2000). At the moment, there is no substantial information on the functional significance of exercise-induced protein oxidation. Whether protein oxidation may have suppressive effects on exercise performance via inhibition of enzymes such as glycerinaldehyde phosphate-dehydrogenase or mitochondrial ATP synthetase (Cochrane 1992) remains speculative and awaits further research.

Exercise-Induced DNA Damage

There is growing evidence that certain regimens of exercise are also capable of inducing oxidative modifications of DNA. Principally, four different endpoints have been used to investigate the effects of exercise on DNA damage: (1) analysis of DNA strand breaks in peripheral leukocytes using the Comet assay, (2) the excretion rate of oxidized bases and nucleosides from DNA as a measure of the total amount of damage in the body, (3) the level of oxidized bases in DNA within specific cells, and (4) analysis of exercise-related alterations of DNA on the chromosomal level.

First indices of exercise-induced oxidative DNA damage were revealed by a study that detected elevated excretion rates of 8-OHdG in the urine of trained athletes directly after a marathon race (Alessio & Cutler 1990). Further investigations could not confirm these initial findings after a triathlon competition (Hartmann et al. 1998), after 6 min of maximal rowing exercise (Nielsen et al. 1995), or after a single training run (Inoue et al. 1993). In contrast to findings from such single bouts of exercise, measurable increases of urinary 8-OHdG excretion have been shown after 30 days of intense exercise (Poulsen et al. 1996), in marines engaged in 14-day field training at moderate altitude (Chao et al. 1999), after a runners' training camp (Okamura et al. 1997), and during a four-day supermarathon race (Radák et al. 2000). Thus, it appears that accumulation of oxidative DNA damage may occur only in response to prolonged or repeated severe exercise. Importantly, analysis of urinary excretion of 8-OHdG does not allow any conclusion regarding the cells, where DNA damage occurs.

Using the Comet assay, Hartmann et al. (1994) were the first to demonstrate the occurrence of DNA strand breaks in leukocytes 24 h after intensive anaerobic exercise. The Comet assay is a sensitive gel electrophoretic method to determine single strand breaks and alkali-labile sites in the DNA of single cells (McKelvey-Martin et al. 1993). Other studies confirmed these delayed DNA effects in the Comet assay in response to a half-marathon (Niess et al. 1998) and triathlon competition (Hartmann et al. 1998), whereas no effects were observed after 45 min of moderate aerobic or eccentric exercise (Niess et al. 1999b). Indirect evidence that the results obtained by the Comet assay reflect oxidative DNA damage is derived from findings of a lower extent of exercise-induced DNA effects in leukocytes under administration of α-tocopherol (Hartmann et al. 1995) and a more pronounced occurrence of DNA damage in the neutrophils compared to mononuclear cells after a half-marathon race (Hartmann & Niess 2000). The latter observation could explain in part results from Inoue et al. (1993), who could not observe any evidence of exercise-induced oxidative DNA damage by measuring 8-OHdG in isolated human lymphocytes. More recently, Tsai et al. (2001) revealed that pyrimidines are major targets for oxidative modification in peripheral leukocytes in response to prolonged and vigorous running exercise. They used a modified protocol of the Comet assay, in which the cells were additionally treated with lesion-specific endonucleases.

With respect to the delayed time course of leukocyte DNA strand breaks in response to exercise, it is likely that the interaction of lipid peroxidation products and DNA, or other secondary auto-oxidative mechanisms including the formation of ·NO due to activation of the cells, are involved (Hartmann & Niess 2000). As reflected by augmented levels of 8-OHdG in muscle 24 h after eccentric muscle exercise (Radak et al. 1999b), exercise-related DNA damage does not seem to be restricted to immunocompetent cells.

The functional significance of exercise-induced DNA damage is not completely clear and has to be further elucidated. Importantly, exercise-induced DNA effects in leukocytes are not paralleled by elevated frequency of micronuclei, suggesting the absence of chromosomal damage in response to exercise (Hartmann et al. 1998, 1994). This indicates adequate repair of DNA lesions by corresponding endonucleases.

Exercise-Induced Changes in Immune Function—Possible Roles of RONS

In vitro studies have shown that the influence of ROS leads to changes in the function of neutrophils. Studies by Baehner et al. (1977) documented a reduction in both the migration and the phagocytosis capacity under the influence of H_2O_2. Other studies confirmed the inhibitory impact of ROS on locomotor and microbicidal function of phagocytes (Anderson 1982). Via its blocking effect on NADPH oxidase, ·NO exerts a suppressive effect on $\cdot O_2^-$ production and thus on the oxidative burst of neutrophils (Lincoln et al. 1997). Thus, exercise-induced generation of RONS could explain a reduced capacity of neutrophils to produce $\cdot O_2^-$ after strenuous physical exertion (Peake 2002). As could be shown in the rat model, reduced glutathione (GSH) exerts a stabilizing effect on mobilization and oxidative burst of neutrophils after exhaustive exercise (Atalay et al. 1996). Oxidants are also potent inhibitors of lymphocyte functions. The capacity of lymphocytes to respond to a mitogenic stimulus, as well as cell viability, was reduced after exposure to ROS (Marini et al. 1996). Heavy physical exercise has been shown to decrease the proliferative response of lymphocytes (Nieman et al. 1995). However, the precise involvement of RONS formation in exercise-related changes in immune function awaits further research (Niess et al. 1999a).

Exercise-Induced Muscle Damage

Damage to skeletal muscles has been described after intense, prolonged, or eccentric exercise or a combination of these (Clarkson &Tremblay 1988), and is followed by the delayed onset of muscle soreness. The mechanisms leading to this phenomenon are not completely clear as yet. A relationship could be disclosed between the extent of immigrating neutrophils and muscle damage (Fielding et al. 1993). The formation of RONS by the leukocytes invading muscle tissue must be anticipated in connection with manifest inflammation. In the animal model, vigorous exercise has been shown to be followed by an increased free radical signal measured by ESR (Jackson et al. 1985). Secondary indices of increased oxidative stress, such as elevated parameters of lipid peroxidation and protein oxidation, could also be detected in muscle tissue after heavy exercise. In humans, a delayed increase of TBARS in plasma after eccentric running exercise (Maughan et al. 1989) was postulated to reflect an ROS-associated mechanism, which influences the onset of muscle damage. In addition to the mitochondrial respiration chain or xanthine oxidase, invading neutrophils and monocytes are assumed to be essential sources of ROS and causes of oxidative stress in the muscle during and after physical exertion. Using muscle glutathione status as a marker of oxidative stress, in the mouse model, Duarte et al. (1994) showed that inhibition of the function of neutrophils invading the muscle by colchicine reduces the extent of oxidative stress and damage to muscle. In contrast, antioxidant supplementation failed to prevent ultrastructural muscle injury induced by eccentric exercise (Warren et al. 1992). Thus, clear indicative findings regarding the causal role of oxidative stress in eccentric muscle damage are not available as yet.

Cellular Redox State and Contractile Force of Skeletal Muscle

Exercise-induced generation of RONS can exert modulating effects on skeletal muscle contractility (Reid 2001). Under basal conditions, contractility of unfatigued skeletal muscle is enhanced by a moderate shift of the redox homeostasis toward a more pronounced pro-oxidant state. Alternatively, antioxidants can diminish force production in unfatigued muscle fibers (Coombes et al. 2001). At higher concentrations, ROS exert suppressive effects on force production, a scenario that seems to be involved in exercise-related fatigue of skeletal muscle and can be reversed in part by antioxidants (Reid et al. 1994). Thus, a biphasic relation between the cellular redox state and contractile function has been proposed (figure 9.5).

In contrast to ROS, ·NO depresses force production in both fatigued and unfatigued muscle (Reid 2001). This effect is quite similar to the effect that ·NO exerts on vascular smooth muscle. Treatment with NOS inhibitors has been shown to increase force production during twitch and submaximal contractions in exercising diaphragma, while ·NO donors have the opposite effect (Kobzik et al. 1994).

The mechanisms by which RONS affect force production in skeletal muscle are only still partly defined. Suppressive effects of RONS result, in part, from modifications of redox-sensitive proteins including myofilaments and components of the sarcoplasmic reticulum. In this context, RONS may contribute to perturbations in calcium homeostasis, with the onset of muscle fatigue as a result. Actions of ·NO on muscle contractility further include suppressive effects on key enzymes of the energy metabolism and an increase in cyclic guanosine monophosphate (cGMP) concentrations (Reid 2001).

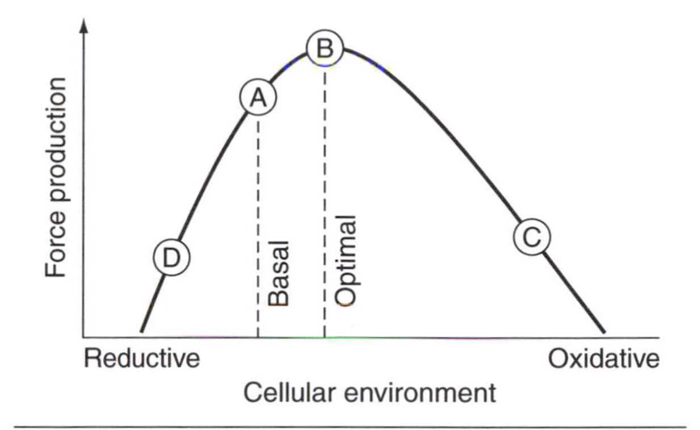

Figure 9.5 Influence of the cellular redox state on contractile force in nonfatigued skeletal muscle. Model as proposed by Reid (2001) and used with permission. A, Basal conditions in unfatigued muscle; B, addition of low doses of reactive oxygen species (ROS); C, addition of large doses of ROS; D, addition of antioxidants.

Adapted, by permission, from M.B. Reid, 2001, "Invited Review: redox modulation of skeletal muscle contraction: What we know and what we don't," *Journal of Applied Physiology* 724-731.

Redox Sensitivity of Transcription Factor Pathways

Recent research provides evidence that acute physical exertion can activate redox-sensitive intracellular signaling cascades. In a well-conducted study, Hollander et al. (2001) investigated the impact of exhaustive treadmill running lasting 1 h on NF-κB and AP-1 binding activity in rat vastus lateralis muscle. NF-κB binding activity increased 2 h postexercise and remained elevated up to 48 h after exercise. AP-1 binding was also increased but returned to baseline within a few hours. Importantly, MnSOD mRNA increased in parallel, which may point to an important role of NF-κB and AP-1 in modulation of the antioxidant defense in response to exercise. In humans, increased NF-κB binding activity has been detected in mononuclear cells up to 6 h after a 60-min treadmill run performed at 80% of $\dot{V}O_2$max (Vider et al. 2001; Weiss et al. 2002). These changes were paralleled by augmented levels of lipid peroxidation products and a decreased GSH/GSSG ratio, respectively. Whether an augmented NF-κB binding activity in immunocompetent cells may represent an underlying mechanism in the acute phase response to exercise remains to be determined.

MAPKs control the differentiation and adaptation of various cell types including myocytes. Based on structural differences, MAPKs are represented by four parallel cascades: stress-activated protein kinase p38 ($MAPK^{p38}$), c-jun N-terminal kinases ($MAPK^{jnk}$), extracellular signal-regulated kinase 1 and 2 ($MAPK^{erk1/2}$), and the extracellular signal-regulated kinase 5 ($MAPK^{erk5}$) (Martindale & Holbrook 2002). MAPK activation can be triggered by lowered pH, hormonal changes, or mechanical and oxidative stress, the occurrence of which is also seen in the context of physical exertion. Indeed, studies on human skeletal muscle revealed that acute exercise can activate MAPKs such as $MAPK^{erk1/2}$ (Aronson et al. 1997; Widegren et al. 1998). As shown recently in isolated rat skeletal muscle, concentric exercise induces phosphorylation of $MAPK^{erk1/2}$ but not $MAPK^{p38}$, whereas eccentric contractions activate both kinases (Wretman et al. 2001). Interestingly, antioxidant treatment reduced contraction-induced phosphorylation of $MAPK^{erk1/2}$ from a fivefold to only a 1.5-fold increase after the concentric protocol. Thus, exercise-induced activation of $MAPK^{erk1/2}$ seems to be modulated in part by changes of the cellular redox state. Further research is needed to elucidate the potential role of ROS in the induction of training-adaptive changes via activation of redox-sensitive transcription factors.

Acute Exercise and Antioxidant Systems

Among the three major antioxidant enzymes, MnSOD and GPX are of mitochondrial location, a predominant site of ROS formation during exercise. Similar to CAT, SOD and GPX activities are highest in oxidative type 1 muscle and lowest in fast-twitch fibers (Powers & Hamilton 1999). Based on the paradigm that enzymatic antioxidant systems such as SOD, CAT, and GPX provide first-line defense against ROS, it is expected that exercise may have inducing effects on these protective mechanisms. Indeed, acute strenuous exercise has been shown to increase SOD activity in various rodent tissues such as skeletal muscle, heart, lung, liver, erythrocytes, and platelets. Current data also suggest that exercise exerts elevating effects on CuZnSOD rather than on its manganese isoform (Ji 1998). The few available data from studies in humans show conflicting results, with no acute effects in muscle after a marathon run but an elevated erythrocyte SOD activity in response to a 20-day bicycle race (Mena et al. 1991).

More recent research revealed increased MnSOD mRNA abundance in rat muscle after 60-min treadmill running at a moderate intensity of approximately 65% of $\dot{V}O_2max$. However, only a small concomitant increase of MnSOD protein content, with no change in enzyme activity, was observed (Hollander et al. 2001). Thus, it appears that substantial up-regulation of the SOD system requires cumulative stimulation as occurs during chronic exercise training.

Reports about the influence of acute exercise on GPX activity have been inconsistent. These discrepancies may be partially related to the different exercise regimens. Interestingly, GPX transcription was found to be decreased in deep and superficial vastus lateralis muscle of rats after acute exercise (Ji 1998). Largely consistent findings exist regarding the failure of CAT activity to be up-regulated in response to one bout of exercise.

Essig et al. (1997) were the first to demonstrate an increased expression of HO-1 at the RNA level in muscles of mice stressed by acute exercise. Rapid activation of HO-1 transcription in working muscle has also been shown in humans subjected to 60- to 90-min one-legged knee extensor exercise or 4-h cycling ergometry performed at 50% to 60% of $\dot{V}O_2max$ (Pilegaard et al. 2000). However, with respect to the more pronounced increase in HO-1 mRNA elicited by the knee extensor protocol, the authors concluded that induction of this stress protein is more related to exercise intensity. In immunocompetent cells, up-regulation of HO-1 on protein level seems to be restricted to vigorous endurance exercise, such as competing in a half-marathon run (Fehrenbach et al. 2003; Niess et al. 1999c; Niess et al. 2000b) (figure 9.6).

Compared to data on HO-1, there are considerably more data regarding the inducing effects of acute exercise on HSP70 expression (detailed in chapter 10). However, findings regarding a potential contribution of ROS to exercise-related up-regulation of HSP70 are still controversial (Kelly et al. 1996; Niess et al. 2002b; Paroo et al. 1999, 2002). It is most likely that ROS exert stimulating effects via cellular protein damage, which in turn provides signals for HSP expression. Up-regulation of GSH-increasing HSP27 has been shown after a single bout of eccentric exercise in human skeletal muscle (Thompson et al. 2001) and in leukocytes

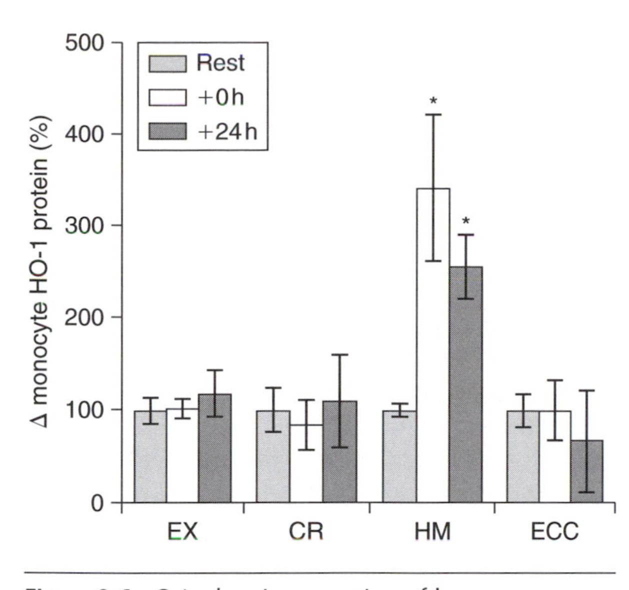

Figure 9.6 Cytoplasmic expression of heme oxygenase-1 (HO-1) in monocytes at rest, 0 h, and 24 h after different exercise procedures. The flow cytometric results (mean fluorescence intensity) are presented in percentage changes of monocyte HO-1 after exercise. Pre-exercise values are normalized to 100%; values are means (columns) and corresponding 95% confidence intervals (bars). EX, exhaustive treadmill run (total duration 30 min on average); CR, treadmill run (60 min) at 75%$\dot{V}O_2$max; HM, competitive half-marathon run (duration 90 min on average); ECC, eccentric exercise (M. quadriceps femoris). * Significantly different from resting conditions (5% level).

of runners after endurance exercise (Fehrenbach et al. 2000).

Effects of Regular Training on RONS and Associated Mechanisms

Adaptation to regular training is the basis for augmented tolerance to exercise. A training-induced limitation of exercise-induced oxidative stress may be the result of a lower formation of RONS but also of an upregulation of the antioxidative systems.

Modulating Effects of Exercise Training on RONS Formation

Initial results showing that adaptation to regular training includes modulation of RONS formation are derived from studies on human leukocytes. Moderate endurance training augments oxidative burst capacity upon stimulation (Smith et al. 1992), which may reflect a beneficial effect on immunity. Conversely, training with high volume or intensity, or both, exerts suppressive effects on stimulated $\cdot O_2^-$ formation in neutrophils—a finding that could be demonstrated in cross-sectional studies (Lewicki et al. 1987; Smith et al. 1990) as well as in longitudinal trials (Miyazaki et al. 2001; Pyne et al. 1995). Importantly, training-induced changes of oxidative burst activity were not related to susceptibility to infectious episodes (Pyne et al. 1995). From a hypothetical point of view, a lower oxidative burst activity may rather reflect a protective effect by lowering the inflammatory response to repeated vigorous exercise (Hoffman-Goetz & Pedersen 1994).

Reduced basal levels of ·OH in plasma and soleus muscle of endurance-trained rats (Itoh et al. 1998) also point to an attenuating effect of chronic exercise on the formation of ROS, whereas the underlying mechanisms are still unclear. Sjödin et al. (1990) postulated that there is a training-induced reduction in ROS formation in contracting muscle. In this context, a more tightly coupled electron transport chain could decrease endogenous ROS formation at rest and during exercise. However, there is no substantial evidence as yet to suggest a more efficient mitochondrial energy transfer in response to regular training (Tonkonogi et al. 2000).

Reactive Nitrogen Species

As shown by Jungersten et al. (1997), regular physical exercise goes along with higher plasma levels of nitrate at rest. However, data from this study did not permit any conclusion as to whether augmented formation of ·NO at rest would be of vascular or other origin. Studies in rodents addressed this issue and revealed an increased expression of the eNOS gene in the aortic endothelium (Yang et al. 2002) and of the corresponding protein in the coronary arterial tree and in skeletal muscle (Balon & Nadler 1997; Laughlin et al. 2001) after chronic exercise. In humans, evidence was provided that endurance training may increase the amount eNOS in microvessels within skeletal muscle, while the expression of nNOS was not affected (Frandsen et al. 2000). Thus, it is premature to draw final conclusions regarding beneficial training effects on eNOS and nNOS expression with regard to improved blood flow or augmented glucose uptake in skeletal muscle.

Young mice subjected to chronic treadmill training exhibited an augmented expression of iNOS mRNA and protein in stimulated macrophages, which in turn showed enhanced in vitro tumor cell killing (Woods et al. 2000). Similarly, regular endurance training exerts a moderate but significant priming effect on iNOS mRNA expression in stimulated human blood mononuclear cells, while no change in the corresponding protein content was observed (Niess, Fehrenbach, Vogel et al. 2002c). In contrast, basal levels of iNOS mRNA in nonstimulated total leukocytes were lower in endurance-trained athletes than in sedentary subjects (Niess, Fehrenbach, Schlotz, Northoff et al. 2002a). This finding points to an attenuating effect of regular training on basal iNOS expression, which may be of therapeutic value in diseases, accompanying a chronic elevated iNOS expression. Indeed, recent research has shown that a long-term training program is associated with reduced local expression of inflammatory cytokines and iNOS in skeletal muscle of patients with chronic heart failure (Schulze et al. 2002). Similar to that of acute exercise, the effect of regular training on NOS expression appears to be highly complex and seems to vary between isoforms, investigated cell type, and the stimulation procedure used.

Adaptation of Antioxidant Systems to Exercise Training

There is growing evidence that beneficial effects of regular physical activity include the enhanced functional capacity of the antioxidant network. High-intensity and long-duration exercise seems to be capable of increasing muscle GSH content in the animal model (Powers & Sen 2000). In contrast, eight weeks of moderate endurance training consisting of 35-min cycling, three times a week, failed to affect the GSH/GSSG ratio in human vastus lateralis muscle (Tiidus et al. 1996), whereas a training program of similar duration was capable of enhancing GSH content in human lymphocytes (Hack et al. 1997).

Data regarding the stimulating impact of regular training on GPX activity in skeletal muscle are reasonably consistent (Ji 1998; Powers & Sen 2000). Furthermore, it could be shown that up-regulation of GPX activity is related to training volume. However, it appears that increased GPX activity due to regular exercise is limited to oxidative skeletal muscles and that the response of mitochondrial GPX is more pronounced than that of its cytosolic fraction (Powers & Sen 2000).

Regular endurance training can induce both activity and protein levels of SOD in rodent skeletal muscle (Powers & Sen 2000; Suzuki, Ohno et al. 2000). However, this training effect seems to be restricted to oxidative muscle fibers and is not associated with increased steady state mRNA levels (Hollander et al. 1999). Thus, exercise-induced activation of SOD gene expression may be a transient process limited to the postexercise period (Hollander et al. 2001). In humans, SOD activity has been shown to be enhanced in skeletal muscle and red blood cells in response to a sufficient training stimulus (Mena et al. 1991; Ortenblad et al. 1997; Suzuki, Ohno et al. 2000), while adaptational effects on SOD protein have been shown in only one study for its manganese isoform in plasma (Ohno et al. 1992). Initial results further suggest that physical activity on a regular basis may be sufficient to augment expression of CU/ZnSOD and GPX genes by approximately 100% in skeletal muscle of patients with chronic heart failure (Ennezat et al. 2001).

Compared to activity of SOD and GPX, CAT activity does not seem to respond to regular training and even exhibited a down-regulation in some studies (Powers & Sen 2000). The latter phenomenon has also been observed for the antioxidant stress protein HO-1. In a cross-sectional study, basal expression of HO-1 protein in leukocytes was lower in endurance-trained subjects compared to untrained controls (Niess, Passek et al. 1999c). This may point to a distinctive reaction of the antioxidant network to regular training. As postulated by Essig and Nossek (1997), antioxidant stress protein expression may be closely related to the capacity of other antioxidant systems. Thus, a lower pro-oxidant state of the cell by adaptation of systems like GPX or SOD may reduce the necessity for expression of stress proteins under basal conditions.

Effects of Regular Training on Susceptibility to Oxidative Damage

If adaptational mechanisms of the antioxidant network occur in response to exercise training, a lower extent of exercise-induced stress could be the result. Indeed, exercise training can attenuate susceptibility of rodent skeletal muscle to spontaneous and ROS-stimulated lipid peroxidation (Pansarasa et al. 2002). Similarly, regular training seems to be sufficient to decrease exercise-induced lipid peroxidation (Miyazaki et al. 2001; Vincent et al. 1999, 2002). With regard to the prevention of arteriosclerosis, a reduced susceptibility of low-density lipoproteins to oxidation after a regular exercise program (Vasankari et al. 1998) merits special interest. Radák et al. (1999a) found that nine weeks of swimming training reduced basal reactive carbonyl content and 8-OHdG in gastrocnemius of rats while corresponding measures of lipid peroxidation did not respond to exercise intervention. More recent work from the same group revealed increased activity of DNA repair enzymes as one causal mechanism involved in lower 8-OHdG content in skeletal muscle of trained rats (Radák et al. 2002). Similarly, a lower extent of exercise-induced DNA strand breaks in peripheral leukocytes of long distance runners compared to sedentary controls (Niess et al. 1996) appears to result from such adaptational effects of regular training.

Conclusion

Research over the past 25 years has established the important role of RONS in exercise physiology. However, the picture is still incomplete. Actual knowledge provides a broad basis for future studies on the cellular and molecular mechanisms of RONS action. Such future research will provide more information regarding the important question of whether oxidative stress is merely a by-product of the stress response to vigorous exercise or whether it plays an active role in exercise-related features such as changes in immune function, muscle damage, or overtraining.

More insight into redox mechanisms in skeletal muscle will yield a clearer picture of contractile regulation and will broaden our understanding of the process of muscle fatigue as it occurs under physiological and pathophysiological conditions. Similarly, initial evidence of a potential involvement of redox-regulated signaling cascades in the process of training adaptation awaits further confirmation.

Future research in this area will also contribute to a better understanding of the beneficial effects of physical exercise in prevention and therapy not only of lifestyle-associated diseases. In the case of several chronic disorders, a down-regulation of augmented RONS formation together with the adaptation of antioxidant systems may be a desired response to exercise training. Whatever the goal, modulation of pathways involved in mediating the cellular response to oxidative injury offers unique opportunities for therapeutic interventions aimed at diseases or conditions in which oxidative stress has proven to be an important factor.

Chapter 10

Cellular Responses to Environmental Stress

Elvira Fehrenbach

Exposure to environmental stress may disturb the cellular equilibrium. These changes impair or alter normal functions and may possibly have lethal consequences. The extent of impairment is dependent on type, intensity, and duration of the stress. Environmental stress includes hyperthermia, hypoxia, oxidative stress, heavy metals, tissue damage, and infections. The ability of cells to respond rapidly and appropriately to stress is essential for surviving in an ever changing environment. The activation of a cellular stress response helps the organism to cope with stress. The proteins expressed by this response are the so-called stress or heat shock proteins.

Physical activity and exercise under special conditions involve one or more of these stress conditions and also induce perturbations of homeostasis on the cellular level. Because exercise is a common stress, a better understanding of the mechanisms and limits is important for general health and for enhancing performance. Hyperthermia and hypoxia, in particular, are two main environmental factors that play a role in the exercise-induced stress response. Furthermore, they are important modulators of exercise performance. This chapter focuses on the cellular reaction to hyperthermia and hypoxia.

Hyperthermia

Exposure of cells to environmental stress like heat shock, oxidative stress, heavy metals, hypoxia, tissue damage, infections, and glucose starvation results in the induction of heat shock protein (HSP) expression. These HSP function as molecular chaperones, by either physical stabilization or proteolysis of certain proteins, or as mediators of an immune response.

HSP are among the most highly conserved of all proteins examined to date, which implies a vital, universal function for these proteins (Lindquist & Craig 1988; Kaufmann 1990). HSP are synthesized by cells of all organisms in response to a sudden increase in temperature. After Ritossa's description of a new chromosomal puffing pattern in *Drosophila* larvae after heat shock (Ritossa 1962), the universal nature of this response was gradually uncovered. The isolated puffs were determined to be sites of intensive transcriptional activity. It became apparent that heat shock was activating a set of genes that led to mRNA accumulation and subsequent synthesis of a group of proteins. Initially it was thought that HSP were effective in preventing the unfolding of proteins only at high temperature (Lindquist & Craig 1988). In addition to heat, many other stressors are capable of inducing HSP synthesis (Locke & Noble 1995; Locke 1997). Stress-inducing agents often affect the redox state and hydration of a cell. This, in turn, causes increased levels of misfolded proteins that may be deleterious by virtue of their altered biological activities. Exercise is included in the group of stimulators capable of inducing the HSP response (Locke et al. 1990; Salo et al. 1991; Ryan et al. 1991).

Several subsets of homologous but differentially regulated stress proteins, a term that is often used interchangeably with HSP, include the HSP and the glucose-regulated proteins (GRP) (Pelham 1986). HSP are induced by heat shock and other stressors whereas GRP are not induced by heat shock. GRP are induced by glucose starvation, anoxia, calcium ionophores, and other agents that disrupt the N-linked glycosylation of nascent proteins.

Molecular Mechanisms of HSP Induction

Sequence analysis of the promoter regions of HSP from several organisms has revealed a highly conserved *cis*-acting element, termed heat shock element (HSE) (Bienz & Pelham 1986; Amin et al. 1988). The presence of a functional HSE specifies an HSP gene. The HSE is the binding site for the heat shock transcription factor (HSF) (for review see Wu 1995; Morimoto 1998).

Multiple HSF genes have been identified (Sarge et al. 1991). Fruit fly and yeast possess a single HSF (Sorger & Pelham 1988). Two HSF genes, HSF-1 and -2, have been isolated from mouse (Sarge et al. 1991); HSF-1, -2, and -4 from the human (Schuetz et al. 1991; Nakai et al. 1997), and HSF-1, -2, and -3 from chicken (Nakai & Morimoto 1993).

Stress-induced regulation of HSP is principally mediated via HSF-1 binding to HSE. A regulatory domain on HSF-1 consisting of 20 amino acids confers heat responsiveness (Newton et al. 1996). In mammals, the expression of HSF-1 is constitutive and not stress inducible (Sarge et al. 1993). Alternatively spliced transcripts for HSF-1, -2, and -4 have been reported (Goodson & Sarge 1995; Nakai et al. 1997; Tanabe et al. 1999).

HSF-1 exists in the cytoplasm of unstressed cells as a monomer complexed with HSP70. It has been postulated that HSP70 is involved in the regulation of HSF-1 activation as a potential autoregulatory factor (Baler et al. 1992; Mosser et al. 1993). In stress situations, malfolded proteins capture HSP70 from HSF binding and thus activate HSF and HSP synthesis (Welch et al. 1991; Baler et al. 1992). On the other hand, HSP70 overexpression causes a reduced HSF activation (Mosser et al. 1993). By this mechanism a free pool of HSP70 might also regulate the expression of other HSP during stress (Craig & Gross 1991).

Uncomplexed HSF-1 converts to a trimeric state and thereby acquires DNA binding activity (Baler et al. 1993; Sarge et al. 1993). The process of trimerization and DNA binding is termed HSF activation (figure 10.1). The achievement of transactivation competence (Wu 1995) and the attenuation of HSF-1 activity (Abravaya et al. 1991) are also involved in the regulation of HSF-1 activity. Additionally, cellular regulators such as HSP70 and 90 (Baler et al. 1992; Mosser et al. 1993; Zou et al. 1998; Shi et al. 1998), the stress-induced translocation of HSF-1 from the cytoplasm into the nucleus (Baler et al. 1993; Sarge et al. 1993; Morimoto 1998), and phosphorylation (Knauf et al. 1996; Kline & Morimoto 1997) may participate in these events. HSF-1 is activated within minutes by heat shock

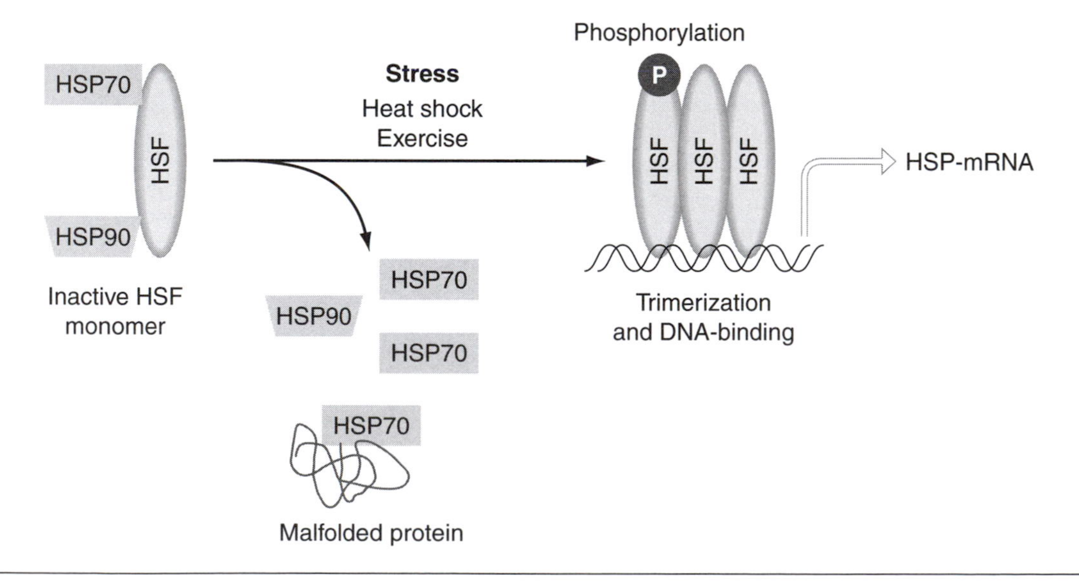

Figure 10.1 Gene expression of heat shock protein (HSP) is regulated by activation of heat shock factor-1 (HSF-1), which includes capturing of HSP by malfolded proteins, HSF trimerization, DNA binding, and transactivation (see "Molecular Mechanisms of HSP Induction").

(Kim et al. 1995; Locke et al. 1995a), hypoxia (Beckmann et al. 1990; Giaccia et al. 1992), ATP depletion (Benjamin et al. 1992), changes in pH (Petronini et al. 1995), metabolic inhibitors (Benjamin et al. 1992), and exercise (Locke et al. 1995a).

Much less is known about the other HSF. They seem to play a role in developmental and differentiation processes and seem to be tissue specific. The existence of more than one isoform of HSF suggests distinct ways in which HSP expression can be regulated (Sarge et al. 1991). HSF-2 is activated by hemin (Sarge et al. 1993).

Exercise-induced HSF activation has been observed in rat hearts; this was similar to HSF activation in hearts from heat-shocked animals (Locke et al. 1995a), suggesting that exercise and heat shock may induce HSP by the same general mechanism. In some cases, core temperatures of exercising animals were below the temperatures required for HSF activation by heat shock. This finding suggests that factors other than heat may also contribute to exercise-induced HSF activation (Locke & Noble 1995).

Although HSP are primarily regulated at the transcriptional level, post-transcriptional mechanisms may also be important. HSP70 mRNA is relatively unstable under normal physiologic conditions (Petersen & Lindquist 1988) and is nearly undetectable in unstressed muscle using Northern blot analysis (Locke et al. 1995a). Thus, when the physiological status returns to normal, the quantity of viable HSP70 mRNA is reduced, leading to a lower translation level. The accumulation of HSP70 proteins may regulate translation via a negative feedback loop (DiDomenico et al. 1982). A rapid accumulation of HSP mRNA after heat shock or exercise and its subsequent decay have been documented (Salo et al. 1991; Locke et al. 1995a; Fehrenbach et al. 2000b). Regulation via the adenine/uracil-rich sequence in the 3' untranslated region of the human HSP70 gene by heat shock has already been investigated (Moseley et al. 1993).

In addition, the physical state of the membranes may control the expression of HSP genes (Vigh et al. 1998). Heat shock increases fluidity, decreases the molecular order, and denatures proteins in the membrane, resulting in HSP induction. Accumulation of HSP causes a rigidification of the heat-fluidized membrane (Torok et al. 1997).

The HSP Family

HSP are classified by their molecular masses. Although variable, depending on the method used to determine the molecular mass and the species examined, the most common HSP demonstrate molecular masses of 110, 100, 90, 70, and 60 kDa. There are also a number of low molecular weight HSP (15-45 kDa) and ubiquitin (8.5 kDa) (Subjeck & Shyy 1986; Welch et al. 1991; Morimoto et al. 1994). Within each gene family are members that are constitutively expressed, inducibly regulated, targeted to different compartments, or a combination of these. Many stress proteins have also been identified in mammalian cells, but only very few studies deal with HSP and exercise. The following sections introduce and discuss several stress proteins (HSP27, 32, 60, 70, 90) with respect to exercise (table 10.1).

Ubiquitin

Ubiquitin is a small molecular weight protein (8.5 kDa). It is part of the ubiquitin-proteasome pathway that is responsible for most of the non-lysosomal protein degradation in mammals. Proteins for degradation by the proteasome in the cytosol are targeted by covalent binding of a polyubiquitin chain (Ciechanover et al. 2000). On the other hand, ubiquitin may also exert a positive effect on protein turnover by modulating accessibility to the DNA transcription complex (Mizzen & Allis 2000) and by influencing the viability of cytokine and proto-oncogene mRNAs (Laroia et al. 1999). Exercise-induced changes of ubiquitin are described in association with muscle protein turnover (Kee et al. 2002; Willoughby et al. 2002; Thompson & Scordilis 1994; Podhorska-Okolow et al. 1998).

HSP27

HSP27 is localized in the cytoplasm of unstressed mammalian cells and is found around or inside the nucleus during stress (Arrigo & Welch 1987). Compared to other HSP, HSP27 accumulates with slower kinetics and is synthesized for a longer time after stress (Arrigo & Welch 1987; Landry et al. 1991). Although the exact function of HSP27 remains unknown, it has been shown to be involved in signal transduction, growth, development, differentiation, and transformation processes. HSP27 appears to inhibit apoptosis mediated through Fas and other receptor-mediated pathways (Mehlen et al. 1996). It restrains specifically the cytochrome C- and ATP-triggered activity of caspase-9 on the apoptotic pathway (Garrido et al. 1999).

A number of agents that activate signal transduction pathways, such as cytokines and growth factors, have been shown to cause a rapid phosphorylation of HSP27, resulting in up

Table 10.1 Exercise-Related HSPs and Their Functions

Name	Synonym	Localization	Comments
Ubiquitin	8.5 kDa	Cytoplasm, nucleus	Ubiquitinylation, ATP-dependent protein degradation
HSP27	27 kDa	Cytoplasm, nucleus	Signal transduction, development, differentiation, apoptosis, microfilament stability, thermoresistance, chaperone
HSP60	60 kDa	Mitochondria	Chaperone activity, association with inflammation
HSP72 and HSP73	Inducible HSP70 and constitutive HSC70	Cytoplasm, nucleus, nucleolus, near ribosomes	Heat-inducible, constitutive functions, chaperone activity, protection, translocation, renaturing and reactivating malformed proteins, membrane stabilization, antigen presentation, cytoprotection
GRP78 and GRP75		Sarcoplasmic/ endoplasmic reticulum, mitochondria	Glucose- but not heat-regulated cytoprotection, chaperone
HSP90	HSP90α, HSP90β	Cytoplasm, nucleus	High basal expression level, signal transduction, steroid hormone receptor inactivation, protein folding, ATP dependent
Heme oxygenase-1	HO-1, HSP32	Cytoplasm	Oxidative degradation of heme, three isoforms, nonfunctional HSE in humans, protection against oxidative stress

to four isoelectric variants (Arrigo & Welch 1987; Landry et al. 1991). The regulation of expression and phosphorylation of HSP27 by inflammatory cytokines such as tumor necrosis factor-α (TNF-α) and interleukin-1 (IL-1) may be highly relevant for the stress response to intensive exercise (Guesdon et al. 1993). Cytokine-related stress due to heavy exercise may in part be responsible for the increased HSP27 expression in peripheral leukocytes after intensive endurance exercise (Fehrenbach et al. 2000a,b). The transient overexpression of HSP27 seems to be essential for preventing cells from undergoing apoptosis, a switch that may be redox regulated (Arrigo & Welch 1987; Mehlen et al. 1996). Increased HSP27 expression associated with attenuated DNA damage in response to exercise at high ambient temperature may support its protective, antioxidative, and antiapoptotic function (Fehrenbach et al. 2003).

HSP27 is thought to regulate apoptosis by maintaining the redox equilibrium of the cell. It may neutralize the toxic effects of oxidized proteins by its ATP-independent chaperone-like activity (Dillmann 1999). Inhibition of apoptosis by increasing the intracellular level of the antioxidant glutathione by HSP27 has been demonstrated (Mehlen et al. 1996; Creagh et al. 2000). An additional protective mechanism involves the modulation of microfilament stability by monomeric, nonphosphorylated HSP27 (Welsh & Gaestel 1998; Creagh et al. 2000). Although HSP27 is phosphorylated, phosphorylation does not appear to be necessary for thermoresistance or chaperone activity but may play a role in dissociating the large oligomers (Freeman et al. 1999). The intracellular localization, level of oligomerization, and phosphorylation status of HSP27 are involved in the regulation of its biological activity (Arrigo et al. 1988).

HSP60

Mammalian HSP60 apparently functions to facilitate the appropriate oligomeric assembly of proteins within the matrix of mitochondria (Welch et al. 1991) and to stabilize preexisting proteins under stress conditions (Martin et al. 1992). This chaperone-like activity results in designating HSP60 a molecular chaperone, or chaperonin (Ellis 1987).

HSP60 is strongly associated with inflammation and autoimmunity (Kleinau et al. 1991; Ferm et al. 1992). An inflammatory response is also induced by the physiological stimulus of extensive exercise (Northoff et al. 1995; Pedersen 1996). An increased expression of HSP60 in mono- and granulocytes detected after a half-marathon competition may be allied with the inflammatory reaction caused by this run (Fehrenbach et al. 2000b). In muscle cells, HSP60 is constitutively expressed in proportion to their mitochondrial content and increases after chronic electrical stimulation (Ornatsky et al.

1995). One of the major effects of regular exercise training or conditioning is an increased mitochondrial biogenesis in which the mitochondrial content of muscle can actually be doubled (Davies et al. 1981). Recently published results indicate that human skeletal muscle responds to a single bout of non-damaging exercise by increasing HSP60 and HSP70 (Khassaf et al. 2001).

HSP70

Four main isoforms of the HSP70 family have been identified in mammalian cells (Beckmann et al. 1990; Welch et al. 1991; Locke 1997). A cognate isoform, termed the heat shock cognate (HSC73 = HSP73 = HSC70), is constitutively synthesized in most cells and is only slightly stress inducible (Pelham 1986; Locke et al. 1990; Welch et al. 1991; Locke 1997). HSC73 is normally located in the cytoplasm, but during heat shock it migrates to the nucleus and nucleolus where it may bind with denaturing or unfolding pre-ribosomes, possibly facilitating renaturation (Welch & Suhan 1986).

A second isoform, HSP72, is closely related to HSC73 and is referred to as the inducible isoform of the HSP70 family (Pelham 1986; Locke et al. 1990). Following heat shock, HSP72 is rapidly synthesized in the cytoplasm and migrates to the heat-sensitive nucleoli, where it may bind to proteins or other structures (Pelham 1986; Welch & Suhan 1986). HSP72 is the most thoroughly investigated HSP in relation to exercise (see "Exercise-Induced HSP Response").

The two other isoforms of the HSP70 family are GRP78 and GRP75, located in the sarcoplasmic/endoplasmic reticulum and mitochondria, respectively (Pelham 1986; Mizzen et al. 1989; Locke 1997). They are glucose regulated but not heat inducible. Only two studies indicate exercise-induced GRP changes in rat myocardial and skeletal muscle (Gonzalez et al. 2000; Harris & Starnes 2001).

HSP90

There are two closely related cytoplasmic isoforms, termed HSP90α and HSP90β (Welch et al. 1991). HSP90 is a very abundant protein in all cells grown under normal conditions, and its synthesis increases three- to fivefold after heat shock. HSP90 is a specialized but essential protein-folding tool. The majority of its substrates are signal transduction proteins such as steroid hormone receptors and protein kinases (for review see Young et al. 2001; Richter & Buchner 2001). HSP90 is thought to bind to the unoccupied steroid hormone receptor in the function of a chaperone (Pratt 1987) and conserves it in an inactive form (Dalman et al. 1991). It functions in conjunction with HSP70, as well as other HSP and non-HSP, in the receptor maturation process (Pratt 1987). On hormone presentation, the receptor-HSP90 complex dissociates, and the receptor is capable of binding to DNA.

In cells deprived of glucose or oxygen, or treated with agents that perturb calcium homeostasis, synthesis of HSP90 declines concomitantly with an increased synthesis of GRP and HSP70 (Welch et al. 1991). HSP90 in the eukaryotic cytosol interacts with a variety of co-chaperone proteins (e.g., p23, Cdc37, PA28, HSC70, Hop) that assemble into a multi-chaperone complex and regulate the function of HSP90 and HSP70. An ATPase cycle is expected to regulate the interaction of HSP90 with substrate polypeptides via an ATP-mediated clamping of the substrate by HSP90 that can adopt a circular structure. The ATP-dependent function of HSP90 has been demonstrated in steroid receptor maturation structure (Young et al. 2001; Richter & Buchner 2001).

An increase in HSP90 synthesis has been shown in lymphocytes, spleen cells, and soleus muscles of untrained rats after an acute bout of exercise (Locke et al. 1990). The constantly high level of HSP90 in leukocytes of human subjects after heavy exercise (Fehrenbach et al. 2000b; Shastry et al. 2002) may indicate multiple mechanisms playing a role in exercise-related HSP expression. Stimulation by high temperature and a decline caused by glucose and oxygen deprivation might eventually explain a fairly constant level of HSP90 (Welch et al. 1991).

Heme Oxygenase (HO)

Heme oxygenase (HO), a ubiquitous enzyme in higher eukaryotes, catalyzes the initial and rate-limiting step in the oxidative degradation of heme to the antioxidant bilirubin (Tenhunen et al. 1970; Abraham et al. 1988; Camhi et al. 1995). Three isoforms of HO, HO-1 (32 kDa = HSP32), HO-2 (36 kDa), and HO-3 (33 kDa), have been shown to be the products of distinct genes (Maines 1988). While HO-2 is a constitutive enzyme distributed throughout the body, HO-1 is highly inducible. Inducers of HO-1 expression are heme, heavy metals, cytokines, hypoxia, hormones, bacterial toxins, sulfhydryl reagents, and heat shock (Yoshida et al. 1988; Abraham et al. 1991; Lutton et al. 1992; Mitani et al. 1992; Rizzardini et al. 1993). Thus, HO-1 is established as a stress protein; in particular, rat HO-1 is a heat shock protein (HSP32) (Shibahara et al. 1989). However, the inducibility of HO-1 by heat shock is controversially discussed, depending on the cell type investigated (Shibahara et al. 1989; Okinaga et al. 1996). The promoter region of the human HO-1 gene contains a putative HSE, which is potentially functional but can be repressed in vivo and in certain cell culture systems (Okinaga et al. 1996).

The third heme oxygenase protein, HO-3, differs from both HO-1 and HO-2 but is closely related to HO-2 (McCoubrey et al. 1997). The HO-3 transcript is found in the spleen, liver, thymus, prostate, heart, kidney, brain, and testis and is the product of a single-copy gene. It is a poor heme catalyst and displays hemoprotein spectral characteristics. A potential regulatory role in cellular heme-dependent processes is suggested.

It is suggested that HO-1 may play a role as a defense system against oxidative stress in response to exercise (Stocker 1990; Camhi et al. 1995; Vogt et al. 1995; Essig et al. 1997). HO-1 expression has been described as significantly increased after strenuous endurance exercise in human leukocytes (Niess et al. 1999) and in rat muscle after repetitive contractions (Essig et al. 1997). The increased HO-1 expression in response to various oxidative stresses, including exercise (Applegate et al. 1991; Essig et al. 1997; Niess et al. 1999, 2000), does not appear to involve typical chaperone functions in which specific proteins are stabilized or transported. Rather, HO-1 may provide protection through the elimination of heme with coincident production of the antioxidants bilirubin or biliverdin (Okinaga et al. 1996), as well as the production of specific cellular messengers (Maines 1997). The secondary enzymatic product of HO-1 is carbon monoxide, which may directly cause a decrease in blood pressure or may function in conjunction with nitric oxide (Maines 1997; Vesely et al. 1998).

Function

HSP may have different constitutive and inducible functions that are described in the following chapter. The specific roles of HSP depend on the localization in different compartments in the cell, the cellular membrane, and outside of the cell. Also, the distribution in various tissues and the type of HSP influence the functions.

Protective and Housekeeping Functions of HSP

HSP collectively function to maintain cellular protein conformation during stressful proteotoxic insults (Essig & Nosek 1997). Their main function is to guarantee intracellular protein homeostasis, thus preserving the cells' viability in the face of denaturing agents. HSP27 (Landry et al. 1989) and, to a lesser extent HSP90 (Bansal et al. 1991), have been associated with cellular protection; but the vast majority of evidence implicates HSP70 members, particularly HSP72, as cytoprotective proteins responsible for conferring protection to cells (Li 1985; Johnston & Kucey 1988; Angelidis et al. 1991; Plumier et al. 1995). Protective effects of exercise were also associated with a synergistic expression of HSP72. The prevention of an ethanol-induced fatty liver in rats by exercise was paralleled by an increase of HSP72 in the liver (Trudell et al. 1995). Protective functions against exercise-induced DNA damage are suggested for heat-induced HSP27 (Fehrenbach et al. 2003).

The regulation of HSP72 synthesis is tissue specific at high physiological temperatures. Hyperthermia resulted in an increase in HSP72 in the liver, small intestine, and kidney, but not in the brain or quadriceps muscles of rats (Flanagan et al. 1995). HSP72 expression may identify a critical target tissue susceptible to thermal damage. The cells' ability to synthesize HSP72 could be used as a predictor of survival during heat stress in rat fibroblasts (Li 1985) and in human lymphocytes (Ryan et al. 1991; Locke & Noble 1995). Thus, measurement of HSP content may indicate the exercise-related thermal history of certain tissues or cells.

Although HSP synthesis is induced in response to physiological perturbations, HSP70 proteins are also normal cell constituents. They are present at low concentrations even when stress is absent, demonstrating survival and housekeeping functions, depending on the environment (Kilgore et al. 1998). A really important function of HSP70 during exercise is the ability to renature and reactivate denatured or malformed proteins (Kilgore et al. 1998). This particular function may be relevant for several physiological processes, and it may directly affect enzyme function after stress. HSP may facilitate proper folding, packaging, and transporting of enzyme precursors. The activity of several enzymes, such as phospholipase A2 (Jaattela 1993), protein kinase C (Ritz et al. 1993), and citrate synthase (Locke 1997), relates to HSP70 concentration under certain conditions. It was demonstrated that the oxidative capacity in skeletal muscle, represented by citrate synthase activity, positively correlated to HSP75 expression, the mitochondrial form of HSP70 (Kilgore et al. 1998). Phosphorylation of HSP72 seems to be an early event in the stress response of skeletal muscle to exercise stress (Hernando & Manso 1997).

Another interesting area is the proposed translocase function of HSP proteins. Proteins normally destined for the endoplasmic reticulum or mitochondria accumulated abnormally in the cytoplasm of cells depleted of HSP70 (Deshaies et al. 1988). HSP70 transiently interacts with nascent peptides during protein synthesis and releases the protein when translation is completed and the new protein folds into the native state (Beckmann et al. 1990; Kilgore et al. 1998). This chaperone function may be important when increased protein synthesis follows exercise stress. The studies described later support the assumption of a relationship between exercise-induced skeletal muscle hypertrophy and the chaperone function of HSP72. An association of stimulated HSP expression with an increase and new synthesis of contractile proteins in human muscle during rowing training has been described (Liu et al. 1999). Hypertrophying skeletal muscle had higher HSP72 concentrations compared to nonhypertrophied control skeletal muscle (Kilgore et al. 1998), and HSP72 content was related to fiber type profile shifts during skeletal muscle hypertrophy (Locke 1997).

A different function may be the stress-induced accumulation of HSP at or in the cell membrane, causing the heat-fluidized membrane to rigidify to the consistency that existed prior to heat shock (Vigh et al. 1998). Thus, changes in membrane fluidity of muscle cells or peripheral leukocytes may also be induced by exercise-related hyperthermia.

The presence of HSP on the surface or in the plasma of the peripheral blood may indicate a role for HSP in antigen presentation and immune regulation as covered in the following section.

Immunological Function of HSP

In addition to the protective effects in the cell, HSP play a functional role in the activation of immune cells, their antigen processing, and presentation by major histocompatibility complex (MHC) I (Guzhova et al. 1998; Sondermann et al. 2000; Binder et al. 2001). Soluble HSP may also function as extracellular signals to activate the immune response (Moseley 2000a). The ability of HSP to activate antigen-presenting cells (APC) provides a unified mechanism for the response to internal and external stimuli (Basu et al. 2000). Activation of pro-monocytes/macrophages or dendritic cells by soluble HSP70 and gp96 to express differentiation markers and antigen-presenting and co-stimulatory molecules, as well as to produce cytokines, has been described (Asea et al. 2000; Basu et al. 2000; Kuppner et al. 2001). Recombinant HSP70 stimulated cytokine production from monocytes and enhanced NK cell proliferation and cytotoxicity. Moreover, HSP was able to induce maturation of immature dendritic cells (Kuppner et al. 2001). Evidence was provided that a proteinaceous receptor exists on the surface of macrophages and monocytes that is specific for mammalian HSP70 (Sondermann et al. 2000). Subsequent to binding, the differentiation markers CD11c and CD23 were up-regulated and HSP70 was taken up by endocytosis (Guzhova et al. 1998; Sondermann et al. 2000). Furthermore it was shown that exogenous HSP72 bonded specifically to the cell surface of human monocytes and activated the production of the proinflammatory cytokine IL-6, IL-1β, and TNF-α via a CD14-dependent pathway (Asea et al. 2000). It has been suggested that HSP can be released from cells into the extracellular milieu to bind to membranes of other cells (Hightower & Guidon 1989; Child et al. 1995; Asea et al. 2000). The induction of the cytokine production may be mediated by interaction with the NF-κB pathway (Basu et al. 2000).

Besides the presence in plasma and the intracellular expression, the localization of HSP on the surface of several cells in free form, or in the context of MHC class I molecules, or both, has been described (Multhoff & Botzler 1998). HSP are expressed selectively on the surface of virally or bacterially infected cells, on cells of patients with autoimmune disease, or on tumor cells, but not on the surface of vital normal cells (Erkeller et al. 1992). It is hypothesized that alterations in the calcium and pH level, hypoxia, and nutrient depletion in the cells are responsible for conformational changes of HSP that result in cell surface localization. The HSP expression in stressed cells is stimulated in the cytoplasm and may subsequently be presented on the cell surface. The mechanism by which the HSP, especially members of the HSP70 family, are displayed on the surface is unclear because HSP do not possess transmembrane domains (Gunther & Walter 1994). Surface expression on stressed cells enables recognition by NK cells or cytotoxic T-lymphocytes, which can subsequently eliminate them (Kaufmann 1990; Multhoff & Botzler 1998).

So far there is only one study monitoring HSP expression on the cell surface following exercise (Fehrenbach et al. 2000b). HSP on the surface was nearly undetectable on the majority of leukocytes of athletes at rest as well as of untrained persons.

Immediately after a half-marathon, however, an increase of monocytes and granulocytes expressing HSP27, HSP60, or HSP70 on their surface was detected in some athletes. The rise of HSP-positive granulocytes was even significant 24 h after the competition. It may be hypothesized that maximally stressed cells due to the half-marathon also reacted with HSP expression on their surface to be recognized by other immunocompetent cells. Alternatively, HSP expression on the cell's surface is relevant for stabilization of membrane fluidity.

Soluble HSP72 may also play an important role in the exercise-induced inflammatory response. It may function as an extracellular signal to activate the immune system, also in response to exercise (Moseley 2000a). Exercise resulted in an increase in serum HSP72 in humans (Walsh et al. 2001), which was paralleled by a significant increase of the cytokine IL-6 in plasma after endurance exercise (figure 10.2). This observation may support a regulatory function of HSP72 in the exercise-induced IL-6 response (Asea et al. 2000).

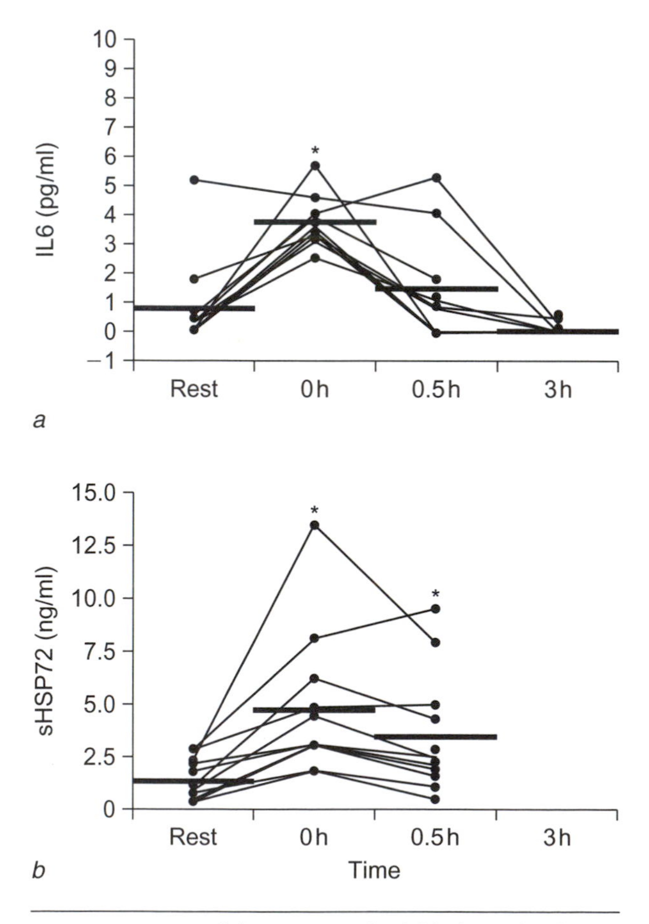

Figure 10.2 Parallel increases of soluble heat shock protein (HSP)72 and interleukin-6 (IL-6) in plasma after exercise. Ten subjects performed extensive interval training (10 × 1,000 m) with an intensity of 88% $\dot{V}O_2$max. Horizontal bars represent the mean, $*p < 0.05$.

The tissue of origin of circulating HSP72 is not clear. Non-damaged muscle did not release HSP72 into the circulation, although an exercise-induced HSP72 increase in this muscle was detected (Febbraio et al. 2002). On the other hand, exercise-induced muscle damage may result in a release of HSP72 in the blood circulation. Necrotic cells released HSP70 and delivered a maturation signal to dendritic cells and activated the NF-κB pathway (Basu et al. 2000). However, the increase of serum HSP72 preceded any HSP72 increase in exercising muscle, suggesting that it was released from other tissues or organs (Walsh et al. 2001). Exercise-induced increases have been shown in leukocytes (Ryan et al. 1991; Fehrenbach et al. 2000a), heart (Salo et al. 1991; Skidmore et al. 1995), liver (Salo et al. 1991), and brain (Walters et al. 1998).

According to the danger theory of Matzinger et al. (Moseley 2000b; Matzinger 2002), one may suggest that circulating HSP72 is partly responsible for the priming and maturation of antigen-presenting cells after exercise that then can respond in the event of infection or inflammation. One may hypothesize that soluble HSP72 in the circulation plays a regulatory role in the prevention or protection of exercise-induced pathophysiological processes involving immune recognition and activation of immune cells.

Association of HSP and Exercise

The metabolic changes caused by exercise are manifold and are similar to those known to modulate HSP synthesis. The effects of acute exercise and chronic training on HSP expression are differentiated and will be discussed with respect to cancer, aging, and gender.

Exercise-Induced HSP Response

Exercise is a sufficient stimulus to induce or enhance the synthesis of HSP in mammalian cells and tissues (Locke et al. 1990; Salo et al. 1991; Ryan et al. 1991; Flanagan et al. 1995).

Hyperthermia, to some degree, is the inevitable consequence of physical activity. Physical exercise can elevate core temperature to 44° C and muscle temperatures up to 45° C (Salo et al. 1991). While most researchers have found that exercise increased HSP concentrations in various tissues, Hammond et al. (1982) failed to demonstrate increased HSP70 concentrations after 1 h of swimming. Water submersion during exercise in these experiments may, however, blunt the temperature increases that are normally seen with exercise. Cold environmental temperature prevented the stimulating effect of chronic exercise on HSP70 expression in myocardial and skeletal muscle in comparison to normal temperature conditions (Harris & Starnes 2001). Also, passive exercise at lower temperature (26-31° C) in an automatic round treadmill had no effect on HSP72 in rat lymphocytes. Induction of HSP72 protein could be detected only in lymphocytes of rats exercising at high temperature (36-37° C) for more than 60 min and when core temperature rose above 41° C (Chen et al. 1995). Elevated ambient temperature has an additional impact on the exercise-induced HSP response in human leukocytes (Fehrenbach et al. 2001) and rat myocard and skeletal muscle (Harris & Starnes 2001). These findings support the idea that abrupt temperature increases cause the cell membranes to undergo a rapid decrease in molecular order, which may be an important factor in determining HSP gene expression (Vigh et al. 1998).

Exercise, aside from increasing body temperature, effects other physiological changes that may also elicit the HSP response. An exercise-induced HSP response is not solely linked to thermal stress but may also be induced by oxidative, mechanical, metabolic, and cytokine stress (Delcayre et al. 1988; Iwaki et al. 1993; Kukreja et al. 1994). Strenuous endurance exercise causes metabolic alterations that include changes of intracellular pH and calcium.

Furthermore, sustained physical activity results in the progressive depletion of glucose and glycogen stores, a phenomenon that is highly correlated with fatigue (Essig & Nosek 1997). In human muscle the exercise-induced HSP expression at the protein and mRNA level is glycogen dependent (Febbraio et al. 2002).

Exercise also causes oxidative stress via increased generation of reactive oxygen intermediates (ROI) and nitrogen species (RNI) (Davies et al. 1982) in the muscle as well as the hematopoietic system. Exercise-induced hyperthermia may actually cause oxidative stress. Along this line, Salo et al. (1991) found that muscle mitochondria undergo progressive uncoupling and increased ROI generation with increasing temperatures. HSP70 mRNA levels were increased in skeletal and cardiac muscle by exercise and by both heat shock and oxidative stress (Salo et al. 1991). A role for ROI, primarily produced as a consequence of elevated rate of mitochondrial respiration, is postulated in skeletal muscle fatigue. Given the potential of ROI to damage intracellular proteins during consecutive series of muscle contractions, the capacity of preexisting antioxidant pathways may be complemented by the synthesis of HSP. Especially, increases in HO-1 mRNA may underlie an inducible antioxidant pathway in skeletal muscle responsiveness to metabolic stresses associated with repeated muscle contractions (Essig & Nosek 1997). Also nitric oxide itself may be involved in HO-1 gene stimulation in skeletal muscle cells (Vesely et al. 1998).

Activation of blood neutrophils described after exercise and an increase of lipid peroxidation products suggest that oxidative stress plays a role in exercise-induced changes in hematopoietic cells (Davies et al. 1982; Smith et al. 1990; Fehrenbach et al. 2000b). In particular, the antioxidative stress protein HO-1 is induced by hypoxia and oxidative stress. A redox-sensitive pathway mediates, at least in part, the hypoxic induction of the HO-1 gene in cardiac myocytes (Borger & Essig 1998). A recently detected induction of HO-1 in leukocytes after long-lasting strenuous endurance exercise (figure 10.3) correlated with increased plasma IL-8, an indicator of ROI occurring in vivo (Niess et al. 1999). Furthermore, oxyblot analysis revealed oxidative modifications of proteins in protein lysates of human leukocytes after endurance exercise (Martel et al. 2000). The role of free radicals in exercise-induced expression of HSP in leukocytes is indirectly supported by the slightly attenuating effect of the antioxidant α-tocopherol on these markers after exercise (Niess et al. 2002). These results account for oxidative stress as one potential stimulus for HSP expression due to exercise.

Different types of exercise result in activation of immunocompetent cells and an acute phase response. Thus, exercise-induced increases of IL-1, -6, -8, tumor necrosis factor (TNF-α), and interferon-γ (IFN-γ) have been demonstrated (Northoff et al. 1995; Fehrenbach et al. 2000b). Several cytokines have already been described as positive regulators of HSP expression (Arrigo & Welch 1987; Landry et al. 1991). This indicates cytokine stress as one possible inducer of the exercise-related HSP response. Parallel increases of several HSP and TNF-α, IL-6, and IL-8 after strenuous endurance exercise in human leukocytes (Niess et al. 1999; Fehrenbach et al. 2000b) suggest that cytokines may be involved in HSP regulation. On the other hand, exercise-induced extracellular HSP72 (Walsh et al. 2001), which was shown to bind specifically

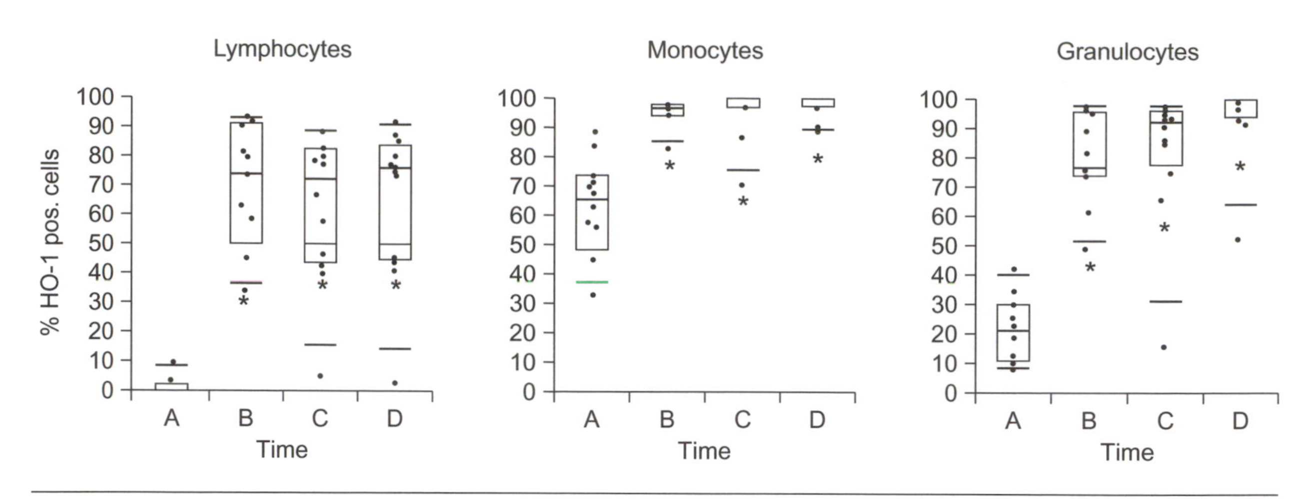

Figure 10.3 Expression of heme oxygenase-1 (HO-1) before and 0 h, 3 h, and 24 h after a half-marathon (A-D). Percent HO-1-positive cells was analyzed by flow cytometry. $*p < 0.05$.

to the cell surface of human monocytes (Schneider et al. 2002), may activate the production of the proinflammatory cytokine IL-6 via a CD14-dependent pathway (Asea et al. 2000).

Finally, several important factors additionally influence the exercise-induced HSP response. Muscular stress, the duration of strenuous exercise, and the competitive situation may be included. Passive treadmill running of rats without muscular stress (no CK increase) or exercise at low temperature resulting in rectal temperature <40° C, or both, revealed no or only a small increase of HSP72 in leukocytes (Ryan et al. 1991; Chen et al. 1995). Non-damaged muscles did not secrete HSP72 in the circulation (Febbraio et al. 2002). Furthermore, eccentric exercise or brief strenuous exercise alone was not sufficient to stimulate HO-1 expression in human peripheral leukocytes (Fehrenbach et al. 2002). In contrast, significant increases of HSP27, HSP60, HSP70, and HO-1 were detected in human leukocytes after a competitive half-marathon at the protein and transcriptional level (figures 10.3 and 10.4) (Fehrenbach et al. 2000b). The stimulated HSP expression was accompanied by significant increases in leukocyte counts, CK, myeloperoxidase, uric acid, and cytokines, confirming that the competitive half-marathon was a strenuous athletic event (Davies et al.

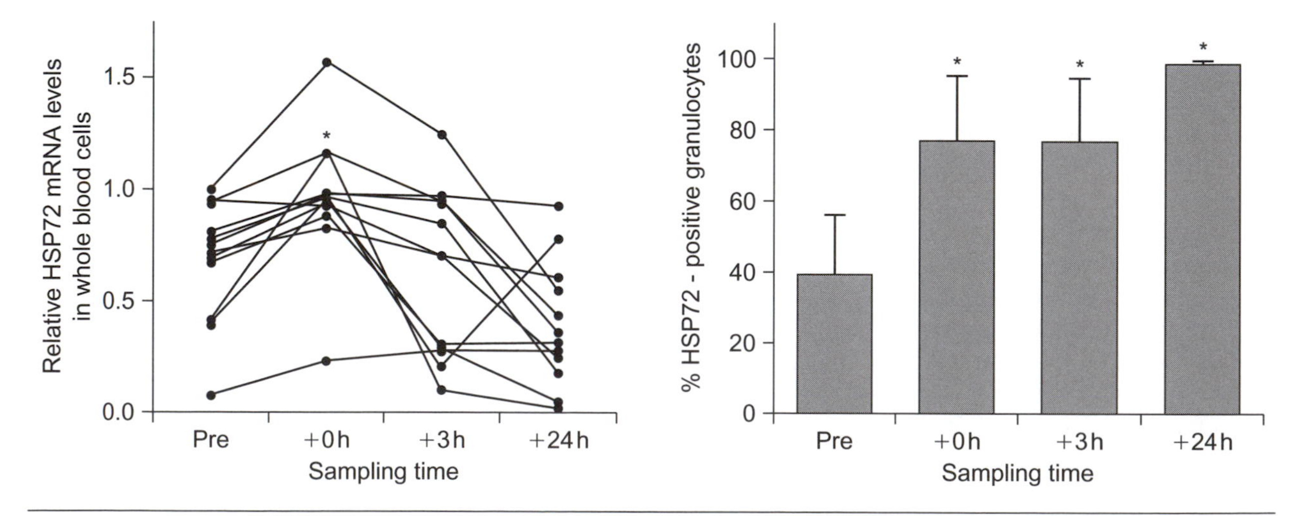

Figure 10.4 Increase of HSP72-expression after a competitive half marathon at mRNA and protein level analyzed by RT-PCR and flow cytometry, respectively. The specific mRNA values of whole blood cells are described in relative units normalized to transcript levels of beta–actin. Each curve represents a single subject. Granulocytes were gated in the scattergram and intracellular protein expression is presented as percent positive cells. Columns represent mean values ± SD (n=12), $*p < 0.05$.

1982; Locke et al. 1990; Smith et al. 1990; Flanagan et al. 1995; Bury & Pirnay 1995; Northoff et al. 1995; Fehrenbach et al. 2000a).

In summary, one may conclude that acute exercise causes the synthesis of HSP through some unique pathway, but it is more probable that a common mechanism is shared among several exercise-related stressors. Moreover, differential regulation may occur at the protein as well as the mRNA level.

Thermotolerance and Adaptation to Training

An elevated level of HSP has been associated with a phenomenon termed acquired thermotolerance, defined as the transient resistance to the cytotoxic effects of a subsequent lethal shock induced by short exposure to a nonlethal heat treatment (Li 1985; Mizzen & Welch 1988). During recovery from the first heat shock, HSP accumulates, thus conferring protection against a subsequent stress. In addition, when HSP are induced by stressors other than heat, thermo- or stress tolerance is still conferred and protection against other stresses is also supplied (Li 1983). The concept of acquired thermo-/stress tolerance may have important implications for the protection conferred to cells and tissues after exercise-induced HSP production. Recovery from exercise is associated with specific transient changes in the expression of immediate early and stress protein genes, including c-fos, alpha B-crystallin, citrate synthase RNA, myoglobin, and HSP70, suggesting that the products of these genes may have specific roles in the remodeling process evoked by repeated periods of contractile activity (Neufer et al. 1998). A single session of eccentric exercise has been shown to increase HSP synthesis in muscle (Locke et al. 1990; Salo et al. 1991) and can provide protection against future injury (Schwane & Armstrong 1983). Thus, considering the repair- and protection-related functions of certain HSP, it is conceivable that induction of HSP by exercise may provide protection against injury from future exercise. An elevated HSP70 expression has been shown to protect muscle cells against ischemic stress (Iwaki et al. 1993). Induction of HSP by exercise can protect the heart from ischemic stress (Locke et al. 1995b).

While acute intensive exercise uniformly leads to HSP up-regulation in leukocytes as well as muscle cells, the situation in trained athletes at rest seems to be different. Regulation of protection mechanisms after regular training seems to be involved in the beneficial effects of moderate exercise on the immune system (Pedersen 1996; Nieman & Pedersen 1999). There are several redox-related mechanisms in the body that react to regular exercise training and thereby may contribute to the overall health benefit from chronic moderate exercise. Of course, there is adaptation at the muscular level (for review see Radak et al. 2001). In the muscle cells, training was paralleled by increased HSP expression (Salo et al. 1991; Locke & Noble 1995; Puntschart et al. 1996; Essig & Nosek 1997; Hernando & Manso 1997; Essig et al. 1997; Kilgore et al. 1998; Liu et al. 1999). Long-term rowing training induced HSP72 protein in highly trained human muscle, which was interpreted as adaptation to training (Liu et al. 1999). An HSP72 increase indicated that treadmill run training for eight weeks stresses the exercising rat hindlimb muscle (Kelly et al. 1996). In a rat model, exercise training provided protection by increased HSP72 against myocardial lipid peroxidation induced by short-term ischemia and reperfusion (Demirel et al. 1998). Western blot analyses for HSP72 showed a significantly elevated myocardial HSP content in the heart of both heat-shocked animals and animals exposed to three bouts of exercise, compared to those of either nonexercised controls or animals subjected to only one session of exercise (Locke et al. 1995b).

In leukocytes, however, moderately trained persons showed a down-regulated HSP level (Niess et al. 1999; Fehrenbach et al. 2000a,b; Shastry et al. 2002). Some other functions seem also to be diminished due to regular training. The ROS production in neutrophils of highly trained athletes was reduced by endurance training over one month (Kumae et al. 1999). Neutrophil-killing capacity is reduced in elite athletes compared to untrained controls (Gabriel et al. 1994; Nieman & Pedersen 1999). Similarly, a suppressed ability of blood neutrophils to produce ROI in trained versus untrained individuals (Smith et al. 1990), and a lower superoxide anion generation in trained athletes, have been reported (Benoni et al. 1995). Trained athletes also showed less DNA damage in their leukocytes than untrained controls after exhaustive exercise (Niess et al. 1997; Asami et al. 1998). Besides increased activity of the antioxidative systems (Giuliani & Cestaro 1997; Powers & Lennon 1999; Jenkins & Beard 2000), up-regulation of DNA repair systems (Asami et al. 1998) or further repair mechanisms (Janssen et al. 1993) probably contributes to the effect. Platelets from trained human subjects and isolated rat hearts have an increased activity of antioxidant enzymes (Somani et al. 1995; Kedziora et al. 1995).

First-line antioxidant systems such as superoxide dismutase, catalase, or glutathione peroxidase that are stimulated during training (Kretzschmar et al. 1991; Criswell et al. 1993; Powers et al. 1999) may be efficient in down-regulating the oxidative potential. Moderate regular training induces so much first-line protection against ROS that secondary repair systems like HSP can be partly shut down. The lower pro-oxidant state due to training results in less activated immunocompetent cells (Pyne 1994).

The heat shock response of leukocytes in vitro seems to be training dependent (Fehrenbach et al. 2000a). Just two bouts of exercise induced an adaptive HSP response at intramuscular and intra-leukocyte levels (Fehrenbach et al. 2001; Thompson et al. 2002). The regulation in different tissues and expression levels is highly complex. A cautious interpretation may be that indiscriminate up-regulation of this potent protection system may not be advantageous. It is noteworthy that the capability of leukocytes to react with a fast heat shock response in vitro seems to be dependent on the training-regulated basal HSP level. Variable basal synthesis of stress proteins contributed to differential inducibility (Boshoff et al. 2000).

The high baseline expression of HSP70 mRNA in moderately trained individuals (Fehrenbach et al. 2000a) may be attributed to a unique characteristic of HSP70 mRNA. It is unstable under normal physiologic conditions and stabilizes under stress conditions (Petersen & Lindquist 1988). The resulting accumulation of HSP70 proteins themselves may regulate translation via a negative feedback loop if they are not used for an immediate stress response (DiDomenico et al. 1982; Welch et al. 1991).

HSP and Cancer

Exercise and training interfere with HSP in a complex way that includes up-regulation through acute exercise and sophisticated regulation through regular training. The extent to which this relates to the beneficial effects of exercise on the immune system or the occurrence of certain kinds of cancer is still open. However, no proven knowledge exists about the exact mechanisms whereby exercise influences cancer development or tumor growth (McTiernan et al. 1998). One possible mechanism may be the activation of the protective, antioxidative HSP system as a kind of adaptive reaction due to regular exercise (Fehrenbach & Northoff 2001).

In relation to cancer, HSP may function as a double-edged sword with potentially opposite effects (for recent reviews see Vayssier et al. 1998; Wang et al. 2000; Fehrenbach & Northoff 2001; Parmiani et al. 2002; Neckers 2002; Srivastava 2002; Manjili et al. 2002). On the one hand, HSP have been shown to participate in the anti-tumor T-cell response by chaperoning antigen processing. Tumor-derived peptides associated with HSP can be recognized by T-cells, and vaccination with HSP from tumor tissue has been shown to elicit an anti-tumor response (Srivastava 1994; Wang et al. 2000). Expression of peptide-loaded HSP on the cell surface may contribute to immune recognition (Singh et al. 2000).

On the other hand, induction of HSP may have unwanted side effects. HSP can save stressed cells from death or provoke a shift from necrosis to apoptosis (Fuller et al. 1994; Vayssier et al. 1998). Thus, up-regulated HSP in tumor cells may save them from necrosis or shift them from immunogenic necrosis to tolerogenic apoptosis, ultimately resulting in tumor survival. Particularly, it has been suggested that small HSP allow cancerous cells to escape immunosurveillance and make them resistant to therapy (Arrigo 2000). However, once the tumor cell is necrotic, HSP will be released and can effectively immunize against tumor cells by presenting peptides (Rammensee 1996; Multhoff & Botzler 1998; Singh et al. 2000) or inducing cytokines (Hall 1994).

HSP and Aging

Passive or exertional heat stress both resulted in liver HSP72 accumulation in mature rats. In contrast, senescent rats demonstrated no significant increase in HSP72 with heating but a large increase with exercise. The observed blunted HSP response to heating observed with aging is not a result of the inability to produce HSP72, because the senescent rats had a significant response to exercise-induced hyperthermia (Kregel & Moseley 1996).

One might speculate that HSP induction is more efficiently up-regulated by mechanisms leading to reactive oxygen species. Similar to the differential HSP responsiveness in tissues, aging may alter efficiency of defined sets of transcription factors.

HSP and Gender

A gender-specific regulation of HSP70 with exercise has been described. Following an acute bout of exercise, males demonstrated a greater content of HSP70 than gonadally intact females in skeletal and cardiac muscle and other tissues like liver and lung (Paroo & Noble 2002). Removal of the ovaries resulted in HSP induction similar to that observed for males, and estrogen treatment of ovariectomized animals reversed this effect similarly in these tissues (Paroo et al. 2002). Thus, the sex-specific HSP response to exercise is mediated by the ovarian hormone estrogen. But this effect is not regulated in a receptor-dependent manner, because tamoxifen, an estrogen receptor antagonist, did not change the HSP response in intact females. Furthermore, 17γ-estradiol, a synthetic, receptor-inactive stereoisomer, and tamoxifen treatment to ovariectomized animals inhibited HSP70 expression, similar to the effect of 17β-estradiol (Paroo & Noble 2002). Estrogens are lipophilic agents that reduce membrane fluidity and attenuate lipid peroxidation, and have been classified as antioxidant membrane-stabilizing molecules (Wiseman et al. 1993). Through this indirect antioxidant, potential estrogens maintain the integrity of cell structure and influence the exercise-induced HSP response. The antioxidant, membrane-stabilizing mechanism of action is supported by exercise-induced production of the antioxidative stress protein HO-1 in ovariectomized animals (Wiseman et al. 1993; Bar & Amelink 1997).

An estrogen response element in the promoter region of HSP27, besides several stress-inducible heat shock elements, provides responsiveness and protective effects of HSP27 to estrogen treatment (Locke 1997).

Summary—HSP

Strenuous exercise creates physiological stress, disturbs cellular homeostasis, and ultimately induces cellular adaptations. HSP are considered to play an essential role in protecting cells from stress, preparing them to survive new environmental challenges, and inducing an immune response (for overview see figure 10.5). Exercise-induced changes such as heat shock, oxidative, metabolic, muscular, and cytokine stress seem to be responsible for the HSP response to exercise. Acute exercise is obviously able to stimulate the expression of certain HSP in different cells and tissues including muscle, heart, liver, brain, leukocytes, and plasma. The HSP response due to exercise has been documented in animal as well as the human species.

The extent of the HSP response is dependent on type, intensity, and duration of exercise; environmental conditions; and cell type examined. Training status and gender also have an impact on the HSP response. A large individual variability in induction of HSP expression and the particular HSP being examined have to be considered.

Repeated exercise bouts, regular training, and high ambient temperature induce an adaptation of the HSP response in muscle and immune cells. Measurement of HSP content seems to be an indicator of exercise-related stress situations that may possibly allow conclusions concerning adaptation to training or may even represent a marker to warn one of overtraining in the near future.

In conclusion, exercise and training interfere with HSP in a complex way, which includes upregulation through acute exercise and sophisticated regulation through regular training. The extent to which this relates to the beneficial effects of exercise on the immune system or the occurrence of certain kinds of cancer remains open to debate.

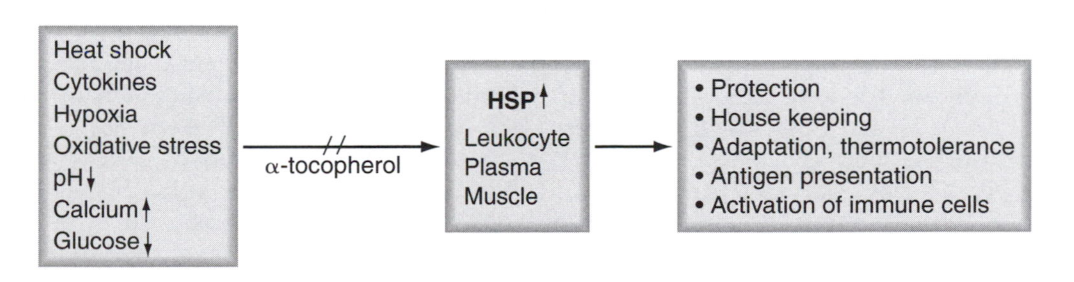

Figure 10.5 Summary of the exercise-related changes and inducers of heat shock protein (HSP) expression and their functions (modified according to Fehrenbach & Northoff 2001; for detailed information see text).

Hypoxia

Hypoxia is the natural consequence of some stressful environmental conditions such as high altitude or diving. Many species live with hypoxia as an everyday occurrence, and they express adaptive responses at the cellular, molecular, and systemic level to minimize the injurious effects of hypoxia. Hypoxia-induced alterations include hematological adaptations; changes in heart rate and ventilation; muscular modifications; and changes in blood lactate, oxygen consumption, and oxygen-sensitive proteins. Short-term exposure to altitude or to simulated hypoxia during exercise training is widely used with endurance athletes with the aim to improve athletic performance during competition at sea level. It represents an additional stimulus in the adaptive response to endurance training. However, these beneficial effects of hypoxia are individually different and are controversially discussed. This section deals with underlying cellular and molecular mechanisms.

Regulation of Hypoxia Sensitive-Proteins

Even slight hypoxia induces a group of physiologically important genes at the cellular level that are involved in the regulation of oxygen homeostasis. Hypoxia-sensitive proteins are erythropoietin (EPO), vascular endothelial growth factor (VEGF, angiogenesis), transferrin (tf) and transferrin receptor (iron transport and uptake), glucose transporter 4 (GLUT4), heme oxygenase-1 (HO-1, CO production), and inducible nitric oxide synthase (iNOS, NO production) (Wenger 2002). These target genes are transcriptionally up-regulated by the hypoxia-inducible factor-1 (HIF-1), a global regulator that belongs to the bHLH-PAS family (Gu et al. 2000). This protein family contains a basic helix–loop–helix (bHLH) binding domain that is necessary for DNA binding (Jiang et al. 1996). Furthermore, there is a common domain called PAS (an acronym for the first three members of this family, PER, ARNT, and SM), which is required for dimerization (Lindebro et al. 1995) and target gene specificity (Zelzer et al. 1997) (figure 10.6). In addition to the ubiquitously expressed HIF-1, two other members of this family, HIF-2 and HIF-3, were identified that show a more restricted tissue expression pattern (Wenger 2002).

HIF-1 is a heterodimer composed of the constitutively expressed HIF-1β (91-94 kDa) and the rate-limiting, hypoxia-sensitive factor HIF-1α (120 kDa) (Wang & Semenza 1995; Semenza 2000). HIF-1β is identical to the aryl hydrocarbon nuclear translocator (ARNT) shown to heterodimerize with the aryl hydrocarbon receptor (AHR) (Hoffman et al. 1991). HIF-1α is a short-lived protein and nearly undetectable in normoxia. Its levels are tightly regulated by the oxygen concentration via the ubiquitin-proteasome system (Salceda & Caro 1997; Huang et al. 1998).

Hypoxia stabilizes HIF-1α by abrogating the ubiquitin-proteasome degradation via the oxygen-dependent degradation domain (ODD) within HIF-1α (Huang et al. 1998) (figure 10.6). In addition to hypoxic stabilization, phosphorylation of HIF-1α by mitogen-activated protein kinases (MAPK) or redox-dependent processes might also trigger HIF-1α function (Wenger 2002). Hypoxia-induced nuclear translocation of HIF-1α is a further means by which the protein escapes from proteasomal degradation (Kallio et al. 1998). Within 2 min of anoxic/hypoxic exposure, the HIF-1α protein accumulates in the nucleus where it autonomously heterodimerizes with HIF-1β (figure 10.6). The functional HIF-1α/β complex binds to an HBS (HIF-1 binding site) located in the HRE (hypoxia response element) and thereby induces transcription (transactivation) of the hypoxia-sensitive target genes (Jewell et al. 2001) (figure 10.6). Transactivation (TA) occurs via modifications (hydroxylation, phosphorylation) of the TA domains in HIF-1α (Wenger 2002). The TA domains confer transcriptional activation of target genes mainly by the recruitment of transcriptional cofactors including CBP/P300 or SRC-1 (Ebert & Bunn 1998).

Apart from the relatively well-characterized mechanisms of hypoxic HIF-1α subunit stabilization, many growth factors, cytokines (Wenger 2002), heat (Katschinski et al. 2002), nitric oxide (Palmer et al. 2000), and insulin (Jelkmann & Hellwig-Burgel 2001) are known to stabilize HIF-1α under normoxic conditions. In some aspects there seems to be synergy in the cellular responses to hypoxia, glucose deficiency, and inflammation. A possible role of HIF-1α in inflammatory processes has been suggested (Hellwig-Burgel et al. 1999; Jelkmann & Hellwig-Burgel 2001).

Upon reoxygenation within 16 min (Jewell et al. 2001) and during normoxia, HIF proteins undergo efficient proteasomal degradation (figure 10.6). The oxygen-dependent enzymatic hydroxylation of the ODD and TA domains confers rapid degradation of HIF-1α mediated by the product of the von Hippel-Lindau tumor suppressor gene (pVHL). pVHL binds to the hydroxylated ODD domain and

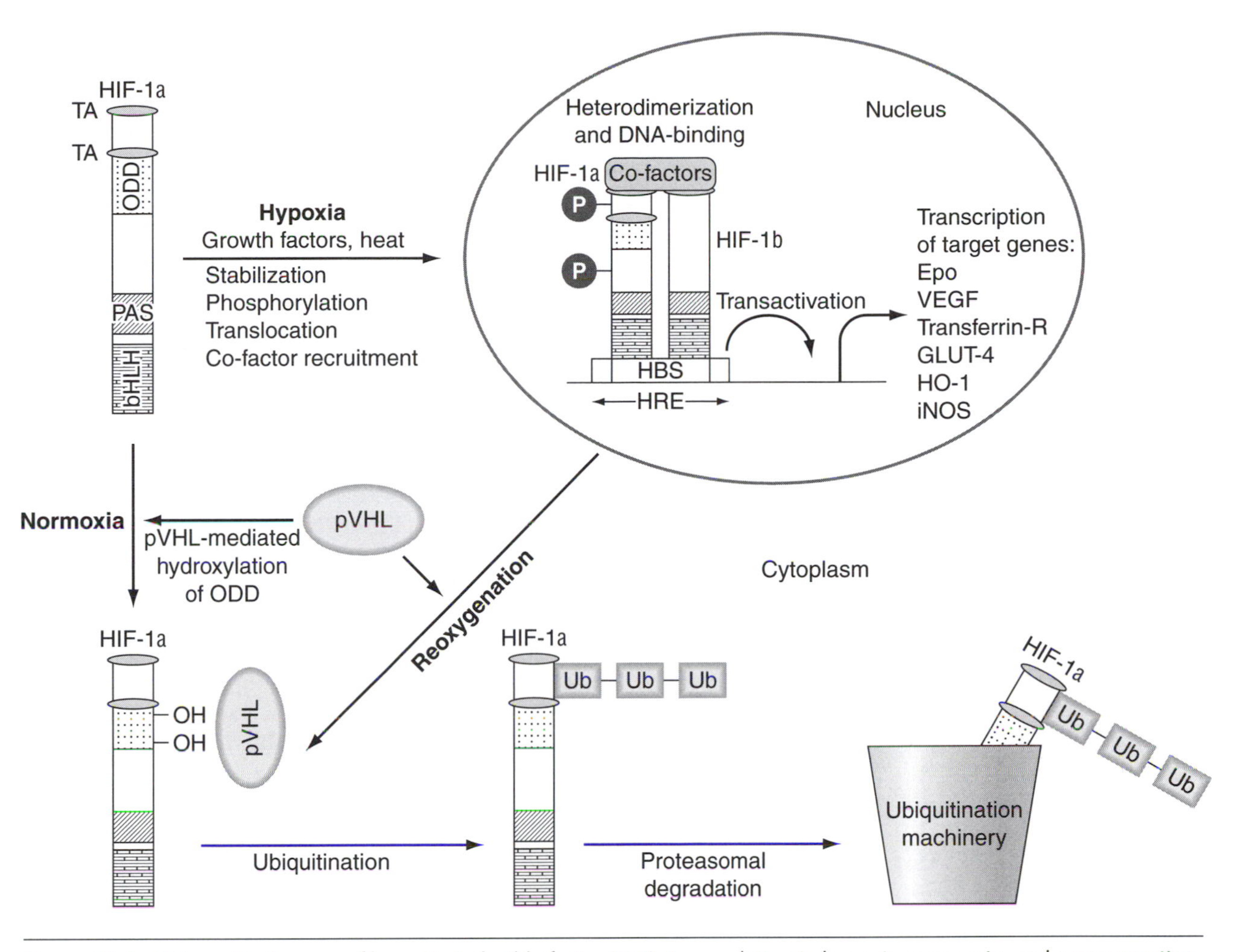

Figure 10.6 Schematic diagram of hypoxia-inducible factor (HIF)-1α regulation in hypoxia, normoxia, and reoxygenation. Details are explained in the text.

recruits HIF-1α to the ubiquitination machinery to be degraded (Maxwell et al. 1999). The association of pVHL with HIF-1α has been reported to take place even in hypoxic conditions. The key enzyme controlling this oxygen-dependent step is a specific HIF-1α-proline hydroxylase (Jaakkola et al. 2001).

Von Hippel-Lindau (VHL) disease is a hereditary cancer syndrome caused by germline mutations of the VHL tumor suppressor gene. pVHL represents the adaptor unit of a protein complex that targets specific proteins including HIF-1 and HIF-2 for ubiquitination and proteolysis. Tumor cells lacking pVHL overproduce the products of HIF target genes (for review see Clifford & Maher 2001; Kondo & Kaelin 2001).

The HIF family

Erythropoietin (EPO) and its receptor function as primary mediators of the normal physiologic response to hypoxia. With increasing altitude, atmospheric pressure decreases, resulting in a decline of oxygen partial pressure (pO_2), arterial pO_2, and oxygen saturation. This decline is detected by a sensor within the kidney and results in increased levels of the hormone EPO. EPO is produced by the kidney in the adult and by the liver in the fetus. The 30-kDa glycoprotein increases red cell mass and maximal oxygen consumption to improve tissue oxygenation, which may be beneficial in exercise physiology (see "Hypoxia and Exercise"). Hypoxia or the acclimatization to altitude may improve performance. The beneficial effect of altitude exposure implies that the increase in serum EPO causes an increase in reticulocyte release from the bone marrow and, subsequently, an increase of the oxygen-carrying capacity of the blood (Bailey & Davies 1997; Boning 1997). On the other hand, a chronic overexpression of EPO inducing severe erythrocytosis may have detrimental effects such as cardiac dysfunction, reduced exercise performance, and markedly reduced life span (Wagner et al. 2001).

EPO acts primarily to rescue erythroid cells from apoptosis to increase their survival. It acts synergistically with several growth factors (e.g., GM-CSF, IGF-1) to cause maturation and proliferation of erythroid progenitor cells. Other effects of EPO include a hematocrit-independent, vasoconstriction-dependent hypertension; increased endothelin production; regulation of tissue renin; change in vascular tissue prostaglandin production; stimulation of angiogenesis; and stimulation of endothelial and vascular smooth muscle cell proliferation (Fisher 2003).

The oxygen-dependent regulation of EPO gene expression is mediated primarily by HIF-1. Cytokines are additional stimulators of hypoxia-induced HIF-1 activity. Nevertheless, EPO synthesis is impaired in inflammatory diseases. Redox-sensitive transcription factors (GATA-2, NF-κB) seem to be involved in the suppression of EPO gene expression by cytokines (IL-1β, TNF-α) and may be responsible for impaired EPO synthesis in inflammatory diseases (La Ferla et al. 2002). This cytokine-induced inhibition of EPO production is not mediated by impairment of HIF-1 function (Hellwig-Burgel et al. 1999).

EPO and EPO receptor gene polymorphisms (Sistonen et al. 1993; Sokol & Prchal 1994; Zeng et al. 2001) seem to be responsible for the individual sensitivity to EPO and hypoxia, familial erythrocytosis, and differences in hematocrit between males and females (Arcasoy et al. 1997; Percy et al. 1998). They may be associated with individual differences in endurance performance (Zeng et al. 2001; Rankinen et al. 2002).

Vascular endothelial growth factor (VEGF) is the most prominent HIF-1 target gene involved in vascular biology. VEGF is mostly expressed by endothelial cells and is important in blood vessel formation. A hypoxia-responsive element represents the binding site for HIF-1 and is responsible for the hypoxia-sensitive regulation of this gene (Ferrara 2000).

VEGF may also play a role in ischemia processes and individual responses to hypoxia, hypoxic training, and acute mountain sickness (Semenza 2000; Maloney et al. 2000; Siren et al. 2001). The well-known, strong individual variability of VEGF concentrations in plasma seems to be determined by common mutations in the VEGF gene (Renner et al. 2000) and is associated with the incidence of acute mountain sickness (Maloney et al. 2000).

VEGF also functions as mediator in exercise-induced capillary growth in skeletal muscle (Gustafsson et al. 1999). A considerable fall of local muscle oxygen tension (3.1-2.1 mmHg intramyocellular) during exercise (Richardson et al. 1995) was responsible for the increase of VEGF mRNA in human (Gustafsson et al. 1999) and rat skeletal muscle (Minchenko et al. 1994; Breen et al. 1996). The VEGF mRNA expression was further augmented when the exercise was performed under hypoxia (Breen et al. 1996). Increased serum levels of VEGF were observed during high-altitude swimming training in humans (Asano et al. 1998). The increase in VEGF mRNA correlated to HIF-1 mRNA levels, which may indicate that HIF-1 influences the exercise-induced VEGF gene expression (Gustafsson et al. 1999). Intensive exercise training in normobaric hypoxia induced HIF-1α and its target gene VEGF in human muscle, in contrast to identical training intensities in normoxia (Vogt et al. 2001). HSP70, which is not HIF regulated, increased under both conditions. Prolonged strenuous endurance exercise at moderate altitude induced a significant long-lasting increase in serum VEGF and EPO that was accompanied by an activation of the immune system (Schobersberger et al. 2000).

Glutamine transporter 4 (GLUT4) is the predominant mammalian facultative glucose transporter isoform expressed in insulin-sensitive tissues including skeletal muscle, adipose tissue, and heart (Charron et al. 1989). Muscle contraction and hypoxia are well-established stimuli of glucose transport activity in skeletal muscle (Ploug et al. 1984; Cartee et al. 1991). A rapid increase in GLUT4 expression is an early adaptive response of muscle to exercise. This adaptation appears to be mediated by pretranslational mechanisms (Ren et al. 1994). GLUT4 plays an essential role in the regulation of muscle glucose uptake and transport in response to hypoxia and to exercise (Zierath et al. 1998).

Iron is required for heme formation and is the most common limiting factor in erythropoiesis. Hypoxia was found to increase the expression of transferrin, probably to enhance the iron transport to erythroid tissues (Rolfs et al. 1997). The transferrin receptor is an HIF-1 target gene, enabling cellular transferrin uptake (Tacchini et al. 1999).

Hypoxia and Exercise

Exposure to hypoxia has specific biological effects that may be important in exercise physiology (Vogt et al. 2001; Hoppeler & Vogt 2001). This hypoxia may represent an additional training stimulus, or the acclimatization to altitude may improve performance; or both may be the case.

To achieve these beneficial effects, there are several concepts of hypoxic training. The classical form of altitude training is "living high and training high." Besides adaptive effects to altitude per se, which improve oxygen transport, utilization, or both, there is also adaptation to the more pronounced stress stimulus during hypoxic exercise. Nevertheless, controlled studies of "typical" altitude training, involving both altitude acclimatization and hypoxic exercise, have never demonstrated improved sea level performance (Levine & Stray-Gundersen 1997). This failure has been attributed to reduced training loads at altitude. If exercise is performed at the same absolute workload in hypoxia, the relative workload is increased as a result of a lower $\dot{V}O_2max$ during hypoxia as compared to normoxia (Maher et al. 1974).

During the training type "living low—training high," the oxygen deficiency represents the additional stimulus. This strategy has been proposed by Hoppeler's group (Hoppeler & Vogt 2001). Hypobaric chambers are often used for this training. Living low—training high clearly improves endurance performance in athletes of all abilities. This improvement is primarily mediated by an increase in erythropoietin leading to increased red cell mass, $\dot{V}O_2max$, and running performance (Roskamm et al. 1969; Terrados et al. 1988; Melissa et al. 1997). Although improved endurance performance can be achieved under altitude conditions, altitude training may evoke some problems. Acute mountain sickness, problems with acclimatization, appetite suppression, inhibition of protein synthesis, muscle wasting, excessive ventilatory work, and detraining due to decreased intensity are believed to influence the effectiveness of altitude training (Boning 1997; Chapman et al. 1998; Levine & Stray-Gundersen 2001).

In the model "living high—training low," adaptive responses to hypoxia in the training-free intervals are expected (Levine & Stray-Gundersen 1997). Training is performed in normoxia with normal volume and intensity. Some athletes sleep and live at moderate altitudes but train near sea level to evoke an increase in the hemoglobin concentration without incurring the deleterious effects of altitude exposure (Hahn et al. 2001). The concept of "sleep high—train low" can be realized relatively simply via simulation of different altitudes in normobaric hypoxia chambers.

The unaccustomed hypoxia in altitude induces several reactions in the body. A reduction of the plasma volume in the first days increases hemoglobin concentration and hematocrit, resulting in an elevation of the oxygen transport capacity. Three weeks of training at 2,300 m induced a 6% decline of plasma volume that persisted after the training session.

The oxygen transport capacity is also induced by stimulated erythropoiesis. The hormone EPO is primarily responsible for stimulating erythropoiesis and thereby maintaining a constant red blood cell mass. The EPO concentration in blood is increased 15 min to 2 h after the start of hypoxic exposure. Kinetic studies revealed that there is a delay of three to four days between increases in serum EPO and reticulocyte release (Major et al. 1994; Breymann et al. 1996). Different methods to measure the EPO-induced red blood cell increase are used. The total hemoglobin mass (Hb_{mass}) is analyzed with the CO-rebreathing technique (Ashenden et al. 1999), and the red cell volume (RCV) is measured using the Evans blue method (Levine & Stray-Gundersen 2001). The literature presents inconsistent results, which may depend on duration of exposure, altitude, methods of measurement, time course of measurement, or individual conditions of the athletes. Figure 10.7 graphically presents the variety of relative changes in Hb_{mass} and RCV as acquired in different studies (Levine & Stray-Gundersen 1997; Hahn et al. 2001; Hoppeler & Vogt 2000). Continuous residence at moderate heights (2,000-2,500 m) improves the oxygen transport capacity by an erythropoietin-induced increase in the hemoglobin (Bunn & Poyton 1996). Nevertheless, it is controversially discussed if exposure to moderate altitudes is sufficient to elicit these hematological adaptations. Ashenden's group found that normobaric hypoxia (2,650-3,000 m) for up to 23 days did not explicitly induce positive effects on red blood cell mass, hemoglobin, maximal oxygen consumption, and reticulocyte counts with the exception of an early increase of serum EPO (Ashenden et al. 1999). Also, exercise alone did not exert a stimulating effect on serum EPO (Schmidt et al. 1991).

In contrast, Levine's group has shown that four weeks of living high—training low (2,500 m/1,250 m) improves sea level running performance in practiced runners because of increases in red cell mass and maximal oxygen consumption, whereas living high—training high or living low—training low for similar periods elicits no such improvement in running performance (Levine &

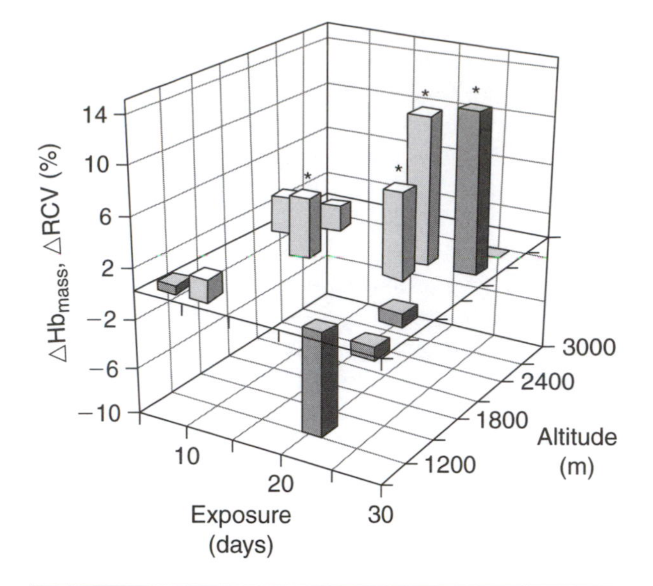

Figure 10.7 Overview of the relative changes (Δ) of hemoglobin mass (Hb_{mass}) and red cell volume (RCV) after hypoxic exercise. Depending on exposure time, altitude, and methods of analysis the different studies revealed highly variable results. Black columns represent studies with continuous residence at the respective altitude; gray columns, studies using the "sleep high—train low" concept. *Significant increases.

Stray-Gundersen 1997). The authors postulated that it is necessary to stay at least three weeks at at least 2,500 m to induce a relevant increase in erythrocyte mass. Individually different responses to altitude and to hypoxic training with respect to the hypoxic ventilatory response, arterial oxygen saturation, and hypoxia-induced EPO and VEGF response of the athlete may be causally involved (Chapman et al. 1998).

Furthermore, repeatedly intermittent exposure to hypoxia may induce a decline in EPO formation (Eckardt et al. 1990). In another study it was observed that intermittent exposure to moderate normobaric hypoxia 12 h daily for one week induced a stimulation of erythropoiesis similar to that with continuous exposure (Koistinen et al. 2000).

Improvement of performance is not likely due to increased hematological parameters alone (Hahn et al. 2001). It is proposed that EPO levels do not respond to tissue hypoxia alone but that hypoxia-induced changes in the hormonal milieu (Bouissou et al. 1988) may also play a role in the EPO response to moderate altitude (Fisher 1993; Saltin 1996). Additionally, effects of hypoxia on the exercise-induced inflammatory stress response are described (Gabriel et al. 1993; Mazzeo et al. 2001; Pedersen & Steensberg 2002). A training session (10 × 1,000 m) at the moderate altitude of 1,800 m above sea level also induced higher increases of adrenaline, noradrenaline, IL-6, blood glucose, cortisol, and growth hormone compared with an almost identical session near sea level (figure 10.8) (Niess et al. 2003). Neuroendocrinological factors such as catecholamines, growth hormone, and cortisol are suggested to play a role in the hypoxia- and exercise-induced changes of leukocyte subpopulations (Pedersen & Steensberg 2002).

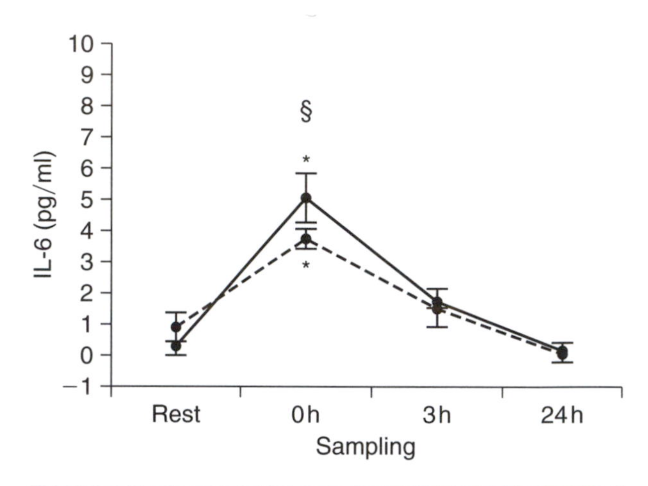

Figure 10.8 Increases of IL-6 in plasma after interval training were significantly different (§) when training was performed at moderate altitude (1,800 m, dotted line) compared to sea level (bold line). To attain similar blood lactate levels at altitude, running velocity was reduced from 112% to 107% of the velocity at the individual anaerobic threshold. Data are presented as means ± SE ($n = 12$), $*p < 0.05$.

Furthermore, one has to consider the effect of hypoxia on muscle composition and mitochondrial biogenesis (Hoppeler et al. 2003). Permanent or long-term exposure to severe environmental hypoxia decreases the mitochondrial content of muscle fibers. As a consequence of hypoxia-inducible gene expression mediated by HIF, glucose utilization and intramyocellular lipid substrate stores are increased. On the other hand, strenuous training in hypoxia while living near sea level leads to muscle adaptations that compensate for the reduced availability of oxygen by improving conditions for transportation and utilization of oxygen in the exercising muscle. Several studies suggest that exercise under hypoxic conditions could possibly induce muscular adaptations that are either absent or found to a lesser degree after training under normoxic conditions. Hypoxic training induced increased mitochondrial densities, capillary-to-fiber ratios, and fiber cross-sectional areas (Desplanches et al. 1993); significant increases in the activities of oxidative enzymes and in capillary density (Green et al. 1999); and elevated activity of citrate synthase and significantly higher myoglobin protein content (Terrados et al. 1990).

In summary, although improvement of performance following exposure to altitude, hypoxia, or hypoxic training may be achieved, there are still conflicting results. Further controlled studies are necessary to explore the molecular responses to hypoxia and to characterize the individual pheno- and genotype of the athlete.

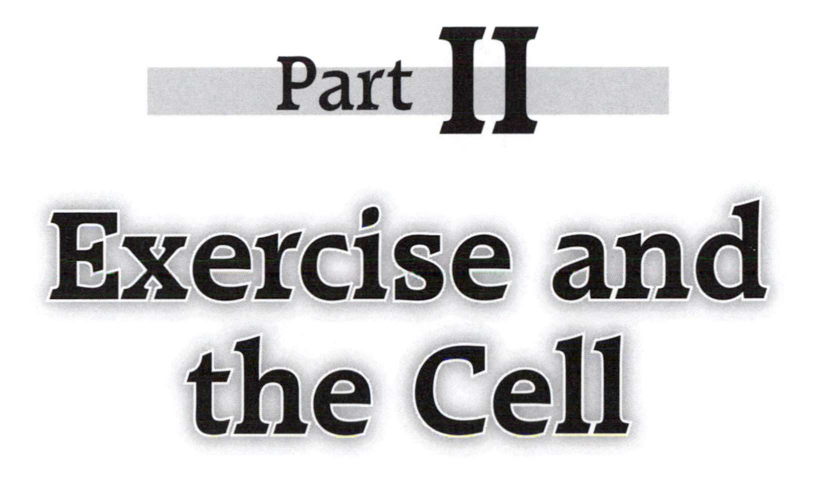

Part II

Exercise and the Cell

Chapter 11

Exercise and the Cardiac Myocyte

Ulrik Wisløff, Jan Pål Loennechen, and Øyvind Ellingsen

During exercise, higher metabolic activity requires increased cardiac output. Over the short term this is accomplished by immediate regulation of filling pressure, heart rate, myocardial contractility, and peripheral vascular resistance. These regulatory mechanisms are well characterized and extensively described in standard physiology textbooks. Over time, exercise improves cardiac function by changing the cellular and molecular phenotype of the cardiac myocyte, including adaptive changes in cell size, contractile function, calcium handling, and resistance to metabolic challenges. Many of these changes have only recently been characterized or are currently being investigated. The purpose of this chapter is to describe some of the long-term adaptations to exercise in the cardiac myocyte, to review potential intracellular signaling pathways and other molecular mechanisms, and to relate these changes to integrated performance and health effects. First we briefly cover some essential cellular and molecular features of the cardiac myocyte.

Structure of the Cardiac Myocyte

The structure of the cardiac myocyte is strongly characterized by its functional properties. Its main function is mechanical work, and the contractile apparatus occupies most of the cellular volume. Contraction and relaxation are highly energy-demanding processes. Myocytes contain more energy-producing mitochondria than any other type of cell. Electrical excitation and excitation–contraction coupling are due to fast and complex ion fluxes. Consequently, cardiomyocyte membranes contain a large number of different ion-transporting proteins. In order to respond promptly to changing circulatory demands, cardiomyocytes contain several regulatory receptors connected to an advanced intracellular signaling system.

Organelles and Membranes

Cardiomyocytes constitute about 75% of the total volume of the myocardium, but only one third of all its cells (Braunwald 1997). The cardiomyocytes are about 10 to 30 μm in diameter and 50 to 180 μm in length. Cell size varies substantially within the same heart and depends on whether the heart is normal, exercise trained, or failing. Each cardiomyocyte is surrounded by a complex cell membrane (sarcolemma), which invaginates the myocyte and forms an extensive tubular network (T-tubules) that extends the extracellular space into the interior of the cell. As with most biological membranes, the sarcolemma is a lipid bilayer virtually impermeable to charged molecules. A large number of transmembrane receptors and ion-transporting proteins constitute properties necessary for normal cardiac function. Ion channels and exchangers use the energy from transmembrane chemical or electrical gradients, or both, for ion transport. Ion pumps transport ions across membranes against gradients utilizing energy from hydrolysis of ATP. Important cardiomyocyte ion-transporting proteins of the sarcolemma and sarcoplasmic reticulum are listed in table 11.1. These proteins consist of one or more subunits with different numbers of membrane-spanning segments. The cation pumps typically consist of 10 transmembrane segments with specific sites for phosphorylation and for binding of ATP and the cation itself. The sodium-calcium exchanger has 11 transmembrane α-helical segments and transports three sodium ions for each calcium ion in the opposite direction. The sodium-hydrogen exchanger extrudes one proton from the cell in exchange for one sodium ion and is composed of 12 membrane-spanning α-helices. The voltage-sensitive potassium, sodium, and calcium channels consist of two or more subunits each with six membrane-spanning α-helices. Cardiomyocytes receive external signals via plasma membrane receptors, intracellular receptors, and adhesion molecules. A large number of membrane receptors have been characterized, and the major receptor groups and their typical ligands are listed in table 11.2. Enzyme-linked receptors activate intracellular enzymes, usually a protein kinase, upon ligand binding. G-protein-coupled receptors are characterized by seven α-helical membrane-spanning segments. When stimulated

Table 11.1 Typical Cardiomyocyte Ion Transporting Proteins

Classes of ion transporting proteins	Important ion transporting proteins
Ion channels	
Sarcolemma	Ca^{2+} channels, K^+ channels, Na^+ channels
Sarcoplasmic reticulum	Ca^{2+} release channel
Ion exchangers	
Sarcolemma	Na^+/Ca^{2+} exchanger, Na^+/H^+exchanger
Ion pumps	
Sarcolemma	Ca^{2+}-ATPase, Na^+ K^+-ATPase
Sarcoplasmic reticulum	SR Ca^{2+}-ATPase

Table 11.2 Cardiac Plasma Membrane Receptor Classes and Ligands

Receptor classes	Ligands
Enzyme-linked receptors	
Tyrosine kinase receptors	IGF-1, TGFβ, FGF, PDGF
Cytokine receptors	TNFα, IL-1, IL-6, cardiotrophin-1
G-protein-coupled receptors	
α- and β-adrenergic receptors	Epinephrine, norepinephrine
Peptide binding receptors	Angiotensin II, endothelin
Muscarinic receptors	Acetylcholine
Ion channel receptors	
Calcium channels	Calcium channel blockers

IGF, insulin-like growth factor; TGF, transforming growth factor; FGF, fibroblast growth factor; PDGF, platelet-derived growth factor; TNF, tumor necrosis factor; IL, interleukin.

they interact with GTP binding proteins (G-proteins), which mediate the intracellular signal. G protein-coupled receptors are important in the regulation of heart rate, cardiomyocyte contractility, and hypertrophy. Ion channel receptors seem not to be important in cardiomyocytes. However, Ca channels bind calcium channel blockers but have no known physiological ligand.

Cardiomyocytes are either mono- or binucleated. The nucleus, which contains almost all of the cell's genetic information, is centrally located. Interspersed between the myofibrils and immediately beneath the sarcolemma are numerous mitochondria, the site of oxidative phosphorylation and the tricarboxylic acid cycle. These organelles occupy about 25% to 35% of the volume of mammalian ventricular cardiomyocytes. Cardiomyocytes have a well-developed cytoskeleton including microtubules and microfilaments. The cytoskeleton has major impact on the cell shape and is involved in altering cardiac mechanical properties (Zile et al. 1999; Tagawa et al. 1998), as well as the intracellular calcium-handling and myocyte contractile function (Gomez et al. 2000; Kerfant et al. 2001).

The sarcoplasmic reticulum (SR) is an extensive membranous organelle that serves important functions in contracting cells. Its lipid bilayer is quite similar to that of the sarcolemma, but the membrane proteins are different. Part of the SR is located close to the T-tubules, where it expands into bulbous hollow swellings, located along the inner surface of the sarcolemma or around the T-tubules. These expanded regions of the SR are often called sarcolemmal cisternae or junctional SR (Braunwald 1997). Their function is to release calcium through the calcium release channels (the ryanodine receptors) to initiate the contractile cycle. The tubular part of the SR is involved in the reabsorption of the calcium, causing cardiomyocyte relaxation. This calcium uptake is performed by an ATP-requiring pump, the sarco-endoplasmic reticulum ATPase (SERCA2). SR calcium is stored at a high concentration in the storage protein calsequestrin.

Contractile Apparatus

Cardiac muscle cells are characterized by a tight organization of contractile proteins into sarcomeres, the functional unit of the muscle, which is bounded by Z-lines. One of its main components is the motor protein myosin, which causes muscle contraction when activated by actin. Myosin is a relatively thick filament (10-15 nm) and consists of about 300 individual myosin molecules, each ending in a bi-lobed head. This head constitutes the cross-bridges that interact with actin to generate contraction. Actin is the main component of a thinner filament that also contains the regulatory troponin and tropomyosin proteins. Calcium binding to troponin C removes the inhibitory effect of troponin I on the interaction between actin and myosin heads. Troponin T links the whole troponin complex to tropomyosin. Cytoskeletal proteins in the myofilaments are related to the thick and thin filaments and the Z-line. The giant protein titin, which extends from the Z-band into the thick filament, is connected to myosin near the center of the thick filament by myosin binding protein C. The parts of the titin molecule that lie within the A-band are quite rigid, whereas the regions found in the I-band are more elastic. Several proteins make up the M-bands, which are transverse structures that link the centers of the thick filament; these include M-protein, myomesin, and the MM isoform of the enzyme creatine phosphokinase (CK-MM). Proteins that support the thin filaments include nebulette, a protein related to skeletal muscle nebulin, which connects the ends of the thin filaments to the Z-band, and tropomodulin, which caps the ends of the thin filaments. Proteins that connect the thin filaments to the Z-line include α-actinin and cap Z (β-actinin).

The cardiomyocytes form a cellular syncytium because of their tight intracellular coupling by intercalated discs. These connectors serve at least three functions: (1) they connect neighbor cells via desmosomes; (2) they connect actin filaments of adjacent cells; and (3) they contain gap junctions that allow coupling of the electrical excitation.

Contractile Cycle and Excitation–Contraction Coupling

Excitation-contraction coupling requires the delivery of calcium to the contractile proteins. The diffusion of an activator from cell surface is not rapid enough to account for the abrupt onset of the active state in the large contracting cardiomyocyte. The following section describes the process of initiating the contractile cycle in cardiomyocytes.

The Action Potential

Cardiac function depends on the appropriate timing of contraction in various regions, as well as appropriate heart rate. The initiating event in cardiac excitation–contraction coupling is the action potential (AP). The AP is the membrane potential waveform that is determined by a complex interplay of many ion transporters and the calcium transient itself. The AP is also responsible for the propagation of excitation information from cell to cell through gap junctions, and allows the heart to function as an electrical and mechanical syncytium. The normal cardiac impulse originates in a group of cardiac pacemaker cells located in the sinoatrial (SA) node and propagates through the atria to reach the atrioventricular node. From the atrioventricular node, electrical activity passes rapidly through the cable-like His-Purkinje system to reach the ventricles, triggering myocyte excitation–contraction coupling and cardiac pumping. The cells in the SA node have normally the fastest intrinsic pacemaker activity. The SA node consists of cells with very few contractile elements and a relatively simple action potential. As the pacemaker cells controlling the rate and the rhythm of the beating heart, SA cells have ever changing membrane potentials. In contrast, ventricular cells are packed with contractile elements and have more complex, triggered action potentials.

The SA node is paced by the i_f current ("funny" Na currents; hyperpolarization activated) and by decreasing opposition to depolarization by the inward rectifier, i_k, current. i_f is a nonselective inward current allowing sodium, potassium, and some calcium ions to enter the cell between –35 and –100 mV of membrane potential. This inward current drives the membrane potential (Em) toward E_f (–20 mV). i_f is opposed by the outward, inwardly rectifying K^+ current, i_{k1}, resulting in a slow but carefully timed pacemaker depolarization to –40 mV. At –40 mV, Ca^{2+} channels begin to open, more rapidly depolarizing the membrane. The slow upstroke of the action potential in nodal cells results from a relative lack of Na^+ channels and dependence of repolarization on the fewer, slower inward Ca^{2+} channels. The first Ca^{2+} channel to be activated is the transient (T type) Ca^{2+} current, i_{CaT}. This current drives Em toward E_{Ca} and in the process triggers the activation of the L-type voltage-activated Ca^{2+} current (i_{CaL}) at –30 mV. Almost at the same time these inward currents are activated, competing outwardly conducting delayed rectifying K^+ currents (i_k) are triggered at Em depolarized to –40 mV. The result is again a tug of war between the inward conductance's (i_{CaT}, i_{CaL}, and $i_{Na/Ca}$; collectively the forces of depolarization) and the outward hyperpolarizing K^+ currents, i_k. The balance is reached at a peak depolarization of +10 mV before the outward K^+ current, i_k, slowly overcomes the inward currents, is joined by i_{k1}, and repolarizes the cell back toward E_k. A schematic presentation of individual ionic currents responsible for depolarization of a pacemaker cell in the SA node is presented in figure 11.1. In the resting heart, a high level of parasympathetic tone slows spontaneous depolarization of the SA node. Blocking of both parasympathetic and sympathetic activity increases the heart rate to about 100 beats

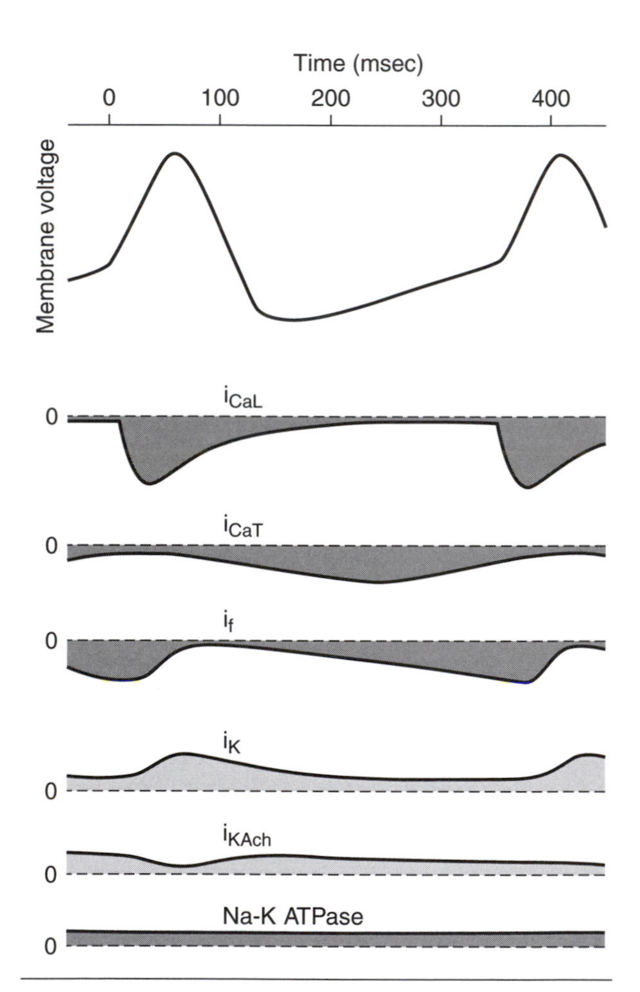

Figure 11.1 Individual ionic currents responsible for depolarization of a pacemaker cell in the SA node. Upper tracing: voltage changes. Diastolic depolarization is caused by declines in three inward currents shown as solid lines (i_{CaL}, i_{CaT}, i_f) and increases in two outward currents shown as dotted lines (i_k and $i_{K.Ach}$). A background outward current generated by the Na-K ATPase (dashed line) also participates in pacemaker activity. For description and definition of the individual ionic currents, see text.

Adapted from A.M. Katz, 2001, *Physiology of the heart* (3rd ed.). (Philadelphia, PA: Lippincott, Williams, and Wilkins).

per minute, which represents the intrinsic or true resting heart rate. Sympathetic activity increases substantially during acute endurance exercise; during strenuous activity in this situation, the heart rate increases up to the individual's maximum, which ranges from about 150 to 230 beats per minute in healthy humans. A signature of the endurance-trained state is resting and submaximal bradycardia. This is due to increased parasympathetic and reduced sympathetic activity. The effects of vagal and sympathetic stimulation on the SA node are largely mediated through G-protein-coupled receptors ($G\alpha_i$ and $G\alpha_s$) (Yatani et al. 1990). The ability of sympathetic stimulation to increase heart rate occurs when $G\alpha_s$ acts directly on the i_f channels. $G\alpha_s$ also stimulates adenylyl cyclase, which activates protein kinase A to phosphorylate calcium channels.

Parasympathetic slowing of heart rate is mediated by $G\alpha_i$, which directly activates the inward rectifier current $i_{K.ACh}$ (inward-rectifying acetylcholine-activated potassium channels) and inhibits cAMP production; both responses hyperpolarize the cells of the SA node (Katz 2001).

In the atrioventricular (AV) node, the AP resembles that in the SA node. In atrial and ventricular muscle cells, the resting membrane potential is about –80 mV and the AP has a very fast upstroke attributable to a Na^+ current, reaching a peak at +30 to 50 mV. Repolarization is much faster in atria than in ventricular myocytes and Purkinje fibers. Thus in ventricular cells there is a more prominent plateau. Moreover, the ventricular AP duration is shortest in epicardial cells, longer in endocardial cells, and longest in midmyocardial cells, partly due to differences in ion channel expression. The long AP duration in ventricular myocytes serves two functions. First, it prevents electrical re-excitation by keeping the membrane depolarized and thus Na^+ and Ca^{2+} channels inactivated. Second, it allows contraction to relax before the next beat, since the AP duration is almost as long as the calcium transient and contraction (Bers 2002). This also prevents tetanization of cardiac muscle. Figure 11.2 summarizes the key features of 10 principal currents involved in the generation of the ventricular myocardial action potential. The action potential is initiated by a rapid depolarization: The fast upstroke in ventricular cells is accomplished by Na^+ channels (i_{Na}). Once the threshold potential is reached, Na^+ channels are activated, resulting in an enormous but brief (<2 msec) inward Na^+ current driving the cell toward E_{Na}. The fast Na^+ channels open as a function of time and voltage, and inactivation causes the current to shut down almost as quickly as it turns on. The threshold-dependent activation of i_{Na} quickly depolarizes the membrane to the levels of activation of both inward Ca^{2+} currents and outward

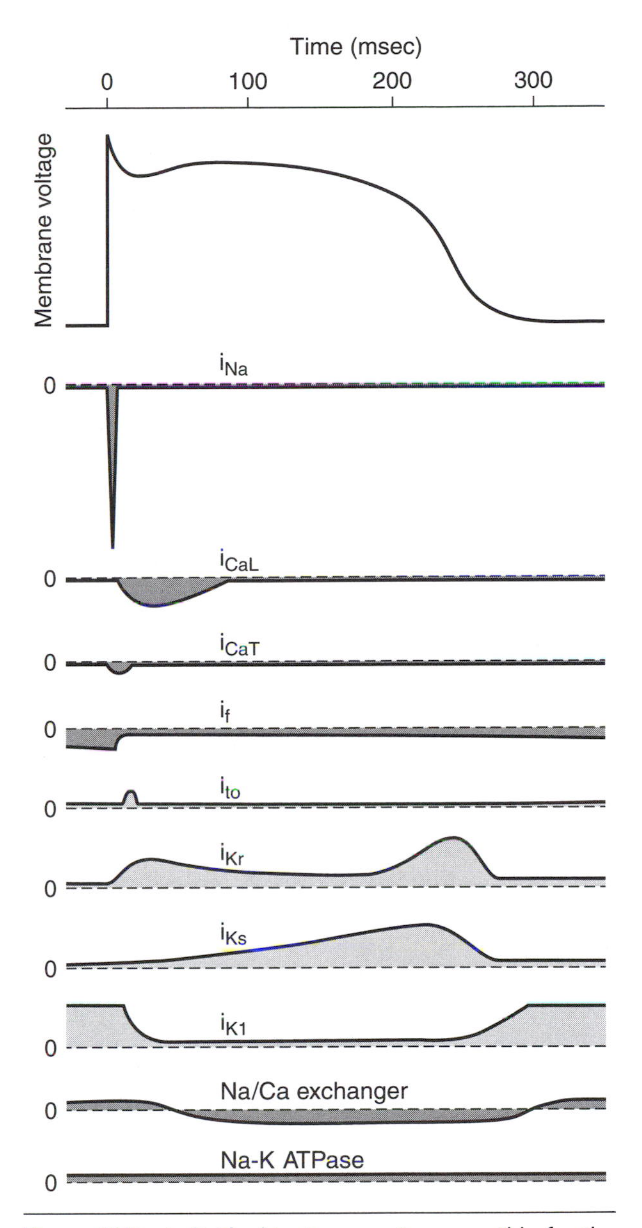

Figure 11.2 Individual ionic currents responsible for the cardiomyocyte action potential. Upper tracing: voltage changes. Lower tracings show 10 different currents: Inward currents are downward; outward currents are upward. Four depolarizing ionic currents (solid lines at the top) include i_{Na}, i_{CaL}, i_{CaT}, i_f. Four repolarization ionic currents (dotted lines in the middle) include three outward rectifiers (i_{to}, i_{kr}, and i_{Ks}) and the inward rectifier (i_{K1}) that inactivates during the plateau. Currents generated by the Na^+-Ca^{2+} exchanger and Na^+, K-ATPase (dashed lines) are shown at the bottom. For description and definition of the individual ionic currents, see text.

Adapted from A.M. Katz, 2001, *Physiology of the heart* (3rd ed.). (Philadelphia, PA: Lippincott, Williams, and Wilkins).

K^+ currents. The Ca^{2+} currents are substantially smaller than the fully activated I_{Na} but also pull the membrane potential to positive potentials. The final result is that the upstroke terminates at +47 mV before reaching E_{Na}. This is followed by the early/rapid phase of repolarization as a consequence of the rapid voltage-dependent inactivation of i_{Na}; the activation of the transient outward K^+ current (i_{to} or $i_{k.4AP}$ or i_{Kto1}); and the activation of a second type of transient outward current (i_{to2}), the Ca^{2+}-activated chloride current, $i_{Cl.Ca}$. Then a plateau phase follows, which is unique to cardiac cell electrophysiology with strikingly few channels open. The plateau phase is maintained by a fine-tuned balance between two types of inward Ca^{2+} currents and at least four types of outward K^+ currents; and the Em is driven back toward E_k. During this phase Ca^{2+} enters the cardiac myocyte via i_{CaT} and i_{CaL} to initiate contraction, with the latter dominating. Thereafter there is a phase of rapid repolarization: As the time-dependent inward currents are inactivated, the outward K^+ currents rapidly drive the membrane potential toward E_K, thus repolarizing the cell. These channels are unable to drive the cell back to E_K because they are inactivated at membrane potentials more negative than –40 mV. Delayed rectifying K^+ currents, i_K, consist of at least three distinct populations of K^+ channels: the rapid, K_r; the slow, i_{Ks}; and the ultra-rapid, i_{Kur} (the most recently described human cardiac i_K subtype). These drive the cell back to resting membrane potential and diastolic depolarization. i_{K1} does not inactivate with time and continues to repolarize the cell. During most of the repolarization phase, the heart cannot be triggered to fire another action potential because Na^+ channels are still inactivated; only hyperpolarization below –70 mV can reprime them for activation. This period of time is called the absolute refractory period. Once i_k repolarizes the cell back to –40 mV, i_{k1} begins to contribute outward currents and drive the cell toward E_k. i_{k1} conducts inward current better than outward current; hence the name inward rectifier.

Calcium Handling

In cardiac muscle, the force of contraction depends on the peak intracellular calcium concentration during systole, the sarcomere length, and the myofilaments' responsiveness to calcium. To understand the basic physiology of heart function it is important to appreciate some details concerning how calcium is moved around the various organelles of the myocyte in order to bring about excitation–contraction coupling. Failure of normal Ca^{2+} handling is a central cause of both contractile dysfunction and arrhythmias in pathophysiological conditions (Bers 2002). The brief increase in cytoplasmic calcium concentration (often called the calcium transient) allows Ca^{2+} to bind to the myofilament protein troponin C, which activates the myofilaments and transduces the chemical signal and energy (ATP) into cardiomyocyte shortening in a calcium-dependent manner.

As already described, during the action potential Ca^{2+} ions enter the cells via voltage-activated Ca^{2+} channels as an inward Ca^{2+} current ($i_{Ca}{}^{2+}$), which contributes to the action potential plateau. In addition, cardiomyocytes exhibit two types of voltage-dependent Ca^{2+} channels (L type and T type). As T-type $i_{Ca}{}^{2+}$ is negligible in most ventricular myocytes, $i_{Ca}{}^{2+}$ generally refers to L-type $i_{Ca}{}^{2+}$ (dihydropyridine receptors). L-type Ca^{2+} channels are located primarily at sarcolemmal-SR junctions where the SR Ca^{2+} release channels (the ryanodine receptors) exist. In addition, the Na^+-Ca^{2+} exchanger contributes to calcium influx and efflux with a stoichiometry of three Na^+ ions to one Ca^{2+} ion that produce an ionic current either inward (forward mode: during high intracellular Ca^{2+} concentrations) or outward (reverse mode: during positive membrane potentials and high intracellular Na^+). The calcium entering the myocyte from the outside contributes directly only to a minor degree to myofilament activation, and its main effect is to stimulate calcium release from the intracellular pool of calcium: the SR. This is normally termed calcium-induced calcium release. Contributions from other pathways have also been proposed, such as the T-type Ca^{2+} current and one through Na^+-Ca^{2+} exchange; but both of these have been shown to be much less effective and slower than the L-type current (Bers 2002). Despite overwhelming evidence that Ca^{2+} influx is essential for cardiac excitation–contraction coupling, a few studies have suggested a voltage-dependent Ca^{2+} release that does not require Ca^{2+} influx (Ferrier & Howlett 2001). Several major concerns have strongly challenged this hypothesis, and at this point it is not convincing (Bers 2002). Inositol (1,4,5)-triphosphate ($InsP_3$) can also trigger Ca^{2+} release in cardiac myocytes, but the rate and extent of Ca^{2+} are very much lower than for calcium-induced calcium release, and action potentials are not known to stimulate $InsP_3$ production (Bers 2002).

A high load of calcium in the SR directly increases the amount of calcium available for release, but also enhances the fraction of SR Ca^{2+} that is released for a given i_{Ca}^{2+} trigger. This increased sensitivity of the ryanodine receptors to calcium at high SR Ca^{2+} concentration means that what is often referred to as spontaneous SR Ca^{2+} release at high cellular SR Ca^{2+} content might be considered mechanistically to be triggered by high Ca^{2+} in SR. This is the basis of after-contractions. Transient inward current and delayed after-depolarizations can trigger arrhythmias (Bers 2001). The ryanodine receptors (RyR) are both the SR Ca^{2+} release channels and a scaffold protein that localizes numerous key regulatory proteins to the junctional complex. These include calmodulin (which can exert Ca^{2+}-dependent modulation of RyR function), FK-506 binding protein (which may stabilize RyR gating and also couple the gating of both individual and adjacent RyR tetramers), PKA (which can alter RyR and i_{Ca}^{2+} gating), phosphatases 1 and 2A, and sorcin (which binds to RyR and L-type Ca^{2+} channels). RyRs are also coupled to other proteins at the luminal surface of SR (triadin, junctin, and calsequestrin), which are involved in intra-SR Ca^{2+} buffering and modulation of the Ca^{2+} release process. RyRs are arranged in large organized arrays (up to 200 nm in diameter with more than 100 RyRs) at the junction between the SR and sarcolemma beneath the L-type Ca^{2+} channels (on the surface and in the T-tubules) (Bers 2002). These arrays constitute a large functional Ca^{2+} release complex at the junction (or the couplon). This local unit concept is supported by observations of spontaneous local Ca^{2+} transients called Ca^{2+} sparks. Ca^{2+} sparks reflect the nearly synchronous activation of a cluster of about 6 to 20 RyRs at a single junction. Ca^{2+} sparks are the fundamental units of SR Ca^{2+} release both at rest and during excitation–contraction coupling. During excitation–contraction coupling, several thousand Ca^{2+} sparks in each cell are synchronized in time by the action potential, such that the local rises in calcium are completely overlapping in time and space, making the calcium transient appear spatially uniform.

For relaxation and filling of the heart to occur, the intracellular Ca^{2+} concentration must decline. This requires Ca^{2+} transport out of the cytosol by four pathways involving SR Ca^{2+}-ATPase, sarcolemmal Na^+-Ca^{2+} exchange, sarcolemmal Ca^{2+}-ATPase, and mitochondrial Ca^{2+} uniport (Bers 2001). The SR Ca^{2+}-ATPase and Na^+-Ca^{2+} exchange are most important quantitatively. In rabbit ventricular myocytes, the SR Ca^{2+}-ATPase removes about 70% of the activator Ca^{2+} from the cytosol, whereas the Na^+-Ca^{2+} exchange removes 28%, with only about 1% each for the sarcolemmal Ca^{2+}-ATPase and mitochondrial Ca^{2+} uniporter. In rat ventricle, the SR Ca^{2+}-ATPase activity is higher, presumably due to more pump molecules per unit cell volume, and Ca^{2+} via Na^+-Ca^{2+} exchange is less, resulting in a balance of 92:7:1 for SR Ca^{2+}-ATPase, Na^+-Ca^{2+} exchange, and mitochondrial Ca^{2+} uniporter/sarcolemmal Ca^{2+}-ATPase, respectively. In the mouse ventricle, the situation is quantitatively similar to that in the rat, whereas the balance of Ca^{2+} fluxes in the human ventricle is similar to that in the rabbit.

In heart failure the expression of SERCA2 is normally reduced and Na^+-Ca^{2+} exchange increased, and the systems contribute equally to the decline in intracellular Ca^{2+} in diastole. Both changes tend to reduce the Ca^{2+} content in SR, limiting SR Ca^{2+} release, which may be a central cause of systolic deficit in heart failure (Bers 2002). SR Ca^{2+} content can be raised by increasing Ca^{2+} influx, decreasing efflux, or enhancing Ca^{2+} uptake into SR by, for example, adrenergic stimulation or an increase in stimulation frequency, action potential duration, i_{Ca}^{2+}, or intracellular Na^+. Phospholamban is an endogenous inhibitor of SERCA2. Phosphorylation of phospholamban by cyclic AMP-dependent or calmodulin-dependent protein kinases relieves this inhibition, allowing faster twitch relaxation and decline of Ca^{2+} in diastole. Targeted gene knockout of phospholamban results in animals with hyperdynamic hearts, with little apparent negative consequence (Brittsan & Kranias 2000). Furthermore, del Monte et al. (1999) showed that overexpression of SERCA2 by adenoviral gene transfer restores contractile function in cardiomyocytes from failing human hearts. Figure 11.3 shows the main mechanisms that contribute to the excitation–contraction coupling and removal of calcium from the cytosol after contraction.

Calcium Sensitivity

Calcium sensitivity of the cardiomyocyte is defined as the degree of tension developed at a given calcium concentration. Since calcium activates the cardiac myofilaments in a graded manner, the relationship between calcium and force development is of fundamental importance. Several factors are known to modify this relationship. Both cooling and shortening of the sarcomere decrease the calcium sensitivity and the maximum force generated by the myofilaments. Similarly, acidosis and high phosphate concentrations decrease both the calcium sensitivity and maximum force. This effect is important during pathological conditions such as hypoxia or ischemia. The molecular and cellular

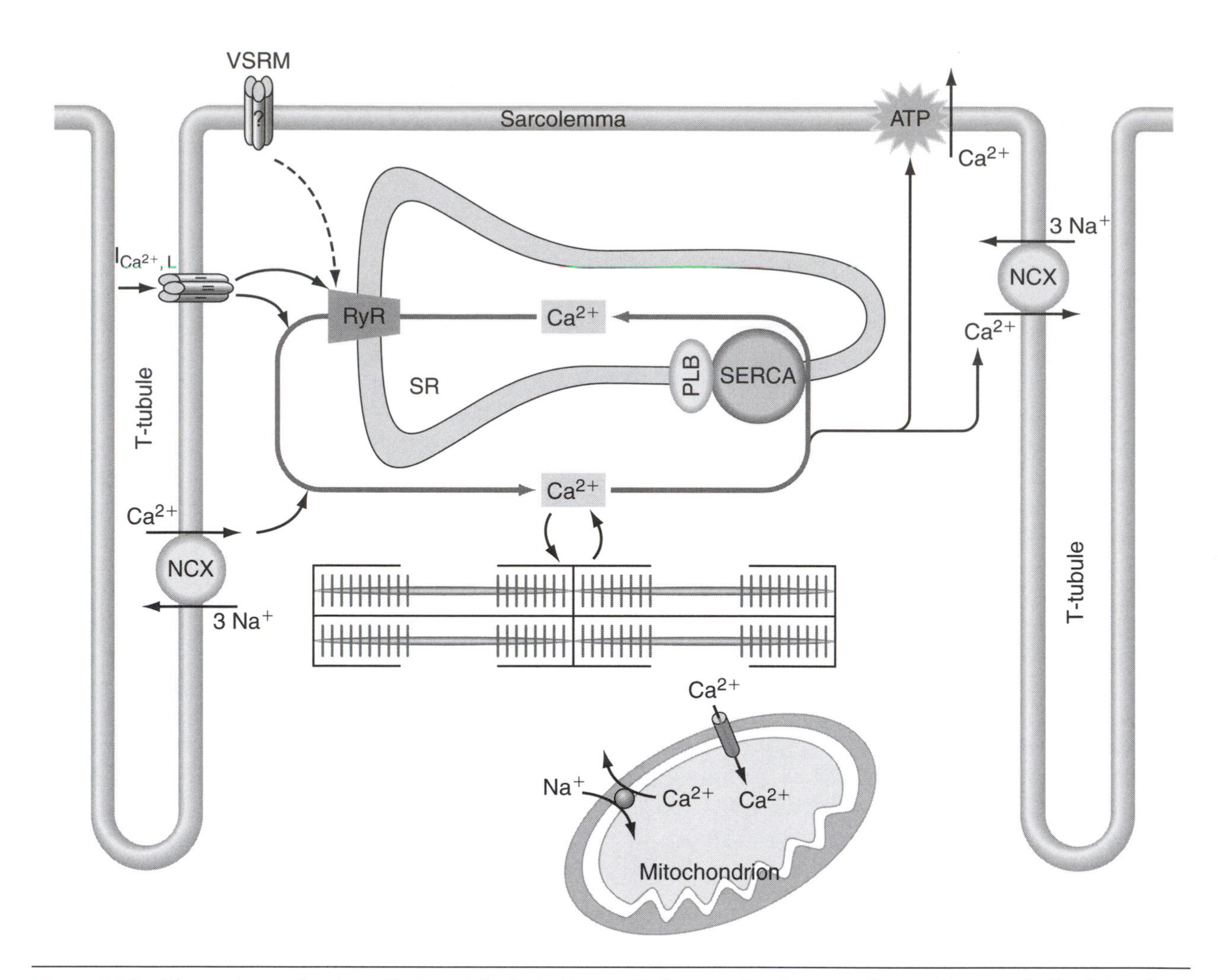

Figure 11.3 The main mechanisms that contribute to the excitation–contraction coupling and removal of calcium from the cytosol after contraction. VSRM, voltage-sensitive release mechanism; ATP, adenosine triphosphate; NCX, sodium-calcium exchanger; SERCA, sarcoplasmic reticulum calcium ATPase; $i_{Ca^{2+}L}$, inward calcium flux via L-type channels; PLB, phospholamban; RyR, ryanodine receptor. For details, see text.

Adapted from Bers 2001.

basis for long-term changes in calcium sensitivity is not fully understood. Multiple biochemical alterations of the contractile proteins have been suggested, including isoform switching of troponin T (Anderson et al. 1995) and troponin I, as well as suppressed α- and increased β-myosin heavy chain expression (Nakao et al. 1997).

Contractile Cycles

The mechanism of muscle contraction is explained by the sliding filament model, in which the actin and myosin filaments move past one another. The cardiac muscle contraction is the result of the following cycle.

In the relaxed state, the myosin heads are attached to actin. Upon binding to ATP, the heads detach themselves from actin. ATP is hydrolyzed to ADP and Pi, but does not release the two reaction products. ATP cleavage results in an allosteric change that builds up tension in the myosin head. When present, calcium binds to troponin C and causes a conformational change in troponin that distorts tropomyosin and allows the myosin to bind to actin. Actin facilitates the release of Pi and shortly thereafter ADP. This converts the allosteric tension in the myosin head into a conformational change that acts like a rowing stroke. The cycle continues as long as ATP and calcium are available. At the end of each contraction, the calcium level drops as a result of reuptake in the SR and transport out of the cell. Calcium is released from troponin C, and tropomyosin returns to its original position on the actin molecule, blocking further cycling.

Adaptive Hypertrophy and Growth Signaling

Regular exercise improves performance of the cardiac myocyte. Recent studies link training-induced increases in calcium-handling proteins with enhanced cardiomyocyte contractile function and higher work capacity, both in healthy individuals and in heart failure (Moore et al. 1993; Diffee et al. 2001, 2003; Wisløff et al. 2002). Exercise also induces adaptive cardiac myocyte growth regulation. In a well-trained athlete's heart, cell elongation contributes to enlarged chambers, increased stroke volume, and higher aerobic capacity (Ferguson et al. 2001; Wiebe et al. 1998). In heart failure, exercise amends excessive cardiac hypertrophy. After experimental myocardial infarction, endurance training reduces cardiac myocyte dimensions and gene expression of atrial natriuretic peptide (ANP), a marker of the hypertrophic response (Wisløff et al. 2002). These and other cardiac effects of exercise are described in the sections that follow.

Adaptive Hypertrophy in Healthy Hearts

In healthy rats there is a time-dependent increase in cardiac mass and cell size due to endurance training (figure 11.4, *a* through *c*). It is apparent that longitudinal cardiomyocyte growth by sarcomeres added in series is sufficient to explain the effect of training on myocardial mass studies (e.g., Moore et al. 1993; Wisløff et al. 2001, 2002). This provides a cellular mechanism to the eccentric ventricular hypertrophy that is often elicited by programs of aerobic exercise in humans and animal models of exercise. In healthy individuals there is a positive correlation between $\dot{V}O_2$max and myocardial hypertrophy (figure 11.4). Figure 11.4*c* shows that this relationship has an exponential form, which fits with

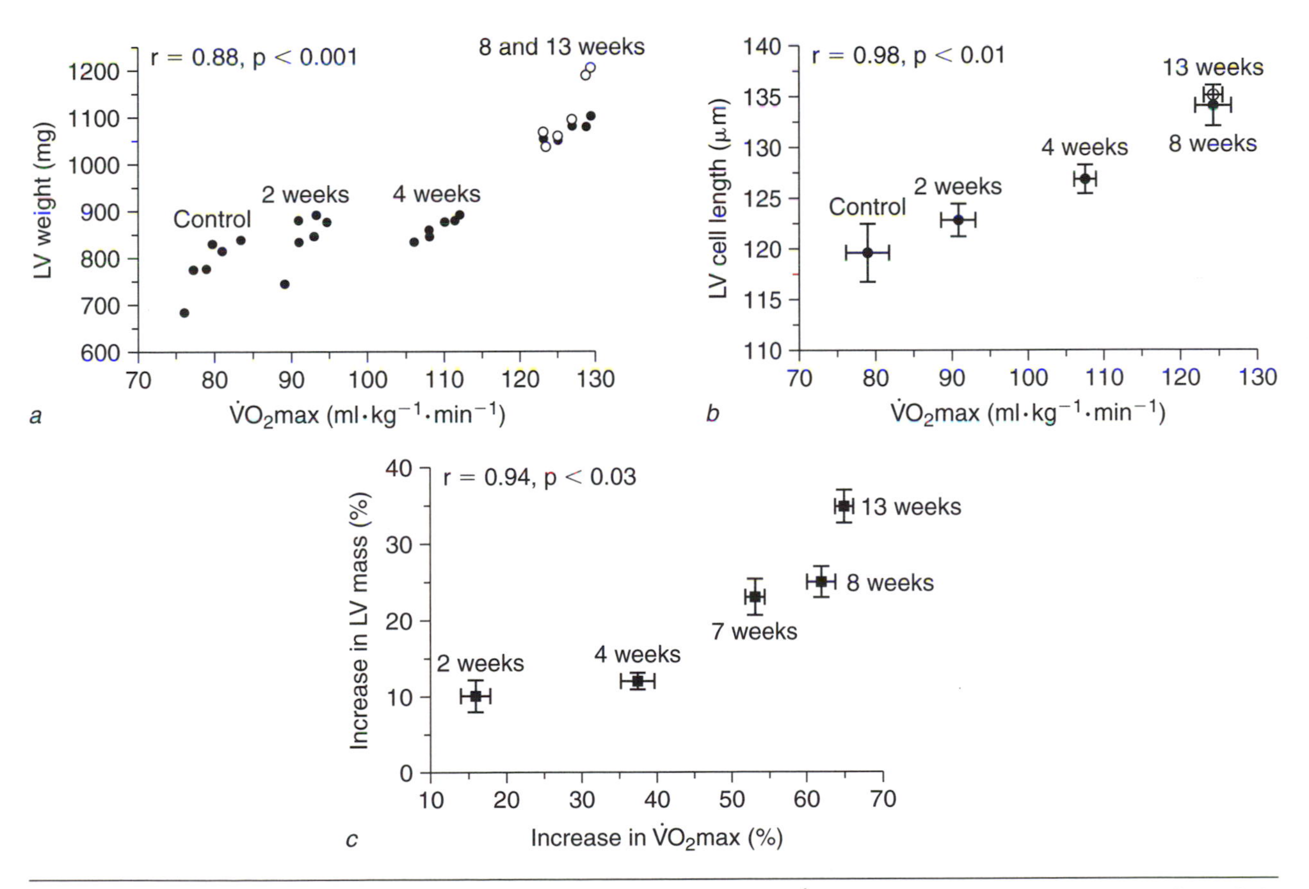

Figure 11.4 *(a)* Relationship between individual maximal oxygen uptake ($\dot{V}O_2$max) and left ventricle (LV) weight and *(b)* LV cell length. Open circles represent animals trained for 13 weeks. *(c)* Relationship between increases in $\dot{V}O_2$max (%) and left ventricular hypertrophy (%). Data are presented as individual values in *a* and as mean ± SD in *b* and *c* from about 180 cells from each of the same individuals, a total of 12,875 cells.

Data from Wisløff (2001).

the hypothesis that untrained subjects are demand limited and that improvement in $\dot{V}O_2max$ early in the training period is due to peripheral factors (demand limited, Wagner 2000). Furthermore, fit subjects seem to be supply limited; that is, improvement in $\dot{V}O_2max$ is mostly due to increased maximal cardiac output. Adaptive hypertrophy seems to be a very dynamic process. Most of the cardiac cell elongation occurs within the first five to seven weeks of regular exercise training, whereas size regresses back to sedentary baseline levels after two to four weeks when training is stopped (Kemi et al. 2004).

Reduced Hypertrophy in Heart Failure

Cardiomyocytes from chronically failing hearts are characterized by hypertrophy. At the cellular level, hypertrophy is due to new sarcomeres, added either in series (elongation) or in parallel (widening), or both. The pattern of hypertrophy depends on the character of the inducing mechanical and humoral load, which differs between models of heart failure. Experimental and human heart failure studies support the notion that diastolic load mainly induces longer myocytes and eccentric hypertrophy of the left ventricle, whereas systolic load induces wider myocytes and concentric left ventricular hypertrophy. In contrast to findings in healthy rats, increased $\dot{V}O_2max$ in post-MI rats is associated with a reduction in ventricular weights and cardiomyocyte dimensions (figure 11.5) (Zhang et al. 1998; Wisløff et al. 2002). By directly reducing cell length in post-infarction myocytes, aerobic interval training minimizes ventricular remodeling and is thus likely to prevent heart failure progression (Orenstein et al. 1995; Jain et al. 2000). Reduced ventricular hypertrophy in trained

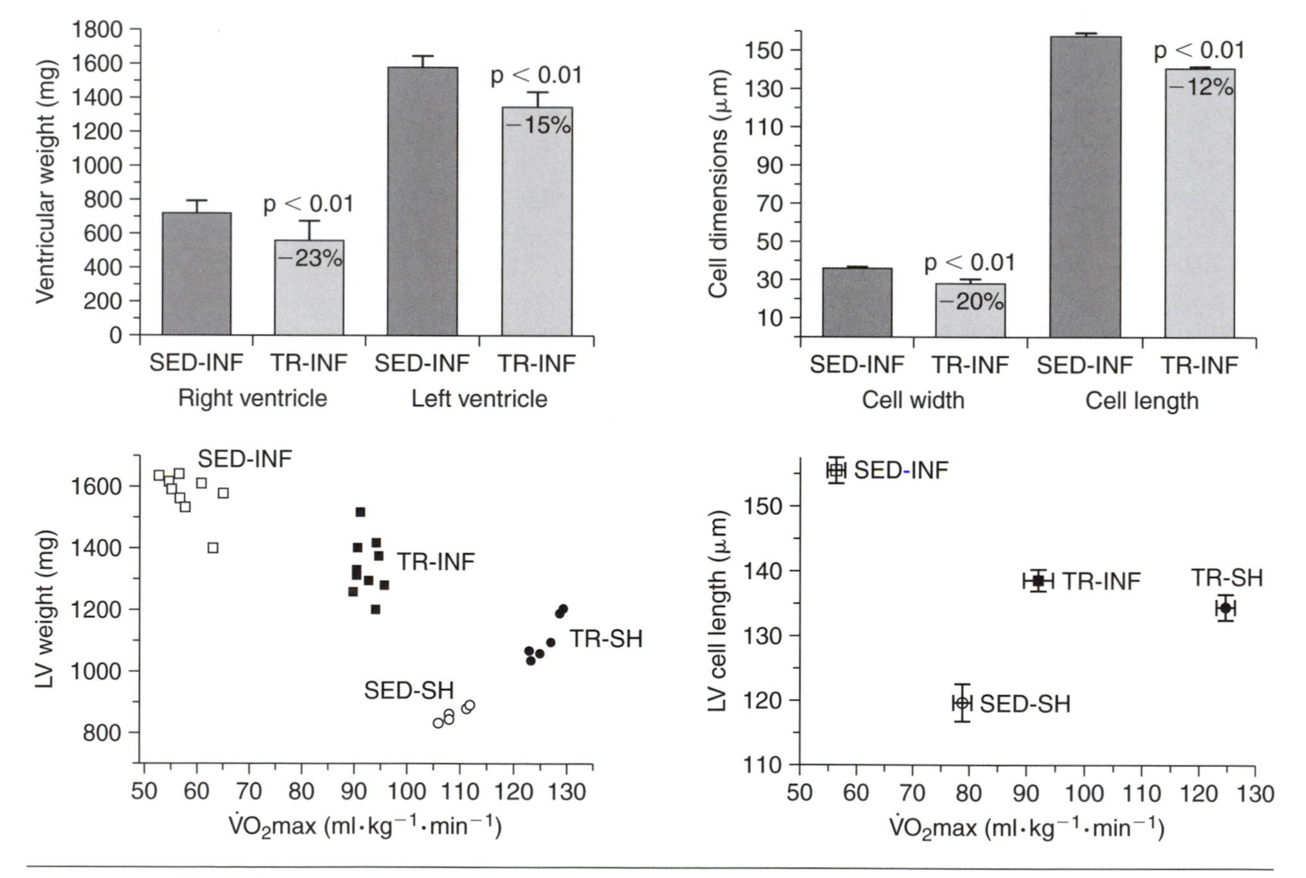

Figure 11.5 Effects of intensity-controlled endurance training on ventricular weights and cardiomyocyte dimensions after eight weeks (upper panel). Data are presented as mean ± SD from 6 rats in each group of sham rats, and 10 and 11 animals in the trained infarction group (TR-INF) and sedentary infarction group (SED-INF), respectively. Cellular dimensions were calculated from 200 ± 12 myocytes from each of the same animals, a total of about 6,500 cells. Relationship between maximal oxygen uptake ($\dot{V}O_2max$) and left ventricular (LV) weight and increases in cardiomyocyte cell length after eight weeks of endurance training (lower panel).

Data from Wisløff 2001 and Wisløff et al. 2002.

rats after myocardial infarction is consistent with reduced expression of atrial natriuretic peptide (ANP), a known marker of cardiac hypertrophy. The exercise-induced reduction of ANP in rats after myocardial infarction is substantial (Wisløff et al. 2002), and is comparable with the response in attenuated ventricular hypertrophy after heart failure treatment with the angiotensin II receptor blocker losartan (Loennechen et al. 2002).

Adaptive Signals and Molecular Pathways

Both the adaptations in cardiac myocyte growth previously discussed and the alterations in contractility and calcium handling described in the following sections depend on changes in gene expression. During the last 10 to 15 years, detailed studies of transcriptional, translational, and post-translational regulation have characterized a host of molecular mechanisms and signaling pathways associated with regulation of growth, contractility and other cell functions. In cardiac myocytes several growth factors, transcription factors, and other regulators of the hypertrophic response have been identified. For many years, investigations focused on cardiomyocyte growth during pathological conditions, in which hypertrophy is typically associated with reduced function (pathological hypertrophy). However, several recent studies indicate that distinctive regulatory pathways are activated in physiological hypertrophy (adaptive cardiomyocyte growth with increased or neutral function) during normal development, in response to exercise training, and in compensatory phases of pressure and volume overload (McMullen et al. 2003, Wilkins et al. 2004, O'Connell et al. 2003). We anticipate that current and future investigations combining mechanistic studies in genetically modified experimental models and human data from controlled clinical studies will clarify specific roles of the molecular pathways regulating exercise-induced cardiac phenotypes.

Adaptive Stimuli in Heart Muscle

Exercise induces many changes in the internal and external milieu that may serve as stimuli for adaptive changes in gene expression of the cardiac myocyte. These may be categorized into five main classes: (1) external humoral factors such as hormones, peptide growth factors, and neurotransmitters; (2) mechanical load; (3) changes in intracellular calcium concentration related to contractile activity; (4) hypoxia; and (5) cellular redox state. At present, most of the evidence suggests that humoral factors, mechanical load, and intracellular calcium are important regulators of myocardial growth, whereas data regarding putative roles for hypoxia and redox state in myocardial hypertrophy are sparse (reviewed in Copeland et al. 2002; Consitt et al. 2001; Baar et al. 1999). The following sections cover the main factors implicated in cardiac hypertrophy in general, with special emphasis on observations more directly related to exercise.

Exercise-Induced Endocrine Factors

Aerobic exercise increases blood levels of many hormones, neurotransmitters, cytokines, and peptide growth factors, for example epinephrine, norepinephrine, cortisol, growth hormone, thyroxine, glucagon, renin, angiotensin, aldosterone, antidiuretic hormone, atrial natriuretic peptide, and interleukin-6, some of which are often termed "stress hormones" because of their correlation with biological stress. Although there are important differences in function, extent, and pattern of secretion, their concentrations all go up during exercise and return to normal levels during rest. Thus, the mode of receptor stimulation and downstream effects is transient or cyclic, in contrast to the tonic or chronic exposure often encountered in disease states like pressure overload and heart failure. This difference may have important implications for long-term downstream effects, such as gene regulation, molecular phenotype, and cell structure. For many of the endocrine factors, the magnitude of secretion corresponds to the level of relative exercise intensity. Thus, a regularly exercising person will have peaks of stimulation exceeding those of a more sedentary person, whereas the average levels during daily life activities are lower. Some of the salutary effects of physical activity are consistent with the notion that chronically elevated levels of stress hormones and endocrine factors may be detrimental, whereas transient peaks are neutral or beneficial.

Sympatho-Adrenal Neurohormones

Adrenergic stimulation is by far the most prominent regulator of acute cardiac myocyte function. During exercise, sympathetic nerve endings and the adrenal glands secrete noradrenaline and adrenaline in an intensity-dependent manner. At low to moderate exercise intensities, noradrenaline from sympathetic nerve terminals is the dominant component; at higher intensities and prolonged exercise, adrenaline from the adrenal medulla contributes increasingly. Abundant cardiac sympathetic innervation may selectively increase local catecholamine levels within the heart far above circulating blood concentrations (Ellingsen et al. 1987). As with many other stress hormones, plasma levels of adrenaline and noradrenaline are related to relative exercise intensity. This means that a well-trained person will have lower adrenergic stimulation than an untrained person when the two perform work at the same exercise intensity. It usually also means that a regularly exercising person will expose him- or herself to higher peaks and probably higher average levels of adrenergic stimulation than one who leads a more sedentary life. E.g., male soccer athletes had slightly higher noradrenaline concentrations in venous blood from the heart muscle at rest, than sedentary age-matched healthy controls (Neri Serneri et al. 2001). An important exception is heart failure patients with a low exercise capacity, who may approach their maximal oxygen consumption during daily life activities, like walking, dressing, and household chores. For patients who increase their $\dot{V}O_2max$ by regular exercise training, average adrenergic and other stress hormone levels may actually be significantly reduced.

In contrast to the well-defined acute changes, the chronic effects of sympathetic stimulation on cardiac myocyte function are less clear. Both adrenaline and noradrenaline bind to alpha-1 and beta-adrenergic receptors on the myocardial cell membrane and activate downstream intracellular signaling pathways described below. Recent observations from mice where the main cardiac subtypes of alpha-1 adrenergic receptors ($\alpha_{1A/C}$ and α_{1B}) were genetically ablated, suggest a sigificant role for alpha-1 adrenergic stimulation in normal heart growth and adaptation to exercise (O'Connell et al. 2003). Although an extensive number of other effects on gene regulation and subcellular structure and function have been described in experimental models, the exact mechanisms and their contribution to long-term effects in the context of exercise training need further investigation. In contrast, both experimental and clinical studies have demonstrated beyond doubt that beta-blockers (i.e., drugs that competitively inhibit binding of adrenaline and noradrenaline to beta-adrenergic receptors) significantly reduce the negative consequences of neurohormonal and hemodynamic loading on heart structure and function during congestive heart failure.

Renin, Angiotensin, Aldosterone

Although the renin-angiotensin-aldosterone hormonal axis is transiently stimulated during endurance activities, its significance for long-term effects of training remains to be determined. The important regulation of electrolyte and water balance by renin, angiotensin, and aldosterone had been known for many years when the profound influence of these hormones on myocardial growth and gene expression was identified. Stimulation of the G-protein-coupled AT_1 receptors and downstream pathways by angiotensin II causes cardiac myocytes to hypertrophy both in pressure overload and in congestive heart failure, and also changes the molecular phenotype to a fetal pattern of gene expression, with increased levels of atrial natriuretic peptide, reduced sarcoplasmic reticulum ATPase, and a lower ratio of alpha versus beta isoforms of myosin heavy chain proteins. Although potentially useful in the early phase of compensatory hypertrophy following myocardial infarction, long-term stimulation of the renin-angiotensin-aldosterone pathway is associated with pathological myocardial enlargement, reduced contractile function, and increased mortality. Blockade of the pathway by angiotensin-converting enzyme inhibitors, AT_1 receptor blockers, and aldosterone antagonists markedly prevents and ameliorates the detrimental health effects and is therefore a basic component in drug therapy of heart failure and other manifestations of cardiovascular disease. During exercise, increased sympathetic nerve activity reduces renal blood flow and induces renin secretion, which leads to higher levels of angiotensin and aldosterone in circulating blood during and immediately after physical activity. However, resting levels are similar between trained and untrained individuals (Copeland et al. 2002; Consitt et al. 2001).

Growth Hormone and Insulin-Like Growth Factor

A large body of evidence indicates that many of the favorable effects of exercise on the cardiac myocyte are consistent with stimulation by growth hormone and insulin-like growth factor IGF-1 (reviewed in Copeland et al. 2002; Consitt et al. 2001). Growth hormone is a peptide secreted from the anterior pituitary gland in a pulsatile pattern regulated by hypothalamic hormones. It exerts many of its metabolic effects by increasing the secretion of IGF-1 locally in target organs, among which liver accounts for the majority of circulating blood levels. The cardiac myocyte has receptors for both growth hormone and IGF-1 and responds to stimulation by either hormone. IGF-1 may act as a paracrine hormone within the myocardium. Its expression responds to stretching of the cardiac myocyte and is elevated in heart failure, probably because of increased wall tension. Growth hormone increases during endurance exercise, after a brief latent period, and may remain elevated postexercise. In the resting state, levels usually revert to normal or even subnormal. However, exercise training may increase the pulsatile secretion of growth hormone at rest. This may be particularly advantageous since pulsatile delivery enhances tissue responses (Copeland et al. 2002). In male soccer athletes, increased levels of IGF-1 correlated with physiological cardiac hypertrophy (Neri Serneri et al. 2001). Decline in the amplitude of pulsatile growth hormone secretion and the resulting decrease in plasma levels of IGF-1 are likely to be important features of aging (Ueki et al. 2002). Binding of IGF-1 to the type 1 IGF receptor (IGF-R1) initiates autophosphorylation of the receptor, phosphorylation of insulin receptor substrate (IRS-1), and tyrosine phosphorylation. Downstream effects are mediated via Ras/Raf, PI3-kinase (PI3K), and MAP kinases as described in subsequent sections. IGF-1 mimics many of the beneficial long-term effects of exercise on cardiac function. It induces adaptive growth of the cardiac myocyte and improves contractile function in heart failure, probably by increasing the myofilament sensitivity to changes in intracellular calcium (Cittadini et al. 1998). Tajima and coworkers (1999) found that restored contractility by growth hormone correlated with increased levels of sarcoplasmic reticulum calcium ATPase (SERCA2) in the myocardium.

Thyroid Hormones

The significance of training-induced changes in thyroid hormone concentrations has not yet been established (Copeland et al. 2002; Consitt et al. 2001). Nevertheless, changes in secretion are potentially interesting, because stimulation of the cardiac myocyte by triiodothyronine (T3) mimics several effects of exercise training. It causes hypertrophy, improves contractility, and induces increased levels of SERCA2 and an increased alpha-/beta-myosin heavy chain ratio. During exercise, an increased amount of thyroid-stimulating hormone (TSH) is secreted from the anterior pituitary. This presumably enhances production of the effector hormones thyroxine and triiodothyronine from the thyroid gland. Since turnover of these hormones is small, acute changes are hard to detect. Moreover, endurance training seems to have little effect on basal and postexercise levels of TSH (Copeland et al. 2002; Consitt et al. 2001).

Cytokines

Cytokines are peptides that were originally characterized as paracrine or endocrine cell-to-cell signaling molecules involved in immune responses. Several of them, including interleukin-1 (IL-1), leukemia inhibitory factor (LIF), tumor necrosis factor-α (TNF-α), and cardiotrophin-1 (CT-1), may induce or modulate growth in cardiac myocytes (Chien 2004). Although their significance in exercise-induced adaptive hypertrophy has not been established, the effects of CT-1, for example, may be of interest. Cardiotrophin-1 induces a hypertrophic phenotype of elongated cardiac myocytes with minimal change in cell width. In some respects the pattern of gene expression differs from the standard fetal program induced by most hypertrophic stimuli. These features may be related to a distinctive signaling pathway, which involves activation of the gp130 transmembrane receptor, activation of Janus kinases (JAK), and downstream signal transducers and activators of transcription (STATs). Interestingly, stimulation of the pathway also activates a family of negative regulators, called suppressors of cytokine signaling or SOCS (Yasukawa et al. 2001). So far, the in vivo studies of cytokine signaling involving cardiotropin-1, LIF receptor, gp130, JAK, STAT, and SOCS have been carried out in a model of pathological myocardial hypertrophy induced by mechanical overload by thoracic aorta constriction. Studies of exercise-induced hypertrophy are required to determine whether positive and negative molecular regulators in this pathway participate in physiological and adaptive growth of the heart.

Mechanical Stress

Evidence from models of cardiovascular disease indicates that cardiac myocytes respond to increased mechanical load by adaptive changes in growth and gene expression. In pressure overload by aortic constriction, cardiac myocytes contract against increased resistance and hypertrophy mainly by increasing their cross-sectional area. In volume overload, as in post-infarction heart failure, for example, myocardial hypertrophy mainly occurs by cardiac myocyte lengthening. Both types of increased mechanical stress are accompanied by induction of immediate early genes (e.g., c-fos, c-jun, egr-1, c-myc) and up-regulation of fetal genes (e.g., atrial natriuretic factor, beta-myosin heavy chain, skeletal alpha-actin) and growth factors (e.g., IGF-1, endothelin, FGF-2). Several of the cellular and molecular mechanisms evoked by mechanical stress have been investigated in isolated cardiac myocytes. Even though there are important differences in the molecular and functional phenotype of pathological hypertrophy and exercise-induced adaptive hypertrophy, some of the same putative mechano-sensors and intracellular signaling pathways may be involved (reviewed in van Wamel et al. 2002; Baar et al. 1999).

Three molecular principles have been proposed as mechano-sensors in the cardiac myocyte. The first, and perhaps most relevant, is signaling via a family of cell surface receptors called integrins. These molecules serve as a mechanical link between the milieu externe and the interior of the cell when their extracellular domain binds to proteins in the extracellular matrix or to counter-receptors on other cells. Downstream signaling may occur either by mechanical transmission or by biochemical pathways. According to the mechanical hypothesis, stress is first received by integrins and then transmitted by interlinked actin microfilaments in concert with microtubules and intermediate filaments. Modulation of gene regulation is hypothesized to occur by interaction with cell chromatin. In the biochemical perspective, integrins function more like classic transmembrane receptors by co-localizing signaling molecules in focal adhesion complexes at particular locations of the cell membrane. Upon clustering at these focal adhesion sites, integrins may activate focal adhesion kinase (FAK) and facilitate signal transduction in mitogen-activated and stress-activated protein kinase pathways (reviewed in, e.g., Ruwhof & van der Laarse 2000).

A second interesting mechano-sensing principle is activation of nonreceptor tyrosine kinases (such as the Src family) by stretch-induced conformational changes. This is one of the earliest responses to stress activation of cardiac myocytes; downstream tyrosine phosphorylation of cellular signaling proteins occurs within 5 sec of stretching of the cell membrane by hypotonic swelling. Although the relevance of hypotonic swelling in isolated cardiac myocytes has not been established for exercise-induced changes in the athletic heart, these experiments suggest that tyrosine kinases may induce immediate early genes (e.g., c-fos) independent of angiotensin receptors and activation of phospholipase C and protein kinase C (Sadoshima & Izumo 1997).

The third putative principle of mechano-sensors in cardiac myocytes is stretch-activated ion channels that induce increased Ca^{2+} influx, which may in turn activate downstream pathways and serve as an intracellular signal of hypertrophy and gene regulation. The caveat is that gadolinium ions, which block the stretch-activated elevation of myoplasmic Ca^{2+} concentration, do not inhibit up-regulation of early immediate genes and protein synthesis (Ruwhof & van der Laarse 2000). Although these observations do not preclude other stretch-activated ion channels as mechano-sensors of growth signaling, the previously described possibilities seem more likely.

An interesting body of experiments suggests that autocrine and paracrine stimulation of cardiac myocytes may be an important mechanism for induction of hypertrophy and gene regulation by mechanical stress (Sadoshima & Izumo 1997). Mechanical stretching of neonatal rat cardiac myocytes cultured on an elastic membrane increases the concentration of angiotensin II and endothelin-1 in the conditioning medium and stimulates protein synthesis and gene expression associated with mitogen-activated protein kinase activity. These responses can be mimicked by exposing nonstretched myocytes to conditioning medium from stretched cells; and most, but not the full extent, of the responses are inhibited by blockers of angiotensin and endothelin receptors. Intracellular abundance of mRNA for precursors of angiotensin II and endothelin-1 is increased upon mechanical stretching, indicating de novo synthesis. Detectable increments of the peptides occur in conditioning medium within a few minutes of stretching, suggesting secretion by a mechanism yet to be determined. The pattern of downstream changes in gene expression and intracellular signaling patterns is indistinguishable from that of stimulation of the corresponding receptors (Sadoshima & Izumo 1997). Secretion of growth factors may also occur in response to contractile activity. Electrical stimulation of cultured adult rat cardiac myocytes causes secretion of basic fibroblast growth factor (FGF-2) and induces a hypertrophic response, as indicated by increased phenylalanine uptake, elevated protein content, and cell enlargement. These responses to increased mechanical activity could be mimicked by recombinant FGF-2 and inhibited by neutralizing anti-FGF antibodies (Kaye et al. 1996). Thus, several lines of evidence indicate that growth factors secreted by myocytes and other cell types within the myocardium may modulate growth and gene expression by autocrine and paracrine mechanisms in response to mechanical stimulation.

Intracellular Calcium

Evidence accumulated over decades has established that changes in intracellular calcium cycling integrate regulatory stimuli from sources that control contraction and relaxation of the cardiac myocyte in the acute, minute-to-minute time frame. Recent and current research is uncovering calcium-dependent intracellular signaling pathways that regulate growth and gene regulation of the cardiac myocyte and render the heart responsive to short-term functional demands by long-term changes in organ size and molecular phenotype. A pathway including the Ca^{2+}-calmodulin-dependent phosphatase calcineurin and the nuclear factor of activated T-cells (NFAT) may play a central role (figure 11.6), either alone or in interaction with mitogen-activated protein kinase pathways (e.g., Crabtree & Olson 2002; Wilkins & Molkentin 2002; Frey et al. 2000). Thus, variations in intracellular calcium may be equally important for long-term regulation of the functional cardiac phenotype as for acute control of performance. Many actions of calcium are mediated through its interaction with intracellular calmodulin (CaM), which acts as an intracellular sensor for calcium ions. Exactly how CaM is able to specifically target downstream enzymes, ion channels, and transcription factors is incompletely understood. It is likely that both subcellular localization and pattern of calcium oscillations in terms of timing and extent are important. Differential sensitivities to calcium of downstream target molecules may also provide specificity. Calcineurin has a particularly high sensitivity to calcium and responds to sustained low-amplitude calcium changes by phosphatase

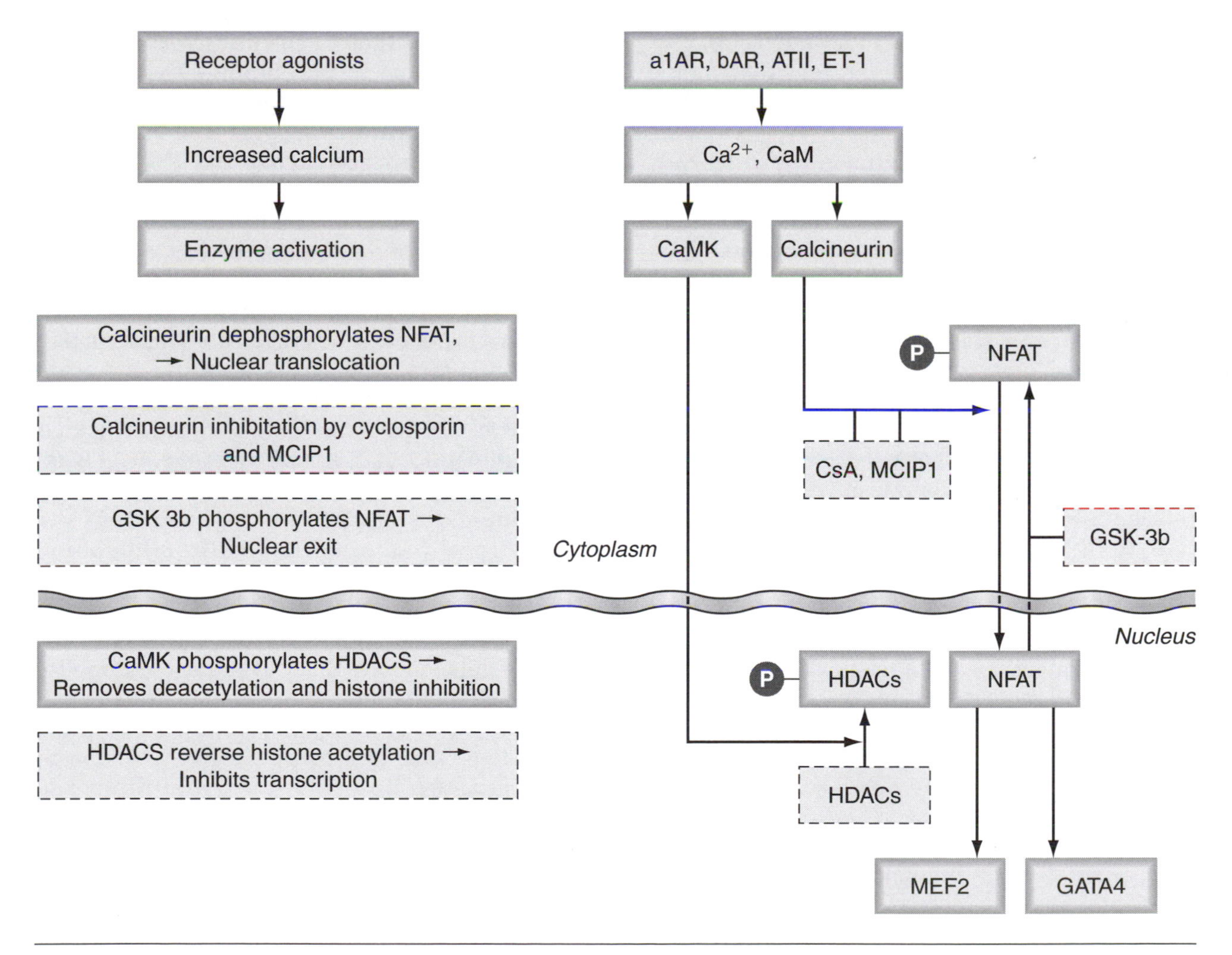

Figure 11.6 Calcium- and calmodulin-dependent signaling pathways of myocardial hypertrophy. Abbreviations are explained in text.

activity. Upon activation, calcineurin dephosphorylates the transcription factor NFAT and permits its translocation to the nucleus, where it associates with the transcription factor GATA-4 to induce myocardial hypertrophy and fetal gene expression. For comparison, the Ca^{2+}-CaM-dependent protein kinase, CaM kinase II (CaMKII), has about 10-fold lower sensitivity for calcium than has calcineurin. CaMKII responds to transient high-amplitude calcium spikes by phosphorylation of L-type calcium channels in the sarcolemma, calcium release channels in the sarcoplasmic reticulum, and phospholamban (which regulates the sarcoplasmic reticulum CaATPase, SERCA2), thus up-regulating the activity of several calcium-cycling proteins. The CaM-activated myosin light chain kinase (MLCK) is associated with actomyosin and has the regulatory myosin light chain subunit MLC-v2 as its main substrate. Phosphorylation of MLC-v2 accompanies sarcomere organization and may facilitate incorporation of additional actin/myosin filaments into the sarcomere. In addition, phosphorylation of MLC-v2 by MLCK increases the calcium sensitivity of the myofilaments, thereby increasing contractility of the cardiac myocyte.

A number of elegant experiments in genetically engineered mice indicate that activation of the calcineurin/NFAT pathway may be both a sufficient and necessary mechanism of pathological hypertrophy, as well as providing further development into heart failure and premature death (reviewed in Frey et al. 2000; Wilkins & Molkentin 2002), whereas physiological hypertrophy in response to exercise training does not seem to involve this pathway (Wilkins et al. 2004). Stimulation by constitutively active mutated forms of either calcineurin or NFAT induces severe hypertrophy in mice. Pharmacological blockade of calcineurin activation by cyclosporin A (CsA) inhibits hypertrophy induced by catecholamines, angiotensin, and endothelin, both in isolated cells and in vivo. It is likely that activation of hypertrophy by the calcineurin/NFAT pathway is counterbalanced by endogenous negative regulators. Genetic overexpression of the endogenous myocyte-enriched calcineurin-interactive protein 1 (MCIP1) inhibits cardiac hypertrophy induced by catecholamines, pressure overload, and exercise training. Glycogen synthase kinase 3β (GSK 3β) specifically phosphorylates NFAT and facilitates translocation out of the nucleus, and thus inhibits the hypertrophy response and associated fetal gene expression. Activity of GSK 3β is high in the normal state and is inhibited by mitogen stimulation. Cardiac expression of a constitutively active mutated form of GSK 3β rendered the mice resistant to cardiac hypertrophy during pressure overload and catecholamine stimulation (Antos et al. 2002). In addition to activation of NFAT, calcium may also induce transcription by phosphorylating the transcription factor MEF-2 (see following discussion). Although these observations open new perspectives in pathogenesis of myocardial hypertrophy and remodeling, failure, and future therapeutic targets, there are still controversies and unanswered questions regarding the role of the pathway in the normal state and in exercise-induced adaptive growth and gene regulation. A recent study in NFAT-luciferase reporter transgenic mice excluded calcineurin/NFAT signaling in physiological hypertrophy induced by either exercise training or GH/IGF-1 infusion. Interestingly, these interventions did not induce the fetal gene expression pattern associated with pathological hypertrophy, whereas pressure-overloaded animals displayed both significant NFAT-luciferase activity and fetal gene expression, suggesting that calcineurin/NFAT signaling may be restricted to maladaptive hypertrophy and heart failure (Wilkins et al. 2004).

Recent observations suggest that CaM-kinase and the transcription factor MEF-2 (myocyte-enhancer factor 2) can mediate calcium-induced hypertrophic signals (reviewed in Frey et al. 2002). Expression of a constitutively active isoform of CaM-kinase in transgenic mice leads to cardiac hypertrophy that progresses to heart failure, but it is not clear whether CaM-kinase signaling is initiated by specific extracellular receptors or by mechanical stress. MEF-2-dependent gene activation seems to be involved, not through direct phosphorylation but by removal of transcriptional repressors. Upon phosphorylation by CaM-kinase, histone acetylase dissociates from MEF-2, thereby relieving the gene from inhibition of transcription. Acetylation of histones relaxes binding to the gene; thus removal of histone deacetylases facilitates transcription. Recent experiments indicate that CaM-kinase can modulate the effect of other mitogen-activated protein kinase (MAPK) pathways. Histone deacetylases prevent MEF-2 activation by the p38 MAP kinase, which may also be involved in hypertrophy signaling. Thus de-repression by CaM-kinase may be necessary for the p38 MAPK pathway to regulate the molecular phenotype of the cardiac myocyte. It is also possible that CamKII contributes to exercise-induced adaptation in cardiomyocytes by activating Akt independent of PIK3K (e.g. Latronica et al. 2004).

Mitogen- and Stress-Activated Signaling

A large body of observations indicates that growth signals from the cardiac cell membrane are processed by a complex system of mitogen- and stress-activated protein kinases before they are transmitted to the nucleus to regulate gene transcription. Experiments involving genetic and pharmacologic stimulation and inhibition demonstrate that activation of some of the key molecules is both necessary and sufficient to induce changes in cardiac growth and gene expression. The role of hypertrophic signaling in the heart by MAPKs and other intracellular pathways is being reviewed frequently (e.g., Chien 2004; Sugden 2001; Katz 2001).

MAPK signaling includes activation of three families of protein kinases with both distinctive and overlapping functions. The extracellular receptor kinases (ERKs), often termed classic or mitogenic MAPKs (mMAPKs) because they are typically activated by extracellular receptors, constitute the main signaling pathway for mitogenic substances and other growth stimuli (e.g., mechanical activation) and lead to hypertrophic growth and gene expression in the cardiac myocyte. Recent observations suggest that although ERK1/2 may be involved in heart growth during normal development (O'Connell et al. 2003), physiological hypertrophy in response to exercise training occurs without significant activation (McMullen 2003). Two other protein kinase families, p38 MAP kinases (p38) and c-jun protein kinases (JNKs), are often termed stress-activated protein kinases (SAPKs) because they are responsive to cellular stress (heat shock, ischemia–reperfusion, viruses, toxic substances, and mechanical activation) and may induce programmed cell death (apoptosis). Interestingly, JNK is activated in an intensity-dependent way during initial stage of exercise training (Boluyt et al. 2003), whereas p38 remains unchanged. Upstream activators of JNK and downstream events have not been identified. Although potentially involved in several forms of myocardial growth and gene expression, the role of SAPK signaling in cardiac hypertrophy is less clear than the classic or "true" mitogenic MAPK pathway.

The general principles of MAPK signal transduction and the three main pathways are outlined in figure 11.7. Stimulated receptors or cellular stress activates different types of GTP binding proteins (G-proteins), which in turn activate a series of three protein kinases, including either of the MAP kinases ERK, JNK, or p38 as the final step. For the ERK pathway, members of the Ras family of monomeric small G-proteins activate the first kinase, Raf-1. Upon activation, Raf-1 phosphorylates the second level of kinases, one of the isoenzymes MEK-1 or MEK-2, which in turn activates the third and final kinase, ERK. The activated MAP kinases translocate to the nucleus of the cardiac cell and activate immediate early genes and other transcription factors by phosphorylation.

Somewhat confusingly, the protein kinases upstream of the MAP kinases ERK, p38, and JNK (e.g., MEK-1) are often referred to by generic names (either MAPK kinases, MAPKKs, MKKs, or MEKs). Accordingly, the protein kinases upstream of the MAPK kinases (e.g., Raf-1) are called either MAPK kinase kinases, MAPKKKs, MKKKs, or MEKKs). This is partly because the identity of some of the upstream kinases in the p38 and JNK pathways has been uncertain. The multistep cascade format allows for signal amplification and modulation by interacting pathways at every step. A variety of protein kinase isoforms at different levels plus modulating interaction between signaling pathways probably provides both diversity and specificity of different types of growth signals.

Cardiac Transcription Factors

Fascinating possibilities for therapeutic interventions are related to identification of the specific and ubiquitous transcription factors that regulate the molecular phenotype of the cardiac myocyte in response to exercise training and in common diseases. One such possibility is that transfection of cardiac fibroblasts with genes that orchestrate the cardiac gene program (analogous to the MyoD helix–loop–helix protein family in skeletal muscle) could transform the development of scar tissue after myocardial infarction into reparative myocardial myogenesis. Drug therapies to inhibit pathological growth and myocardial dysfunction, or to promote the athletic molecular phenotype, could be achieved by pharmacological interventions targeting specific stimulatory and inhibitory transcription factors that regulate the associated patterns of gene expression. So far, our understanding of how transcription factors regulate growth and functioning of the cardiac myocyte is in an exploratory phase (e.g., Chien 2004). As also indicated in preceding sections, induction of the ubiquitous immediate early genes c-jun, c-fos, c-myc, and egr-1 is required for activation of cardiac genes involved in stimulation by growth factors and hypertrophy induced by mechanical stress. A

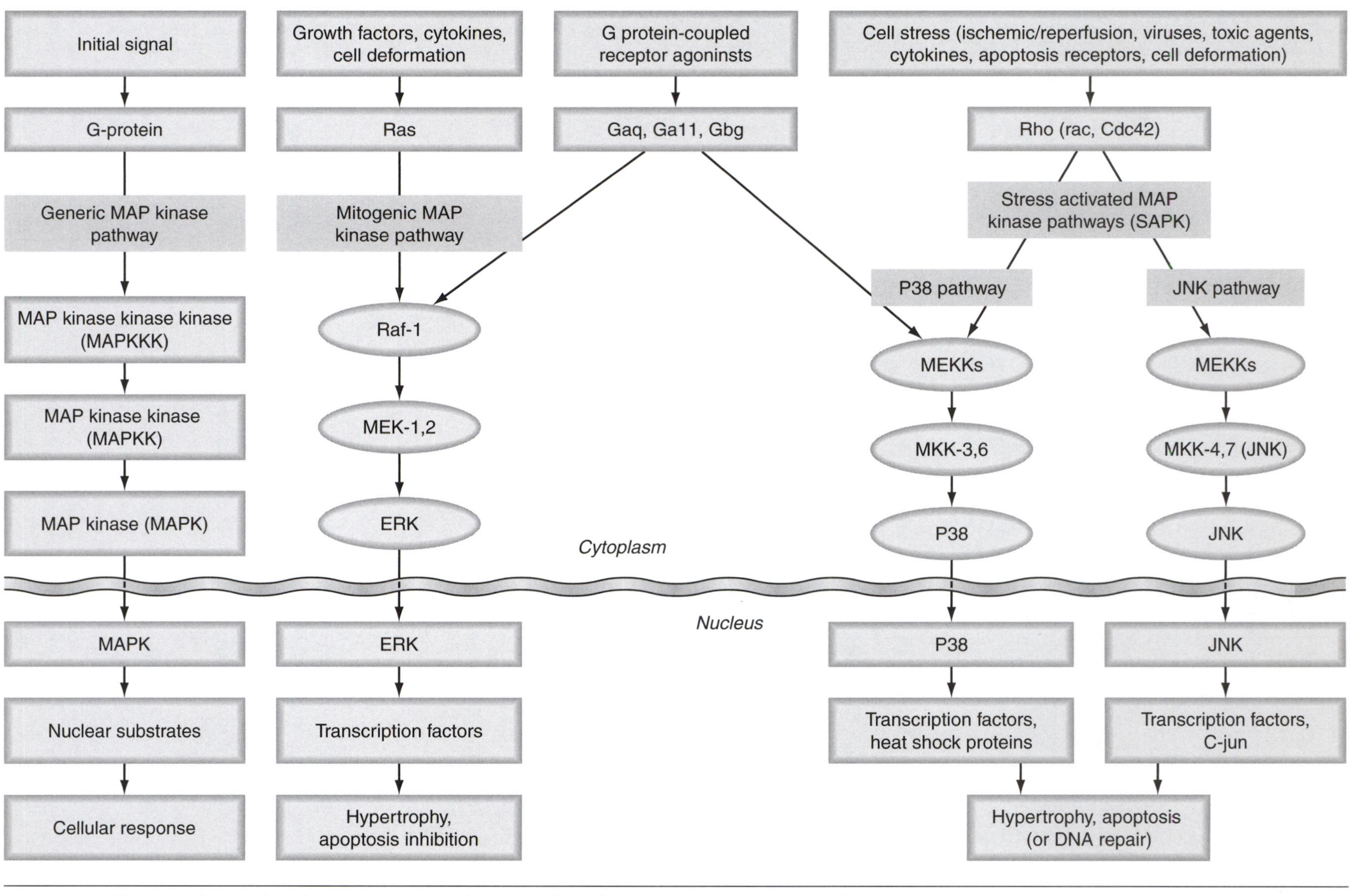

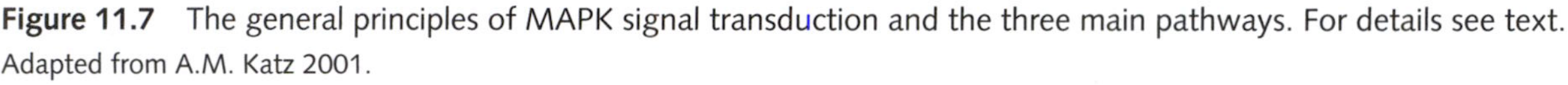
Figure 11.7 The general principles of MAPK signal transduction and the three main pathways. For details see text.
Adapted from A.M. Katz 2001.

family of myocyte-specific enhancer factors (MEF-2) regulates transcription of cardiac muscle genes in response to mechanical overload and growth factor stimulation by forming dimers with other proteins in the same class and by associating with other transcription factors. The cardiac muscle-specific transcription factor GATA-4 regulates cardiac-specific gene expression and is involved in hypertrophic signaling. Most of the molecular mechanisms described previously have been investigated in the context of pathological hypertrophy, either by chronic growth factor stimulation or by permanent mechanical overload. So far, little is known of how these may actually participate in exercise-induced adaptive growth regulation.

Exercise and Gene Expression

Even though the cardiac myocyte is subject to many of the same stimuli during exercise as during cardiovascular diseases involving volume and pressure overload, the athletic molecular phenotype seems to be distinctly different from that of pathological hypertrophy. So far, most of the characteristic changes in gene expression during pathological growth of the heart have not been observed in exercise-induced hypertrophy; in fact, some of the prototypic patterns are reversed. In the athletic phenotype, sarcomeric gene expression associated with increased contractile function is induced. α-myosin heavy chain (α-MHC) and atrial myosin light chain 1 (aMLC-1) are up-regulated in the trained state (Jin et al. 2000; Diffee et al. 2002), while the β-MHC isoform is increased in response to overload hypertrophy (Chien et al. 1991).

Some of the main proteins that regulate myocardial calcium handling and contractile function are reciprocally regulated by exercise and pathological hypertrophy. SR calcium ATPase (SERCA2a) is induced by exercise, both in normal hearts and in the context of heart failure (Wisløff et al. 2001, 2002; Lu et al. 2002). Phospholamban is up-regulated to the same extent as SERCA2a (Wisløff et al. 2001). Regulation of the sodium calcium exchanger (NCX-1) seems to be species dependent. NCX-1 is up-regulated in failing dog (and human) hearts and down-regulated by exercise training (Lu et al. 2002), whereas in rat, NCX-1 is down-regulated in heart failure and up-regulated by exercise both in normal hearts and in heart failure (Wisløff et al. 2002).

Exactly how exercise affects the natriuretic peptides ANP and BNP is not quite clear, as either increased (Buttrick et al. 1994; Allen 2001; Iemitsu et al. 2002), unchanged (Jin et al. 2000; Calderone et al. 2001; Iemitsu et al. 2001; Wisløff et al. 2002), or reduced (Diffee et al. 2003) levels have been reported in normal hearts. Of note is that the observed changes were small (less than 2-fold) compared to the marked increments in pathologic hypertrophy (more than 5- to 10-fold). In failing hearts, exercise training attenuates the up-regulation of ANP and improves cardiac myocyte contractility to the same extent as drug therapy with the angiotensin antagonist losartan (Wisløff et al. 2002; Loennechen et al. 2002). Some of the studies that report increased natriuretic peptide expression also demonstrate levels of endothelin-1 and transforming growth factor β_3 that were increased compared to those in sedentary controls (+50% and +35%, respectively) but markedly less than in pathological hypertrophy (Calderone et al. 2001; Iemitsu et al. 2002). The growth-inhibitory peptide adrenomedullin had a different expression pattern, with similar levels in controls and pathological hypertrophy and more than 50% reduction after exercise training (Iemitsu et al. 2001).

Emerging evidence suggests that heat shock proteins (HSP) may be involved in induction of the athletic molecular phenotype of the cardiac myocyte. High-intensity exercise training induces a significant increase in myocardial HSP70 protein and improves myocardial contractile function (Paroo et al. 2002). HSP synthesis is induced by binding of the HSP transcription factor 1 (HSF-1) to specific heat shock elements in the heat shock gene. Stimulation of HSP70 synthesis requires exercise at a high intensity (Milne & Noble 2002). A gender-specific attenuation of the response related to estrogen has been demonstrated in female rats (Paroo et al. 2002), but it is not known whether inhibition is absolute or if estrogen markedly elevates the threshold for induction. Although gender-dependent differences regarding mechanisms of physiological hypertrophy have been reported in mice (O'Connell et al. 2003), the relevance for human physiology is not known. Recent experiments suggest that mechanical stress like increased preload of the heart, but not increased afterload, activates HSF-1 (Nishizawa et al. 2002), whereas HSPs may not be persistently elevated in compensatory myocardial hypertrophy (e.g., Snoeckx et al. 2001).

Within the next years, functional genomics, in which gene expression patterns are linked to exercise-induced changes in cardiac myocyte function, is likely to identify novel molecular mechanisms of the athletic phenotype. In addition to control of exercise intensity, timing of sampling may be a critical factor for detecting relevant changes in gene expression. For example, the level of sodium-hydrogen exchanger-1 mRNA (NHX-1) was markedly elevated immediately after a single bout of exercise, but not detectably changed 24 h after a 13-week training period (Wisløff et al. 2001). It is now well documented that the pattern of gene expression varies substantially during the first 48 h following mechanical overload by aortic constriction (Hoshijima & Chien 2002). Temporal variations are probably even more marked in association with cyclic loading of the heart by regular exercise.

Translational and Post-translational Regulation by PI3Kα and Akt

A host of recent studies links physiological hypertrophy in normal development and during exercise training to activation of phoshophoinositol 3-kinases (PI3Ks), protein kinase B/Akt and downstream targets, which in turn leads to more effective formation of the translational initiation complex, a rate-limiting step in eukaryotic protein synthesis. The result is increased protein synthesis and cardiomyocyte growth with little induction or even reversal of fetal gene transcription (reviewed by e.g. Chien 2004, Latronico et al. 2004, Hardt & Sadoshima 2002; Sugden & Clerk 1998). Although not definitively proven, GH/IGF-1 is currently the most likely candidate to initiate the cascade of signaling events. Systemic and local levels of IGF-1 are elevated in response to exercise, and its administration mimics exercise-induced effects from signaling pathway activation to adaptive cardiomyocyte growth including increased function (Wilkins et al. 2004). Genetic ablation of the PI3Ka isoform leads to smaller hearts in unchallenged mice, significantly reduces activation of Akt and other downstream targets, and blunts exercise-induced cardiomyocyte hypertrophy and resistance to pressure overload (McMullen et al. 2003). Activation of Akt increases the rate of protein synthesis by removing the inhibiting effect of the glycogen synthase kinase-3β (GSK-3β). Phosphorylation by this enzyme negatively regulates several downstream processes, including binding of the eukaryotic translation initiation factor 2 (eIF2) to the activated initiator tRNA as well as binding of several transcription factors to DNA. GSK-3β is active in unstimulated cells, and its main regulation is inactivation by phosphorylation. Inhibition of GSK-3β in failing, hypertrophied hearts suggests relevance in human biology; it has been suggested that it may play a role in hypertrophy where MAPKs are peripherally engaged, for example in response to beta-adrenergic stimulation. In addition to its effect on protein synthesis, GSK-3β reduces binding and attenuates the effect of a number of transcription factors, including NFAT, GATA-4, HSF-1, and STAT (Hardt & Sadoshima 2002), and may thus inhibit or modulate cardiac myocyte growth by interacting with other hypertrophic pathways. Other substrates downstream of PI3Kα include the mammalian target of rapamycin (mTor), which increases protein synthesis by activating a protein kinase that specifically acts on the small ribosomal unit S6 (p70 S6K).

Contractile Function and Calcium Handling

A number of studies in humans, and in all mammalian species examined, have shown that exercise training induces resting and submaximal bradycardia, increased maximal stroke volume, increased left ventricular end-diastolic dimension, improved myocardial contractile function, and subtle to moderate increase in myocardial mass (reviewed in Moore & Korzick 1995). The following section focuses on adaptation occurring at the single-myocyte level that underlies the training-induced changes in heart function. At slow stimulation frequencies (0.067-0.2 Hz) and low temperatures (23-29°C) there is little evidence of training-induced improvement in the cardiac myocytes' shortening characteristics (Laughlin et al. 1992; Moore et al. 1993; Palmer et al. 1998). However, training-induced adaptations, such as increased degree of fractional shortening and reduced re-lengthening time, become more evident as both the stimulation frequency and temperatures approach in vivo conditions (Zhang et al. 1999, 2001, 2002). For the rat this is 300 to 600 beats per minute at 37°C. There seems to be a progressive increase in cardiomyocyte contractility in response to regular exercise training until a limit of training effects has been reached. This coincides with the maximal increase in $\dot{V}O_2max$ and cardiomyocyte hypertrophy. The relationship between cardiomyocyte cell length and maximal extent of shortening is presented in figure 11.8. The figure demonstrates that increased cell length does not necessarily improve maximal extent of shortening, as was the case in healthy rats. In contrast to values in healthy rats, increased $\dot{V}O_2max$ in post-MI rats is associated with reduced cell dimensions and improved maximal extent of cardiomyocyte shortening. Similarly, sprint training in post-MI rats was associated with reduced cardiomyocyte dimensions and improved contractile function (Zhang et al 2000a). Thus, different types of hypertrophic stimuli (training vs. myocardial infarction) elicit qualitatively different adaptations in cardiomyocyte shortening, despite marked hypertrophy in both situations. Exercise training seems to induce beneficial effects on cardiomyocyte hypertrophy and contractility comparable to those of angiotensin antagonists. The effect of combined exercise and inhibition of the renin-angiotensin system is still unsettled. Note the reduced hypertrophy and improved contractile function by losartan and exercise training after myocardial infarction, whereas cariporide affected only contractility. Also note that training in SHAM induced both hypertrophy and contractility (figure 11.8).

In experimental and human end-stage heart failure, most investigators demonstrate both impaired shortening and prolonged relaxation time of cardiomyocytes at physiological stimulation rates. It is therefore believed that reduced cardiomyocyte contractile function significantly contributes to heart failure development (e.g.,

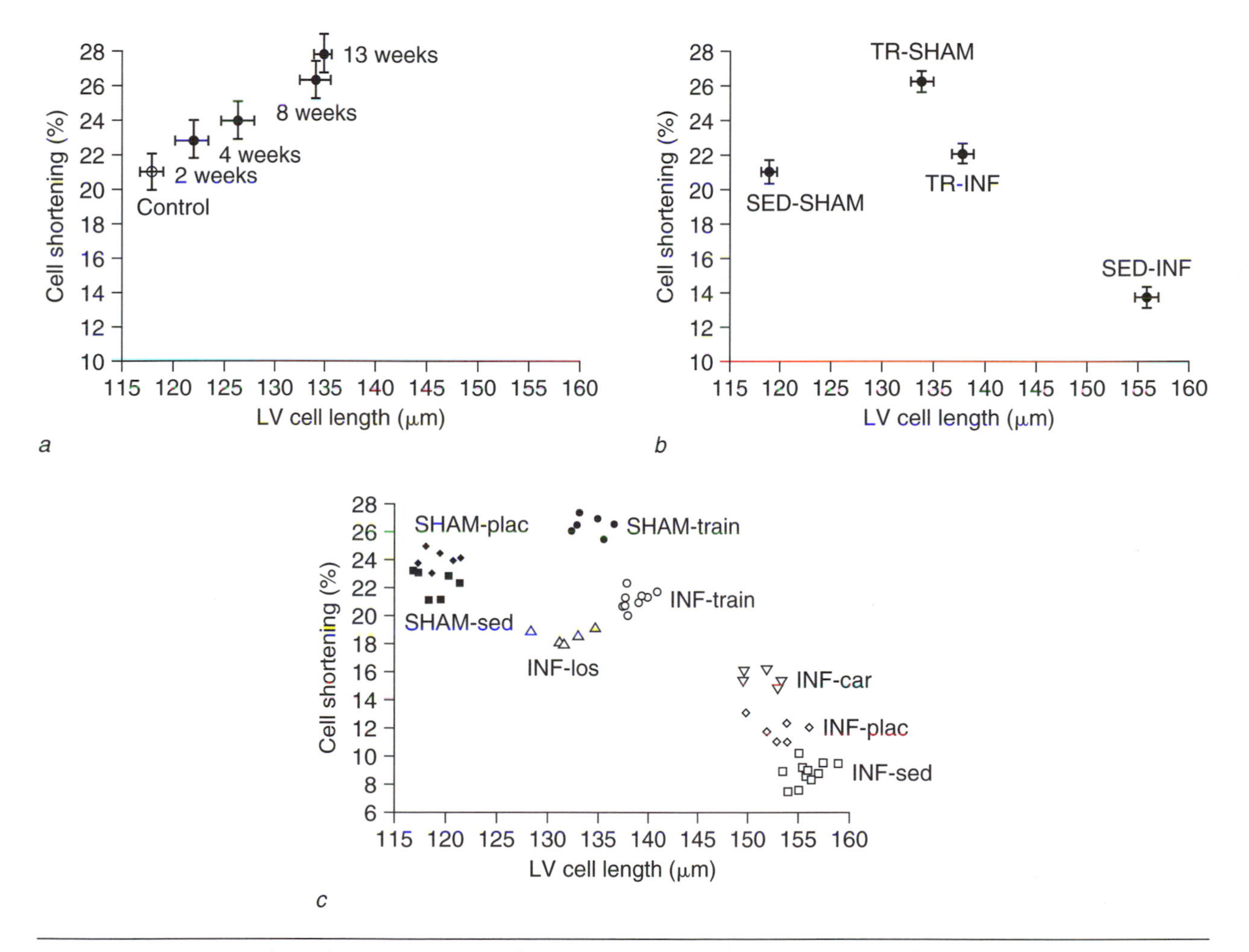

Figure 11.8 *(a)* Time-dependent increase in cardiomyocyte length and maximal extent of shortening in cardiomyocytes isolated from endurance-trained and sedentary rats. Each data point represents mean ± SD of 60 cells, 9 ± 3 in each rat (n = 6). In each cell, data were calculated as the mean of 10 consecutive contractions after stabilization at 7 Hz. *(b)* The relationship between cardiomyocyte hypertrophy and maximal extent of shortening at 7 Hz in trained (TR) and sedentary (SED) sham and rats with post-infarction heart failure (LV cells from the remote area). Data are presented as mean ± SD from 6 rats in each group of sham rats, and 10 and 11 animals in the trained infarction group (TR-INF) and sedentary infarction group (SED-INF), respectively. Cellular dimensions were calculated from 200 ± 12 myocytes from each of the same animals, a total of about 6,500 cells. *(c)* Resting length and relative shortening (7 Hz) of cardiomyocytes isolated remote from large myocardial infarction (INF) after treatment with losartan (los), cariporide (car), placebo (plac), exercise training (train), or sedentary observation (sed) and from left ventricle of sham-operated controls (SHAM).

Data from Loennechen 2002, Wisløff 2001, and Wisløff et al. 2002.

Davies et al. 1995; del Monte et al. 1999; Gomez et al. 2001; Loennechen et al. 2001).

Training-induced elongation of LV myocytes occurs in the absence of changes in sarcomere length (Moore et al. 1993), and the changes in cardiac contractile function induced by endurance training are due in part to myocyte length-independent changes in contractile function. Several lines of evidence support this notion. Schaible and Scheuer (1981) demonstrated that treadmill training increased end-diastolic volume, stroke work, ejection fraction, and midwall fractional shortening in the absence of changes in end-diastolic wall stress in perfused working rat hearts. Furthermore, isometric force development by rat LV papillary muscle maintained at optimal length is increased by endurance training (Mole 1978; Tibbits et al. 1978, 1981a). Recently Diffee and Chung (2003) showed that training increased the velocity of loaded shortening and increased peak power output in the single permeabilized myocyte preparation. Figure 11.9 summarizes the effects of

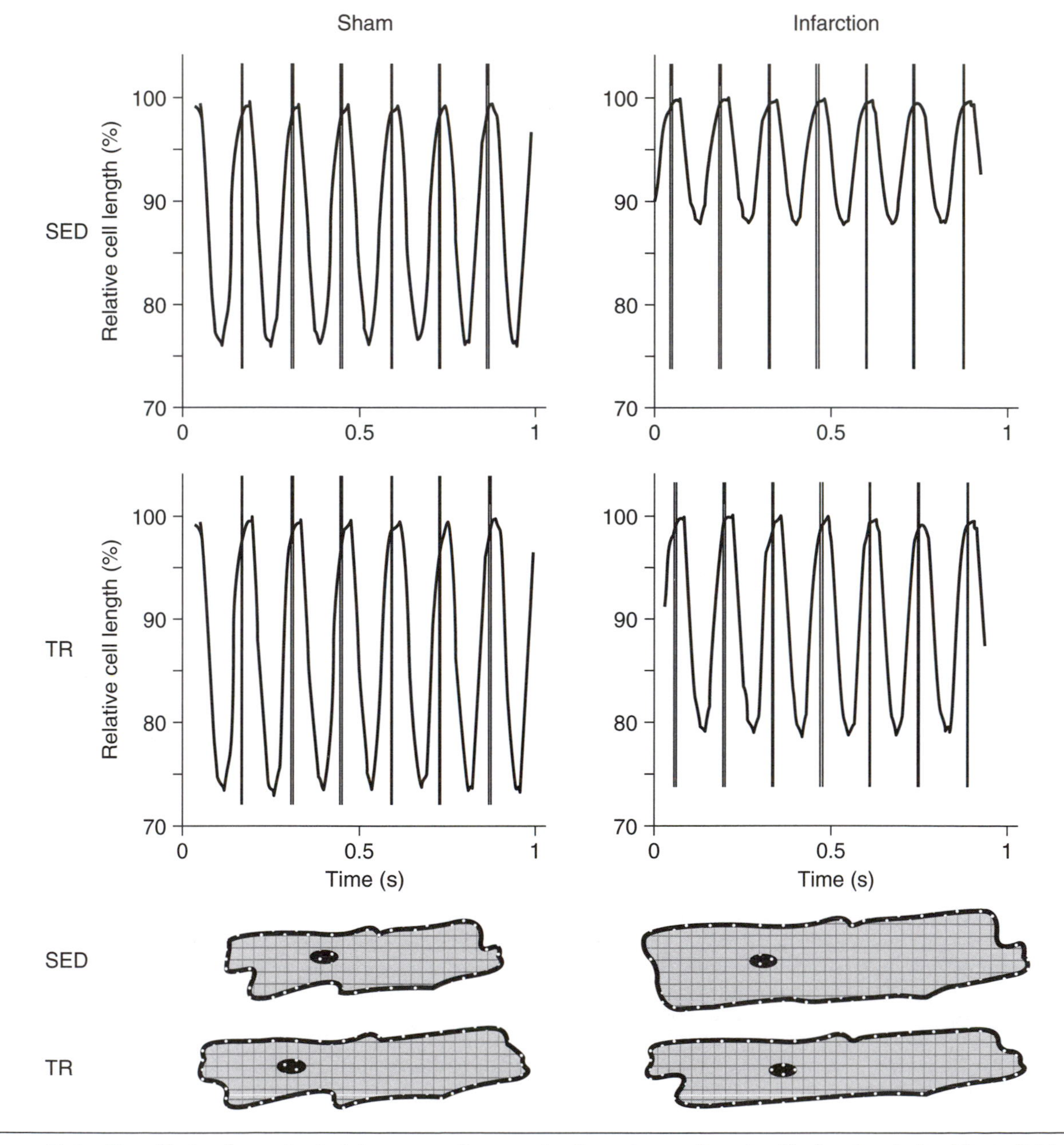

Figure 11.9 The effects of exercise training on cardiomyocyte dimension and contractile function in healthy individuals and in those with post-infarction heart failure. SED: sedentary; TR: trained. Vertical lines represent electrical stimulation of the isolated cardiomyocyte.

exercise training on cardiomyocyte dimension and contractile function in healthy and failing hearts. Briefly, interval running improves contractile function (degree of shortening, time to peak shortening normalized to fractional shortening, and the time to 50% re-lengthening time) in left ventricular myocytes from normal and failing hearts. In contrast to observations in healthy rats, the dimensions in cardiomyocytes from failing hearts decrease in response to exercise training.

Improved Calcium Handling

Changes in cardiomyocyte shortening and relaxation during heart failure, endurance training, and drug treatment are associated with changes in calcium handling. Cardiomyocyte shortening in healthy endurance-trained rats is associated with lower peak systolic and diastolic intracellular calcium (Moore et al. 1993; Libonati et al. 1997; Wisløff et al. 2001, 2002) and reduced time for the calcium decay from systole (Moore et al. 1993; Wisløff et al. 2001, 2002). The reduction in peak systolic calcium concentration in myocytes from trained hearts may be due to (1) reduced calcium released into cytosol via sarcolemma and SR, (2) dilution of released calcium into the sarcoplasm due to increased average myocyte volume, or (3) increased intracellular calcium buffering capacity. The first two possibilities are unlikely since reduced calcium influx or diluted cytosolic calcium would reduce the calcium binding to myofilaments and reduce contractility. The final possibility is feasible since only a small fraction of calcium that is released into and removed from the sarcoplasm during an excitation–contraction coupling cycle exists as free Ca^{2+} (Sipido & Wier 1991). This adaptation to training is consistent with lower diastolic and systolic calcium in TR myocytes. Tibbits et al. (1981b) demonstrated that calcium binding sites increased by about 65% in papillary muscle from trained rats. Penpargkul et al. (1977) reported enhanced calcium binding by cardiac SR from trained rats. Lower diastolic calcium in TR myocytes could also result from enhanced SL ATP-dependent calcium extrusion (Pierce et al. 1989), mitochondrial metabolism (Beyer et al. 1984), or both, thus effectively lowering the set point for calcium regulation (Cheung et al. 1986) in trained myocytes. Gene analysis demonstrates a marked up-regulation of SERCA2 and Na^+-Ca^{2+} exchanger in trained hearts (table 11.3) (e.g., Tibbits et al. 1989; Tate et al. 1990, 1996; Wisløff et al. 2001, 2002). Chronically raised Na^+-Ca^{2+} exchanger levels are known to reduce systolic calcium (Terracciano et al. 1998) and may contribute to the reduced peak systolic calcium observed in myocytes from endurance-trained rats. Furthermore, increased calcium uptake capacity of the SR due to increased SERCA2 expression could account for the increased rate of decay of the calcium transient.

In cardiomyocytes from failing hearts, several alterations in cardiomyocyte calcium handling have been demonstrated that might explain the impaired cardiomyocyte shortening. According to the calcium-induced calcium release theory, reduced calcium release from SR in failing cardiomyocytes might be due to impaired detection of sarcolemmal calcium influx during depolarization (Gomez et al. 1996). Furthermore, reduced SR calcium loading might reduce systolic calcium-induced calcium release (Currie & Smith 1999; Neary et al. 1998). Several lines of evidence point toward changes in SERCA2 as an important mechanism of reduced cardiomyocyte contractile function in chronic heart failure. Reduced SERCA2 expression and function have consistently been shown in heart failure (e.g., Hasenfuss et al. 1994). Impaired SERCA2 function probably causes delayed reuptake of calcium into SR, which impairs cardiomyocyte relaxation. Furthermore, reduced SR calcium loading might impair cardiomyocyte contraction due to reduced calcium release in systole. Gene transfer of SERCA2 restores contractile function in cardiomyocytes from patients with end-stage heart failure and in mice with heart failure due to aortic banding (del Monte et al. 1999).

A few studies have addressed the effects of exercise training on altered calcium handling in heart failure. Intense anaerobic training post-MI significantly restored cell length (Zhang et al. 1998) and systolic [Ca^{2+}] but further decreased SERCA2 and phospholamban levels (Zhang et al. 2000b) in conjunction with improved cardiomyocyte contractility (Zhang et al. 2000a). This contrasts with findings from a study using high-intensity aerobic endurance training showing restoration of cell length, reduced calcium transients, and increased SERCA2 and Na^+-Ca^{2+} exchanger levels in cardiomyocytes from failing hearts, in conjunction with improved cardiomyocyte contractility (Wisløff et al. 2002). It seems reasonable to speculate that these differences are due to different training regimens used in the studies. However, a recent study showed that anaerobic training post-MI increased both the expression of SERCA2 and Na^+-Ca^{2+} exchanger levels in conjunction with

Table 11.3 Exercise and Gene Expression

Gene	Change (%)	Method	Species	State	Exercise	Duration	Intensity	Sampling	Reference
PKB	No change	Western	Rat	N	Swim	13 wk	low	>48 h	Lajoie et al. 2004
PKB	No change	Western	Rat	DT-2	Swim	13 wk	low	>48 h	Lajoie et al. 2004
pPKB Ser^{473}	No change	Western	Rat	N	Swim	13 wk	low	>48 h	Lajoie et al. 2004
pPKB Thr^{308}	-41%	Western	Rat	N	Swim	13 wk	low	>48 h	Lajoie et al. 2004
pPKB/Akt Thr^{308} Ser^{473}	Normalized	Western	Rat	DT-2	Swim	13 wk	low	>48 h	Lajoie et al. 2004
pPKB Ser^{473}	+ 166%	Western	Rat	HT	Treadmill	8 wk	Medium	>48 h	Lajoie et al. 2002
pPKB Thr^{308}	+ 120%	Western	Rat	HT	Treadmill	8 wk	Medium	>48 h	Lajoie et al. 2002
GSK-3β	No change	Western	Rat	N	Swim	13 wk	low	>48 h	Lajoie et al. 2004
GSK-3β	No change	Western	Rat	DT-2	Swim	13 wk	low	>48 h	Lajoie et al. 2004
pGSK-3α	Normalized	Western	Rat	DT-2	Swim	13 wk	low	>48 h	Lajoie et al. 2004
pGSK-3β	+68%	Western	Rat	N	Swim	13 wk	low	>48 h	Lajoie et al. 2004
MYOSINS									
α-MHC	+30%	PCR	Rat	N	Treadmill	13 wk	Medium	?	Jin et al. 2000
	+ 54%	Western	Rat	PI	Treadmill	8 wk	High	?	Zhang et al. 2002
	+230%	PCR	Rat	N	Swim	8 wk	Medium	24 h	Iemitsu et al. 2004
	+40%	PCR	Rat	N	Swim	15 wk	Medium	>24 h	Iemitsu et al. 2001
β-MHC	No change	PCR	Rat	N	Treadmill	13 wk	Medium	?	Jin et al. 2000
	No change	Array	Rat	N	Treadmill	11 wk	Medium/low	>72 h	Diffee et al. 2003
	No change	PCR	Rat	N	Swim	15 wk	Medium	>24 h	Iemitsu et al. 2001
	No change	PCR	Rat	N	Swim	8 wk	Medium	24 h	Iemitsu et al. 2004
MLC-1	No change	Northern	Rat	N	Swim	6 wk	Medium ?	?	Buttrick et al. 1994
	No change	Northern	Rat	HT	Swim	6 wk	Medium ?	?	Buttrick et al. 1994
	+270%	Array	Rat	N	Treadmill	11 wk	Medium/low	>72 h	Diffee et al. 2003
MLC-2	No change	PCR	Rat	N	Treadmill	13 wk	Medium	?	Jin et al. 2000
ACTINS									
α-CA	No change	PCR	Rat	N	Treadmill	13 wk	Medium	?	Jin et al. 2000
α-SA	No change	PCR	Rat	N	Treadmill	13 wk	Medium	?	Jin et al. 2000
α-SMA	No change	PCR	Rat	N	Treadmill	13 wk	Medium	?	Jin et al. 2000
β-actin	+220%	Array	Rat	N	Treadmill	11 wk	Medium/low	>72 h	Diffee et al. 2003
α-SMA	No change	PCR	Rat	N	Treadmill	13 wk	Medium	?	Jin et al. 2000

TROPONIN									
TnI	+32%	Northern	Rat	N	Swim	6 wk	Medium ?	?	Buttrick et al. 1994
	+10%	Northern	Rat	HT	Swim	6 wk	Medium ?	?	Buttrick et al. 1994
CA^{2+} HANDLING									
SERCA-2	+12%	Northern	Rat	N	Swim	6 wk	Medium ?	?	Buttrick et al. 1994
	+12%	Northern	Rat	HT	Swim	6 wk	Medium ?	?	Buttrick et al. 1994
	No change	Western	Rat	N	Treadmill	8 wk	High	?	Zhang et al. 2002
	+81%	Western	Rat	N	Treadmill	8 wk	85-90% $\dot{V}O_2max$	24 h	Wisløff et al. 2001
	+35%	Western	Rat	PIHF	Treadmill	8 wk	85-90% $\dot{V}O_2max$	24 h	Wisløff et al. 2002
	54%	Western	Rat	PI	Treadmill	6-8 wk	High	?	Zhang et al. 2000b
	+300%	PCR	Rat	N	Swim	8 wk	Medium	24 h	Iemitsu et al. 2004
	+175%	Western	Rat	N	Swim	8 wk	Medium	24 h	Iemitsu et al. 2004
	No change	PCR	Rat	N	Treadmill	13 wk	Medium	?	Jin et al. 2000
	+33%	Western	Dog	HF	Treadmill	4 wk	Medium	?	Lu et al. 2002
NCX	No change	Western	Rat	PI	Treadmill	6-8 wk	High	?	Zhang et al. 1998
	+31%	Western	Rat	PIHF	Treadmill	8 wk	85-90% $\dot{V}O_2max$	24 h	Wisløff et al. 2002
	300%	Western	Rat	N	Treadmill	6-8 wk	High	?	Zhang et al. 2002
	+27%	Western	Rat	N	Treadmill	8 wk	85-90% $\dot{V}O_2max$	24 h	Wisløff et al. 2002
	38%	Western	Dog	HF	Treadmill	4 wk	Medium	?	Lu et al. 2002
PLB	+45%	Western	Rat	N	Treadmill	13 wk	High	24 h	Wisløff et al. 2001
	24%	Western	Rat	PI	Treadmill	6-8 wk	High	?	Zhang et al. 2000b
	No change	PCR	Rat	N	Treadmill	13 wk	Medium	?	Jin et al. 2000
Calsequestrin	No change	Western	Rat	PI	Treadmill	6-8 wk	High	?	Zhang et al. 2002
Ryanodine	No change	Western	Dog	HF	Treadmill	4 wk	Medium	?	Lu et al. 2002
	+ 33%	Western	Rat	N	Treadmill	24 wk	Medium	48	Morán et al. 2003

(continued)

Table 11.3 *(continued)*

CA^{2+} HANDLING									
ANP	+50%	Northern	Mouse	N	Wheel	8-10 wk	Medium ?	?	Allen et al. 2001
	44%	PCR	Rat	PIHF	Treadmill	8 wk	85-90% $\dot{V}O_2$max of $\dot{V}O_2$max	24 h	Wisløff et al. 2002
	No change	PCR	Rat	N	Treadmill	8 wk	85-90% $\dot{V}O_2$max	24 h	Wisløff et al. 2002
	220%	Array	Rat	N	Treadmill	11 wk	Medium/low	>72 h	Diffee et al. 2003
	No change	PCR	Rat	N	Treadmill	13 wk	Medium	?	Jin et al. 2000
	+45%	PCR	Rat	N	Swim	8 wk	Medium	24 h	Iemitsu et al. 2002
	No change	Northern	Rat	N	Wheel	6 wk	Medium/low	?	Calderone et al. 2001
BNP	+50%	PCR	Rat	N	Swim	8 wk	Medium	24 h	Iemitsu et al. 2002a
	No change	PCR	Rat	N	Swim	15 wk	Medium	>24 h	Iemitsu et al. 2001
ET-1	+130%	PCR	Rat	N	Treadmill	Acute	80% $\dot{V}O_2$max	<10 min	Maeda et al. 1998
	No change	PCR	Rat	N	Treadmill	8 wk	85-90% $\dot{V}O_2$max	24 h	Wisløff et al. 2002
	No change	PCR	Rat	PIHF	Treadmill	8 wk	85-90% $\dot{V}O_2$max	24 h	Wisløff et al. 2002
	No change	PCR	Rat	N	Swim	15 wk	Medium	>24 h	Iemitsu et al. 2001
	+50%	PCR	Rat	N	Swim	8 wk	Medium	24 h	Iemitsu et al. 2002a
IGF-1	No change	PCR	Rat	PIHF	Treadmill	8 wk	85-90% $\dot{V}O_2$max	24 h	Wisløff et al. 2002
	No change	PCR	Rat	N	Treadmill	8 wk	85-90% $\dot{V}O_2$max	24 h	Wisløff et al. 2002
TGFβ_1	+51%	Northern	Rat	N	Wheel	6 wk	Medium/low	?	Calderone et al. 2001
TGFβ_3	No change	Northern	Rat	N	Wheel	6 wk	Medium/low	?	Calderone et al. 2001
HSP	+500%	PCR	Rat	N	Treadmill	10-12 wk	65-75% $\dot{V}O_2$max		Demirel et al. 1998
	+500%	Western	Rat	N	Treadmill	Acute	High	24 h	Milne and Noble 2002
	No Change	Western	Rat	N	Treadmill	Acute	Low	24 h	Milne and Noble 2002
	+ 51%	Western	Rat	DT-2	Swim	13 wk	Low	>48 h	Lajoie et al. 2004
	+ 82%	Western	Rat	N	Swim	13 wk	Low	>48 h	Lajoie et al. 2004
	+ 73%	Western	Rat	HT	Treadmill	8 wk	Medium	>48 h	Lajoie et al. 2002

ACE	No change	PCR	Rat	N	Swim	15 wk	Medium	>24 h	Iemitsu et al. 2001
Adrenomedullin	-65%	PCR	Rat	N	Swim	15 wk	Medium	>24 h	Iemitsu et al. 2001
β-adrenergic receptor	+65%	PCR	Rat	N	Swim	15 wk	Medium	>24 h	Iemitsu et al. 2001
β-adrenergic receptor kinase	No change	PCR	Rat	N	Swim	15 wk	Medium	>24 h	Iemitsu et al. 2001
Muscarine M2-receptor	No change	PCR	Rat	N	Swim	15 wk	Medium	>24 h	Iemitsu et al. 2001
PPAR-α	+150% +100%	PCR Western	Rat	N	Swim	8 wk	Medium	24 h	Iemitsu et al. 2002b
HAD	+100%	PCR	Rat	N	Swim	8 wk	Medium	24 h	Iemitsu et al. 2002b
CPT-I	+80%	PCR	Rat	N	Swim	8 wk	Medium	24 h	Iemitsu et al. 2002b
CD36	No change	PCR	Rat	N	Swim	15 wk	Medium	>24 h	Iemitsu et al. 2003
Acyl CoA synthase	No change	PCR	Rat	N	Swim	15 wk	Medium	>24 h	Iemitsu et al. 2003
CPT-I	No change	PCR	Rat	N	Swim	15 wk	Medium	>24 h	Iemitsu et al. 2003
CPT-II	+63%	PCR	Rat	N	Swim	15 wk	Medium	>24 h	Iemitsu et al. 2003
Isocitrate dehydrogenase	No change	PCR	Rat	N	Swim	15 wk	Medium	>24 h	Iemitsu et al. 2003
LDH	No change	PCR	Rat	N	Swim	15 wk	Medium	>24 h	Iemitsu et al. 2003
phosphofructokinase	No change	PCR	Rat	N	Swim	15 wk	Medium	>24 h	Iemitsu et al. 2003
Thyroid hormone receptor-α	+125%	Western	Rat	N	Swim	8 wk	Medium	24 h	Iemitsu et al. 2004
Thyroid hormone receptor-β	+100%	Western	Rat	N	Swim	8 wk	Medium	24 h	Iemitsu et al. 2004
c-jun	+100%	Northern	Rat	N	Treadmill	Acute	Medium	0-24 h	Boluyt et al., 2003
JNK activity	+50% (NS)	Western/IP	Rat	N	Treadmill	Acute	Low ~60% of $\dot{V}O_2max$	0-24 h	Boluyt et al., 2003
JNK activity	+100%	Western/IP	Rat	N	Treadmill	Acute	Medium ~94% of $\dot{V}O_2max$	0-24 h	Boluyt et al., 2003
JNK activity	+150%	Western/IP	Rat	N	Treadmill	Acute	High > 100% of $\dot{V}O_2max$	0-24 h	Boluyt et al., 2003
JNK activity	No change	Western/IP	Rat	TR	Treadmill	6 wk/ Acute	Low	0-24 h	Boluyt et al., 2003
SEK phosphorylation	No change (trend)	Western/IP	Rat	N	Treadmill	Acute	High > 100% of $\dot{V}O_2max$	0-24 h	Boluyt et al., 2003

State: N, normal; DT-2, *Type 2* diabetes; PI, post-infarction; HT, hypertensive; PIFH, post-infarction heart failure. Sampling: the time of tissue collection after the last exercise training. For additional genes see Diffee et al. (2003).

improved contractile function in cardiomyocytes from post-MI rats (Song et al. 2004). The covariation of SERCA2 expression and cardiomyocyte calcium handling, shortening, and relaxation in trained and untrained heart failure supports the notion that SERCA2 function is important in reduction of cardiomyocyte function during heart failure and improvement after exercise training.

It should be mentioned that changes in calcium transients and calcium regulatory proteins are not uniform across studies, probably due to experimental conditions. In contrast to most studies (e.g., Gwathmey et al. 1987; Vahl et al. 1994; Kleiman & Houser 1988), there are studies indicating an increased peak systolic calcium transient in cardiomyocytes from rats with myocardial infarction (Wisløff et al. 2002; Bing et al. 1991; Heller 1979).

Interestingly, a recent study shows contrasting changes in peak systolic calcium transients in myocytes from different regions of failing myocardium (McIntosh et al. 2000). Subendocardial cells from the heart failure group displayed an increased peak systolic calcium concentration in contrast to the decreased values observed in subepicardial cells and M-cells (from the midmyocardial region of the ventricle). The results from McIntosh et al. (2000) might explain the previous range of disparate results in the literature, including both a decreased (Gwathmey et al. 1987; Vahl et al. 1994; Kleiman & Houser 1988) and an enhanced (Bing et al. 1991; Heller 1979) peak systolic calcium concentration.

Increased Calcium Sensitivity

An additional mechanism for the increased contractile force in the cardiomyocyte is that exercise training may result in an increase in the sensitivity of the myofilaments to activation by calcium. An increase in calcium sensitivity of tension would result in a greater level of isometric tension generation at the same intracellular calcium level. In healthy rats, treadmill running induces an increased cardiomyocyte sensitivity to calcium, both in intact cardiomyocytes (Moore et al. 1993; Wisløff et al. 2001) and in permeabilized myocytes (figure 11.10) (Diffee et al. 2001, 2003; Wisløff et al. 2001), with more pronounced changes in endocardial compared to epicardial cells (Diffee & Nagle 2003). Recently Bupha-Intr & Wattanapermpool (2004) demonstrated improved calcium sensitivity in isolated myofibrils from ovariectomized rats. There are also indications that permeabilized cardiomyocytes from trained hearts are less affected by low pH at constant calcium than sedentary counterparts (Wisløff et al. 2001, 2002). As previously reported (reviewed in Allen & Kentish 1985), low pH decreases and alkaline pH increases myofilament shortening in myocytes from sedentary and trained cardiomyocytes. In an analogous way to intracellular calcium, this indicates that a component of the enhanced myocyte contractility could be attributed to the more alkaline intracellular pH in the trained myocytes at high stimulus frequencies.

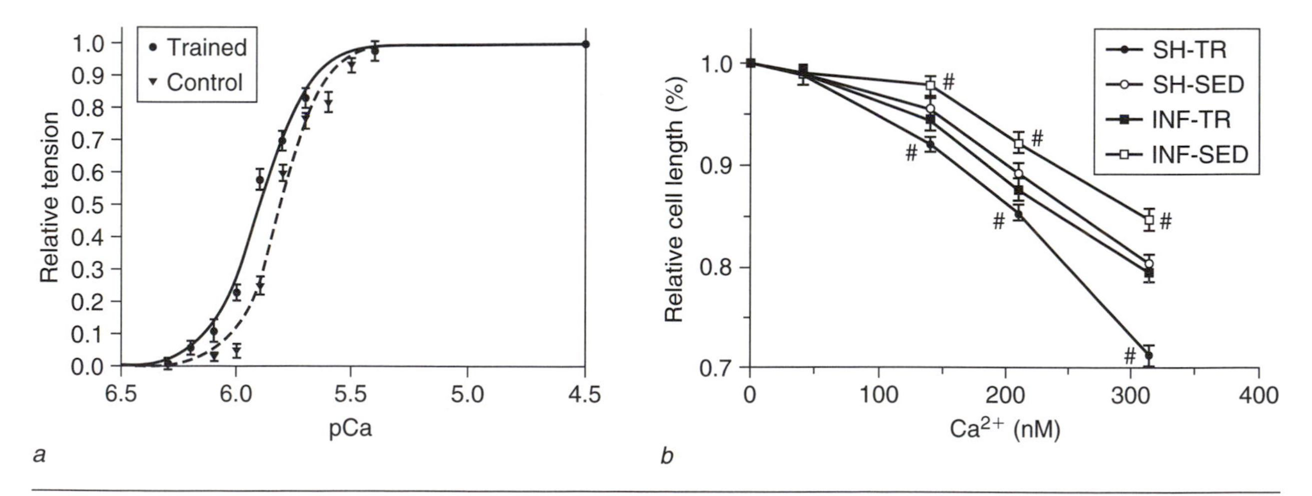

Figure 11.10 *(a)* Relationship between relative tension and pCa in skinned myocytes. Data were compiled from 70 control myocytes and 70 trained myocytes. Relative tension data at each pCa were averaged from all myocytes in the group and are presented as means ± SD. *(b)* After eight weeks of training, skinned myocytes responded more to changes in Ca^{2+}. There was no difference between INF-TR and SH-SED. SH, sham; INF, post-infarction heart failure; TR, trained; SED, sedentary.

Adapted, by permission, from G.M. Diffee, et al., 2001, "Exercise training increases the Ca2+ sensitivity of tension in rat cardiac myocytes," *Journal of Applied Physiology 91: 311.*

In human heart failure, myofilament calcium responsiveness has been reported to be both decreased (Hajjar et al. 1988; D'Agnolo et al. 1992) and unchanged (Schwinger et al. 1994). The discrepancies might be due to whether the concentration of substances affecting calcium sensitivity such as protons, ADP, cAMP, and inorganic phosphate had been held constant or not, in addition to different temperatures used in previous experiments. Recent studies in human heart failure (Hajjar et al. 2000) suggest that changes in the myofilaments' responsiveness to calcium are a central cause of reduced contractile function in heart failure, and that changes in calcium cycling are compensatory rather than causative. Demonstration of reduced calcium sensitivity in cardiomyocytes from patients and animals with heart failure has led to the development of calcium sensitizers, that is, drugs that improve the responsiveness to calcium, which improves function in human heart failure. Endurance training in post-infarction heart failure increases the contractile machinery's sensitivity to calcium, shown indirectly in intact cardiomyocytes by increased maximal extent of shortening despite lower peak systolic calcium. This has been shown more directly in skinned cardiomyocytes, with cardiomyocytes from trained rats responding more to changes in calcium than sedentary controls (Wisløff et al. 2002) (figure 11.10). Trained myocytes exhibited a leftward shift in the tension–pCa relationship compared with control myocytes. The pCa50 of these composite curves was 5.83 for control and 5.89 for trained myocytes. Whereas myofilament calcium sensitivity is depressed in cardiomyocytes from sedentary post-infarction rats (Hajjar et al. 1988, 2000; Perez et al. 1999), this effect was not evident in myocytes from endurance-trained post-infarction rats. Thus, training appears to normalize myofilament calcium sensitivity (figure 11.10).

The cellular basis for improved calcium sensitivity is not known, but multiple biochemical alterations of the contractile proteins have been described in heart failure, including isoform switching of troponin T (Anderson et al. 1995) and suppression of α- and increased β-myosin heavy chain expression (Nakao et al. 1997). Responsiveness of myofilaments to calcium is also affected by the levels of protons, inorganic phosphate, cAMP, and cGMP (Hajjar et al. 2000). Little is known about changes in cardiac myofilaments following aerobic exercise training, although stress-induced changes in the expression of troponin I isoforms have been noted (Wattanapermpool et al. 1995). It is also known that treadmill training in rat might induce a further increase in the α-myosin heavy chain expression (Jin et al. 2000; Scheuer & Tipton 1977), which is associated with high ATPase activity and increased contractility. It is clear, however, that changes in ventricular ATPase activity or isoform composition are not obligatory for improved ventricular performance secondary to chronic exercise (Buttrick et al. 1994; Ritter et al. 1999). Recently Diffee et al. (2003) found an upregulation of atrial myosin light chain 1 (aMLC-1) in left ventricle of endurance-trained rats. Atrial myosin light chain 1 has previously been shown to increase in human cardiac hypertrophy and has been associated with increased Ca^{2+} sensitivity to tension and increased power output.

Changes in AP Duration

The two most consistent electrophysiological abnormalities associated with pathological myocyte hypertrophy are the prolongation of action potential duration (APD) (McIntosh et al. 2000; Rozanski and Xu 2002) and reduction of transient outward current (i_{to}) density (Lue & Boyden 1992; Qin et al. 1996). The decreased density of the outward current appears to be caused by reduced expression of the genes for the delayed rectifier $K_v4.2$ and $K_v4.3$ and may be caused in part by an inhibitory effect of α-adrenoreceptor stimulation. Prolongation of APD might be an important compensatory mechanism to increase Ca^{2+} influx in post-MI myocytes during the action potential. Through prolongation of the plateau phase of the action potential, more Ca^{2+} may enter the cell via reverse mode Na^+-Ca^{2+} exchanger and thus partially compensate for the reduced Ca^{2+} (both directly and via Ca^{2+}-induced calcium release from SR) available for the contractile elements normally observed in post-MI myocytes. On the other hand, prolongation of APD may be one of the mechanisms for development of tachyarrythmias in the hypertrophied ventricle (Qin et al. 1996). Thus prolonged APD in post-MI myocytes may be regarded as a maladaptation with potentially fatal consequences. There are few data on the effect of endurance exercise on APD in post-MI myocytes. However, Zhang et al. (2001) showed that anaerobic treadmill running in post-MI rats normalized the APD and that this was due to increased fast and slow component of the i_{to}. Whether this resulted from increased expression of $K_v4.2$ and $K_v4.3$ is not known. Recently Natali et al. (2002) showed that training prolonged the action potential of subepicardial myocytes from healthy rats, reducing the transmural gradient in APD. This observation may be important for understanding the cellular causes of T-wave abnormalities found in the electrocardiograms of some athletes.

Cell Metabolism

Training-induced elongation of cardiomyocytes is accompanied by angiogenesis and adaptation in the mechanisms regulating blood flow and O_2 delivery to hypertrophied cardiomyocytes, increasing the heart's ability to avoid energetic limitations during periods of increased mechanical activity. Endurance training increases arterial size so that it is commensurate with or is greater than the relative increase in cardiac mass (reviewed in Moore & Korzick 1995). There is also evidence of arteriolar proliferation and longitudinal development (Breisch et al. 1986) in addition to a pronounced capillary angiogenesis (Hudlicka 1991) that is similar or more marked than the increase in cardiac size.

Due to the high intrinsic heart rate (350 beats per minute), the activities of the cardiomyocytes' mitochondria are about three times higher than those observed in skeletal muscle, indicating a robust mechanism for matching energy supply with demand. High-intensity aerobic endurance training (with high heart rates) seems to have more impact on mitochondrial volume and enzyme activity than low-intensity training (reviewed in Moore & Korzick 1995).

Cardioprotection: Preconditioning

Exercise training has been associated with improved recovery of the heart from ischemic–reperfusion (I/R) insult (Libonati et al. 1997; Spencer et al. 1997). Recent studies have given more insight into the cellular basis for this training-induced cardioprotection. Several studies (e.g., Demirel et al. 1998) suggest that training-induced cardioprotection against I/R injury is associated with increased HSP72 expression and cellular oxidative stress, both acting via ATP-sensitive K^+ (K_{ATP}) channels (Joyeux et al. 1998). In light of the evidence that HSP72 and cellular oxidative stress (both of which can be profoundly influenced by training) can modulate K_{ATP} channel activity, it seems reasonable that training might alter the responsiveness of sarcolemmal K_{ATP} channels to I/R stress. This speculation is consistent with findings by Jew and Moore (2001) demonstrating that glibenclamide, a K_{ATP} channel blocker, improved the recovery of contractile function of I/R challenge much faster in hearts isolated from endurance-trained rats. These results were confirmed more recently (Jew & Moore 2002) in an investigation showing that, in response to an anoxic challenge, the onset of glibenclamide-sensitive outward current was markedly delayed and peak current density reduced in cardiomyocytes from endurance-trained rats. This provides evidence that training markedly blunts the responsiveness of sarcolemmal K_{ATP} channels to metabolic stress in single, isolated cardiomyocytes. This in turn raises the possibility that increased HSP72 and reduced reactive oxygen species in trained cardiomyocytes may represent adaptations to suppress K_{ATP} channel opening, and that this is an adaptation designed to preserve tissue mechanical function and myocardial oxygen supply (Jew & Moore 2001, 2002). This suggestion is consistent with earlier observations that administration of a K_{ATP} channel blocker can normalize Ca^{2+} transients and cardiomyocyte shortening under conditions of metabolic stress (Lederer et al. 1989; Stern et al. 1988), and with the finding of Jew and Moore (2001) that administration of a K_{ATP} channel blocker during reperfusion is conducive to the functional recovery of the myocardium after ischemic insult, particularly in hearts from trained rats.

Integrated Function and Health Effects

For patients and athletes to fully benefit from exercise training, it is important to know the basal mechanism of training effects in health and disease. Substantial knowledge linking the cardiac myocyte to clinical effects of exercise training has emerged over the last decade. In moderately and highly trained healthy persons, further improvement of aerobic capacity mainly depends on increased cardiac output (Richardson et al. 1999; Wagner 2000). As maximal heart rate remains constant even in highly trained athletes, improved cardiac output depends on increased LV stroke volume achieved by increased diastolic LV volume and increased myocardial contractility (Wiebe et al. 1999; Ferguson 2001). These changes are due to increased cardiomyocyte length, mass, and contractility. As presented in this chapter, several molecular mechanisms and signaling pathways for these changes are being uncovered and form the basis for modern exercise physiology and directions of future research.

In heart failure, which is a common end stage of coronary heart disease, hypertension, cardiomyopathy, and valvular heart disease, there is accumulating evidence of beneficial effects from training on quality of life and survival (Belardinelli et al. 1999; Wielenga 1997; Myers et al. 2002). LV dilation and reduced contractility, associated with death and severity of disease, significantly improve after training (Hambrecht et al. 2000; Gianuzzi et al. 1997). These effects are directly related to cellular changes, reduced cardiomyocyte hypertrophy, and improved cardiomyocyte contractility. Improper relaxation of the cardiac muscle frequently induces dyspnea and fatigue during exercise due to increased pulmonary capillary pressure. Prolonged myocardial relaxation is a frequent finding in hypertension, in hypertrophic cardiomyopathy, in the elderly, and in most stages of heart failure and is a strong predictor of mortality. Exercise training consistently improves myocardial relaxation, probably because of enhanced diastolic absorption of calcium into SR due to increased SERCA function. Physical exercise may therefore be an efficient tool to improve symptoms in cardiac patients, especially the elderly.

Cardiac arrhythmias seem to be a more frequent cause of death than previously supposed. In more than 60% of cardiac deaths, the patient does not live long enough to reach hospital treatment. Although sudden death occurs more frequently during or immediately after exercise, habitual exercise markedly diminishes the overall risk of sudden cardiac death (Albert et al. 2000). The mechanism of this reduced risk of sudden death is not settled but is probably due to antiischemic and antiarrythmic effects. It has been shown that training shortens the prolonged cardiomyocyte action potential in heart failure and thereby might reduce electrical instability and sudden cardiac death.

As exercise physiology and molecular biology are rapidly expanding sciences, new knowledge will certainly explain many effects of exercise training on both cardiac myocytes and other cells during the coming years. One of the future challenges will be to implement the increasing knowledge of basal physiological mechanisms into everyday practice of sport and medicine.

Exercise and Endothelium

Klaus Völker

For a long time the endothelium has been regarded simply as an inert membrane that lines the circulatory system, with its primary function to maintain flow and vessel wall permeability. However, during the last 20 years of research, opinion of the endothelium as a passive structure has completely changed. Rather early the endothelium was regarded as a dynamic, heterogeneous, disseminated organ that processes vital secretory, synthetic, metabolic, and immunologic functions (Fishman 1982). This opinion also represents the current view.

Structurally the endothelium consists of a flattened monolayer of interlinked cells. The endothelium organ contains about 1 to 6 $\times$ 10^{13} endothelium cells, weighs nearly 1 kg, and covers a surface area of 1 to 7 m^2 (Augustin et al. 1994). Endothelium-lined vessels can be found in every organ system, regulating the blood flow and acting as a barrier between the intravascular and the interstitial space. Thereby endothelial cells regulate the transport of nutrient substances and diverse biologically active molecules, regulate the adhesion/diapedesis of blood cells, and are involved in hemostasis control (Cines et al. 1998). While the endothelium organ is able to act autonomously, it often functions in close interaction with the vascular smooth muscle. Both structures are essential to maintain function and structure of the vessel wall. Figure 12.1 summarizes the main regulatory tasks endothelium cells have to fulfill.

The purpose of this chapter is to give a brief overview of endothelium participation in these multiple biological processes, followed by a focus on the interaction between endothelial cell function and exercise.

Endothelial Cell Function

Of the main regulatory tasks the autonomous organ endothelium has to fulfill, the regulation of vascular tone, the induction of alteration in vascular structure, and the interaction with blood cells are of special interest in the context of physical activity.

Regulation of Vascular Tone

The endothelium plays a pivotal role in blood pressure and blood flow regulation. Thereby vascular tone is under control of local and systemic factors and is the result of a balance between vasodilatating and vasoconstricting agonists. Locally, autoregulation of vascular tone occurs especially in response to mechanical stimulation. The two mechanical forces acting on the blood vessels in response to exercise are flow-associated augmentations in pulsatile stretch and shear stress. Experimental data especially on epicardial coronary vessels suggest that the exercise-induced increase of shear stress in endothelial shear stress leads to a proportional vasodilatation whereas changes in blood pressure that are responsible for the pulsatile stretch show no effect (Hilton 1959; Lie et al. 1970). The mechanism by which the mechanical stimulus of flow leads to flow-induced vasodilatation is still a field of research. A shear stress-induced deformation in cytoskeletal formation seems to be involved in this process (Olesen et al. 1988).

The shear stress-induced effects are mediated by release of vasodilatators such as nitric oxide (NO) and prostacyclin (PGI_2) on the one hand and vasoconstrictors such as endothelin (ET) and platelet-activating factor ((PAF) on the other hand (Cines et al. 1998).

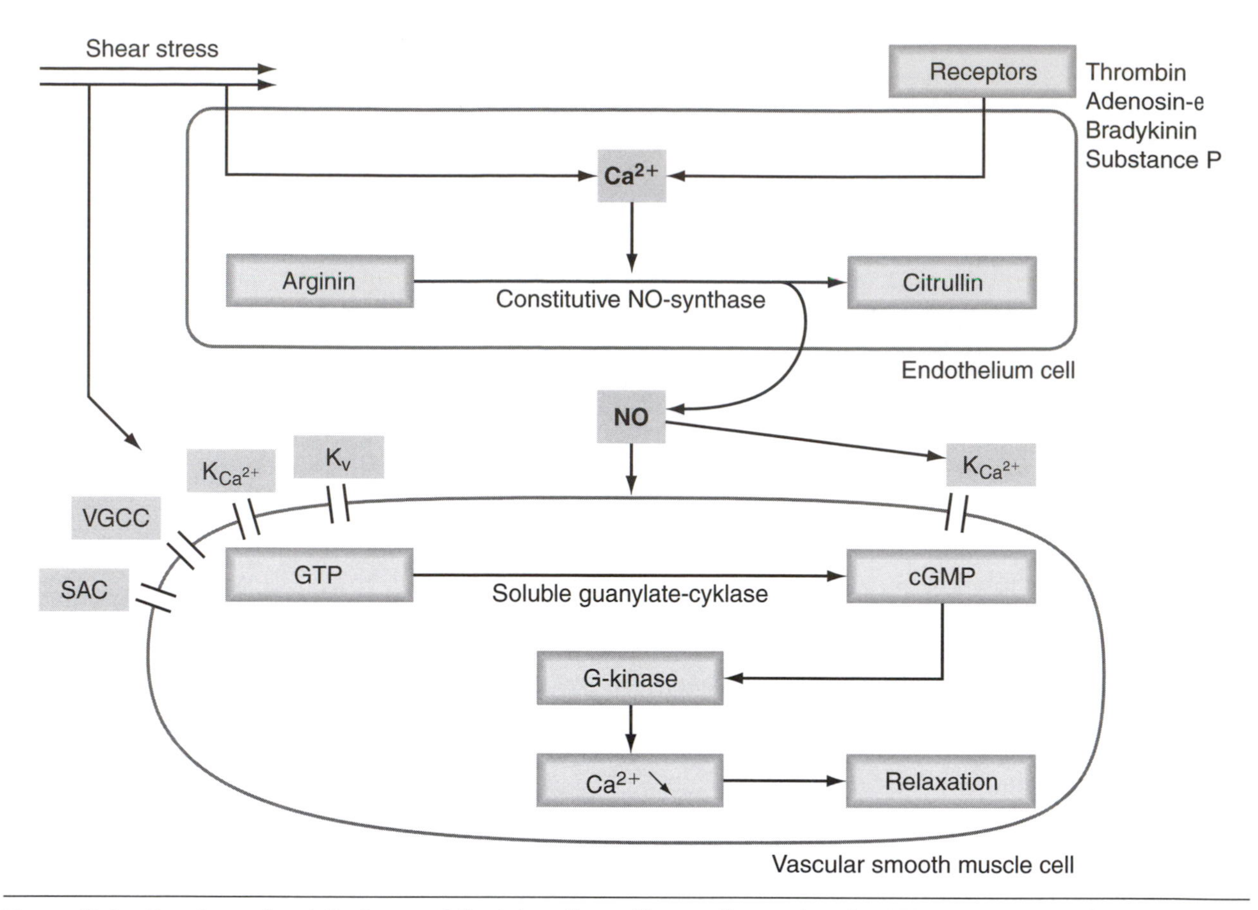

Figure 12.1 Functions of endothelium and the interaction with smooth muscle cells. GTP, guanosine trisphospate; cGMP, cyclic guanosine monophosphate; NO, nitric oxid; VGCC, L-type voltage-gated calcium channel; SAC, stretch-activated channel; Ca^{2+}, activated K^+ channel (K_{ca2+}); Kv, voltage-gated potassium channel.

The release of these chemically diverse substances is regulated by the localization of specific receptors on vascular cells, through their rapid metabolism or at the level of gene transcription. Because of their continuous secretion there is no evidence for intracellular storage of these substances. The continuous release of NO by the endothelium cells is modulated by a number of exogenous chemical and physical stimuli. The other mediators (PGI_2, ET, and PAF) are synthesized primarily in response to changes in the external environment (Cines et al. 1998).

Vasodilatory Agonists

NO, first described as endothelium-derived relaxing factor (EDRF) in 1980 by Furchgott and Zawadzki (1980), is a heterodiatomic free radical generated by oxidation of L-arginine into L-citrulline by NO synthases (NOS). There are three isoforms of NOS:

- NOS I (nNOS), which is expressed in the nerve tissue
- NOS II (iNOS), which is inducible in many cell types by factors like LPS or cytokines but is normally not expressed or extremely rarely expressed
- NOS III (eNOS), which is mainly expressed in endothelium cells but also in cardiomyocytes and platelets

In contrast to iNOS, eNOS is constitutively active in endothelium cells. It also can be stimulated by receptor-dependent agonists that increase intracellular calcium (like thrombin, adenosine 5'-disphosphate, bradykinin, substance P, and muscarinic agonists) and additionally through perturbation of the plasma membrane phospholipid symmetry (Venema et al. 1995) such as shear stress (Topper et al. 1996) and cyclic strain (Awolesi et al. 1994). NO exerts its relaxing effect on vascular smooth muscle by activating soluble guanylate cycles to produce cyclic guanosine monophosphate. Additionally, NO may stimulate a calcium (Ca^{2+})-dependent potassium channel, which leads to hyperpolarization and relaxation of vascular smooth muscle (figure 12.2; Bolotina et al. 1994).

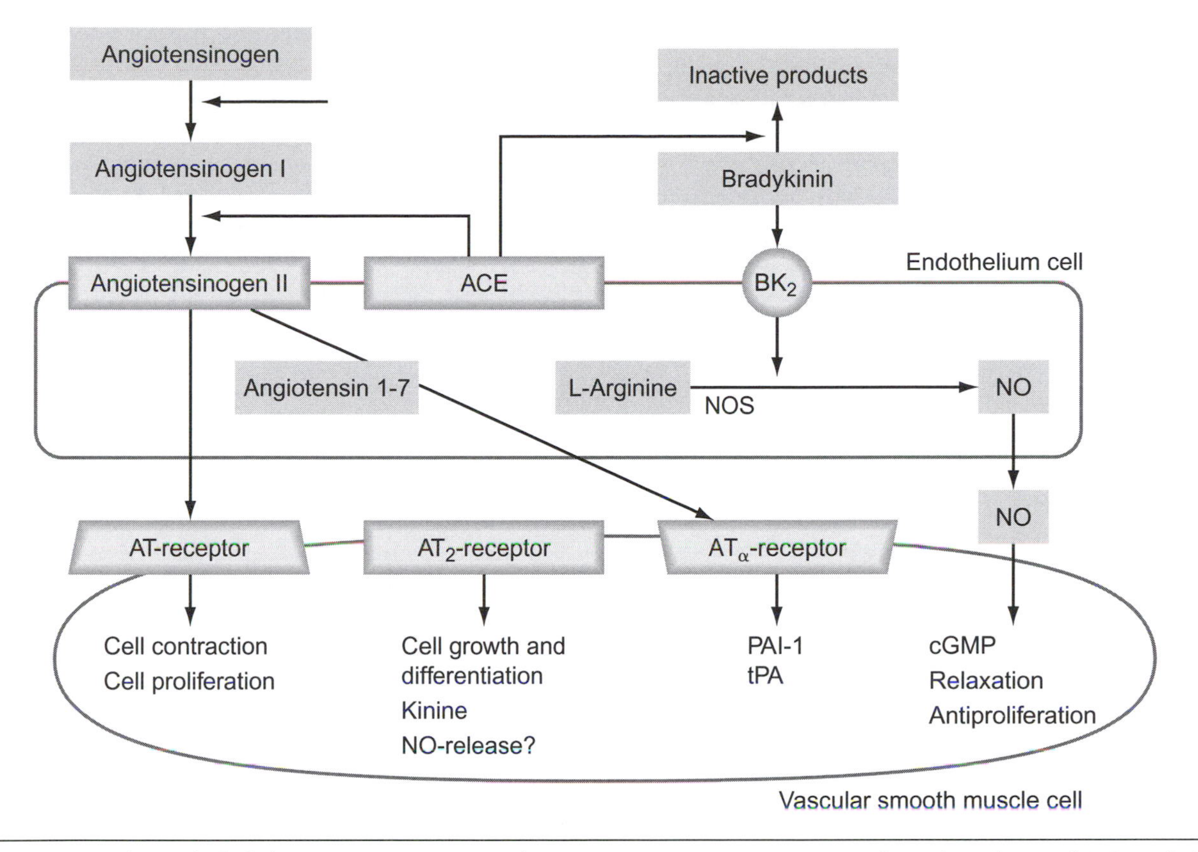

Figure 12.2 The endothelial arginine/NO system. Shear stress or receptor agonists induce the release of NO and the activation of different channels, leading to relaxation of smooth muscle cells. ACE, angiotensin converting enzyme; NO, nitric oxid; NOS, nitric oxid synthase; PAI-I, plasminogen activator inhibitor-1; tPA, tissue plasminogen activator; cGMP, cyclic guanosine monophosphate; BK2, bradykinin receptor.

The findings of Ayajiki et al. (1996) indicate that the flow-induced release of NO has a biphasic nature. The initial Ca^{2+}-dependent phase of NO release is followed by a phase of sustained NO secretion that is Ca^{2+} independent. There is recent evidence for two signal transduction pathways. The Ca^{2+}-dependent pathway involves the increase of intracellular Ca^{2+} and the activation of phospholipase C (Geiger et al. 1992; Shen et al. 1992); the Ca^{2+}-independent pathway seems to require the activity of the sodium-hydrogen exchanger and the activation of protein kinase C and tyrosine kinase (Ayajiki et al. 1996). Additionally, some studies indicate that there may be an autocrine mechanism that contributes to the flow-mediated vasodilatation. The release of bradykinin from endothelium cells is regarded as induced by flow. It is postulated that the binding of bradykinin to its respective receptor B2 induces the release of NO (Groves et al. 1995).

Besides NO synthesis, chronic flow augmentation may enhance prostacyclin synthesis as well. A continuous release of NO and prostacyclin from endothelium cells was observed in response to shear stress (Hecker et al. 1993); also, an enhanced dilator response of gracilis muscle arterioles in a rat model after short-term daily exercise (Koller et al. 1995) was attributed to the release of both relaxation factors. Prostacyclin may thereby play a larger role in flow-mediated relaxation of skeletal muscle microvessels. This may counteract the neurogenic and myogenic vasoconstriction stimuli induced by exercise (Niebauer & Cooke 1996).

Flow-mediated vasodilatation is furthermore mediated by the release of other agonists such as ATP, substance P, prostacyclin, or endothelium-derived hyperpolarization factor (EDHF) (Koller et al. 1995; Feletou & Vanhoutte 1988). ATP and substance P, which are released from endothelium cells by shear stress, stimulate specific receptors and cause the release of NO. Another source of ATP may be the circulating ATP. The increased flow overwhelms the degradation of the circulating ATP by ectopyrase and leads to greater delivery to the endothelial surface (Dull & Davies 1991; Mo et al. 1991). A burst of prostacyclin produced by endothelium cells could be observed after a sudden increase of shear stress, indicating that the gradient is more likely to be the physiologic stimulus than the absolute level of shear stress (Frangos et al. 1985). The endothelium-derived hyperpolarization factor (EDHF) is a less well described product of endothelium cells. Stimulation with muscarine agonists leads to a release of EDHF causing a transient hyperpolarization of the cell membrane. Two ways to achieve its vascular effects are discussed: activating ATP-sensitive potassium channels, smooth muscle sodium-potassium ATPase, or both (Feletou & Vanhoutte 1988). Recent studies indicate that cytochrome P450 2C may serve as an EDHF synthase (Kessler et al. 1999). However, further studies are required to explain the role of prostacyclin and EDHF in vascular physiology.

Vasoconstrictory Agonists

Important vasoconstrictory agonists are endothelin-1 and angiotensin II, the latter of which is not secreted by endothelium cells but produced by the angiotensin converting enzyme (ACE) at the surface of the endothelium.

Endothelin-1 (ET-1) (figure 12.3) is one of the most potent peptide vasoconstrictors identified to date. Endothelin is a member of the 21-amino acid peptide family that is produced by many different cell types. There are no intracellular granules for ET-1 (Nakamura et al. 1990). ET-1 is formed after transcription of the gene encoding preproendothelin-1, which is the inactive precursor of active ET-1. Besides shear stress, hypoxia and ischemia are stimulating factors. ET-1 binds to the G-protein-coupled ET-A receptor, which is abundantly expressed on vascular smooth muscle cells, resulting in an increase of muscular calcium concentration and leading to an augmentation of vascular smooth muscle tone (Simonson 1990). The effect of ET-1 on intracellular calcium seems to persist even after the hormone dissociates from the receptor. This longer-lived effect can be shortened by NO through acceleration of the restoration of intracellular calcium (Goligorsky et al. 1994). The release of ET-1 in response to shear stress is complex. Physiologic levels of shear stress induce a transient increase of peptide release and an upregulation of the expression of ET-1 messenger ribonucleic acid (mRNA), which is followed by a mark suppression (Yoshizumi et al. 1989; Malek et al. 1993). The release of ET-1 in human umbilical

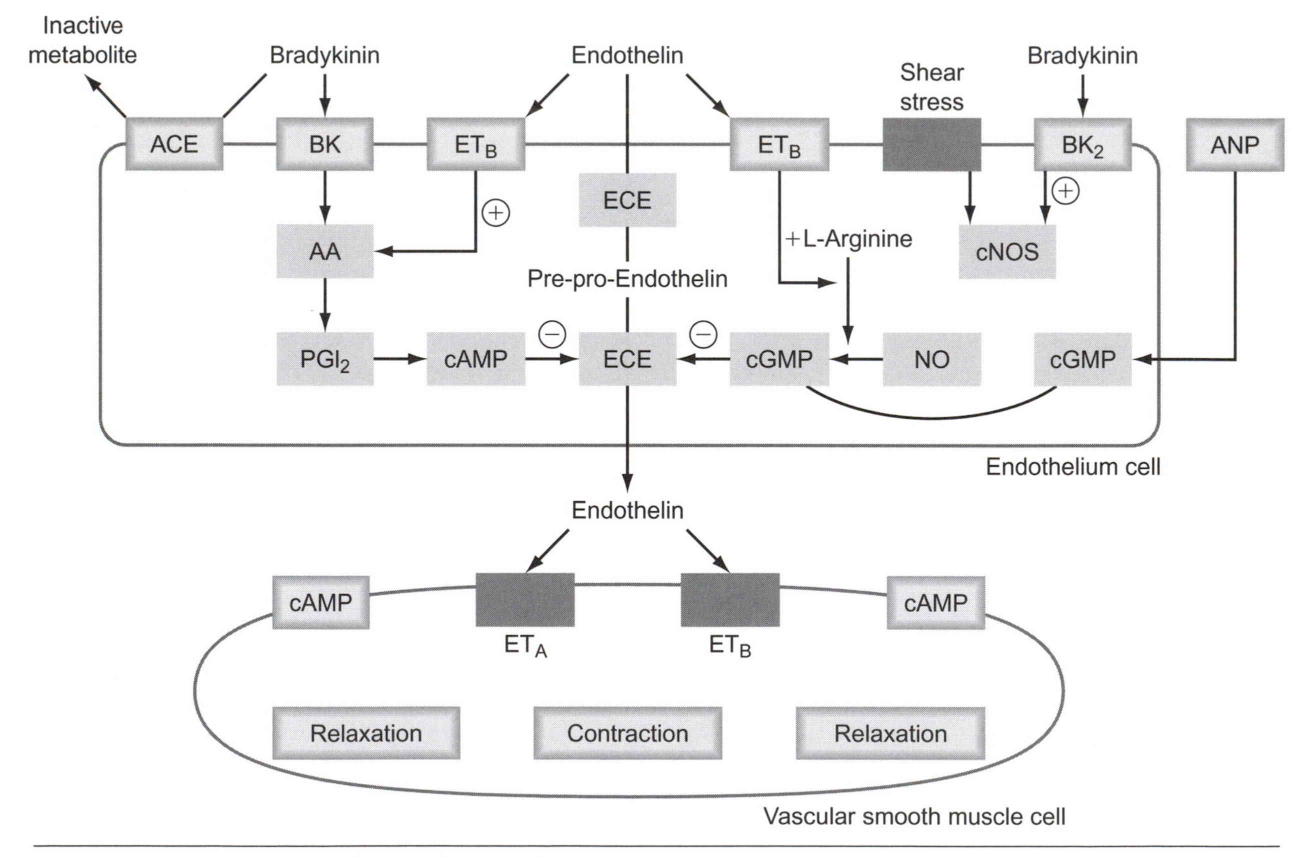

Figure 12.3 The interaction of NO and endothelin in endothelium and smooth muscle cells. ACE, angiotensin converting enzyme; NO, nitric oxid; cNOS, constitutive nitric oxid synthase; cGMP, cyclic guanosine monophosphate; cAMP, cyclic adenosine monophosphate; BK2, bradykinin receptor; ANP, atrial natriuretic peptide; ECE, endothelin-converting enzyme; ETA/ETB, endothelin A/B receptor; AA, arachidonic acid; PG12, prostacyclin.

endothelium cells is augmented with low shear stress but inhibited with high stress (Kuchan & Frangos 1993).

Angiotensin II as a potent vasoconstrictor, as well as the whole renin-angiotensin system, shows interactions to endothelium-derived NO (Linz et al. 1999). ACE on the surface of the endothelium not only is responsible for converting angiotensin I to angiotensin II but also leads to an inactivation of bradykinin and diminishes the bradykinin-induced release of NO from endothelium (Linz et al. 1999). In contrast, some data indicate that angiotensin II itself can lead to a kinine-induced release of NO by stimulating the AT_2 receptor (figure 12.4).

Endothelium-Independent Regulation of Vascular Tone

Besides the endothelium-dependent regulation of vascular tone by the release of vasodilatory and vasoconstrictory substances, there is also an endothelium-independent mechanism caused by mechanical stretch. The following model describes the possible signaling pathway (Davis et al. 1992; Meininger & Davis 1992).

The following cascade was modeled to explain the reaction (Meininger & Davis 1992; Nelson & Quayle 1995). Stretch leads to an activation of a nonselective stretch-activated cation (SAC) channel in the plasma membrane, resulting in cell depolarization. The depolarization in turn activates dihydropyridine-sensitive, voltage-gated Ca^{2+} channels (VGCC) promoting a secondary Ca^{2+} influx in addition to the Ca^{2+} entry through SAC channels, leading to smooth muscle contraction. Conversely, membrane depolarization and Ca^{2+} influx also activate K^+ channels (voltage-dependent K^+ [K_v] channel and Ca^{2+}-activated [K_{Ca}] channel). The activation of the potassium channel leads to K^+ outflux and thereby maintains, through hyperpolarization, the electrochemical gradient necessary for Ca^{2+} influx, which has been observed by different groups (Dull & Davies 1991; Mo et al. 1991). This mechanism may act as a negative feedback mechanism to limit depolarization, VGCC activation, and thereby smooth muscle contraction (Brayden & Nelson 1992; Wellner & Isenberg 1994).

The flow-dependent Ca^{2+} influx may be modulated by vasodilatory agents such as ATP (Mo et al. 1991) and regulated by the activity of kinases (e.g., mitogen-activated protein [MAP] kinase). MAP kinases are associated with cytoskeletal elements, which react to mechanical deformation such as stretch (Powell et al. 1987; Sadoshima & Izumo 1993) or physical forces such as osmotic stress (Brewster et al. 1993).

Vascular Control in Skeletal Muscle at Rest and During Exercise

For a long time (Hilton et al. 1970) it has been known that vascular control in the skeletal muscle vascular bed is controlled by the sympathetic nervous system and codetermined by the endothelium. This control may vary depending

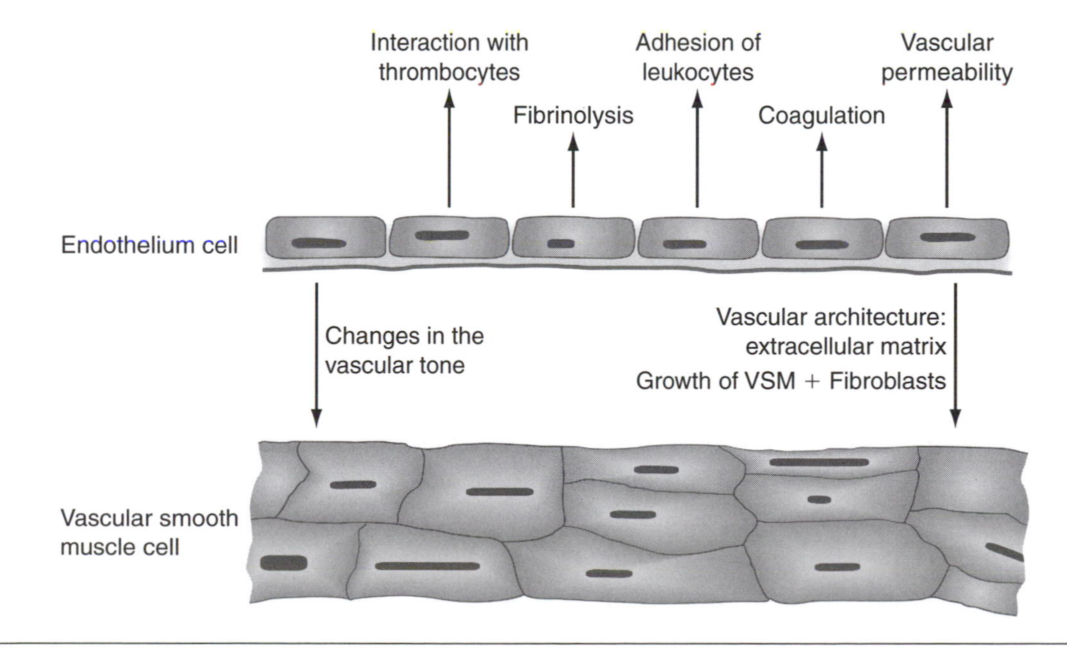

Figure 12.4 The interaction between the renin-angiotensin system and the arginine/NO system. VSM, vascular smooth muscle.

on muscle fiber composition and regional specificity of the vascular tree within muscle tissue (Mellander & Bjornberg 1992; Hester et al. 1993). Endothelium-derived NO is a significant regulator of basal vascular tone in resting muscle (Ekelund & Mellander 1990).

The contribution of endothelium-derived NO to exercise-induced vasorelaxation in active skeletal muscle is less clear (Delp & Laughlin 1998). Some data indicate that the endothelium-derived NO may be responsible for the initial (1 min) hyperemic response mediating the dilatation of first- and second-order arteries, but not third-order arterioles (Hester et al. 1993). The inhibition of endothelium cell formation in rats caused an attenuation of exercise-induced hyperemia that differed between different fiber compositions, indicating a linear correlation to the content of oxidative capacity (Delp & Laughlin 1998). On the other hand, blockade of endothelium-derived NO did not alter the exercise-induced hyperemia in skeletal muscles of cat (Ekelund et al. 1992) or rabbit (Persson et al. 1990).

In humans, intrabrachial infusion of NO inhibitors produced a decrease of forearm blood flow at rest as well as during rhythmic and static forearm exercise. In magnitude the decrease in blood flow during exercise was similar to the reduction at rest (Endo et al. 1994; Gilligan et al. 1994; Dyke et al. 1995). Because of the same magnitude of reduction in blood flow at rest and during exercise, these findings suggest a lack of additional NO-mediated control of muscle perfusion during exercise. Whether other endothelium-derived vasodilatators are involved in skeletal blood flow regulation remains to be investigated (Delp & Laughlin 1998).

Flow-Induced Alterations in Vascular Structure

The endothelium-derived factors involved in the regulation of vascular tone are also responsible for induction of changes in vascular structure. The same molecular pathway leading to flow-mediated vasorelaxation is involved in the flow-induced transcription of transforming growth factor-beta$_1$ (TGF-beta$_1$) (Ohno et al. 1995). On the other hand, the vasoconstrictor ET-1 is a potent mitogenic agent that can be potentiated by other growth-stimulating factors like angiotensin II, platelet-derived growth factor-beta (PDGF-beta), and insulin (Dubin et al. 1989; Hirata et al. 1989; Yang et al. 1990). This leads to the hypothesis that the initial response (vasorelaxation) to an increase in flow and the later adaptation (change in vascular structure) may share common pathways (Niebauer & Cooke 1996).

In response to long-term change in flow, vessels have the capability to remodel significantly. In animal models, the increased flow induced by endurance training enlarges the vessel diameter of the carotid artery or the coronary artery independently of the reaction of the end organ (Leon & Bloor 1968; Kramsch et al. 1981; Snell et al. 1987). In contrast, over the long term a decrease in flow diminishes the response to structural modification of the vessel wall (Langille & O'Donnell 1986). The fact that vascular remodeling can be abolished through removal of the endothelium provides evidence that flow-induced vascular remodeling is as endothelium dependent as flow-mediated vasorelaxation. Although the mechanism mediating these structural changes has still to be elucidated, the findings of various studies imply that the endothelium may serve as a transducer of the flow signal by generating the paracrine effectors that mediate vascular remodeling. Identified mediators involved in cell growth, extracellular matrix production, and proteolysis are, for example, ET-1, NO, PDGF-beta, prostacyclin, TGF-beta$_1$, and tissue plasminogen activator (Diamond et al. 1989; Cooke et al. 1990; Kuchan & Frangos 1993; Mitsumata et al. 1993; Koller et al. 1995; Ohno et al. 1995).

Flow-Induced Effects on Hemostasis/Thrombolysis/ Fibrinolysis

Besides its role in vasodilatation, endothelium-derived NO has remarkable effects on hemostasis. NO, which is also produced by the platelets' own NO synthase III, inhibits platelet adhesion, activation, secretion, and aggregation and also promotes platelet disaggregation. The mechanism inducing these effects is partly cyclic GMP dependent (Mendelsohn et al. 1990). NO also inhibits other steps in the platelet activation cascade and interacts there synergistically with PGI_2, which alone has no effect on platelet adhesion (Radomski et al. 1987; Stamler et al. 1989). The agonist-dependent increase in platelets' calcium concentration is inhibited by NO (Mendelsohn et al. 1990), leading to a down-regulation of P-selectin expression on the platelets' surface. Furthermore, the calcium-sensitive conformational change in the heterodimeric integrin glycoprotein $\alpha_{IIb}\beta_3$ (GP IIb-IIIa), which is a precondition for fibrinogen binding (Michelson et al. 1996), is suppressed. The activity of phosphoinositide 3-kinase, which supports conformational changes in integrin glycoprotein by making the association with fibrinogen completely irreversible, is impaired by NO. This effect indirectly supports platelet disaggregation.

Regarding fibrinolyis, mechanical shear stress has been shown to be a potent stimulator of t-PA transcription and protein production in vitro (Diamond et al. 1989). Since an increase in cytoplasmic calcium can induce acute t-PA release, it may be hypothesized that the augmentation of intracellular calcium concentration in response to regular exercise also contributes to a greater capacity of t-PA release.

NO inhibits the interaction of leucocytes with the vessel wall either by inhibiting the activation of leucocytes or the expression of adhesion molecules (Kubes et al. 1991; De Caterina et al. 1995).

Effects of Physical Training on Vascular Reactivity

Physical training is a well known modulator of blood pressure emphasizing the important role of sports in therapeutic regimens of hypertonicity. Recent investigations documented that the vascular wall is an important target for exercise and adapts to chronic physical activity in several ways.

Molecular Aspects of Vascular Training Adaptation

Besides the effects of acute increased blood flow, evidence is accumulating that the chronic augmentation of blood flow, as induced by regular exercise training, exerts beneficial effects on vascular reactivity. These adaptations have been generalized as an overall enhanced vasodilatation and attenuated vasoconstriction to vasoactive agents (Laughlin & McAllister 1992; Parker et al. 1994).

Receptor/Channel/Signaling Level

The effects of training that can be seen in enhancement in vasoreactivity have their correlates in the expression of receptors, in the contribution of channels, and on the signaling level.

A single bout of exercise, examined in an animal model with Wistar rats, can enhance vascular responsiveness to ACh (Cheng et al. 1999). This study showed an up-regulation of M_3 endothelial receptors, their high-affinity binding sites increasing in number but not in affinity. In contrast, the α_2 receptor binding sites decreased in number but increased in affinity. It was concluded that the enhanced receptor-mediated vasodilatation with a single bout of exercise is partly due to the regulation of either the endothelium receptor number or the receptor affinity.

An increased K^+ channel contribution, which is responsible for the myogenic basal tone, was found in an endurance training animal model for porcine coronary resistance arteries. Consistent with the greater number of K^+ channels was an increase in myogenic tone (Brayden & Nelson 1992; Meininger & Davis 1992; Nelson et al. 1995; Bowles et al. 1998). The increased tone in resistance and conduit coronary arteries (Bowles et al. 1998) seems to contrast with the overall pattern of enhanced vasodilatation and reduced vasoconstriction found after exercise training (Laughlin & McAllister 1992; Parker et al. 1994; Bowles et al. 1998). The high tone induced by training may be the precondition for enhanced vasodilatation (Bowles et al. 1998).

The results of a study with humans support this model. Comparison of the proximal coronary artery diameter using coronary autograph in ultradistance runners with that of sedentary controls showed a 2.2-fold greater increase of diameter in runners after the application of nitroglycerin (Haskell et al. 1993). Consistent with the heterogeneous training response of vessels of different size, the activation of potassium channels is different. After eight weeks of training, Wistar-Kyoto rat mesenteric arteries showed on EDHF a more pronounced reaction than in the larger conduit artery thoracic aorta. This can be explained by activation of large conductance calcium-activated K^+ (BK_{CA}) channels (Chen et al. 2001).

Acute exercise enhances vasodilatation by modulating endothelial calcium signaling. Aortic endothelium cells of Wistar rats were analyzed in situ by Ca^{2+}_i imaging method. There was a logarithmic correlation between the intracellular calcium elevation and vasorelaxation. Acute exercise enhanced ACh-induced calcium concentration and vasorelaxation without altering their relationship. The augmentation in stimulus-induced $[Ca^{2+}]_i$ elevation in response to acute exercise is largely due to an increase in calcium influx (Jen et al. 2002).

There is strong evidence that the enhanced endothelium-dependent vasodilatation in response to exercise training is due to a change in endothelium Ca^{2+} signaling. Aortic endothelium of male Wistar rats after 10 weeks of training showed an increased response of intracellular Ca^{2+} to ACh and ATP induction. Consistent with other findings, the Ca^{2+} elevations in endothelium cells of different individuals, especially on ACh stimulation, were different. Chronic exercise seems to facilitate Ca^{2+} influx (Chu et al. 2000).

Transcriptional Level

The training-induced long-term increase in blood flow modulates the expression of NO synthase. In cultured endothelium cells exposed to long-term shear stress, the expression of mRNA for NO synthase is up-regulated (Sessa et al. 1992). The mechanism by which shear stress induces the transcriptional activation leading to an enhanced expression of NO synthase requires shear stress responsive elements (SSRE) within the promoter region of the Nos3 gene. The consensus sequence for the SSRE in NO synthase promoter is GAGACC. This transcription factor binding site may, in addition to certain other cis-acting transcriptional regulatory elements, also serve as a promoter region for a number of other endothelial genes involved in the response to shear stress. Among these are tissue plasminogen activator, intracellular adhesion molecule, TGF-$beta_1$, PDGF-beta, and ET-1 (Resnick et al. 1993; Resnick & Gimbrone 1995).

Factors of Influence

There are several factors that are important for the linkage between exercise and vascular adaptation. Those factors are discussed here.

Magnitude and Nature of Flow

In vitro experiments using cultured human umbilical vein endothelium cells suggest that the magnitude and the nature of flow as well as the shear stress may have significant influence on endothelium-derived NO production. Whereas laminar flow induces a dose-dependent increase in NO synthesis, no effect of turbulent flow on NO synthesis has been observed (Noris et al. 1995). Moreover, the salutary changes of endothelium NO release in vitro are extracted from experiments using sustained constant flow. However, blood flow caused by exercise is more intermittent. Therefore the question arises whether a long-term intermittent increase in flow could change the endothelium NO synthesis in a similar way. Different in vivo exercise studies with animal models (coronary heart arteries of dogs after exercise training, Hornig et al. 1996), as well as training studies with patients (brachial artery of patients with congestive heart failure after exercise training, Hornig et al. 1996), showed a consistent enhancement in NO-mediated vascular responsiveness.

Vessel Size

Some observations suggest that the vasodilatory reaction depends on the vessel size. Moreover the training-induced adaptations are not uniform but heterogeneous (Laughlin & McAllister 1992; Parker et al. 1994). The results are rather different, and there are conflicting data. Enhanced endothelium-dependent vasodilatation in coronary arteries after exercise training was shown in resistance arteries, but not in conduit coronary arteries. In contrast, enhanced vascular responsiveness to adenosine could be observed in conduit and small arteries but was absent in smaller resistance arteries (Oltman et al. 1992; Muller et al. 1994; Parker et al. 1994). After 7 to 10 days using the same intensive training protocol, an enhanced vasodilatation in large conduit coronary arteries of dogs was found but not in coronary resistance arterioles (Wang et al. 1993; Sessa et al. 1994). In adult miniature swine, conduit coronary arteries exhibited no change either in endothelium-dependent relaxation or in eNOS protein content after 16- to 20-week training (Oltman et al. 1995; Laughlin et al. 2001). In contrast, coronary arterioles exhibited enhanced endothelium-dependent vasodilatation, as well as increased eNOS mRNA and eNOS protein content (Muller et al. 1994; Woodman et al. 1997; Laughlin et al. 2001). The assumption that the differences between these results may be caused by species-specific differences was counteracted by contrary findings even in the same species, for example in dogs (Rogers et al. 1991).

Frequency of Exercise

Another question that must be discussed in this field is whether short-term exercise or even a single bout of exercise—or only long-term exercise such as physical training—may enhance vascular reactivity. For endothelium-mediated vasodilatation in coronary arteries, the findings of animal studies indicate that short-term as well as long-term exercise (Wang et al. 1993; Sessa et al. 1994; Koller et al. 1995) increases the mRNA expression of NO synthase, leading to augmented NO activity and thereby enhancing vascular reactivity.

Type of Exercise

Physical activity represents a broad field of different types of exercise with different characteristics.

The question that arises is whether different types of exercise may influence the vascular reactivity in different ways.

The influence of a 10-week combined aerobic and anaerobic exercise program on brachial arterial responsiveness in 25 healthy male military recruits was measured using a high-resolution ultrasonic vessel wall tracking device. In comparison to findings in nonactive controls, the endothelium-dependent vasodilatation, verified by reactive hyperemia, was enhanced. Endothelium-independent vasodilatation after a sublingual application of glyceryl trinitrate was unchanged in the trained recruits (Clarkson et al. 1999).

Consistent with these findings, four-week cycle training in healthy subjects improved the endothelium-dependent reaction in forearm resistance vessels (Kingwell et al. 1997). Four-week forearm handgrip training with a comparable group failed to confirm the enhanced vasodilatation either at rest or during exercise (Green et al. 1994; Franke et al. 1998).

Also, moderate-intensity circuit training that improves functional capacity, body composition, and strength in middle-aged subjects showed no influence on forearm resistance vessel reactivity. These findings contrast with previous studies indicating that endothelial dysfunction may be improved with comparable interventions (O'Driscoll et al. 1999; Maiorana et al. 2000a- b, 2001). A higher training level seems to be necessary when baseline vascular function is not impaired (Maiorana et al. 2001).

Relatively intense running training for three months with healthy young men decreased the endothelium-dependent forearm vasorelaxation induced by intrabrachial infusion of ACh while leaving the endothelium-nondependent relaxation (as indicated by the infusion of nitroprusside) unchanged. Besides the impairment of endothelium function, a decrease in most antioxidant concentrations except for ascorbate could be observed. There was a significant correlation between the decrease in uric acid, an antioxidant predominately localized in endothelium cells, and the degree of endothelium dysfunction. These results led to the hypothesis that changes in antioxidant defense may influence endothelium function (Bergholm et al. 1999).

Time Point of Exercise Training Adaptation

Diametrically opposed findings for coronary arteries of the same vessel size, even in the same animal model, support the hypothesis that the kinetics of the training adaptation may be involved. It is well known that training induces time-dependently progressive and different adaptations. Evidence from the available literature is consistent with the following concept (Laughlin et al. 2003). Endothelium-dependent diameter control of conduit coronary arteries is enhanced in the early (7-19 days) exercise training process. In the later training period this value returns to normal levels, because of structural adaptations leading to an increased artery diameter and thereby reducing the coronary shear stress, especially in comparison to the shear stress of the first training bouts (Haskell et al. 1993; Laughlin 1995). In a study with female Yucatan miniature swine, seven-day short-term training increased the receptor-induced dilatation in coronary conduit arteries, but not in coronary arterioles. The eNOS protein level in aortic endothelium was augmented while the expression of eNOS or SOD-1 synthase was not altered (Laughlin et al. 2003). For coronary arterial endothelium, exercise-induced adaptations are highly dependent on artery size (Laughlin et al. 2001) and the time point (one week or some months) (Wang et al. 1993; Sessa et al. 1994) at which the study was conducted during the adaptive training process.

The Effect of Detraining

The effect of eight-week aerobic exercise training (bicycle ergometer) and eight-week detraining on the plasma level of endothelium-derived factors, NO and ET-1, was examined in seven young men. After the training, the plasma concentration of NO_x increased significantly while the plasma level of ET-1 decreased significantly. The analysis was repeated after four and eight weeks of detraining. The increase of NO_x and the decrease of ET-1 level lasted through the fourth week after cessation of exercise training but returned to the pre-exercise level after the eighth week. There was a significant negative correlation between plasma NO_x and plasma ET-1 concentration (Maeda et al. 2001).

Age

Several studies with healthy volunteers indicate that aging is associated with a progressive decline in endothelium-dependent dilatation. Using ultrasound techniques, the reactive hyperemia in the brachial artery, indicating the vasodilatator response, in male subjects younger than 40 years was 70% higher than in older ones. Such differences were less pronounced when female subjects were examined (Wang et al. 1993). However, another study demonstrated a progressive impairment in endothelium function in the aging process for both genders. An explanation for the contradictory findings might be that the process occurs in men and women at different ages. While a remarkable decline commences for men at the age of about 40 years, for women the decline may start one decade later (Celermajer et al. 1994). Similar findings have been reported with other methods for inducing vasodilatation such as receptor stimulation with ACh in the brachial artery (Taddei et al. 1995) or in coronary arteries (Chauhan et al. 1996).

The influence of endurance training in older men on endothelium-dependent and -independent vasorelaxation was investigated in a cross-sectional study (Rywik et al. 1999). Twelve endurance athletes were compared to sedentary controls; all subjects were between 61 and 83 years of age. Vasoreactivity in the brachial artery of older endurance-trained men was enhanced, showing about 50% higher ischemia-induced endothelium-dependent dilatation in comparison to the value in sedentary controls. For endothelium-independent relaxation after sublingual application of nitroglycerin, there was only a slight trend toward a higher vasodilator response. Multiple linear regression analysis showed that aerobic capacity, as measured by $\dot{V}O_2max$, was an independent predictor for the endothelium-dependent as well as endothelium-independent vasodilatation (Rywik et al. 1999). Another study comparing 39- to 66-year-old distance runners to sedentary controls showed a 120% enhanced endothelium-independent vasodilatation in coronary arteries (Haskell et al. 1994).

In a cross-sectional study that included endurance-trained and sedentary healthy men (aged 22-35 or 50-76 years) using strain gauge plethysmography, forearm blood flow response to intra-arterial infusion of ACh and sodium nitroprusside was investigated, thereby quantifying the endothelium-dependent and endothelium-independent vasorelaxation. There was no age-related decline in endothelium-dependent vasodilatation in endurance-trained men in comparison with their sedentary peers. Middle-aged and older men regularly practicing aerobic exercise showed a more pronounced ACh-mediated vasodilatation than their sedentary peers. Regular aerobic training can restore the loss of endothelium-dependent vasodilatation in previously sedentary middle-aged and older men. There were no differences in flow to sodium nitroprusside between the groups in relation to either age or training status (DeSouza et al. 2000).

Another study, using reactive hyperemia and sublingual nitroglycerin in order to distinguish between endothelium-dependent and nondependent relaxation in long-term endurance-trained older men (68.5 ± 2.3 years old) and healthy sedentary peers at the same age (64.7 ± 1.4 years), confirmed the results (Rinder et al. 2000). Longitudinal enhancement of non-endothelium-dependent vascular reactivity was demonstrated in a prior study in men and women aged 64 ± 3 years who performed aerobic exercises regularly for more than six months (Martin et al. 1990). Together these data suggest that training can restore the age-related loss of endothelium-dependent vasorelaxation, but the data for the endothelium-independent relaxation are inconsistent.

Besides endothelium-dependent vasoreactivity, the capacity of the endothelium to release t-PA antigen also declines significantly in the aging process in healthy sedentary men. In men regularly performing endurance exercise, endothelial t-PA antigen release is well preserved with age in contrast to sedentary peers. Even for previously sedentary older men, a rather brief period (13 weeks) of aerobic exercise training can reverse the age-related decline in capacity of the endothelium to release t-PA antigen (Smith et al. 2003).

Conclusion

Finally, it is important to keep in mind that all the findings presented in this chapter have many limitations. Most conclusions drawn from these studies are based on either in vitro or animal models, with only a few derived from human studies. Besides a broad variation in investigation techniques, the endothelium examined was derived from different vessels, which differ in size and function. Most studies focus on the coronary artery, fewer on the aorta or on extremity vessels. Exercise training programs differ in type, intensity, and duration. Most studies have been cross-sectional, and there is a lack of longitudinal studies, especially with humans. Although data are conflicting, there is an overall trend toward an improvement of endothelium function caused by exercise. Based on the presumption that endothelium dysfunction is systemic in nature, one might assume that regular, especially aerobic, exercise produces favorable global endothelial adaptations. To complete the inchoate puzzle of existing data, further investigation is necessary.

Chapter 13

Activity-Dependent Adaptive Responses of Skeletal Muscle Fibers

Dirk Pette

Skeletal muscle is a complex tissue composed of a large variety of functionally diverse fiber types. In addition, postmitotic, fully differentiated muscle fibers represent highly versatile entities capable of responding to altered functional demands and a variety of exogenous and endogenous signals by changing their metabolic and phenotypic profiles. This adaptive responsiveness, also termed muscle plasticity (Pette 2001), encompasses reversible fiber type transitions. It is based on the existence of multiple isoforms of most sarcomeric proteins and their expression in various combinatorial patterns. This chapter is an attempt to explain adaptive changes in muscle fiber properties as the result of qualitative and quantitative alterations in the expression levels of metabolic enzymes and myofibrillar and other sarcomeric proteins. Special attention is given to myosin and its isoforms as key markers of fiber types and fiber type transitions. For additional information, the reader is referred to previous reviews (Bottinelli & Reggiani 2000; Pette & Staron 1990, 1997; Schiaffino & Reggiani 1996).

The Multiplicity of Sarcomeric Protein Isoforms

Isoforms of a given sarcomeric protein have similar but not identical properties. Protein isoforms thus represent elements of functional diversity. In many but not all cases, their expression patterns are fiber type specific. The number of isoforms expressed varies between different sarcomeric proteins (for reviews see Pette & Staron 1990; Schiaffino & Reggiani 1996). In adult skeletal muscles, it may be restricted to fast and slow isoforms, as in the case of some proteins involved in excitation–contraction (EC) coupling or Ca^{2+} sequestration, such as the voltage-sensing α_1 subunit of the dihydropyridine receptor (DHPR) or the sarcoplasmic reticulum Ca^{2+}-ATPase (SERCA).

Most proteins of the thin and thick filaments are composed of two or more subunits, each existing as various isoforms (table 13.1). For example, the dimeric tropomyosin (TM) molecule may exist in up to six different species composed of homo- or heterodimers of three different subunit isoforms, that is, $TM\alpha_f$, $TM\alpha_s$, and $TM\beta$. Troponin (Tn), the major regulatory protein of the thin filament, is composed of three subunits, each existing as two or more isoforms. The inhibitory subunit TnI and the Ca^{2+} binding subunit TnC exist as fast and slow isoforms (TnI_f, TnI_s and TnC_f, TnC_s), whereas the tropomyosin binding subunit TnT exists as at least four fast and two slow isoforms (TnT_{1f}, TnT_{2f}, TnT_{3f}, TnT_{4f} and TnT_{1s}, TnT_{2s}). Four different fast troponin molecules may thus be formed by the assembly of one of the four different fast TnT subunits with TnI_f and TnC_f. Similarly, the assembly of TnI_s and TnC_s with either one of the two slow TnT isoforms generates two different slow troponin molecules.

Table 13.1 Major Myofibrillar Protein Isoforms in Slow (Type I) and Fast (Type II) Fibers of Limb Muscles in Small Mammals

Filament	Slow	Fast
Thin filament		
Actin	α-skeletal	α-skeletal
Tropomyosin	$TM\alpha_s < TM\beta$	$TM\alpha_f > TM\beta$
Troponin		
Troponin T	TnT_{1s}, TnT_{2s}	TnT_{1f}, TnT_{2f}, TnT_{3f}, TnT_{4f}
Troponin I	TnI_s	TnI_f
Troponin C	TnC_s	TnC_f
Thick filament		
Myosin heavy chains (MHC)	MHCI	MHCIIa, MHCIId(x), MHCIIb
Myosin light chains		
Essential (ELC)	ELC1sa, ELC1sb	ELC1f, ELC3f
Regulatory (RLC)	RLC2s	RLC2f

Myosin, a hexameric protein, is composed of two heavy chains (MHC), two essential (ELC), and two regulatory light chains (RLC). Essential and regulatory light chains exist as several fast and slow isoforms, whereas at least 10 different MHC isoforms have been identified in extrafusal fibers (Pette & Staron 2000; Schiaffino & Reggiani 1996; Weiss et al. 1999) (table 13.1). Some MHC isoforms are expressed in a muscle-specific manner (e.g., extraocular, laryngeal, masticatory muscles), while others, namely the slow MHCI and the three fast MHCIIa, MHCIId(x), and MHCIIb isoforms, are widely distributed in a variety of fully differentiated skeletal muscles.

Functional myosin molecules are formed by the assembly of light and heavy chains according to the formula $[(ELC)_2(RLC)_2(MHC)_2]$. The RLC complement of native fast (FM) or slow myosins (SM) consists of two identical fast or slow isoforms. The ELC complement, however, exists in three possible combinations, namely in fast myosins as ELC1f homodimer, ELC1f/ELC3f heterodimer, and H ELC3f homodimer, and in slow myosins as ELC1sa homodimer, ELC1sa/ELC1sb heterodimer, and ELC1sb homodimer. Combinations of the three ELC couples with the RLC homodimer and an MHC homodimer thus generate three isomyosins for each MHC isoform: three MHCI-based slow isomyosins and three fast isomyosins each for MHCIIa, MHCIId(x), and MHCIIb. An even greater number of different isomyosins results from combinations of MHCIIa and MHCI with both fast and slow light chain isoforms (Pette & Staron 1990).

Myofibrillar Protein Isoforms and Fiber Diversity

MHC isoforms represent the best markers for delineating specific fiber types in adult muscles (Pette & Staron 1990, 2000; Schiaffino & Reggiani 1996). Immunohistochemistry and single-fiber biochemistry have delineated fibers that contain either a single MHC isoform (pure fiber types) or two or more MHC isoforms (hybrid fiber types). Four pure fiber types exist in adult skeletal muscles of small mammals: slow type I fibers with MHCI and three fast fiber (type II) subtypes—namely type IIA with MHCIIa, type IID/X with MHCIId(x), and type IIB with MHCIIb. The slow MHCI is thought to be identical with the β-cardiac MHC isoform (Lompré et al. 1984), although this has been questioned (Galler et al. 2002). MHCIId and fiber type IID are considered equivalent to MHCIIx and fiber type IIX, respectively (LaFramboise et al. 1990; Vikstrom et al. 1997; Weiss & Leinwand 1996). The two nomenclatures coexist, but the designation MHCIId(x) and type IID(X) fiber is also used, as in this chapter. Furthermore, it should be noticed that fibers previously classified as type IIB in human muscle have been renamed type IID(X) according to their MHC complement, which resembles the MHCIId(x) isoform of rat (Ennion et al. 1995; Smerdu et al. 1994).

Hybrid fibers with coexisting pairs of different MHC isoforms can be classified according to their predominant MHC isoform: type I/IIA (MHCI > MHCIIa); type IIA/I (MHCIIa > MHCI); type IIA/D (MHCIIa > MHCIId); type IID/A (MHCIId > MHCIIa); type IID/B (MHCIId > MHCIIb), and type IIB/D (MHCIIb > MHCIId). Fiber types I/IIA and IIA/I correspond to types IC and IIC, a nomenclature often used when fibers are classified by myofibrillar adenosinetriphosphatase (mATPase) histochemistry (Pette & Staron 1990).

MHC-based fiber type and maximum unloaded shortening velocity (V_{max}) are correlated (Bottinelli et al. 1991; Galler et al. 1994; Larsson & Moss 1993; Pellegrino et al. 2003). Velocity is lowest in type I fibers and highest in type IIB fibers. Types IID(X) and IIA display V_{max} values similar to each other, but lower than for type IIB (figure 13.1). Overlapping V_{max} values of the pure fiber types most likely result from specific MLC combinations (Bottinelli et al. 1994a) that have been shown to modulate the V_{max} properties determined by the MHC complement (Lowey et al. 1993).

Stretch activation characteristics of single fibers also correlate with the MHC complement and represent another parameter for delineating functionally diverse fiber types. On this basis, pure and hybrid fibers have been separated without overlaps in a spectrum extending in small mammals between type I, as the slowest, and type IIB, as the fastest (Galler et al. 1994). MHCIIb is not expressed at the protein level in the human (Horton et al. 2001). According to measurements of V_{max} (Larsson & Moss 1993) and stretch activation kinetics (Hilber et al. 1999) (figure 13.2), type IID(X) fibers are the fastest in human skeletal muscles.

The differences in contractile velocities of the various fiber types correspond to differences in their myofibrillar ATPase activities and tension costs (figure 13.3). Type IIB fibers display the highest ATPase activity and tension cost, type I fibers the lowest, and types IID(X) and IIA intermediate values (Bottinelli et al. 1994b). Likewise, single-fiber measurements of the $[ATP]/[ADP_{free}]$ ratio, a measure of the cellular energy potential, demonstrate a similar gradient between type IIB and type I fibers (Conjard et al. 1998) (figure 13.4).

In addition to qualitative differences in their isoform patterns, fiber types exhibit quantitative differences in protein levels. Such differences may be either graded or all-or-none. For example, the fast Ca^{2+}-ATPase isoform, SERCA1, is expressed at

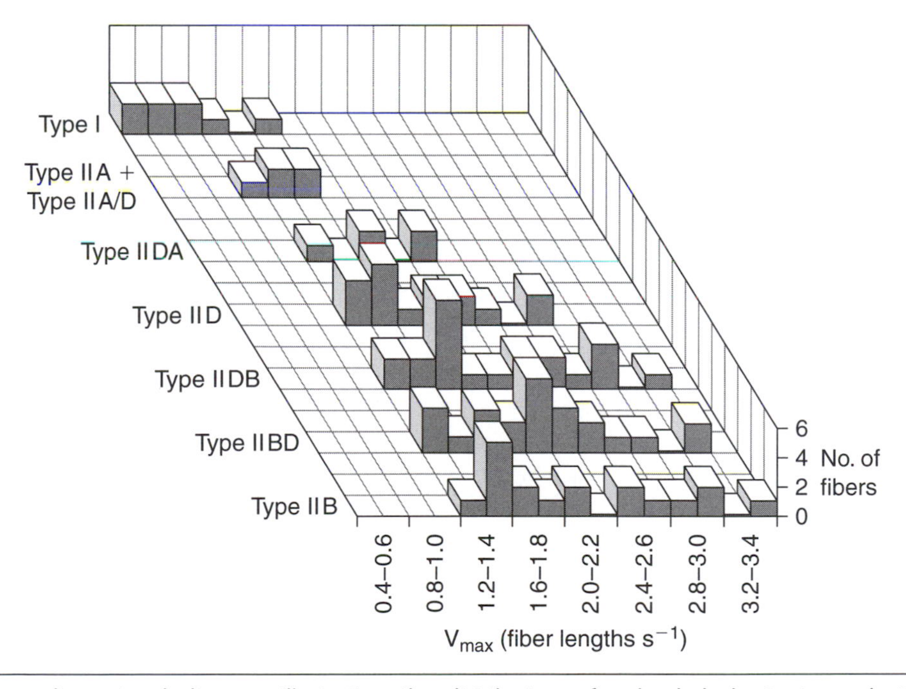

Figure 13.1 Three-dimensional diagram illustrating the distribution of unloaded shortening velocity (V_{max}) values (expressed as fiber lengths^{-1}) among different pure and hybrid myosin heavy chain (MHC)-based fiber types from rat skeletal muscle. Data are means ± SD. The large overlaps between the different fiber types most likely result from various ELC combinations.

Reprinted, by permission, from S. Galler, T.L. Schmitt, D. Pette, 1994, "Stretch activation, unloaded shortening velocity, and myosin heavy chain isoforms of rat skeletal muscle fibres," *The Journal of Physiology* 478: 513-521.

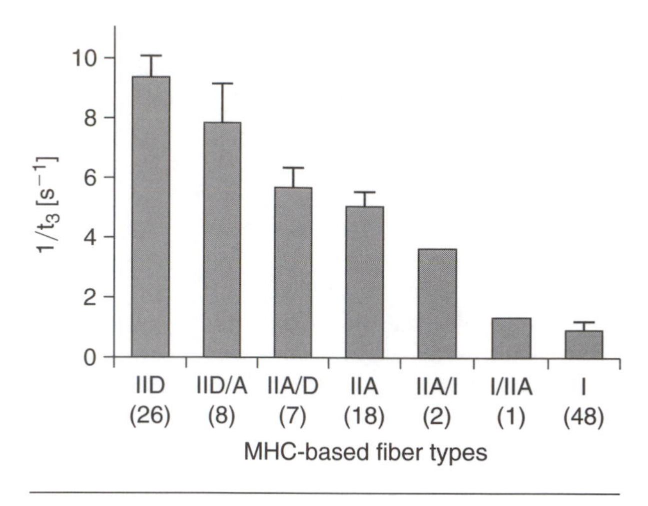

Figure 13.2 Stretch activation properties of pure and hybrid human muscle fibers. Stretch activation was measured on maximally Ca^{2+}-activated skinned fibers by determining t_3, that is, the time elapsed between the beginning of the stretch and the peak value of the delayed force increase. Bars represent reciprocal t_3 values (means ± SD) of different fiber types classified according to their electrophoretically determined myosin heavy chain (MHC) complement. Numbers of the different fiber types analyzed are given in parentheses. Note that pure and hybrid fibers are clearly separated without overlaps.

Adapted from K. Hilber, S. Galler, B. Gohlsch, and D. Pette, 1999, "Kinetic properties of myosin heavy chain isoforms in single fibers of human skeletal muscle" *FEBS Letters* 455: 267-270.

higher levels in type IIB and IID(X) fibers than in type IIA fibers (Krenács et al. 1989; Leberer & Pette 1986). Phospholamban, a regulatory protein of the sarcoplasmic reticulum Ca^{2+}-ATPase, is expressed in type I but not in type II fibers (Jorgensen & Jones 1986). Conversely, parvalbumin, a cytosolic Ca^{2+}-buffering protein, is present at high levels in type IIB and IID(X) fibers, hardly detectable in type IIA, and undetectable in type I fibers (Schmitt & Pette 1991). Parvalbumin plays an essential role in Ca^{2+} sequestration of skeletal muscles with high relaxation speeds, such as fast-twitch muscles of small mammals, where its concentrations are highest. Parvalbumin concentrations are low in muscles with relatively low relaxation speeds, such as fast muscles of larger mammals and the human (Heizmann et al. 1982).

Large variations also exist in enzyme levels of anaerobic and aerobic metabolic pathways. Fibers with low mitochondrial content base their energy supply mainly on glycogen breakdown and glycolysis and thus have been termed "glycolytic," whereas "oxidative" fibers characterized by high mitochondrial contents base their energy metabolism preferentially on aerobic-oxidative

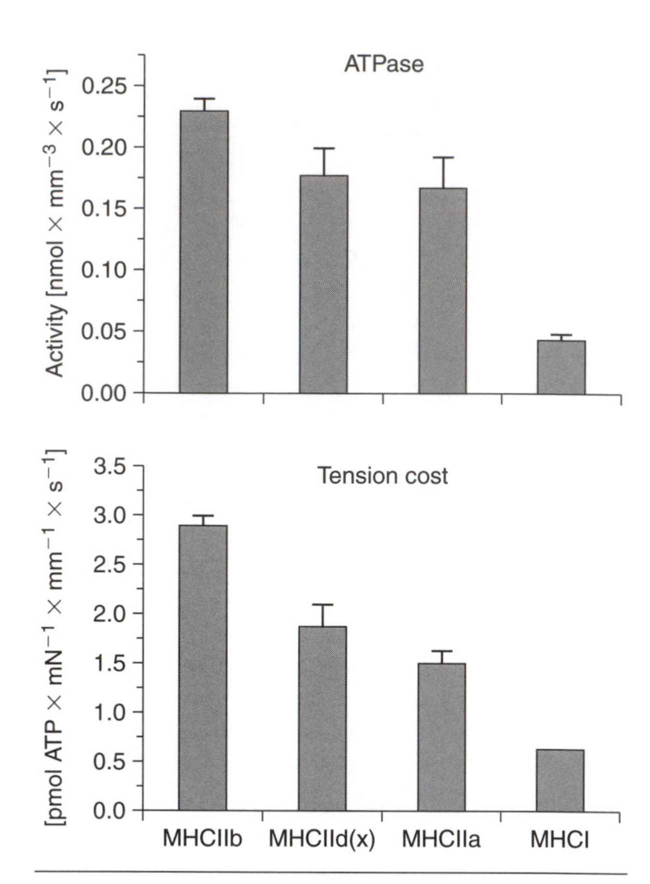

Figure 13.3 Myofibrillar ATPase activity and tension cost, that is, the ratio between ATPase activity and isometric tension of single myosin heavy chain (MHC)-based rat muscle fiber types. Plots were drawn according to the data (means ± SEM) from Bottinelli et al. 1994b.

Reprinted, by permission, from R. Bottinelli et al., 1994b, "Myofibrillar ATPase activity during isometric contraction and isomyosin composition in rat single skinned muscle fibres" *The Journal of Physiology* 481: 663-675.

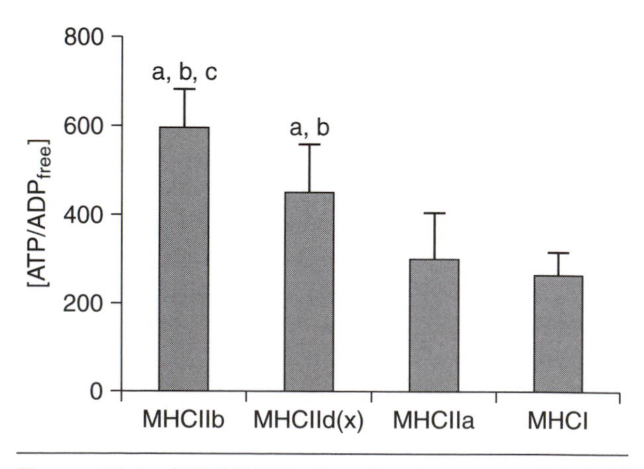

Figure 13.4 [ATP]/[ADP_{free}] ratios in pure and hybrid fiber types from normal rabbit muscles. Values are means ± SD. [a]Different from myosin heavy chain (MHC)I; [b]different from MHCIIa; [c]different from MHCIId(x).

Adapted, by permission, from A. Conjard et al., 1998, "Energy state and myosin chain isoforms in singe fibres of normal and transforming rabbit muscles" *Pflügers Archiv* 436:962-969.

pathways. Combining histochemically assessed metabolic enzyme activities with myofibrillar ATPase histochemistry and contractile properties, Peter and coworkers (1972) distinguished slow-twitch oxidative (SO), fast-twitch glycolytic oxidative (FOG), and fast-twitch glycolytic (FG) fiber types. The SO fibers contain MHCI and therefore correspond to type I fibers. FG and FOG fibers represent metabolically distinct fast fiber subtypes without stringent reference to the MHC-based fast subtypes IIA, IID(X), and IIB (Pette & Staron 1990).

Metabolic Adaptations of Muscle Fibers to Altered Functional Demands

Muscle fibers are dynamic structures capable of changing their phenotype and functional properties in response to altered functional demands. Persistent changes in neuromuscular activity or mechanical load, or both, evoke a multitude of adaptive responses at the levels of molecular and cellular organization. The range of the induced responses depends on the type, intensity, and duration of the change in function. However, as discussed here, it also depends on the fiber type. In general, moderate changes in neuromuscular activity remain restricted to metabolic adaptations, while greater and sustained changes in neuromuscular activity lead to fiber type transitions in addition.

The energy supply for the chemomechanical transformation underlying the contractile process varies considerably depending on the recruitment and performance of different motor units. Energy supply for short-term, high-intensity performance is based on phosphocreatine breakdown, glycogenolysis, and glycolysis, especially of type II fibers, while aerobic carbohydrate and fatty acid oxidation predominate in muscle fibers destined for sustained activity, such as type I fibers with their anti-gravitational function. Persistently enhanced neuromuscular activity, therefore, leads to metabolic adaptations toward the use of fuels and metabolic pathways with higher efficiency in energy-rich phosphate supply. Because of their intrinsically high aerobic-oxidative potential, type I fibers are less involved in this type of metabolic adaptation than the more glycolytically oriented type II fibers, especially types IIB and IID(X). Reduced neuromuscular activity induces changes in the opposite direction, that is, a shift toward a more glycolytically based energy metabolism. Also under these conditions, fast fibers seem to be more involved in metabolic remodeling than slow fibers. Microgravity is an exception, however, as it releases slow fibers from their anti-gravitational function.

Exercise-induced increases in the degree of coupling between mitochondrial oxidation and phosphorylation, as well as shifts in fuel utilization, for example reduced glycogenolysis and lactate formation, are early adaptive events of prolonged exercise, preceding increases in mitochondrial content and mitochondrial enzyme activities (Green et al. 1992b; Phillips et al. 1996; Zoll et al. 2002). Enhanced contractile activity has an immediate stimulatory effect upon glucose uptake of the working muscle by enhanced incorporation of GLUT4, the muscle-specific glucose transporter, into the sarcolemmal membrane and its transverse tubules (Etgen et al. 1993; Roy et al. 1997). Further increases in glucose uptake capacity result from enhanced transcription and synthesis of GLUT4 (Dohm 2002). Also, increases in the catalytic activity of hexokinase II, the muscle-specific isozyme, contribute to enhanced glucose utilization. The gain in catalytic activity results from reversible binding of the enzyme to the external mitochondrial membrane (Parra et al. 1997; van Houten et al. 1992). In addition, exercise has a stimulatory effect on hexokinase II transcription and translation (Hofmann & Pette 1994; Odoherty et al. 1996). Sustained contractile activity enhances fatty acid uptake by elevated levels of sarcolemmal fatty acid translocase. Increases in translocase activity result from its intracellular redistribution and its enhanced synthesis (Bonen et al. 1999, 2001). Some of the immediate organismic responses to exercise have been attributed to cytokine release from working muscle, such as interleukin-6, which, by its effect on liver and adipose tissue, contributes to the maintenance of glucose homeostasis and mediates exercise-induced lipolysis (Pedersen et al. 2001).

In addition to these early metabolic responses, skeletal muscle reacts to persistently elevated contractile activity with long-term metabolic adaptations. The pioneering studies of Holloszy (1967) and Gollnick (1969) and coworkers have shown that the enzyme activity profiles of energy metabolism in muscle depend on its training state. According to these and numerous other studies, endurance training leads to increases in mitochondrial content, in parallel with elevations in enzyme levels of aerobic-oxidative energy metabolism, such as fatty acid and amino acid oxidation, ketone body utilization, the citric acid cycle, and respiratory chain (Booth & Baldwin 1996; Henriksson 1992; Holloszy & Coyle 1984; Saltin & Gollnick 1983; Turner et al. 1997; Williams & Neufer 1996). In other words, skeletal muscle responds to persistently enhanced contractile activity by switching its energy supply to fuels and metabolic pathways with maximum ATP yield. The time courses of these changes at the protein level vary between days and weeks. They are preceded by changes in the tissue levels of specific mRNAs. A single bout of exhaustive exercise is sufficient to induce in human muscle severalfold increases in transcriptional activities of specific mitochondrial enzymes (Barash et al. 2004; Pilegaard et al. 2000). These increases are transient, and therefore, cumulative effects of repetitive increases in transcriptional activities seem to be necessary for persistent elevations also of translation, ultimately resulting in substantial increases in mitochondrial enzyme levels.

Fiber Type Transitions

Persistently altered neuromuscular activity or mechanical load (or both) evokes changes in gene expression that, due to qualitatively and quantitatively altered isoform profiles of myofibrillar and other sarcomeric proteins, result in fiber type transitions. Experimental protocols and major conditions for establishing long-term changes in neuromuscular activity are summarized schematically in figure 13.5. In general, enhanced neuromuscular activity leads to shifts in the fiber population toward slower and metabolically more oxidative phenotypes, whereas fiber type transitions in the opposite direction result from long-lasting decreases in neuromuscular activity (Baldwin & Haddad 2001; Loughna et al. 1990; Pette & Staron 1997; Talmadge 2000).

Increased neuromuscular activity/overloading
(e.g., endurance exercise, low-frequency stimulation, stretch)

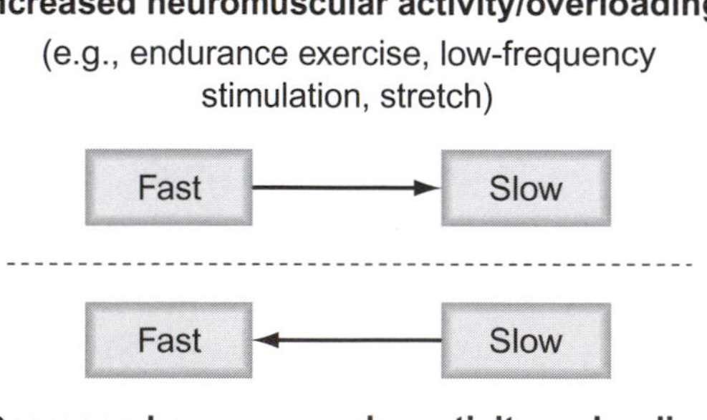

Decreased neuromuscular activity, unloading
(e.g., detraining, immobilization, microgravity)

Figure 13.5 Schematic representation of reversible phenotype transitions in response to altered neuromuscular activity.

Fast-to-Slow Transitions

Endurance training induces fiber type transitions with shifts toward slower phenotypes, such as moderate increases in type IIA and eventually also type I fibers. These transitions are accompanied by increases in hybrid fibers co-expressing MHCIIa and MHCI (Green et al. 1984; Howald et al. 1985; Jansson & Kaijser 1977; Luginbuhl & Dudley 1984; Schantz et al. 1982; Schantz & Henriksson 1983; Simoneau et al. 1985).

Fast-to-slow transitions are most efficiently induced by chronic electrostimulation of fast-twitch muscles with tonic, low-frequency (10-20 Hz) impulse patterns (Salmons & Vrbová 1969). Chronic low-frequency stimulation (CLFS) offers several advantages as compared to exercise training (Pette & Vrbová 1999): It is a standardized and reproducible regimen that activates all motor units of the stimulated muscle and imposes higher levels of activity over time than any other exercise regimen. CLFS, therefore, challenges the adaptive potential of the target muscle to its maximal limits. Because high levels of activity can be applied with stimulation onset, time-dependent changes in molecular, structural, and functional properties can be followed from the beginning. Finally, CLFS offers the possibility to investigate dose-response relationships between increases in neuromuscular activity and adaptive responses of the various cellular elements.

It is not surprising, therefore, that CLFS has provided the most extensive insights into the molecular, metabolic, and structural changes related to activity-induced fiber type transitions. The CLFS-induced changes in fiber type composition, sarcomeric protein isoforms, and metabolic properties exceed by far the changes induced by endurance exercise. The greater efficiency of CLFS in triggering fiber type transitions is due to the fact that, contrary to exercise training, forced contractile activity by electrical stimulation can be performed for up to 24 h per day.

When exposed to CLFS, fast-twitch, fatigable muscles turn into slower-contracting, less fatigable muscles (for reviews see Ohlendieck 2000; Pette & Vrbová 1992, 1999; Salmons 1994). The changes in contractile properties (e.g., reduced speed of contraction, increases in time-to-peak of isometric twitch contraction and half-relaxation time) result from fast-to-slow transitions in the isoform patterns of myofibrillar and Ca^{2+}-regulatory proteins, including down-regulation of parvalbumin and up-regulation of phospholamban (Ausoni et al. 1990; Gundersen et al. 1988; Leeuw & Pette 1993; Ohlendieck et al. 1999; Pette & Vrbová 1992).

The improvement in fatigue resistance primarily results from metabolic adaptations, such as increases in mitochondrial content (Eisenberg & Salmons 1981; Reichmann et al. 1985), myoglobin (Michel et al. 1994), and capillarization (Egginton & Hudlická 1999; Hudlická et al. 1977; Mathieu-Costello et al. 1996; Škorjanc et al. 1998). The changes in the enzyme activity profile encompass elevations in mitochondrial creatine kinase and mitochondrial enzymes involved in aerobic-oxidative metabolism (transport, activation, and oxidation of fatty acids; ketone utilization; amino acid oxidation; the citric acid cycle; the respiratory chain) concomitant with decreases in enzyme activities of glycogenolysis, glycolysis, and the glycerolphosphate cycle (Pette & Vrbová 1992).

Up to sixfold increases in mitochondrial content and enzyme activities of terminal substrate oxidation have been observed in rabbit fast-twitch muscle exposed to CLFS for 28 days. The mitochondrial volume density of the stimulated muscle increased from approximately 3% to 20%, which is far beyond the mitochondrial content of the slow-twitch soleus muscle (~8%) (Hood 2001; Hood & Pette 1989; Reichmann et al. 1985). On the other hand, CLFS of the slow-oxidative soleus muscle does not lead to significant elevations in mitochondrial enzyme activities (Pette et al. 1975). These different responses indicate that the range of metabolic adaptations to enhanced activity is determined by the "actual" metabolic profile of the muscle. In other words, a muscle that, according to its enzyme profile, preferentially generates energy-rich phosphate from anaerobic glycogen breakdown and glycolysis responds to persistently enhanced contractile activity by elevating its aerobic-oxidative potential. Conversely, a muscle rich in mitochondria is metabolically prepared to meet the energetic requirements for sustained activity and, therefore, may not need to increase its aerobic-oxidative potential.

CLFS-induced transitions in myofibrillar protein isoform expression follow specific time courses. Figure 13.6 illustrates time-dependent fast-to-slow transitions of MHC isoforms in low frequency-stimulated tibialis anterior muscle of the rabbit (Leeuw & Pette 1993). The relative concentration of MHCIId(x), the fastest and predominant isoform in this muscle, decreases in parallel with an increase in MHCIIa. With prolonged stimulation (>35 days), MHCIIa is partially replaced by MHCI. Changes in MHC isoform composition of low frequency-stimulated rabbit muscle have also been studied at the mRNA level (Brownson et al. 1992; Kirschbaum et al. 1990a). Thus, the changes in the amounts of the MHC mRNA isoforms precede the corresponding changes at the protein level. A time course study on rat extensor digitorum longus muscle revealed decreases in MHCIIb mRNA already one day after the onset of CLFS. The decay of the MHCIIb mRNA followed an exponential time course with an apparent half-life of ~60 h (Jaschinski et al. 1998). A rapid up-regulation of MHCIIb mRNA occurred during recovery when CLFS was interrupted. Significant increases in the amount of MHCIIb mRNA, which was reduced to ~4% of its normal level in 15-day-stimulated rat tibialis anterior muscle, were detected already after 21 h of recovery (Kirschbaum et al. 1990b).

The stimulation-induced transitions in MHC expression increase the fraction of hybrid fibers (Conjard et al. 1998; Peuker et al. 1998; Staron et al. 1987). As also shown in studies using other conditions or experimental protocols, coexistence of different MHC isoforms is a characteristic of transforming muscle fibers (Andersen et al. 1996, 1999; Caiozzo et al. 1998; Klitgaard et al. 1990; Li & Larsson 1997; Staron & Hikida 1992). Changes in MHC isoform expression in transforming fibers are accompanied by changes in the pattern of myosin

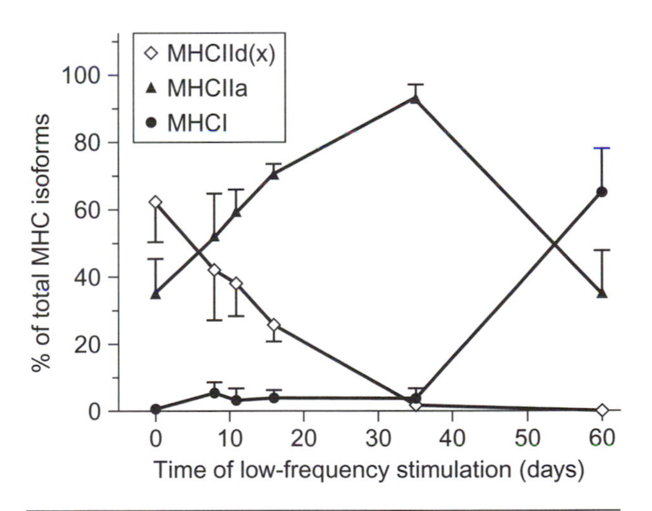

Figure 13.6 Sequential changes in the myosin heavy chain (MHC) isoform profile of low frequency-stimulated tibialis anterior muscle of the rabbit. Values are means ± SD.

Reprinted, by permission, from T. Leeuw and D. Pette 1993, "Coordinate changes in the expression of troponin subunit and myosin heavy-chain isoforms during fast-to-slow transition of low-frequency-stimulated rabbit muscle" *European Journal of Biochemistry* 213: 1039-1046.

light chain (MLC) isoforms. Additional isomyosins, therefore, may be formed by combinations of fast MLC with slow MHC isoforms and slow MLC with fast MHC isoforms (Hämäläinen & Pette 1997).

Other experimental models, such as mechanical overload, also induce fast-to-slow fiber type transitions. For example, stretch overload of rabbit tibialis anterior muscle by immobilization in a lengthened position causes fast-to-slow transitions in MHC isoform expression (Loughna et al. 1990). According to mRNA analyses, the reprogramming of the MHC isoform profile proceeds within a few days (Goldspink et al. 1992). An increase in slow fibers was shown in another study, but in comparison with the changes at the mRNA level, fiber type transitions extended over several weeks (Pattullo et al. 1992).

Slow-to-Fast Transitions

Reduced neuromuscular activity or mechanical unloading or both, through reduction in weight-bearing activity, immobilization of the muscle in a shortened position, or exposure to microgravity, cause muscle atrophy and slow-to-fast fiber type transitions (Baldwin & Haddad 2001; Caiozzo et al. 1996; Edgerton & Roy 1996; Fitts et al. 2000; Loughna et al. 1990; Pette & Staron 1997; Staron et al. 1998; Talmadge 2000). Muscle atrophy mainly results from reduced protein synthesis. In addition to reduced protein synthesis, enhanced proteolysis via the proteasome-ubiquitin pathway seems to be the major mechanism implicated in muscle atrophy (Jagoe & Goldberg 2001). Under microgravity conditions, myofibrillar proteins are lost at a faster rate than other cellular proteins, and the thin filaments are lost disproportionately to the thick filament (Fitts et al. 2000).

The changes in muscle fiber phenotypes resulting from inactivity or mechanical unloading correspond to slow-to-fast transitions in the isoform patterns of thick- and thin-filament and Ca^{2+}-regulatory proteins. Alterations in protein composition are preceded by corresponding changes at the mRNA level. For example, mRNA encoding MHCIIb normally not expressed in rabbit soleus muscle is up-regulated after only one day if the muscle is unloaded by immobilization in a shortened position (Loughna et al. 1990). The induction of all fast MHC mRNA isoforms was observed in rat soleus muscle after only four days of unweighting by hindlimb suspension (Stevens et al. 1999a).

Also slow-to-fast transitions seem to follow a regular and sequential order. Unweighting of the slow-twitch rat soleus muscle results in coordinated slow-to-fast transitions in MHC and troponin subunit isoforms (Stevens et al. 1999b, 2002). MHCI, the predominant isoform in normal rat soleus muscle, initially decreases in parallel with a moderate up-regulation of MHCIIa. These changes are followed by elevations both in MHCIId(x) and in MHCIIb. However, the effects of unweighting are small in comparison with the fast-to-slow transitions in low frequency-stimulated fast-twitch muscles (Stevens et al. 1999a,b). Interestingly, a similar sequence of slow-to-fast MHC isoform transitions was observed on leg muscles of paraplegic patients (Burnham et al. 1997). After a progressive decrease in MHCI-containing fibers, there was an increase in the number of fibers co-expressing slow and fast MHC isoforms. The hybrid fibers were eventually replaced by fibers almost exclusively expressing fast myosin.

The question whether or not slow-to-fast fiber type transitions can also be induced by specific regimens of exercise training is still open. The literature on this point is restricted to a few studies on the human. A moderate shift from histochemically identified type I to type IIA fibers was observed in the vastus lateralis muscle of young adults after four to six weeks of high-intensity sprint training (Jansson et al. 1990). Similar observations were made in a study using single-fiber electrophoresis for MHC-based fiber classification. Sprint training has been reported to increase the fraction of fibers expressing MHCIIa at the expense of type I fibers (Andersen et al. 1994).

Do Fiber Type Transitions Occur in a Sequential Order?

A large body of evidence suggests that fiber type transitions generally occur in a sequential order. Immunohistochemical studies and biochemical analyses of single fibers support the notion that fast-to-slow transitions in MHC isoform expression occur in the following order (Pette & Staron 1997):

$$\text{MHCIIb} \rightarrow \text{MHCIId(x)} \rightarrow \text{MHCIIa} \rightarrow \text{MHCI}$$

An inverse order has been deduced from the temporal changes in MHC isoform expression during slow-to-fast transitions:

$$\text{MHCI} \rightarrow \text{MHCIIa} \rightarrow \text{MHCIId(x)} \rightarrow \text{MHCIIb}$$

The notion of sequential transitions in both directions (figure 13.7) assumes that fiber type conversions occur according to a "nearest-neighbor rule" (Pette & Staron 1997), following gradients of contractile speed (figure 13.1), stretch activation kinetics (figure 13.2), myofibrillar ATPase activity, tension cost (figure 13.3), and cellular energy potential (figure 13.4). The nearest-neighbor rule suggests that responses to altered functional demands generally occur stepwise and not in an all-or-none manner jumping from one extreme to the other. Gradual transitions in MHC expression translate into sets of isomyosins finely tuned by various MLC/MHC combinations, thus resulting in smooth and gradual fiber type conversions instead of producing abrupt changes in contractile properties. Also, metabolic adaptations to enhanced contractile activity do not occur in an all-or-none manner, but display dose-response relationships between elevations in mitochondrial content and both the intensity and duration of the imposed workload (Pette & Vrbová 1992; Sutherland et al. 1998).

The scheme of the fiber type spectrum in figure 13.7 suggests that the ranges of possible fast-to-slow or slow-to-fast transitions depend on the position of a given fiber type in this spectrum. Furthermore, it suggests that, with the exception of the fastest and slowest types, muscle fibers are capable of transforming in both directions. A similar conclusion may be drawn with regard to metabolic adaptations that seem to be predestined by the actual metabolic type of the fiber. Muscle-specific and species-specific differences in adaptive responses to altered neuromuscular activity may thus be explained by differences in fiber type composition and specific conversion ranges of the different fiber types.

Signaling Pathways Related to Fiber Type Transitions

Differential gene expression requires control by specific signals and regulatory elements. Regulatory elements involved in the expression of individual members of the MHC gene family have been identified in upstream promoter sequences (e.g., Allen et al. 2001). Differences in their responsiveness to transcription factors generated as end products of specific signaling pathways seem to underlie fiber type-specific expression patterns and their modulation by various physiological stimuli.

The signaling pathways involved in fiber type-specific gene expression are beginning to be elucidated (figure 13.8). Some, but not all, appear to be Ca^{2+} mediated due to alterations in myoplasmic Ca^{2+} concentration. Several directly or indirectly Ca^{2+}-related pathways have been identified, including the calcineurin/NFAT pathway, the Ca^{2+}/calmodulin protein kinase pathway, a protein kinase C pathway, and a pathway involving the Ras/mitogen-activated protein kinase (MAPK/ERK) system.

The Ca^{2+}/Calcineurin Pathway

A key role in the up-regulation of slow fiber-specific genes has been proposed for calcineurin (CaN), a Ca^{2+}/calmodulin-dependent, serine/threonine protein phosphatase (Bassel-Duby & Olson 2003; Chin et al. 1998; Olson & Williams 2000a). Ca^{2+}-activated calcineurin dephosphorylates members of the NFAT (nuclear factors of activated T-cells)

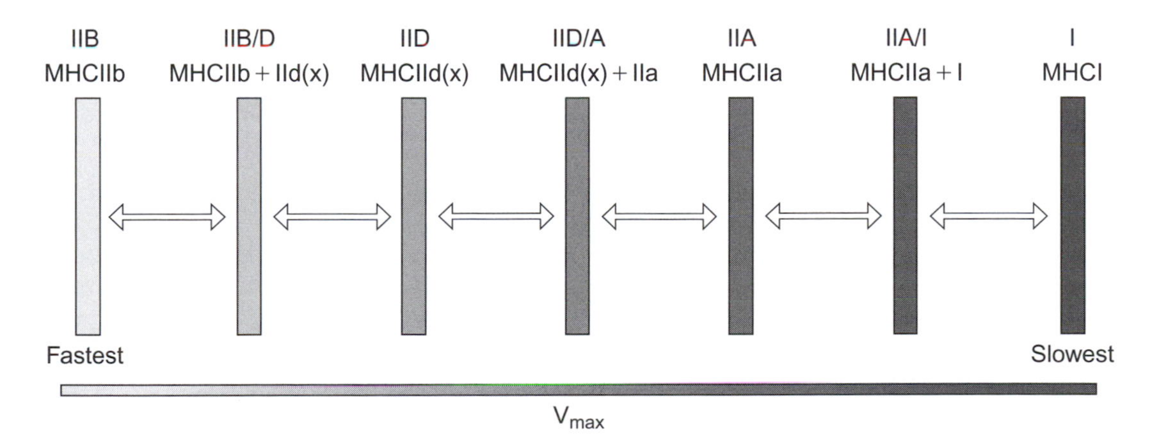

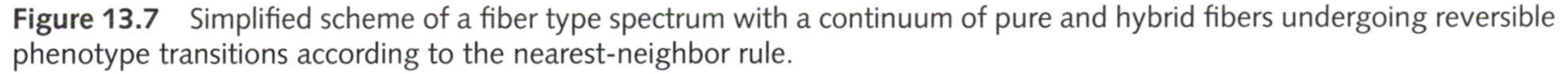

Figure 13.7 Simplified scheme of a fiber type spectrum with a continuum of pure and hybrid fibers undergoing reversible phenotype transitions according to the nearest-neighbor rule.

Reprinted, by permission, from D. Pette, 2002, "The adaptive potential of skeletal muscle fibers," *Canadian Journal of Applied Physiology* 27:423-448.

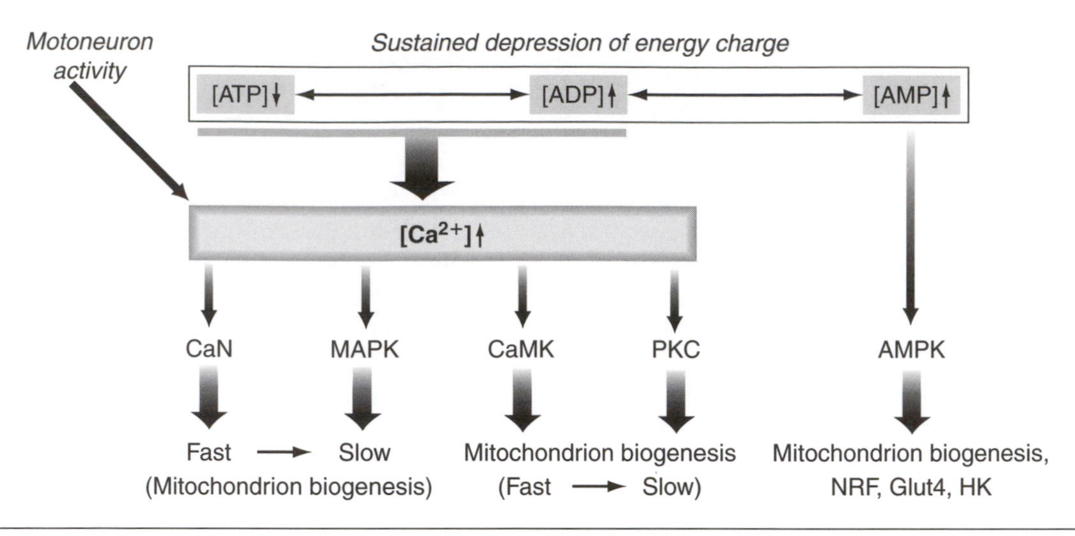

Figure 13.8 Schematic representation of major signals and signaling pathways thought to be involved in muscle fiber type differentiation/transformation and metabolic adaptation. Gray arrows indicate activation; upright or downward black arrows indicate elevated or reduced concentrations, respectively. AMPK, 5'AMP-activated protein kinase; $[Ca^{2+}]$, free myoplasmic calcium concentration; CaMK, Ca^{2+}/calmodulin-dependent protein kinase; CaN, calcineurin; MAPK, Ras/mitogen-activated protein kinase; GLUT4, glucose transporter-4; HK, hexokinase; NRF, nuclear respiratory factors; PKC, protein kinase C. For further explanations, see text.

isoform family, thereby unmasking nuclear signals that enhance intranuclear translocation (Beals et al. 2002). Intranuclear NFATs bind to specific consensus sequences of upstream promoter and enhancer elements, thus activating transcription of genes encoding fiber type-specific proteins (Chin et al. 1998; Karasseva et al. 2003). Chin and coworkers based their hypothesis on the observation that a constitutively active form of calcineurin stimulates the promoters of slow troponin I (TnI_s) and myoglobin in myogenic cell cultures (Chin et al. 1998). Similarly, transgenic mice expressing constitutively active calcineurin in their skeletal muscles exhibited increases in slow fibers in parallel with increases in TnI_s and myoglobin (Naya et al. 2000). However, the use of the slow TnI subunit and myoglobin as slow fiber-specific markers may be misleading because TnI_s and myoglobin do not exclusively exist in type I fibers, but may also be expressed in type IIA fibers. That the calcineurin pathway is not restricted to the up-regulation of slow fiber type-specific proteins, such as MHCI, but also activates the MHCIIa gene has been demonstrated by a significant Ca^{2+}/calcineurin-mediated activation of the MHCIIa promoter (Allen et al. 2001; Allen & Leinwand 2002).

The role of calcineurin in the expression of fiber type-specific genes has been substantiated by studies on adult rats treated for several weeks with cyclosporin A, an inhibitor of calcineurin. The slow-twitch soleus muscle of the treated rats displayed slow-to-fast fiber type transitions, namely decreases in both MHCI and the slow sarcoplasmic reticulum Ca^{2+}-ATPase (SERCA2) concomitant with increases both in MHCIIa and the fast Ca^{2+}-ATPase (SERCA1) (Bigard et al. 2000; Chin et al. 1998; Serrano et al. 2001). The involvement of calcineurin in the induction of slow myosin by low-frequency stimulation follows from studies on regenerating denervated rat soleus muscle (Serrano et al. 2001). In the absence of its nerve, regenerating soleus muscle predominantly expresses fast MHCIId(x) and MHCIIb. However, MHCI is induced when the denervated, regenerating soleus muscle is exposed to CLFS. However, the CLFS-induced up-regulation of MHCI is counteracted by treatment with the calcineurin inhibitor cyclosporin A (Serrano et al. 2001).

The calcineurin-mediated effects on the expression of fiber type-specific genes may be relevant with regard to specific, neurally induced oscillations in the concentration of free myoplasmic Ca^{2+} in fast and slow fibers (Olson & Williams 2000b). The activation by specific patterns of motor neuron firing, that is, tonic low-frequency in slow and phasic high-frequency impulses in fast fibers (Hennig & Lömo 1985), generates different temporal patterns of free myoplasmic Ca^{2+} that may exert differential effects on gene expression (Dolmetsch et al. 1997). Such effects may be of special relevance with regard to fiber type transitions, especially related to enhanced neuromuscular activity by endurance exercise or CLFS. Significant increases in free myoplasmic Ca^{2+} were observed in single mouse muscle fibers during fatiguing stimulation (Westerblad & Allen 1991). According to single-fiber studies on fast-twitch muscles of rat (Carroll et al. 1999) and rabbit (Sréter et al. 1987), low-frequency stimulation results in severalfold increases in the concentration of free myoplasmic Ca^{2+}.

Sustained elevations in free myoplasmic Ca^{2+} may trigger fast-to-slow switches in gene expression via the calcineurin-dependent pathway, other Ca^{2+}-mediated pathways, or both. Experimental evidence exists that CLFS-induced increases in myoplasmic Ca^{2+} are not solely due to the neurally transmitted tonic impulse pattern, but additionally result from impaired Ca^{2+} sequestration as reflected by incomplete muscle relaxation (Hicks et al. 1997). The function of the sarcoplasmic reticulum Ca^{2+}-pumping ATPase depends on a sufficiently high energy charge or ATP phosphorylation potential, as its activity decreases at depressed $[ATP]/[ADP_{free}]$ ratios (Läuger 1991). In fact, the $[ATP]/[ADP_{free}]$ ratio is drastically and persistently depressed by CLFS as demonstrated by measurements on whole muscles (Green et al. 1992a) and single fibers (Conjard et al. 1998).

The role of increased Ca^{2+} as a signal for the expression of slow fiber-specific genes has been further corroborated by studies on rabbit myotube cultures. Direct stimulation of these myotubes at low frequencies led to an up-regulation of MHCI. Significant increases in MHCI mRNA and MHCI protein could also be induced by treatment of the myotubes with the Ca^{2+} ionophore A23187, which markedly elevated (up to 10-fold) the intracellular Ca^{2+} concentration. These effects were prevented by the calcineurin antagonist cyclosporin A (Meissner et al. 2000, 2001).

Results highly relevant to the specificity of tonic and phasic impulse patterns in CLFS-induced fiber type transitions were obtained in a study on the activity-dependent translocation of NFATc in skeletal muscle fibers from adult mice (Liu et al. 2001). The authors investigated the effects of various frequency patterns on NFAT translocation in isolated muscle fibers. Stimulation protocols using 1 Hz and 50 Hz were inefficient, whereas 10-Hz stimulus trains caused nuclear translocation and characteristic intranuclear NFAT patterns (Liu et al. 2001). These results form a basis for understanding the specific action of neural impulse patterns on gene expression in muscle. The activation of slow fiber-specific genes might thus be linked to the sensing of specific Ca^{2+} pulses generated by specific neural impulse patterns (Dolmetsch et al. 1997).

Pathways Mediated by Protein Kinase C and Ca^{2+}/Calmodulin Kinase

Additional Ca^{2+}-dependent signal chains seem to be involved in the control of muscle fiber phenotypes. For example, Ca^{2+} has been implicated as a signal in the up-regulation of nuclear genes encoding mitochondrial proteins via a protein kinase C-dependent pathway (Freyssenet et al. 1999). Protein kinase C was also suggested to be involved in the control of slow myosin expression in chick muscle (DiMario 2001).

Another Ca^{2+}-dependent signal chain has been deduced from studies on transgenic mice expressing a constitutively active Ca^{2+}/calmodulin-dependent protein kinase IV (CaMK) (Wu et al. 2002). Skeletal muscles of these animals were characterized by moderate increases in slow fibers, enhanced mitochondrial DNA replication and mitochondrial biogenesis resulting in elevated levels of mitochondrion enzymes. CaMK IV is an important factor regulating PGC-1α (peroxisome-proliferator-activated receptor co-activator-1) expression (Lin et al. 2002). PGC-1α stimulates the expression of nuclear respiratory factors (NRFs) and mitochondrial transcriptional factor A (mtTFA).

The Ras/MAPK Pathway

An involvement of Ras/MAPK(ERK) pathways in the up-regulation of slow fiber-specific genes was suggested by Murgia and coworkers (2000). Denervated, regenerating rat soleus muscle expresses only fast MHC isoforms. MHCI can be induced, however, by transfection of constitutively active Ras or a Ras mutant that selectively activates the MAPK(ERK) pathway. Activation of the MAPK pathway, therefore, mimicks the effects of slow motor neuron innervation. Finally, severalfold increases in MAPK activity were observed when denervated, regenerating soleus muscle was exposed to low-frequency stimulation (Murgia et al. 2000).

Studies on the effects of exercise in human muscle revealed that several kinases related to parallel MAPK pathways, namely ERK1, ERK2, and to a lesser extent also p38 MAPK, are rapidly phosphorylated after the onset of contractile activity. These observations support the notion that activation of the MAPK pathway is important in the regulation of specific transcriptional events (Hayashi et al. 1999; Widegren et al. 1998).

Ras can be activated by Ca^{2+}, and temporal patterns of Ca^{2+} transients have been shown to exert specific effects on gene expression (Dolmetsch et al. 1997; Fields et al. 1997; Finkbeiner & Greenberg 1996). It is interesting that MAPK signaling has been implicated, in addition to the calcineurin and CaMK pathways, in the transcriptional control of myogenin (Xu et al. 2002), a muscle regulatory factor involved in skeletal muscle differentiation (Arnold & Winter 1998) and possibly also in the formation of the slow phenotype (Hughes et al. 1993).

The AMPK Pathway

A separate, Ca^{2+}-independent signaling pathway that has been implicated in exercise-induced metabolic adaptations relates to the function of 5'AMP-activated protein kinase (AMPK). AMPK is interconverted into its active form by phosphorylation through an AMPK kinase. In addition, AMPK is allosterically activated by 5'AMP and inhibited by phosphocreatine (Hardie 2004; Winder 2001). In other words, AMPK is activated in response to ATP depletion. Its activation has been suggested to initiate a signal chain triggering exercise-induced metabolic adaptations, such as enhanced expression of GLUT4, increases in hexokinase, and mitochondrial enzyme activities (Goodyear 2000; Holmes et al. 1999; Winder et al. 2000; Winder 2001; Wright at al. 2004; Zheng et al. 2001).

Noticeably, chronic activation of AMPK in vivo by administration of AICAR (5-aminoimidazole-4-carboxamide-1-beta-D-ribofuranoside) increases mitochondrial enzyme activities in rat muscle without changes in MHC isoforms or fiber type transitions (Putman et al. 2003). These observations indicate that metabolic adaptations are under separate control and may, at least partially, occur independently of alterations in myofibrillar protein composition.

Exercise has been shown to increase AMPK activity in muscles of rodents (Winder & Hardie 1996) and humans (Fujii et al. 2000; Stephens et al. 2002; Wojtaszewski et al. 2000). Increases in AMPK activity have been elicited in rat muscle also by electrical stimulation (Hutber et al. 1997). Finally, the role of AMPK in long-term metabolic adaptations has been established in a study in which AMPK was activated by depleting the phosphocreatine and ATP stores through administration of β-guanidinopropionic acid (GPA), a metabolic competitor of creatine (Bergeron et al. 2001). GPA administration stimulated mitochondrion biogenesis and increased mitochondrial density in skeletal muscle. However, GPA-treatment of transgenic mice expressing a dominant-negative mutant of AMPK in muscle had no effect on mitochondrial content (Zong et al. 2002).

Both the activation by 5'AMP and inhibition by phosphocreatine connect AMPK activity directly to the phosphorylation potential or the energy charge of the muscle fiber. Sustained depressions of the energy charge with elevated 5'AMP levels thus might be directly transmitted by AMPK to signaling pathways involved in adaptive metabolic responses, for example fuel uptake, enzyme levels, and mitochondrial content. In other words, the energy charge of the muscle fiber appears to be an overall sensor in the control of metabolic adaptations (Conjard et al. 1998; Green et al. 1992a). This control may be exerted directly by AMPK and indirectly by changes in myoplasmic Ca^{2+} via signaling chains that act on different target proteins of the transcription apparatus.

Conclusion

Skeletal muscles contain a spectrum of functionally different pure and hybrid fiber types extending between type IIB and type I, the fastest and slowest fibers, respectively. This spectrum reflects the dynamic nature and versatility of the muscle fibers, which change their phenotypic properties in response to sustained changes in functional demands. Adaptive responses to altered neuromuscular activity or mechanical loading encompass a multitude of qualitative and quantitative changes in gene expression, ultimately resulting in reversible fiber type transitions. In general, increases in neuromuscular activity evoke fast-to-slow transitions, whereas reduced neuromuscular activity results in slow-to-fast transitions. The ranges of fiber type transitions depend on both the extent and duration of the changes in functional demands, but also depend on the fiber type itself.

The metabolic adaptations to sustained alterations in neuromuscular activity seem to obey a rule similar to that underlying the adaptive changes in myofibrillar protein isoform expression. The transitions between anaerobic-glycolytic and aerobic-oxidative pathways of energy metabolism depend on the "actual" metabolic profile of the fiber. In other words, "glycolytic" fibers respond to enhanced recruitment with greater increases in aerobic-oxidative potential than "oxidative" fibers. Oxidative fibers are metabolically prepared to meet the energetic demands of sustained activity and, thus, exhibit a relatively small adaptive range for increasing their aerobic-oxidative potential. Conversely, reduced neuromuscular activity hardly affects the metabolic profile of glycolytic fibers, but shifts that of oxidative fibers toward the glycolytic phenotype.

The signaling pathways involved in adaptive changes of gene expression are beginning to be elucidated. Their multiplicity most likely relates to different targets and the possibility of their independent regulation. Changes in free myoplasmic Ca^{2+} and changes in the cellular phosphorylation potential seem to be major signals for triggering calcineurin-, CaMK-, PKC-, and Ras/MAPK-mediated signaling pathways involved in the control of fiber type-specific expression of myofibrillar and other sarcomeric protein isoforms. AMPK-mediated signaling pathways, directly related to the energy charge of the cell, seem to be primarily implicated in adaptive responses adjusting energy supply to energy demand.

Chapter 14

Exercise and the Alveolar and Bronchial Epithelial Cell

Heimo Mairbäurl

Respiratory epithelial cells form the barrier that separates gas space from fluid phases in the entire lung. Their functions are crucial for effective exchange of respiratory gases in the alveoli, that is, elimination of CO_2 formed in cell metabolism and uptake of O_2 from the environmental air. Two basic structures are required to fulfill these functions: the conductive airways, which distribute the respiratory gases within the lung and guide them to the alveoli, and the alveoli, where the exchange between alveolar gas and blood occurs (figure 14.1). Airway epithelial cells actively secrete Cl^- and reabsorb Na^+ to generate the osmotic gradient for water transport in the direction of net ion movement required for moistening the airways. In the adult lung, alveolar epithelial cells actively

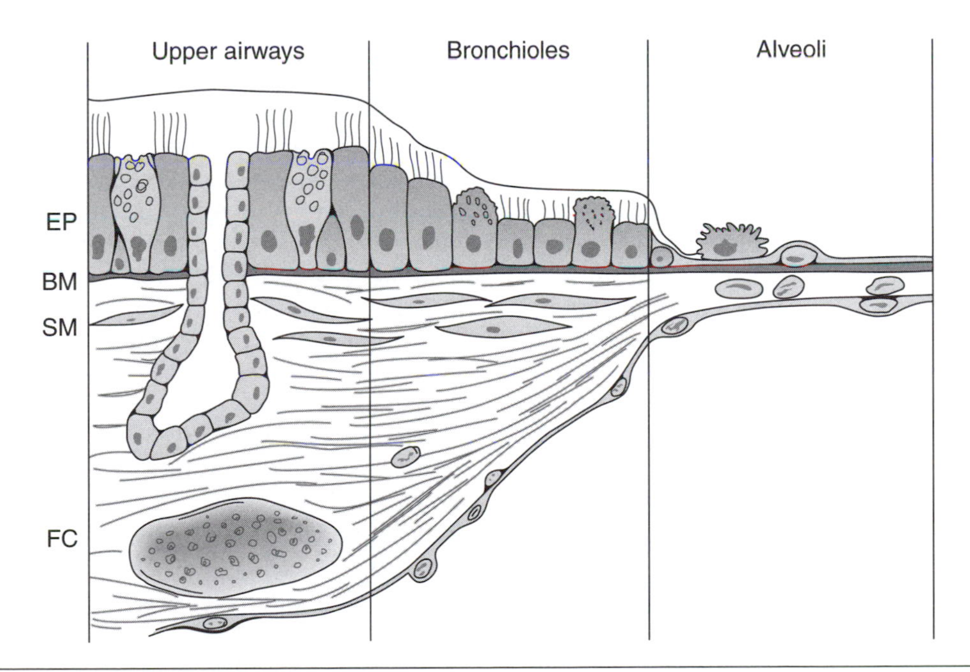

Figure 14.1 Change in structure of the airway and alveolar wall. The epithelial layer (EP) changes from pseudostratified in upper airways to cuboidal and squamous in the alveoli. Cilia-containing cells, secretory cells, and smooth muscle cells (SM) are not found in the alveoli. The fibrous coat (FC) containing cartilage is present only in the trachea and major bronchi. A basal membrane (BM) separates the epithelial cell layer from deeper structures.

Reprinted, by permission, from E.R. Weibel, 1984, *The pathway for oxygen,* (Cambridge, MA: Harvard University Press).

reabsorb Na^+ and water to prevent alveolar fluid accumulation, whereas in the fetal lung, alveolar epithelial cells secrete the fluid that fills the lungs. Regarding their function, cell types differ along the airways in a kind of hierarchical order. This chapter describes the basic ion transport functions of airway and alveolar epithelia. The secretion of mucus, airway smooth muscle function, and pulmonary blood flow are not covered.

Airways

Trachea and bronchi are the conducting airways. Their major function is to transfer inspired air to the alveolar space of the lung and to exhale CO_2-rich, O_2-deprived alveolar gas. Inspired air needs to be cleared from particles, humidified, and brought to body temperature.

Histology of the Airway Epithelium

Airways are covered by a superficial layer of mucus and a thin layer of fluid that lines the immediate epithelial cell surface underneath the mucus. Together, mucus, lining fluid, and the movement of cilia form an important defense system that traps inhaled bacteria and particles contained in the inspired air and clears the mucus from the airways. The cellular barrier at the air–liquid interface is formed by polarized epithelial cells (EP in figure 14.1). Their basal side is anchored to a basal membrane (BM); the apical side faces the airspace and carries various specialties like cilia and microvilli. The lateral sides of the cells provide junctions for cell-to-cell contacts and a seal of the intercellular cleft that also separates interstitial and lining fluid (tight junctions). There are also desmosomes for mechanical coupling of the cells and gap junctions that allow a selective exchange of ions and small molecules between cells (Weibel 1984).

Various cell types contribute to the formation of the airway epithelium (figure 14.1), their relative number changing along the airways. In general the airway epithelium is a pseudostratified columnar epithelium that contains mainly ciliated cells, goblet cells, and basal cells in the trachea. In smaller bronchi the composition is similar, although other cell types such as Clara cells are also present but in much lower proportion. Clara cells are nonsquamous, cuboidal to columnar, nonciliated, secretory, multifunctional cells that are found mostly in distal airways. Ciliated cells contain kinocilia, which are organelles that beat rhythmically in a coordinate fashion such that the mucus blanket that forms a thin layer on top of the cilia is moved up the airways in a steady stream. When the production of mucus is enhanced or when large amounts of dust are inhaled, mucus movement by cilia is enforced by coughing. Coughing alone seems not to be sufficient for mucus transport. Factors released from inflammatory cells seem to inhibit ciliary motility (Robbins & Rennard 1997). Some of the secretory cells actually seem to form small glands in the upper airways. Goblet cells reach the apical side of the airways and contain immature secretory granules. In distal bronchi, Clara cells contribute to secretion. Basal cells are thought to serve as precursors of ciliated cells and goblet cells.

Ion Transport Across the Airway Epithelium

Control of osmolarity and ionic composition of the airway epithelial lining fluid by secretory and reabsorptive processes is vital for cilia movement and defense against bacterial colonization (Guggino 1999). Epithelial ion transport properties are different along the airways. A transepithelial potential is generated due to secretory and reabsorptive processes as discussed further on. The highest values of transepithelial potential and transepithelial resistance (a measure of the tightness of the epithelial barrier) have been found in the trachea; values were lower in main stem and terminal bronchi, whereas short-circuit currents (ISC; a measure of the magnitude of vectorial ionic movement across the epithelium), which are in the range of 30 to 100 $\mu A/cm^2$, seem not to be different in different areas of the airways (Welsh 1986). However, ion fluxes differ markedly along the airways. In tracheal epithelium, ISC is mainly the result of net secretion of Cl^-, which is about two to four times higher than Na^+ reabsorption. In contrast, in distal bronchial epithelium, Na^+ reabsorption seems to dominate. Cl^- secretion and Na^+ reabsorption can reside in the same cells. The basic mechanisms shown in figure 14.2 are described in the subsections that follow. Both secretion and reabsorption require active Na^+/K^+ pumps that keep the intracellular Na^+ concentration low. K^+ channels and electrogenic Na^+/K^+ pumps contribute to the cell-inside negative membrane potential.

Secretion

Cl^- secretion is an electrogenic process that generates the osmotic gradient required to drive the secretion of water. Secretion across the airway

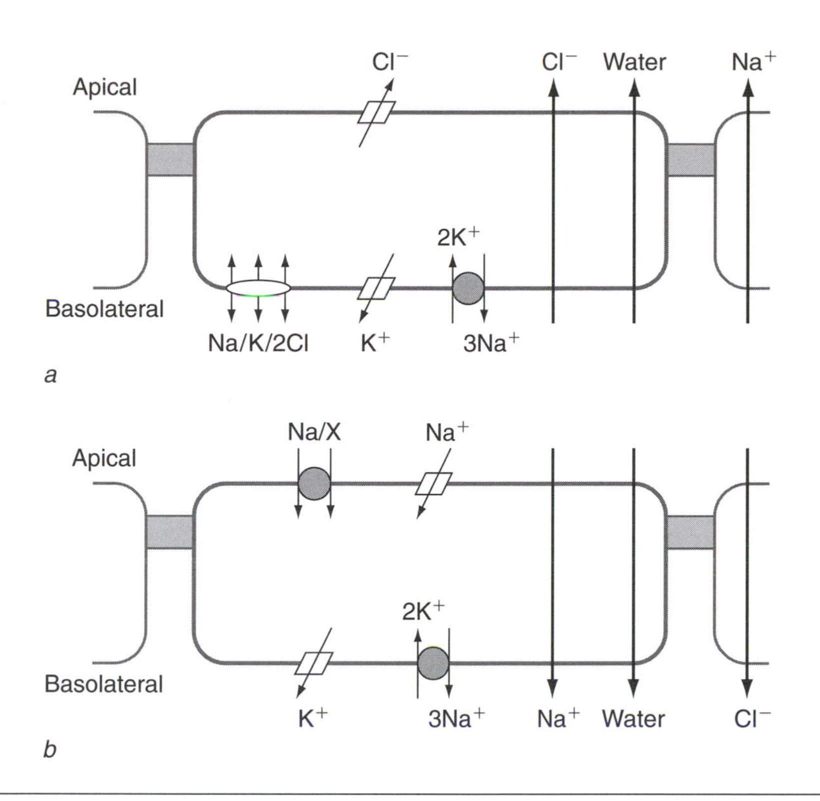

Figure 14.2 Airway secretion and reabsorption. *(a)* Secretion is mediated by basolateral $Na^+/K^+/2Cl^-$ cotransport, which causes uptake of Na^+, K^+, and Cl^- into the cell and Cl^- extrusion via apical Cl^- channels. *(b)* In reabsorptive epithelia, the low intracellular Na^+ concentration allows apical Na^+ entry via Na^+-coupled transport (Na/X) and via Na^+ channels. Transepithelial Na^+ transport is electrically balanced by Cl^- reabsorption, thus generating the osmotic driving force for water reabsorption. For simplicity, the pathways that mediate Na^+ (in *a*), Cl^- (in *b*), and water transport (in *a* and *b*) are not specified.

epithelium is similar other types of secretory epithelia (figure 14.2*a*). The model proposes an electroneutral Na^+-dependent entry of Cl^- across the basolateral side of the cell. The driving force for the uptake of Na^+Cl^- is generated by the Na^+/K^+ pump. If basolateral Na entry is mediated by $Na^+/K^+/2Cl^-$ cotransport, the K^+ that entered the cell together with Na^+ and Cl^- leaves across basolateral K^+ channels. These channels also recycle the K^+ taken up by the Na^+/K^+ pumps. The increased intracellular Cl^- concentration and the cell's inside negative membrane potential drive Cl^- out of the cell. Cl^- exit is mediated by Cl^- channels, mainly the cystic fibrosis transmembrane conductance regulator (CFTR). Furosemide prevents basolateral Cl^- entry and thus reduces the electromotive force across the apical membrane.

Reabsorption

Reabsorption of fluid from the apical side of the epithelium is driven by active reabsorption of Na^+ (figure 14.2*b*). Various transporters mediate apical Na^+ entry, although most entry seems to be mediated by amiloride-inhibitable Na^+ channels. Other sources of Na^+ uptake are Na^+-coupled transporters such as Na/glucose or Na/amino acid transporters and Na/H exchange. In any case, Na^+ uptake is driven by the Na^+ gradient across the apical membrane generated by the activity of Na^+/K^+ pumps, which extrude Na^+ in exchange for K^+ across the basolateral membrane. Basolateral K^+ channels are required for recycling of K^+.

Transporters Involved in Secretion and Reabsorption

As outlined previously in this chapter, secretion and reabsorption of fluid is coupled to the transport of solutes that generate the osmotic driving force to move water across the epithelium. Ion transport is mediated by distinct transporters, the most important of which are described here.

Apical Membrane

The major Cl^- transporting protein in the apical membrane of airway epithelial cells is CFTR, which is activated already under baseline conditions. CFTR is a nonrectifying Cl^- channel with a unit conductance of about 10 pS. Increasing intracellular cAMP causes almost no change in the Cl^- conductance and in ISC in human tracheal epithelium, whereas in cultured cells expressing CFTR, an elevation of cAMP causes considerable stimulation of transport. An increase in intracellular Ca^{2+} also stimulates Cl^- secretion by CFTR.

The CFTR protein belongs to the family of ABC transporters—transport proteins that contain nucleotide binding sites (ATP binding cassette). These latter domains are required for phosphorylation of the CFTR protein and for ATP binding and hydrolysis (Ko & Pedersen 1997). There is evidence that CFTR, besides transporting Cl^-, also mediates the release of cellular ATP, which might then serve as an extracellular signal to regulate the secretion of mucus and water via specific receptors and an increase in intracellular calcium.

There are also outwardly rectifying Cl^- channels that are stimulated by cAMP. These channels have a unit conductance of 50 pS and mediate Cl^- secretion also in the airway epithelium of CFTR-deficient mice (Gabriel et al. 1993). Increased levels of intracellular Ca^{2+} stimulate these channels as well.

Na^+ channels mediate the uptake of Na^+ across the apical plasma membrane of airway epithelial cells. Most of the transport is inhibited by amiloride. The predominant form of Na^+ channels seems to be ENaC (epithelial Na^+ channels), a channel highly selective for Na^+. It consists of three subunits. The α subunit of ENaC is required for transport activity, whereas the presence of β and γ subunits stimulates transport activity (Voilley et al. 1997). However, nonselective cation channels are also present, which are also inhibited by amiloride. Control of Na^+ channel activity seems to be coordinated with the activity of CFTR in that a high activity (or level of expression) of CFTR inhibits ENaC (König et al. 2001).

Basolateral Membrane

Cl^- enters the cell across the basolateral membrane mainly by $Na^+/K^+/2Cl^-$ cotransport (NKCC). In vitro experiments also indicate the presence of Na^+/Cl^- cotransport. Transport is inhibited by loop diuretics such as furosemide and also depends on the presence of external Na^+ (and K^+) (Widdicombe 1997). Loop diuretic-sensitive transport is stimulated by α-adrenergic agents and increased intracellular Ca^{2+} but not by cAMP (Haas & Forbush 2000). NKCC stimulation by these agents might be secondary to Cl^- channel activation, Cl^- efflux, and subsequent cell shrinkage. Isoproterenol and cAMP affect the number of copies of NKCC in the basolateral plasma membrane of nasal epithelium (Haas & McBrayer 1994).

Na^+/K^+ pumps are located exclusively in the basolateral membrane, where they mediate the extrusion of Na and the uptake of K in a 3:2 ratio (Lingrel 1992). ATP is utilized to drive transport against concentration gradients. There are no differences in the sensitivity to ions and ATP between Na^+/K^+ pumps in airway epithelia and those in other tissues. The major determinant of the activity of plasma membrane-inserted pumps is the intracellular Na^+ concentration, whereas regulation by second messengers seems not to play a role in the regulation of the airway Na^+/K^+ pump.

Basolateral K channels determine the membrane potential. Their main function is the recycling of K^+ that has entered the cell via NKCC and the Na^+/K^+ pump. The cell negative potential also generates the electromotive force to drive Cl^- out of the cell across apical Cl^- channels and supports the uptake of Na^+. Thus, K^+ channel activity increases upon stimulation of Cl^- secretion to prevent K^+ accumulation. The inhibition of K^+ recycling by Ba^{2+} and TEA indicates that this function is mediated by maxi-K^+ channels (Widdicombe 1997). Other types of K^+ channels in airway epithelial cells seem to be sensitive to cAMP, epinephrine, changes in cell volume, and stretch (Widdicombe 1997). A functional role for K^+_{ATP} channels remains to be demonstrated.

Control of Secretion and Reabsorption Across Airway Epithelium

Cl^- secretion across airway epithelium is regulated by a variety of neurohumoral mediators that act via intracellular cAMP or Ca^{2+}, whereas Na^+ reabsorption shows only little control by β-adrenergic agents. Regulation of ion transport by cAMP seems not to be of great importance in humans, although there is some evidence that β-adrenergic agonists can stimulate CFTR-related Cl^- secretion in healthy subjects. In contrast, cAMP stimulation of CFTR is well documented in cultured cells. It involves the action of PKA. An increase in intracellular Ca^{2+} is the most import stimulator of Cl^- secretion in human airway epithelium. Ca^{2+} can affect transporters directly but can also act via Ca^{2+}-dependent protein kinases. Many mediators such as bradykinin and extracellular ATP exert their effects by increasing intracellular Ca^{2+} in response to increased IP_3, whereas effects of cAMP seem to be independent of IP_3. Cyclic AMP might also cause an increase in intracellular Ca^{2+}. Both increased intracellular cAMP and Ca^{2+} control transport activity by stimulation of the phosphorylation of transport proteins.

There is evidence that transport activity might be regulated by the intracellular Cl^- concentration. A drop in intracellular Cl^- activates secretion by stimulation of the $Na^+/K^+/2Cl^-$ cotransport and of K^+ channels in the basolateral membrane (Haas et al. 1993).

A direct control of the composition of the airway lining fluid seems to be mediated by cross-talk between CFTR and ENaC first shown by Stutts et al. (1995). An increase in the Cl^- conductance, for example by overexpression of CFTR, causes a decrease in ENaC-mediated Na^+ transport, thus allowing a switch back and forth between secretion and reabsorption. A possible mediator of this cross-talk appears to be a change in the intracellular Cl^- concentration upon application of secretagogues (König et al. 2001). Possible other mechanisms include electrical phenomena and effects on the expression of transport proteins.

Transepithelial Nasal Potential Difference

The measurement of the ion transport activity of airway epithelial cells provides important information on malfunction of certain transporters such as CFTR. Transport can be measured through determination of the transepithelial potential difference across the nasal mucosa (Knowles et al. 1981). A typical experimental setup is shown in figure 14.3. The nasal mucosa is superfused with various media that may contain inhibitors of different transporters located in the apical membrane of airway epithelial cells; the measuring electrode is placed in this line. A reference electrode might be placed either on the surface of the skin or into an intravenous catheter. Switching between different media that contain inhibitors or stimulators allows one to distinguish between certain transport pathways. Figure 14.3 also shows the tracing of a typical experiment indicating an apical-side negative transepithelial potential of about −15 mV. When Na transport is inhibited (e.g., with amiloride), the transepithelial potential decreases and the transepithelial potential becomes more positive. On the other hand, when a Cl^- gradient from the blood to the epithelial surface is applied by superfusion with a medium of low Cl^- concentration, Cl^- is forced to follow down that gradient toward the epithelial surface. This causes a hyperpolarization of the transepithelial potential. Lack of hyperpolarization might point to a defect in the Cl^- conductance such as in cystic fibrosis (Knowles et al. 1981).

Exercise

Exercise is a particular challenge to the airways since an increased volume of inspired air is transported to the alveoli associated with increased gas flow rates by increasing tidal volume and

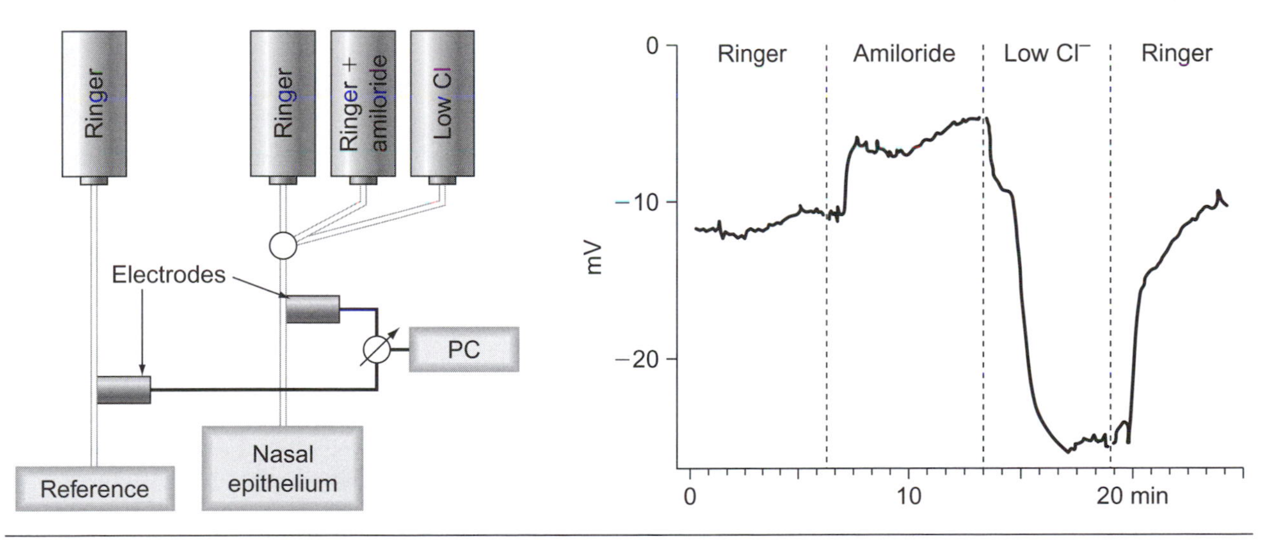

Figure 14.3 Transepithelial nasal potential measurement. A reference electrode is inserted into an intravenous line; the measuring electrode is inserted into a catheter placed on the surface of nasal epithelium. Both lines are perfused at a slow rate. A manifold allows switching between media. The electrical potential (mV) between the electrodes is recorded on line. The graph shows the tracing of a typical experiment. After switching to amiloride-containing Ringer, the potential decreases due to Na^+ channel inhibition. The capacity of Cl^- channels can be estimated by with a medium low in Cl^- concentration. Isoproterenol stimulates Cl^- transport to maximal capacity, which causes the transepithelial potential to become more negative. Switching back to Ringer restores the original potential.

frequency of breathing, as well as by dilation of the airways. The latter is caused by increased sympathetic activity. An increased ventilation, however, bears the risk of airway cooling and dehydration, as well as the uptake of increased amounts of particles and bacteria, a major factor that is discussed as a cause of exercise-induced asthma (see the next section).

Catecholamines stimulate airway epithelial Cl^- secretion via activation of CFTR. Therefore one would expect a shift of the transepithelial potential to more negative values during exercise. The response of airway epithelial ion transport has been evaluated through measurement of the nasal potential difference during cycling in a semi-supine position. The change in transport activity seems to depend on exercise intensity. Exercise below the anaerobic threshold was found to cause a decrease in transepithelial potential (Hebestreit et al. 2001). This low-intensity exercise also decreases the amiloride-sensitive portion of the transepithelial potential, which indicates an inhibition of epithelial Na^+ channels. Therefore Cl^- secretion should dominate and favor secretion into the airways, which improves humidification of inspired air. The mechanism of inhibition of Na^+ transport is not clear. Inhibition by negative feedback with Cl^- conductance (stimulation of Cl^- secretion inhibits Na^+ transport and vice versa) can be ruled out since the potential change induced by imposing a transepithelial Cl^- gradient was not altered by this type of exercise (Hebestreit et al. 2001). These results indicate a mechanism that specifically acts on Na^+ channels. Less is known about airway epithelial transport at high-intensity exercise. In a different study, exercise above the anaerobic threshold was shown to shift the transepithelial potential toward more negative values, which fits the expectation of a stimulation of Cl^- transport by elevated levels of catecholamines. However, the contribution of individual transport pathways has not been further evaluated. Although the exact mechanisms are not understood, these results indicate that airway secretion is adjusted to exercise intensity and to the rate of gas flow through the airways, thus avoiding airway dryness.

Disturbances of Airway Function

A variety of dysfunctions of airways are known that concern airway conductance and defense against bacterial infections. Here two aspects are discussed that deal with these functions, namely exercise-induced airway asthma and cystic fibrosis.

Exercise-Induced Asthma (EIA)

EIA refers to narrowing of the airways following intense exercise. It is thought to be related to drying or cooling, or both, of the airway epithelium and surrounding cells, causing reactive hyperemia, edema, and airway smooth muscle contraction (Andersen & Daviskas 2000). Mechanisms are summarized in figure 14.4. As large volumes of air pass by the airway epithelium during exercise, water evaporates, thus increasing the osmolarity of the airway epithelial lining fluid. Adrenergic

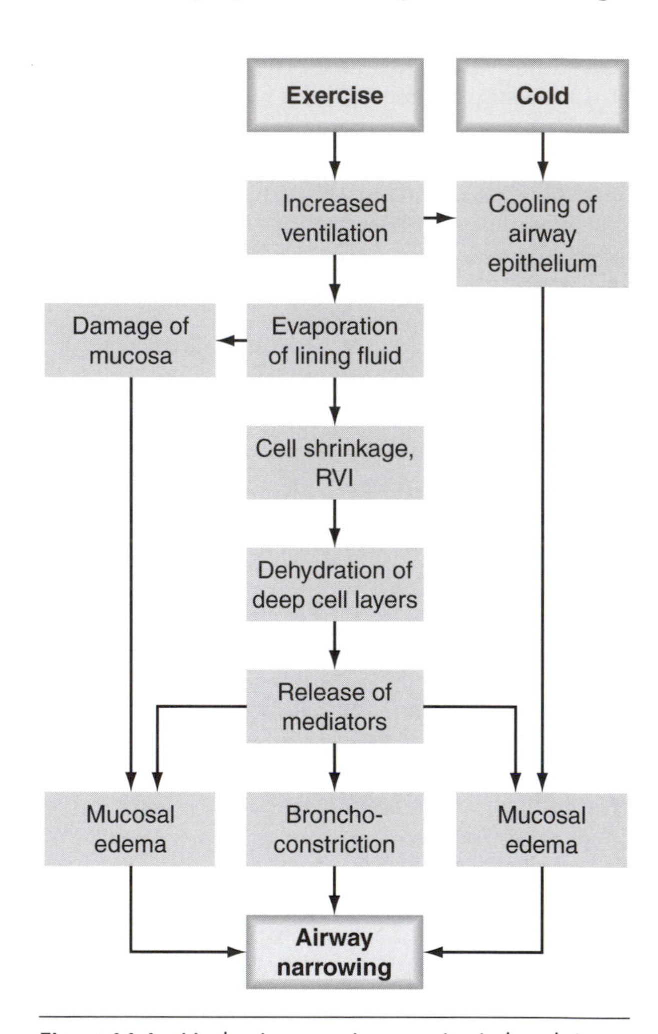

Figure 14.4 Mechanisms causing exercise-induced airway obstruction. Increased ventilation during exercise causes airway cooling and drying. Airway lining fluid evaporates and increases extracellular osmolarity. Both cooling and increased osmolarity can cause mucosal damage and edema. Airway epithelial cells shrink, which induces regulatory volume increase (RVI) and withdraws water from surrounding interstitium and causes cells in deeper layers to shrink also. Cell shrinkage and damage lead to release of prostaglandins, leukotrienes, and histamine causing hyperemia, airway edema, airway smooth muscle constriction, and narrowing of the airways.

stimulation of Cl^- and water secretion seem not to be able to match this loss of water. Cells shrink in response to the hyperosmolar environment. Regulatory volume increase of airway epithelial cells withdraws water from interstitial space and cells lying underneath the epithelial surface and causes a spreading of cell volume decrease into deeper cell layers of the airways. Drying of airway epithelium in response to dry air can cause mucosal injury. Cell injury, change in cell volume, regulatory volume increase, and an increase in intracellular Ca^{2+} and IP_3 are all known to mediate the release of mediators such as histamine, prostaglandins, and leukotrienes by a variety of cell types present in the airways such as mast cells and to cause bronchial smooth muscle contraction (Anderson & Holzer 2000). Therefore, inhibitors of transporters such as Cl^- channels and Na/K/2Cl cotransport that are involved in cell volume regulation are also effective in preventing airway narrowing upon exercise and inhalation of hyperosmolar aerosols (Anderson & Daviskas 2000). Also β-agonists prevent airway narrowing by their action on airway smooth muscle cells.

Mechanisms increasing the susceptibility to EIA are unclear. They might involve existing (chronic) inflammation of the airways, presence of inflammatory cells, and altered cell volume regulation, which, as outlined previously, can all cause bronchoconstriction and edematous airway narrowing. Airway inflammatory cells seem to be increased in athletes, both at rest and during exercise, with some variability among sports that is most likely associated with the environmental exposure (Bonsignore et al. 2003). However, they seem not to be activated (Bonsignore et al. 2003). The aspect of defective cell volume regulation might be of particular importance since it has been found that airway epithelium can also produce bronchodilator substances such as PGE_2 in response to increased osmolarity and hyperpnea with dry air (Anderson & Daviskas 2000).

Cystic Fibrosis

Cystic fibrosis (CF) is an inborn metabolic error in children and young adults mostly found in countries with Caucasian populations. CF is not rare (~0.1% of all live births). It is characterized by a deficiency in Cl^- and fluid secretion not only in the airways but also in other secretory epithelia such as in pancreas, intestine, and sweat glands.

Several hundred mutations of the CFTR gene associated with various degrees of the disease have been reported. Most prominent is a deletion of the amino acid phenylalanine in position 508 (ΔF508). $CFTR_{\Delta F508}$ is retained in the endoplasmic reticulum, where it is actually a functioning Cl^- channel, but is not processed to the apical surface of the cell. This disease therefore resembles a disturbance of the trafficking of newly synthesized CFTR protein. Thus the capacity for Cl^- secretion is decreased while at the same time Na^+ reabsorption is stimulated (for review see Ko & Pedersen 1997).

In the airways, CF is associated with impaired clearance of mucus, inflammation, and chronic bacterial infections of the respiratory tract. The decreased rate of Cl^--dependent secretion and a relative increase in the rate of Na^+ reabsorption are responsible for the decreased amount of airway epithelial lining fluid and its altered ion composition (Guggino 1999). The decreased volume of lining fluid impairs mucociliary clearance, whereas the altered ional composition of the airway lining fluid seems to favor bacterial colonization (Guggino & Guggino 2000). However, inhaled bacteria and viruses might also affect transport proteins directly.

Because Na^+ reabsorption determines the transepithelial potential when the Cl^- secretion is diminished, the resting transepithelial potential across the nasal mucosa is more negative in persons with CF than in healthy controls. Exercise-induced sympatho-adrenergic activity is a known stimulator for secretion in airway epithelial cells. In CF no such increase in secretion can be found, which makes these patients even more vulnerable to airway infections when they breathe large volumes of air. Other exercise-related problems of CF patients include the diminished thermoregulation due to decreased sweat production as well as metabolic disturbances.

The Alveolar Epithelium

In the adult lung, the alveolar epithelium covers the side of the alveolar barrier that is exposed to air. It mediates diffusional gas exchange between the respiratory gases and the blood. Alveolar epithelial cells secrete surfactant and, at the same time, maintain a thin layer of alveolar lining fluid.

Histology of the Alveolar Barrier

The alveolar barrier is formed from alveolar type I and type II cells (figures 14.1 and 14.5). More than 90% of the surface area is covered by alveolar type I cells (ATI cells), though these cells represent only about 30% of all alveolar epithelial cells (Weibel 1984). ATI cells are flat, with small nuclei and a few mitochondria centered around the nucleus. In a cross section of more distal parts of the cell, the apical and basal membrane appear to be attached to each other (Weibel 1984). The thickness of this sheet is about 0.1 μm. ATI cells are attached to a basal lamina that also contacts capillary endothelial cells.

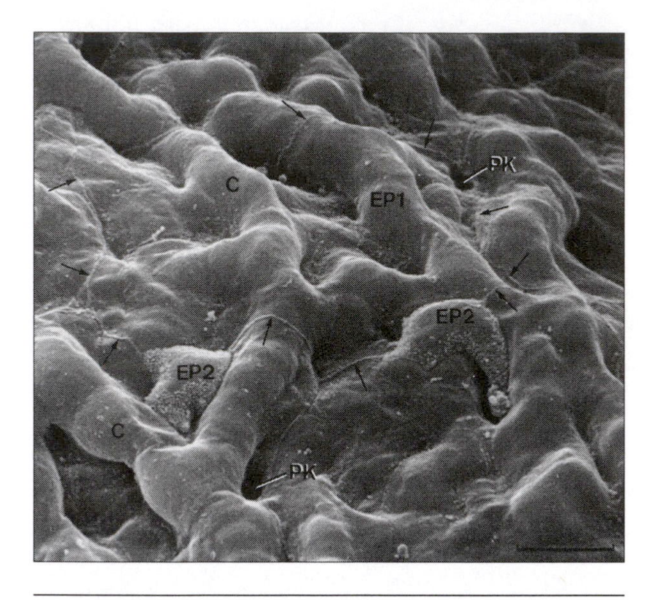

Figure 14.5 Scanning electron micrograph of the surface of the alveolar wall of the human lung. The alveolar surface is formed by alveolar type I (EP1) and type II (EP2) cells. Arrows indicate the boundary between adjacent type I cells forming a leaflet that covers capillaries (C). Scale marker: 10 μm.

Reprinted, by permission, from E.R. Weibel, 1984, *The pathway for oxygen,* (Cambridge, MA: Harvard University Press).

Alveolar type II cells (ATII cells) are found in the corners of alveoli. They are cuboidal and heavily structured. Microvilli, which usually serve to increase the surface area for optimal secretion and reabsorption, are found on the apical surface. ATII cells have numerous mitochondria, endoplasmic reticulum, and a well-developed Golgi apparatus. The most prominent organelles are lamellar bodies, which are in charge of producing and secreting surfactant.

Surfactant Secretion

Surfactant is a mixture of phospholipids and protein (Mason et al. 1998) that is highly surface active. It lines the alveolar surface of the lung and reduces the surface tension at the air–liquid interface and thus prevents alveoli from collapsing. It is synthesized in the fetal and adult lung. Lack or dysfunction of surfactant causes respiratory distress.

Surfactant is a mixture of phospholipids and protein, mostly plasma proteins. There are also unique surfactant proteins (SP) such as SP-A, SP-B, SP-C, and SP-D, which bind lipids and carbohydrates and interact with specific cell surface receptors. The major lipid components are phosphatidylcholine and phosphatidylglycine. The saturated phosphatidylcholine is almost entirely dipalmitoylphosphatidylcholine. This substance contributes most to the decrease in surface tension.

Surfactant is synthesized in the Golgi apparatus of ATII cells and packaged into secretory granula, which in a first stage are composite granula but then become lamellar bodies (figure 14.6). Ca^{2+}- and

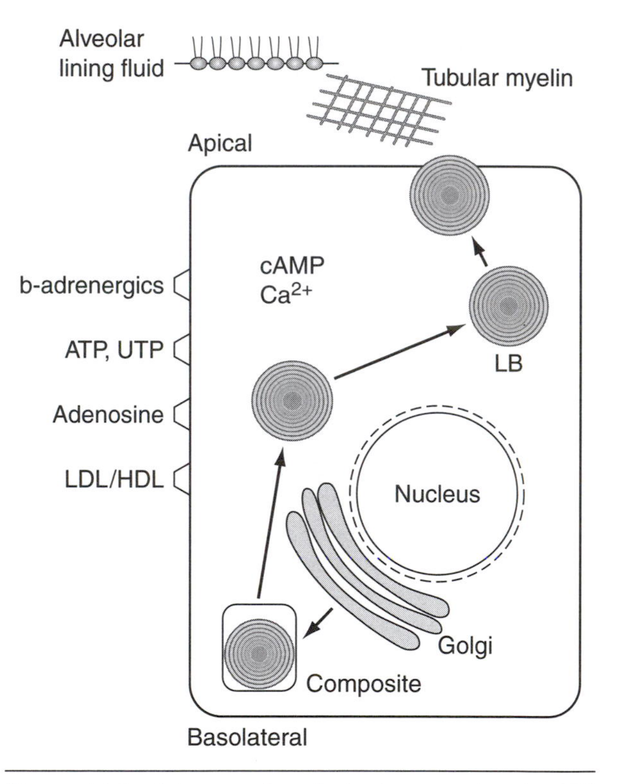

Figure 14.6 Scheme of surfactant processing in ATII cells. After synthesis in the endoplasmic reticulum (not shown), surfactant components are processed in the Golgi apparatus and moved into composite and, from there, into lamellar bodies (LB). Upon intra- or extracellular stimuli, or both, LB fuse with the plasma membrane to release surfactant. Tubular myelin is formed, and surfactant proteins (not shown) and phospholipids separate. Whereas surfactant proteins are taken up via receptor-mediated pathways, phospholipids form a monolayer to cover the lining fluid at the air–liquid interface. Known stimuli for secretion are ATP (adenosinetriphosphate), UTP (uridinetriphosphate), and low- and high-density lipoproteins (LDL and HDL, respectively).

cAMP-dependent mechanisms that involve PKA- and PKC-dependent phosphorylation processes are engaged in the control of exocytosis (Rooney et al. 1994). The most important stimulus seems to be mechanical stretch; less effective stimuli are extracellular ATP, thought to be released during extensive lung inflation, and hormones such as catecholamines. Once at the alveolar surface, the material forms tubular myelin, which probably represents a state of separation of surfactant protein and lipid as the lipid forms a monolayer at the air–liquid interface. This process seems to be triggered by surfactant proteins. Training significantly increases the ratio between surfactant proteins and phospholipids contained in surfactant, which might indicate improved surfactant secretion (Doyle et al. 2000). In racehorses, however, a functional decrease in surfactant activity was found at highly intense exercise (Morrison et al. 1999). Most of the surfactant is recycled by reuptake into ATII cells by endocytosis through coated vesicles for insertion into lamellar bodies and reuse. Some surfactant is removed by alveolar macrophages; some is also removed via the airways and circulation (Rooney et al. 1994).

Alveolar Fluid Balance

During fetal and perinatal development the function of the alveolar epithelium changes considerably. Whereas in the fetal lung the alveolar epithelium actively secretes fluid, the perinatal and the adult alveolar epithelium reabsorbs the fluid that has entered the alveolar space. The mechanisms that control the change in function around the time of birth are not fully understood. In the adult lung, ATI and ATII cells are involved in the Na^+-dependent reabsorption of excessive alveolar fluid, which is important to maximize alveolar gas exchange (Matthay et al. 2002).

Changes During Lung Development

Alveoli of the *fetal lung* need to be filled with fluid in order to allow proper lung development. Fluid secretion is in accordance with the classical models of secretory epithelia (see figure 14.2*a*). It is dependent on the active secretion of Cl^-, which provides the osmotic gradient for the transepithelial movement of water. Days before birth, the epithelium begins to prepare itself for the switch from secretion to reabsorption. The expression of Na^+ channels, mainly ENaC, increases. Stress hormones during delivery and increased levels of oxygenation augment this process (O'Brodovich 1991). Rafii et al. (1999) have shown that oxygenation of cultured fetal alveolar epithelial cells increases activity and expression of Na^+ transporters and that heme proteins and increased levels of reactive oxygen species might be involved in upregulation (Rafii et al. 2000).

In the *adult lung* the alveolar epithelium forms the barrier at the air–liquid interface. Its function is the maintenance of a large alveolar surface area that needs to be covered with a thin layer of alveolar lining fluid. Both are an absolute requirement for optimized alveolar gas exchange. ATII cells produce surfactant to reduce the surface tension of the lining fluid. Both ATI and ATII cells are involved in removal of excessive alveolar fluid. ATII cells also are responsible for repair of the alveolar epithelium upon damage. They are capable of defense against bacteria by respiratory bursts, a role they share with alveolar macrophages. On the other hand, ATII cells also have a high activity of enzymes to inactivate free oxygen radicals.

Na^+ and Water Reabsorption

Fluid enters the alveolar space through occasional leaks driven by hydrostatic pressure gradients and by secretion in distal bronchioles. A small amount of this fluid is required to allow physical solution of respiratory gases at the air–liquid interface before diffusion across the alveolar barrier. The time required for the diffusion of gases from the air into the capillary increases if this layer of lining fluid is thickened as in alveolar edema. Active reabsorption is therefore necessary to limit the thickness of the layer of alveolar lining fluid and to optimize gas diffusion.

Active reabsorption of water from the alveolar (apical) to the basolateral side of cultured alveolar epithelial cells was discovered when dome formation of cells grown on impermeable support was observed. Domes are thought to be fluid-filled spaces under the basolateral side of confluent cell monolayers of primary rat alveolar epithelial cells that are generated by the reabsorption of Na^+ and water, since dome formation was not observed in the presence of inhibitors of active Na^+ transport (Goodman & Crandall 1982). These results indicated that the reabsorption of water from the alveolar space was coupled to the reabsorption of Na^+ and Cl^- from the alveolar surface to the basolateral side of the epithelium. Results on cultured alveolar epithelial cells were confirmed with experiments on the intact, fluid-filled lung. In this system, isotonic media containing marker molecules that are not reabsorbed are instilled into the lung. Aspired instilled fluid had an increased concentration of these marker molecules, which is indicative of net water reabsorption from the alveolar space. Addition of amiloride to block apical Na^+ entry via Na^+ channels inhibited water reabsorption, which indicates that also in vivo, the reabsorption of water depends on the net reabsorption of Na^+.

Recent literature indicates that both ATI and ATII cells are involved in Na^+ reabsorption, although the body of information on the control of Na^+ transport comes from experiments on cultured ATII cells, as well as from measurements of the rate of reabsorption of fluid instilled into lungs of experimental animals. The mechanism of Na^+ transport is similar to that described earlier for the airway epithelium (figure 14.2*b*): Na^+ enters the cells on their apical side via Na^+ channels; Na^+-coupled transporters such as Na^+/glucose, Na^+/amino acid transport, and Na^+/phosphate cotransport, and Na^+/H^+ and Na^+/Ca^{2+} exchange. Cultured alveolar epithelial cells express both highly Na^+-selective and nonselective cation channels that both transport Na^+ and are both inhibited by amiloride. About 50% to 75% of transepithelial transport and lung fluid reabsorption has been shown to be amiloride sensitive. The most important cation channel is the epithelial Na^+ channel (ENaC), which is highly selective for Na^+. Its significance has been demonstrated in knockout mice that lack the α subunit of this channel. These mice die within 40 h after birth due to the lack of reabsorption of the fluid that is contained in the fetal lung. Partial gene recovery that restores the presence of this subunit, but at a reduced level of ENaC expression, is sufficient for the clearance of alveolar fluid after birth (Hummler et al. 1996), whereas the other subunits of ENaC seem not to be a basic requirement for basal alveolar fluid balance (Rossier et al. 2002).

The Na^+ gradient into the cell that allows Na^+ uptake is maintained by basolateral Na^+/K^+ pumps. Overexpression of the β1 subunit of the Na^+/K^+ pump also increases the rate of transepithelial Na^+ transport (Factor et al. 1998). Interestingly, overexpression of the α1 subunit did not alter fluid reabsorption.

To maintain electroneutrality, Cl^- follows the Na^+ transported across the epithelium. The pathways for Cl^- transport are not well understood. Recent literature indicates a possible involvement of very distal bronchi in lung fluid reabsorption. However, a physiological role for CFTR in alveolar epithelial cells has not been demonstrated. At birth, CFTR knockout mice are able to clear the water contained in the lung and maintain an adequate alveolar fluid balance in adult life (Matthay et al. 2002).

Cellular and paracellular pathways mediate water movement across the epithelium in the direction of net Na^+ and Cl^- transport. The alveolar epithelium has a very high water conductance (Verkman et al. 2000). Alveolar epithelial cells express aquaporins, which are water channels. Aquaporin 5 is found in ATI cells, whereas ATII cells express aquaporin 3 and 4. However, no disturbance of the alveolar water balance at birth and in the adult lung has been observed in mice when the genes for these water channels have been knocked out although the water conductivity of the alveolar epithelium was considerably reduced in these mice (Verkman et al. 2000).

Regulation of Alveolar Reabsorption

Stimulation of alveolar fluid clearance is required when the filtration of fluid into the alveolar space is increased. Both in primary cultured ATII cells and in fluid-filled lungs, reabsorption is stimulated by β-adrenergic agents, cAMP, or both. In ATII cells these agents stimulate the reabsorption of Na^+ by activation of amiloride-sensitive pathways, most prominently by activation of ENaC (Matthay et al. 2002). There is evidence that the action of β-adrenergics is caused by increase in the number of Na^+ channel molecules in the plasma membrane due to insertion of endogenous ENaC present in endocytic vesicles. It is unclear whether the channel conductance is also modulated. Increased levels of cAMP cause an increase in intracellular Ca^{2+}, which increases the open probability of nonselective cation channels independent of the cAMP effects on ENaC.

It has been postulated that the stimulation of Na^+ transport by β-adrenergic agents requires preceding activation of Cl^- transport (O´Grady et al. 2000). Recently it has been shown that the stimulation of water reabsorption by terbutaline required the stimulation of CFTR-mediated Cl^- transport, since the stimulated portion of water reabsorption (but not the basal rate) was drastically reduced in CFTR knockout mice. As the presence of CFTR in alveolar epithelium is not well documented, it appears likely that this result indicates reabsorption occurring in distal bronchiolar epithelium (Matthay et al. 2002).

Stimulation by β-adrenergic agents also increases the activity of the Na^+/K^+ pump. The increase seems to be independent of increased apical Na^+ entry. It occurs rapidly (~15 min) and is associated with an increased pump capacity and an increased number of ouabain binding sites indicative of an increased number of pump molecules in the basolateral plasma membrane. As with Na^+ channels, long-term exposure with β-adrenergics of alveolar epithelial cells also seems to affect the rate of expression of Na^+/K^+ pumps (Matthay et al. 2002).

Although there is evidence that alveolar epithelial cells lack the mineral corticoid receptor, both glucocorticoids and aldosterone stimulate alveolar epithelial Na^+ transport. The main effect seems to be an increased expression of both ENaC and Na^+/K^+ pumps. In vivo, normal levels of glucocorticoids are important for maintaining a normal alveolar fluid clearance (Matthay et al. 2002).

Growth factors are involved in tissue development and repair. They also stimulate alveolar fluid clearance. Application of growth factors such as epidermal growth factor (EGF), transforming growth factor (TGF-α), and keratinocyte growth factor (KGF), independent of changes in cAMP, seems to stimulate the expression of both Na^+ channels and Na^+/K^+ pumps. Since these growth factors also stimulate the proliferation of ATII cells, it is unclear whether the effects are related (Matthay et al. 2002).

Atrial natriuretic peptide (ANP) inhibits Na^+ reabsorption in the kidney. Its effects on lung water balance are less clear. In cultured alveolar epithelial cells, ANP seems to inhibit Na^+ reabsorption by decreasing amiloride-sensitive Na^+ transport. Na^+/K^+ pumps are not affected by ANP (Matthay et al. 2002).

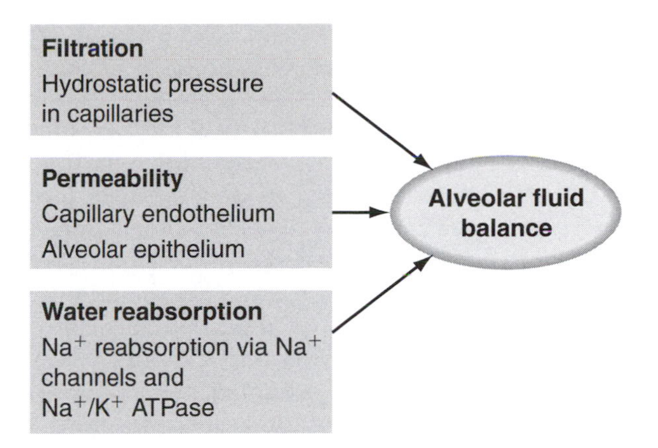

Figure 14.7 Mechanisms that affect the alveolar fluid balance. Water reabsorption is required to minimize the amount of alveolar lining fluid. An increased alveolar permeability increases the rate of fluid filtration into the interstitial and alveolar space. Increased alveolar capillary pressure and inflammatory processes at the alveolar wall can cause (transient) leaks of the barrier, increase the rate of filtration, and cause interstitial and alveolar edema. An inhibition of alveolar Na^+ reabsorption blunts the reabsorption of fluid from the alveolar surface. All three mechanisms have been shown to occur at high altitude (Schoene 1987; Swenson et al. 2002; Höschele & Mairbäurl 2003). Increased filtration into the alveolar space has also been shown to occur during high-intensity exercise (West 2000).

Disturbances of Alveolar Fluid Balance

The function of the alveolar barrier can be affected in many ways. Increased alveolar capillary pressure, inflammatory processes, the inhalation of aggressive substances or particles, hypoxia inhibition of the reabsorption of alveolar fluid, and other mechanisms may disturb the integrity of the epithelial surface, impair surfactant secretion, and cause alveolar fluid accumulation—all of which also impair alveolar gas exchange. This section addresses only selected examples of alveolar dysfunction (figure 14.7).

Increased Capillary Pressure

The capillary network supplying the alveoli is particularly sensitive to increased capillary hydrostatic pressure, since distension of the capillary wall is not counteracted by adjacent tissue. It has been shown that perfusion of the lung with high pressure causes capillary stress failure (Costello et al. 1992) characterized by leaks of the capillary endothelium, interstitial edema, and occasional leaks of the alveolar barrier such that even red blood cells can enter the alveolar space. Submicroscopic leaks seem to occur already at lower pressures and cause the filtration of albumin-rich fluid into the alveolar space (Parker et al. 1979). Typical clinical situations that cause interstitial and alveolar edema are pulmonary hypertension and left heart dysfunction. In exercise, pulmonary capillary stress failure with extravasation of large volumes of blood has been observed in racehorses and has been discussed as a factor limiting maximal exercise (West & Mathieu-Costello 1995). Alveolar fluid clearance is ineffective in clearing hydrostatic alveolar edema as long as epithelial leaks persist and when the rate of fluid filtration exceeds the rate of reabsorption (Höschele & Mairbäurl 2003). Alveolar flooding with fluid, blood, or both can be prevented only by decreasing the hydrostatic pressure in alveolar capillaries.

Defective Na^+ Transport

It was indicated earlier that most of the reabsorption of water from the alveolar space depends on the presence and activity of apical Na^+ channels. When the α subunit of ENaC is genetically removed, these knockout mice die within 40 h after birth due to lack of reabsorption of alveolar fluid (Hummler et al. 1996). Loss-of function mutations of the other subunits of ENaC do not impair lung fluid clearance (Rossier et al. 2002). This indicates that the low activity of the Na^+ channel formed by αENaC alone is sufficient to mediate the removal of alveolar fluid after birth, though reabsorption is slower than that in littermates. β and γENaC mutations simulate a disease called pseudohypoaldosteronism (PHA-1). Mutations that cause PHA-1 are distributed all along the ENaC sequences. The severe lethal types found in knockout mice have not been observed in humans. Also, in the human lung, phenotypes manifest later and symptoms are less pronounced than in the mouse model. The role in malfunction of ENaC in different kinds of pulmonary edema including HAPE is currently under investigation.

Hypoxia

When mountaineers ascend rapidly to high altitude, about 50% will suffer from acute mountain sickness with symptoms such as headache, insomnia, anorexia, and nausea (Bärtsch 1997). About 0.2% to 5% will develop high-altitude pulmonary edema (HAPE). The incidence increases to 15% due to lack of acclimatization when the ascent is too fast. It can increase to nearly 70% in mountaineers with previous HAPE. The clinical picture of HAPE shows dyspnea, cough, gurgling in the chest, and pink, frothy sputum. At an altitude of 4,559 m, where the arterial PO_2 is about 45 mmHg (arterial SO_2 above 75%) in healthy mountaineers, in HAPE arterial PO_2 and SO_2 can decrease to 30 mmHg and below 60%, respectively. Systolic pulmonary artery pressures higher than 90 mmHg have been found (Swenson et al. 2002).

HAPE is a noncardiogenic form of edema, in which lung water in the interstitium, the alveolar space, or both is increased due to an increased hydrostatic pressure in lung capillaries subsequent to (inhomogeneous) arterial or venous hypoxic vasoconstriction (or a combination of the two), which augments filtration and can even rupture the alveolar barrier. A contribution of increased permeability of the capillary endothelium or the alveolar epithelium is unlikely. A role of a defective alveolar fluid reabsorption in the pathogenesis of HAPE is unclear (figure 14.7).

Hypoxia of alveolar epithelial cells in culture has been shown to impair the reabsorption of Na^+ across the alveolar epithelium (Mairbäurl et al. 2002). Transport inhibition is brought about by inhibition of apical entry of Na^+ via ENaC and by inhibition of basolateral Na^+/K^+ pumps (Clerici 1998), caused by inhibition of transport activity and by a reduction of the copy number of those transporters in the plasma membrane of alveolar epithelial cells (Wodopia et al. 2000; Planes et al. 1997). The mechanisms that lead to either effect are entirely unclear. They might involve altered cellular levels of reactive oxygen species and effects of hypoxia on the signal transduction pathways that control transport activity and membrane insertion of active transporters.

HAPE can be prevented or treated through decrease in pulmonary capillary pressure when the alveolar PO_2 is increased by descent to low altitudes or application of oxygen. Treatment with nifedipine also releases pulmonary vasoconstriction and improves oxygenation (Bärtsch 1997). Once normoxic, edema is rapidly reabsorbed. Prophylactic treatment has also been attempted through alveolar application of β-adrenergics (Sartori et al. 2002). The idea of this treatment is based on the known stimulation of these agents of alveolar fluid clearance, decrease in pulmonary artery pressure, and tightening of the endothelial barrier. In fact, inhalation of aerosolized salmeterol decreased the occurrence of HAPE by about 25% in HAPE-susceptible mountaineers.

Chapter 15

Exercise and the Liver Cell

Jean-Marc Lavoie

The liver is involved in a multitude of specialized biochemical functions that make it the body's major metabolic processing unit. Among these are the synthesis of blood plasma proteins, the degradation of porphyrins and nucleic acid bases, the storage of iron, and the detoxification of biologically active substances such as drugs, poisons, and hormones by a variety of oxidation, reduction, hydrolysis, conjugation, and methylation reactions. The liver is also the main organ responsible for maintaining the proper levels of nutrients in the blood for use by peripheral tissues including muscles. At rest, the unique situation of the liver contributes to this task since all nutrients absorbed by the intestines except fatty acids are released into the portal vein, which drains directly into the liver. During exercise, this task becomes more complex because physical exercise is a unique situation that requires a tremendous increase in energy production in order for the body to respond to the increased energy demand while maintaining homeostasis.

Prominent liver responses to exercise include high activation of glycogenolysis, the conversion of glucogenic amino acids into pyruvate or citric acid cycle intermediates to supply glucose through gluconeogenesis, and fatty acid degradation to Acetyl-CoA and then to ketone bodies for export via the bloodstream to the peripheral tissues. All of these functions require adaptations of the liver at the cellular level, an understanding of which has recently begun to emerge. The purpose of this chapter is to present the cellular and molecular aspects related to the metabolic function of the liver during exercise and the molecular adaptations of the liver cell following physical training.

Hepatic Carbohydrate Metabolism

The main role of the liver during exercise is to ensure the maintenance of an adequate blood glucose supply in response to an increased glucose utilization by the working muscles. Hepatic glucose production, through hepatic glycogenolysis and gluconeogenesis, increases during prolonged exercise in a way very similar to muscle glucose utilization. As a result, hypoglycemia does not occur, at least until hepatic glycogen content is significantly reduced.

Physiological Aspects

The initial increment in hepatic glucose production during the early stages of exercise is due almost entirely to accelerated glycogenolysis, while gluconeogenesis becomes increasingly more important with prolonged exercise as hepatic glycogen stores decrease. The increase in hepatic glucose production during exercise is due to the actions of several regulatory factors. The decrease in plasma insulin and the increase in glucagon appear to be of primary importance for the increase in hepatic glucose production during exercise (for a review, see Wasserman 1995). There is evidence that sympathetic nervous activity to the liver and circulating norepinephrine do not play an important role in the exercise-induced rise in glucose production, whereas epinephrine has only a minor stimulating effect on hepatic glucose production during intense exercise, and late during prolonged exercise (Kjaer 1998). In spite of this information, the mechanism responsible for the close link between glucose production and muscle glucose utilization during exercise remains unclear. Glucoregulatory mechanisms during exercise seem to be very sensitive to transient or subtle blood glucose changes, and independently of pancreatic hormone and hepatic catecholamine action (Coker et al. 2002). There is also evidence that glucose sensors in the liver may be involved in monitoring liver glycogen breakdown or a related variable through pancreatic hormone responses during exercise (Lavoie et al. 1989). Afferent neural feedback signals from contracting muscle and motor center-related feedforward mechanisms (Kjaer 1992) have also been hypothesized to be involved in glucoregulation during exercise. Finally, the recent finding that transcriptional activation of interleukin 6 (IL-6) in skeletal muscle is higher when glycogen levels become low (Keller et al. 2001) supports the hypothesis that IL-6 may be produced by contracting myofibers as a means of signaling the liver to increase glucose production.

Molecular Aspects

Hormonal regulation of hepatic metabolism in response to exercise occurs through activation of both cyclic adenosine monophosphate (cAMP)-dependent and cAMP-independent mechanisms (Nieto et al. 1996). Cyclic AMP levels increase during exercise, and the magnitude of the increase depends on the duration and intensity of exercise. Hepatic glycogenolysis during exercise is stimulated through allosteric/covalent modification of the enzymes glycogen phosphorylase and glycogen synthase within the liver. Gluconeogenesis, on the other hand, is regulated by mechanisms that influence precursor delivery to the liver, hepatic extraction, and the activity of key gluconeogenic enzymes (Wasserman & Cherrington 1991). The hepatic enzyme phospho*enol*pyruvate carboxykinase (PEPCK) is considered a major control point in the regulation of flux through the gluconeogenic pathway. Evidence that intrahepatic gluconeogenic mechanisms are stimulated by exercise comes from the demonstration that maximal activity of PEPCK continuously increased with exercise duration and was highest at exhaustion (figure 15.1*a;* Dohm et al. 1985). In contrast, phosphofructokinase (PFK) submaximal activity and its activator, fructose-2,6-biphosphate, were decreased by 30 min of exercise and remained at that low activity until exhaustion.

In search of a molecular mechanism that controls the increased hepatic PEPCK activity with exercise, Friedman (1994) reported a 290% increase in the levels of hepatic PEPCK mRNA after 30 min of exercise in mice fed a high-carbohydrate diet (80%). The level of PEPCK mRNA continued

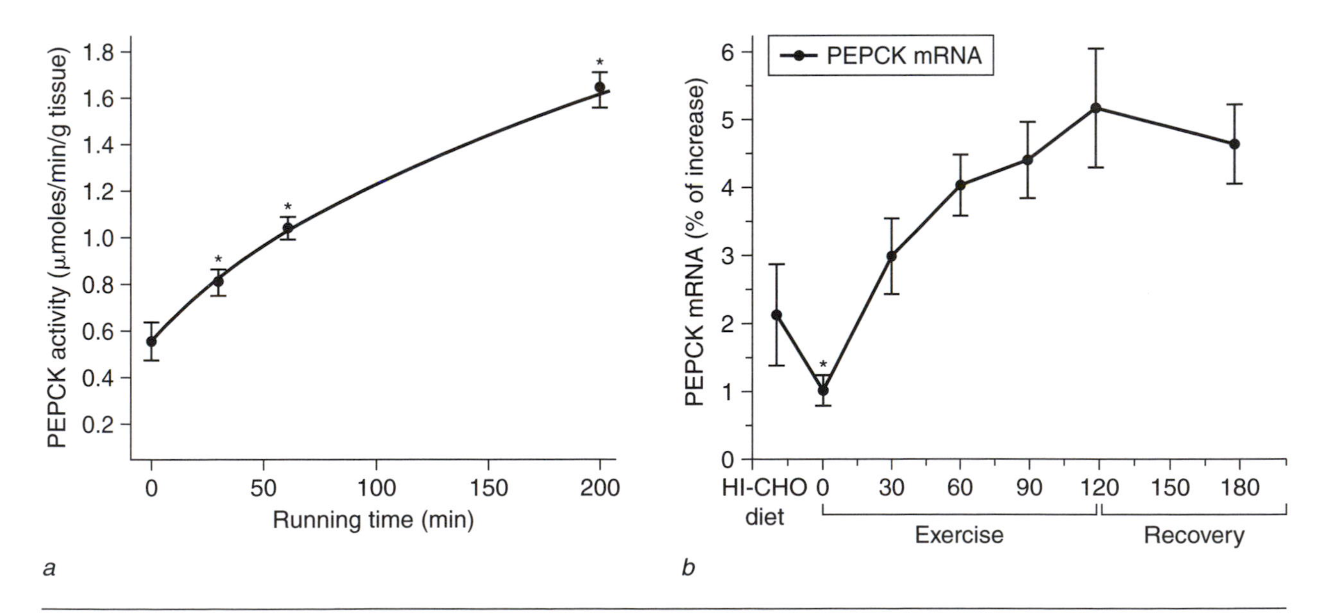

Figure 15.1 Changes in *(a)* hepatic phospho*enol*pyruvate carboxykinase (PEPCK) activity after various running times and *(b)* PEPCK mRNA in mice fed a high-carbohydrate diet and exercised for up to 120 min. *Significantly different than rested control, P < 0.05, in *a* and significantly (P < 0.01) less than all other time points in *b*.

Reprinted, by permission, from G.L. Dohm et al., 1985, "Time course of changes in gluconeogeneic enzyme activities during exercise and recovery," *American Journal of Physiology* 249: E6-E11 and from J.E. Friedman et al. 1994, "Role of glucocorticoids in activation of hepatic PEPCK gene transcription during exercise," *American Journal of Physiology* 266: E560-E566.

to increase progressively during the course of prolonged exercise, reaching 510% after 120 min of exercise, and remained elevated 60 min after exercise (figure 15.1*b*). Nuclear runoff experiments using nuclei isolated from livers of control and exercised mice revealed a 700% increase in PEPCK gene transcription within 30 min, reaching 1,000% after 120 min and returning to control level after 60 min of recovery (Friedman 1994). The author explains the excess transcription rate relative to mRNA by suggesting that initial transcripts may not be fully processed into mature cytosolic RNA molecules or by a shorter mRNA half-life relative to the transcription rate. The transcriptional activation of the PEPCK promoter in cell culture is enhanced by the addition of specific nuclear proteins (transcription factors), which bind to distinct sequences of the promoter regulatory region. The promoter regulatory region of the PEPCK gene contains a cAMP response element (CRE; -87/-74), which also requires the presence of a second binding site (P31; –248/–230) for full responsiveness to cAMP (Patel et al. 1994. Three families of leucine zipper proteins are capable of binding to the CRE or P31 domain of the PEPCK promoter, or both: cAMP response element binding protein (CREB)/activating transcription factor, Fos-Jun/activator protein 1, and CAAAT/enhancer binding protein β (C/EBP-β). It has been found that exercise induces a 12-fold increase in C/EBP-β gene transcription that parallels the induction of the PEPCK gene, while *c-jun* transcription increased by 350% and transcription of *c-fos* was not significantly altered during exercise (Friedman 1994; figure 15.2). In the same study, Friedman (1994) also showed that deletion of the PEPCK glucocorticoid regulatory unit (GRU; –451/–351) abolished the transcriptional response to exercise in transgenic mice. In a following study, Nizielski et al. (1996), using transgenic mice carrying a single mutation in either the CRE or P31 element alone, showed that both sites are required for full induction of PEPCK gene expression by exercise, as well as the GRU site.

The mechanism for the enhancement of PEPCK gene transcription as a result of hormonal changes during exercise is unknown. The rapid appearance of C/EBP-β in the nucleus suggests that the immediate effects of exercise on PEPCK gene transcription may involve phosphorylation of C/EBP-β, which is phosphorylated by calcium/calmodulin-dependent protein kinase II (Wegner et al. 1992), and translocalization from the cytosol to the nucleus. However, Nizielski et al. (1996) were unable to detect the protein in cytoplasmic extracts from either control or exercised mice. Exercise also resulted in an increase in the proportion of total nuclear proteins bound to oligonucleotides corresponding to DNA binding site CRE or P31 of the PEPCK promoter due to C/EBP-β (Nizielski et al.

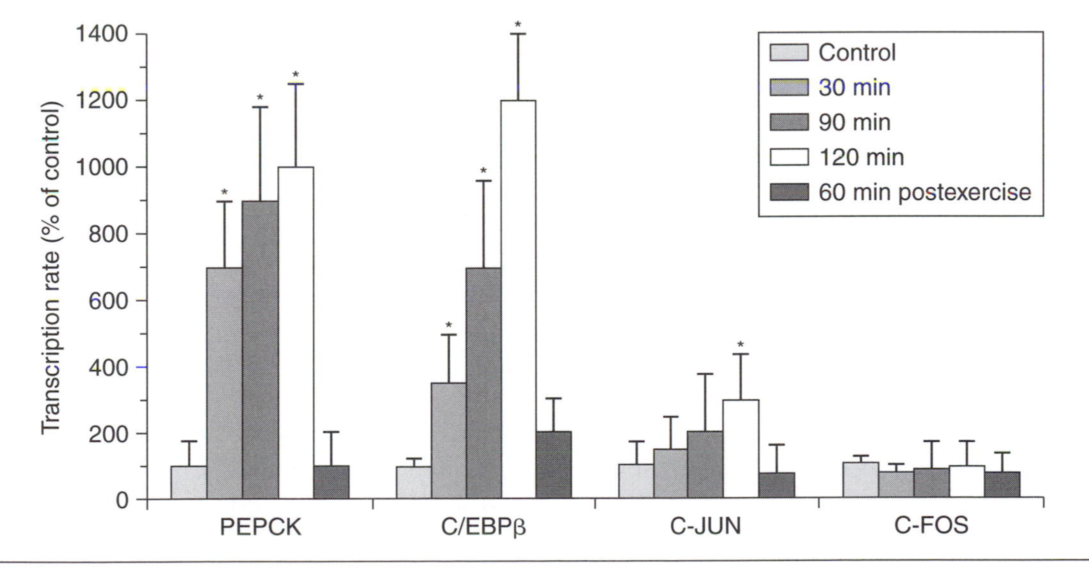

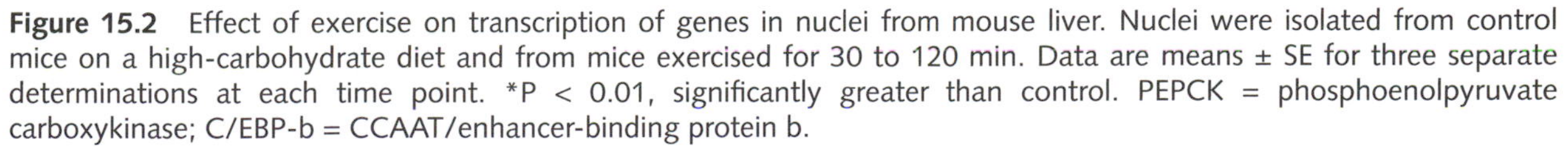

Figure 15.2 Effect of exercise on transcription of genes in nuclei from mouse liver. Nuclei were isolated from control mice on a high-carbohydrate diet and from mice exercised for 30 to 120 min. Data are means ± SE for three separate determinations at each time point. *P < 0.01, significantly greater than control. PEPCK = phosphoenolpyruvate carboxykinase; C/EBP-b = CCAAT/enhancer-binding protein b.

Reprinted, by permission, from J.E. Friedman et al. 1994, "Role of glucocorticoids in activation of hepatic PEPCK gene transcription during exercise," *American Journal of Physiology* 266: E560-E566.

1996). However, this change was small compared with the large increase in PEPCK gene transcription observed during exercise. The authors suggested that elevation of C/EBP-β during exercise may increase transcription, by altering interactions with proteins that normally inhibit PEPCK gene transcription, or that C/EBP-β interacts with other transcription factors to trigger a productive interaction at sites other than CRE and P31. Taken together, these results are compatible with a role for the activation of gene transcription of key gluconeogenic enzymes to prevent hypoglycemia during prolonged exercise. Transcription factors, such as C/EBP-β, may play an important role in stimulating gluconeogenesis at the molecular level.

Hepatic Zonation

Metabolic zonation is a 25-year-old concept proposed by Sasse et al. (1975) on the basis that hepatocytes, depending on their position along the acinus, possess different amounts and activities of enzymes and thus different metabolic capacities. The first zone includes hepatocytes located in proximity to the terminal portal venules or periportal hepatocytes (PP-H), which are perfused with blood rich in O_2, substrates, and hormones. The second zone includes hepatocytes in the vicinity of terminal hepatic venules or perivenous hepatocytes (PV-H), which are perfused with blood containing lower concentrations of O_2, substrates, and hormones and with higher concentrations of CO_2 and other by-products of metabolism. The periportal cells show a greater capacity for gluconeogenesis, and perivenous cells for glycolysis (Jungermann & Thurman 1992). The regional heterogeneity of the hepatocytes that compose the liver causes the liver cells to respond differently to physiological and pathological situations such as starvation, cold adaptation, and diabetes (Jungermann & Kietzmann 1996). With regard to exercise, Désy et al. (2001) examined the effects of a single bout of running exercise on the gluconeogenic activity of isolated PP-H and PV-H in fasted rats. Hepatocytes selectively isolated, using anterograde and retrograde liver perfusions, were incubated with saturating concentrations of alanine or a mixture of lactate and pyruvate to determine the glucose production flux ($J_{glucose}$) in the incubation medium. Exercise resulted in the stimulation of $J_{glucose}$ in PP-H and PV-H. However, the magnitude of this stimulation appears to be greater in PV-H than in PP-H, resulting in the disappearance of the portovenous difference observed in the resting state (figure 15.3). In a subsequent study using the same experimental approach, Bélanger et al. (2003) reported that the increase in the gluconeogenic capacity of the liver following endurance training was first observed in perivenous hepatocytes. These results indicate that, similarly to muscle, liver cells do not respond homogeneously to the exercise stimulus and that this aspect should be taken into account in discussion of changes at the cellular level.

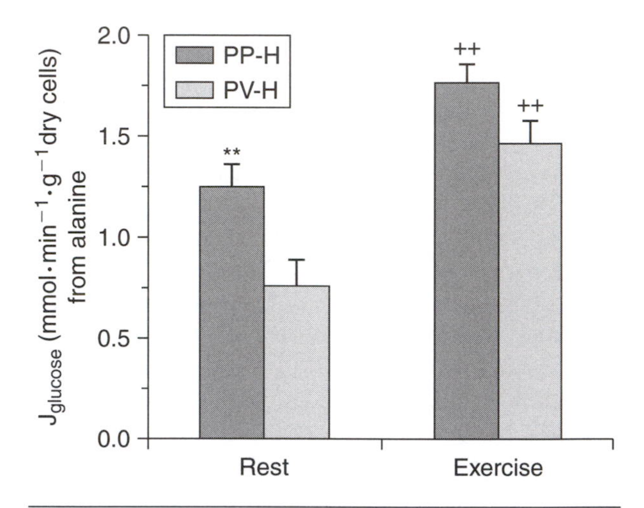

Figure 15.3 Glucose production flux ($J_{glucose}$) from linear accumulation of glucose over the incubation period of periportal (PP-H) and perivenous (PV-H) hepatocytes in the presence of alanine (20 mM) and oleate (2 mM) after correction for the respective endogenous glucose production. Results are means ± SE for $n = 9$ in each group. ++Significantly different ($P < 0.01$) from the corresponding rest group of identical type of hepatocyte preparation. **Significantly different ($P < 0.01$) between PP-H and PV-H of the same activity status.

Reprinted, by permission, from F. Désy et al. 2001, "Effects of acute exercise on the gluconeogenic capacity of periportal and perivenous hepatocytes," *Journal of Applied Physiology* 91: 1099-1104.

Training Effects

Endurance training has long been known to result in a better maintenance of blood glucose level during prolonged exercise, both in humans and in rats. However, the question whether this was due to decreased glucose uptake or to increased glucose production was, until the early 1990s, the subject of some speculation. In vivo studies conducted in rats by Donovan and Sumida (1990) demonstrated that this was ascribed to an increased glucose production attributable to a higher rate of gluconeogenesis. Further evidence of the impact of training on gluconeogenesis was presented by the group of Donovan and collaborators, who reported, in perfused livers, a ~25% increase in the Vmax for hepatic gluconeogenesis at saturating lactate and alanine concentrations (Sumida & Donovan 1996). Podolin, Gleeson et al. (1994) also reported an increase in the rate of [^{14}C]-lactate incorporation into glucose in liver slices of trained animals of different ages. The training effect on gluconeogenesis was thereafter confirmed by Burelle et al. (2000), who reported an increased glucogenic flux in hepatocytes from trained rats following incubation with either alanine (+64%), glycerol (+21%), lactate-pyruvate (+18%), or dihydroxyacetone (+10%). Further experiments conducted on perifused hepatocytes showed that the increased gluconeogenic flux ($J_{glucose}$) from alanine could not be due to adaptations at the level of the phospho*enol*pyruvate-pyruvate cycle, since the relationship between the glucogenic flux and pyruvate was similar for hepatocytes from trained and control rats (Burelle et al. 2000). By plotting together the concentrations of selected intermediates and $J_{glucose}$, Burelle et al. (2000) showed that the increased gluconeogenesis from alanine in trained rats was due to an increased alanine transport capacity at low alanine concentration and to an increased transamination of alanine into pyruvate at high alanine concentration. The increase in $J_{glucose}$ from lactate-pyruvate after training, however, points to an adaptation at the level of the phospho*enol*pyruvate-pyruvate cycle, in spite of the fact that the maximal activity of PEPCK, pyruvate carboxylase, and PEPCK mRNA appears unchanged in response to training (Podolin, Pagliassotti et al. 1994; Horn et al. 1997). It must be mentioned, however, that the increase in maximal gluconeogenic capacity with training is not incompatible with the report that endurance training reduces the contribution of both hepatic glycogenolysis and gluconeogenesis during moderate-intensity exercise performed at the same absolute intensity after, compared with before, training (Coggan et al. 1995).

The increase in gluconeogenic capacity of the liver with exercise training is something of a paradox in view of the decrease in glucoregulatory hormones and gluconeogenic precursors observed during exercise in trained individuals. One possible explanation for this apparent paradox comes from recent studies showing an increase in hepatic hormonal sensitivity following endurance training. Podolin and colleagues published two papers in 1994 in which they showed clearly a decreased capacity for hepatic gluconeogenesis with age (Podolin, Pagliassotti et al. 1994; Podolin, Gleeson et al. 1994). In one of these studies (Podolin, Gleeson et al. 1994), they reported an increased responsiveness of gluconeogenesis to norepinephrine in trained animals. Previously, Bonen et al. (1986) failed to observe any change in insulin binding properties in liver membranes after six weeks of endurance training in rats. Sensitivity of the liver to glucagon, on the other hand, has been shown to increase with exercise training. Nieto et al. (1996) were first to report an increase in adenyl cyclase activity over a wide range of glucagon concentrations. Subsequently, Drouin et al. (1998) found that hepatic glycogenolysis in response to glucagon was faster and greater (53%) in trained compared to untrained subjects. To go one step further, Légaré et al. (2001) studied the effects of endurance training on the density and affinity of glucagon receptors in plasma membrane isolated from the livers of rats submitted to eight weeks of swimming. They observed an increase in the density, but not in the affinity, of the glucagon receptors. The recent finding that hepatic glucose production, in response to hyperglucagonemia, was increased in rats immediately after an acute bout of exercise indicates that the increased sensitivity of the liver to glucagon is a phenomenon that takes place rapidly (Matas Bonjorn et al. 2002). Finally, Podelin et al. (2001) investigated the effects of aging and training on alterations in glucagon receptor-mediated signal transduction in rat livers. They found an increase in stimulatory G-protein content in old but not young or middle-aged rats. Furthermore, training resulted in an increase in cAMP production (adenyl cyclase activity) when the catalytic subunit of adenyl cyclase was directly stimulated by the pharmacological stimulator, forskolin, which bypasses both receptor and G-proteins. In summary, it is clear that the liver is capable of molecular adaptations to chronic exercise. These adaptations would lead to enhanced hepatic hormonal responsiveness and capacity that may be important to improve glucose homeostasis during exercise, but also in other situations such as aging and diabetes.

Ureagenesis

Many studies have indicated that exercise increases serum urea concentration, urinary urea excretion, and sweat loss. Moreover, it has been shown that ureagenesis increases during exercise (Wasserman et al. 1991). Uptake of amino acids by the liver increases during exercise in dogs, suggesting that these amino acids are used for augmented gluconeogenesis in the liver (Wasserman et al. 1991). In agreement with these findings, it was recently reported that glucose administered with amino acids before and during exercise decreases hepatic ureagenesis due to reduced hepatic gluconeogenesis, in comparison to administration of amino acids alone (Hamada et al. 1998). In addition to increased gluconeogenesis, urea formation also needs to be increased to process ammonia released from working muscles. In livers perfused with ammonia to determine the capacity of the liver to uptake ammonia and produce urea, it was found that both of these parameters were increased by acute exercise (Ferreira et al. 1998). The authors pointed out that the increased ammonia concentration during exercise is a consequence of the limited hepatic blood flux during exercise but not due to a limited capacity of the liver to uptake ammonia and produce urea. Ferreira et al. (1998) also suggested that the changes in ureagenesis induced by exercise were mediated by cortisol (ammonia intake) and by glucagon (urea production). Indeed, glucagon and glucocorticoids have been shown to elicit a synergistic effect raising the mRNA of the five urea-cycle enzymes in cultured hepatocytes (Ulbright & Snodgrass 1993). In the study of Ferreira et al. (1998), acute exercise also resulted in higher urea production from L-glutamine. The amino acid glutamine is one of the two amino acids, with alanine, that are of primary importance for nitrogen delivery from peripheral tissue to liver. In exercising dogs, glutamine uptake increases by twofold as a result of an increase in hepatic fractional extraction (Halseth et al. 1998). In a subsequent study, Krishna et al. (2000) showed that the exercise-induced increase in glucagon was essential for the increase in hepatic glutamine uptake and fractional extraction and was required for the full increment in ureagenesis. Consistent with these findings, glucagon has been reported to increase glutaminase activity (Brosnan et al. 1995) in the isolated perfused rat liver. All in all, it seems that the increase in ureagenesis during exercise is in large part related to enhanced gluconeogenesis from amino acids and that the capacity of the liver to produce urea does not constitute a metabolic limitation.

Hepatic Glucose Uptake

Physical exercise depletes glycogen stores in skeletal muscle and in the liver. It is well known that muscle glucose uptake is greatly enhanced during exercise and remains elevated in the postexercise state to replenish muscle glycogen stores. On the other hand, liver produces glucose during exercise, and hepatic glucose production continues in the recovery period. When glucose is administered peripherally after exercise, the liver stores glycogen through direct and indirect pathways (Johnson & Bagby 1988). It was not known until recently whether exercise could affect the disposal of exogenous glucose in the liver. The first piece of evidence that supported this idea came from Matsuhisa et al. (1998), who observed that prior contraction of the rabbit hindlimb caused an increase in the rate of liver deposition of the glucose analog 3-fluoro-3-deoxy-D-glucose during a glucose load. These results were confirmed by Galassatti, Coker et al. (1999), who reported that prior exercise led to a 74% greater net hepatic glucose uptake of an intraportal glucose load in conscious dogs in which several factors known to stimulate net hepatic glucose uptake were controlled. It was calculated that 28% of the whole-body glucose uptake could be accounted for by the net hepatic glucose uptake. At the cellular level, glucose transport across the hepatocyte membrane is primarily mediated by the GLUT2 isoform of the glucose transporter family. It is the phosphorylation step of glucose by the enzyme glucokinase, however, that is thought to be rate limiting for glucose uptake in the liver. An increase in glucokinase activity has been reported after exercise in rats (Dohm & Newsholme 1983). It is not known, however, how exercise specifically affects the mechanism (i.e., translocalization, inhibitory protein) that regulates glucokinase activity.

In resting conditions, hepatic glucose uptake is greater if glucose is administered intraportally than if it is given peripherally. The intraportal infusion of glucose generates a negative arterial-portal venous gradient as in the case of normal glucose feeding. This condition is believed to create a portal signal capable of not only activating hepatic glucose uptake (Myers et al. 1991) but also inhibiting glucose uptake by rested skeletal muscle (Galassetti et al. 1998). In the postexercise state, it was found that the portal signal still stimulated hepatic glucose uptake but that its ability to inhibit muscle glucose uptake was overridden by the stimulatory effect of prior exercise (Galassetti, Koyama et al. 1999). In a recent study (Burcelin et al. 2000), however, it was reported that in mouse, activation of the portal signal induced hypoglycemia by stimulating glucose utilization by a subset of tissues.

The postulated mechanism of action of the portal signal implies that the liver is afferently able to contribute to glucose homeostasis. There is considerable evidence that the liver may act as a metabolic sensor to provide signals for the control of food intake (for a review, see Friedman & Tordoff 1996). There is also growing evidence that the liver is afferently able to contribute to metabolic regulation of exercise. The hypothetical construct underlying the liver's afferent contribution to metabolic regulation during exercise is that a decrease in liver glycogen or a related metabolic intermediate is sensed by the liver and the signal is transduced to the central nervous system, most likely through the afferent activity of the hepatic vagus nerve, where it contributes to the orchestration of the metabolic and hormonal responses of exercise. Support of this construct comes mainly from the demonstration that sectioning of the hepatic vagus nerve attenuates the normal hormonal response to exercise (Lavoie et al. 1989). In other respects, the hepatic mechanism responsible for linking metabolic activity in the liver to an afferent signal capable of regulating the metabolic response to exercise remains speculative. Substrates or derivatives of substrate oxidation, energy-related compounds (Pi/ATP), or changes in cell volume may all be related to changes in transmembrane potential in the liver cell, which according to the "potentiostatic" theory would determine the afferent vagal activity (for a recent review, see Lavoie 2002).

Hepatic Lipid Metabolism

In an adult, the liver represents only 15 to 20 g/kg body weight but accounts for 20% to 25% of the resting energy expenditure, suggesting a high organ metabolic rate (Müller 1995). Under basal aerobic conditions the liver primarily selects fatty acids as its fuel. In the fed state, liver fuel mix changes, and amino acids and pyruvate predominate as fuel whereas most of the carbohydrate is stored as glycogen (Müller 1995). Starvation is associated with increased rates of hepatic gluconeogenesis, fatty acid oxidation, and ketogenesis. In the rat, hepatic ketone body production may account for up to 68% (after 24 h of starvation) of hepatic free fatty acid uptake (Remesy & Demigne 1983). As already mentioned, liver glycogenolysis and gluconeogenesis gain increasing importance with increasing duration of exercise. Splanchnic blood flow decreases in response to heavy exercise (Wahren et al. 1975). However, hepatic free fatty acid uptake and ketogenesis increase with prolonged exercise (Wasserman & Cherrington 1991). In addition to hepatic free fatty acid uptake, the mobilization of intrahepatic fat stores may also add to hepatic fat oxidation and ketogenesis. During prolonged exercise, hepatic ketogenic efficiency may reach 40% of hepatic free fatty acid uptake (Wasserman & Cherrington 1991). To summarize, during prolonged exercise, free fatty acids become more important both as a fuel and as determinants of increased gluconeogenesis. Data presented by Yamatani et al. (1992), using an inhibitor of free fatty acid oxidation, suggest that increased free fatty acid oxidation enhances hepatic gluconeogenesis.

Insulin-Like Growth Factor Binding Protein-1 (IGFBP-1)

Increased visceral adiposity is an important risk factor for insulin resistance and diabetes. One hypothesis proposed to explain this link is that visceral fat results in hepatic insulin resistance via a portal effect of free fatty acids and glycerol released by increased omental fat (Bjorntorp 1990).

Insulin-like growth factor-I (IGF-I) is a small polypeptide whose actions in vitro are either acute anabolic effects on protein and carbohydrate metabolism or longer-term effects on cell replication and differentiation. IGBBP-1 is one of the IGF binding proteins that modulate IGF's actions through IGF bioavailability and bioactivity. The present interest in IGFBP-1 stems from the fact that most circulating IGFBP-1 is synthetized in the liver. In addition, of all the IGF binding proteins, IGFBP-1 is unique in that its levels show rapid modulation by metabolic circumstances (Lee et al. 1993). In addition, while IGFBP-1 is not directly involved in the regulation of glucose production, a common motif has been characterized for the insulin response element of PEPCK, glucose-6-phosphatase, and IGFBP-1 genes (O'Brien et al. 1995). The marked similarities between the IGFBP-1 and PEPCK promoters have spurred interest in the use of this circulating protein as a noninvasive index of the liver-specific transcriptional action of insulin in vivo. The interest of such an index is illustrated by the report that surgical removal of visceral fat results in markedly decreased concentration of IGFBP-1, suggesting an increased hepatic sensitivity, which was supported by the decrease in gene expression of both glucose-6-phosphatase and PEPCK (Barzilai et al. 1999). With regard to exercise, a large increase in IGFBP-1 has been repeatedly observed in humans and rats (Anthony et al. 2001). Most recently, our group (Lavoie et al. 2002) has reported data showing that IGFBP-1 response to exercise was not always linked to a decrease in plasma glucose and that the decrease in liver glycogen content may be related to the increase in IGFBP-1 during exercise (figure 15.4). These data suggest that the increase in hepatic fatty acid uptake and the decrease in liver glycogen during exercise may be associated with a transient increase in hepatic insulin resistance, as indicated by the increased exercising levels of IGFBP-1.

Hepatic Lipogenic Enzymes

The hormonal milieu during exercise that favors lipolysis and oxidation of fatty acids, the increased gluconeogenesis, and the increased muscle glucose uptake and plasma free fatty acids—all of these are expected to antagonize hepatic lipogenesis. Accordingly, high-fructose diet induction of key hepatic lipogenic enzymes (fatty acid synthetase, pyruvate kinase) is inhibited by an acute bout of exercise (Griffiths et al. 1995). In a following study, the same group (Griffiths et al. 1996) reported that an exhaustive bout of prolonged exercise inhibited dietary induction of hepatic fatty acid synthetase activity and mRNA abundance, suggesting that exercise may influence fatty acid synthetase transcription, mRNA stability, or both. Both the decrease in insulin and the increase in glucagon during exercise may cause a decrease in the rate of transcription of the fatty acid synthetase gene (Paulauskis & Sul 1989). The effects of exercise on the down-regulation of carbohydrate-induced fatty acid synthetase was, thereafter, examined at the gene level. Fiebig et al. (1999) showed that compared to fasted rats, fructose-re-fed rats had twofold higher liver nuclear protein binding (upstream stimulatory factors 1 and 2) to oligonucleotides corresponding to the insulin-responsive sequence (–71/–50) and carbohydrate response element (+283/+303) on the fatty acid synthetase promoter. Exercise attenuated this binding in liver nuclear extracts to the levels seen in fasted rats. Although a causal relationship could not be established, the authors postulated that the altered nuclear protein binding might be a primary reason for the observed fatty acid synthetase down-regulation with exercise. Reduced liver nuclear protein binding might, on the other hand, be associated with the exercise-increased liver cAMP levels. In a follow-up study, Fiebig et al. (2001) used streptozotocin-treated rats to mimic diminished plasma insulin status due to exercise and to examine the role of insulin in the exercise down-regulation of fatty acid synthetase. They found that, although insulin status had a great influence on fatty acid synthetase gene expression, exercise-induced down-regulation of fatty acid synthetase mRNA was not mediated by altered insulin response sequence binding but primarily by increased inverted CCAAT-box

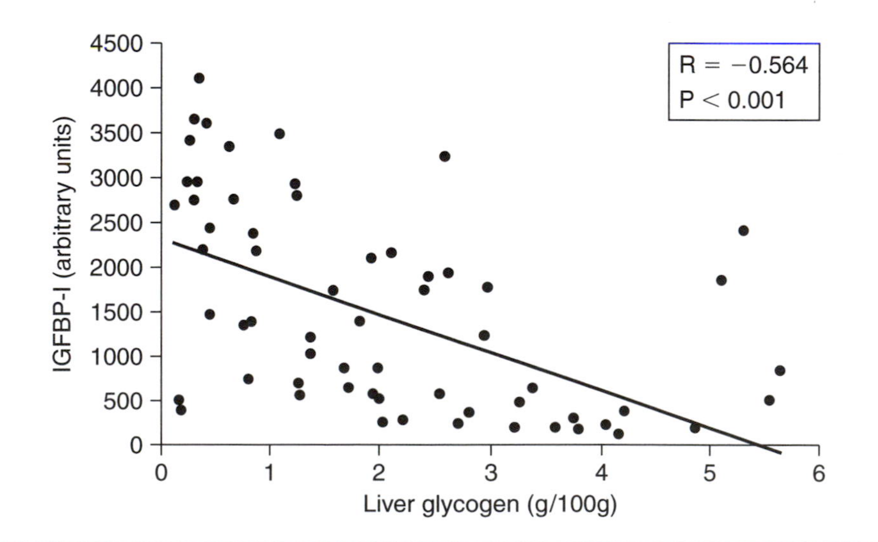

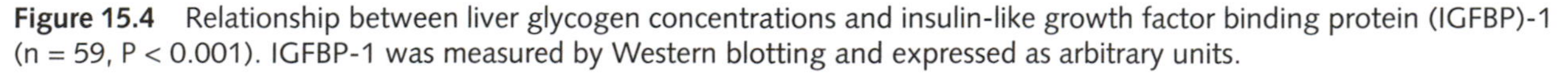

Figure 15.4 Relationship between liver glycogen concentrations and insulin-like growth factor binding protein (IGFBP)-1 (n = 59, $P < 0.001$). IGFBP-1 was measured by Western blotting and expressed as arbitrary units.

Reprinted, by permission, from J.M. Lavoie et al. 2002, "Evidence that the decrease in liver glycogen is associated with the exercise-induced increase in IGFBP-1," *Journal of Applied Physiology* 93: 789-804.

element (ICE) binding to fatty acid synthetase promoter. Diminished hepatic glycolytic intermediates during exercise might also play a role in destabilizing fatty acid synthetase mRNA. ICE is located adjacent to the insulin response sequence in the promoter region of the fatty acid synthetase gene. This region can bind the transcription factor NF-Y in response to cAMP, and its occupancy can attenuate up-regulation of fatty acid synthetase by insulin in vitro (Roder et al. 1997).

Together, these studies provide strong evidence that physical exercise is a powerful means to reduce the lipogenic potential in liver of rats fed a carbohydrate-rich diet. Hepatic lipogenic enzyme induction by high-carbohydrate meal feeding may be inhibited by exercise training (Fiebig et al. 1998). The benefit of exercise on inhibiting lipogenesis in the liver is especially important when the diet is high in monosaccharides, such as fructose, that promote lipogenic enzyme induction and obesity.

Malonyl-CoA and AMP-Activated Protein Kinase

It is well established that exercise induces a decline in liver malonyl-CoA (Beattie & Winder 1985). Malonyl-CoA is the first committed intermediate in the lipogenic pathway and an inhibitor of carnitine palmitoyltransferase-1 (CPT-1). CPT-1 activity can be limiting for fatty acid oxidation and ketogenesis (McGarry & Brown 1997). The decline in liver malonyl-CoA has been postulated to be responsible for the increased ketogenesis during and after exercise (Beattie & Winder 1985). Malonyl-CoA is synthetized by Acetyl-CoA carboxylase (ACC). Recently, Carlson and Winder (1999) reported that, contrary to what occurs in skeletal muscle, the decrease in malonyl-CoA in the liver during exercise in rats at a low rate for 2 h could not be attributed to phosphorylation and inactivation of ACC by either protein kinase A or 5'-AMP-activated protein kinase (AMPK). They postulated that an increased level of palmitoyl-CoA, an allosteric inhibitor of ACC, and a diminished substrate supply (fructose-2,6-biphosphate) caused the decreased malonyl-CoA in liver during prolonged exercise rather than an allosteric and covalent regulation of ACC. During intense short-term (32 m/min; 10 min) exercise in rats, however, the decrease in malonyl-CoA was accompanied by an increase in AMPK activity and changes in kinetic properties (K_a and V_{max}) that provided indirect evidence that ACC was phosphorylated (Carlson & Winder 1999). The decreased liver content of fructose-2,6-biphosphate during exercise of high intensity could also contribute. In agreement with these observations, Park et al. (2002) most recently reported that intense exercise (21 m/min up a 12% grade for 30 min) in rats resulted in a decrease in liver malonyl-CoA levels associated with diminished activities of ACC and glycerol-3-phosphate acyltransferase (GPAT) and an increase in malonyl-CoA decarboxylase activity, an enzyme responsible for malonyl-CoA catabolism. AMPK activity in liver was also increased by exercise, and all of these effects could be reproduced by injection of 5-amino-4-imidazole-carboxamide (AICAR), an AMPK activator.

The mechanism of activation of liver AMPK during high-intensity exercise is not known (Winder & Hardie 1999). The AMPK system responds to the AMP/ATP ratio (Hardie & Carling 1997). There is some evidence that liver ATP level is decreased by exercise (Ghanbari-Niaki et al. 1999). Hardie and Carling (1997) have proposed that AMPK protects cells against stresses that deplete ATP by inhibiting energy-requiring biosynthetic pathways. In addition to fatty acid oxidation, the overall effects of AMPK activation in the liver would, therefore, be inhibition of the energy-requiring biosynthetic pathways of fatty acids and their esterification, cholesterol, and glycogen synthesis through inactivation of liver ACC and GPAT, 3-hydroxy-methylglutaryl-CoA, and glycogen synthetase, respectively (Hardie & Carling 1997; Park et al. 2002). Finally, the fact that the exercise-induced increase in AMPK activity in liver reported by Park et al. (2002) was measured 30 min postexercise raises the possibility that a prolonged increase in AMPK activity following exercise leads to changes in the expression of genes encoding key enzymes of lipid partitioning.

Liver Lipid Infiltration

The increase in the prevalence of obesity is often ascribed to the changing lifestyle in westernized societies. Among these changes, the consumption of high-fat diets is a major concern. The capacity of the human body to respond to an increased fat intake with an increased fat oxidation is limited. High-fat diets, thus, would preferentially lead to deposition of dietary fat into adipocytes, but also into muscle and liver. Fatty liver, or steatosis, refers to a histopathological condition characterized by the excess accumulation of lipids within the hepatocytes. Although it is generally thought to be a benign process, a subset of patients go on to develop hepatitis (resembling alcohol induced), which may then progress to fibrosis, cirrhosis, and liver failure (Teli et al. 1995). Besides these liver complications, the recent interest in pure fatty liver is its strong association with features of the metabolic syndrome. The role of central obesity and increased visceral fat turnover with respect to hepatic fat deposition and insulin resistance deserves special attention. Most recently, Saltiel and Kahn (2001) specifically mentioned that the link between increased circulating free fatty acids and insulin resistance might involve an accumulation of triglycerides in muscle and liver.

Although regular physical exercise is an integral part of the treatment of obesity, the role of exercise training on liver lipid infiltration has received very little attention. Straczkowski et al. (2001) submitted rats to a diet containing a high percentage of fat (60% of calories as fat) for three weeks while rats were trained for four weeks. The authors found a large increase in liver triglyceride levels with the high-fat diet, which did not change with training. Recent data from our laboratory, however, indicate that in female Sprague-Dawley rats submitted to a high-fat (42% of calories as fat) diet for eight weeks, regular physical training largely attenuated liver lipid infiltration induced by the high-fat diet, as measured by biochemical and histological approaches (Gauthier et al. 2003). Exercise and restricted diet have also been reported to improve liver histology and biochemical markers in obese patients with fatty liver (Ueno et al. 1997). Taken together, these data suggest that exercise may play a role in controlling liver lipid infiltration related to obesity and may constitute a means by which exercise counteracts the development of insulin resistance.

Conclusion

Metabolically, the liver is the most versatile organ of the body; about 25 compounds are either taken up or released in substantial amounts by the liver (Müller 1995). During exercise, the study of the role of the liver has, to a large extent, been limited to its function of glucose production. In recent years, the cellular and molecular mechanisms by which the liver accomplishes this function have received increased attention. Data on the sensitivity of the liver to hormonal stimulation, mechanisms of activation or inhibition of metabolic signaling, and factors contributing to transcriptional activation of key regulating enzymes have started to emerge. How mechanistically hepatic glucose production during or after exercise is influenced by external factors such as the composition of the diet, training status, gender, or aging is also under intensive investigation. From a broad perspective, more attention will have to be devoted to the study of the impact of exercise on liver metabolic disturbances and their associations with pathological states such as obesity, diabetes, or cirrhosis. The increasing proportion of overweight people has caused a concomitant surge in the prevalence of fatty liver. In addition to problems related to glucose metabolism such as hepatic insulin resistance, increased hepatic triglyceride levels are also associated with increased hepatic lipid peroxidation and increased production of reactive oxygen species by hepatic mitochondria (Yang et al. 2000). Severe fasting is not recommended, since it may cause glutathione depletion, which enhances lipid peroxidation. Instead, it seems that physical exercise in combination with a moderately hypocaloric diet can progressively decrease adipocyte fat stores and improve liver tests (Pessayre et al. 2002). Finally, there is some evidence that exercise attenuates ethanol-induced fatty liver (Trudell et al. 1995). This is consistent with reports that exercise also reduces some of the harmful effects of ethanol on liver mitochondria and increases ethanol clearance (Ardies et al. 1989).

Chapter 16

Exercise and the Adipocyte

Gale B. Carey

The adipocyte is traditionally viewed as an energy storehouse—storing triacylglycerol in times of energy excess and mobilizing triacylglycerol in times of energy need. More recently, the adipocyte has been recognized as an endocrine cell, releasing over three dozen signaling molecules that have local as well as systemic effects. The aim of this chapter is to examine the cellular and molecular effects of exercise on adipocyte size, function, and signaling molecules. The first section reviews morphological characteristics of adipose tissue and the adipocyte and effects of exercise on adipocyte energy storage, energy release, and endocrine production. The second section examines the influence of exercise on the number and binding affinity of various adipocyte plasma membrane receptors, and on intracellular molecules that may be critical conduits for translating the physiological exercise signal into an adipocyte response.

The Adipocyte: Characteristics and Functions

The processes by which energy is stored in and released from the adipocyte are complex and influenced by many factors. These factors include the location of the adipose tissue in the body, the hormones to which the adipose tissue is exposed, and the sensitivity of the adipose tissue to its own endocrine secretions. The following section describes how exercise combines with these factors to affect the processes of energy storage and release.

Morphology

Nearly 90% of adipocyte volume is dedicated to triacylglycerol storage, and the remaining 10% of volume—which includes cytoplasm, intracellular membrane systems, mitochondrion, and nuclei—is relegated to the periphery of the cell (see figure 16.1). One of the unique features of the adipocyte is its variable size: Not only can the diameter of an adipocyte range from 20 to 200 μm, but the diameter changes some 20-fold in response to physiological factors (Fruhbeck et al. 2001). One factor known to influence adipocyte size is exercise.

Exercise causes mobilization of stored energy that is subsequently used by active muscles. If chronic exercise is accompanied by a negative energy balance, adipocyte size will diminish over time. This has been demonstrated in studies using rats (Askew & Hecker 1976), swine (Carey & Sidmore 1994; Carey 2000), and humans (Crampes et al. 1986).

Regionality

Unlike amphibians, fast-moving reptiles, and insects, which have one intra-abdominal adipose tissue depot, most mammals and birds have a dozen or more discrete adipose tissue depots scattered throughout their bodies (Pond 1992). These depots are structurally integrated with local neighboring skin, muscle, or organs. Their parenchymal cell, the adipocyte, is suspended in a connective tissue matrix. In humans, factors such as energy balance, age, gender, adipose tissue

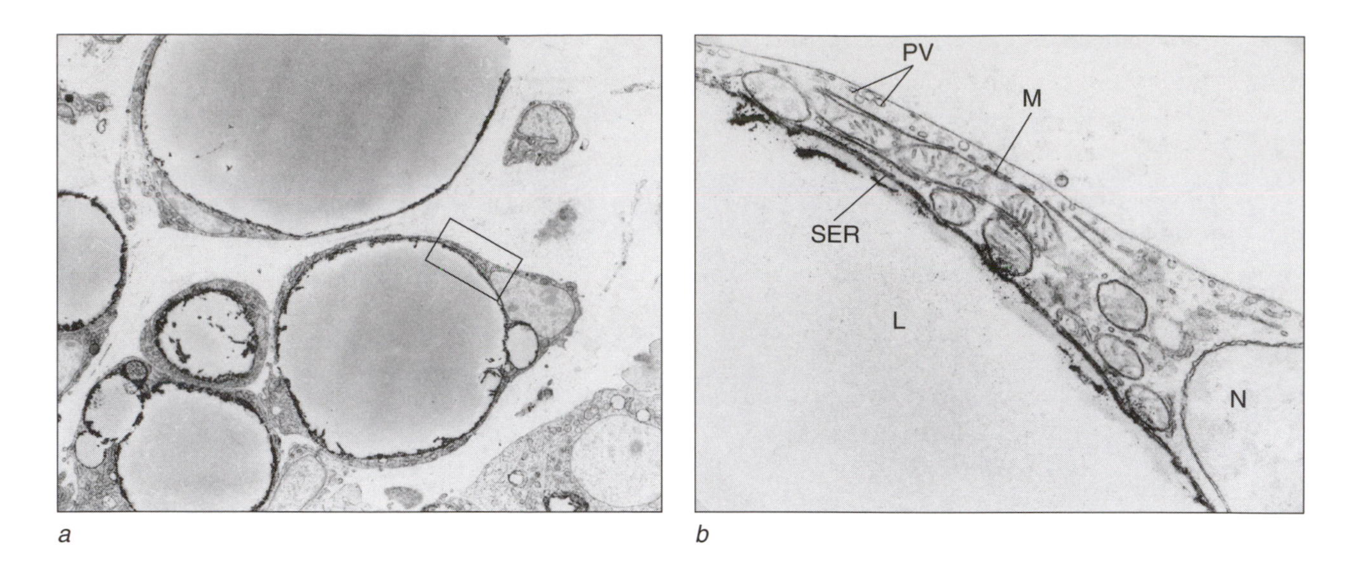

Figure 16.1 Transmission electron microscopy of adipose tissue. *(a)* Epididymal white adipose tissue of a 12-day-old rat. Magnification ×3,300. *(b)* Enlargement of the framed area in *a,* showing the cytoplasm of a white adipocyte containing few and small mitochondria with randomly oriented cristae. M, mitochondria; PV, pinocytosis vesicles; SER, smooth endoplasmic reticulum; L, lipid droplet; N, nucleus. Magnification ×32,000.

Reprinted with permission from S. Cinti (2001), "The adipose organ: morphological perspectives of adipose tissues," *Proceedings of the Nutrition Society* 60:319-328.

biology, sex steroids, and glucocorticoids play a role in determining an individual's regional adipose tissue distribution (Bouchard et al. 1993).

Regional distribution of adipose tissue correlates with risk of disease. For example, upper body adiposity (more common in males than females) is associated with an increased risk for diabetes and cardiovascular disease (Bouchard et al. 1993). Regional adipose depots vary widely in their metabolic characteristics such as lipolytic sensitivity; fatty acid/triacylglycerol cycling; leptin secretion; gene expression; and activities of enzymes such as lipoprotein lipase, phosphofructokinase, and hexokinase (Pond 1992; Vidal 2001). Because they differ in their metabolic characteristics, adipose depots vary in their response to exercise. For example, endurance-trained women mobilized fat preferentially from the subcutaneous abdominal depot rather than the femoral depot (Mauriege et al. 1997).

Energy Uptake and Storage

The adipocyte takes up glucose from extracellular fluid primarily via the facilitative GLUT4 transporter. Under basal conditions, this transporter resides intracellularly in the Golgi complex and endosomes. Within minutes after exposure to insulin, GLUT4 are translocated to the plasma membrane and dramatically increase the capacity of the adipocyte to take up glucose. The details of this insulin-stimulated translocation process are the subject of intense study, particularly in muscle tissue (Bryant et al. 2002), yet it is still speculative whether or not the translocation mechanisms discovered in muscle universally apply to other tissues, such as adipose tissue (Ploug & Ralston 2002).

Adipocyte glucose uptake also is stimulated by exercise training (Hirshman et al. 1989; Goodyear & Kahn 1998). This increased ability to remove glucose from the blood serves two functions for the adipocyte. First, it provides a fuel source. Second, it provides a precursor for glycerol-3-phosphate, the backbone for synthesizing triacylglycerol. This latter function is crucial because, under most but not all conditions (Guan et al. 2002), the adipocyte lacks glycerol kinase to phosphorylate glycerol directly.

The fatty acids required for triacylglycerol synthesis can originate from three sources: non-esterified fatty acids from the adipocyte, non-esterified fatty acids from plasma, or lipoproteins from plasma. A primary player in the hydrolysis of plasma lipoproteins is lipoprotein lipase. Although this glycoprotein enzyme is synthesized and glycosylated within the adipocyte, it is packaged into vesicles, secreted from the adipocyte, and bound to glycosaminoglycans on the luminal side of capillary endothelial cells. Here, the enzyme has access to blood-borne, triacylglycerol-rich lipoproteins. Once activated by apolipoprotein CII, a component of chylomicrons and very low-density lipoproteins, lipoprotein lipase hydrolyzes the lipoprotein-lipid, and the resulting fatty acids enter the adipocyte and are re-esterified to glycerol-3-phosphate, forming triacylglycerol (Arner & Eckel 1998).

Exercise training has been shown to cause either no change or a reduction in adipose tissue lipoprotein lipase activity and message. Five to thirteen consecutive days of supervised exercise training of 32 sedentary men caused no change in adipose tissue lipoprotein lipase mRNA and lipoprotein lipase activity, but did increase muscle lipoprotein lipase message and enzyme activity (Seip et al. 1995). Likewise, after six weeks of treadmill training of male rats, there was no change in adipose tissue lipoprotein lipase activity or mRNA levels compared to values in sedentary rats (Ong et al. 1995). However, detraining can influence lipoprotein lipase. Two weeks of detraining caused an 86% increase in lipoprotein lipase activity and a 100% increase in lipoprotein lipase mass in adipose tissue of runners, but no change in lipoprotein lipase synthetic rate or mRNA level (Simsolo et al. 1993).

The response of lipoprotein lipase to acute exercise appears to differ with species. An acute, 2-h swim of male rats caused a 43% decrease in adipose tissue lipoprotein lipase activity and a 42% reduction in lipoprotein lipase message; 24 h later, lipoprotein lipase activity had returned to normal while message was 23% below control. In contrast, a 90-min exercise bout in humans caused an acute increase in lipoprotein lipase activity (Savard & Bouchard 1990). One factor complicating these findings could be a lack of control over food ingestion prior to exercise: Food ingestion significantly increases lipoprotein lipase activity in adipose tissue.

Energy Mobilization: Lipolysis

The lipolytic cascade is a well-known metabolic pathway in which activated β-adrenergic receptor subtypes (β_1, β_2, or β_3AR) couple with heterotrimeric G-stimulatory proteins composed of three subunits (α, β, and γ), to activate adenylyl cyclase and increase intracellular cyclic AMP levels (see figure 16.2). Cyclic AMP activates protein kinase A (PKA), which is a serine/threonine kinase with two catalytic subunits and two regulatory subunits. When two molecules of cyclic AMP bind to the two protein kinase A regulatory subunits, autoinhibition is relieved so that protein kinase A catalytic subunits dissociate and phosphorylate target substrates, which include the β_1 and β_2 receptors, perilipins, GLUT4, cyclic AMP phosphodiesterase, and hormone-sensitive lipase (Michel & Scott 2002).

When activated, hormone-sensitive lipase catalyzes the hydrolysis of fatty acids from the triacylglycerol droplet, eventually releasing free fatty acids and glycerol. Thus, protein kinase A activity is directly related to lipolysis rate (Honnor et al. 1985). Recently, a non-cyclic AMP-related lipolytic pathway specific to humans has been identified. This pathway is stimulated by natriuretic peptides (Sengenes et al. 2000). These peptides bind to specific natriuretic peptide receptors that stimulate cyclic GMP production and fail to influence cyclic AMP. However, the details and effects of exercise on this pathway are unknown at present.

Lipolytic stimulation can be suppressed by at least three biochemicals that initiate a lowering of intracellular cyclic AMP levels, albeit via distinct mechanisms. One is adenosine that binds to the A_1 adenosine receptor. The A_1 adenosine receptor couples with G-inhibitory protein and inhibits adenylyl cyclase activity, thus reducing cyclic AMP synthesis. A second is insulin, which binds to the insulin receptor. The insulin receptor, acting via insulin receptor substrate-1 and phosphatidylinositol 3-kinase, will trigger the phosphorylation and activation of phosphodiesterase 3B, thereby lowering cyclic AMP levels and decreasing lipolysis. A third is epinephrine or norepinephrine binding to the α_2-adrenergic receptor, which, similarly to the A_1 adenosine receptor, couples with G-inhibitory proteins. At low concentrations, epinephrine binds with high affinity to the α_2-adrenergic receptor; at higher concentrations, epinephrine activates the β-adrenergic receptor. Thus, lipolytic response to epinephrine represents a balance between activated α_2 and β receptors.

An acute bout of exercise increases adipocyte lipolytic sensitivity in humans (Wahrenberg et al. 1987). In normal-weight male subjects, this increase involves activation of both the α_2- and β-adrenergic pathways. However, the stimulatory signal is stronger, as indicated by a significant increase in extracellular glycerol measured using microdialysis, a technique in which a probe covered with a semipermeable membrane is perfused with buffer while it sits in the adipose

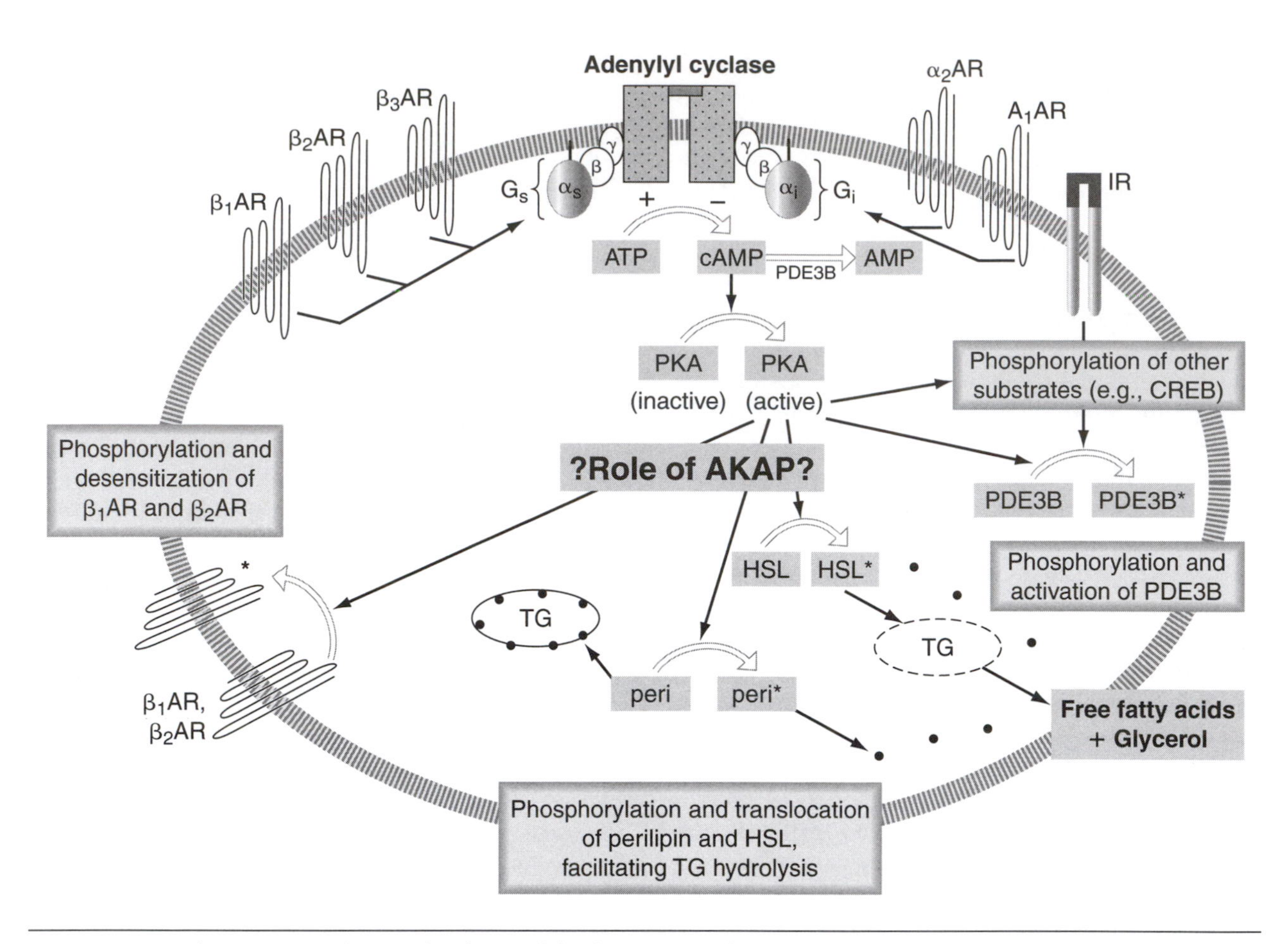

Figure 16.2 Adipocyte signaling molecules and lipolysis. AR, adrenergic receptor; A_1AR, adenosine A_1-adrenergic receptor; IR, insulin receptor; G_s, G-stimulatory protein; G_i, G-inhibitory protein; cAMP, cyclic AMP; PKA, protein kinase A; AKAP, A-kinase anchoring protein; PDE3B, cyclic GMP-inhibited phosphodiesterase 3B; CREB, cyclic AMP-responsive element binding protein; HSL, hormone-sensitive lipase; TG, triglyceride; peri, perilipin.

Adapted, by permission, from M. Lafontan and Michel Berlan, 1995, "Fat cell alpha2-adrenoceptors: The regulation of fat cell function and lipolysis," *Endocrine Reviews* 16(6): 716-738. Copyright © 1995 The Endocrine Society.

tissue (Stich et al. 1999). In contrast, when obese male subjects were acutely exercised, glycerol release into extracellular fluid was only 20% that of lean subjects (Stich et al. 2000). The exercise-induced glycerol release was markedly increased when phentolamine, an α_2 receptor antagonist, was infused into the microdialysis probe in the obese subjects, but not in the lean. This indicates physiological stimulation of the α_2-adrenergic pathway during exercise in obese, but not lean, men (Stich et al. 2000).

Exercise training clearly increases the sensitivity of isolated adipocytes to lipolytic stimulation by isoproterenol (synthetic catecholamine) and epinephrine. This is observed in both humans (Crampes et al. 1989) and animals (Askew & Hecker 1976; Izawa et al. 1991). A study by Crampes et al. (1989) suggests that an increase in β-adrenergic receptor pathway efficiency is responsible for the increase in lipolysis in adipocytes from exercise-trained females, while in males the response may be due to both an increase in β-adrenergic efficiency and a decrease in α_2-adrenergic pathway efficiency. These in vitro findings were confirmed by Hellstrom et al. (1996) using microdialysis. In addition, exercise training is reported to decrease lipolytic sensitivity of swine adipocytes to adenosine (Carey & Sidmore 1994) and of obese human adipocytes to α_2-agonists and insulin (de Glisezinski et al. 1998). Adipocytes from exercise-trained rats, however, show an increased sensitivity to the anti-lipolytic effects of insulin, compared to adipocytes from sedentary controls (Suda et al. 1993). This increased sensitivity with exercise may be the result of increased adipocyte cyclic AMP phosphodiesterase activity (Kenno et al. 1986).

Recent research reveals that the lipolytic story is more complex than originally believed. A number of signaling molecules may play important roles in controlling the lipolytic response, such as perilipin, protein kinase A anchoring proteins, cyclic AMP, and calmodulin. (These molecules are discussed on pages 305-306.)

Substrate Cycling

Another contributor to fatty acid appearance with exercise is the fatty acid-triacylglycerol substrate cycle. Like all substrate cycles, this cycle requires energy and generates heat but produces no net product. In the fatty acid-triacylglycerol cycle, fatty acids are released via lipolysis but are not oxidized; they are simply re-esterified to reform triacylglycerol. This re-esterification may take place within the adipocyte or in another tissue such as liver. Research shows that, at rest, nearly two-thirds of the fatty acids released by the adipocyte are immediately re-esterified to triacylglycerol. During exercise, however, re-esterification drops to 25% while lipolysis increases threefold, allowing for a sixfold increase in fatty acid availability (Wolfe et al. 1990). Moreover, at rest, exercise-trained humans have four times the amount of fatty acid-triacylglycerol cycling than sedentary controls (Romijn et al. 1993): This may allow for the rapid fatty acid availability that occurs at the start of exercise in trained athletes. Thus, rates of fatty acid re-esterification and lipolysis may work in concert to regulate the exercise-induced appearance of fatty acids into the plasma.

Endocrine Function

The traditional view, that the primary role of the adipocyte is as a passive receptacle for storing and releasing energy, has changed. Beginning with the discovery in 1994 that the adipocyte secretes the signaling protein leptin (Zhang et al. 1994), it has become well accepted that the adipocyte also is an endocrine cell. The number of products secreted by the adipocyte is well over three dozen. These products include prostaglandins, adipsin, tumor necrosis factor-α, (TNF-α), acylation stimulating protein, interleukin-6, adenosine, steroids, adiponectin, angiotensin, leptin, and resistin (Trayhurn & Beattie 2001). These products can, in turn, have both local and systemic effects. Local effects include regulating adipocyte lipolysis, triacylglycerol synthesis, and adipocyte differentiation while systemic effects include signaling the brain regarding appetite and energy status, influencing the immune system, altering peripheral insulin sensitivity, and influencing the reproductive system (Fried & Russell 1998; Fruhbeck et al. 1998; Pond & Mattacks 1998; Mohamed-Ali et al. 1998; Halle et al. 1998).

Two adipocyte products whose secretion is affected by exercise are leptin and TNF-α. Both of these proteins play a role in energy balance and insulin sensitivity. Leptin is a 16-kDa peptide hormone with receptors in nearly all tissues of the body. It is produced by adipocytes, and thus plasma leptin levels are directly proportional to body fat content. Because exercise training has the potential to reduce fat mass, exercise can reduce leptin levels. This reduction can affect a variety of physiological properties including food intake, reproduction, hematopoiesis, angiogenesis, immune response, blood pressure control, and bone formation (Fruhbeck 2001). Leptin also affects adipose tissue directly: It inhibits adipocyte lipogenesis by decreasing acetyl-CoA carboxylase gene expression, fatty acid synthesis, and lipid synthesis (Bai et al. 1996). It also stimulates lipolysis, perhaps via a nitric oxide-mediated mechanism (Fruhbeck 2001).

TNF-α is a cytokine, originally identified as a product of macrophages in response to inflammation. It is associated with wasting of body protein and is known to inhibit muscle protein synthesis while stimulating muscle protein degradation. TNF-α has three adipocyte-specific effects that are consistent with its role as a catabolic signal: It inhibits fatty acid uptake, decreases expression of regulated enzymes of lipogenesis, and stimulates lipolysis.

Animal studies have measured adipose tissue levels of leptin and TNF-α in response to exercise. Using the sucrose-fed rat as a model for insulin resistance, Baba et al. (2000) found that four weeks of voluntary wheel running significantly reduced mesenteric and subcutaneous adipose tissue leptin levels while it increased TNF-α levels compared to those in sedentary sucrose-fed rats. Friedman et al. (1997) used a rat model of type II diabetes and insulin resistance and found that 8 to 12 weeks of treadmill running of lean and obese rats reduced subcutaneous adipose tissue leptin mRNA levels by 85% and 50%, compared to levels in lean and obese sedentary rats, respectively.

In humans, the effects of exercise on leptin and TNF-α levels have been measured in plasma only. In a study of pre- and postmenopausal women, 10 weeks of aqua exercise training tended to reduce body fat, increase plasma TNF-α levels, and reduce plasma leptin levels (Hayase et al. 2002). Because leptin production directly correlates with fat mass, a reduction in fat mass due to exercise training often leads to a reduction in plasma leptin levels; hence circulating leptin levels are low in trained athletes. However, another study in which six weeks of walking and cycle ergometer exercise caused no change in body fat mass showed a decrease in plasma leptin compared to levels in the sedentary group (Ishii et al. 2001). The effect of exercise on adipocyte leptin production, independent of changes in fat mass, remains to be fully explored (Hickey & Calsbeek 2001).

Adipocyte Receptors, Signaling Molecules, and Exercise

The adipocyte is home to a variety of proteins and signaling molecules that respond to an exercise signal. These include cell surface receptors; intracellular second messengers; anchoring and docking proteins; and enzymes. The following section describes our current understanding in this rapidly evolving field of how exercise alters the content and activity of these adipocyte molecules.

Stimulatory Receptors

Three distinct genes code for three β-adrenergic receptors: β_1, β_2, and β_3 (see figure 16.2). These gene products belong to the superfamily of G-protein-coupled receptors, characterized by seven transmembrane spans, an extracellular amino terminus, and an intracellular carboxy terminus (Carey 1998). The extracellular domain is responsible for ligand binding, while the intracellular domain is involved in G-stimulatory protein activation. The β_1 and β_2 receptors have a higher affinity for catecholamines than β_3, and there is species specificity in the adipose tissue expression of these proteins.

Much of the research on the increase in lipolysis due to exercise has focused on stimulatory receptors, especially β_1. In general, however, β receptor number or binding does not change with acute or chronic exercise (see table 16.1). This is in contrast to the exercise-induced decrease in β receptor number seen in lymphocytes and cardiac cells. The adipocyte β_1 and β_2 receptors, but not β_3, can be rapidly desensitized when phosphorylated by protein kinase A and β-adrenergic receptor kinase (β-ARK), which decreases the receptor's ability to transmit signal (Carey 1998); the effects of exercise on β-ARK activity have not been examined.

Table 16.1 Adipocyte Receptor Response to Exercise

			RESPONSE		
Receptor type	Species	Exercise type	# Receptors	Binding affinity	Reference
β-adrenergic	Humans	Acute	No Δ	No Δ	Wahrenberg et al. 1987
	Humans	Acute	No Δ	Unknown	Wahrenberg et al. 1991
	Rats	Trained	No Δ	No Δ	Williams & Bishop 1982
	Rats	Trained	↓	Unknown	Suda et al. 1993
	Rats	Trained	No Δ	No Δ	Bukowiecki et al. 1980
	Rats	Trained	No Δ	No Δ	Shepherd et al. 1986
	Rats	Trained	↑	No Δ	Nieto et al. 1996
α_2-adrenergic	Humans	Acute	No Δ	Unknown	Wahrenberg et al. 1991
	Rats	Acute	No Δ	Unknown	Wahrenberg et al. 1987
Insulin	Humans	Acute	No Δ	Unknown	Wahrenberg et al. 1987
A_1 adenosine	Swine	Trained	↓	No Δ	Carey 2000; Dong et al. 1994

Two mechanisms have been proposed to explain the exercise-induced increase in lipolysis. One mechanism is enhanced coupling of the β receptor to G-stimulatory proteins or to adenylyl cyclase (Williams & Bishop 1982). A second mechanism is downstream of the adenylyl cyclase signal. Dibutyryl cyclic AMP, a cell-permeable cyclic AMP mimic, enhanced lipolysis in adipocytes from exercise-trained but not sedentary rats (Bukowiecki et al. 1980). This suggests a role for intracellular players in the exercise-enhanced lipolytic response.

Inhibitory Receptors

The α_2-adrenergic receptor is also a member of the G-protein-coupled receptor family. It has high binding affinity for norepinephrine and epinephrine. However, when activated, it couples to G-inhibitory protein and inhibits adenylyl cyclase activity. Two studies failed to show a change in α_2-adrenoreceptor ligand binding after an acute bout of exercise (Wahrenberg et al. 1987, 1991). Izawa et al. (1988) showed a 40% decrease in G_i proteins with exercise training, which would reduce the potency of the inhibitory pathway.

Activation of the 43-kDa A_1 adenosine receptor also elicits an inhibitory response on lipolysis by coupling to G-inhibitory protein and reducing adenylyl cyclase activity (Carey 1998). Carey and Sidmore (1994) and Shinoda et al. (1989) found a reduction in adenosine sensitivity of adipocytes from exercise-trained swine and rats, respectively. Carey and Sidmore postulated that a reduction in A_1 adenosine receptor number would facilitate exercise-induced lipolysis, and explored this possibility using exercise-trained miniature swine. Twelve weeks of treadmill training caused a 50% reduction in A_1 adenosine receptor number in adipocyte plasma membranes with no change in binding affinity, compared to control (Carey 2000; Dong et al. 1994). This response would facilitate lipolysis at rest and during exercise.

Cyclic AMP Levels

Cyclic AMP and one of its targets, protein kinase A, have the potential to dramatically influence rates of lipolysis by influencing the activity of hormone-sensitive lipase, a key enzyme in lipolysis. Because exercise training increases lipolysis, one might expect intracellular cyclic AMP levels to be higher in adipose tissue of exercise-trained animals compared to controls. However, trained rats actually have lower cyclic AMP levels than sedentary rats (Askew et al. 1978).

Intracellular cyclic AMP levels reflect the rate at which cyclic AMP is synthesized by adenylyl cyclase and degraded by phosphodiesterase 3B. With exercise training, adipocyte adenylyl cyclase activity remains unchanged but phosphodiesterase 3B activity is increased (Askew et al. 1978; Shepherd et al. 1981; Kenno et al. 1986). Thus cyclic AMP levels are attenuated by exercise training. However, because adipocytes from trained rats have a greater lipolytic sensitivity and glycerol release than adipocytes from sedentary rats, this suggests a shift in lipolytic sensitivity to cyclic AMP. This was confirmed by Izawa et al. (1991), who reported that the concentration of cyclic AMP required for half-maximal lipolytic response in adipocytes from trained rats was nearly one-half that needed in adipocytes from controls. Clearly, there must be other players involved.

Protein Kinase A and Anchoring Proteins

Protein kinase A maximal activity was reported to be unchanged or increased (Nomura et al. 2002) by exercise training. Protein kinase A is rather unusual in being tethered to A-kinase anchoring protein, or AKAP (Michel & Scott 2002). AKAPs are a structurally diverse family of proteins comprising over 70 members, each of which contains a unique subcellular targeting domain. AKAPs bind not only protein kinase A, but other protein kinases and phosphatases. Interestingly, the 250-kDa AKAP, gravin, binds to the β_2-adrenergic receptor in epidermoid carcinoma cells and is believed to act as a scaffold protein by organizing protein kinase A, protein kinase C, protein phosphatase 2B, and G-protein-linked receptor kinase 2 with the β_2-adrenergic receptor (Fan et al. 2001). Thus, AKAPs may be important as tethering or scaffolding proteins, or both, conferring spatial specificity in the protein kinase A response. AKAP could play a role in regulating lipolysis, particularly during exercise. This possibility was recently validated by Nomura et al (2002), who found that expression of the 150-kDa AKAP was greater in the 40,000 × g membrane fraction of adipocytes from exercise-trained, compared to sedentary, rats.

Hormone-Sensitive Lipase and Perilipin

Hormone-sensitive lipase (HSL) is a neutral lipase, 82 to 88 kDa in size, found in adipocytes of mammals and is traditionally viewed as the rate-limiting enzyme of lipolysis. It catalyzes triacylglycerol hydrolysis and is acutely regulated by protein kinase A phosphorylation. There are four phosphorylation sites on the enzyme: Ser 563, 565, 659, and 660. In vitro activation of hormone-sensitive lipase occurs when Ser 563, 659, and 660 are phosphorylated by protein kinase A. Ser 565 (also termed the basal site) is phosphorylated in unstimulated cells by several kinases including AMP-activated protein kinase, Ca^{+2}/calmodulin-dependent kinase II, and glycogen synthase kinase-4, but not by protein kinase A (Londos et al. 1999).

At present, it is unknown which phosphorylation sites on hormone-sensitive lipase are physiologically relevant in vivo. This is, in part, due to the observation that homogenates of unstimulated cells, which contain nonphosphorylated hormone-sensitive lipase, exhibit lipase activity equal to that of intact, stimulated cells (Lafontan & Berlan 1993; Okuda et al. 1992). It is conjectured that in the intact adipocyte, hormone-sensitive lipase has restricted access to its substrate and that once the cells are homogenized, this access increases dramatically (Londos et al. 1999). This conjecture is supported by evidence of hormone-sensitive lipase translocation from the cytosol to the lipid droplet surface when adipocytes are stimulated by lipolytic hormones or cyclic AMP analogs (Egan et al.1992; Brasaemle et al. 2000).

A key player in hormone-sensitive lipase translocation and subsequent lipolysis may be perilipin. This family of four proteins (A, B, C, and D) are 47 to 56 kDa, heavily phosphorylated by protein kinase A, and dephosphorylated by insulin (Londos et al. 1996; Greenberg et al. 1991). Perilipin A is the most abundant perilipin in adipocytes (Londos et al. 1996) and coats the lipid droplet of these cells. It has been proposed that this coating of nonphosphorylated perilipin serves as a barrier for hormone-sensitive lipase access to substrate and that once phosphorylated, perlipins disperse and reveal the triacylglycerol substrate, making it vulnerable to lipase action (Londos et al. 1999). This proposal is supported by two pieces of data from perilipin null mice: Adipose tissue mass of null mice is reduced 30% compared to values in wild type, and isolated adipocytes from null mice have elevated basal lipolysis compared to adipocytes from wild-type mice (Tansey et al. 2001).

HSL activity is increased by acute exercise: It increases sixfold above baseline at 5 min, but gradually returns to baseline over the remaining 25 min of exercise (Petridou & Mougios 2002). HSL is reported to be increased (Nomura et al. 2002), unchanged (Shepherd et al. 1981; de Glisezinski et al. 1998), or decreased (Shepherd et al. 1981) by exercise training. Shepherd et al. (1981) found that hormone-sensitive lipase activity was lower in homogenates of adipose tissue from exercise-trained rats compared to controls if the rats were killed at rest. However, if the rats were exercised just prior to killing, there was no difference in hormone-sensitive lipase activity. The effect of exercise on perilipin expression and phosphorylation is unknown.

Lastly, a hormone-sensitive lipase docking protein has been identified, called lipotransin (Syu & Saltiel 1999). This protein is expressed in 3T3-L1 cultured adipocytes and binds with hormone-sensitive lipase in the presence of insulin. Much remains to be learned about the biochemical mechanism of action and physiological role of this protein.

Ca-Calmodulin

Optimal calcium levels are important for lipolysis. Adipocyte lipolysis is reduced by 30% to 40% when adipocytes are incubated in calcium-free medium with isoproterenol or dibutyryl cyclic AMP (Allen & Beck 1986). Likewise, higher than normal calcium levels activate phosphodiesterase and reduce intracellular cyclic AMP, leading to suppressed lipolysis (Xue et al. 2001).

Research shows that adipocytes from trained rats had higher intracellular calcium levels under basal and sustained stimulated conditions compared to adipocytes from untrained rats. In addition, inhibition of lipolysis by the calmodulin inhibitor N-(6-aminohexyl)-5-chloro-1-naphthalene sulfonamide (W-7) was more pronounced in adipocytes from trained, compared to untrained, rats (Izawa & Komabayashi 1994). Thus, cyclic AMP regulation of protein kinase A in adipocytes from trained rats appears to depend more on the calcium-calmodulin complex than in adipocytes from sedentary rats.

Conclusion

The perception of adipocyte function has shifted from that of an energy repository to that of a metabolically active endocrine tissue. Likewise, our understanding of how exercise impacts cellular and molecular events in the adipocyte has grown. The dated observation that exercise training enhances adipocyte glucose uptake and lipolysis is now accompanied by details about the many players and processes involved: transporters, receptors, coupling proteins, translocation protein, anchoring proteins, and ions. Continued research at the cellular and molecular levels will lead to a deeper understanding of how these and other molecules may be involved in the exercise response.

Chapter 17
Erythrocytes

Walter F.J. Schmidt, Dieter Böning, and Norbert Maassen

The red blood cell is one of the best-investigated systems in the organism because of its accessibility and stability under in vitro conditions as well as its relatively simple structure. It contains no nucleus and no organelles; therefore there is no replacement of essential components restricting the survival to 110 to 130 days. Adaptation to increased exercise load can occur only during maturation in the bone marrow.

Essential parts of the erythrocyte are a large membrane (in relation to its volume) with channels (e.g., anion transporters for chloride and lactate) and pumps, supported by a flexible cell skeleton consisting largely of spectrin allowing deformation important for flow, and the cytoplasmic fluid with dissolved nonpermeating substances like hemoglobin, enzymes, and organic phosphates. These large negatively charged molecules produce an osmotic drag and an inequality in permeating ion distribution, resulting in a membrane potential (approximately +10 mV outside, Juel 1997); the volume constancy is maintained by pumping cations outside.

Functions, besides gas transport and buffering, are antioxidative defense and, as new aspects, influences on vessel diameter and ventilation. Anaerobic glycolysis with its end-product lactic acid supplies ATP and 2,3-biphosphoglycerate (2,3-BPG) as allosteric effector for Hb-oxygen affinity, and the pentose phosphate cycle NADPH for antioxidative defense.

In this chapter, after a description of red cell production, the three major functions of the erythrocyte, which all depend on each other, are reviewed: (1) oxygen transport by the hemoglobin, (2) CO_2 transport and buffering, and (3) red cell volume and osmotic regulation. In each section a general description of the system is followed by information about the influences of exercise and training.

Red Cell Production

As one important step of adaptation to increased O_2 utilization of the muscle or reduced O_2-supply to the muscle, the mass of erythrocytes can be modulated by mostly hormonal inflences.

Hormonal Regulation

The red cell production in the bone marrow is under the control of the glycoprotein erythropoietin (EPO) and, to lesser extent, of other growth-promoting hormones such as testosterone, growth hormone, and insulin-like growth factor-I (IGF-1). EPO is produced and secreted in peritubular cells of the kidney. It increases the proliferation of precursor cells in the bone marrow beginning in the early stage of differentiation (i.e., colony forming [CFU-E] and burst forming units-erythroid [BFU-E]) and regulates iron incorporation into the proerythroblasts and erythroblasts.

Erythropoietic regulation is generally modulated by the degree of oxygen availability that is sensed in the kidney. Similar to all other cells, peritubular cells possess an oxygen sensor, which was recently discovered by Jaakkola et al. (2001) as an HIF-1 (hypoxia-inducible factor-1)α prolyl-hydroxylase (HIF-PH), regulating the interaction between a specific domain of the HIF-1α subunit and the von Hippel-Lindau tumor suppressor E3 complex (pVHL). Hypoxia suppresses the pVHL effects and thereby stabilizes HIF-1α, which is immediately degraded under normoxia, leading the signal "hypoxia" into the nucleus of the cell. In the nucleus, HIF-1α binds together with HIF-1β as HIF-1 complex to the hypoxia-responsive element (HRE) of the oxygen-regulated genes and induces an adequate production of erythropoietin and also of VEGF, transferrin, endothelin-1, and other oxygen-dependent factors. From the HIF-1 complex, only the α subunit shows an oxygen-dependent expression, while HIF-1β is always present in relatively high concentrations. Because HIF-1α is not regulated on the transcription level but via protein stabilization, it is immediately available in case accelerated erythropoiesis is necessary (Semenza 2000).

Under normal environmental conditions, plasma EPO activity of a healthy subject is in the range between 5 and 15 mU/ml. To increase EPO production and secretion, a hypoxic stimulus for at least 80 min is necessary. Using such a stimulus, plasma EPO reaches peak values 4 to 5 h after the onset of hypoxia (Eckardt et al. 1989; Schmidt et al. 1991). On the molecular level, Stroka et al. (2001) showed maximum HIF-1α levels after 60-min hypoxia, which returned to normal values some hours later. The hypoxic stimuli are a low inspiratory PO_2 (e.g., due to altitude exposure), anemia, blood loss, cardiopulmonary disorders, and altered Hb-O_2 affinity. Previously, it was hypothesized that physical exercise also leads to low oxygen availability, increasing renal EPO production and, by this mechanism, the erythropoietic rate.

This, however, is not the case. Although short intensive or longer-lasting submaximal bouts of exercise have been shown to induce a remarkable reticulocytosis, an increase in plasma EPO concentration has never been found up to 5 h after exercise (e.g., Schmidt et al. 1991). Other mechanisms such as changed EPO receptor affinity or effects of other growth-promoting factors (growth hormone, paracrine-secreted IGF-1) have to be taken into consideration.

Red Blood Cell Mass in Trained Subjects

Physical exercise increases red cell production as proven by increased expression of the transferrin receptor (TFR) on membranes of erythroblasts after strenuous exercise (Qian et al. 1999) and by higher reticulocyte number during training periods (Schmidt et al. 1988). The influence of training on red cell mass, however, is differently described, ranging from no effects (Shoemaker et al. 1996) to a daily renewal rate of 1.2% of the total red cell mass (Schmidt et al. 1988). Cross-sectional studies, however, consistently show remarkable differences between trained and untrained subjects. As early as 1949, Kjellberg et al. demonstrated a total hemoglobin mass (tHb) of 11.5 g/kg in untrained, 13.6 g/kg in moderately trained, and 15.7 g/kg in highly trained athletes. In comparison to untrained subjects we found 40% increased tHb values in athletes from different endurance disciplines and 13% increased values in anaerobically trained subjects (Heinicke et al. 2001).

The effects of training on red cell mass (RCM) can be augmented by living at altitude. A longer stay or permanent living at moderate altitude (2,600 m) is associated with about 10% higher RCM (Böning et al. 2001). Cyclists native to 2,600 m have the highest RCM values ever found in endurance sports (see figure 17.1). In comparison to sea level athletes, they increase their RCM mass by 12% (51.7 ml/kg and 46.3 ml/kg, respectively; Schmidt et al. 2002).

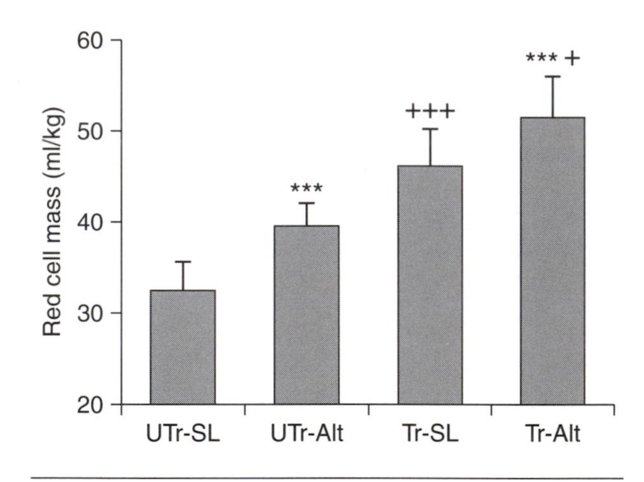

Figure 17.1 Red cell mass calculated for kilogram body mass in highly trained cyclists from altitude (Alt, 2,600 m) and from sea level (SL) and their corresponding control groups (according to Schmidt et al. 2002). UTr, untrained subjects; Tr, trained athletes (cyclists). *Altitude effects; +training effects; + = $p < 0.05$, +++,*** = $p < 0.001$.

The effects of high tHb (or erythrocyte mass) on aerobic performance can be estimated by the strong relationship between tHb and $\dot{V}O_2max$ (r = 0.722, Heinicke et al. 2001). It is therefore not surprising that in some endurance disciplines, doping with blood or EPO has become very common and is considered as a serious danger for sports.

Principles of O_2 Transport by the Hemoglobin Molecule

About 1% of the transported O_2 is dissolved in the plasma water, the remaining 99% has to be transported by the hemoglobin molecule. Hemoglobin was the first large protein investigated in the 1960s and 1970s.

As the main respiratory protein, hemoglobin transports oxygen from the lung to the tissue and facilitates CO_2 transport. One gram of hemoglobin binds 1.39 ml O_2 (Hüfner´s number); thus a person with a normal hemoglobin concentration of 15 g/dl possesses an O_2 transport capacity of 20.85 ml/dl blood. Human hemoglobin has a molecular weight of about 64.500 and consists of two pairs of two identical polypeptide chains with one heme molecule each. The polypeptide chains of hemoglobin A, which amounts to 94% of normal adult hemoglobin, consists of 141 amino acids (α-chain) and 146 amino acids (β-chain). The different structure of the chains with their hydrophil and hydrophobe parts, as well as electrostatic forces, produces eight helix structures of each chain and special interactions between them. These are the reasons for preventing an irreversible binding of the central iron with ligands, that is, mostly with oxygen. The interaction of the four subunits, which is the basis for the cooperative binding of O_2 molecules, is described by Hill's equation (formula 1):

$$SO_2 = k(PO_2)^n / [1 + k(PO_2)^n]$$

where k is the dissociation constant and n is number of heme units contributing to the cooperativity. Before the binding of the first O_2 molecule, the Hb-molecule is characterized by a low O_2 affinity. Thereafter, this changes into a high-affinity state to bind the next two O_2 molecules, which is again followed by a low affinity state. This flexibility allows an optimal binding of O_2 in the lungs and a high O_2 supply to the tissue by maintaining a high PO_2 gradient between the capillaries and the surrounding tissue during the capillary passage.

The second mechanism adapting oxygen transport by the hemoglobin molecule to physiological conditions is the influence of temperature and allosteric effectors on the position of the oxygen dissociation curve (ODC), which is characterized by the PO_2 at half-saturation of the blood with oxygen, that is, the P_{50}. High temperatures decrease the Hb-O_2 affinity, while low temperatures increase it. Because the core temperature is relatively constant at 37° C, the temperature effect is low under normal resting conditions. The human hemoglobin molecule is characterized by a high intrinsic O_2 affinity; but the allosteric effectors, that is, H^+, CO_2, and 2,3-BPG, decrease this affinity at physiological conditions and move the ODC into their normal in vivo position.

Allosteric Effectors

The Hb-O_2 affinity directly depends on red cell pH (Bohr effect). When pH is in the range of 6.5 to 9, O_2 affinity decreases with increasing H^+-ion concentration (alkaline Bohr effect). The Bohr effect can be quantified at constant O_2 saturation by calculating the Bohr coefficient (BC) (formula 2):

$$BC = \Delta \log PO_2 / \Delta pH$$

The extent of proton binding or release induced by changes in Hb-O_2 saturation can be expressed as the Haldane coefficient (HC), indicated by formula 3:

$$HC = H^+ / \Delta Hb\text{-}O_2$$

The molecular mechanism of the Bohr effect is based on the change of the pK value of some weak bases that form salt bridges to negatively charged groups in the deoxygenated, but not the oxygenated, state. Depending on the binding of the allosteric factors we distinguish between intrinsic and nonintrinsic Bohr groups, which require further cofactors such as Cl^- and 2,3-BPG. The most important intrinsic Bohr group is His β-146 forming a salt bridge to Asp β-94 in the deoxy state by changing the pK from 7.1 to 8.0, which is responsible for 40% to 50% of the whole Bohr effect.

The most effective non-intrinsic Bohr groups stabilizing the deoxygented state are Val α-1, which increases its pK when binding a Cl^- ion, and the three groups of Val β-1, His β-2, and His β-143, which all change their pK when 2,3-BPG is bound. The Bohr effect therefore does not depend solely on changes in hydrogen ion concentrations, but also on the presence and concentration of other allosteric factors such as 2,3-BPG, CO_2, and Cl^-. When pH is lowered by a fixed acid, for example physiologically by lactic acid, the Bohr coefficient has its maximum at half-saturation of the hemoglobin molecule. In the case of acidification with CO_2, the highest Bohr coefficients are measured at low SO_2 because the carbamate formed attenuates Hb-O_2 affinity (see Bohr-coefficients for lactic acid and CO_2 in table 17.1).

Table 17.1 Changes of Muscle Capillary PO_2 at Fixed O_2 Saturation (20%) in Untrained and Trained Male Subjects Caused by Acidosis and Temperature Effects on Oxygen Affinity During Exercise

	UNTRAINED		TRAINED	
	Coefficient	PO_2 (mmHg)	Coefficient	PO_2 (mmHg)
Standard values (pH 7.4, PCO_2 40 mmHg, 37° C)	—	15.0	—	17.0
PCO_2 + 35 mmHg – 0.2 pH units	–0.54	19.2	– 0.73	23.8
+7 mmol/L lactate – 0.1 pH unit	–0.38	21.0	–0.48	26.6
In vivo Bohr effect – 0.3 pH units	–0.09	22.3	–0.14	29.3
+4° C	+0.015	25.7	+0.015	33.6

Data from Böning et al 1978, 1982; Braumann et al. 1979, 1982.

The 2,3-BPG molecule is considered the main allosteric factor of hemoglobin. Its binding occurs in a stochiometric ratio of 1:1 at the positively charged groups of Val-1, His-2, and His-143 of both β-chains and of the α amino group of one Lys β-82. Normal concentrations of 2,3-BPG in the blood tighten the deoxygenated conformation and increase the P_{50} from 13.2 mmHg in a hemoglobin solution (37° C, pH 7.2, 0.15 mol/L Cl^-) to 27 mmHg under standardized conditions (37° C, pH 7.4, PCO_2 40 mmHg) (Baumann et al. 1984). Under physiological conditions P_{50} is shifted to the right side by 1 mmHg when 2,3-BPG increases by 0.4 mmol/L erythrocyte (Bellingham 1971).

Acute Exercise Effects

The oxygen saturation of the hemoglobin in venous blood rapidly decreases after the beginning of exercise until minimum values of 5% to 10%. The PO_2, however, remains relatively stable in the range of 20 mmHg while lactic acid concentration increases. The stabilizing of the blood PO_2 by Bohr and temperature effects prevents a decrease in O_2 diffusion by maintaining a high PO_2 gradient between erythrocyte and mitochondrium (see table 17.1). The temperature of the exercising muscle and therefore the capillary blood may increase considerably, by 3° to 4° C up to 40° C, contributing to the right shift of the ODC. The acidification of the capillary blood is first brought about by the increase in tissue PCO_2 due to accelerated aerobic metabolism and at higher exercise intensity also by the production of lactic acid. At the first glance acidification with lactic acid seems to be less effective compared with the effect of CO_2, because (1) the BC is much higher for CO_2 at low oxygen saturation (Böning et al. 1982) and (2) the diffusion of lactic acid across the erythrocyte membrane is delayed (Juel et al. 1990). There exists, however, a mechanism similar to a gas pump that completely compensates for these disadvantages (Böning et al. 1991). When lactic acid is buffered by bicarbonate, the resulting H_2CO_3 immediately dissociates into CO_2 and H_2O. CO_2 diffuses into the erythrocyte and rapidly changes again into H_2CO_3 via the effects of the enzyme carboanhydrase. Especially at low O_2 saturation, H_2CO_3 possesses a high Bohr coefficient and produces a strong Bohr effect. Because of its good diffusion properties, the additionally liberated CO_2 rapidly disappears in the lungs; this is accompanied by a fall in capillary PO_2 facilitating the diffusion of O_2 from the alveoli into the plasma.

In addition to the well-known Bohr effect under in vitro conditions, a further shift of the ODC to the right was observed in vivo (Braumann et al. 1982) that cannot yet be explained but may be due to changing cellular hemoglobin and Cl^- concentrations during intensive exercise.

Adaptation Due to Training

Regular endurance training accelerates the erythrocyte turnover (Schmidt et al. 1988), leading to a reduction of its life span from about 120 days to 70 days (Reefsum et al. 1976). The blood of an endurance athlete is therefore characterized by the O_2 binding properties of young erythrocytes (P_{50} and Hill's n in old cells, 22.5 mmHg and 2.38; in young cells, 28.6 mmHg and 2.67; Schmidt et al. 1987). Because of higher 2,3-BPG concentration, the ODC is right shifted (P_{50} +2-3 mmHg) and the cooperativity of the hemoglobin tetramers is increased (n +0.3 units) in athletes (Böning et al. 1975; Braumann et al. 1979; Schmidt et al. 1988). In addition to the training effects on the ODC, which are mostly due to a younger red cell population, the Bohr effect determined under in vitro conditions, and even more the effects of acidification obtained during physical exercise (in vivo), are considerably increased compared to the effects observed in untrained subjects (see figure 17.2; Böning et al. 1975; Braumann et al. 1982). All of these influences on the ODC seem small; but because of the logarithmic nature of these coefficients, their effects are not additive—rather they must be multiplied with each other. Therefore even small changes contribute to remarkable effects as shown in table 17.1.

Altitude Effects

At altitude, the ODC is shifted to the right, due to increased 2,3-BPG. The production of this substance is due to alkalosis occurring during the first days at altitude because of hyperventilation, and to a younger erythrocyte population after some weeks at altitude because of effective erythropoiesis. When endurance athletes live and train at altitude, both effects are present, shifting the standard ODC (P_{50}) from 28.5 mmHg (untrained lowlanders) to 32.3 mmHg in trained highlanders from 2,600 m (for comparison: P_{50} in trained lowlanders, 31.0 mmHg; in untrained highlanders, 29.6 mmHg; Schmidt et al. 1991). To date, research on the altitude training effect has mostly focused on red cell expansion, as total hemoglobin mass is about 11% higher in altitude than in sea level athletes (Schmidt et al. 2002). How long the right shift of the ODC remains after descent from altitude and whether this may have any positive effect on physical performance need to be investigated.

A right-shifted ODC is generally beneficial at low altitudes, leading to higher oxygen extraction from the blood. At higher altitude the point at which a right shift becomes disadvantageous is at about 5,000 m at rest because of lower saturation of the hemoglobin molecule in the lungs (Bencowitz et al. 1982). During heavy anaerobic exercise, this point markedly decreases to 3,500 m or lower. For altitude expeditions above 8,000 m, therefore, the

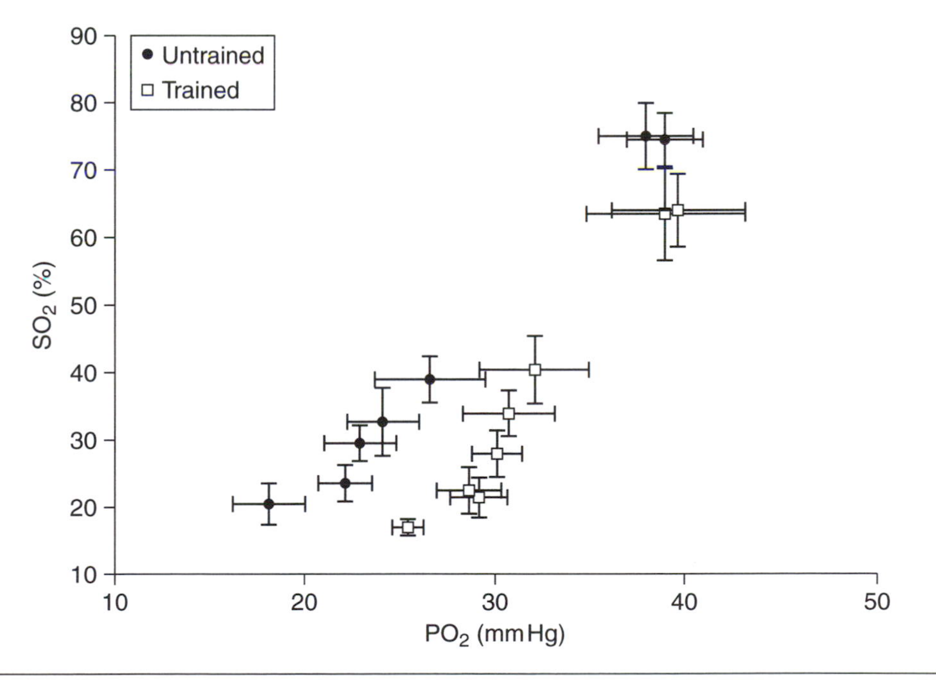

Figure 17.2 Relationship between O_2 saturation and PO_2 in femoral venous blood during rest and cycle ergometer exercise in untrained (point) and endurance-trained subjects (unfilled squares). Presented are mean values and standard errors of the mean.

Reprinted, by permission, from D. Böning et al., 1975, "Influences of exercise and endurance training on the oxygen dissociation curve of blood under 'in vivo' and 'in vitro' conditions," *European Journal of Applied Physiology* 34: 1-10.

noncompensated alkalosis (pH > 7.6, PCO_2 at the top of Mount Everest is 7.5 mmHg; West et al. 1983) is a beneficial adaptation for success at that extreme altitude.

Principles of CO_2 Transport in Blood

In contrast to O_2, the polarized CO_2 molecule is well dissolved in water and diffuses 20 times more rapid across the body fluids. But a pure physical solution is not sufficient for transport in blood, because this would require high pressure differences between venous and arterial blood accompanied by a marked acidosis resulting from the formation of carbonic acid (H_2CO_3). Buffering of this acid in the form of bicarbonate (chemical binding) is the main mechanism for the uptake of CO_2 (approximately 0.5 L in 1 L blood) and its transport without remarkable gas pressure changes.

The red cells play an indispensable role in these functions (for a more extended description see Greger & Windhorst [1996] and Geers & Gros [2000]). The most time-consuming step is the following reaction (formula 4):

$$CO_2 + H_2O \leftrightarrow H_2CO_3$$

which is markedly accelerated by the erythrocytic enzyme carboanhydrase (type carboanhydrase II) leading to an equilibrium within milliseconds instead of minutes; prerequisite is diffusion of the CO_2 into the cells.

The next step is buffering of H^+ after immediate dissociation of carbonic acid by hemoglobin (especially histidine groups) and phosphates (2,3-biphosphoglycerate, ATP, etc.) The nonbicarbonate buffer capacity in the erythrocytes (approximately 60 mmol/L per pH unit) is markedly greater than in plasma (plasma proteins and phosphates). The bicarbonate formed rapidly leaves the cell by electroneutral exchange with Cl^- (Hamburger shift) using anion transporters (band 3 proteins). Since the membrane is slightly charged (approximately +10 to 20 mV outside, resulting from a Donnan equilibrium imposed by nonpermeating proteins and organic phosphates [Juel 1997]), an uphill transport against the higher extracellular [HCO_3^-] (approximately 26 vs. 17 mmol/kg H_2O) is possible.

The decrease of pH by the bicarbonate loss favors the uptake of H^+ ions by the buffers, making them more effective. The rapid exchange of CO_2 and HCO_3^- between red cells and plasma leads to complete equilibrium within seconds. Because of this and the presence of carboanhydrase in endothelial and other cell membranes, the enzyme is not necessary in plasma.

In addition to dissolved CO_2 and bicarbonate, a third form of transport is binding of CO_2 to four amino groups of valines at the ends of the α and β Hb chains as carbamate ($R\text{-}N\text{-}H_2 + CO_2 \leftrightarrow R\text{-}NHCOOH \leftrightarrow R\text{-}NH\text{-}COO^- + H^+$, R Hb chain). The reaction is, however, attenuated at low pH. At rest it accounts for approximately 10% of the total CO_2 transport.

The CO_2 transport is additionally promoted by the Haldane effect, which corresponds to the Bohr effect for O_2 transport. The conformational changes of the Hb tetramer when binding O_2 favor the uptake of H^+ by the so-called Bohr groups (especially imidazoles and amino groups in histidines and valines) and the carbamate formation at the β-chains. The whole Haldane effect changes the CO_2 content of blood by maximally 3 mmol/L.

The total amount of CO_2 in blood (erythrocytes + plasma) in dependence on PCO_2 is illustrated in the form of the CO_2 dissociation curves (figure 17.3). In contrast to the O_2 dissociation curve, these are similar to a parabola, flattening at high gas pressures without reaching a saturation point.

The slope depends on the buffer concentration, that is, mainly on [Hb]. At rest, 2 mmol/L CO_2 is taken up in the tissue capillaries because of the Haldane effect; the increase of PCO_2 amounts to only 6 to 8 mmHg (arrow R in figure 17.3).

Exercise and CO_2 Transport

During aerobic exercise, the Haldane effect increases markedly because of enlarged oxygen extraction; additionally PCO_2 in mixed venous blood rises to approximately 60 mmHg while the arterial value remains nearly constant. Thus, the binding properties of blood alone allow a threefold increase of CO_2 transport. The moderate acidosis in capillaries and extracellular fluid is tolerated by the organism.

With a marked rise of lactic acid concentration at intense exercise, some problems appear. Because of buffering, the bicarbonate concentration and the slope of the CO_2 dissociation curve are reduced.

In addition, carbamate formation is reduced by the acidosis. These drawbacks are partly compensated for by increased capillary PCO_2 and oxygen extraction, as well as by a small rise of [Hb] because of a water shift to the extracel-

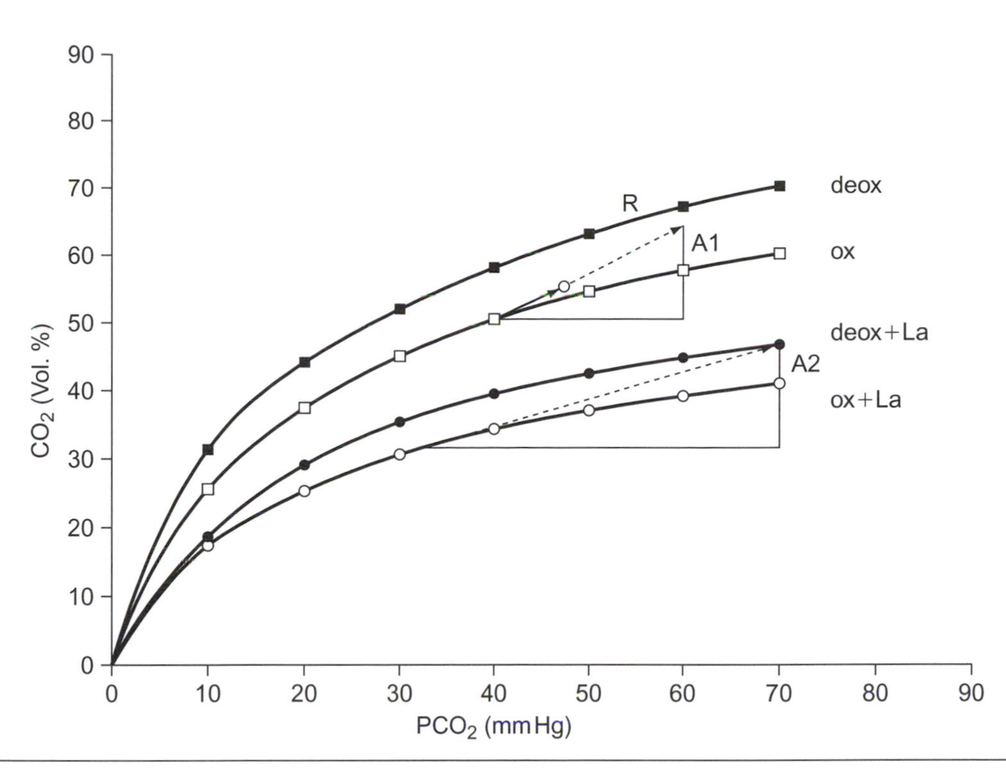

Figure 17.3 CO_2 dissociation curves of blood (schematically). Ox = 100% O_2 saturation; deoxy = 0% O_2 saturation; arrows indicate arteriovenous changes at rest (R), during aerobic exercise (A1), and during exercise with a 10 mmol/L increase in lactic acid concentration (A2, lower pair of curves). In the latter case the arteriovenous difference, because of the flat curves, is sufficiently large only if venous PCO_2 increases and arterial PCO_2 decreases.

lular fluid and working muscle cells. Also the now-beginning hyperventilation reduces the alveolar PCO_2, shifting the arterial point to the steep initial part of the CO_2 dissociation curve. Because of the hyperventilation, CO_2 excretion is larger than O_2 uptake during heavy exercise. This not only attenuates the lactic acidosis but also buffers the increase of osmolality caused by lactate and other metabolites (maximally approximately 40 mosmol/kg H_2O), since bicarbonate molecules leave the blood and tissues (Böning & Maassen 1983; Maassen 1984).

Concerning lactic acid there are two additional effects or special points. First, the large molecule does not fit the band 3 protein anion transporter in the red cell membranes but uses a much slower lactate-proton cotransporter. At 37° C the half-time is 1.2 min; the time needed for complete equilibration is much longer (Juel et al. 1990). Thus, at first glance the use of red cell buffers seems to be retarded. This is, however, not the case. Lactic acid liberates CO_2 from plasma bicarbonates; the CO_2 immediately enters the erythrocytes and reacts with nonbicarbonate buffers (Böning et al. 1991). The additional implications for O_2 transport are described in the preceding section.

The time delay of equilibration, as well as the fact that because of the Donnan effect the intra-erythrocytic lactate concentration is rather low, is not generally known and often leads to confusion in comparison of measurements of lactate performed in whole blood or plasma/serum (Juel 1997; Böning 2001).

The second additional effect is that lactate competes with CO_2 at the valines of Hb (Böning et al. 1993). According to in vitro experiments, this might change the direction of carbamate transport now going from the lung to tissue. However, this effect decreases with increasing temperature and should not play an important role after warm-up and during longer-lasting exercise (Böning et al. 2000).

Training Effects on CO_2 Transport

Physical training does not cause remarkable changes of blood acid–base equilibrium at rest; pH, PCO_2, and $[HCO_3^-]$ are not significantly different between athletes and sedentary people. Of course the reduction of lactic acid concentration at a given power in endurance-trained subjects improves CO_2 transport. An additional positive effect is the larger proportion of young erythrocytes in this group: On the one hand, these cells contain more biphosphoglycerate, and on the other hand the buffer properties of the Hb molecule decrease with time, perhaps by chemical modifications like glycation (Böning et al. 1999). Also the increase of RCM enlarges the total amount of buffers in trained subjects.

Altitude Effects

At altitude, the shift of arterial PCO_2 to lower values (i.e., to the steeper part of the CO_2 dissociation curve) and the increased Hb concentration in combination with younger erythrocytes (at least initially) facilitate the CO_2 transport. The increase in nonbicarbonate buffer capacity, which is further increased by the rise in 2,3-BPG, possibly surpasses the loss of bicarbonate through renal compensation of respiratory acidosis (Böning et al. 2002).

An interesting point in this context is the effect of blockers for carboanhydrase (e.g., acetazolamide) often used for prevention or treatment of acute mountain sickness. Bicarbonate formation in blood during passage through tissues is markedly retarded (Swenson & Maren 1978). Under these conditions, venous PCO_2 rises while arterial PCO_2 drops. CO_2 output and peak O_2 uptake are little influenced, but the endurance time is markedly reduced (Stager et al. 1990). This might be of importance for long-lasting tours in the mountains.

Regulation of Red Cell Volume

Red cells experience severe mechanical stress during the transitions of capillaries. This is increased during exercise as the muscles compress capillaries during contraction. The mechanical stress is exaggerated when red cells pass through tissues with osmolalities differing from isotonicity, for example renal medulla or heavily exercising muscles. To sustain that stress, the cytoskeleton (especially spectrin) and the remarkable constancy of the red cell volume are important. Under resting conditions the latter is achieved by regulation of the intracellular osmolality.

Factors Contributing to Intracellular Osmolality

Under resting conditions the volume of the erythrocyte is mainly regulated by the following transport systems determining the intracellular electrolyte content:

Na^+/K^+ ATPase

Na^+-K^+-Cl^- cotransporter

Na^+/H^+ antiport

KCl cotransport

The transporters can be activated by different stimuli. The Na^+/K^+ pump is stimulated to a small extent by extracellular K^+ and, more efficiently, by intracellular Na^+ (Sachs 1970). Additionally this ATPase is activated by catecholamines.

The Na^+-K^+-Cl^- cotransporter is also stimulated by the factors mentioned. Further potent activators are osmolality, cell shrinkage, or both, and pH (acidosis). pH may have an additional indirect effect by stimulating the Na/H exchange and thus increasing the intracellular Na concentration. The KCl cotransport is dependent on the cell volume and is activated by cell swelling (Parker et al. 1995). The element of cytoskeleton spectrin seems to be involved in this activation (Sachs & Martin 1999).

The concentration or activity of these proteins in the membrane is in part dependent on cell age. In rat red cells, the highest activity of Na^+/K^+ATPase was found in reticulocytes. The concentration of this enzyme rapidly decreases during the first days of cell maturation. Concomitantly Na^+-K^+-Cl^- transport is activated by cell shrinkage (Mairbaurl et al. 2000). Cell shrinkage continues during the whole life span of the red cell. This is accompanied by increased intracellular Na^+ concentration, falling K^+ concentration, and an increase of MCHC (cell density) (Schmidt et al. 1987). With cell shrinkage, an additional factor comes into play: the osmotic coefficient of the hemoglobin, which depends on its concentration (Wittmann & Gros 1981). For electrolytes like bicarbonate, Na^+, or lactate this coefficient is in the range of 0.93 (Böning & Maassen 1983; Maassen & Böning 1987), meaning that for an increase (or decrease) of 1 mmol/kg, osmolality increases by 0.93 mosmol/kg H_2O. For normal red cell hemoglobin concentration of 330 g/L (corresponding to 6.9 mmol/kg H_2O), this coefficient is about 3. Thus, hemoglobin contributes with about 21 mosmol/kg H_2O to the total osmolality. In young red cells, hemoglobin concentration is about 5.7 mmol/kg H_2O; in old red cells it is 7.3 mmol/kg H_2O (Schmidt et al. 1987). Due to an increase of the osmotic coefficient from about 2.6 to 3.4 (Wittmann & Gros 1981), the contribution of hemoglobin to the intracellular osmolality therefore increases from 15 to 25 mosmol, compensating in part for the reduction of transport capacity for cations. This mechanism also works when red cell volume is changed rapidly, and is enhanced by changes in the ionic strength (Wittmann & Gros 1981; Gary-Bobo & Solomon 1968). The effect is particularly large with hemoglobin concentrations and ionic strength higher than the normal mean values. Both of these conditions are present during passage through the renal medulla and through heavily exercising muscles.

Water Exchange

To let these transport mechanisms work with regard to volume regulation, a prerequisite has to be fulfilled. The red cell membrane has to be permeable to water. This is accomplished through two different ways of exchange water between the intracellular compartment and the surrounding fluid (plasma). The first is pure diffusion across the lipid bilayer. The second is the rapid shift of water through pores due to osmotic gradients (Lee et al. 1997). These pores are now identified as aquaporin1, a membrane-spanning protein (28 kDa), which is present not only in red cells but at least also in renal proximal tubules. Diffusional and osmotic water permeability of red cell membranes differ markedly. Diffusional water permeability ($3 \cdot 10^{-3}$ cm/sec, measured at 25° C; Brahm 1982) is about 14% of osmotic water permeability ($20 \cdot 10^{-3}$ cm/sec, measured at 25° C; Moura et al. 1984). Neither permeability can be attributed to either the bilayer or the pores exclusively. Aquaporin1 contributes about 64% of the diffusional permeability and >85% of the total osmotic permeability (Mathai et al. 1996). The large osmotic water permeability results in a rapid osmotic equilibrium when red cells are exposed to hypertonic medium. It takes less than 1 sec to achieve the respective volume when the external osmolality is doubled suddenly. Stated simply, this means that under normal conditions red cells are always near the osmotic equilibrium.

Whether other solutes are exchanged through the aquaporin1 channels is still a matter of debate, since it was shown that urea is transported by a separate protein across the red cell membrane. If it occurs at all, this phenomenon is restricted to very small molecules. But it seems to be established that a part of the CO_2 transport is enabled by these water channels (Cooper et al. 1998).

Whereas the structure is known and the function seems to be clear at least in large part, the importance of this system is a matter of question, because people having aquaporin1 deficiency (Colton null phenotype) do not experience any obvious clinical problem (Preston et al. 1994). Red cells of these subjects show some functional differences compared with cells from healthy people. Osmotic water permeability is 13% of that of normal red cells, and the diffusional permeability is about 36% of the normal value. Membrane surface area is slightly reduced; the ratio surface area to volume is smaller as well; and the life span of the red cells seems to be reduced (Mathai et al. 1996). The latter might be a result of a decreased mechanical stability. In red cells from Colton null subjects, CO_2 permeability is reduced by about 85% compared to normal (Endeward personal communication). Because the Colton null phenotype is very rare, possible limitations during exercise, especially during exercise of high intensity, have not been investigated.

Red Cell Volume During Exercise

There are two major challenges for volume regulation during exercise. The first is the increase in extracellular (plasma) osmolality, and the second is the occurrence of acidosis. Both are pronounced during exercise of high intensity. Thus, it is not surprising that red cell volume is almost unchanged during exercise of low or medium intensity.

During exercise of high intensity, when sweat loss can be neglected because of the short duration, or during exercise with a small muscle group, plasma osmolality may increase by about 40 mosmol/kg H_2O. The cause of the osmolality increase is not increase in the concentration of electrolytes like Na^+, K^+, or Cl^-. These concentrations change because of exchanges across a membrane or because of water shifts. Movement of ions is accompanied by water movements, as the water permeability of muscle cells (like red cells) is very high (Sejersted et al. 1982). Since ions are followed by water, osmolality remains constant, but the volume of the compartments may change. The cause for the increase in osmolality is the production of metabolites inside the muscle cells. A portion of the metabolites (lactate and the products of phosphocreatine splitting) accumulate inside the cells, causing a water shift from the interstitial space and the blood into the muscle cells (Maassen 1984). This water shift should result in red cell shrinkage, as the red cell membranes are easily permeable to water.

The conditions present during intensive exercise increase the activity of the transporters responsible for potassium influx. Increase in plasma [K^+] enhances the Na^+/K^+ ATPase activity; the pH decrease influences the Na^+/H^+ exchanger, thus increasing the intracellular Na^+ concentration. This is in turn an additional stimulator for the Na^+/K^+ ATPase and the Na^+-K^+-Cl^- cotransporter. During exercise of a large muscle group as in cycling, the additionally increasing catecholamine concentration enhances the activity of the Na^+/K^+ ATPase. The influx of K^+ into the red cell is augmented by about 25% (Maassen et al. 1998; Lindinger et al. 1999). Resting values are between 1.2 mmol · L^{-1} · h^{-1} and 2.1 mmol · L^{-1} · h^{-1}, and end exercise values are 1.5 and 2.5 mmol · L^{-1} · h^{-1}, respectively. This increase results in an uptake of 0.02 mmol of K^+/L of red cells within 3 min. During this period of time, tissue osmolality increases by about 35 mosmol/kg H_2O during intense exercise. The potassium uptake is far too small to compensate for any volume challenges. Changes in red cell [K^+] during heavy exercise therefore are not caused by changes in the activity of the transporters. They are the consequence of water shifts (Juel et al. 1999; Maassen et al. 1998).

The driving force for the water shift out of the red cells is high tissue osmolality. But this osmotic effect is minimized, as the metabolites liberated from the muscle can be easily taken up by the red cells. The main metabolites are lactate and HCO_3^-. Their enhanced concentration increases the intracellular content of osmoles and thus reduces the water loss from the red cells. The uptake of these substances is further enhanced as the Donnan equilibrium changes because of the falling pH, further supporting the distribution of lactate and HCO_3^- (and chloride) into the red cells. As the activation of transporters is of minor importance, the red cell volume is the result of these two factors: a water shift from the red cells due to tissue osmolality, and a water shift into the red cells due the acidosis occurring. Under normal conditions the two factors cancel each other out. Therefore red cell volume undergoes only small changes.

These interrelationships show that the red cells are involved in transport of lactate and CO_2 from the working muscles to the central circulation, but not in the transport of K^+. One additional role of the described mechanisms might be to prevent excessive water loss from the blood to the heavily working muscle during exercise.

Red Cell Influence on Circulation and Respiration

A very interesting aspect is the reaction of Hb with NO produced by the endothelial cells. There are different forms of binding either to Fe^{++}, partly producing methemoglobin (Fe^{+++}), favored in the deoxygenated state, or to cysteine β-93, favored in the oxygenated molecule (McMahon et al. 2000).

Whereas the physiological importance of changes in oxygen affinity because of low concentrations of NO may be in doubt, the varying fixation, especially to cysteine, seems to be of interest for circulation and ventilation. Liberation of NO from Hb with increased oxygen extraction during exercise may cause vasodilation; entrance of deoxyhemoglobin-derived S-nitrosothiols into the brain (nucleus tractus solitarius) increases ventilation (Lipton et al. 2001). Another fascinating idea is that release of ATP from red cells at low PO_2 and high PCO_2 in the venules stimulates endothelial cyclooxygenase, leading to production of prostacyclin; the latter might diffuse into the accompanying arterioles and cause upstream vasodilation during exercise that is as yet unexplained (Hester & Choi 2002).

Erythrocyte and Radicals

A high concentration of iron and contact with high oxygen pressure in the lungs make the red cell very vulnerable to oxidative stress. The Fe^{++} is protected by the surrounding peptide chains, but continuous production of Fe^{+++} (methemoglobin) accompanied by the superoxide anion O_2^- is inevitable. The enzyme methemoglobin reductase catalyzes the restoration of Hb-Fe^{++} using mainly NADH + H^+ supported by glycolysis. The pentose phosphate pathway is the source of NADPH used for regeneration of reduced glutathione GSH (Stryer 1995). The concentration of this substance is rather high in the erythrocyte (approximately 2 mmol/L). It serves for defense of the functionally important SH groups in Hb and other proteins, and in addition contributes to acidification of the cell interior and influences oxygen affinity. The membrane is protected by vitamin E bound to specific proteins.

Radicals produced in the muscles during exercise should have an impact on the erythrocyte after entering the plasma; intravascular hemolysis has been partly ascribed to this mechanism (e.g.,

Sentürk et al. 2001). The influence of acute and chronic exercise on antiradical defense in red cells has been investigated, with inconsistent results. Measurement of glutathione is dependent on oxygen saturation, which usually has not been considered (Hütler et al. 2000). The erythrocytic content of vitamin E seems to be well regulated during various forms of exercise (Hütler et al. 2001). Increases in the activities of superoxide dismutase, peroxidases, and catalase immediately after exercise (enzyme activation), and also of glutathione reductase in trained animals and human subjects (increased enzyme production in new erythrocytes), have been described (Ohno et al. 1988; Mena et al. 1991; Tauler et al. 1999; Sentürk et al. 2001; Wozniak et al. 2001). To our knowledge, methemoglobin reductase has never been investigated with respect to exercise. In conclusion, a better antiradical defense should be expected after endurance training because of the increasing proportion of young red cells with highly active enzymes and without previous membrane damage (Schmidt et al.1987, 1988).

Excessive increase in young cells as occurs during EPO doping should be accompanied by very high enzyme activity. Therefore, measuring age-dependent enzyme activity (e.g., ASAT) might be used as a screening method to detect EPO abuse.

Conclusion

Mass and properties of the erythrocytes are important for physical exercise and are frequently assumed to be performance-limiting factors. Both are changed by training influences and adapt to the increased need for oxygen. Public interest is focused on quantitative aspects, because these are easy to be understand and manipulate. Special training forms and methods such as hypoxia and altitude training have been developed to increase total erythrocyte mass. Doping with blood or EPO also seems to be widespread and is difficult to detect. No special forms of training or manipulation, however, have been developed thus far to improve the qualitative properties of erythrocytes.

Chapter 18

Leukocytes

Bente Klarlund Pedersen

This chapter focuses on exercise-induced changes in leukocyte subpopulations. Circulating leukocyte subpopulations can be divided into granulocytes, lymphocytes, and monocytes. Granulocytes are classified as neutrophils, eosinophils, or basophils on the basis of cellular morphology and cytoplasmic staining characteristics. The dominant granulocyte is the neutrophil, which has a granulated cytoplasm and is often called a polymorphonuclear leukocyte because of its multi-lobulated nucleus. The neutrophils are produced in the bone marrow and released into the circulation. They appear generally to be the first cell to arrive at a site of inflammation. They are active phagocytic cells. Lymphocytes are mononuclear cells, the so-called white blood cells, responsible for the specific immune response. They can be divided in numerous ways, and their characteristics account for the immune system's attributes of diversity, memory, specificity, and self-/non-self-recognition. Monocytes are phagocytic mononuclear cells, which after a few hours in the circulation migrate into the tissue and differentiate into macrophages.

Over the past 20 years, a variety of studies have demonstrated that exercise induces considerable changes in circulating leukocyte subpopulations. The interactions between exercise stress and the immune system provide a unique opportunity to link basic and clinical physiology, and to evaluate the role of underlying stress and immunophysiological mechanisms. It has been suggested that exercise represents a quantifiable model of physical stress (Hoffman-Goetz & Pedersen 1994; Pedersen & Hoffman-Goetz 2000), as many clinical physical stressors (e.g., surgery, trauma, burn, sepsis) induce a pattern of hormonal and immunological responses that has similarities to that of exercise.

Acute Exercise and Leukocytes

Responses of blood leukocyte subpopulations to an episode of acute exercise are highly stereotyped. However, the sensitivity to the exercise stimulus varies between subpopulations.

Leukocyte Subpopulations

Neutrophil concentrations increase during and postexercise (figure 18.1*a*), whereas lymphocyte concentrations increase during exercise and fall below pre-values after long-duration physical work (figure 18.1*b*) (McCarthy & Dale 1988). Monocytes do not change during exercise, but increase in number in the postexercise period. Several reports describe exercise-induced changes in subsets of blood mononuclear cells (BMNC) (Brines et al. 1996; Hoffman-Goetz 1996; Keast et al. 1988; Keast & Morton 1992; Mackinnon 1994; Nieman 1994; Pedersen 1997; Pedersen & Nieman 1998; Pedersen et al. 1998). Increased lymphocyte concentration is likely due to the recruitment of all lymphocyte subpopulations to the vascular compartment: T-cells, B-cells, and natural killer (NK) cells. NK cells are more sensitive to exercise than T-cells, which are more sensitive than B-cells (figure 18.2*a*). Lymphocyte subpopulations are identified by a so-called cluster of differentiation (CD) number. During exercise the CD4/CD8 ratio decreases, reflecting the greater increase in CD8+ cytotoxic lymphocytes than in CD4+ helper lymphocytes (figure 18.2*b*). CD4+ and CD8+ cells can be further divided into memory cells, which are activated cells, and naive cells or virgin cells, which have not yet met an antigen. CD45RO+ memory and CD45RA+ naive cells are identified

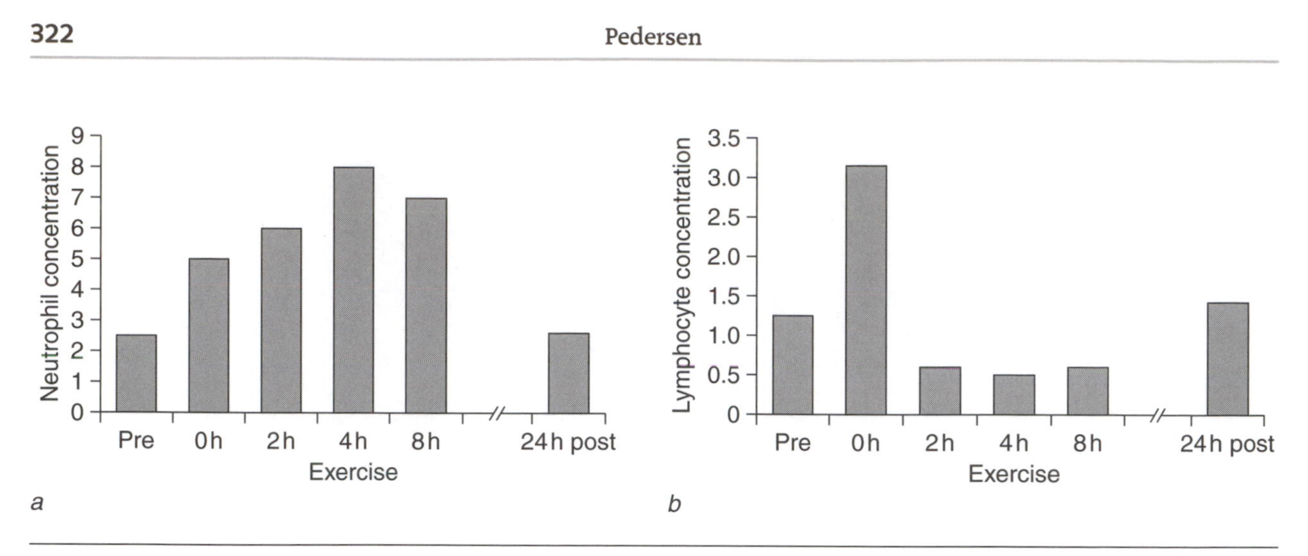

Figure 18.1 Schematic presentation of the changes in neutrophil number *(a)* and lymphocyte number *(b)* in relation to an acute bout of exercise.

by the absence of 45RO and the presence of CD62L (figure 18.2*c*). Data show that the recruitment is primarily of CD45RO+ lymphocytes (Gabriel et al. 1993). Recent studies indicate that the concentrations of CD45RO+ and CD45RO-CD62L increase during exercise, suggesting that memory, but not naive, lymphocytes are rapidly mobilized to the blood in response to physical exercise (Bruunsgaard et al. 1999) (figure 18.2*d*).

The CD4+ T helper (Th) and the CD8+ T cytotoxic (Tc) cells can be further divided into type 1 (Th1 and Tc1) and type 2 (Th2 and Tc2) cells according to their cytokine profile. Type 1 T-cells produce interferon (IFN)- and interleukin (IL)-2, whereas type 2 T-cells produce IL-4, IL-5, IL-6, and IL-10. Type 1 T-cell responses are stimulated by IL-12 and have been shown to protect against intracellular pathogens such as several viruses. IL-6 has been shown to induce Th2 polarization by stimulating the initial production of IL-4. Type 1 T-cells mediate protection against intracellular microorganisms such as virus, whereas type 2 T-cells are important in the defense against extracellular parasites such as several helminths (figure 18.2*e*). Two recent studies (Ibfelt et al. 2002; Steensberg, Toft, Bruunsgaard et al. 2001) demonstrate that the postexercise decrease in T-lymphocyte number is accompanied by a more pronounced decrease in type 1 T-cells, which may be linked to high plasma epinephrine. The relatively more pronounced decrease in type 1 compared with type 2 T-cells in the recovery period may explain the increased sensitivity to infections following strenuous exercise, as these infections are often caused by viruses.

To obtain information about lymphocyte turnover in cells recruited during exercise, we recently analyzed telomeric terminal restriction fragment (TRF) length. Telomeres are the extreme ends of chromosomes that consist of TTAGGG repeats. After each round of cell division telomeric sequence was lost because of the inability of DNA polymerase to fully replicate the 5 prime end of the chromosome. Telomere lengths have been used as a marker for replication history and the proliferation potential of the cells. In response to exercise, telomere lengths in CD4+ and CD8+ lymphocytes were significantly shorter than in cells isolated at rest, indicating that mature cells with a long history of replication are mobilized to the blood during exercise (Bruunsgaard et al. 1999).

Thus, the initial increase in CD4+ and CD8+ cells after exercise appears not to be due to repopulation by newly generated cells but may be a redistribution of activated cells.

Lymphocyte Proliferation

A commonly used method for evaluating the functional capacity of lymphocytes is the assessment of mitogen- or antigen-induced proliferative responses by measuring incorporation of [^{3}H]thymidine in DNA. Studies in humans indicate that the lymphocyte responses to the T-cell mitogens, phytohemagglutinin (PHA) and concanavalin A (ConA), decline during and for up to several hours after exercise (Nielsen & Pedersen 1997). This is at least partly due to the increase in NK cells in circulation and the relative decline of CD4+ cells in in vitro assays (Fry et al. 1992; Keast et al. 1988). In contrast, lymphocyte proliferation to B-cell mitogens, pokeweed mitogen (PWM), and lipopolysaccharide (LPS) increases or remains unchanged after exercise (Field et al.

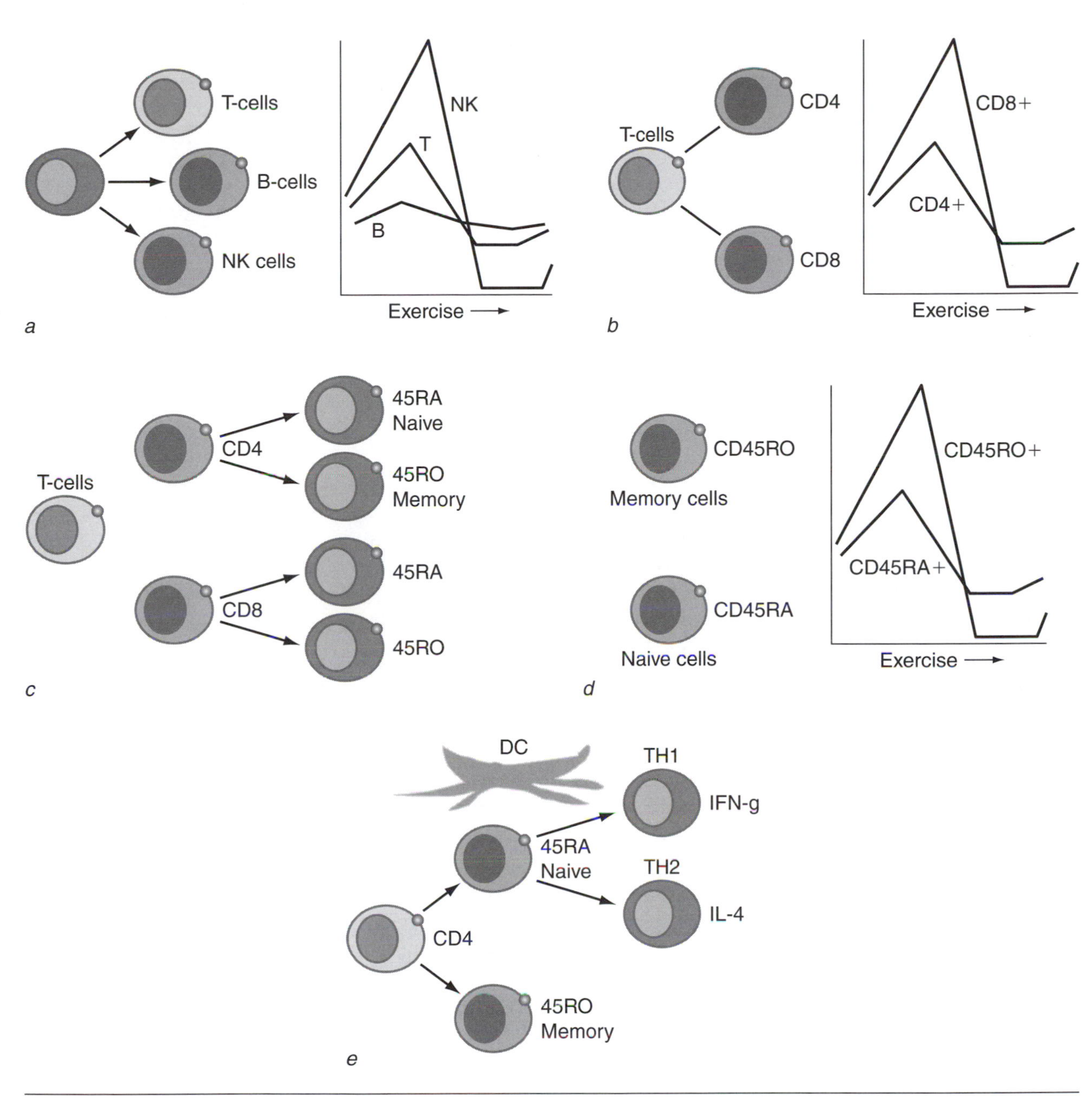

Figure 18.2 *(a)* Schematic of NK, T-, and B-cells; *(b)* CD4+ and CD8+ T-cells; *(c)* CD45+ naive cells; *(d)* CD45RO+ memory cells; and *(e)* type 1 and type 2 cytokine-producing cells. The effect of exercise is demonstrated in *a*, *b*, and *d*. Following exercise, type 1 T-cells disappear from the circulation more than type 2 T-cells. Thus a shift in the type 1/type 2 T-cell balance occurs, with type 2 T-cell dominance.

1991). The large increase in NK cells, relative to T-cells, following intense exercise was the most likely cause of the reduced mitogen response of total lymphocyte cultures (Hinton et al. 1997). It is important to bear in mind that during exercise more lymphocytes are recruited to the blood, and on a per cell basis the lymphocyte proliferation response is not actually suppressed. Thus, the lower responses to PHA and ConA during exercise simply reflect the proportional changes in lymphocyte subsets and the decline in the percentage of T-cells (Fry et al. 1992; Nielsen & Pedersen 1997; Tvede, Pedersen et al. 1989). Following exercise, the total lymphocyte concentration declines and the proliferation response is unchanged from values obtained prior to exercise. Consequently, the total in vivo lymphocyte function in the blood can be considered "suppressed" after exercise.

NK Cell Activity

NK cells mediate non-major histocompatibility complex (MHC)-restricted cytotoxicity, with potential resistance to viral infections (Welsh & Vargas-Cortes 1992) and cytolysis of some malignant cells (O'Shea & Ortaldo 1992). The method frequently applied to measure NK cell activity is the ^{51}Cr release assay, for which the percentage of lysed target cells is the endpoint. The cytolytic activity of NK cells is enhanced by interferon (IFN) (Ortaldo et al. 1983) and IL-2 (O'Shea & Ortaldo 1992), whereas certain prostaglandins (PG) (Brunda et al. 1980) and immune complexes (Pedersen et al. 1986) down-regulate the function of NK cells. Exercise of various types, durations, and intensities induces recruitment to the blood of cells expressing characteristic NK cell markers (Mackinnon 1989; Pedersen & Ullum 1994). NK cell activity (lysis per fixed number of BMNC) increases consequent to the increased proportions of cells mediating non-MHC-restricted cytotoxicity. Following exercise, the NK cell activity is suppressed on a per NK cell basis (Nielsen, Secher, Kappel et al. 1996) if the exercise has been of high intensity and long duration (more than 45 min) (figure 18.3). Maximal reduction in NK cell concentrations, and hence the lower NK cell activity, occur 2 to 4 h after exercise. In one study, the reduction in NK cell activity was due to down-regulation of NK cell activity by prostaglandins (PG) (Pedersen et al. 1990).

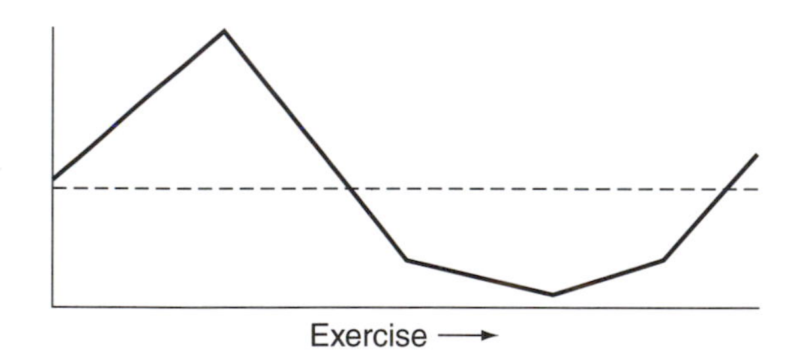

Figure 18.3 The natural killer (NK) cell activity is suppressed below pre-values following intense exercise of longer duration.

Generally, NK cell activity is increased when measured immediately after or during either moderate and intense exercise of a few minutes. The intensity, more than the duration of exercise, is responsible for the degree of increment in the number of NK cells. If the exercise has lasted for a long period and has been very intense (e.g., a triathlon race), only a modest increase in NK cells is found postexercise (Rohde et al. 1996). NK cell count and the NK cell activity are markedly lower only following intense exercise of at least 1-h duration. The definitive study to map the time course in terms of postexercise NK cell immune impairment has not been done. Initial fitness level or sex does not appear to influence the magnitude of exercise-induced changes in NK cells (Brahmi et al. 1985; Kendall et al. 1990).

Cell Function and Antibody Production

The secretory immune system of mucosal tissues such as the upper respiratory tract is considered by many clinical immunologists to be the first barrier to its colonization by pathogenic microorganisms (Mackinnon & Hooper 1994). Although IgA constitutes only 10% to 15% of the total immunoglobulin in serum, it is the predominant immunoglobulin class in mucosal secretions, and the level of IgA in mucosal fluids correlates more closely with resistance to upper respiratory tract infections than serum antibodies (Liew et al. 1984). Lower concentrations of the salivary IgA have been reported in cross-country skiers after a race (Tomasi et al. 1982). This finding was confirmed by a 70% decrease in salivary IgA, which persisted for several hours after completion of intense, long-duration ergometer cycling (Mackinnon et al. 1987). Decreased salivary IgA was found after intense swimming (Gleeson et al. 1995; Tharp & Barnes 1990), after running (Steerenberg et al. 1997), and after incremental treadmill running to exhaustion (McDowell et al. 1992). Submaximal exercise had no effect on salivary IgA (Housh et al. 1991; McDowell et al. 1992). The percentage of B-cells among BMNC does not change in relation to exercise. This finding suggests that the suppression of immunoglobulin-secreting cells (plaque-forming cells) is not due to changes in numbers of B-cells. Purified B-cells produce plaques only after stimulation with Epstein-Barr virus, and in these cultures no exercise-induced suppression was found. The addition of indomethacin to IL-2-stimulated cultures of BMNC partly reversed the postexercise suppressed B-cell function. Therefore, exercise-induced suppression of the plaque-forming cell response may be mediated by monocytes or their cytokines (Tvede, Heilmann et al. 1989).

In Vivo Immunological Response

There are only a few studies that document immune system responses in vivo in relation to exercise. In vivo impairment of cell-mediated immunity and specific antibody production could be demonstrated after intense exercise of long duration (triathlon race) (Bruunsgaard et al. 1997). The cellular immune system was evaluated as a skin test response to seven recall antigens, whereas the humoral immune system was evaluated as the antibody response to pneumococcal polysaccharide vaccine (this vaccine is generally considered to be T-cell independent) and tetanus and diphtheria toxoids (both of which are T-cell dependent). The skin test response was significantly lower in the group who performed a triathlon race compared to triathlete controls and untrained controls who did not participate in the triathlon. No differences in specific antibody titers were found between the groups. The latter result is in accordance with findings by others (Eskola et al. 1978).

Neutrophil Function

Neutrophils represent 50% to 60% of the total circulating leukocyte pool. These cells are part of the innate immune system, are essential for host defense, and are involved in the pathology of various inflammatory conditions. This inflammatory involvement reflects tissue peroxidation resulting from incomplete phagocytosis. One of the more pronounced features of physical activity on immune parameters is the prolonged neutrocytosis following acute long-term exercise (McCarthy & Dale 1988).

The increase in neutrophils following exercise is a biphasic response. Thus, there is an immediate transient increase within the first 30 min immediately postexercise, followed by a delayed elevation of neutrophil numbers several hours later (Peake et al. 2002). The first increase is believed to represent neutrophils demarginated from endothelial walls through the actions of epinephrine, whereas the second phase represents immature neutrophils released from the bone marrow by glucocorticoids. The late increase also represents inhibition by glucocorticoid of the ability of neutrophils to migrate into tissues.

Exercise promotes neutrophil degranulation, as measured by the increase of elastase and myeloperoxidase (Belcastro et al. 1996; Tiidus & Bombardier 1999). A number of reports show that exercise triggers a series of changes in the neutrophil population and may affect certain subpopulations differentially. A reduction in the expression of L-selectin (CD62L) immediately after exercise, followed by an increase during recovery, has been reported (Kurokawa et al. 1995). There were no concomitant changes in CD11a or CD11b expression. In contrast, however, increased expression of CD11b in response to exercise was found (Smith et al. 1996). Increased expression of the cell adhesion molecules following exercise may contribute to neutrophil extravasation into damaged tissue, including skeletal muscle. Regarding the function of neutrophils, exercise has both short- and long-term effects. The neutrophil responses to infection include adherence, chemotaxis, phagocytosis, oxidative burst, degranulation, and microbial killing. In general, moderate exercise boosts neutrophil functions, including chemotaxis, phagocytosis, and oxidative burst activity. Extreme exercise, on the other hand, reduces these functions, with the exception of chemotaxis and degranulation, which are not affected (Brines et al. 1996; Ortega et al. 1993; Smith et al. 1990, 1992).

Repetitive Bouts of Exercise

In contrast to a vast number of investigations on immune cell responses to a single bout of exercise, there is limited information on how repeated bouts of exercise on the same day affect the immune system. The few studies that have measured changes in concentration and function of leukocytes associated with repeated bouts of exercise have all used different exercise and recovery protocols, as well as subjects with varying training status (Field et al. 1991; Hoffman-Goetz et al. 1990; Nielsen, Secher, & Pedersen 1996). A methodological limitation that applies to all of the previous investigations is the lack of control for diurnal variations in various blood constituents, as the response to a first bout of exercise in the morning has been compared with that to a second or third bout later on the same day. In addition, the exercise and recovery protocols used in these studies hardly reflect the daily exercise and recovery regime practiced by most elite endurance athletes. A recent study used a design that eliminated the effect of diurnal variations. This study (Ronsen et al. 2001) compared leukocyte counts and lymphocyte responsiveness during and after a second bout of high-intensity endurance exercise on the same day with the response to a similar but single bout of exercise. Athletes participated in three 24-h trials: (1) rest in bed (Rest); (2) one bout of exercise (One); and (3) two bouts of exercise (Two). All bouts consisted of 75 min at ~75% of maximal O_2 uptake on a cycle ergometer. The second bout of exercise in the Two trial was associated with significantly increased concentrations of total leukocytes, neutrophils, lymphocytes, CD4+, CD8+, and CD56+ NK cells. These differences suggest a "carryover" effect in the immune system from a first to a second bout of exercise on the same day, whereas immune changes from five repetitive bouts of cycling over five days (each separated by a rest of 24 h) were not different from those elicited by the first bout (Hoffman-Goetz et al. 1990).

Mechanisms of Action

Acute, intense muscular exercise increases the concentrations of a number of stress hormones in the blood, including adrenaline, noradrenaline, growth hormone, beta-endorphins, and cortisol, whereas the concentration of insulin slightly decreases (Galbo 1983; Kjaer & Dela 1996; Volek et al. 1997). This section presents the evidence for exercise-induced changes in neuro-immune interactions.

Catecholamines

During exercise, adrenaline is released from the adrenal medulla, and noradrenaline is released from the sympathetic nerve terminals. Arterial plasma concentrations of adrenaline and noradrenaline increased almost linearly with duration of dynamic exercise and exponentially with intensity, when expressed relative to the individual's maximal oxygen uptake (Kjaer & Dela 1996). The expression of beta-adrenoceptors on T-, B-, and NK cells, macrophages, and neutrophils in numerous species provides the molecular basis for these cells to be targets for catecholamine signaling (Madden & Felten 1995). The numbers of adrenergic receptors on the individual lymphocyte subpopulations may determine the degree to which the cells are mobilized in response to catecholamines. In accordance with this hypothesis, it has been shown that different subpopulations of BMNC have different numbers of beta-adrenergic receptors (Khan et al. 1986; Maisel et al. 1990; Rabin et al. 1996; van Tits et al. 1991). NK cells contain the highest number of beta-adrenergic receptors, with CD4+ lymphocytes having the lowest number. B-lymphocytes and CD8+ lymphocytes are intermediate between NK cells and CD4+ lymphocytes (Maisel et al. 1990). Dynamic exercise up-regulates the beta-adrenergic density, but only on NK cells (Maisel et al. 1990). Interestingly, NK cells are more responsive to exercise and other stressors than any other subpopulation. CD4+ cells are less sensitive, and CD8+ cells and B-cells are intermediate (Hoffman-Goetz et al. 1994). Thus, a correlation exists between numbers of beta-adrenergic receptors on lymphocyte subpopulations and their responsiveness to exercise. Selective administration of adrenaline to obtain plasma concentrations comparable to those obtained during concentric cycling for 1 h at 75% of $\dot{V}O_2max$ mimicked the exercise-induced effect on BMNC subsets, NK cell activity, lymphokine-activated killer (LAK) cell activity, and the lymphocyte proliferative response (Kappel et al. 1991; Steensberg, Toft, Halkjaer-Kristensen et al. 2001; Tonnesen et al. 1987; Tvede et al. 1994). However, adrenaline infusion caused either a very small or no increase in neutrophil concentrations compared to that observed following exercise (Kappel et al. 1991; Tvede et al. 1994). Thus, adrenaline seems to be responsible for the recruitment of lymphocytes, and in particular NK cells, to the blood during physical exercise.

Growth Hormone

Plasma levels of pituitary hormones increase in response to exercise both with duration and with intensity. Growth hormone responses are more related to the peak exercise intensity than to duration of exercise or total work output (Kjaer & Dela 1996). An intravenous bolus injection of growth hormone at blood concentrations comparable to those observed during exercise had no effect on BMNC subsets, NK cell activity, cytokine production, or lymphocyte function but induced a highly significant neutrocytosis (Kappel et al. 1993).

Cortisol

The plasma concentrations of cortisol increase only in relation to exercise of long duration (Galbo 1983). Thus, short-term exercise does not increase the cortisol concentration in plasma, and only minor changes in the concentrations of plasma cortisol were described in relation to acute time-limited exercise stress of 1 h (Galbo 1983). It is well documented that corticosteroids given intravenously to humans cause lymphocytopenia, monocytopenia, eosinopenia, and neutrophilia, which reach their maximum 4 h after administration (Rabin et al. 1996). The increase in cortisol during exercise is mediated by IL-6 (Steensberg, Toft, Bruunsgaard et al. 2001). The link between exercise-induced lymphocyte changes and an effect of IL-6 on cortisol production is further supported by several studies demonstrating that carbohydrate loading during exercise attenuates both the exercise-induced increase in circulating IL-6 and the exercise effect on lymphocyte number and function (Nehlsen-Canarella et al. 1997; Nieman, Fagoaga et al. 1997; Nieman, Henson et al. 1997).

Glutamine

Skeletal muscle is the major tissue involved in glutamine production and known to release glutamine into the bloodstream at a high rate. Therefore, skeletal muscle may play a vital role in maintenance of the key process of glutamine utilization in the immune cells. Consequently, the activity of the skeletal muscle may directly influence the immune system. It has been hypothesized (the so-called glutamine hypothesis) that under intense physical exercise, or in relation to surgery, trauma, burn, and sepsis, the demands on muscle and other organs for glutamine is such that the lymphoid system may be forced into a glutamine debt, which temporarily affects its function. Thus, factors that directly or indirectly influence glutamine synthesis or release could theoretically influence the function of lymphocytes and monocytes (Newsholme 1990, 1994). Following intense long-term exercise and in other physical stress disorders, the glutamine concentration in plasma declines (Essen et al. 1992; Keast et al. 1995; Lehmann et al. 1993; Parry Billings et al. 1992). In four placebo-controlled glutamine intervention studies (Krzywkowski et al. 2001a,b; Rohde, Asp et al. 1998; Rohde, MacLean et al. 1998), it was found that glutamine abolished the postexercise decline in plasma glutamine without influencing postexercise impairment of NK and LAK cell function, or mitogen-induced proliferative responses or salivary IgA. Thus, contrary to the glutamine hypothesis, the latter studies did not support the idea that postexercise decline in immune function is caused by a decrease in the plasma glutamine concentration (Hiscock & Pedersen 2002).

Other Factors

Exercise induces numerous physiological changes, including hyperthermia, and if the exercise is very intense it may also induce oxygen desaturation. These changes may partly be responsible for exercise-induced immune changes (Pedersen 1997).

Furthermore, as a consequence of the catecholamine- and growth hormone-induced immediate changes in leukocyte subsets, the relative proportions of these subsets change and activated leukocyte subpopulations may be mobilized to the blood. Free oxygen radicals and PG released by the elevated number of neutrophils and monocytes may influence the function of lymphocytes and contribute to the impaired function of the latter cells.

To summarize, adrenaline, and to a lesser extent, noradrenaline, contribute to the acute effects on lymphocyte subpopulations. The increase in growth hormone mediates the acute effects on neutrophils. The exercise-induced increase in cortisol is mediated by IL-6. Cortisol exerts its effects within a time lag of at least 2 h and contributes to the maintenance of lymphopenia and neutrocytosis after prolonged exercise. Plasma glutamine has no effect on the acute exercise-induced changes (figure 18.4).

Chronic Exercise and Leukocyte Subpopulations

In contrast to knowledge based on the large number of studies on the immune response to acute exercise, much less is known concerning the effect of physical conditioning or training on immune function. This is largely due to the difficulties in separating fitness effects from the actual physical exercise and the long-duration studies that need to be performed. Thus, the changes induced by intense physical exercise may last at least 24 h, and even moderate acute exercise induces significant immune changes for several hours. In brief, a number of human studies indicate that the NK cell activity measured at rest is elevated in trained versus untrained humans (Nieman, Buckley et al. 1995; Pedersen et al. 1989; Tvede et al. 1991) and animals (Hoffman-Goetz et al. 1992, 1994; Jonsdottir et al. 1997; MacNeil & Hoffman-Goetz 1993a,b). In contrast, the lymphocyte proliferative responses have been described as decreased (Papa et al. 1989), elevated (Baj et al. 1994; Nieman et al. 1993), or unchanged (Nieman, Brendle et al. 1995; Nieman, Buckley et al. 1995; Oshida et al. 1988; Pedersen et al. 1989; Tvede et al. 1991) in comparisons of athletes and nonathletes. Neutrophil function is either suppressed (Lewicki et al. 1988; Pyne 1994) or not significantly influenced by exercise training

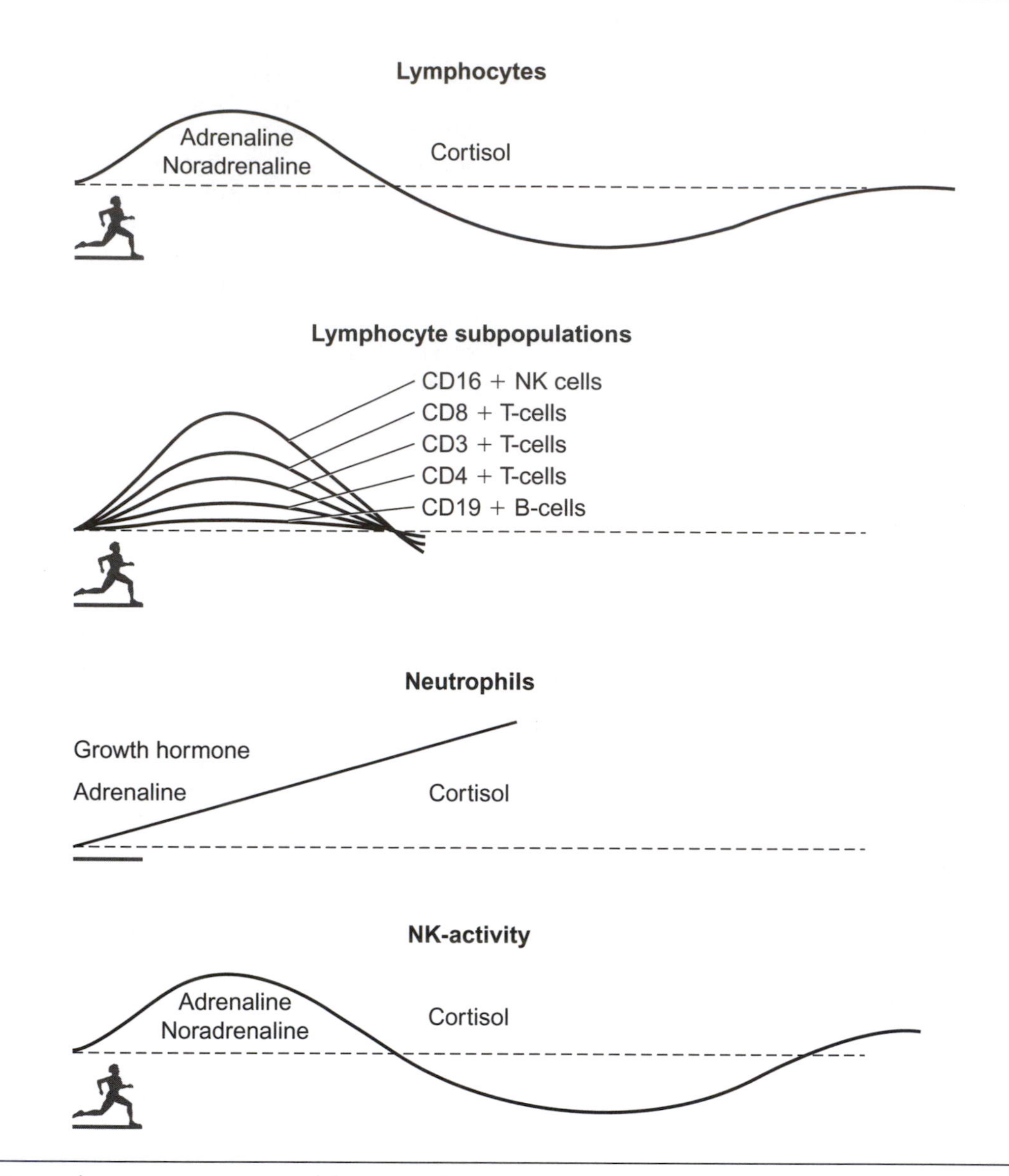

Figure 18.4 A schematic presentation of the classical hormones involved in exercise-induced lymphocyte and neutrophil cell trafficking.

(Green et al. 1981; Hack et al. 1992). Neutrophil function was unchanged in athletes during a low-training period but decreased during periods of high-intensity training (Baj et al. 1994; Hack et al. 1994).

Exercise and Infections

Without doubt, exercise and training influence the concentration of immunocompetent cells in the circulating pool, the proportional distribution of lymphocyte subpopulations, and the function of these cells. An important question is, however, the degree to which these cellular changes are of clinical significance, especially with respect to resistance to infectious diseases. From animal studies (Cannon & Kluger 1984; Ilback et al. 1991; Nicholls & Spaeth 1922) it is clear that effects of exercise on disease lethality vary with the type and time that it is performed. In general, exercise or training before infection has either no effect or decreases morbidity and mortality. Exercise during the incubation period of the infection appears either to have no effect or to increase severity of infection. In contrast to the limited experimental evidence, there are several epidemiological studies on exercise and upper respiratory tract infections (URTI). These studies are based on self-reported symptoms rather than clinical verification. In general, increased numbers of URTI symptoms have been reported in the days following strenuous exercise (e.g., a marathon race) (Heath et al. 1991; Kendall et al. 1990; Nieman et al. 1989; Nieman, Johanssen et al. 1990), whereas it has been claimed that moderate training reduces the number of symptoms (Nieman et al. 1993; Nieman, Nehlsen-Canarella et al. 1990) (figure 18.5). However, in neither strenuous nor moderate exercise have these symptoms been causally linked to exercise-induced changes in immune function.

Conclusion

Exercise influences immunocyte dynamics and possibly immune function. These effects are mediated by diverse factors including exercise-induced release of cytokines and classical stress hormones. In general, intense exercise of long duration induces immunosuppression and increased susceptibility to infectious diseases, whereas moderate exercise seems to enhance resistance to infections.

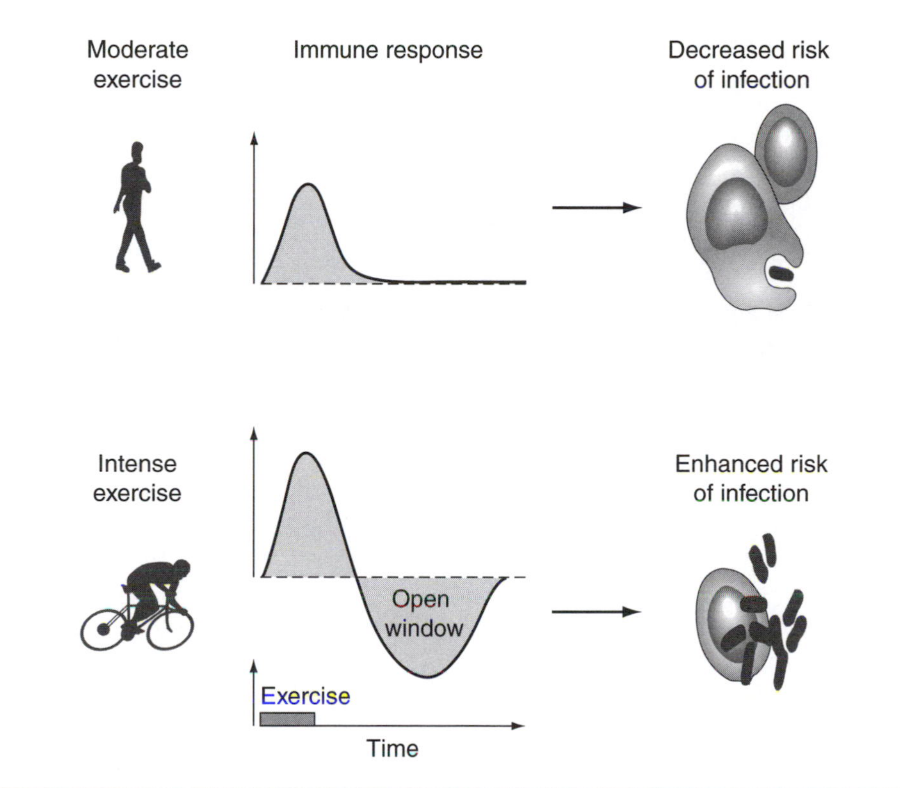

Figure 18.5 A current hypothesis is that moderate exercise boosts the immune system and offers protection against infections. In contrast, intense exercise induces temporary immune impairment, which may allow microorganisms to invade the body and establish as infections.

Chapter 19

Exercise and the Brain

Carl W. Cotman, Nicole C. Berchtold, Paul A. Adlard, and Victoria M. Perreau

Until recently, the effects of exercise on the brain have been largely unexplored. However, current research in this area reveals that the brain is remarkably responsive to exercise, evoking changes at anatomical, cellular, and molecular levels. Intriguingly, many of these changes occur in areas of the brain critical to learning and memory, as well as higher cognitive functioning. While these cognitive effects are becoming increasingly documented, an appreciation of the underlying mechanisms has only just begun.

In this chapter we review recent data generated from human and animal studies suggesting that exercise and increased activity or fitness helps preserve cognitive function and the structural integrity of the brain, particularly with aging. Recent studies at the cellular and molecular levels show that exercise affects brain plasticity by inducing specific molecules that subserve learning and memory, such as brain-derived neurotrophic factor and a number of other genes that modulate synaptic plasticity.

Exercise Improves Cognitive Function in Humans and Prevents Age-Related Brain Atrophy

It has long been suspected that moderate exercise improves cognition in people, and recently, a number of human studies have provided strong evidence that exercise benefits brain health and function (figure 19.1). Exercise participation and aerobic fitness have consistently emerged as predictors of superior performance on cognitive and behavioral measures in a number of cross-sectional studies, in both young and old adults (Blomquist & Danner 1987; Rogers et al. 1990; Berkman et al. 1993; Hill et al. 1993; Colcombe & Kramer 2003).

Improved cognitive function—executive function, memory (Colcombe & Kramer 2003)

Protection from depression (DiLorenzo et al. 1999; Lawlor & Hopker 2001)

Prevention of age-related declines in cerebral perfusion (Rogers et al. 1990)

Prevention of age-related brain tissue loss (Colcombe et al. 2003)

Decreased risk and incidence of Alzheimer's disease and general dementia (Friedland et al. 2001; Laurin et al. 2001)

Figure 19.1 Recent literature documenting the benefits of exercise on human brain health and function. IGF-1 = insulin-like growth factor; NARP = neuronal activity-regulated pentraxin; COX-2 = Cyclooxygenase-2.

Interestingly, the most robust positive effects have emerged in studies of the older population. Retrospective analysis has revealed that behavioral stimulation and physical activity reduce the risk of developing Alzheimer's disease, an age-related neurodegenerative disease that affects 20 to 30 million individuals worldwide (Friedland et al. 2001). Bolstering these findings, a large five-year prospective study of older individuals (65-plus) demonstrated that physical activity is associated with lower risks of Alzheimer's disease, cognitive impairment, and in fact dementia of any type (Laurin et al. 2001). Certain aspects of cognitive function benefit more than others from increased aerobic fitness (Colcombe & Kramer 2003). The most robust enhancement is seen for processes that show age-related decrements in performance such as executive-control processes (scheduling, planning, multitasking, inhibition, working memory) and visuo-spatial processing (Colcombe & Kramer 2003). A key recent study using magnetic resonance imaging (MRI) has provided clues to the underlying mechanisms, revealing that exercise actually protects against the age-related loss of brain tissue in the frontal, parietal, and temporal cortices—brain areas critical for higher cognitive processes such as executive-control functions (Colcombe et al. 2003). While these studies clearly demonstrate a beneficial role of exercise in humans, they do not address the underlying biological changes and mechanisms that drive these effects.

Animal Models to Study the Effects of Exercise on Brain Function

Researchers have turned to animal models to study the anatomical, cellular, and molecular changes that occur in the brain in response to exercise. Mice and rats are principally used in exercise paradigms such as treadmill exercise (forced activity), voluntary wheel running activity, or "environmental enrichment." In environmental enrichment, animals are exposed to a number of different stimulating conditions, such as access to running wheels, group housing (providing social interaction), and a complex environment containing toys, tunnels, and frequent changes in food location. Of these paradigms, wheel running may be the most ideal for studying the brain response to exercise in that it isolates exercise as the critical variable. In addition, wheel running is quantifiable, avoids the stress introduced in forced activity paradigms, and allows the animal to choose how much to run, thus paralleling human patterns of exercise.

Wheel running and environmental enrichment induce a number of changes in the anatomy, neurochemistry, and electrophysiological activity of the brain, demonstrating that the adult brain is remarkably responsive to exercise and behavioral stimulation (van Praag et al. 2000; Cotman & Berchtold 2002). Such plasticity is thought to underlie the positive effects of exercise on cognition and brain health. A number of anatomical changes are observed in the brain after wheel running or environmental enrichment, including increased neuron number (resulting from increased neurogenesis and enhanced neuron survival) and increased number and length of dendrites and dendritic spines (van Praag et al. 2000). Because dendrites and spines are the sites of synapses and communication between neurons, these structural changes promote synaptic complexity and thus greater potential for information processing. In addition to such anatomical changes, exercise evokes changes on a neurochemical level in certain brain regions, inducing the release of neurotransmitters (acetylcholine, serotonin, and noradrenaline) and neuropeptides (substance P, neurokinin A, neuropeptide Y) (Diamond 2001). Such neurochemical alterations provide one mechanism for exercise-induced changes in electrophysiological activity in the brain, including the hippocampus, a brain structure critical for learning and memory. For example, long-term potentiation (LTP)—a synaptic analogue of learning and memory—is enhanced in the dentate gyrus of the hippocampus following exercise (van Praag, Christie et al. 1999). LTP is thought to enable encoding of new information, suggesting that exercise can drive mechanisms that facilitate learning. Exercise has also been shown to improve neuronal survival and increase resistance to brain insults (Stummer et al. 1994; Carro et al. 2001). The mechanisms underlying these changes in brain function and health are currently incompletely understood. However, recent discoveries of target molecules regulated by exercise are providing clues to the biology that mediates the beneficial effects of exercise on the brain.

Exercise Up-Regulates Brain-Derived Neurotrophic Factor

One molecule that has emerged as critical for both neuronal health and learning and memory is brain-derived neurotrophic factor (BDNF). This growth factor promotes the survival and health of a variety of types of neurons and is also an important modulator of plasticity (Barde 1994; Lu & Chow 1999; McAllister et al. 1999; Tyler et al. 2002; Vicario-Abejon et al. 2002). BDNF is produced by neurons, particularly neurons in the hippocampus and cortex, areas involved in learning and memory. Neuronal activity (such as occurs during encoding of information) stimulates BDNF gene regulation and protein release. In turn, BDNF release at synapses enhances synaptic transmission and neuronal excitability (Figurov et al. 1996). In addition, behavioral stimulation and learning both increase BDNF gene expression in the brain. A critical role for BDNF in learning was demonstrated using BDNF-deficient transgenic mice. These animals show impairments in both learning and LTP; and importantly, restoring BDNF reverses both the electrophysiological and learning deficits (Levine et al. 1995; Korte et al. 1996; Patterson et al. 1996). Mice deficient in BDNF also show decreased synaptic innervation and reduced levels of synaptic vesicle proteins (Martinez et al. 1998; Pozzo-Miller et al. 1999), indicating that BDNF is important for normal synaptic signaling (Martinez et al. 1998).

Given the key role that BDNF has in learning and memory and neuronal survival—processes also affected by exercise—we postulated that physical activity may increase levels of BDNF and other growth factors in the brain. In 1995 we first reported that voluntary wheel running activity increases the gene expression of a number of growth factors in the rat brain, such as nerve growth factor (NGF) and fibroblast growth factor 2 (FGF-2) (Neeper et al. 1996; Gomez-Pinilla et al. 1998). The largest induction, however, was seen for BDNF, which was increased in a number of different brain regions (Neeper et al. 1995, 1996). Initially it was predicted that a neurotrophin response to exercise would likely be restricted to motor-sensory systems of the brain such as the cerebellum, primary cortical areas, or basal ganglia. While BDNF was moderately increased in the cerebellum, cortex, and spinal cord (but not the striatum) after wheel running, unexpectedly, the most robust and sustained response was in the hippocampus, a highly plastic structure critical for higher cognitive function, particularly learning and memory (Neeper et al. 1996; Gomez-Pinilla et al. 2001).

In the hippocampus, exercise increases BDNF mRNA in neurons, particularly those of the dentate gyrus, hilus, and CA3 subfields (figure 19.2). The gene up-regulation appears rapidly (after several hours of exercise) (Oliff et al. 1998), occurs in both male (Berchtold et al. 2002; Neeper et al. 1996) and female (Berchtold et al. 2001) rats and mice, is sustained even after several months of exercise (Russo-Neustadt et al. 1999), and is paralleled by increased amounts of BDNF protein (figure 19.2). Once induced, protein levels remain elevated for several days after exercise has stopped, then decay back to baseline (figure 19.3). For example, after 28 days of running, BDNF protein levels continue to rise for three days after exercise has ceased but return to near baseline levels following one week of sedentary lifestyle. Interestingly, alternating days of exercise is as effective as daily

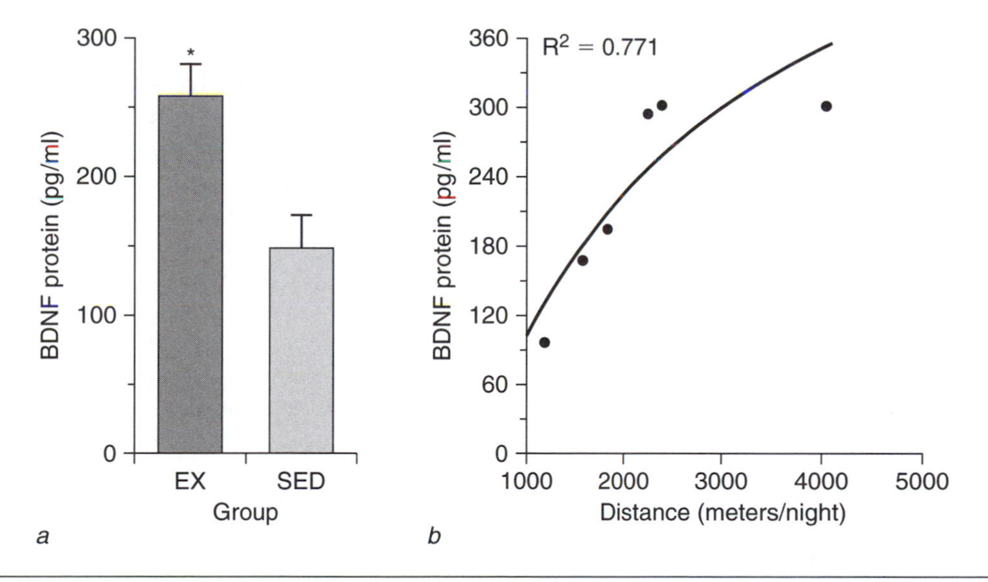

Figure 19.2 Effects of exercise on hippocampal BDNF mRNA and protein levels. In situ hybridization shows that BDNF mRNA levels are increased after seven days of exercise *(a)* in the rate dentate gyrus (DG), hilus, CA3-CA1, and cortex compared to *(b)* sedentary animals.

activity for increasing hippocampal BDNF protein levels (Berchtold et al. submitted) (figure 19.3). The practical significance of this is that even intermittent exercise is a powerful enough stimulus to activate the molecular machinery that drives synaptic plasticity.

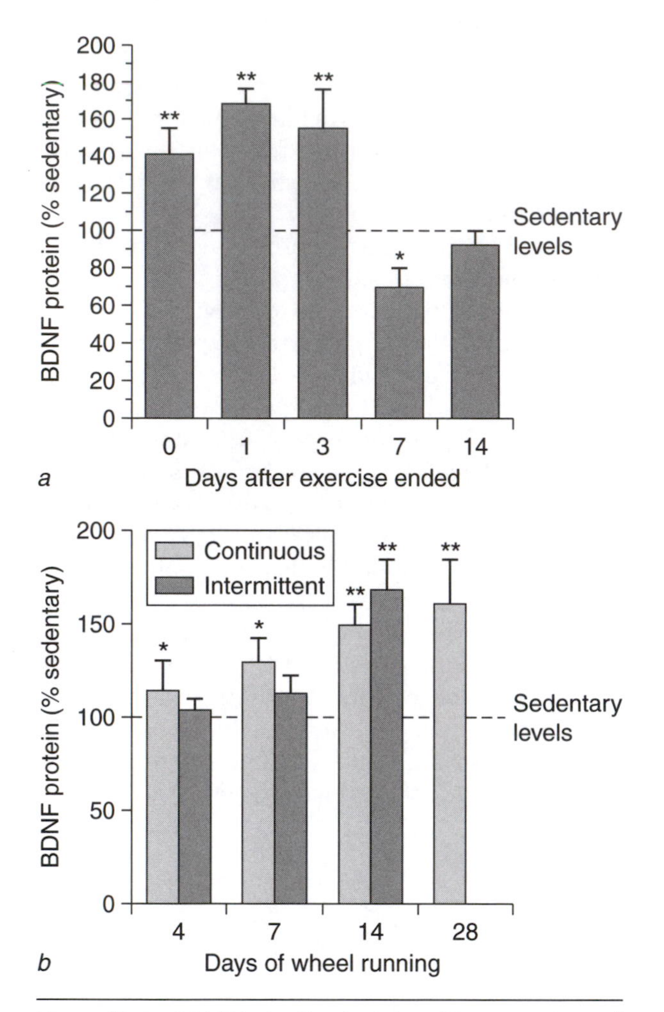

Figure 19.3 BDNF induction by intermittent exercise and rate of decay after continuous exercise. *(a)* Time course of decay of BDNF protein levels in the hippocampus after exercise has ceased. Animals exercised for four weeks in running wheels, and then wheels were removed from the cages. Levels of BDNF were measured at 0, 1, 3, 7, and 14 days after the end of running. BDNF was significantly increased over baseline at 0, 1, and 3 days after running ceased. *(b)* BDNF protein induction in the hippocampus following continuous or intermittent wheel running exposure. Animals underwent 4, 7, or 14 days of exercise of either daily running or alternating days of running. By 14 days of activity, intermittent exercise was as effective as daily activity in increasing BDNF levels. Values represent averages ± SEM, normalized to sedentary levels (100%). $^*p < .05$, $^{**}p < .01$.

Part *a* reprinted from *Trends in Neuroscience,* Vol. 25, C.W. Cotman and N.C. Berchtold, "Exercise: a behavioral intervention to enhance brain health and plasticity," 295-301, Copyright (2002), with permission of Elsevier.

Investigation of brain molecules regulated by exercise initially focused on BDNF and other growth factors. However, it is likely that many other genes in the brain are responsive to exercise as well. The recent development of gene microarrays has made feasible the large-scale investigation of gene expression profiles in response to stimuli such as exercise.

Gene Microarray Analysis Reveals Other Genes That Are Regulated by Exercise

Microarray analysis of the hippocampus demonstrated altered expression of 130 genes in response to three weeks of exercise (Tong et al. 2001). These genes can be broadly grouped into four main categories of function: plasticity, metabolism, anti-aging, and immune related. Interestingly, the majority of the genes are associated with supporting synaptic plasticity. A few of these plasticity-related genes are highlighted here.

Neuronal activity-regulated pentraxin (NARP), homer-1a, and cyclooxygenase 2 (COX-2) are examples of three plasticity-related genes that are up-regulated in the hippocampus in response to exercise. NARP and homer-1 have been recently identified as key molecules in regulating synaptic plasticity through controlling the trafficking of receptors for glutamate. Glutamate is the principal excitatory neurotransmitter in the brain and is a major modulator of plasticity mechanisms. Glutamate acts by way of specific glutamate receptors, particularly the alpha-amino-3-hydroxy-5-methyl-isoxazole-4-propionate (AMPA) receptor, the group I metabotropic glutamate receptor (mGluR1), and the n-methyl-d-aspartate (NMDA) receptor. NARP and homer-1a are involved in clustering and insertion of glutamate receptors (AMPA, mGluR1) into the synaptic membrane of dendrites and dendritic spines (Tu et al. 1998; O'Brien et al. 2002). Because dendrites and dendritic spines are the sites of synapses and neuronal communication, regulating the amount of glutamate receptors available for transmission is a key process in plasticity-related mechanisms (e.g., LTP). Like BDNF, NARP is rapidly induced in hippocampal and cortical neurons by synaptic activity (Tsui et al. 1996). Interestingly, BDNF also stimulates NARP expression. Thus, regulation of NARP availability may be one of the downstream consequences of BDNF induction by exercise.

Another gene shown to be up-regulated in response to exercise is COX-2. While COX-2 is best known for its role in prostaglandin synthesis and immune function, in the CNS it has an additional role in plasticity. In the CNS, COX-2 gene expression is up-regulated by synaptic activity and BDNF (Yamagata et al. 1993; Adams et al. 1996; Marcheselli & Bazan 1996; Tu & Bazan 2003). In addition, COX-2 is localized in dendritic spines (Kaufmann et al. 1996), where it modulates postsynaptic excitability through the regulation of prostaglandin-E2 signaling. Specifically, inhibition of COX-2 decreases postsynaptic excitability, resulting in decreased calcium accumulation in dendrites and suppression of LTP induction (Chen et al. 2002). These effects are reversed by the addition of prostaglandin-E2, indicating a physiological role for COX-2 in the regulation of plasticity. Thus, the effect of COX-2 induction by exercise is likely to increase postsynaptic excitability in the hippocampus, with the result that the postsynaptic neuron is more responsive to incoming synaptic activity. Functionally, such effects potentially translate into more efficient encoding of information and facilitated learning.

Microarray analysis has identified a number of genes associated with synaptic plasticity that are up-regulated in the hippocampus by exercise. NARP, COX-2, and BDNF are all increased with synaptic activity, and in turn support and strengthen synaptic activity. Moreover, BDNF regulates NARP and COX-2 expression, suggesting that BDNF is regulating select downstream gene products and may be one of the initiating molecules underlying some of the beneficial effects of exercise. Along with NARP, COX-2, homer, and BDNF, exercise increases a number of other synaptic-related genes in the hippocampus such as synapsin I and synaptotagmin (Chen et al. 1998; Molteni et al. 2002). The up-regulation of these genes by exercise would be predicted to have effects on hippocampal function, including improved learning and memory.

Exercise Enhancement of Learning and BDNF

A key molecule involved in learning and memory is BDNF. Infusion of BDNF enhances learning (Alonso et al. 2002), and deficiencies in BDNF impair learning. The fact that exercise increases BDNF suggests that exercise has the potential to enhance learning. Specifically, we hypothesized that exercise would increase the rate of learning as a result of increased availability of BDNF, in addition to other synaptic plasticity gene products. Here we explore the relationship between BDNF and learning.

In tests of learning and memory in rodents, exercise facilitates learning, and the improvement is correlated with hippocampal BDNF levels. For example, after exercise, animals require fewer trials to learn spatial memory tasks (such as the Morris water maze and 8-arm maze; Anderson et al. 2000). In the case of the water maze, the enhanced learning correlated with significantly elevated levels of BDNF protein within the hippocampus of the exercised animals. However, once the task had been learned to criterion, there were no differences in escape latencies, and there was also no difference in the BDNF protein levels between the groups (Adlard, Perreau et al. 2004). This suggests that the acquisition of new information is initially enhanced in exercised animals (figure 19.4), potentially as a result of elevated BDNF levels. In addition, results from a passive avoidance task indicate that exercise helps memory. That is, animals allowed to exercise prior to a passive avoidance task had significantly improved short- and long-term memory as compared to control animals (Radak et al. 2001).

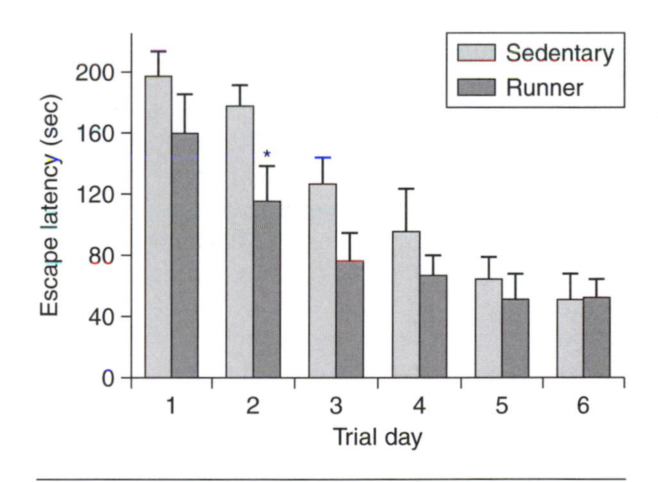

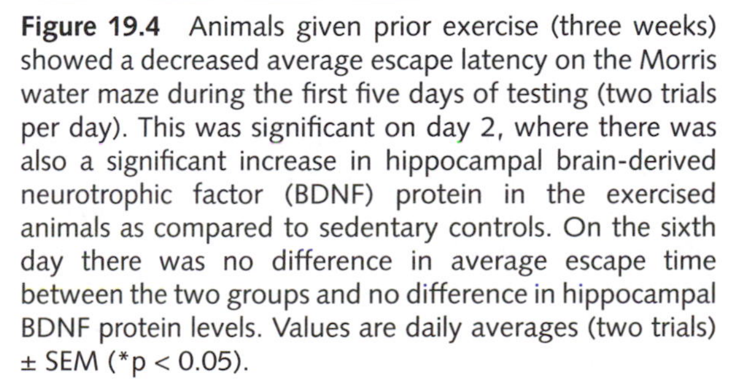

Figure 19.4 Animals given prior exercise (three weeks) showed a decreased average escape latency on the Morris water maze during the first five days of testing (two trials per day). This was significant on day 2, where there was also a significant increase in hippocampal brain-derived neurotrophic factor (BDNF) protein in the exercised animals as compared to sedentary controls. On the sixth day there was no difference in average escape time between the two groups and no difference in hippocampal BDNF protein levels. Values are daily averages (two trials) ± SEM (*$p < 0.05$).

While we hypothesize that the exercise-induced increase in BDNF is one of the critical mechanisms in exercise enhancement of learning, it is also possible that exercise is modulating other factors involved in the acquisition of new information. Hippocampal neurogenesis, for example, is associated with improvements in spatial memory performance and has been shown to result from exposure of animals to exercise or enriched environments (van Praag, Christie et al. 1999; van Praag, Kempermann et al. 1999). The modulation of central processes by physical activity may thus be crucial not only for learning, but also for the maintenance of the anatomical integrity of the hippocampus and CNS, as implied from the recent imaging studies in humans described earlier.

Exercise and Depression

In addition to a key role in learning and memory, the hippocampus is involved in the etiology of mood disorders such as depression. Recent investigations also implicate BDNF as having an important role in such conditions. For example, major classes of antidepressants, while acting through diverse mechanisms, all appear to converge on BDNF, and up-regulate BDNF gene expression in the hippocampus (Nibuya et al. 1995; Duman et al. 1997; Russo-Neustadt et al. 1999; Fujimaki et al. 2000). Further, direct hippocampal infusion of BDNF protein produces antidepressant effects in animal models of depression, such as learned helplessness (Karege et al. 2002; Shirayama et al. 2002). Given the role of BDNF in depression and the up-regulation of this molecule by exercise, it would be predicted that exercise may be a simple nonpharmacological means to prevent and treat depression.

A number of laboratory studies have examined the effect of exercise on the induction of learned helplessness, an animal model of behavioral depression. Chronic wheel running (9-12 weeks) in young rats has been reported to offset the induction of depression, as evidenced by a decreased escape latency in a learned helplessness model. In addition, this activity was associated with alterations in monoamine levels, including 5HT, in brain regions such as the dorsal raphe and hippocampus (Dishman et al. 1997). Similarly, a recent report by Greenwood and colleagues (Greenwood et al. 2003) demonstrated that six weeks of free-wheel-running in rats was sufficient to prevent behavioral depression, also perhaps via modulation of the activity of serotonergic neurons in the dorsal raphe nucleus. Shorter periods of exercise (four weeks) have also been shown to prevent stress-induced behavioral depression in young rats (Moraska & Fleshner 2001). Recent studies in our own laboratory have further indicated that one week of exercise prior to the induction of learned helplessness improves escape latency in both young and old mice, suggesting that the effects of exercise on behavioral depression may be applicable across the life span (unpublished observations).

Thus, data from animal models suggests that exercise can serve as an intervention to prevent and treat depression. In addition, the fact that exercise and antidepressants both act on BDNF suggests that a combination of the two interventions may be more effective than either alone. Indeed, exercise and antidepressant treatment interact to increase the rate and extent of BDNF induction in the hippocampus (Russo-Neustadt et al. 1999, 2001). Thus, antidepressants and exercise appear to share a common additive mechanism that converges on BDNF regulation.

While exercise has been shown to decrease depressive symptoms in animal models, it is important to determine whether or not exercise can serve a similar function in humans. In support of a possible role for exercise as an intervention for depression, a number of cross-sectional and longitudinal human studies have demonstrated a positive correlation between increased physical activity and lower levels of depression (Byrne & Byrne 1993; Blumenthal et al. 1999; Hassmen et al. 2000; Lawlor & Hopker 2001; Pollock 2001). In addition, exercise has been shown to be protective against recurrent depression. In one-, five-, and eight-year follow-up studies, people who maintained or increased their activity level had fewer depressive symptoms and showed more physiological and psychological benefits than those who decreased their activity level (DiLorenzo et al. 1999; Lampinen et al. 2000; Strawbridge et al. 2002). Meta-analyses of randomized clinical trials demonstrate that, on the whole, exercise reduces symptoms of depression and has psychological benefits on mood, mental performance, concentration, and confidence (DiLorenzo et al. 1999; Lawlor & Hopker 2001). Interestingly, BDNF is a possible candidate molecule that could play a fundamental role in the benefits of exercise for depression. In support of a possible central role of BDNF for depression in humans, a decrease in BDNF protein, which correlated with the severity of depression, was noted in the serum of major-depressed patients (Dunn et al. 2002). Further, postmortem studies indicate that BDNF is up-regulated in the hippocampus of antidepressant-treated depressive patients (Dunn et al. 2002). The up-regulation of BDNF may then affect a number of different systems and processes within the brain to help promote mental health.

The optimal level and duration of exercise/activity is an area of importance in human studies. That is, how much exercise and of what type is necessary to confer protection against disorders such as depression? As little as 30 min of daily treadmill activity for 10 days has been demonstrated to decrease depression (Dimeo et al. 2001). Most studies, however, typically use longer training periods with less frequent exercise (two to three times per week); and in fact, less frequent exercise may be better than daily training, based on depression scores (Hassmen et al. 2000). Interestingly, the data emerging from animal models, as discussed earlier (figure 19.3*b*), suggests that the induction of BDNF within the hippocampus is similar, but perhaps occurs at a different rate, with differing frequencies of exercise (daily vs. intermittent). The questions of optimal level and duration of activity may, therefore, have significant clinical relevance as predicted from both animal and human studies. Exercise as a "treatment" for depression therefore needs to be evaluated in placebo-controlled trials that take into account the stimulatory effect of a diverse range of behavioral activities on the induction of BDNF.

Definitive Role for BDNF in Human Cognition

While rodent studies demonstrate that BDNF is an important molecule in cognitive performance, recent genetic data underscore the pivotal role of BDNF for humans. Indeed, recent genetic studies highlighting a series of polymorphisms in the BDNF gene have established a decisive role for BDNF in human cognition. An amino acid substitution in the coding region of the BDNF gene (val/met) results in impaired processing and release of BDNF (Egan et al. 2003). Genetic linkage studies of the allelic distribution of this polymorphism have demonstrated that individuals carrying the met-BDNF allele have poorer memory function and abnormal hippocampal activation. Remarkably, these cognitive effects were shown in a cohort of normal young adults who manifested no overt signs of cognitive or neurological abnormalities (25-45 years old, 641 individuals [Egan et al. 2003]). Other studies have additionally demonstrated that met-BDNF and a separate BDNF polymorphism are risk factors for Alzheimer's disease (Kunugi et al. 2001; Ventriglia et al. 2002; Egan et al. 2003). These studies make it clear that when BDNF is not functioning properly, cognition suffers in the long term. The fact that deficiencies in BDNF can result in such cognitive deficits even in young individuals underscores the need to understand the mechanisms that regulate activity of this gene.

In summary, exercise has anatomical, cellular, molecular, and functional consequences for the brain. Exercise appears to protect from age-related brain atrophy in humans. In animal models, exercise increases dendritic complexity and alters neurotransmitters and synaptic activity in the brain. In addition, a number of exercise-induced molecular changes support plasticity. Functional consequences of exercise include improvements in learning and memory and protection from depression and other neurological disorders. BDNF is emerging as a key molecule in modulating these effects and is also currently the best understood.

CNS and Peripheral Regulatory Mechanisms of Exercise Effect on BDNF

The data summarized in the preceding section highlight the need for additional animal and human studies to understand the basic mechanisms by which exercise and BDNF regulation can serve as a simple nonpharmacological intervention to improve brain function. In the next section we summarize current information on the central neurotransmitter regulatory mechanisms and the peripheral neuroendocrine mechanisms that converge on the overall effect of exercise on brain and BDNF regulation.

Recent data show that the regulation of BDNF protein is controlled by expression of the BDNF gene in neurons, and the gene expression levels are modulated by a convergence of neurotransmitter interactions. In addition to CNS neurotransmitters, recent evidence indicates that peripheral factors, such as circulating hormone levels (e.g., estrogen, corticosteroids, and insulin-like growth factor-1) also impact BDNF regulation/expression in the brain. Here we explore both these central and peripheral regulatory mechanisms.

CNS (Neurotransmitter) Mechanisms

BDNF gene expression is regulated by neuronal activity. In the CNS, glutamate is the predominant excitatory neurotransmitter driving neuronal activity. Glutamate signaling is central to mediating up-regulation of BDNF levels in the hippocampus, since BDNF is expressed by glutamatergic neurons. BDNF gene regulation is additionally sensitive to a number of other neurotransmitter systems, which converge to modulate glutamatergic neuronal activity in the hippocampus. Neurotransmitter systems thus far explored for their role in mediating exercise-dependent regulation of BDNF in the hippocampus are the acetylcholine, GABA, serotonin, and norepinephrine systems.

A major source of acetylcholine and GABA input to the hippocampus comes from a connected brain structure called the medial septum, an important modulator of normal hippocampal function and excitability. Interestingly, during physical activity, the septo-hippocampal circuit is activated, and cholinergic and GABA activity in the medial septum drives a rhythmic neuronal firing activity in the hippocampus (Vanderwolf 1969; Lawson & Bland 1993; Lee et al. 1994). In addition, exercise causes levels of acetylcholine to increase in the hippocampus (Dudar et al. 1979; Nilsson et al. 1990; Mizuno et al. 1991). Because acetylcholine regulates BDNF gene expression and is increased by physical activity, acetylcholine appeared to be a good candidate to mediate exercise-induced increases in BDNF mRNA (Lapchak et al. 1993; Knipper et al. 1994; Ferencz et al. 1997). Surprisingly, this is not the case. Lesion studies demonstrate that although acetylcholine from the medial septum provides tonic regulation of baseline hippocampal BDNF gene expression, it is not a key regulator in the activity-dependent state (Berchtold et al. 2002). The cholinergic neurons of the medial septum can be selectively lesioned, resulting in complete loss of cholinergic input/terminals to the hippocampus. Complete cholinergic lesion, despite reducing baseline (resting) levels of BDNF mRNA, does not impair exercise induction of BDNF. By contrast, when partial loss of cholinergic afferents is combined with loss of medial septal GABAergic neurons, exercise-dependent BDNF regulation is disrupted in the hippocampus (Berchtold et al. 2002). Thus, there is a strong involvement of the medial septum in activity-dependent regulation of BDNF gene expression, and it appears to involve either non-cholinergic-mediated signaling or a combination of neurotransmitter systems.

In addition to cholinergic and GABAergic modulation, monoamine neurotransmitter signaling may also contribute to BDNF gene regulation. This idea is based on the observation that agents that increase transmission at monoaminergic synapses, such as antidepressants, also increase BDNF gene expression in the hippocampus. Antidepressants act to raise the levels of the monoamine neurotransmitters serotonin and norepinephrine (Nibuya et al. 1995; Fujimaki et al. 2000). In addition, exercise increases norepinephrine levels in several brain regions including the hippocampus, and may also increase serotonin neurotransmission. As described previously, antidepressant treatment in combination with exercise further enhances exercise-dependent BDNF up-regulation, suggesting that exercise and antidepressants may be working through synergistic systems to induce BDNF changes, such as additive effects on brain monoaminergic neurotransmission (Russo-Neustadt et al. 1999, 2001).

Studies investigating whether monoaminergic neurotransmission participates in exercise-dependent BDNF regulation demonstrate that norepinephrine is involved (Ivy et al. 2003). Neurotransmitter-specific toxins can be used to selectively destroy noradrenergic, or serotonergic, neurotransmission throughout the brain. After noradrenergic lesion, exercise fails to increase BDNF gene expression in the hippocampus. However, baseline (resting) levels of BDNF are not affected by noradrenergic lesion. Serotonergic lesions have little, if any, effect on BDNF (Garcia et al. in press). These results demonstrate that neither norepinephrine nor serotonin provides tonic regulation of BDNF gene expression in the hippocampus, while noradrenaline is important in the activity-dependent regulation of BDNF.

Peripheral Regulatory Mechanisms

Although CNS activity-dependent mechanisms are pivotal in mediating the effects of exercise on the brain, the concept is now emerging that peripheral influences are also important. Components contributing to this peripheral control include estrogen, corticosterone, and insulin-like growth factor (IGF-1). Interestingly, some of these same regulatory mechanisms not only control BDNF but also modulate basal neurogenesis, as well as the effect of exercise on neurogenesis.

Estrogen

There are a number of possible mechanisms by which estrogen regulates effects of exercise on BDNF expression. Estrogen can have *direct* molecular effects on gene expression, as well as *indirect* effects via stimulatory effects on physical activity levels.

Some of the beneficial effects of estrogen in the brain may be mediated by BDNF, as estrogen regulates BDNF gene expression to increase availability of this trophic factor (Singh et al. 1995). Interestingly, in females, the presence of estrogen appears necessary in order for exercise regulation of BDNF to occur (Berchtold et al. 2001). After two months of estrogen deprivation, exercise fails to increase BDNF mRNA or protein in the rat hippocampus. By contrast, when exercise is combined with estrogen replacement, BDNF protein levels show a greater increase than in response to estrogen replacement alone. The presence of estrogen in females might thus be a permissive factor necessary for exercise-induced regulation of BDNF availability (Berchtold et al. 2001).

Interestingly, levels of voluntary physical activity also depend on estrogen status. In the absence of estrogen, animals are less active, while estrogen replacement restores activity to normal levels (Berchtold et al. 2001). This result is paralleled in human studies, showing an association between use of hormone replacement therapy and physical activity levels; hormone replacement users report greater exercise participation than nonusers (Matthews et al. 1996; Persson et al. 1997). This effect of estrogen raises the interesting possibility that some of the health benefits associated with hormone replacement in women could be related to increased exercise participation.

Corticosterone/Stress

Corticosteroids constitute another neuroendocrine regulatory mechanism that can affect BDNF levels and may modulate the effects of exercise on the brain. These hormones, released from the adrenal gland in response to stressful events, can enter the brain and bind to select glucocorticoid receptors to modify gene expression. The concentration of glucocorticoid receptors within the hippocampus makes this region of the brain particularly vulnerable to the effects of stress and glucocorticoids.

Prolonged stress increases the vulnerability of hippocampal neurons to injury from neuronal insult (McIntosh & Sapolsky 1996; Sapolsky 1999), and is both a predisposing and a precipitating factor in the onset of certain types of depression (Altar 1999). In addition, glucocorticoids decrease neuronal excitability and suppress hippocampal LTP (Schaaf et al. 1998), contributing to stress-induced impairments in learning and memory (Luine et al. 1996; Smith & Cizza 1996; Holscher 1999; Zhou et al. 2000; Bowman et al. 2001). Thus, interventions that counteract the effects of stress are important to delineate.

Exercise is thought to offset/protect from the negative effects of stress exposure. One mechanism might be through the regulation of BDNF, as elevated glucocorticoid levels decrease hippocampal BDNF mRNA and protein expression (Schaaf et al. 1998; Nitta et al. 1999; Zhou et al. 2000). A commonly used behavioral paradigm to elicit stress in rodent models is immobilization stress, which raises circulating glucocorticoid levels and reduces hippocampal BDNF mRNA and protein. Importantly, three weeks of exercise prior to immobilization stress protects against the stress-induced reduction in BDNF protein (Adlard, Cotman et al. 2003). Because exercise can offset the negative effects of stress on both a behavioral and a neurochemical level, BDNF may be a key mechanism involved in protection from stress-related processes. Voluntary exercise may, therefore, represent a therapeutic intervention to protect from stress-related impairments such as depression, compromised learning and memory, and potential long-term neuronal loss/damage.

Growth Hormone (GH) and Insulin-Derived Growth Factor (IGF-1)

A third peripheral mechanism that may mediate the effects of exercise and BDNF in the brain is the GH/IGF-1 axis. Exercise increases circulating GH, which is the major stimulus for IGF-1 production. While the majority of circulating IGF-1 is derived from the liver, many different tissues can produce IGF-1, including the brain. IGF-1 has multiple biological effects on processes including neurogenesis, learning, memory, cognition, amyloid processing, and other systemic effects (Sonntag et al. 2001; Carro et al. 2002; Carter et al. 2002; Holzenberger et al. 2003). IGF-1 may also be involved in a number of different neurodegenerative diseases (Busiguina et al. 2000); in addition, the IGF-1 receptor is even reported to regulate mammalian life span (Holzenberger et al. 2003).

Increased IGF-1 has also been suggested as one of the mechanisms responsible for the beneficial effects of exercise on the brain, as blocking the entry of peripheral IGF to the brain prevents exercise stimulation of neurogenesis and the protection from select brain insults (Carro et al. 2000, 2001). Interestingly, this suggests that a peripherally derived peptide may have significant central effects. Further, the cognitive impairment, memory loss, and decreases in neurogenesis associated with advanced age can be rescued by the exogenous administration of IGF-1 (Markowska et al. 1998; Lichtenwalner et al. 2001). Additionally, there may be an interplay between IGF and BDNF because peripheral IGF administration induces BDNF in the hippocampus (Carro et al. 2000).

Thus the idea is now emerging that IGF-1 may be a key mediator of many cortical processes. IGF-1 and BDNF act primarily through different signaling pathways, suggesting that they may act individually or in concert to promote and maintain brain health and plasticity. There are many other regulatory mechanisms that may contribute to the exercise up-regulation of growth factors such as BDNF (see figure 19.5).

Conclusion

In summary, basic science research and animal models provide a solid foundation documenting that exercise can promote changes in the brain associated with enhanced brain health, increased resistance to adverse environmental conditions, and facilitated learning and other adaptive responses to the environment (figure 19.5). Exercise induces BDNF mRNA in glutamate neurons in the hippocampus and select other brain regions within a few days of voluntary running. BDNF protein level follows mRNA induction after a lag time of several days, and the protein levels are maintained for months with continued exercise. In addition to BDNF, exercise induces the expression of a number of other genes that also support synaptic plasticity. Interestingly, expression of some of these plasticity-related genes, such as NARP and COX-2, can be induced by BDNF. The synthesis of BDNF is modulated by converging neurotransmitter systems and peripheral neuroendocrine mechanisms. The induction of BDNF, in conjunction with downstream molecular responses, predicts that behavioral functions associated with BDNF such as learning and memory and depression may improve with exercise. Animals that have exercised show an enhanced learning rate, as would be predicted from the increased levels of BDNF and downstream synaptic plasticity gene products. Exercise also prevents the development of depression in animal models. Thus current data on animal models suggest that exercise is benefi-

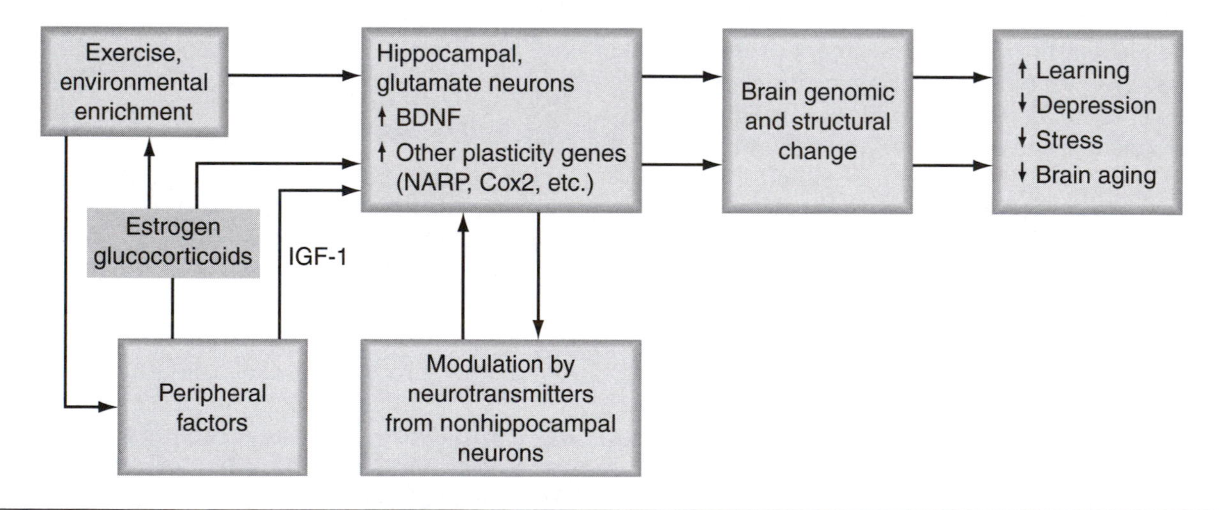

Figure 19.5 Mechanisms by which exercise and environmental enrichment prime the brain to enhance brain function by improving resistance to injury and age-related decline. Exercise and enrichment act on glutamate neurons in the hippocampus to increase levels of protective and plasticity factors such as BDNF. BDNF up-regulates NARP and COX-2, suggesting that BDNF may be one of the initiating molecules regulating some of the beneficial effects of exercise. Multiple factors control BDNF expression in the hippocampus, including neurotransmitters from nonhippocampal neurons and peripheral circulating factors such as estrogen, glucocorticoids, and IGF-1 (which is itself increased by exercise). Regulation of protective and plasticity factors results in further downstream genomic changes and structural changes in the brain that ultimately improve brain function.

cial to brain function; however, human studies are still at an early stage.

From the human literature it is now clear that BDNF regulation can indeed interact with cognitive function, because the polymorphism in BDNF—associated with processing and release of the protein—is associated with reduced cognitive functioning and activation. This is an important finding as it suggests that animal studies will translate to human studies. In the human literature there is a paucity of studies examining exercise effects that can be classified as clinical trials containing the appropriate placebo controls. Most of the published exercise studies are only descriptive, being based on retrospective analysis of data sets. Nonetheless, the current human literature suggests that exercise/increased aerobic fitness is one of the variables that consistently emerges as a predictor of higher cognitive functioning and is also associated with less depression. Exercise in turn may be associated with other variables that are critical synergistic or surrogate markers. The exciting conclusion is that exercise is a readily practiced, inexpensive benefit to brain health and aging that, as it becomes increasingly understood and better documented, can be directly translated into practices for improving the quality of brain function.

List of Abbreviations

8-OhdG = 8-hydroxydeoxy-2'-deoxyguanosine
AA = arachidonic acid
ACC = coenzyme A carboxylase
ACE = angiotensin converting enzyme
ACTH = adrenocorticotropyic hormone
ADP = adenosine diphosphate
aFGF = acidic fibroblastic growth factor
AICAR = 5-aminoimidazole-4-carboxamide ribonucleoside
Akt = protein kinase B
AMP = adenosine monophosphate
AMPK = 5' AMP-activated protein kinase
ANP = atrial natriuretic peptide
AP-1 = activator protein-1
Apaf-1 = apoptosis activating factor-1
Asp = aspartic acid
ATP = adenosine triphosphate
[a-v] = arteriovenous
Bcl-2 = B-cell lymphoma
bFGF = basic fibroblast growth factor
BK2 = bradykinin receptor
BMNC = blood mononuclear cell
bp = base pair
Ca^{2+} = calcium
CaMK = Ca^{2+} calnodulin-dependent protein kinase
cAMP = cyclic adenosine monophosphate
CaN = calcineurin
CAT = catalase
CD = cluster of differentiation
cdks = cyclin-dependent kinases
cGMP = cyclic guanine monophosphate
CHO = carbohydrate
CLFS = chronic low-frequency stimulation
cNOS = constitutive nitric oxid synthase
Cip = cyclin inhibitor protein
DD = death domain
DHPR = dihydropyridine receptor
DNA = deoxyribonucleic acid
d.w. = dry weight
EC = excitation-contraction coupling
ECE = endothelin-converting enzyme
ECL = essential myosin light chain
ECM = extracellular matrix
EDRF = endothelium-derived relaxing factor
EGF = epithelial growth factor
EGFR = epithelial growth factor receptor
ELISA = enzyme linked immunosorbant assay
ENOS = endothelial nitric oxide synthase
ER = endoplasmic reticulum
ERK 1/2 = extracellular regulated kinase
ESAF = endothelial cell-stimulating angiogenic factor
ESR = electron spin resonance spectroscopy
ETA/ETB = endothelin A/B receptor
FA = fatty acid
FasL = Fas ligand
FFA = free fatty acid
FG = fast-twitch glycolytic
FGF-1 = fibroblast growth factor-1
FGF-2 = fibroblast growth factor-2
FGFR-4 = fibroblast growth factor receptor 4
FM = fast isomyosin
FOG = fast-twitch oxidative glycolytic
GC = glucocorticoid
GLUT = glucose transporter protein
GPA = β-guanidinopropionic acid
GPX = glutathione peroxidase
GSH = glutathione (reduced)
GSSG = glutathione (oxidized)
GTP = guanosine trisphosphate
H_2O_2 = hydrogen peroxide
HIF-1 = hypoxia-inducible factor 1
HIF-1α = hypoxemia-inducible factor 1α
HIF-1β = hypoxemia-inducible factor 1β
HK = hexokinase

HO-1 = heme oxygenase-1
HOCl = hypochlorous acid
HPA = hypothalamic-pituitary-adrenal axis
HSF-1 = heat shock transcriptional factor-1
HSP = heat shock protein
I-κB = inhibitory protein-κB
IFN-α = interferon-α
IFN-β = interferon-β
IFN-γ = interferon-γ
IGF-1 = insulin-like growth factor-1
IL = interleukin
IL-2 = interleukin-2
IL-12 = interleukin-12
IMP = inosine monophosphate
INK4 = inhibitors of cyclin-dependent kinase 4
iNOS = inducible NO-synthase
IRS = insulin receptor substrate
Jak = Janus kinase
JNK = c-Jun NH_2-terminal kinase
K_{Ca2+} = Ca^{2+}-activated K^+ channel
Kip = kinase inhibitor protein
KO = knockout
Kv = voltage-gated potassium channel
LPS = lipopolysaccharide
MAPK = mitogen-activated protein kinase
MAPKAP-K = MAPK-activated protein kinase
MDA = malondialdehyde
MEF2 = myocyte enhancer factor 2
MHC = myosin heavy chain
MLC = myosin light chain
MMP = matrix metalloproteinase
MPO = myeloperoxidase
mRNA = messenger ribonucleic acid
MSK = mitogen- and stress-activated kinase
mtDNA = mitochondrial DNA
MTP = mitochondrial permeability transition pore
mtTFA = mitochondrial transcriptional factor A
NAD = nicotinamide adenine dinucleotide
NADP = nicotinamide adenine dinucleotide phosphate
nDNA = nuclear DNA
NFAT = nuclear factors of activated T-cells
nNOS = neuronal nitric oxide synthase
NO = nitric oxide
NOS = nitric oxide synthase
NRF = nuclear respiratory factor
O_2 = oxygen
O_2^- = superoxide radical
OH = hydroxyl radical
$ONOO^-$ = peroxynitrite
PAI-1 = plasminogen activator inhibitor-1
PBMC = peripheral blood mononuclear cells
PD-ECGF = platelet-derived endothelial cell growth factor
PDGF = platelet-derived growth factor
PDGFR = platelet-derived growth factor receptor
PF4 = platelet factor 4
PGI2 = prostacyclin
PGC-1 = peroxisome-proliferator-activated receptor-γ coactivator 1
PHOS = glycogen phosphorylase
PI = propidium iodide
PI3 kinase = phosphatidylinositol-3 kinase
PI3K = phosphatidylinositol-3 kinase
PKB/Akt = protein kinase B
PKC = protein kinase C
PS = phosphatidylserine
PUFA = polyunsaturated fatty acid
Rb = retinoblastoma
RCD = reactive carbonyl derivative
RCL = regulatory myosin light chain
RNA = ribonucleic acid
RNS = reactive nitrogen species
RONS = reactive oxygen and nitrogen species
ROS = reactive oxygen species
RU-486 = mifepristone
SAC = stretch-activated channel
SERCA = sarcoplasmic reticulum adenosine triphosphatase
SM = slow isomyosin
SO = slow-twitch oxidative
SOD = superoxide dismutase
SODD = silencer of death domain
STAT = signal transducer and activator of transcription

TBARS = thiobarbituric acid reactive substance

TCA = tricarboxylic acid

TGF-β = transforming growth factor-β

TGF-BR I = transforming growth factor beta receptor 1

TGF-BR II = transforming growth factor beta receptor 2

TGF-BR III = transforming growth factor beta receptor 3

TIMPs = tissue inhibitors of metalloproteinase

TM = tropomyosin

Tn = troponin

TnC = troponin C (CA^{2+} binding subunit)

TNF-α = tumor necrosis factor-α

$TNFR_1$ = tumor necrosis factor receptor 1

TNFR-55 = tumor necrosis factor receptor 55

TnI = troponin I (inhibitory subunit)

TnT = troponin T (tropomyosin binding subunit)

TP = thymidine phosphorylase

t-PA = tissue-type plasminogen activator

TRAIL = tumor necrosis factor-related apoptosis-inducing ligand

TRF = terminal restriction fragment length

TRF1 = telomeric repeat binding factor 1

TRF2 = telomeric repeat binding factor 2

TRX = thioredoxin

TSP-1 = thrombospondin-1

TUNEL = terminal deoxynucleotidyl transferase-mediated dUTP nick end labeling

u-PA = urokinase-type plasminogen activator

VGCC = L-type voltage-gated calcium channel

VEGF = vascular endothelial growth factor

V_{max} = maximum unloaded shortening velocity

$\dot{V}O_2max$ = maximal oxygen uptake

[$\dot{V}O_2max$] = maximal aerobic power

VSM = vascular smooth muscle

XD = xanthine dehydrogenase

XO = xanthine oxidase

References

Chapter 1

Ali, M.H. & P.T. Schumacker. 2002. Endothelial responses to mechanical stress: where is the mechanosensor? *Crit. Care Med.* 30(5 Suppl): S198-206.

Allan, V.J., H.M. Thompson & M.A. McNiven. 2002. Motoring around the Golgi. *Nat. Cell Biol.* 4: 236-42.

Allen, D.L. & L.A. Leinwand. 2002. Intracellular calcium and myosin isoform transitions. Calcineurin and calcium-calmodulin kinase pathways regulate preferential activation of the IIa myosin heavy chain promoter. *J. Biol. Chem.* 277(47): 45323-30.

Andersson, A., A. Sjodin, R. Olsson et al. 1998. Effects of physical exercise on phospholipid fatty acid composition in skeletal muscle. *Am. J. Physiol.* 274(3 Pt 1): E432-38.

Apodaca, G. 2002. Modulation of membrane traffic by mechanical stimuli. *Am. J. Physiol.* 282(2): F179-90.

Armstrong, R.B. 1990. Initial events in exercise-induced muscular injury. *Med. Sci. Sports Exerc.* 22(4): 429-35.

Azenabor, A.A. & L. Hoffman-Goetz. 2000. Effect of exhaustive exercise on membrane estradiol concentration, intracellular calcium, and oxidative damage in mouse thymic lymphocytes. *Free Radic. Biol. Med.* 28 (1): 84-90.

Baar, K., A.R. Wende, T.E. Jones et al. 2002. Adaptations of skeletal muscle to exercise: rapid increase in the transcriptional coactivator PGC-1. *FASEB J.* 16(14): 1879-86.

Balnave, C.D. & D.G. Allen. 1995. Intracellular calcium and force in single mouse muscle fibres following repeated contractions with stretch. *J. Physiol.* 488 (Pt. 1): 25-36.

Barakat, A.I. 1999. Responsiveness of vascular endothelium to shear stress: potential role of ion channels and cellular cytoskeleton. *Int. J. Mol. Med.* 4(4): 323-32.

Belcastro, A.N. 1993. Skeletal muscle calcium-activated neutral protease (calpain) with exercise. *J. Appl. Physiol.* 74(3): 1381-86.

Bergeron, R., J.M. Ren, K.S. Cadman et al. 2001. Chronic activation of AMP kinase results in NRF-1 activation and mitochondrial biogenesis. *Am. J. Physiol.* 281(6): E1340-46.

Bitbol, M. & P.F. Devaux. 1988. Measurement of outward translocation of phospholipids across human erythrocyte membrane. *Proc. Natl. Acad. Sci.* 85(18): 6783-87.

Bizeau, M.E., W.T. Willis & J.R. Hazel. 1998. Differential responses to endurance training in subsarcolemmal and intermyofibrillar mitochondria. *J. Appl. Physiol.* 85(4): 1279-84.

Bonifacino, J.S. & A.M. Weissman. 1998. Ubiquitin and the control of protein fate in the secretory and endocytic pathways. *Annu. Rev. Cell Dev. Biol.* 14: 19-57.

Bretscher, M.S. 1972. Asymmetrical lipid bilayer structure for biological membranes. *Nat. New Biol.* 236(61): 11-12.

Brun, J.F. 2002. Exercise hemorheology as a three acts play with metabolic actors: is it of clinical relevance? *Clin. Hemorheol. Microcirc.* 26(3): 155-74.

Caimi, G., P. Assennato, B. Canino et al. 1997. Exercise test: trend of the leukocyte flow properties, polymorphonuclear membrane fluidity and cytosolic Ca2+ content in normals, in subjects with previous acute myocardial infarction and in subjects with aortocoronary by-pass. *Clin. Hemorheol. Microcirc.* 17(2): 127-35.

Child, R.B., D.M. Wilkinson, J.L. Fallowfield et al. 1998. Elevated serum antioxidant capacity and plasma malondialdehyde concentration in response to a simulated half-marathon run. *Med. Sci. Sports Exerc.* 30(11): 1603-07.

Chilibeck, P.D., G.J. Bell, R.P. Farrar et al. 1998. Higher mitochondrial fatty acid oxidation following intermittent versus continuous endurance exercise training. *Can. J. Appl. Physiol.* 76(9): 891-94.

Chilibeck, P.D., D.G. Syrotuik & G.J. Bell. 2002. The effect of concurrent endurance and strength training on quantitative estimates of subsarcolemmal and intermyofibrillar mitochondria. *Int. J. Sports Med.* 23(1): 33-39.

Chin, E.R., R.W. Grange, F. Viau et al. 2003. Alterations in slow-twitch muscle phenotype in transgenic mice overexpressing the Ca2+ buffering protein parvalbumin. *J. Physiol.* 547: 649-63.

Chin, E.R., E.N. Olson, J.A. Richardson et al. 1998. A calcineurin-dependent transcriptional pathway controls skeletal muscle fiber type. *Genes Dev.* 12(16): 2499-509.

Clague, M.J. 1998. Molecular aspects of the endocytic pathway. *Biochem. J.* 336: 271-82.

Connor, M.K., O. Bezborodova, C.P. Escobar et al. 2000. Effect of contractile activity on protein turnover in skeletal muscle mitochondrial subfractions. *J. Appl. Physiol.* 88(5): 1601-06.

Cooke, J.P. 2003. Flow, NO, and atherogenesis. *Proc. Natl. Acad. Sci.* 100(3): 768-70.

Cooper, C.E., N.B. Vollaard & T. Choueiri. 2002. Exercise, free radicals and oxidative stress. *Biochem. Soc. Transact.* 30(2): 280-85.

Coulombe, P.A., O. Bousquet, L. Ma et al. 2000. The "ins" and "outs" of intermediate filament organization. *Trends Cell Biol.* 10(10): 420-28.

Dai, J. & M.P. Sheetz. 1999. Membrane tether formation from blebbing cells. *Biophys. J.* 77(6): 3363-70.

Davies, K.J., L. Packer & G.A. Brooks. 1981. Biochemical adaptation of mitochondria, muscle, and whole-animal respiration to endurance training. *Arch. Biochem. Biophys.* (2): 539-54.

Davies, P.F. 1995. Flow-mediated endothelial mechanotransduction. *Physiol. Rev.* 75(3): 519-60.

Denker, S.P. & D.L. Barber. 2002. Ion transport proteins anchor and regulate the cytoskeleton. *Curr. Op. Cell Biol.* 14(2): 214-20.

Diaz, C. & A.J. Schroit. 1996. Role of translocases in the generation of phosphatidylserine asymmetry. *J. Membr. Biol.* 151(1): 1-9.

Dillard, C.J., R.E. Litov, W.M. Savin et al. 1978. Effects of exercise, vitamin E, and ozone on pulmonary function and lipid peroxidation. *J. Appl. Physiol.* 45(6): 927-32.

Dohm, G.L., H. Barakat, T.P. Stephenson et al. 1975. Changes in muscle mitochondrial lipid composition resulting from training and exhaustive exercise. *Life Sci.* 17(7): 1075-80.

Duncan, C.J. 1988. The role of phospholipase A2 in calcium-induced damage in cardiac and skeletal muscle. *Cell Tiss. Res.* 253: 457-62.

Durante, P.E., K.J. Mustard, S.H. Park et al. 2002. Effects of endurance training on activity and expression of AMP-activated protein kinase isoforms in rat muscles. *Am. J. Physiol.* 283(1): E178-86.

Fisher, A.B., S. Chien, A.I. Barakat et al. 2001. Endothelial cellular response to altered shear stress. *Am. J. Physiol.* 281(3): L529-33.

Foretz, M., D. Carling, C. Guichard et al. 1998. AMP-activated protein kinase inhibits the glucose-activated expression of fatty acid synthase gene in rat hepatocytes. *J. Biol. Chem.* 273(24): 14767-71.

Frey, T.G. & C.A. Mannella. 2000. The internal structure of mitochondria. *Trends Biochem. Sci.* 25(7): 319-24.

Freyssenet, D., P. Berthon, A. Geyssant et al. 1994. ATP synthesis kinetic properties of mitochondria isolated from the rat extensor digitorum longus muscle depleted of creatine with beta-guanidinopropionic acid. *Biochim. Biophys. Acta* 1186(3): 232-36.

Freyssenet, D., M.K. Connor, M. Takahashi et al. 1999. Cytochrome c transcriptional activation and mRNA stability during contractile activity in skeletal muscle. *Am. J. Physiol.* 277(1 Pt 1): E26-32.

Friden, J. & R.L. Lieber. 2001. Eccentric exercise-induced injuries to contractile and cytoskeletal muscle fibre components. *Acta Physiol. Scand.* 171(3): 321-26.

Frixione, E. 2000. Recurring views on the structure and function of the cytoskeleton: a 300-year epic. *Cell Mot. Cytoskeleton.* 46(2): 73-94.

Fuchs, E. & K. Weber. 1994. Intermediate filaments: structure, dynamics, function, and disease. *Annu. Rev. Biochem.* 63: 345-82.

Galbraith, C.G., R. Skalak & S. Chien. 1998. Shear stress induces spatial reorganization of the endothelial cell cytoskeleton. *Cell Mot. Cytoskeleton* 40(4): 317-30.

Gibala, M.J., J.D. MacDougall, M.A. Tarnopolsky et al. 1995. Changes in human skeletal muscle ultrastructure and force production after acute resistance exercise. *J. Appl. Physiol.* 78(2): 702-08.

Gruenberg, J. 2001. The endocytic pathway: a mosaic of domains. *Nature Rev. Mol. Cell Biol.* 2(10): 721-30.

Gundersen, G.G. & T.A. Cook. 1999. Microtubules and signal transduction. *Curr. Op. Cell Biol.* 11(1): 81-94.

Hamill, O.P. & B. Martinac. 2001. Molecular basis of mechanotransduction in living cells. *Physiol. Rev.* 81(2): 685-740.

Helge, J.W., K.J. Ayre, A.J. Hulbert et al. 1999. Regular exercise modulates muscle membrane phospholipid profile in rats. *J. Nutr.* 129(9): 1636-42.

Hood, D.A. 2001. Invited review: contractile activity-induced mitochondrial biogenesis in skeletal muscle. *J. Appl. Physiol.* 90(3): 1137-57.

Hood, D.A., A. Balaban, M.K. Connor et al. 1994. Mitochondrial biogenesis in striated muscle. *Can. J. Appl. Physiol.* 19(1): 12-48.

Hood, D.A., R. Zak & D. Pette. 1989. Chronic stimulation of rat skeletal muscle induces coordinate increases in mitochondrial and nuclear mRNAs of cytochrome-c-oxidase subunits. *Eur. J. Biochem.* 179(2): 275-80.

Hoppeler, H. & M. Fluck. 2003. Plasticity of skeletal muscle mitochondria: structure and function. *Med. Sci. Sports Exerc.* 35(1): 95-104.

Hoppeler, H., P. Luthi, H. Claassen et al. 1973. Ultrastructure of normal human skeletal muscle; a morphometric analysis in controls and men trained in long-distance running. *Hoppe-Seyler's Z. Physiol. Med.* 354(3): 229-30.

Hunziker, W. & H.J. Geuze. 1996. Intracellular trafficking of lysosomal membrane proteins. *Bioessays* 18(5): 379-89.

Iwai, K., M. Miyao, Y. Wadano et al. 2003. Dynamic changes of deleted mitochondrial DNA in human leucocytes after endurance exercise. *Eur. J. Appl. Physiol.* 88(6): 515-19.

Jackson, M.J., D.A. Jones & R.H.T. Edwards. 1984. Experimental skeletal muscle damage: the nature of the calcium-activated degenerative processes. *Eur. J. Clin. Invest.* 14: 369-74.

Jouaville, L.S., P. Pinton, C. Bastianutto et al. 1999. Regulation of mitochondrial ATP synthesis by calcium: evidence for a long-term metabolic priming. *Proc. Natl. Acad. Sci.* 96(24): 13807-12.

Katzmann, D.J., G. Odorizzi & S.D. Emr. 2002. Receptor downregulation and multivesicular-body sorting. *Nature Rev. Mol. Cell Biol.* 3(12): 893-905.

King, S. 2000. The dynein microtubule motor. *Biochim. Biophys. Acta* 1496: 60-75.

Koehler, C.M. 2000. Protein translocation pathways of the mitochondrion. *FEBS Lett.* 476(1-2): 27-31.

Krieger, D.A., C.A. Tate, J. McMillin-Wood et al. 1980. Populations of rat skeletal muscle mitochondria after exercise and immobilization. *J. Appl. Physiol.* 48(1): 23-28.

Lai, M.M. & F.W. Booth. 1990. Cytochrome c mRNA and alpha-actin mRNA in muscles of rats fed beta-GPA. *J. Appl. Physiol.* 69(3): 843-48.

Li, Y.S., J.Y. Shyy, S. Li et al. 1996. The Ras-JNK pathway is involved in shear-induced gene expression. *Mol. Cell Biol.* 16(11): 5947-54.

Lithgow, T. 2000. Targeting of proteins to mitochondria. *FEBS Lett.* 476(1-2): 22-26.

Manno, S., Y. Takakuwa & N. Mohandas. 2002. Identification of a functional role for lipid asymmetry in biological membranes: phosphatidylserine-skeletal protein interactions modulate membrane stability. *Proc. Natl. Acad. Sci.* 99(4): 1943-48.

McNiven, M.A. & K.J. Marlowe. 1999. Contributions of molecular motor enzymes to vesicle-based protein transport in gastrointestinal epithelial cells. *Gastroenterology* 116: 438-51.

Mitchison, T.J. & L.P. Cramer. 1996. Actin-based cell motility and cell locomotion. *Cell* 84(3): 371-79.

Mokelke, E.A., B.M. Palmer, J.Y. Cheung et al. 1997. Endurance training does not affect intrinsic calcium current characteristics in rat myocardium. *Am. J. Physiol.* 273(3 Pt 2): 1193-97.

Mooren, F.C., A. Lechtermann, A. Fromme et al. 2001. Alterations in intracellular calcium signaling of lymphocytes after exhaustive exercise. *Med. Sci. Sports Exerc.* 33(2): 242-48.

Morris, C.E. & U. Homann. 2001. Cell surface area regulation and membrane tension. *J. Membr. Biol.* 179(2): 79-102.

Mukherjee, S., R.N. Ghosh & F.R. Maxfield. 1997. Endocytosis. *Physiol. Rev.* 77: 759-803.

Murakami, T., Y. Shimomura, N. Fujitsuka et al. 1994. Enzymatic and genetic adaptation of soleus muscle mitochondria to physical training in rats. *Am. J. Physiol.* 267(3 Pt 1): E388-95.

Nakano, T., Y. Wada & S. Matsumura. 2001. Membrane lipid components associated with increased filterability of erythrocytes from long-distance runners. *Clin. Hemorheol. Microcirc.* 24(2): 85-92.

Nunoi, H., T. Yamazaki & S. Kanegasaki. 2001. Neutrophil cytoskeletal disease. *Int. J. Hemotol.* 74(2): 119-24.

Ojuka, E.O., L.A. Nolte & J.O. Holloszy. 2000. Increased expression of GLUT-4 and hexokinase in rat epitrochlearis muscles exposed to AICAR in vitro. *J. Appl. Physiol.* 88(3): 1072-75.

Overgaard, K., T. Lindstrom, T. Ingemann-Hansen et al. 2002. Membrane leakage and increased content of Na+-K+ pumps and Ca2+ in human muscle after a 100-km run. *J. Appl. Physiol.* 92(5): 1891-98.

Pilegaard, H., B. Saltin & P.D. Neufer. 2003. Exercise induces transient transcriptional activation of the PGC-1alpha gene in human skeletal muscle. *J. Physiol.* 546: 851-58.

Proske, U. & D.L. Morgan. 2001. Muscle damage from eccentric exercise: mechanism, mechanical signs, adaptation and clinical applications. *J. Physiol.* 537(Pt 2): 333-45.

Rizo, J. & T.C. Südhof. 1998. Mechanics of membrane fusion. *Nature Struct. Biol.* 5(10): 839-42.

Rizzuto, R., P. Pinton, W. Carrington et al. 1998. Close contacts with the endoplasmic reticulum as determinants of mitochondrial Ca2+ responses. *Science* 280(5370): 1763-66.

Rizzuto, R., A.W. Simpson, M. Brini et al. 1992. Rapid changes of mitochondrial Ca2+ revealed by specifically targeted recombinant aequorin. *Nature* 358(6384): 325-27.

Roth, T.F. & K.R. Porter. 1964. Yolk protein uptake in the oocyte of the mosquito Aedes Aegypti L. *J. Cell Biol.* 20: 313-32.

Rutter, G.A., C. Fasolato & R. Rizzuto. 1998. Calcium and organelles: a two-sided story. *Biochem. Biophys. Res. Com.* 253(3): 549-57.

Sachs, F. & M. Sokabe. 1990. Stretch-activated ion channels and membrane mechanics. *Neurosci. Res. Suppl.* 12: S1-4.

Sadoshima, J. & S. Izumo. 1997. The cellular and molecular response of cardiac myocytes to mechanical stress. *Annu. Rev. Physiol.* 59: 551-71.

Samelman, T.R., L.J. Shiry & D.F. Cameron. 2000. Endurance training increases the expression of mitochondrial and nuclear encoded cytochrome c oxidase subunits and heat shock proteins in rat skeletal muscle. *Eur. J. Appl. Physiol.* 83(1): 22-27.

Schlame, M. & D. Haldar. 1993. Cardiolipin is synthesized on the matrix side of the inner membrane in rat liver mitochondria. *J. Biol. Chem.* 268(1): 74-79.

Senturk, U.K., F. Gunduz, O. Kuru et al. 2001. Exercise-induced oxidative stress affects erythrocytes in sedentary rats but not exercise-trained rats. *J. Appl. Physiol.* 91(5): 1999-2004.

Shyy, J.Y. & S. Chien. 2002. Role of integrins in endothelial mechanosensing of shear stress. *Circ. Res.* 91(9): 769-75.

Szygula, Z. 1990. Erythrocytic system under the influence of physical exercise and training. *Sports Med.* 10(3): 181-97.

Takahashi, M., A. Chesley, D. Freyssenet et al. 1998. Contractile activity-induced adaptations in the mitochondrial protein import system. *Am. J. Physiol.* 274(5 Pt 1): C1380-87.

Takahashi, M. & D.A. Hood. 1993. Chronic stimulation-induced changes in mitochondria and performance in rat skeletal muscle. *J. Appl. Physiol.* 74(2): 934-41.

Takahashi, M. & D.A. Hood. 1996. Protein import into subsarcolemmal and intermyofibrillar skeletal muscle mitochondria. Differential import regulation in distinct subcellular regions. *J. Biol. Chem.* 271(44): 27285-91.

Tang, X., M.S. Halleck, R.A. Schlegel et al. 1996. A subfamily of P-type ATPases with aminophospholipid transporting activity. *Science* 272(5267): 1495-97.

Teasdale, R.D. & M.R. Jackson. 1996. Signal-mediated sorting of membrane proteins between the endoplasmic reticulum and the golgi apparatus. *Annu. Rev. Cell Dev. Biol.* 12: 7-54.

Temiz, A., O.K. Baskurt, C. Pekcetin et al. 2000. Leukocyte activation, oxidant stress and red blood cell properties after acute, exhausting exercise in rats. *Clin. Hemorheol. Microcirc.* 22(4): 253-59.

Thyberg, J. & S. Moskalewski. 1999. Role of microtubules in the organization of the Golgi complex. *Exp. Cell Res.* 246: 263-76.

Tinel, H., J.M. Cancel, H. Mogami et al. 1999. Active mitochondria surrounding the pancreatic acinar granule region prevent spreading of inositol trisphosphate-evoked local cytosolic Ca(2+) signals. *EMBO J.* 18(18): 4999-5008.

Venditti, P. & S. Di Meo. 1997. Effect of training on antioxidant capacity, tissue damage, and endurance of adult male rats. *Int. J. Sports Med.* 18(7): 497-502.

Vina, J., A. Gimeno, J. Sastre et al. 2002. Mechanism of free radical production in exhaustive exercise in humans and rats; role of xanthine oxidase and protection by allopurinol. *IUBMB Life* 49(6): 539-44.

Vincent, H.K., S.K. Powers, H.A. Demirel et al. 1999. Exercise training protects against contraction-induced lipid peroxidation in the diaphragm. *Eur. J. Appl. Physiol.* 79(3): 268-73.

Warren, G.L., C.P. Ingalls, D.A. Lowe et al. 2001. Excitation-contraction uncoupling: major role in contraction-induced muscle injury. *Exerc. Sport Sci. Rev.* 29(2): 82-87.

Warren, G.L., D.A. Lowe, D.A. Hayes et al. 1993. Excitation failure in eccentric contraction-induced injury of mouse soleus muscle. *J. Physiol.* 468: 487-99.

Weis, K. 2003. Regulating access to the genome: nucleocytoplasmic transport throughout the cell cycle. *Cell* 112(4): 441-51.

Winder, W.W., B.F. Holmes, D.S. Rubink et al. 2000. Activation of AMP-activated protein kinase increases mitochondrial enzymes in skeletal muscle. *J. Appl. Physiol.* 88(6): 2219-26.

Wu, H., S.B. Kanatous, F.A. Thurmond et al. 2002. Regulation of mitochondrial biogenesis in skeletal muscle by CaMK. *Science* 296(5566): 349-52.

Wu, H., B. Rothermel, S. Kanatous et al. 2001. Activation of MEF2 by muscle activity is mediated through a calcineurin-dependent pathway. *EMBO J.* 20(22): 6414-23.

Wu, Z., P. Puigserver, U. Andersson et al. 1999. Mechanisms controlling mitochondrial biogenesis and respiration through the thermogenic coactivator PGC-1. *Cell* 98(1): 115-24.

Yalcin, O., M. Bor-Kucukatay, U.K. Senturk et al. 2000. Effects of swimming exercise on red blood cell rheology in trained and untrained rats. *J. Appl. Physiol.* 88(6): 2074-80.

Chapter 2

Aigner, T. 2002. Apoptosis, necrosis, or whatever: how to find out what really happens? *J. Pathol.* 198: 1-4.

Allen, D.L., J.K. Linderman, R.R. Roy et al. 1997. Apoptosis: a mechanism contributing to remodeling of skeletal muscle in response to hindlimb unweighting. *Am. J. Physiol.* 273: C579-C587.

Allsopp, R.C., H. Vaziri, C. Patterson et al. 1992. Telomere length predicts replicative capacity of human fibroblasts. *Proc. Natl. Acad. Sci.* 89: 10114-18.

Amaral, S.A., P.E. Papanek & A.S. Greene. 2001. Angiotensin II and VEGF are involved in angiogenesis induced by short-term exercise training. *Am. J. Physiol.* 88: H1163-H1169.

Annex, B.H., C.E. Torgan, P. Lin et al. 1998. Induction and maintenance of increased VEGF protein by chronic motor nerve stimulation in skeletal muscle. *Am. J. Physiol.* 274: H860-H867.

Asano, M., K. Kaneoka, T. Nomura et al. 1998. Increase in serum vascular endothelial growth factor level during high altitude training. *Acta Physiol. Scand.* 162: 455-59.

Azenabor, A.A. & L. Hoffman-Goetz. 1999. Intrathymic and intrasplenic oxidative stress mediates thymocyte and splenocyte damage in acutely exercised mice. *J. Appl. Physiol.* 86: 1823-27.

Azenabor, A.A. & L. Hoffman-Goetz. 2000. Effect of exhaustive exercise on membrane estradiol concentration, intracellular calcium, and oxidative damage in mouse thymic lymphocytes. *Free Radic. Biol. Med.* 28: 84-90.

Balomenos, D. & C. Martinez-A. 2000. Cell-cycle regulation in immunity, tolerance and autoimmunity. *Immunol. Today* 21: 551-55.

Bidere, M. & A. Senik. 2001. Caspase-independent apoptotic pathways in T lymphocytes: a minireview. *Apoptosis* 6: 371-75.

Biral, D., A. Jakubiec-Puka, I. Ciechomska et al. 2000. Loss of dystrophin and some dystrophin-associated proteins with concomitant signs of apoptosis in rat leg muscle overworked in extension. *Acta Neuropathol.* 100: 618-26.

Blackburn, E.H. 1991. Structure and function of telomeres. *Nature* 350: 569-73.

Bodine-Fowler, S. 1994. Skeletal muscle regeneration after injury: an overview. *J. Voice* 8: 53-62.

Bolosover, S.R., J.S. Hyams, S. Jones et al. 1997. *From Genetics to Cells.* New York: Wiley-Liss.

Breen, E.C., E.C. Johnson, H. Wagner et al. 1996. Angiogenic growth factor mRNA responses in muscle to a single bout of exercise. *J. Appl. Physiol.* 81: 355-61.

Brown, M.D., M. Milkiewicz & O. Hudlicka. 2001. Multiple sources and roles for VEGF in activity-induced angiogenesis in skeletal muscle. *J. Physiol.* 12: 536-38.

Bruunsgaard, H., M.S. Jensen, P. Schjerling et al. 1999. Exercise induces recruitment of lymphocytes with an activated phenotype and short telomeres in young and elderly humans. *Life Sci.* 65: 2623-33.

Buchkovich, K.J. 1996. Telomeres, telomerase, and the cell cycle. *Prog. Cell Cycle Res.* 2: 187-95.

Budd, R.C. 2001. Activation-induced cell death. *Curr. Op. Immunol.* 13: 356-62.

Cain, K. & C. Freathy. 2001. Liver toxicity and apoptosis: role of TGF-β1, cytochrome c and the apoptosome. *Toxicol. Lett.* 120: 307-15.

Carraro, U. & C. Franceschi. 1997. Apoptosis of skeletal and cardiac muscles and physical exercise. *Aging* 9: 19-34.

Chautan, M., G. Chazal, F. Cecconi et al. 1999. Interdigital cell death can occur through a necrotic and caspase-independent pathway. *Curr. Biol.* 9: 967-70.

Chang, E. & C.B. Harley. 1995. Telomere length and replicative aging in human vascular tissues. *Proc. Natl. Acad. Sci.* 92: 11190-94.

Cheng, W., B. Li, J. Kajstura, P. Li et al. 1995. Stretch induced programmed myocyte cell death. *J. Clin. Invest.* 96: 2247-59.

Chitko-McKown, C.G. & J.F. Modiano. 1997. Clues to immune function and oncogenesis provided by events that activate the cell cycle machinery in normal human T cells. *J. Leukoc. Biol.* 62: 430-37.

Chiu, C.P., W. Dragowska, N.W. Kim et al. 1996. Differential expression of telomerase activity in hematopoietic progenitors from adult human bone marrow. *Stem Cells* 14: 239-48.

Chua, B.T., K. Guo & P. Li. 2000. Direct cleavage by the calcium-activated protease calpain can lead to inactivation of caspases. *J. Biol. Chem.* 275: 5131-35.

Cohen, G.M. 1997. Caspases: the executioners of apoptosis. *Biochem. J.* 326: 1-16.

Collins, K. 2000. Mammalian telomeres and telomerase. *Curr. Op. Cell Biol.* 12: 378-83.

Counter, C.M. 1996. The roles of telomeres and telomerase in cell life span. *Mut. Res.* 366: 45-63.

Concordet, J.P. & A. Ferry. 1993. Physiological programmed cell death in thymocytes is induced by physical stress (exercise). *Am. J. Physiol.* 265: C626-C629.

Cross, M. & M.T. Dexter. 1991. Growth factors in development, transformation and tumorigenesis. *Cell* 64: 2171-80.

Decary, S., C.B. Hamida, V. Mouly et al. 2000. Shorter telomeres in dystrophic muscle consistent with extensive regeneration in young children. *Neuromusc. Dis.* 10: 113-20.

Denecker, G., D. Vercammen, W. Declercq et al. 2001. Apoptotic and necrotic cell death induced by death domain receptors. *Cell. Mol. Life Sci.* 58: 356-70.

Di Donna, S., V. Renault, C. Forestier et al. 2000. Regenerative capacity of human satellite cells: the mitotic clock in cell transplantation. *Neurol. Sci.* 2: S943-S951.

Evan, G.I., L. Brown, M. Whyte et al. 1995. Apoptosis and the cell cycle. *Curr. Op. Cell Biol.* 7: 825-34.

Fan, G., X. Ma, B.T. Kren et al. 1996. The retinoblastoma gene product inhibits TGF-beta1 induced apoptosis in primary rat hepatocytes and human HuH-7 hepatoma cells. *Oncogene* 12: 1909-19.

Fellstrom, B. & L. Zezina. 2001. Apoptosis: friend or foe? *Transplant. Proc.* 33: 2414-16.

Farrara, N. 2001. Role of vascular endothelial growth factor in regulation of physiological angiogenesis. *Am. J. Physiol.* 280: C1358-C1366.

Folkman, J. 1997. Angiogenesis and angiogenesis inhibition: an overview. *EXS* 79: 1-8.

Forsythe, J.A., B.-H. Jiang, N.V. Iyer et al. 1996. Activation of vascular endothelial growth factor gene transcription by hypoxia-inducible factor 1. *Mol. Cell. Biol.* 16: 4604-13.

Foster, J.S., D.C. Henley, S. Ahamed et al. 2001. Estrogens and cell-cycle regulation in breast cancer. *Trends Endocrinol. Metab.* 12: 320-27.

Franco, S., I. Segura, H.H. Riese et al. 2002. Decreased B16F10 melanoma growth and impaired vascularization in telomerase-deficient mice with critically short telomeres. *Canc. Res.* 62: 552-59.

Friedrich, U., E. Griese, M. Schwab et al. 2000. Telomere length in different tissues of elderly patients. *Mech. Ageing Dev.* 119: 89-99.

Gavin, T.P., D.A. Spector, H. Wagner et al. 2000a. Effect of catopril on skeletal muscle angiogenic growth factor responses to exercise. *J. Appl. Physiol.* 88: 1690-97.

Gavin, T.P., D.A. Spector, H. Wagner et al. 2000b. Nitric oxide synthase inhibition attenuates the skeletal muscle VEGF mRNA response to exercise. *J. Appl. Physiol.* 88: 1192-98.

Goel, S., S. Mani & R. Perez-Soler. 2002. Tyrosine kinase inhibitors: a clinical perspective. *Curr. Oncol. Rep.* 4: 9-19.

Gottlieb, R.A., K.O. Burkson, R.A. Koner et al. 1994. Reperfusion injury induces apoptosis in rabbit cardiomyocytes. *J. Clin. Invest.* 94: 1621-28.

Goyns, M.H. & W.L. Lavery. 2000. Telomerase and mammalian ageing: a critical appraisal. *Mech. Ag. Dev.* 114: 69-77.

Green, K.J., D.G. Rowbottom & L.T. Mackinnon. 2002. Exercise and T-lymphocyte function: a comparison of proliferation in PBMC and NK cell-depleted PBMC culture. *J. Appl. Physiol.* 92: 2390-95.

Griffioen, A.W. & G. Molema. 2000. Angiogenesis: potential for pharmacologic intervention in the treatment of cancer, cardiovascular diseases, and chronic inflammation. *Pharmacol. Rev.* 52: 237-68.

Grogan, S.P., B. Aklin, M. Frenz et al. 2002. In vitro model for the study of necrosis and apoptosis in naïve cartilage. *J. Pathol.* 197: 5-13.

Grünert-Fuchs et al. 1995. Vitamin E prevents exercise-induced DNA damage. *Mutat. Res.* 346: 195-202.

Gustafsson, T., A. Puntschart, L. Kaijser et al. 1999. Exercise-induced expression of angiogenesis-related transcription and growth factors in human skeletal muscle. *Am. J. Physiol.* 45: H679-H685.

Hacker, G. 2000. The morphology of apoptosis. *Cell Tiss. Res.* 301: 5-17.

Haluska, P. & A.A. Adjei. 2001. Receptor tyrosine kinase inhibitors. *Curr. Op. Invest. Drugs* 2: 280-86.

Hang, J., L. Kong, J.-W. Gu et al. 1995. VEGF gene expression is upregulated in electrically stimulated rat skeletal muscle. *Am. J. Physiol.* 269: H1827-H1831.

Hansen-Smith, F., S. Egginton & O. Hudlicka. 1998. Growth of arterioles in chronically stimulated adult rat skeletal muscle. *Microcirculation* 5: 49-59.

Harley, C.B., A.B. Futcher & C.W. Greider. 1990. Telomeres shorten during ageing of human fibroblasts. *Nature* 345: 458-60.

Hartmann, A., U. Plappert, K. Raddatz et al. 1994. Does physical activity induce DNA damage? *Mutagenesis* 9: 269-72.

Hathcock, K.S., N.P. Weng, R. Merica et al. 1998. Cutting edge: antigen-dependent regulation of telomerase activity in murine T cells. *J. Immunol.* 160: 5702-06.

Hirsch, T., P. Marchetti, S.A. Susin et al. 1997. The apoptosis-necrosis paradox. Apoptogenic proteases activated after mitochondrial permeability transition determine the mode of cell death. *Oncogene* 15: 1573-81.

Hiyama, K., Y. Hirai, S. Kyoizumi et al. 1995. Activation of telomerase in human lymphocytes and hematopoietic progenitor cells. *J. Immunol.* 155: 3711-15.

Ho, A. & S.F. Dowdy. 2002. Regulation of G(1) cell-cycle progression by oncogenes and tumor suppressor genes. *Curr. Op. Gen. Dev.* 12: 47-52.

Hoffman-Goetz, L., S. Zajchowski & A. Aldred. 1999. Impact of treadmill exercise on early apoptotic cells in mouse thymus and spleen. *Life Sci.* 64: 191-200.

Hoffman-Goetz, L. & C.L. Fietsch. 2002. Lymphocyte apoptosis in ovariectomized mice given progesterone and voluntary exercise. *J. Sports Med. Phys. Fit.* 42: 481-87.

Hoffman-Goetz, L. & J. Quadrilatero. 2003. Treadmill exercise in mice increases intestinal lymphocyte loss via apoptosis. *Acta Physiol. Scand.* 179: 289-97.

Hoffman-Goetz, L. & S. Zajchowski. 1999. In vitro apoptosis of lymphocytes after exposure to levels of corticosterone observed following submaximal exercise. *J. Sports Med. Phys. Fit.* 39: 269-74.

Hsu, T.G., K.M. Hsu, C.W. Kong et al. 2002. Leukocyte mitochondria alterations after aerobic exercise in trained human subjects. *Med. Sci. Sports Exerc.* 34: 438-42.

Hubbard, S.R. 1999. Structural analysis of receptor tyrosine kinases. *Prog. Biophys. Mol. Biochem.* 71: 343-58.

Hubbard, S.R. & J.H. Till. 2000. Protein tyrosine kinase structure and function. *Annu. Rev. Biochem.* 69: 373-98.

Israels, E.D. & L.G. Israels. 2001. The cell cycle. *Stem Cells* 19: 88-91.

Jin, H., R. Yang, W. Li et al. 2000. Effects of exercise training on cardiac function, gene expression, and apoptosis in rats. *Am. J. Physiol.* 279: H2994-H3002.

Johnson, D.G. & C.L. Walker. 1999. Cyclins and cell cycle checkpoints. *Annu. Rev. Pharmacol. Toxicol.* 39: 295-312.

Jones, S.M. & A. Kazlauskas. 2001. Growth factor-dependent signaling and cell cycle progression. *FEBS Lett.* 490: 10-16.

Kajstura, J., W. Cheng, R. Sarangarajan et al. 1996. Necrotic and apoptotic myocyte cell death in the aging heart of Fischer 344 rats. *Am. J. Physiol.* 271: H1215-H1228.

Kajstura, J., M. Mansukhani, W. Cheng et al. 1995. Programmed cell death and expression of the protoncogene bcl-2 in myocytes during postnatal maturation in the heart. *Exp. Cell Res.* 219: 110-21.

Kamesaki, H. 1998. Mechanisms involved in chemotherapy-induced apoptosis and their implications in cancer chemotherapy. *Int. J. Hemotol.* 68: 29-43.

Kaminska, A.M. & A. Fidzianska. 1996. Experimental induction of apoptosis and necrosis in neonatal rat skeletal muscle. *Basic Appl. Myol.* 6: 251-56.

Kerr, J.F.R. 1971. Shrinkage necrosis: a distinct mode of cellular death. *J. Pathol.* 105: 13-20.

Kerr, J.F.R., A.H. Wyllie & A.R. Currie. 1972. Apoptosis: a basic biological phenomenon with wide-ranging implications in tissue kinetics. *Brit. J. Canc.* 26: 239-57.

King, K.L. & J.A. Cidlowski. 1998. Cell cycle regulation and apoptosis. *Annu. Rev. Physiol.* 60: 601-17.

Kloner, R.A., K. Przyklenk & P. Whittaker. 1989. Deleterious effects of oxygen radicals in ischemia/reperfusion. Resolved and unresolved issues. *Circulation* 80: 1115-27.

Kong, M., E.A. Barnes, V. Ollendorff et al. 2000. Cyclin F regulates the nuclear localization of cyclin B1 through a cyclin-cyclin interaction. *EMBO J.* 19: 1378-88.

Kowaltowski, A.J., R.F. Castilho & A.E. Vercesi. 2001. Mitochondrial permeability transition and oxidative stress. *FEBS Lett.* 495: 12-15.

Kumar, R. & E.B. Thompson. 1999. The structure of the nuclear hormone receptors. *Steroids* 64: 310-19.

Lagranha, C.J., S.M. Senna, T.M. de Lima et al. 2004. Beneficial effect of glutamine on exercise-induced apoptosis of rat neutrophils. *Med. Sci. Sports Exerc.* 36: 210-7.

Leclerc, V. & P. Leopold. 1996. The cyclin C/Cdk8 kinase. *Prog. Cell Cycle Res.* 2: 197-204.

Leist, M., B. Single, A.F. Castoldi et al. 1997. Intracellular adenosine triphosphate (ATP) concentration: a switch in the decision between apoptosis and necrosis. *J. Exp. Med.* 185: 1481-86.

Lemasters, J.J. 1999. Mechanisms of hepatic toxicity. V. Necrapoptosis and the mitochondrial permeability transition: shared pathways to necrosis and apoptosis. *Am. J. Physiol.* 276: G1-G6.

Lin, Y.S., H.L. Kuo, C.F. Kuo et al. 1999. Antioxidant administration inhibits exercise-induced thymocyte apoptosis in rats. *Med Sci Sports Exerc.* 31: 1594-8.

Liu, Y., S.R. Cox, T. Morita et al. 1995. Hypoxia regulates vascular endothelial growth factor gene expression in endothelial cells. *Circ. Res.* 77: 638-43.

Lloyd, P.G., M.B. Prior, H.T. Yang et al. 2002. Angiogenic growth factor expression in rat skeletal muscle in response to exercise training. *Am. J. Physiol.* 281: H2528-H2538.

Loft, S. & H.E. Poulsen. 1996. Cancer risk and oxidative DNA damage in man. *J. Mol. Med.* 74: 297-312.

Los, M., C. Stroh, R.U. Janicke et al. 2001. Caspases: more than just killers? *Trends Immunol.* 22: 31-34.

Lundberg, A.S. & R.A. Weinberg. 1999. Control of the cell cycle and apoptosis. *Eur. J. Canc.* 35: 531-39.

Maisonpierre, P.C., C. Suri, P.F. Jones et al. 1997. Angiopoietin-2, a natural antagonist for Tie2 that disrupts in vivo angiogenesis. *Science* 277: 55-60.

Majno, G. & I. Joris. 1995. Apoptosis, oncosis, and necrosis. An overview of cell death. *Am. J. Pathol.* 146: 3-15.

Mars, M., S. Govender, A. Weston et al. 1998. High intensity exercise: a cause of lymphocyte apoptosis? *Biochem. Biophys. Res. Com.* 249: 366-70.

Martin, S.J. & D.R. Green. 1995. Protease activation during apoptosis: death by a thousand cuts? *Cell* 82: 349-52.

Martinez Arias, A. & A. Stewart. 2002. *Molecular Principles of Animal Development.* New York: Oxford University Press.

Mathieu, J., S. Richard, B. Ballester et al. 1999. Apoptosis and gamma rays. *Ann. Pharmacol. Franc.* 57: 314-23.

Mazzeo, R.S., C. Rajkumar, J. Rolland et al. 1998. Immune response to a single bout of exercise in young and elderly subjects. *Mech. Ag. Dev.* 100: 121-32.

McCarthy, N.J. & G.I. Evan. 1998. Methods for detecting and quantifying apoptosis. *Curr. Top. Dev. Biol.* 36: 259-78.

McCormick, K.M. & D.P. Thomas. 1992. Exercise-induced satellite cell activation in senescent soleus muscle. *J. Appl. Physiol.* 72: 888-93.

McEachern, M.J., A. Krauskopf & E.H. Blackburn. 2000. Telomeres and their control. *Annu. Rev. Gen.* 34: 331-58.

Mooren, F.C., D. Blöming, A. Lechtermann et al. 2002. Lymphocyte apoptosis after exhaustive and moderate exercise. *J. Appl. Physiol.* 93: 147-53.

Musci, M.A., K.M. Latinis & G.A. Koretzky. 1997. Signaling events in T lymphocytes leading to cellular activation or programmed cell death. *Clin. Immunol. Immunpath.* 83: 205-22.

Nakagawa, T., H. Zhu, N. Morishima et al. 2000. Caspase-12 mediates endoplasmic-reticulum-specific apoptosis and cytotoxicity by amyloid-beta. *Nature* 403: 98-103.

Narayan, P., R.M. Mentzer Jr. & R.D. Lasley. 2001. Annexin V staining during reperfusion detects cardiomyocytes with unique properties. *Am. J. Physiol.* 281: H1931-H1937.

Newton, K. & A. Strasser. 2001. Cell death control in lymphocytes. *Adv. Immunol.* 76: 179-226.

Nicoterra, P., M. Leist & E. Ferrando-May. 1998. Intracellular ATP, a switch in the decision between apoptosis and necrosis. *Toxicol. Lett.* 102-03: 139-42.

Nie, D. & K.V. Honn. 2002. Cyclooxygenase, lipoxygenase and tumour angiogenesis. *Cell. Mol. Life Sci.* 59: 799-807.

Nieman, D.C., A.R. Miller, D.A. Henson et al. 1994. Effect of high- versus moderate-intensity exercise on lymphocyte subpopulations and proliferative response. *Int. J. Sports Med.* 15: 199-206.

Niess, A.M., A. Hartmann, M. Grünert-Fuchs et al. 1996. DNA damage after exhaustive treadmill running in trained and untrained men. *Int. J. Sports Med.* 17: 397-403.

Nor, J.E. & P.J. Polverini. 1999. Role of endothelial cell survival and death signals in angiogenesis. *Angiogenesis* 3: 101-16.

Noris, M., M. Morigi & R. Donadelli. 1995. Nitric oxide synthesis by cultured endothelial cells is modulated by flow conditions. *Circ. Res.* 76: 536-43.

Pallini, R., F. Pierconti, M.L. Falchetti et al. 2001. Evidence for telomerase involvement in the angiogenesis of astrocytic tumors: expression of human telomerase reverse transcriptase messenger RNA by vascular endothelial cells. *J. Neurosurg.* 94: 961-71.

Patel, H. & L. Hoffman-Goetz. 2002. Effects of oestrogen and exercise on caspase-3 activity in primary and secondary lymphoid compartments in ovariectomized mice. *Acta Physiol. Scand.* 176: 177-84.

Pedersen, B.K. & L. Hoffman-Goetz. 2000. Exercise and the immune system: regulation, integration, and adaptation. *Physiol. Rev.* 80: 1055-81.

Perrem, K. & R.R. Reddel. 2000. Telomeres and cell division potential. *Prog. Mol. Subcell. Biol.* 24: 173-89.

Phaneuf, S. & C. Leewenburgh. 2001. Apoptosis and exercise. *Med. Sci. Sports Ex.* 33: 393-96.

Pilz, R.B., M. Suhasini, S. Idriss et al. 1995. Nitric oxide and cGMP analogs activate transcription from AP-1-responsive promoters in mammalian cells. *FASEB J.* 9: 552-58.

Podhorska-Okolow, M., M. Sandri, S. Zampieri et al. 1998. Apoptosis of myofibres and satellite cells: exercise-induced damage in skeletal muscle of the mouse. *Neuropath. Appl. Neurobiol.* 24: 518-31.

Price, C.M. 1999. Telomeres and telomerase: broad effects on cell growth. *Curr. Op. Gen. Dev.* 9: 218-24.

Pucci, B., M. Kasten & A. Giordano. 2000. Cell cycle and apoptosis. *Neoplasia* 2: 291-99.

Qian, X., T.N. Wang, V.L. Rothman et al. 1997. Thrombospondin-1 modulates angiogenesis in vitro by up-regulation of matrix metalloproteinase-9 in endothelial cells. *Exp. Cell Res.* 235: 403-12.

Radak, Z., A.W. Taylor, M. Sasvari et al. 2001. Telomerase activity is not altered by regular strenuous exercise in skeletal muscle or by sarcoma in liver of rats. *Redox Rep.* 6: 99-103.

Raidal, S.L., D.N. Love, G.D. Bailey et al. 2000. Effect of single bouts of moderate and high intensity exercise and training on equine peripheral blood neutrophil function. *Res. Vet. Sci.* 68: 141-46.

Richardson, R.S., H. Wagner, S.R.D. Mudaliar et al. 1999. Human VEGF gene expression in skeletal muscle: effect of acute normoxic and hypoxic exercise. *Am. J. Physiol.* 276: H2247-H2252.

Rivilis, I., M. Milkiewicz, P. Boyd et al. 2002. Differential involvement of MMP-2 and VEGF during muscle stretch- versus shear stress-induced angiogenesis. *Am. J. Physiol.* 283: H1430-38.

Roberts, A.B., M.B. Sporn, R.K. Assoian et al. 1986. Transforming growth factor type beta: rapid induction of fibrosis and angiogenesis in vivo and stimulation of collagen formation in vitro. *Proc. Natl. Acad. Sci.* 83: 4167-71.

Ronsen, O., E. Haug, B.K. Pedersen et al. 2001. Increased neuroendocrine response to a repeated bout of endurance exercise. *Med. Sci. Sports Exerc.* 33: 568-78.

Saikumar, P., Z. Dong, V. Mikhailov et al. 1999. Apoptosis: definition, mechanisms, and relevance to disease. *Am. J. Med.* 107: 489-506.

Salvesen, G.S. & V.M. Dixit. 1999. Caspases: intracellular signaling by proteolysis. *Cell* 91: 443-46.

Sandner, P., K. Wolf, U. Bergmaier et al. 1997. Hypoxia and cobalt stimulate vascular endothelial growth factor receptor gene expression in rats. *Pfluegers Arch.* 433: 803-08.

Sarin, A., H. Nakajima & P.A. Henkart. 1995. A protease-dependent TCR-induced death pathway in mature lymphocytes. *J. Immunol.* 154: 5806-12.

Schafer, K.A. 1998. The cell cycle: a review. *Vet. Pathol.* 35: 461-78.

Sheppard, K.E. 2002. Nuclear receptors. II. Intestinal corticosteroid receptors. *Am. J. Physiol.* 282: G742-G746.

Schweichel, J.-U. & H.-J. Merker. 1973. The morphology of various types of cell death in prenatal tissues. *Teratology* 7: 253-66.

Sherr, C. 1996. Cancer cell cycle. *Science* 274: 1672-77.

Singh, N., R.B. Lim & M.A. Sawyer. 2000. The cell cycle. *Hawaii Med. J.* 59: 300-06.

Sjöström, J. & J. Bergh. 2001. How apoptosis is regulated, and what goes wrong in cancer. *Brit. Med. J.* 322: 1538-39.

Smith, H.K., L. Maxwell, C.D. Rodgers et al. 2001. Exercise-enhanced satellite cell proliferation and new myonuclear accretion in rat skeletal muscle. *J. Appl. Physiol.* 90: 1407-14.

Steensberg, A., J. Marrow, A.D. Toft et al. 2002. Prolonged exercise, lymphocyte apoptosis and F2-isoprostanes. *Eur. J. Appl. Physiol.* 87: 38-42.

Stetler Stevenson, W.G. 1999. Matrix metalloproteinases in angiogenesis: a moving target for therapeutic intervention. *J. Clin. Invest.* 103: 1237-41.

Stewart, Z.A. & J.A. Pietenpol. 2001. p53 signaling and cell cycle checkpoints. *Chem. Res. Toxicol.* 14: 243-63.

Squier, M.K., A.C. Miller, A.M. Malkinson et al. 1994. Calpain activation in apoptosis. *J. Cell Physiol.* 159: 229-37.

Suleiman, M.S., A.P. Halestrap & E.J. Griffiths. 2001. Mitochondria: a target for myocardial protection. *Pharmacol. Ther.* 89: 29-46.

Sussman, M.S. & G.B. Bulkley. 1990. Oxygen-derived free radicals in reperfusion injury. *Meth. Enzymol.* 186: 711-23.

Talapatra, S. & C.B. Thompson. 2001. Growth factor signaling in cell survival: implications for cancer treatment. *J. Pharmacol. Exp. Ther.* 298: 873-78.

Tamaki, T., S. Uchiyama, Y. Uchiyama et al. 2000. Limited myogenic response to a single bout of weight-lifting exercise in old rats. *Am. J. Physiol.* 278: C1143-C1152.

Tanaka, M., H. Ito, S. Adachi et al. 1994. Hypoxia induces apoptosis with enhanced expression of Fas antigen m-RNA in cultured neonatal rat cardiomyocytes. *Circ. Res.* 75: 426-33.

Tonini, T., C. Hillson & P.P. Claudio. 2002. Interview with the retinoblastoma family members: do they help each other? *J. Cell. Physiol.* 192: 138-50.

Umansky, S.R., G.M. Cuenco, S.S. Khutzian et al. 1995. Post-ischemic apoptotic death of rat neonatal cardiomyocytes. *Cell Death Diff.* 2: 235-41.

Van Cruchten, S. & W. van den Broeck. 2002. Morphological and biochemical aspects of apoptosis, oncosis and necrosis. *Anat. Histol. Embryol.* 31: 214-23.

Van De Graaff, K.M. & S.I. Fox. 1995. *Concepts of Human Anatomy and Physiology,* 4th ed. Dubuque, IA: Brown.

Vermes, I., C. Haanen, H. Steffens-Nakken & C. Reutelingsperger. 1995. A novel assay for apoptosis. Flow cytometric detection of phosphatidylserine expression on early apoptotic cells using fluorescein labeled annexin V. *J. Immunol. Meth.* 184: 39-51.

Villa, P., S.H. Kaufmann & W.C. Earnshaw. 1997. Caspases and caspase inhibitors. *Trends Biochem. Sci.* 22: 388-93.

Visconti, R.P., C.D. Richardson & T.N. Sato. 2002. Orchestration of angiogenesis and arteriovenous contribution by angiopoietins and vascular endothelial growth factor (VEGF). *Pro. Natl. Acad. Sci.* 99: 8219-24.

Wang, K.K. 2000. Calpain and caspase: can you tell the difference? *Trends Neurosci.* 23: 20-26.

Wang, Q.D., J. Pernow, P.O. Sjoquist & L. Ryden. 2002. Pharmacological possibilities for protection against myocardial reperfusion injury. *Cardiovasc. Res.* 55: 25-37.

Weigel, N.L. 1996. Steroid hormone receptors and their regulation by phosphorylation. *Biochem. J.* 319: 657-67.

Weng, N.P., L. Granger & R.J. Hodes. 1997. Telomere lengthening and telomerase activation during human B cell differentiation. *Proc. Natl. Acad. Sci.* 94: 10827-32.

Weng, N.P., B.L. Levine, C.H. June & R.J. Hodes. 1995. Human naive and memory T lymphocytes differ in telomeric length and replicative potential. *Proc. Natl. Acad. Sci.* 92: 11091-94.

Wyllie, A.H., J.F.R. Kerr & A.R. Currie. 1980. Cell death: the significance of apoptosis. *Int. Rev. Cytol.* 68: 251-306.

Wynford-Thomas, D. 1996. Telomeres, p53 and cellular senescence. *Oncol. Res.* 8: 387-98.

Xu, M., Y. Wang, K. Hirai, A. Ayub et al. 2001. Calcium preconditioning inhibits mitochondrial permeability transition and apoptosis. *Am. J. Physiol.* 280: H899-H908.

Yan, Z. 2000. Skeletal muscle adaptation and cell cycle regulation. *Exerc. Sport Sci. Rev.* 28: 24-26.

Zachsenhaus, E., Z. Jiang, D. Chung et al. 1996. pRb controls proliferation, differentiation, and death of skeletal muscle cells and other lineages during embryogenesis. *Genes Dev.* 10: 3051-64.

Zafonte, B.T., J. Huilt, D.F. Amanatullah et al. 2000. Cell-cycle dysregulation in breast cancer: breast cancer therapies targeting the cell cycle. *Front. Biosci.* 5: 938-61.

Ziche, M. & L. Morbidelli. 2000. Nitric oxide and angiogenesis. *J. Neurooncol.* 50: 139-48.

Chapter 3

Alvarez, R., N. Terrados, R. Ortolano et al. 2000. Genetic variation in the renin-angiotensin system and athletic performance. *Eur. J. Appl. Physiol.* 82: 117-20.

An, P., L. Perusse, T. Rankinen et al. 2003. Familial aggregation of exercise heart rate and blood pressure in response to 20 weeks of endurance training: the HERITAGE Family Study. *Int. J. Sports Med.* 24: 57-62.

An, P., T. Rice, J. Gagnon et al. 2000. Familial aggregation of stroke volume and cardiac output during submaximal exercise: the HERITAGE Family Study. *Int. J. Sports Med.* 21: 566-72.

An, P., T. Rice, L. Perusse et al. 2000. Complex segregation analysis of blood pressure and heart rate measured before and after a 20-week endurance exercise training program: the HERITAGE Family Study. *Am. J. Hypert.* 13: 488-97.

Bey, L., N. Akunuri, P. Zhao et al. 2003. Patterns of global gene expression in rat skeletal muscle during unloading and low-intensity ambulatory activity. *Physiol. Gen.* 11: 11.

Boerwinkle, E., J.E. Hixson & C.L. Hanis. 2000. Peeking under the peaks: following up genome-wide linkage analyses. *Circulation* 102: 1877-78.

Bonnefont, J.P., F. Demaugre, C. Prip-Buus et al. 1999. Carnitine palmitoyltransferase deficiencies. *Mol. Genet. Metab.* 68: 424-40.

Booth, F.W., M.V. Chakravarthy & E.E. Spangenburg. 2002. Exercise and gene expression: physiological regulation of the human genome through physical activity. *J. Physiol.* 543: 399-411.

Bouchard, C. 1995. Individual differences in the response to regular exercise. *Int. J. Obes. Relat. Metab. Disord.* 19: S5-8.

Bouchard, C., P. An, T. Rice et al. 1999. Familial aggregation of VO2max response to exercise training: results from the HERITAGE Family Study. *J. Appl. Physiol.* 87: 1003-08.

Bouchard, C., E.W. Daw, T. Rice et al. 1998. Familial resemblance for VO2max in the sedentary state: the HERITAGE Family Study. *Med. Sci. Sports Exerc.* 30: 252-58.

Bouchard, C., F.T. Dionne, J.A. Simoneau et al. 1992. Genetics of aerobic and anaerobic performances. *Exerc. Sport Sci. Rev.* 20: 27-58.

Bouchard, C., A.S. Leon, D.C. Rao et al. 1995. The HERITAGE Family Study. Aims, design, and measurement protocol. *Med. Sci. Sports Exerc.* 27: 721-29.

Bouchard, C., R. Lesage, G. Lortie et al. 1986. Aerobic performance in brothers, dizygotic and monozygotic twins. *Med. Sci. Sports Exerc.* 18: 639-46.

Bouchard, C. & T. Rankinen. 2001. Individual differences in response to regular physical activity. *Med. Sci. Sports Exerc.* 33: S446-S451.

Bouchard, C., T. Rankinen, Y.C. Chagnon et al. 2000. Genomic scan for maximal oxygen uptake and its response to training in the HERITAGE Family Study. *J. Appl. Physiol.* 88: 551-59.

Bronikowski, A.M., P.A. Carter, T.J. Morgan et al. 2003. Lifelong voluntary exercise in the mouse prevents age-related alterations in gene expression in the heart. *Physiol. Gen.* 12: 129-38.

Bruno, C., G. Manfredi, A.L. Andreu et al. 1998. A splice junction mutation in the alpha(M) gene of phosphorylase kinase in a patient with myopathy. *Biochem. Biophys. Res. Com.* 249: 648-51.

Cardon, L.R. & L.J. Palmer. 2003. Population stratification and spurious allelic association. *Lancet* 361: 598-604.

Cartegni, L., S.L. Chew & A.R. Krainer. 2002. Listening to silence and understanding nonsense: exonic mutations that affect splicing. *Nat. Rev. Genet.* 3: 285-98.

Chagnon, Y.C., T. Rankinen, E.E. Snyder et al. 2003. The human obesity gene map: the 2002 update. *Obes. Res.* 11: 313-67.

Chagnon, Y.C., T. Rice, L. Perusse et al. 2001. Genomic scan for genes affecting body composition before and after training in Caucasians from HERITAGE. *J. Appl. Physiol.* 90: 1777-87.

Comi, G.P., F. Fortunato, S. Lucchiari et al. 2001. Beta-enolase deficiency, a new metabolic myopathy of distal glycolysis. *Ann. Neurol.* 50: 202-07.

Conne, B., A. Stutz & J.D. Vassalli. 2000. The 3' untranslated region of messenger RNA: a molecular "hotspot" for pathology? *Nat. Med.* 6: 637-41.

Davis, M.E., H. Cai & G.R. Drummond. 2001. Shear stress regulates endothelial nitric oxide synthase expression through c-Src by divergent signaling pathways. *Circ. Res.* 89: 1073-80.

Delp, M.D. 1998. Differential effects of training on the control of skeletal muscle perfusion. *Med. Sci. Sports Exerc.* 30: 361-74.

Elston, R.C., S. Buxbaum, K.B. Jacobs et al. 2000. Haseman and Elston revisited. *Gen. Epidemiol.* 19: 1-17.

Fagard, R., E. Bielen & A. Amery. 1991. Heritability of aerobic power and anaerobic energy generation during exercise. *J. Appl. Physiol.* 70: 357-62.

Folland, J., B. Leach, T. Little et al. 2000. Angiotensin-converting enzyme genotype affects the response of human skeletal muscle to functional overload. *Exp. Physiol.* 85: 575-79.

Gayagay, G., B. Yu, B. Hambly et al. 1998. Elite endurance athletes and the ACE I allele—the role of genes in athletic performance. *Hum. Gen.* 103: 48-50.

Hadjigeorgiou, G.M., N. Kawashima, C. Bruno et al. 1999. Manifesting heterozygotes in a Japanese family with a novel mutation in the muscle-specific phosphoglycerate mutase (PGAM-M) gene. *Neuromusc. Dis.* 9: 399-402.

Hagberg, J.M., R.E. Ferrell, S.D. McCole et al. 1998. VO2 max is associated with ACE genotype in postmenopausal women. *J. Appl. Physiol.* 85: 1842-46.

Hamano, T., T. Mutoh, H. Sugie et al. 2000. Phosphoglycerate kinase deficiency: an adult myopathic form with a novel mutation. *Neurology* 54: 1188-90.

Haseman, J.K. & R.C. Elston. 1972. The investigation of linkage between a quantitative trait and a marker locus. *Behav. Genet.* 2: 3-19.

Higashi, Y., S. Sasaki, S. Kurisu et al. 1999. Regular aerobic exercise augments endothelium-dependent vascular relaxation in normotensive as well as hypertensive subjects: role of endothelium-derived nitric oxide. *Circulation* 100: 1194-1202.

Kingwell, B.A., B. Sherrard, G.L. Jennings et al. 1997. Four weeks of cycle training increases basal production of nitric oxide from the forearm. *Am. J. Physiol.* 272: H1070-77.

Laitinen, P.J., K.M. Brown, K. Piippo et al. 2001. Mutations of the cardiac ryanodine receptor (RyR2) gene in familial polymorphic ventricular tachycardia. *Circulation* 103: 485-90.

Lander, E.S., L.M. Linton, B. Birren et al. 2001. Initial sequencing and analysis of the human genome. *Nature* 409: 860-921.

Leon, A.S., T. Rice, S. Mandel et al. 2000. Blood lipid response to 20 weeks of supervised exercise in a large biracial population: the HERITAGE Family Study. *Metabolism* 49: 513-20.

Li, S., M. Kim, Y.L. Hu, S. Jalali et al. 1997. Fluid shear stress activation of focal adhesion kinase. Linking to mitogen-activated protein kinases. *J. Biol. Chem.* 272: 30455-62.

Liu, H.X., S.L. Chew, L. Cartegni et al. 2000. Exonic splicing enhancer motif recognized by human SC35 under splicing conditions. *Mol. Cell. Biol.* 20: 1063-71.

Lopez, A.J. 1998. Alternative splicing of pre-mRNA: developmental consequences and mechanisms of regulation. *Annu. Rev. Gen.* 32: 279-305.

Maekawa, M., K. Sudo, T. Kanno et al. 1990. Molecular characterization of genetic mutation in human lactate dehydrogenase-A (M) deficiency. *Biochem. Biophys. Res. Com.* 168: 677-82.

McArdle, B. 1951. Myopathy due to a defect in muscle glycogen breakdown. *Clin. Sci.* 10: 13-33.

Modrek, B. & C. Lee. 2002. A genomic view of alternative splicing. *Nat. Gen.* 30: 13-19.

Montgomery, H.E., P. Clarkson, C.M. Dollery et al. 1997. Association of angiotensin-converting enzyme gene I/D polymorphism with change in left ventricular mass in response to physical training. *Circulation* 96: 741-47.

Montgomery, H.E., R. Marshall, H. Hemingway et al. 1998. Human gene for physical performance. *Nature* 393: 221-22.

Myerson, S., H. Hemingway, R. Budget et al. 1999. Human angiotensin I-converting enzyme gene and endurance performance. *J. Appl. Physiol.* 87: 1313-16.

Myerson, S.G., H.E. Montgomery, M. Whittingham et al. 2001. Left ventricular hypertrophy with exercise and ACE gene insertion/deletion polymorphism: a randomized controlled trial with losartan. *Circulation* 103: 226-30.

Ookawara, T., V. Dave, P. Willems et al. 1996. Retarded and aberrant splicings caused by single exon mutation in a phosphoglycerate kinase variant. *Arch. Biochem. Biophys.* 327: 35-40.

Perusse, L., J. Gagnon, M.A. Province et al. 2001. Familial aggregation of submaximal aerobic performance in the HERITAGE Family Study. *Med. Sci. Sports Exerc.* 33: 597-604.

Perusse, L., T. Rankinen, R. Rauramaa et al. 2003. The human gene map for performance and health-related fitness phenotypes: the 2002 update. *Med. Sci. Sports Exerc.* 35: 1248-1264.

Perusse, L., T. Rice, M.A. Province et al. 2000. Familial aggregation of amount and distribution of subcutaneous fat and their responses to exercise training in the HERITAGE Family Study. *Obes. Res.* 8: 140-50.

Rankinen, T., P. An, L. Perusse et al. 2002. Genome-wide linkage scan for exercise stroke volume and cardiac output in the HERITAGE Family Study. *Physiol. Gen.* 10: 57-62.

Rankinen, T., P. An, T. Rice et al. 2001. Genomic scan for exercise blood pressure in the Health, Risk Factors, Exercise Training and Genetics (HERITAGE) Family Study. *Hypertension* 38: 30-37.

Rankinen, T., J. Gagnon, L. Perusse et al. 2000. AGT M235T and ACE ID polymorphisms and exercise blood pressure in the HERITAGE Family Study. *Am. J. Physiol.* 279: H368-74.

Rankinen, T., L. Perusse, J. Gagnon et al. 2000. Angiotensin-converting enzyme I/D polymorphism and trainability of the fitness phenotypes. The HERITAGE Family Study. *J. Appl. Physiol.* 88: 1029-35.

Rankinen, T., L. Perusse, R. Rauramaa et al. 2001. The human gene map for performance and health-related fitness phenotypes. *Med. Sci. Sports Exerc.* 33: 855-67.

Rankinen, T., B. Wolfarth, J.A. Simoneau et al. 2000. No association between the angiotensin-converting enzyme ID polymorphism and elite endurance athlete status. *J. Appl. Physiol.* 88: 1571-75.

Rao, D.C. & M.A. Province, eds. 2001. *Genetic Dissection of Complex Traits.* San Diego: Academic Press.

Rauramaa, R., R. Kuhanen, T.A. Lakka et al. 2002. Physical exercise and blood pressure with reference to the angiotensinogen M235T polymorphism. *Physiol. Gen.* 10: 71-77.

Rice, T., P. An, J. Gagnon, A. Leon et al. 2002. Heritability of HR and BP response to exercise training in the HERITAGE Family Study. *Med. Sci. Sports Exerc.* 34: 972-79.

Rice, T., Y.C. Chagnon, L. Perusse et al. 2002. A genome-wide linkage scan for abdominal subcutaneous and visceral fat in black and white families: the HERITAGE Family Study. *Diabetes* 51: 848-55.

Rice, T., J.P. Despres, L. Perusse et al. 2002. Familial aggregation of blood lipid response to exercise training in the Health, Risk Factors, Exercise Training, and Genetics (HERITAGE) Family Study. *Circulation* 105: 1904-08.

Rice, T., Y. Hong, L. Perusse et al. 1999. Total body fat and abdominal visceral fat response to exercise training in the HERITAGE Family Study: evidence for major locus but no multifactorial effects. *Metabolism* 48: 1278-86.

Rice, T., T. Rankinen, Y.C. Chagnon et al. 2002. Genome-wide linkage scan of resting blood pressure: HERITAGE Family Study. Health, risk factors, exercise training, and genetics. *Hypertension* 39: 1037-43.

Rigat, B., C. Hubert, F. Alhenc-Gelas et al. 1990. An insertion/deletion polymorphism in the angiotensin I-converting enzyme gene accounting for half the variance of serum enzyme levels. *J. Clin. Invest.* 86: 1343-46.

Roth, S.M., R.E. Ferrell, D.G. Peters et al. 2002. Influence of age, sex, and strength training on human muscle gene expression determined by microarray. *Physiol. Gen.* 10: 181-90.

Scholte, H.R., R.N. Van Coster, P.C. de Jonge et al. 1999. Myopathy in very-long-chain acyl-CoA dehydrogenase deficiency: clinical and biochemical differences with the fatal cardiac phenotype. *Neuromusc. Dis.* 9: 313-19.

Sherman, J.B., N. Raben, C. Nicastri et al. 1994. Common mutations in the phosphofructokinase-M gene in Ashkenazi Jewish patients with glycogenesis VII—and their population frequency. *Am. J. Hum. Gen.* 55: 305-13.

Skinner, J.S., K.M. Wilmore, J.B. Krasnoff et al. 2000. Adaptation to a standardized training program and changes in fitness in a large, heterogeneous population: the HERITAGE Family Study. *Med. Sci. Sports Exerc.* 32: 157-61.

Smith, C.W. & J. Valcarcel. 2000. Alternative pre-mRNA splicing: the logic of combinatorial control. *Trends Biochem. Sci.* 25: 381-88.

Sonna, L.A., M.A. Sharp, J.J. Knapik et al. 2001. Angiotensin-converting enzyme genotype and physical performance during US Army basic training. *J. Appl. Physiol.* 91: 1355-63.

Sugie, H., Y. Sugie, M. Ito & T. Fukuda. 1998. A novel missense mutation (837T→C) in the phosphoglycerate kinase gene of a patient with a myopathic form of phosphoglycerate kinase deficiency. *J. Child Neurol.* 13: 95-97.

Sundet, J.M., P. Magnus & K. Tambs. 1994. The heritability of maximal aerobic power: a study of Norwegian twins. *Scand. J. Med. Sci. Sports* 4: 181-85.

Swan, H., K. Piippo, M. Viitasalo et al. 1999. Arrhythmic disorder mapped to chromosome 1q42-q43 causes malignant polymorphic ventricular tachycardia in structurally normal hearts. *J. Am. Coll. Cardiol.* 34: 2035-42.

Szathmary, E., F. Jordan & C. Pal. 2001. Molecular biology and evolution. Can genes explain biological complexity? *Science* 292: 1315-16.

Taylor, R.R., C.D. Mamotte, K. Fallonet al. 1999. Elite athletes and the gene for angiotensin-converting enzyme. *J. Appl. Physiol.* 87: 1035-37.

Toscano, A., S. Tsujino, G. Vita et al. 1996. Molecular basis of muscle phosphoglycerate mutase (PGAM-M) deficiency in the Italian kindred. *Muscle Nerve* 19: 1134-37.

Tsujino, S., S. Servidei, P. Tonin et al. 1994. Identification of three novel mutations in non-Ashkenazi Italian patients with muscle phosphofructokinase deficiency. *Am. J. Hum. Gen.* 54: 812-19.

Tsujino, S., S. Shanske, A.K. Brownell et al. 1994. Molecular genetic studies of muscle lactate dehydrogenase deficiency in white patients. *Ann. Neurol.* 36: 661-65.

Tsujino, S., S. Shanske & S. DiMauro. 1993. Molecular genetic heterogeneity of myophosphorylase deficiency (McArdle's disease). *N. Engl. J. Med.* 329: 241-45.

Tsujino, S., S. Shanske, S. Sakoda et al. 1993. The molecular genetic basis of muscle phosphoglycerate mutase (PGAM) deficiency. *Am. J. Hum. Gen.* 52: 472-77.

Tsujino, S., P. Tonin, S. Shanske et al. 1994. A splice junction mutation in a new myopathic variant of phosphoglycerate kinase deficiency (PGK North Carolina). *Ann. Neurol.* 35: 349-53.

Tupler, R., G. Perini & M.R. Green. 2001. Expressing the human genome. *Nature* 409: 832-33.

Vorgerd, M., J. Karitzky, M. Ristow et al. 1996. Muscle phosphofructokinase deficiency in two generations. *J. Neurol. Sci.* 141: 95-99.

Wang, Q., M.E. Curran, I. Splawski et al. 1996. Positional cloning of a novel potassium channel gene: KVLQT1 mutations cause cardiac arrhythmias. *Nat. Gen.* 12: 17-23.

Wehner, M., P.R. Clemens, A.G. Engel et al. 1994. Human muscle glycogenosis due to phosphorylase kinase deficiency associated with a nonsense mutation in the muscle isoform of the alpha subunit. *Hum. Mol. Genet.* 3: 1983-87.

Williams, A.G., M.P. Rayson, M. Jubb et al. 2000. The ACE gene and muscle performance. *Nature* 403: 614.

Wilmore, J.H., P.R. Stanforth, J. Gagnon et al. 2001. Heart rate and blood pressure changes with endurance training: the HERITAGE Family Study. *Med. Sci. Sports Exerc.* 33: 107-16.

Woods, D., M. Hickman, Y. Jamshidi et al. 2001. Elite swimmers and the D allele of the ACE I/D polymorphism. *Hum. Gen.* 108: 230-32.

Woods, D.R., M. World, M.P. Rayson et al. 2002. Endurance enhancement related to the human angiotensin I-converting enzyme I-D polymorphism is not due to differences in the cardiorespiratory response to training. *Eur. J. Appl. Physiol.* 86: 240-44.

Chapter 4

Anthony, J.C., T.G. Anthony, S.R. Kimball et al. 2001. Signaling pathways involved in translational control of protein synthesis in skeletal muscle by leucine. *J. Nutr.* 131: 856S-860S.

Baar, K. & K. Esser. 1999. Phosphorylation of p70(S6k) correlates with increased skeletal muscle mass following resistance exercise. *Am. J. Physiol.* 276: C120-C127.

Ben-Neriah, Y. 2002. Regulatory functions of ubiquitination in the immune system. Nat. Immunol. 3: 20-26.

Boss, O., S. Samec, D. Desplanches et al. 1998. Effect of endurance training on mRNA expression of uncoupling proteins 1, 2, and 3 in the rat. *FASEB J.* 12: 335-39.

Chevet, E., P.H. Cameron, M.F. Pelletier et al. 2001. The endoplasmic reticulum: integration of protein folding, quality control, signaling and degradation. *Curr. Opin. Struct. Biol.* 11: 120-24.

Coux, O., K. Tanaka & A.L. Goldberg. 1996. Structure and functions of the 20S and 26S proteasomes. *Annu. Rev. Biochem.* 65: 801-47.

DeVol, D.L., P. Rotwein, J.L. Sadow et al. 1990. Activation of insulin-like growth factor gene expression during work-induced skeletal muscle growth. *Am. J. Physiol.* 259: E89-E95.

Feasson, L., D. Stockholm, D. Freyssenet et al. 2002. Molecular adaptations of neuromuscular disease-associated proteins in response to eccentric exercise in human skeletal muscle. *J. Physiol.* 543: 297-306.

Fink, C., S. Ergun, D. Kralisch et al. 2000. Chronic stretch of engineered heart tissue induces hypertrophy and functional improvement. *FASEB J.* 14: 669-79.

Friden, J. 1984. Muscle soreness after exercise: implications of morphological changes. *Int. J. Sports Med.* 5: 57-66.

Goldspink, D.F. 1991. Exercise-related changes in protein turnover in mammalian striated muscle. *J. Exp. Biol.* 160: 127-48.

Goto, S., R. Takahashi, A. Kumiyama et al. 2001. Implications of protein degradation in aging. In: *Healthy Aging for Functional Longevity: Molecular and Cellular Interactions in Senescence,* ed. S.C. Park et al.. *Ann. N.Y. Acad. Sci.* 928: 54-64.

Gregory, P., R.B. Low & W.S. Stirewalt. 1987. Fractional synthesis rates in vivo of skeletal-muscle myosin isoenzymes. *Biochem. J.* 245: 133-37.

Groettrup, M., M. van den Broek, K. Schwarz et al. 2001. Structural plasticity of the proteasome and its function in antigen processing. *Crit. Rev. Immunol.* 21: 339-58.

Grune, T., T. Reinheckel & K.J. Davies. 1997. Degradation of oxidized proteins in mammalian cells. *FASEB J.* 11: 526-34.

Hasselgren, P.O. 1999. Glucocorticoids and muscle catabolism. *Curr. Opin. Clin. Nutr. Metab. Care* 2: 201-05.

Hespel, P., E.B. Op't, M. Van Leemputte et al. 2001. Oral creatine supplementation facilitates the rehabilitation of disuse atrophy and alters the expression of muscle myogenic factors in humans. *J. Physiol.* 536: 625-33.

Holloszy, J.O. 1967. Biochemical adaptations in muscle. Effects of exercise on mitochondrial oxygen uptake and respiratory enzyme activity in skeletal muscle. *J. Biol. Chem.* 242: 2278-82.

Hood, D.A., M. Takahashi, M.K. Connor et al. 2000. Assembly of the cellular powerhouse: current issues in muscle mitochondrial biogenesis. *Exerc. Sport Sci. Rev.* 28: 68-73.

Hortobagyi, T., L. Dempsey, D. Fraser et al. 2000. Changes in muscle strength, muscle fibre size and myofibrillar gene expression after immobilization and retraining in humans. *J. Physiol.* 524 (Pt 1): 293-304.

Ikemoto, M., T. Nikawa, S. Takeda et al. 2001. Space shuttle flight (STS-90) enhances degradation of rat myosin heavy chain in association with activation of ubiquitin-proteasome pathway. *FASEB J.* 15: 1279-81.

International Human Genome Sequencing Consortium. 2001. A physical map of the human genome. *Nature* 409: 934-41.

Jacob, R., X. Hu, D. Niederstock et al. 1996. IGF-I stimulation of muscle protein synthesis in the awake rat: permissive role of insulin and amino acids. *Am. J. Physiol.* 270: E60-E66.

Ji, L.L., F.W. Stratman & H.A. Lardy. 1988. Enzymatic down regulation with exercise in rat skeletal muscle. *Arch. Biochem. Biophys.* 263: 137-49.

Jones, T.E., K. Baar, E. Ojuka et al. 2003. Exercise induces an increase in muscle UCP3 as a component of the increase in mitochondrial biogenesis. *Am. J. Physiol.* 284: E96-101.

Kee, A.J., A.J. Taylor, A.R. Carlsson et al. 2002. IGF-I has no effect on postexercise suppression of the ubiquitin-proteasome system in rat skeletal muscle. *J. Appl. Physiol.* 92: 2277-84.

Murakami, T., Y. Shimomura, A. Yoshimura et al. 1998. Induction of nuclear respiratory factor-1 expression by an acute bout of exercise in rat muscle. *Biochim. Biophys. Acta* 1381: 113-22.

Neeper, S.A., F. Gomez-Pinilla, J. Choi et al. 1996. Physical activity increases mRNA for brain-derived neurotrophic factor and nerve growth factor in rat brain. *Brain Res.* 726: 49-56.

Ojuka, E.O., T.E. Jones, D.H. Han et al. 2002. Intermittent increases in cytosolic Ca2+ stimulate mitochondrial biogenesis in muscle cells. *Am. J. Physiol.* 283: E1040-45.

Oliff, H.S., N.C. Berchtold, P. Isackson et al. 1998. Exercise-induced regulation of brain-derived neurotrophic factor (BDNF) transcripts in the rat hippocampus. *Mol. Brain Res.* 61: 147-53.

Park, H., V.K. Kaushik, S. Constant et al. 2002. Coordinate regulation of malonyl-CoA decarboxylase, sn-glycerol-3-phosphate acyltransferase, and acetyl-CoA carboxylase by AMP-activated protein kinase in rat tissues in response to exercise. *J. Biol. Chem.* 277: 32571-77.

Phillips, S.M., K.D. Tipton, A. Aarsland et al. 1997. Mixed muscle protein synthesis and breakdown after resistance exercise in humans. *Am. J. Physiol.* 273: E99-107.

Pilegaard, H., B. Saltin & P.D. Neufer. 2003. Exercise induces transient transcriptional activation of the PGC-1alpha gene in human skeletal muscle. *J. Physiol.* 546: 851-58.

Proud, C.G. 2002. Regulation of mammalian translation factors by nutrients. *Eur. J. Biochem.* 269: 5338-49.

Radák, Z., K. Asano, M. Inoue et al. 1996. Superoxide dismutase derivative prevents oxidative damage in liver and kidney of rats induced by exhausting exercise. *Eur. J. Appl. Physiol.* 72: 189-94.

Radák, Z., T. Kaneko, S. Tahara et al. 1999. The effect of exercise training on oxidative damage of lipids, proteins, and DNA in rat skeletal muscle: evidence for beneficial outcomes. *Free Radic. Biol. Med.* 27: 69-74.

Radák, Z., T. Kaneko, S. Tahara et al. 2001a. Regular exercise improves cognitive function and decreases oxidative damage in rat brain. *Neurochem. Int.* 38: 17-23.

Radák, Z., M. Sasvari, C. Nyakas et al. 2000. Exercise preconditioning against hydrogen peroxide-induced oxidative damage in proteins of rat myocardium. *Arch. Biochem. Biophys.* 376: 248-51.

Radák, Z., A.W. Taylor, H. Ohno et al. 2001b. Adaptation to exercise-induced oxidative stress: from muscle to brain. *Exerc. Immunol. Rev.* 7: 90-107.

Raj, D.A., T.S. Booker & A.N. Belcastro. 1998. Striated muscle calcium-stimulated cysteine protease (calpain-like) activity promotes myeloperoxidase activity with exercise. *Pfluegers Arch.* 435: 804-09.

Rasmussen, B.B., K.D. Tipton, S.L. Miller et al. 2000. An oral essential amino acid-carbohydrate supplement enhances muscle protein anabolism after resistance exercise. *J. Appl. Physiol.* 88: 386-92.

Ricquier, D. & F. Bouillaud. 2000. Mitochondrial uncoupling proteins: from mitochondria to the regulation of energy balance. *J. Physiol.* 529(Pt 1): 3-10.

Rogers, S.W. & M. Rechsteiner. 1988. Degradation of structurally characterized proteins injected into HeLa cells. Effects of intracellular location and the involvement of lysosomes. *J. Biol. Chem.* 263: 19843-49.

Schrauwen, P., V. Hinderling & M.K. Hesselink. 2002. Etomoxir-induced increase in UCP3 supports a role of uncoupling protein 3 as a mitochondrial fatty acid anion exporter. *FASEB J.* 16: 1688-90.

Sorimachi, H., Y. Ono & K. Suzuki. 2000. Skeletal muscle-specific calpain, p94, and connectin/titin: their physiological functions and relationship to limb-girdle muscular dystrophy type 2A. *Adv. Exp. Med. Biol.* 481: 383-95.

Turk, V., B. Turk & D. Turk. 2001. Lysosomal cysteine proteases: facts and opportunities. *EMBO J.* 20: 4629-33.

Venter, J.C. et al. 2001. The sequence of the human genome. *Science* 291:1304-51.

Volarevic, S. & G. Thomas. 2001. Role of S6 phosphorylation and S6 kinase in cell growth. *Prog. Nucleic Acid Res. Mol. Biol.* 65: 101-27.

Wakshlag, J.J., F.A. Kallfelz, S.C. Barr et al. 2002. Effects of exercise on canine skeletal muscle proteolysis: an investigation of the ubiquitin-proteasome pathway and other metabolic markers. *Vet. Ther.* 3: 215-25.

Willoughby, D.S. & M.J. Nelson. 2002. Myosin heavy-chain mRNA expression after a single session of heavy-resistance exercise. *Med. Sci. Sports Exerc.* 34: 1262-69.

Wolfsberg, T.G., J. McEntyre & G.D. Schuler. 2001. Guide to the draft human genome. *Nature* 409: 824-26.

Wu, H., S.B. Kanatous, F.A. Thurmond et al. 2002. Regulation of mitochondrial biogenesis in skeletal muscle by CaMK. *Science* 296: 349-52.

Yan, Z., S. Salmons, Y.I. Dang et al. 1996. Increased contractile activity decreases RNA-protein interaction in the 3'-UTR of cytochrome c mRNA. *Am. J. Physiol.* 271: C1157-C1166.

Chapter 5

Ahtikoski, A.M., S.O.A. Koskinen, P. Virtanen et al. 2001. Regulation of synthesis of fibrillar collagens in rat skeletal muscle during immobilization in shortened and lengthened positions. *Acta Physiol. Scand.* 172: 131-40.

Ahtikoski, A.M., S.O.A. Koskinen, Virtanen et al. 2003. Synthesis and degradation of type IV collagen in rat skeletal muscle during immobilization in shortened and lengthened positions. *Acta Physiol. Scand.* 177: 473-81.

Ameye, L. & M.F. Young. 2002. Mini review. Mice deficient in small leucine-rich proteoglycans: novel in vivo models for osteoporosis, osteoarthritis, Ehlers-Danlos syndrome, muscular dystrophy, and corneal diseases. *Glyobiology* 12:107R-116R.

Aumailley, M. & B. Gayraud. 1998. Structure and biological activity of the extracellular matrix. *J. Mol. Med.* 76:253-65.

Bailey, A.J. 2001. Molecular mechanisms of ageing in connective tissues. *Mech. Ageing Dev.* 122:735-55.

Bailey, A.J. & D.L. Nicholas. 1989. *Connective Tissue in Meat and Meat Products.* London and New York: Elsevier Applied Sciences.

Balcerzak, D., L. Querengesser, W.T. Dixon et al. 2001. Coordinated expression of matrix-degrading proteinases and their activators and inhibitors in bovine skeletal muscle. *Am. Soc. Animal Sci.* 79: 94-107.

Bayne, E.K., M.J. Anderson & D.M. Fambrough. 1984. Extracellular matrix organization in developing muscle: correlation with acetylcholine receptor aggregates. *J. Cell Biol.* 99:1486-1501.

Belkin, A.M., S.F. Retta, O.Y. Pletjushkina et al. 1997. Muscle β1D integrin reinforces the cytoskeleton-matrix link: modulation of integrin adhesive function by alternative splicing. *J. Cell Biol.* 139: 1583-95.

Berg, R.A., W.W. Kao & N.L. Kedersha. 1980. The assembly of tetrameric prolyl hydroxylase in tendon fibroblasts from newly synthesized alpha-subunits and from preformed cross-reacting protein. *Biochem. J.* 189:491-99.

Birkedal-Hansen, H. 1993. Proteolytic remodeling of extracellular matrix. *Curr. Op. Cell Biol.* 7:728-35.

Bishop, J.E. & G. Lindahl. 1999. Regulation of cardiovascular collagen synthesis by mechanical load. *Cardiovasc. Res.* 42:27-44.

Bonaldo, P., P. Braghetta, M. Zanetti et al. 1998. Collagen Vi deficiency induces early onset of myopathy in the mouse: an animal model for Bethlem myopathy. *Hum. Mol. Genet.* 7:2135-40.

Borkakoti, N. Structural studies of matrix metalloproteinases. 2000. *J. Mol. Med.* 78:261-68.

Brew, K., D. Dinakarpandian & H. Nagase. 2000. Tissue inhibitors of metalloproteinases: evolution, structure and function. *Biochim. Biophys. Acta* 1477:267-83.

Brinkmann, J., H. Notbohm, M. Tronnier et al. 1999. Overhydroxylation of lysyl residues is the initial step for altered collagen cross-links and fibril architecture in fibrotic skin. *J. Invest. Dermatol.* 113:617-21.

Cameron-Smith, D. 2002. Exercise and skeletal muscle gene expression. *Clin. Exp. Pharmocol. Physiol.* 29: 209-213.

Carragher, N.O., B. Levkau, R. Ross et al. 1999. Degraded collagen fragments promote rapid disassembly of smooth muscle focal adhesions that correlates with cleavage of $pp125^{FAK}$, paxillin, and talin. *J. Cell Biol.* 147:619-29.

Chiquet, M. 1999. Regulation of extracellular matrix gene expression by mechanical stress. Mini review. *Mat. Biol.* 18:417-28.

Chiquet, M., U. Mumenthaler, M. Wittner et al. 1998. The chick and human collagen $\alpha 1$(XII) gene promotor: activity of highly conserved regions around the first exon and in the first intron. *Eur. J. Biochem.* 257: 362-71.

Chiquet-Ehrismann, R., C. Hagios & K. Matsumoto. 1994. The tenascin gene family. *Persp. Dev. Neurobiol.* 2: 3-7.

Chou, M.Y. & H.C. Li. 2002. Genomic organization and characterization of the human type XXI collagen (COL21A1) gene. *Genomics* 79:395-401.

Christiansen, D.L., E.K. Huang & F.H. Silver. 2000. Assembly of type I collagen: fusion of fibrillar subunits and the influence of fibril diameter on mechanical properties. *Mat. Biol.* 19:409-20.

Colognato, H., D.A. Winkelmann & Y.D. Yurchenco. 1999. Laminin polymerization induces a receptor-cytoskeleton network. *J. Cell Biol.* 145:619-31.

Colognato, H. & P.D. Yurchenco. 2000. Form and function: the laminin family of heterotrimers. *Dev. Dyn.* 218:213-34.

Creemers, L.B., I.D. Jansen, A.J. Docherty et al. 1998a. Gelatinase A (MMP-2) and cysteine proteinases are essential for the degradation of collagen in soft connective tissue. *Mat. Biol.* 17:35-46.

Creemers, L.B., K.A. Hoeben, D.C. Jansen et al. 1998b. Participation of intracellular cysteine proteinases, in particular cathepsin B, in degradation of collagen in periosteal tissue explants. *Mat. Biol.* 16:575-84.

Diab, M., J-J. Wu & D.R. Eyre. 1996. Collagen type IX from human cartilage: a structural profile of intermolecular cross-linking sites. *Biochem. J.* 314:327-32.

Dickinson, C.D., D.A. Gay, J. Parello et al. 1994. Crystals of the cell-binding module of fibronectin obtained from a series of recombinant fragments differing in length. *J. Mol. Biol.* 238:123-27.

Eklund, L. 2001. Genetic studies of collagen types XV and XVIII. Type XV collagen deficiency in mice results in skeletal myopathy and cardiovascular defects, while the homologous endostatin precursor type XVIII collagen is needed for normal development of the eye. *Acta Universitatis Ouluensis, D Medica, 660.* 1-94.

Eklund, L., J. Piuhola, J. Komulainen et al. 2001. Lack of type XV collagen causes a skeletal myopathy and cardiovascular defects in mice. *Proc. Natl. Acad. Sci.* 98:1194-99.

Ellerbroek, S.M., Y.I. Wu, C.M. Overall et al. 2001. Functional interplay between type I collagen and cell surface matrix metalloproteinase activity. *J. Biol. Chem.* 276:24833-42.

Erickson, A.C. & J.R. Couchman. 2000. Still more complexity in mammalian basement membrane. *J. Histochem. Cytochem.* 48:1291-1306.

Eyre, D.R. 1995. The specificity of collagen cross-links as markers of bone and connective tissue degradation. *Dev. Dyn.* (Suppl 266) 66:165-69.

Foidart, M., J-M. Foidart & W.K. Engel. 1981. Collagen localization in normal and fibrotic human skeletal muscle. *Arch. Neurol.* 38:152-57.

Flück, M., V. Tunc-Civelek & M. Chiquet. 2000. Rapid and reciprocal regulation of tenascin-C and tenascin-Y expression by loading of skeletal muscle. *J. Cell Sci.* 113:3583-91.

Flück, M., A. Ziemiecki, R. Billeter et al. 2002. Fibre-type specific concentration of focal adhesion kinase at the sarcolemma: influence of fibre innervation and regeneration. *J. Exp. Biol.* 205:2337-48.

Gordon, S.E., M. Flueck & F.W. Booth. 2001. Plasticity in skeletal, cardiac, and smooth muscle. Selected contribution: skeletal muscle focal adhesion kinase, paxillin, and serum response factor are loading dependent. *J. Appl. Physiol.* 90:1174-83.

Gosselin, L.E., C. Adams, T.A. Cottere et al. 1998. Effect of exercise training on passive stiffness in locomotor skeletal muscle: role of extracellular matrix. *J. Appl. Physiol.* 85:1011-16.

Grounds, M.D., J.K. McGeachie, M.J. Davies et al. 1998. The expression of extracellular matrix during adult skeletal muscle regeneration: how the basement membrane, interstitium and myogenic cells collaborate. *Basic Appl. Myol.* 8(2):129-41.

Halfter, W., S. Dong, B. Schurer et al. 1998. Collagen XVIII is a basement membrane heparan sulfate proteoglycan. *J. Biol. Chem.* 273:25404-12.

Han, X-Y-, W. Wang, J. Komulainen et al. 1999. Increased mRNAs for procollagens and key regulating enzymes in rat skeletal muscle following downhill running. *Eur. J. Physiol.* 437:857-64.

Hanson, D.A. & D.R. Eyre. 1996. Molecular site specificity of pyridinoline and pyrrole cross-links in type I collagen of human bone. *J. Biol. Chem.* 271:26508-16.

Hantai, D., J. Gautron & J. Labbat-Robert. 1983. Immunolocalization of fibronectin and other macromolecules of the intercellular matrix in the striated muscle fiber of the adult rat. *Collag. Rel. Res.* 3:381-91.

Hashimoto, T., T. Wakabayashi, A. Watanabe et al. 2002. CLAC: a novel Alzheimer amyloid plaque component derived from a transmembrane precursor, CLAC-P/ collagen type XXV. *EMBO J.* 21:1524-34.

Heinegard, D.K. & E.R. Pimentel. 1992. Cartilage matrix proteins. In *Articular Cartilage and Osteoarthritis,* eds. K.E. Kuettner, R. Schleyerbach, J.G. Peyron & V.C. Hascall, 95-111. New York: Raven Press.

Heino, J. 2000. The collagen receptor integrins have distinct ligand recognition and signalling functions. *Mat. Biol.* 19:319-23.

Henry, M.D., J.S. Satz, M. Brakebusch et al. 2001. Distinct roles for dystroglycan, beta1 integrin and perlecan in cell surface laminin organization. *J. Cell Sci.* 114. 1137-44.

Huijing, P.A. 1999. Muscle as collagen fiber reinforced composite: a review of force transmission in muscle and whole limb. *J. Biomech.* 32:329-45.

Hägg, P., M. Rehn, P. Huhtala et al. 1998. Type XIII collagen is identified as a plasma membrane protein. *J. Biol. Chem.* 273:15590-97.

Hägg, P., T. Väisänen, A. Tuomisto et al. 2001. Type XIII collagen: a novel cell adhesion component present in a range of cell-matrix adhesions and in the intercalated discs between cardiac muscle cells. *Mat. Biol.* 19:727-42.

Imanaka-Yoshida, K., M. Hiroe, Y. Yasutomi et al. 2002. Tenascin-C is a useful marker for disease activity in myocarditis. *J. Pathol.* 197:388-94.

Ingber, D.E., S.R. Heideman, P. Lamoureux et al. 2000. Opposing views on tensegrity as a structural framework for understanding cell mechanics. *J. Appl. Physiol.* 89:1663-78.

Itoh, Y., A. Ito, K. Iwata et al. 1998. Plasma membrane-bound tissue inhibitor of metalloproteinases (TIMP)-2 specifically inhibits matrix metalloproteinase 2 (gelatinase A) activated on cell surface. *J. Biol. Chem.* 273:24360-67.

Jablecki, C.K., J.E. Heuser & S. Kaufman. 1973. Autoradiographic localization of new RNA synthesis in hypertrophying skeletal muscle. *J. Cell Biol.* 57: 743-59.

Jalkanen, M., S. Jalkanen & M. Bernfield. 1991. Binding of extracellular effector molecules by cell surface proteoglycans. In *Receptors for Extracellular Matrix,* J.A. McDonald & R.P. Mecham, eds., 1-37. San Diego: Academic Press.

Kadi, F. & L-E. Thornell. 2000. Concomitant increases in myonuclear and satellite cell content in female trapezius muscle following strength training. *Histochem. Cell Biol.* 113:99-103.

Kääriäinen, M., T. Liljamo, M. Pelto-Huikko et al. 2001. Regulation of $\alpha 7$ integrin by mechanical stress during skeletal muscle regeneration. *Neuromusc. Dis.* 11:360-69.

Kashtan, C.E. 2000. Alport syndrome: abnormalities of type IV collagen genes and proteins. *Renal Failure* 22:737-49.

Keagle, J.N., W.J. Welch & D.M. Young. 2001. Expression of heat shock proteins in a linear rodent wound. *Wound Repair Regen.* 9:378-85.

Kivirikko, K.I. & R. Myllylä. 1979. Collagen glucosyltransferases. In *International Review in Connective Tissue Research,* eds. D.A. Hall & D.S. Jackson, 8:23-72. New York: Academic Press.

Kivirikko, K.I. & R. Myllylä. 1984. Biosynthesis of collagens. In *Extracellular Matrix Biochemistry,* eds. K.A. Piez & A.H. Reddi, 83-118. New York: Elsevier.

Kivirikko, K.I., R. Myllylä & T. Pihlajaniemi. 1989. Protein hydroxylation: prolyl 4-hydroxylase, an enzyme with four co-substrates and a multifunctional subunit. *FASEB J.* 3:1609-17.

Kivirikko, K.I., R. Myllylä & T. Pihlajaniemi. 1992. Hydroxylation of proline and lysine residues in collagens and other mammalian and plant proteins. In *Posttranslational Modifications of Proteins,* eds. J.J. Harding & M.J.C. Grabble, 1-51. Boca Raton, FL: CRC Press.

Komulainen, J. & V. Vihko. 1998. The course of exercise-induced skeletal muscle fiber injury. In: *Oxidative Stress in Skeletal Muscle,* ed. A.Z. Reznick et al., 55-73. Basel: Birkhäuser Verlag.

Koskinen, S.O.A., A.M. Ahtikoski, J. Komulainen et al. 2002. Short-term effects of forced eccentric contractions on collagen synthesis and degradation in rat skeletal muscle. *Eur. J. Physiol.* 444:59-72.

Koskinen, S.O.A., W. Wang, A.M. Ahtikoski et al. 2001. Acute exercise induced changes in rat skeletal muscle mRNAs and proteins regulating type IV collagen content. *Am. J. Physiol.* 280:R1292-R1300.

Kovanen, V. 1989. Effects of ageing and physical training on rat skeletal muscle. *Acta Physiol. Scand.* 135, Supplementum 577.

Kovanen, V. 2002. Intramuscular extracellular matrix: complex environment of muscle cells. *Exerc. Sport Sci. Rev.* 30:20-25.

Kovanen, V. & H. Suominen. 1988. Effects of age and life-long endurance training on the passive mechanical properties of rat skeletal muscle. *Compr. Gerontol. A* 2:18-23.

Kovanen, V., H. Suominen & E. Heikkinen. 1980. Connective tissue of "fast" and "slow" skeletal muscle in rats—effects of endurance training. *Acta Physiol. Scand.* 108:173-80.

Kovanen, V., H. Suominen & E. Heikkinen. 1984. Mechanical properties of fast and slow skeletal muscle with special reference to collagen and endurance training. *J. Biomech.* 17:10:725-35.

Kühl, U., R. Timpl & K. von der Mark. 1982. Synthesis of type IV collagen and laminin in cultures of skeletal muscle cells and their assembly on the surface of myotubes. *Dev. Biol.* 93:344-54.

Kühl, U., M. Öcalan, R. Timpl et al. 1984. Role of muscle fibroblasts in the deposition of type IV collagen in the basal lamina of myotubes. *Differentiation* 28:164-72.

Kühn, K. 1995. Basement membrane (type IV) collagen. *Mat. Biol.* 14:439-45.

Kvist, A-P., A. Latvanlehto, M. Sund et al. 2001. Lack of cytosolic and transmembrane domains of type XIII collagen results in progressive myopathy. *Am. J. Pathol.* 159:1581-91.

Langberg, H., L. Rosendal & M. Kjaer. 2001. Training-induced changes in peritendinous type I collagen turnover determined by microdialysis in humans. *J. Physiol.* 534:297-02.

Langberg, H., D. Skovgaard, L.J. Petersen et al. 1999. Type I collagen synthesis and degradation in peritendinous tissue after exercise determined by microdialysis in humans. *J. Physiol.* 521:299-306.

Larrain, J., D.J. Carey & E. Brandan. 1998. Syndecan-1 expression inhibits myoblast differentiation through a basic fibroblast growth factor-dependent mechanism. *J. Biol. Chem.* 273:32288-96.

Last, J.A., L.G. Armstrong & K.M. Reiser. 1990. Biosynthesis of collagen crosslinks. *Int. J. Biochem.* 22: 559-64.

Laurent, G.J., R.J. McAnulty & J. Gibson. 1985. Changes in collagen synthesis and degradation during skeletal muscle growth. *Am. J. Physiol.* 249:C352-C355.

Laurent, G.J., M.P. Sparrow, P.C. Bates et al. 1978. Turnover of muscle protein in the fowl. Collagen content and turnover in cardiac and skeletal muscles of the adult fowl and changes during stretch-induced growth. *Biochem. J.* 176:419-27.

Lawrie, R.A. 1979. *Meat Science,* 3rd ed., 75-31. Oxford: Pergamon Press.

Lehto, M. 1983. Collagen and fibronectin in a healing skeletal muscle injury. An experimental study in rats under variable states of physical activity. *Ann. Universitatis Turkuensis, Ser. D Medica-Odontologica* 14. University of Turku.

Li, D., C. Clark & J.C. Myers. 2000. Basement membrane zone type XV collagen is a disulfide-bonded chondroitin sulfate proteoglycan in human tissues and cultured cells. *J. Biol. Chem.* 275:22339-47.

Lieber, R.L., L.-E. Thornell & J. Fridén. 1996. Muscle cytoskeletal disruption occurs within the first 15 min of cyclic eccentric contraction. *J. Appl. Physiol.* 80(1):278-84.

Light, N. & A.E. Champion. 1984. Characterization of muscle epimysium, perimysium and endomysium collagens. *Biochem. J.* 219:1017-26.

Lindahl, G.E., R.C. Chambers, J. Papakrivopoulou et al. 2002. Activation of fibroblasts procollagen α1(I) transcription by mechanical strain is transforming growth factor-β-dependent and involves increased binding of CCAAT-binding factor (CGF/NF-Y) at the proximal promoter. *J. Biol. Chem.* 277:6153-61.

Lindy, S., H. Turto & J. Uitto. 1972. Protocollagen proline hydroxylase activity in rat heart during experimental cardiac hypertrophy. *Circ. Res.* 30:205-09.

Lipton, B.H. 1977. Collagen synthesis by normal and bromodeoxyuridine-modulated cells in myogenic cultures. *Dev. Biol.* 61:153-65.

Marneros, A.G. & B.R. Olsen. 2001. The role of collagen-derived proteolytic fragments in angiogenesis. Mini review. *Mat. Biol.* 20:337-45.

Mayne, R. & R.D. Sanderson. 1985. The extracellular matrix of skeletal muscle. *Collag. Rel. Res.* 5:449-68.

Miner, J.H. & Sanes. 1994. Collagen IV α3, α4 and α5 chains in rodent basal laminae: sequence, distribution, association with laminins, and developmental switches. *J. Cell Biol.* 127:879-91.

Moscatello, D.K., D.M. Santra, D.J. Mann et al. 1998. Decorin suppresses tumor cell growth by activating the epidermal growth factor receptor. *J. Clin. Invest.* 101:406-12.

Myllyharju, J. & K.I. Kivirikko. 2001. Collagens and collagen-related diseases. *Ann. Med.* 33:7-31.

Myllylä, R., A. Salminen, L. Peltonen et al. 1986. Collagen metabolism of mouse skeletal muscle during the repair of exercise injuries. *Pflügers Arch.* 407: 64-70.

Nagase, H. & J.F. Woessner. 1999. Matrix metalloproteinases. Mini review. *J. Biol. Chem.* 274:21491-94.

Nagata, K. 1996. Hsp47: a collagen-specific molecular chaperone. *Trends Biochem. Sci.* 21:23-26.

Nakato, H. & K. Kimata. 2002. Heparan sulfate fine structure and specificity of proteoglycan functions. *Biochim. Biophys. Acta* 1573:312-18.

Neame, P.J., C.J. Kay, D.J. McQuillan et al. 2000. Independent modulation of collagen fibrillogenesis by decorin and lumican. *Cell. Mol. Life Sci.* 57:859-63.

Netzel-Arnett, S., S.K. Mallya, H. Nagase et al. 1991. Sequence specificities of fibroblast and neutrophil collagenases. *J. Biol. Chem.* 266:6747-55.

Nishimura, T., A. Hattori & K. Takahashi. 1994. Ultrastructure of the intramuscular connective tissue in bovine skeletal muscle. A demonstration using the cell-maceration/scanning electron microscope method. *Acta Anat.* 151:250-57.

Nishimura, T., A. Hattori & K. Takahashi. 1996. Arrangement and identification of proteoglycans in basement membrane and intramuscular connective tissue of bovine semitendinosus muscle. *Acta Anat.* 155: 257-65.

Notbohm, H., M. Nokelainen, J. Myllyharju et al. 1999. Recombinant human type II collagens with low and high levels of hydroxylysine and its glycosylated forms show marked differences in fibrillogenesis in vitro. *J. Biol. Chem.* 274:8988-92.

Oldberg, A., P. Antonsson, E. Hedbom et al. 1990. Structure and function of extracellular matrix proteoglycans. *Biochem. Soc. Transact.* 18:789-92.

Orend, G. & R. Chiquet-Ehrismann. 2000. Adhesion modulation by antiadhesive molecules of the extracellular matrix. *Exp. Cell Res.* 261:104-10.

Ortega, N. & Z. Werb. 2002. New functional roles for non-collagenous domains of basement membrane collagens. *J. Cell Sci.* 115:4201-14.

Osses, N. & E. Brandan. 2001. ECM is required for skeletal muscle differentiation independently of muscle regulatory factor expression. *Am. J. Physiol.* 282: C383-C394.

Palokangas, H., V. Kovanen, Duncan et al. 1992. Age-related changes in the concentration of hydroxypyridinium crosslinks in functionally different skeletal muscles. *Matrix* 12:291-96.

Parks, W.C. & R.P. Mecham, eds. 1998. *Matrix Metalloproteinases.* San Diego: Academic Press.

Patterson, M.L., S.J. Atkinson, V. Knäuper et al. 2001. Specific collagenolysis by gelatinase A, MMP-2, is determined by the hemopexin domain and not the fibronectin-like domain. *FEBS Lett.* 503:158-62.

Patton, B.L., J.H. Miner, A.Y. Chiu et al. 1997. Distribution and function of laminins in the neuromuscular system of developing, adult and mutant mice. *J. Cell Biol.* 139:1507-21.

Prockop, D.J. 1990. Mutations that alter the primary structure of type I collagen. The perils of a system for generating large structures by principles of nucleated growth. Minireview. *J. Biol. Chem.* 265: 15349-52.

Purslow, P.P. 2002. Structure and functional significance of variations in the connective tissue within muscle. *Comp. Biochem. Physiol.* Part A 133:947-66.

Purslow, P.P. & V.C. Duance. 1990. Structure and function of intramuscular connective tissue. In: *Connective Tissue Matrix,* vol. 2, ed. D.W.L. Hukins, 127-66. New York: Macmillan.

Purslow, P.P. & J.A. Trotter. 1994. The morphology and mechanical properties of endomysium in series-fibred muscle: variations with muscle length. *J. Musc. Res. Cell Mot.* 15:299-308.

Ravanti, L. & V-M. Kähäri. 2000. Matrix metalloproteinases in wound repair [review]. *Int. J. Mol. Med.* 6: 391-407.

Rehn, M. & T. Pihlajaniemi. 1997. Alpha 1(XVIII), a collagen chain with frequent interruptions in the collagenous sequence, a distinct tissue distribution, and homology with type XV collagen. *Proc. Natl. Acad. Sci.* 91:4234-38.

Riquelme, C., J. Larrain, E. Schönherr et al. 2001. Antisense inhibition of decorin expression in myoblasts decreases cell responsiveness to transforming growth factor beta and accelerates skeletal muscle differentiation. *J. Biol. Chem.* 276:3589-96.

Ruoslahti, E. & A. Vaheri. 1974. Novel human serum protein from fibroblast plasma membrane. *Nature* 248:789-91.

Ruotsalainen, H., L. Sipila, E. Kerkela et al. 1999. Characterization of cDNAs for mouse lysyl hydroxylase 1, 2 and 3, their phylogenetic analysis and tissue-specific expression in the mouse. *Mat. Biol.* 18: 325-29.

Sabatelli, P., P. Bonaldo, G. Lattanti et al. 2001. Collagen VI deficiency affects the organization of fibronectin in the extracellular matrix of cultured fibroblasts. *Mat. Biol.* 20:475-86.

Sado, Y., M. Kagawa, I. Naito et al. 1998. Organization and expression of basement membrane collagen IV genes and their roles in human disorders. *J. Biochem.* 123:767-76.

Sanes, J.R., E. Engvall, R. Butkowski et al. 1990. Molecular heterogeneity of basal laminae: isoforms of laminin and collagen IV at the neuromuscular junction and elsewhere. *J. Cell Biol.* 111:1685-99.

Sasaki, T., H. Larsson, D. Tisi et al. 2000. Endostatins derived from collagens XV and XVIII differ in structural and binding properties, tissue distribution and anti-angiogenic activity. *J. Mol. Biol.* 301:1179-90.

Sasse, J., H. von der Mark, U. Kuehl et al. 1981. Origin of collagen type I, III and V in cultures of avian skeletal muscle. *Dev. Biol.* 83:79-89.

Sato, K., K. Yomogida, T. Wada et al. 2002. Type XXVI collagen, a new member of the collagen gene family, is specifically expressed in the testis and ovary. *J. Biol. Chem.* 277:37678-84.

Sell, D.R., M.A. Lane, W.A. Johnson et al. 1996. Longevity and the genetic determination of collagen glycoxidation kinetics in mammalian senescence. *Proc. Natl. Acad. Sci.* 93(1):485-90.

Simonsen, E.B., H. Klitgaard & F. Bojsen-Moller. 1995. The influence of strength training, swim training and ageing on the Achilles tendon and m. soleus of the rat. *J. Sport Sci.* 13:291-95.

Smyth, N., H.S. Vatansever, P. Murray et al. 1999. Absence of basement membrane after targeting the LAMC1 gene results in embryonic lethality due to failure of endoderm differentiation. *J. Cell Biol.* 144:151-60.

Snellman, A. 2000. Characterization of chain association in collagen XII and XII and other biochemical features of type XIII collagen using baculovirus-directed insect cell expression. *Acta Univ. Ouluensis, D Medica,* 606.

Streuli, C. 1999. Extracellular matrix remodelling and cellular differentiation. *Curr. Op. Cell Biol.* 11:634-40.

Sumiyoshi, H., F. Laub, H. Yoshioka et al. 2001. Embryonic expression of type XIX collagen is transient and confined to muscle cells. *Dev. Dyn.* 220:155-62.

Suominen, H. & E. Heikkinen. 1975. Enzyme activities in muscle and connective tissue of M. vastus lateralis in habitually training and sedentary 33 to 70-year-old men. *Eur. J. Appl. Physiol.* 34:249-54.

Suominen, H., E. Heikkinen & T. Parkatti. 1977. Effect of eight weeks' physical training on muscle and connective tissue of the M. vastus lateralis in 69-year-old men and women. *J. Gerontol.* 32:33-37.

Takala, T.E.S., R. Myllylä & A. Salminen. 1983. Increased activities of prolyl 4-hydroxylase and galactosylhydroxylysyl glucosyltransferase, enzymes of collagen biosynthesis, in skeletal muscle of endurance trained mice. *Pfluegers Arch.* 399:271-74.

Tasab, M., L. Jenkinson & N.J. Bulleid. 2002. Sequence-specific recognition of collagen triple helices by the collagen-specific molecular chaperone HSP47. *J. Biol. Chem.* 277:35007-12.

Tezak, Z., P. Grandini, M. Boscaro et al. 2003. Clinical and molecular study in congenital muscular dystrophy with partial laminin alpha 2 (LAMA2) deficiency. *Hum. Mutation* 21(2):103-11.

Timpl, R. & J.C. Brown. 1996. Supramolecular assembly of basement membranes. *BioEssays* 18:123-32.

Tomono, Y., I. Naito, K. Ando et al. 2002. Epitope-defined monoclonal antibodies against multiplexin collagens demonstrate that type XV and XVIII collagens are expressed in specialized basement membranes. *Cell Struct. Funct.* 27(1):9-20.

Trotter, J.A. 2002. Structure–function considerations of muscle-tendon junctions. *Comp. Biochem. Physiol.* Part A 133:1127-33.

Trotter, J.A., F.J.R. Richmont & P.P. Purslow. 1995. Functional morphology and motor control of series fibred muscles. In: *Exercise and Sport Science Reviews*, vol. 23, ed. J.O. Holloszy, 167-213. Baltimore: Williams & Wilkins.

Trächslin, J., M. Koch, M. Chiquet et al. 1999. Rapid and reversible regulation of collagen XII expression by changes in tensile stress. *Exp. Cell Res.* 247:320-28.

Turto, H., S. Lindy & J. Halme. 1974. Protocollagen proline hydroxylase activity in work-induced hypertrophy of rat muscle. *Am. J. Physiol.* 226:63-65.

Väliaho, J., T. Takala, V. Kovanen et al. 2000. Expression of basement membrane associated genes in adult skeletal muscle fibers. 5th Annual Congress of the European College of Sport Science, Jyväskylä, Finland.

Valtanen, H., K. Lehti, J. Lohi et al. 2000. Expression and purification of soluble and inactive mutant forms of membrane type 1 matrix metalloproteinase. *Prot. Exp. Pur.* 19:66-73.

Velleman, S.G. 1998. The role of the extracellular matrix in skeletal muscle development. *Poultry Sci.* 78:778-84.

Velling, T., J. Risteli, K. Wennerberg et al. 2002. Polymerization of type I and III collagens is dependent on fibronectin and enhanced by integrins $\alpha_{11}\beta_1$ and $\alpha_{12}\beta_1$. *J. Biol. Chem.* 277:37377-81.

Vierck, J., B. O'Reilly, K. Hossner et al. 2000. Satellite cell regulation following myotrauma caused by resistance exercise. *Cell Biol. Int.* 24(5):263-72.

Vuorio, E. & B. Crombrugghe. 1990. The family of collagen genes. *Annu. Rev. Biochem.* 59:837-72.

Wang, C., M. Risteli, J. Heikkinen et al. 2002. Identification of amino acids important for the catalytic activity of the collagen glucosyltransferase associated with the multifunctional lysyl hydroxylase 3, LH3. *J. Biol. Chem.* 277:18568-73.

Zimmerman, S.D., R.J. McCormick, K. Vadlamundi et al. 1993. Age and training alter collagen characteristics in fast- and slow-twitch rat limb muscle. *J. Appl. Physiol.* 75:1670-74.

Zimmerman, S.D., D.P. Thomas, S.G. Velleman et al. 2001. Time course of collagen and decorin changes in rat cardiac and skeletal muscle post-MI. *Am. J. Physiol.* 281:H1816-H1822.

Yamaguchi, Y., D.M. Mann & E. Ruoslahti. 1990. Negative regulation of transforming growth factor-β by the proteoglycan decorin. *Nature* 346:281-84.

Yurchenco, P.D. & J.J. O'Rear. 1994. Basal lamina assembly. *Curr. Op. Cell Biol.* 6:674-81.

Zhang, X., B.G. Hudson & M.P. Sarras Jr. 1994. Hydra cell aggregate development is blocked by selective fragments of fibronectin and type IV collagen. *Dev. Biol.* 164:10-23.

Zhang, Y., J. Li, F.W. Sellke et al. 2003. Syndecan 4 modulates basic fibroblast growth factor (FGF2) signaling in vivo. *Am. J. Physiol.* 284(6):H2078-82.

Zhou, J., T. Mochizuki, H. Smeets et al. 1993. Deletion of the paired alpha 5(IV) and alpha 6(IV) collagen genes in inherited smooth muscle tumors. *Science* 261:1167-69.

Chapter 6

Adragna, N.C. & P.K. Lauf. 1997. Oxidative activation of K-Cl cotransport by diamide in erythrocytes from humans with red cell disorders, and from several other mammalian species. *J. Membr. Biol.* 155(3): 207-17.

Adragna, N.C., R.E. White, S.N. Orlov et al. 2000. K-Cl cotransport in vascular smooth muscle and erythrocytes: possible implication in vasodilation. *Am. J. Physiol.* 278(2): C381-90.

Agus, Z.S., E. Kelepouris, I. Dukes et al. 1989. Cytosolic magnesium modulates calcium channel activity in mammalian ventricular cells. *Am. J. Physiol.* 256(2 Pt 1): C452-55.

Alper, S.L., M.N. Chernova & A.K. Stewart. 2001. Regulation of Na+-independent Cl-/HCO3-exchangers by pH. *J. Pancreas* 2(4 Suppl): 171-75.

Alper, S.L., R.B. Darman, M.N. Chernova et al. 2002. The AE gene family of Cl/HCO3-exchangers. *J. Nephrol.* 15(Suppl 5): S41-53.

Bangsbo, J., C. Juel, Y. Hellsten et al. 1997. Dissociation between lactate and proton exchange in muscle during intense exercise in man. *J. Physiol.* 504(Pt 2): 489-99.

Bara, M., A. Guiet-Bara & J. Durlach. 1993. Regulation of sodium and potassium pathways by magnesium in cell membranes. *Magnesium Res.* 6(2): 167-77.

Basso, L.E., J.B. Ubbink, R. Delport et al. 2000. Erythrocyte magnesium concentration as an index of magnesium status: a perspective from a magnesium supplementation study. *Clin. Chim. Acta* 20;291(1): 1-8.

Bell, G.J. & H.A. Wenger. 1988. The effect of one-legged sprint training on intramuscular pH and nonbicarbonate buffering capacity. *Eur. J. Appl. Physiol.* 58(1-2): 158-64.

Beller, G.A., J.T. Maher, L.H. Hartley et al. 1975. Changes in serum and sweat magnesium levels during work in the heat. *Av. Space Environ. Med.* 46(5): 709-12.

Bize, I. & P.B. Dunham. 1995. H2O2 activates red blood cell K-Cl cotransport via stimulation of a phosphatase. *Am. J. Physiol.* 269(4 Pt 1): C849-55.

Cairns, S.P., W.A. Hing, J.R. Slack et al. 1997. Different effects of raised [K+]o on membrane potential and contraction in mouse fast- and slow-twitch muscle. *Am. J. Physiol.* 273(2 Pt 1): C598-611.

Chan, A., R. Shinde, C.C. Chow et al. 1991. In vivo and in vitro sodium pump activity in subjects with thyrotoxic periodic paralysis. *Brit. Med. J.* 303(6810): 1096-99.

Chen, S.J., C.C. Wu & M.H. Yen. 2001. Exercise training activates large-conductance calcium-activated K(+) channels and enhances nitric oxide production in rat mesenteric artery and thoracic aorta. *J. Biomed. Sci.* 8(3): 248-55.

Choe, S. 2002. Potassium channel structures. *Nature Rev. Neurosci.* 3(2): 115-21.

Clausen, T. 1998. Clinical and therapeutic significance of the Na+,K+ pump. *Clin. Sci.* (London, England, 1979) 95(1): 3-17.

Clausen, T. & J.A. Flatman. 1977. The effect of catecholamines on Na-K transport and membrane potential in rat soleus muscle. *J. Physiol.* 270(2): 383-414.

Cordova, A., J.F. Escanero & M. Gimenez. 1992. Magnesium distribution in rats after maximal exercise in air and under hypoxic conditions. *Magnesium Res.* 5(1): 23-27.

Costa, F., J. Heusinkveld, R. Ballog et al. 2000. Estimation of skeletal muscle interstitial adenosine during forearm dynamic exercise in humans. *Hypertension* 35(5): 1124-28.

Dai, L.J., B. Bapty, G. Ritchie et al. 1998. Glucagon and arginine vasopressin stimulate Mg2+ uptake in mouse distal convoluted tubule cells. *Am. J. Physiol.* 274(2 Pt 2): F328-35.

De Bruijne, A.W., H. Vreeburg & J. van Steveninck. 1985. Alternative-substrate inhibition of L-lactate transport via the monocarboxylate-specific carrier system in human erythrocytes. *Biochim. Biophys. Acta* 812(3): 841-44.

Deuster, P.A., E. Dolev, S.B. Kyle et al. 1987. Magnesium homeostasis during high-intensity anaerobic exercise in men. *J. Appl. Physiol.* 62(2): 545-50.

Dolev, E., R. Burstein, R. Wishnitzer et al. 1991-92. Longitudinal study of magnesium status of Israeli military recruits. *Magnes. Trace Elem.* 10(5-6): 420-26.

Dorup, I. & T. Clausen. 1997. Effects of adrenal steroids on the concentration of Na(+)-K+ pumps in rat skeletal muscle. *J. Endocrinol.* 152(1): 49-57.

Dubouchaud, H., N. Eydoux, P. Granier et al. 1999. Lactate transport activity in rat skeletal muscle sarcolemmal vesicles after acute exhaustive exercise. *J. Appl. Physiol.* 87: 955-61.

Elin, R.J. 1987. Assessment of magnesium status. *Clin. Chem.* 33(11): 1965-70.

Everts, M.E., K. Retterstol & T. Clausen. 1988. Effects of adrenaline on excitation-induced stimulation of the sodium-potassium pump in rat skeletal muscle. *Acta Physiol. Scand.* 134(2): 189-98.

Evertsen, F., J.L. Medbo, E. Jebens et al. 1997. Hard training for 5 mo increases Na(+)-K+ pump concentration in skeletal muscle of cross-country skiers. *Am. J. Physiol.* 272(5 Pt 2): R1417-24.

Eydoux, N., G. Py, K. Lambert et al. 2000. Training does not protect against exhaustive exercise-induced lactate transport capacity alterations. *Am. J. Physiol.* 278(6): E1045-52.

Flatman, P.W. 1991. Mechanisms of magnesium transport. *Annu. Rev. Physiol.* 53: 259-71.

Fliegel, L. 2001. Regulation of myocardial Na+/H+ exchanger activity. *Bas. Res. Cardiol.* 96(4): 301-05.

Frohlich, O. & M. Karmazyn. 1997. The Na-H exchanger revisited: an update on Na-H exchange regulation and the role of the exchanger in hypertension and cardiac function in health and disease. *Cardiovasc. Res.* 36(2): 138-48.

Giebisch, G. 1999. Physiological roles of renal potassium channels. *Sem. Nephrol.* 19(5): 458-71.

Green, H.J., E.R. Chin, M. Ball-Burnett et al. 1993. Increases in human skeletal muscle Na(+)-K(+)-ATPase concentration with short-term training. *Am. J. Physiol.* 264(6 Pt 1): C1538-41.

Green, H., J. MacDougall, M. Tarnopolsky et al. 1999. Downregulation of Na+-K+-ATPase pumps in skeletal muscle with training in normobaric hypoxia. *J. Appl. Physiol.* 86(5): 1745-48.

Green, H., B. Roy, S. Grant et al. 2000. Downregulation in muscle Na(+)-K(+)-ATPase following a 21-day expedition to 6,194 m. *J. Appl. Physiol.* 88(2): 634-40.

Grinstein, S., S. Cohen, J.D. Goetz et al. 1985. Na+/H+ exchange in volume regulation and cytoplasmic pH homeostasis in lymphocytes. *Fed. Proc.* 44(9): 2508-12.

Gueux, E., F. Bussière & Y. Rayaaiguier. 2001. Magnesium content of polymorphnuclear blood cells is not an indicator for magnesium status. In: Y. Rayssiguier, A. Mazur & J. Durlach, eds., *Advances in Magnesium Research in Nutrition and Health,* John Libbey & Company Ltd., Eastleigh, England. 259-261.

Gullestad, L., J. Hallen & O.M. Sejersted. 1995. K+ balance of the quadriceps muscle during dynamic exercise with and without beta-adrenoceptor blockade. *J. Appl. Physiol.* 78(2): 513-23.

Gunther, T. & J. Vormann. 1994. Intracellular Ca(2+)-Mg2+ interactions. *Ren. Physiol. Biochem.* 17(6): 279-86.

Hackam, D.J., S. Grinstein & O.D. Rotstein. 1996. Intracellular pH regulation in leukocytes: mechanisms and functional significance. *Shock* 5(1): 17-21.

Halestrap, A.P. & N.T. Price. 1999. The proton-linked monocarboxylate transporter (MCT) family: structure, function and regulation. *Biochem. J.* 343(Pt 2): 281-99.

Hall, R.A., R.T. Premont, C.W. Chow et al. 1998. The beta2-adrenergic receptor interacts with the Na+/H+-exchanger regulatory factor to control Na+/H+ exchange. *Nature* 392(6676): 626-30.

Halperin, M.L. & K.S. Kamel. 1998. Potassium. *Lancet* 352(9122): 135-40.

Harmer, A.R., M.J. McKenna, J.R. Sutton et al. 2000. Skeletal muscle metabolic and ionic adaptations during intense exercise following sprint training in humans. *J. Appl. Physiol.* 89(5): 1793-803.

Henrotte, J.G. 1988. Genetic regulation of blood and tissue magnesium content in mammals. *Magnesium* 7(5-6): 306-14.

Hespel, P., P. Lijnen, R. Fiocchi et al. 1986. Cationic concentrations and transmembrane fluxes in erythrocytes of humans during exercise. *J. Appl. Physiol.* 61(1): 37-43.

Hoffmann, G. & D. Böhmer. 1988. Magnesiumkonzentration in Vollblut, Serum und Erythrozyten bei trainierten Langläufern vor und nach 25-km-Lauf mit und ohne vierwöchiger oraler Elektrolyteinnahme. *Magnesium Bull.* 10: 56-63.

Hundal, H.S., A. Marette, Y. Mitsumoto et al. 1992. Insulin induces translocation of the alpha 2 and beta 1 subunits of the Na+/K(+)-ATPase from intracellular compartments to the plasma membrane in mammalian skeletal muscle. *J. Biol. Chem.* 267(8): 5040-43.

Hundal, H.S., A. Marette, T. Ramlal et al. 1993. Expression of beta subunit isoforms of the Na+,K(+)-ATPase is muscle type-specific. *FEBS Lett.* 328(3): 253-58.

Hurley, T.W., M.P. Ryan & R.W. Brinck. 1992. Changes of cytosolic Ca2+ interfere with measurements of cytosolic Mg2+ using mag-fura-2. *Am. J. Physiol.* 263(2 Pt 1): C300-07.

Izzard, A.S. & A.M. Heagerty. 1989. Resting intracellular pH in mesenteric resistance arteries from spontaneously hypertensive and Wistar-Kyoto rats: effects of amiloride and 4,4'-diisothiocyanatostilbene-2,2'-disulphonic acid. *J. Hypert. Suppl.* 7(6): S128-29.

Jew, K.N. & R.L. Moore. 2002. Exercise training alters an anoxia-induced, glibenclamide-sensitive current in rat ventricular cardiocytes. *J. Appl. Physiol.* 92(4): 1473-79.

Jew, K.N., M.C. Olsson, E.A. Mokelke et al. 2001. Endurance training alters outward K+ current characteristics in rat cardiocytes. *J. Appl. Physiol.* 90(4): 1327-33.

Joborn, H., G. Akerstrom & S. Ljunghall. 1985. Effects of exogenous catecholamines and exercise on plasma magnesium concentrations. *Clin. Chem.* (Oxford) 23(3): 219-26.

Juel, C. 1988. Intracellular pH recovery and lactate efflux in mouse soleus muscles stimulated in vitro: the involvement of sodium/proton exchange and a lactate carrier. *Acta Physiol. Scand.* 132: 363-71.

Juel, C. 1998a. Muscle pH regulation: role of training. *Acta Physiol. Scand.* 162(3): 359-66.

Juel, C. 1998b. Skeletal muscle Na+/H+ exchange in rats: pH dependency and the effect of training. *Acta Physiol. Scand.* 164(2): 135-40.

Juel, C. 2000. Expression of the Na(+)/H(+) exchanger isoform NHE1 in rat skeletal muscle and effect of training. *Acta Physiol. Scand.* 170(1): 59-63.

Juel, C. 2001. Current aspects of lactate exchange: lactate/H+ transport in human skeletal muscle. *Eur. J. Appl. Physiol.* 86(1): 12-16.

Juel, C. & A.P. Halestrap. 1999. Lactate transport in skeletal muscle—role and regulation of the monocarboxylate transporter. *J. Physiol.* 517 (Pt 3): 633-42.

Juel, C., Y. Hellsten, B. Saltin et al. 1999. Potassium fluxes in contracting human skeletal muscle and red blood cells. *Am. J. Physiol.* 276(1 Pt 2): R184-88.

Juel, C., L. Grunnet, M. Holse et al. 2001. Reversibility of exercise-induced translocation of Na+-K+ pump subunits to the plasma membrane in rat skeletal muscle. *Pfluegers Arch.* 443(2): 212-17.

Juel, C., J.J. Nielsen & J. Bangsbo. 2000. Exercise-induced translocation of Na(+)-K(+) pump subunits to the plasma membrane in human skeletal muscle. *Am. J. Physiol.* 278(4): R1107-10.

Juel, C. & H. Pilegaard. 1998. Lactate/H+ transport kinetics in rat skeletal muscle related to fibre type and changes in transport capacity. *Pfluegers Arch.* 436(4): 560-64.

Khan, F.A. & D.N. Baron. 1987. Ion flux and Na+,K+-ATPase activity of erythrocytes and leucocytes in thyroid disease. *Clin. Sci.* (London, England, 1979) 72(2): 171-79.

Kahn, A.M., E.J. Cragoe Jr., J.C. Allen et al. 1990. Na(+)-H+ and Na(+)-dependent Cl(-)-HCO3-exchange control pHi in vascular smooth muscle. *Am. J. Physiol.* 259: C134-43.

Kjeldsen, K., A. Norgaard, C.O. Gotzsche et al. 1984. Effect of thyroid function on number of Na-K pumps in human skeletal muscle. *Lancet* 2(8393): 8-10.

Laires, M.J., F. Madeira, J. Sergio et al. 1993. Preliminary study of the relationship between plasma and erythrocyte magnesium variations and some circulating pro-oxidant and antioxidant indices in a standardized physical effort. *Magnesium Res.* 6(3): 233-38.

Lauf, P.K. & N.C. Adragna. 2000. K-Cl cotransport: properties and molecular mechanism. *Cell. Physiol. Biochem.* 10(5-6): 341-54.

Lavoie, L., D. Roy, T. Ramlal et al. 1996. Insulin-induced translocation of Na(+)-K(+)-ATPase subunits to the plasma membrane is muscle fiber type specific. *Am. J. Physiol.* 270(5 Pt 1): C1421-29.

Lijnen, P., P. Hespel, R. Fagard et al. 1988. Erythrocyte, plasma and urinary magnesium in men before and after a marathon. *Eur. J. Appl. Physiol.* 58(3): 252-6.

Lindinger, M.I. 1995. Origins of [H+] changes in exercising skeletal muscle. *Can. J. Appl. Physiol.* 20(3): 357-68.

Lindinger, M.I., P.L. Horn & S.P. Grudzien. 1999. Exercise-induced stimulation of K(+) transport in human erythrocytes. *J. Appl. Physiol.* 87(6): 2157-67.

Liu, Y. & D.D. Gutterman. 2002. Oxidative stress and potassium channel function. *Clin. Exp. Pharmacol. Physiol.* 29(4): 305-11.

Maassen, N., M. Foerster & H. Mairbaurl. 1998. Red blood cells do not contribute to removal of K+ released from exhaustively working forearm muscle. *J. Appl. Physiol.* 85(1): 326-32.

McCullagh, K.J., C. Juel, M. O'Brien et al. 1996. Chronic muscle stimulation increases lactate transport in rat skeletal muscle. *Mol. Cell. Biochem.* 156(1): 51-57.

McCutcheon, L.J., R.J. Geor & H. Shen. 1999. Skeletal muscle Na(+)-K(+)-ATPase and K+ homeostasis during exercise: effects of short-term training. *Equ. Vet. J. Suppl.* 30: 303-10.

McKenna, M.J. 1995. Effects of training on potassium homeostasis during exercise. *J. Mol. Cell. Cardiol.* 27(4): 941-49.

McKenna, M.J., T.A. Schmidt, M. Hargreaves et al. 1993. Sprint training increases human skeletal muscle Na(+)-K(+)-ATPase concentration and improves K+ regulation. *J. Appl. Physiol.* 75(1): 173-80.

Medbo, J.I., E. Jebens, H. Vikne et al. 2001. Effect of strenuous strength training on the Na-K pump concentration in skeletal muscle of well-trained men. *Eur. J. Appl. Physiol.* 84(1-2): 148-54.

Montrose, M.H. & H. Murer. 1986. Regulation of intracellular pH in LLC-PK1 cells by Na+/H+ exchange. *J. Membr. Biol.* 93(1): 33-42.

Mooren, F.C., S. Turi, D. Gunzel et al. 2001. Calcium-magnesium interactions in pancreatic acinar cells. *FASEB J.* 15(3): 659-72.

Mudge, G.M. & I.M. Weiner. 1992. Agents affecting volume and composition of body fluids. In: A. Goodman, A. Gilman, T.W. Rall, A.S. Nies, P. Tayler, eds., *Pharmacological Basis of Therapeutics,* vol. 1, 8th ed. New York: McGraw-Hill.

Navas, F.J., J.F. Martin & A. Cordova. 1997. Compartmental shifts of calcium and magnesium as a result of swimming and swimming training in rats. *Med. Sci. Sports Exerc.* 29(7): 882-91.

Nevill, M.E., L.H. Boobis, S. Brooks et al. 1989. Effect of training on muscle metabolism during treadmill sprinting. *J. Appl. Physiol.* 67(6): 2376-82.

Nielsen, O.B. & T. Clausen. 1996. The significance of active Na+,K+ transport in the maintenance of contractility in rat skeletal muscle. *Acta Physiol. Scand.* 157(2): 199-209.

Nielsen, O.B. & A.P. Harrison. 1998. The regulation of the Na+,K+ pump in contracting skeletal muscle. *Acta Physiol. Scand.* 162(3): 191-200.

Nishimuta, M., N. Kodama, H. Takeyama et al. 1997. Magnesium metabolism and exercise in humans. In: Theophanides, T. & Anastassopoulou, J., eds., *Magnesium: Current Status and New Developments*, 109-13. Kluwer Academic Publishers, Dordrecht, Netherlands.

Okada, K., S. Ishikawa & T. Saito. 1992. Cellular mechanisms of vasopressin and endothelin to mobilize [Mg2+]i in vascular smooth muscle cells. *Am. J. Physiol.* 263(4 Pt 1): C873-78.

Panov, A. & A. Scarpa. 1996. Mg2+ control of respiration in isolated rat liver mitochondria. *Biochemistry* 35(39): 12849-56.

Paradiso, A.M., M.C. Townsley, E. Wenzl et al. 1989. Regulation of intracellular pH in resting and in stimulated parietal cells. *Am. J. Physiol.* 257(3 Pt 1): C554-61.

Park, C.O., X.H. Xiao & D.G. Allen. 1999. Changes in intracellular Na+ and pH in rat heart during ischemia: role of Na+/H+ exchanger. *Am. J. Physiol.* 276(5 Pt 2): H1581-90.

Pilegaard, H. & S. Asp. 1998. Effect of prior eccentric contractions on lactate/H+ transport in rat skeletal muscle. *Am. J. Physiol.* 274(3 Pt 1): E554-59.

Pilegaard, H., K. Domino, T. Noland et al. 1999a. Effect of high-intensity exercise training on lactate/H+ transport capacity in human skeletal muscle. *Am. J. Physiol.* 276(2 Pt 1): E255-61.

Pilegaard, H., G. Terzis, A. Halestrap et al. 1999b. Distribution of the lactate/H+ transporter isoforms MCT1 and MCT4 in human skeletal muscle. *Am. J. Physiol.* 276: E843-48.

Poole, R.C., C.E. Sansom & A.P. Halestrap. 1996. Studies of the membrane topology of the rat erythrocyte H+/lactate cotransporter (MCT1). *Biochem. J.* 320: 817-24.

Porta, S., G. Leitner, D. Heidinger et al. 1997. Magnesium während der Alpinausbildung bringt um 30% bessere Energieverwertung. *Magnesium Bull.* 19: 59-61.

Putney, L.K., S.P. Denker & D.L. Barber. 2002. The changing face of the Na+/H+ exchanger, NHE1: structure, regulation, and cellular actions. *Annu. Rev. Pharmacol. Toxicol.* 42: 527-52.

Quamme, G.A. & S.W. Rabkin. 1990. Cytosolic free magnesium in cardiac myocytes: identification of a Mg2+ influx pathway. *Biochem. Biophys. Res. Com.* 167(3): 1406-12.

Rayssiguier, Y., C.Y. Guezennec & J. Durlach. 1990. New experimental and clinical data on the relationship between magnesium and sport. *Magnesium Res.* 3(2): 93-102.

Renaud, J.M. 2002. Modulation of force development by Na+, K+, Na+ K+ pump and KATP channel during muscular activity. *J. Appl. Physiol.* 27(3): 296-315.

Reusch, H.P., R. Reusch, D. Rosskopf, et al. 1993. Na+/H+ exchange in human lymphocytes and platelets in chronic and subacute metabolic acidosis. *J. Clin. Invest.* 92(2):858-65.

Romani, A., C. Marfella & A. Scarpa. 1993. Hormonal stimulation of Mg2+ uptake in hepatocytes. Regulation by plasma membrane and intracellular organelles. *J. Biol. Chem.* 25: 268(21): 15489-95.

Romani, A. & A. Scarpa. 1990. Norepinephrine evokes a marked Mg2+ efflux from liver cells. *FEBS Lett.* 20;269(1): 37-40.

Romero, M.F. 2001. The electrogenic Na+/HCO3-cotransporter, NBC. *J. Pancreas* 2(4 Suppl): 182-91.

Romero, M.F. & W.F. Boron. 1999. Electrogenic Na+/HCO3-cotransporters: cloning and physiology. *Annu. Rev. Physiol.* 61: 699-723.

Roos, A. & W.F. Boron. 1981. Intracellular pH. *Physiol. Rev.* 61(2): 296-434.

Ruff, R.L. 1996. Sodium channel slow inactivation and the distribution of sodium channels on skeletal muscle fibres enable the performance properties of different skeletal muscle fibre types. *Acta Physiol. Scand.* 156(3): 159-68.

Russell, J.M. 2000. Sodium-potassium-chloride cotransport. *Physiol. Rev.* 80(1): 211-76.

Ryan, M.P. 1993. Interrelationships of magnesium and potassium homeostasis. *Min. Electrol. Metab.* 19(4-5): 290-95.

Sahlin, K. & J. Henriksson. 1984. Buffer capacity and lactate accumulation in skeletal muscle of trained and untrained men. *Acta Physiol. Scand.* 122(3): 331-39.

Saur, P., M. Joneleit, H. Tölke et al. 2002. Evaluation des Magnesiumstatus bei Ausdauersportlern. *Dtsch. Zeitschr. Sportmed.* 53(3): 72-78.

Sejersted, O.M. & G. Sjogaard. 2000. Dynamics and consequences of potassium shifts in skeletal muscle and heart during exercise. *Physiol. Rev.* 80(4): 1411-81.

Sharp, R.L., D.L. Costill, W.J. Fink et al. 1986. Effects of eight weeks of bicycle ergometer sprint training on human muscle buffer capacity. *Int. J. Sports Med.* 7(1): 13-17.

Sjogaard, G. 1983. Electrolytes in slow and fast muscle fibers of humans at rest and with dynamic exercise. *Am. J. Physiol.* 245(1): R25-31.

Songu-Mize, E., N. Sevieux, X. Liu et al. 2001. Effect of short-term cyclic stretch on sodium pump activity in aortic smooth muscle cells. *Am. J. Physiol.* 281(5): H2072-78.

Speich, M., B. Bousquet & G. Nicolas. 1981. Reference values for ionised, complexed, and protein-bound plasma magnesium in men and women. *Clin. Chem.* 27(2): 246-49.

Sreter, F.A. 1963. Cell water, sodium and potassium in stimulated red and white mammalian muscle. *Am. J. Physiol.* 205: 1295-98.

Standen, N.B., A.I. Pettit, N.W. Davies et al. 1992. Activation of ATP-dependent K+ currents in intact skeletal muscle fibres by reduced intracellular pH. *Proc. Royal Soc. Biol. Sci.* 247(1320): 195-98.

Stendig-Lindberg, G., Y. Shapiro, Y. Epstein et al. 1987. Changes in serum magnesium concentration after strenuous exercise. *J. Am. Coll. Nutr.* 6(1): 35-40.

Thompson, C.B. & A.A. McDonough. 1996. Skeletal muscle Na,K-ATPase alpha and beta subunit protein levels respond to hypokalemic challenge with isoform and muscle type specificity. *J. Biol. Chem.* 271(51): 32653-58.

Torigoe, T., H. Izumi, T. Ise et al. 2002. Vacuolar H(+)-ATPase: functional mechanisms and potential as a target for cancer chemotherapy. *Anticancer Drugs* 13(3): 237-43.

Touyz, R.M. & E.L. Schiffrin. 1996. Angiotensin II and vasopressin modulate intracellular free magnesium in vascular smooth muscle cells through Na+-dependent protein kinase C pathways. *J. Biol. Chem.* 4;271(40): 24353-58.

Tsakiridis, T., P.P. Wong, Z. Liu et al. 1996. Exercise increases the plasma membrane content of the Na+-K+ pump and its mRNA in rat skeletal muscles. *J. Appl. Physiol.* 80(2): 699-705.

Vormann, J. & T. Gunther. 1993. Magnesium transport mechanisms. In: N.J. Birch, ed., *Magnesium and the Cell.* London: Academic Press, 137-55.

Whyte, K.F., G.J. Addis, R. Whitesmith et al. 1987. Adrenergic control of plasma magnesium in man. *Clin. Sci.* (London, England) 72(1): 135-38.

Williams, R.P.J. 1993. Magnesium—an introduction to its biochemistry. In: N.J. Birch, ed., *Magnesium and the Cell.* London: Academic Press, 15-30.

Wisdom, D.M., G.M. Salido, L.M. Baldwin et al. 1996. The role of magnesium in regulating CCK-8-evoked secretory responses in the exocrine rat pancreas. *Mol. Cell. Biochem.*154(2): 123-32.

Chapter 7

Alessio, H.M. 1993. Exercise-induced oxidative stress. *Med. Sci. Sports Exerc.* 25(2):218-24.

Altin, J.G. & E.K. Sloan. 1997. The role of CD45 and CD45-associated molecules in T cell activation. *Immunol. Cell Biol.* 75(5):430-45.

Anderson, M.T., F.J. Staal, C. Gitler et al. 1994. Separation of oxidant-initiated and redox-regulated steps in the NF-kappa B signal transduction pathway. *Proc. Natl. Acad. Sci.* 91(24):11527-31.

Ando, J. & A. Kamiya. 1993. Blood flow and vascular endothelial cell function. *Front. Med. Biol. Eng.* 4: 245-64.

Angelopoulos, T.J. 2001. β-endorphin immunoreactivity during high-intensity exercise with and without opiate blockade. *Eur. J. Appl. Physiol.* 86:92-96.

Appenzeller, O. & S.C. Wood. 1992. Peptides and exercise at high and low altitudes. *Int. J. Sports Med.* S135-40.

Aronson, D., M.A. Violan, S.D. Dufresne et al. 1997. Exercise stimulates the mitogen-activated protein kinase pathway in human skeletal muscle. *J. Clin. Invest.* 15;99(6):1251-57.

Ashwell, J.D. & R.D. Klusner. 1990. Genetic and mutational analysis of the T-cell antigen receptor. *Annu. Rev. Immunol.* 8:139-67.

Azenabor, A.A. & L. Hoffman-Goetz. 2000. Effect of exhaustive exercise on membrane estradiol concentration, intracellular calcium, and oxidative damage in mouse thymic lymphocytes. *Free Radic. Biol. Med.* 28(1):84-90.

Azuma, M. & L.L. Lanier. 1995. The role of CD28 costimulation in the generation of cytotoxic T lymphocytes. *Curr. Top. Microbiol. Immunol.* 198:59-74.

Baldari, C.T., A. Heguy, M.M. Di Somma et al. 1994. Inhibition of T-cell antigen receptor signaling by overexpression of p120GAP. *Cell Growth Diff.* 5(1):95-98.

Baniyash, M., P. Garcia-Morales, E. Luong et al. 1988. The T cell antigen receptor zeta chain is tyrosine phosphorylated upon activation. *J. Biol. Chem.* 263(34):18225-33.

Barford, D. 1996. Molecular mechanisms of the protein serine/threonine phosphatases. *Trends Biochem. Sci.* 21(11):407-12.

Barletta, H., L. Stefani, R. Del Bene et al. 1998. Effects of exercise on natriuretic peptides and cardiac function in man. *Int. J. Cardiol.* 65:217-25.

Barr, S.M., K.R. Lees, D.D. McBryan et al. 1988. Exercise and platelet intracellular free calcium concentration. *Clin. Sci.* 75(2):221-24.

Barton, B.E. 1997. IL-6: insights into novel biological activities. *Clin. Immunol. Immunpath.* 85(1):16-20.

Beato, M., S. Chavez & M. Truss. 1996. Transcriptional regulation by steroid hormones. *Steroids* 61(4):240-51.

Beebe, S.J. 1994. The cAMP-dependent protein kinases and cAMP signal transduction. *Sem. Canc. Biol.* 5(4): 285-94.

Belcastro, A.N. 1993. Skeletal muscle calcium-activated neutral protease (calpain) with exercise. *J. Appl. Physiol.* 74:1381-86.

Berridge, M.J. 1997. Lymphocyte activation in health and disease. *Crit. Rev. Immunol.* 17(2):155-78.

Berridge, M.J. & R.F. Irvine. 1989. Inositol phosphates and cell signalling. *Nature* 341:197-205.

Bertagna, X. 1994. Proopiomelanocortin-derived peptides. *Endocrinol. Metab. Clin. North Am.* 23:467-85.

Besterman, J.M., V. Duronio & P. Cuatrecasas. 1986. Rapid formation of diacylglycerol from phosphatidylcholine: a pathway for generation of a second messenger. *Proc. Natl. Acad. Sci.* 83(18):6785-89.

Bohm, M., H. Dorner, P. Htun et al. 1993. Effects of exercise on myocardial adenylate cyclase and Gi alpha expression in senescence. *Am. J. Physiol.* 264(3 Pt 2):H805-14.

Bonin, A. & N.A. Khan. 2000. Regulation of calcium signalling by docosahexaenoic acid in human T-cells: implication of CRAC channels. *J. Lipid Res.* 41(2): 277-84.

Boone, J.B., T. Sherraden, K. Pierzchala et al. 1992. Plasma met-enkephalin and catecholamine responses to intense exercise in humans. *J. Appl. Physiol.* 73: 388-92.

Boppart, M.D., S. Asp, J.F.P. Wojtazewski et al. 2000. Marathon running transiently increases c-Jun-NH_2–terminal kinase and p38 activities in human skeletal muscle. *J. Physiol.* 526, 3:663-69.

Bouissou, P., F.X. Galen, J.P. Richalet et al. 1989. Effects of propranolol and pindolol on plasma ANP levels in humans at rest and during exercise. *Am. J. Physiol.* 2(Pt 2):R259-64.

Bourguet, W., P. Germain & H. Gronemeyer. 2000. Nuclear receptor ligand-binding domains: three-dimensional structures, molecular interactions and pharmacological. *Trends Pharmacol. Sci.* 21(10):381-88.

Bowles, D.K., Q. Hu, M.H. Laughlin et al. 1998. Exercise training increases L-type calcium current density in coronary smooth muscle. *Am. J. Physiol.* 275(6 Pt 2):H2159-69.

Breittmayer, J.P., C. Pelassy, J.L. Cousin et al. 1993. The inhibition by fatty acids of receptor-mediated calcium movements in Jurkat T-cells is due to increased calcium extrusion. *J. Biol. Chem.* 268(28):20812-17.

Brenner, B.M., B.J. Ballermann, M.E. Gunning et al. 1990. Diverse biological actions of atrial natriuretic peptide. *Physiol. Rev.* 3:665-99.

Brodde, O.E., A. Daul, A. Wellstein et al. 1988. Differentiation of beta 1- and beta 2-adrenoceptor-mediated effects in humans. *Am. J. Physiol.* 254(2 Pt 2): H199-206.

Brodde, O.E., A. Daul & N. O'Hara. 1984. Beta-adrenoreceptor changes in human lymphocytes, induced by dynamic exercise. *Naunyn Schmiedebergs Arch. Pharmacol.* 325(2):190-92.

Bruunsgaard, H., H. Galbo & J. Halkjaer-Kristensen. 1997. Exercise-induced increase in interleukin-6 is related to muscle damage. *J. Physiol.* 499(3):833-41.

Bunemann, M., K.B. Lee, R. Pals-Rylaarsdam et al. 1999. Desensitization of G-protein-coupled receptors in the cardiovascular system. *Annu. Rev. Physiol.* 61: 169-92.

Bush, J.A., W.J. Kraemer, A.M. Mastro et al. 1999. Exercise and recovery responses of adrenal medullary neurohormones to heavy resistance exercise. *Med. Sci. Sports Exerc.* 31:554-59.

Butler, J., J.G. Kelly, K. O'Malley et al. 1983. Beta-adrenoceptor adaptation to acute exercise. *J. Physiol.* 344:113-17.

Caimi, G., P. Assennato, B. Canino et al. 1997. Exercise test: trend of the leukocyte flow properties, polymorphonuclear membrane fluidity and cytosolic Ca^{2+} content in normals, in subjects with previous acute myocardial infarction and in subjects with aortocoronary by-pass. *Clin. Hemorheol. Microcirc.* 17:127-35.

Cain, B.S., D.R. Meldrum, K.S. Joo et al. 1998. SERCA2a levels correlate inversely with age in senescent human myocardium. *J. Am. Coll. Cardiol.* 32(2): 458-67.

Calder, P.C. 1998. Dietary fatty acids and lymphocyte functions. *Proc. Nutr. Soc.* 57(4):487-502.

Cambier, J.C., M.K. Newell, L.B. Justement et al. 1978. Ia binding ligands and cAMP stimulate nuclear translocation of PKC in B lymphocytes. *Nature* 327(6123): 629-32.

Cancela, J.M., H. Mogami, A.V. Tepikin et al. 1998. Intracellular glucose switches between cyclic ADP-ribose and inositol trisphosphate triggering of cytosolic Ca2+ spiking. *Curr. Biol.* 8(15):865-68.

Carafoli, E. & M. Brini. 2000. Calcium pumps: structural basis for and mechanism of calcium transmembrane transport. *Curr. Op. Chem. Biol.* 4(2):152-61.

Castell, L.M., J.R. Poortmans, R. Leclercq et al. 1997. Some aspects of the acute phase response after a marathon race, and the effects of glutamine supplementation. *Eur. J. Appl. Physiol.* 75(1):47-53.

Castro, M.G. & E. Morrison. 1997. Post-translational processing of proopiomelanocortin in the pituitary and in the brain. *Crit. Rev. Neurobiol.* 11:35-57.

Chakrabarti, R., N.E. Hubbard, D. Lim et al. 1997. Alteration of platelet-activating factor-induced signal transduction in macrophages by n-3 fatty acids. *Cell Immunol.* 175(1):76-84.

Chu, T.F., T.Y. Huang, C.J. Jen et al. 2000. Effects of chronic exercise on calcium signaling in rat vascular endothelium. *Am. J. Physiol.* 279(4):H1441-46.

Chu, Y., E.Y. Lee & K.K. Schlender. 1996. Activation of protein phosphatase 1. Formation of a metalloenzyme. *J. Biol. Chem.* 271(5):2574-77.

Clementi, E., G. Martino, L.M. Grimaldi et al. 1994. Intracellular Ca2+ stores of T lymphocytes: changes induced by in vitro and in vivo activation. *Eur. J. Immunol.* 24(6):1365-71.

Collins, S., M. Bouvier, M.A. Bolanowski et al. 1989. cAMP stimulates transcription of the beta 2-adrenergic receptor gene in response to short-term agonist exposure. *Proc. Natl. Acad. Sci.* 86(13):4853-57.

Comb, M., P.H. Seeburg, J. Adelman et al. 1982. Primary structure of the human met- and leu-enkephalin precursor and its mRNA. *Nature* 295:663-66.

Cosenzi, A., A. Sacerdote, E. Bocin et al. 1996. Neither physical exercise nor alpha 1- and beta-adrenergic blockade affect plasma endothelin concentrations. *Am. J. Hypert.* 8:819-22.

Coyle, E.F. 2000. Physical activity as a metabolic stressor. *Am. J. Clin. Nutr.* 72 (2 Suppl):S512-20.

Croisier, J.L., G. Camus, I. Venneman et al. 1999. Effects of training on exercise-induced muscle damage and interleukin 6 production. *Muscle Nerve* 22(2): 208-12.

Cross, T.G., D. Scheel-Toellner, N.V. Henriquez et al. 2000. Serine/Threonine protein kinases and apoptosis. *Exp. Cell Res.* 256:34-41.

Cumming, D.C. & G.D. Wheeler. 1987. Opioids in exercise physiology. *Semin. Reproduc. Endocrinol.* 5:171-79.

Davies, A.O. 1988. Exercise-induced fall in coupling of human beta 2-adrenergic receptors. *Metabolism* 37(10):916-18.

Davies, E.V., A.K. Campbell, B.D. Williams et al. 1991. Single cell imaging reveals abnormal intracellular calcium signals within rheumatoid synovial neutrophils. *Brit. J. Rheumatol.* 30(6):443-48.

Davies, E.V., B.D. Williams, R.J. Whiston et al. 1994. Altered Ca2+ signalling in human neutrophils from inflammatory sites. *Ann. Rheumatic Dis.* 53(7):446-49.

de Bold, A.J. 1985. Atrial natriuretic factor: a hormone produced by the heart. *Science* 4727:767-70.

De Diego Acosta, A.M., J.C. Garcia, V.J. Fernandez-Pastor et al. 2001. Influence of fitness on the integrated neuroendocrine response to aerobic exercise until exhaustion. *J. Physiol. Biochem.* 57:313-20.

Delitala, G., M. Palermo, P. Tomasi et al. 1991. Adrenergic stimulation of the human pituitary-adrenal axis is attenuated by an analog of met-enkephalin. *Neuroendocrinology* 53:41-46.

de Nucci, G., R. Thomas, P. D'Orleans-Juste et al. 1988. Pressor effects of circulating endothelin are limited by its removal in the pulmonary circulation and by the release of prostacyclin and endothelium-derived relaxing factor. *Proc. Natl. Acad. Sci.* 24:9797-800.

DeRijk, R.H., A. Boelen, F.J. Tilders et al. 1994. Induction of plasma interleukin-6 by circulating adrenaline in the rat. *Psychoneuroendocrinology* 19(2):155-63.

DeSommers, K., J.M. Loots, S.F. Simpson et al. 1990. Circulating met-enkephalin in trained athletes during rest, exhaustive treadmill exercise and marathon running. *Eur. J. Clin. Pharmacol.* 38:391-92.

DeSommers, K., S.F. Simpson, J.M. Loots et al. 1989. Effect of exercise on met-enkephalin in unfit and superfit individuals. *Eur. J. Clin. Pharmacol.* 37:399-400.

Di, Y.P., E.A. Repasky & J.R. Subjeck. 1997. Distribution of HSP70, protein kinase C, and spectrin is altered in lymphocytes during a fever-like hyperthermia exposure. *J. Cell Physiol.* 172(1):44-54.

Dinarello, C.A. 1992. Role of interleukin-1 in infectious diseases. *Immunol. Rev.* 127.

Dixon, J.E. 1996. Protein tyrosine phosphatases: their roles in signal transduction. *Rec. Progr. Horm. Res.* 51:405-14.

Dolmetsch, R.E., R.S. Lewis, C.C. Goodnow et al. 1997. Differential activation of transcription factors induced by Ca2+ response amplitude and duration. *Nature* 386(6627):855-58.

Drenth, J.P., S.H. van Uum, M. van Deuren et al. 1995. Endurance run increases circulating IL-6 and IL-1ra but downregulates ex vivo TNF-alpha and IL-1beta production. *J. Appl. Physiol.* 79:1497-1503.

Dröge, W., V. Hack, R. Breitkreutz et al. 1998. Role of cysteine and glutathione in signal transduction, immunopathology and cachexia. *Biofactors* 8(1-2): 97-102.

Duan, C., M.D. Delp, D.A. Hayes et al. 1990. Rat skeletal muscle mitochondrial [Ca2+] and injury from downhill walking. *J. Appl. Physiol.* 68:1241-51.

Duclos, M., J.B. Corcuff, M. Rashedi et al. 1997. Trained versus untrained men: different immediate post-exercise responses of pituitary adrenal axis. *Eur. J. Appl. Physiol.* 75:343-50.

Dumont, F.J. 2000. FK506, an immunosuppressant targeting calcineurin function. *Curr. Med. Chem.* 7(7): 731-48.

Dunbar, C.C. & M.I. Kalinski. 1994. Cardiac intracellular regulation: exercise effects on the cAMP system and A-kinase. *Med. Sci. Sports Exerc.* 26(12):1459-65.

Duncan, D.D. & D.A. Lawrence. 1989. Differential lymphocyte growth-modifying effects of oxidants: changes in cytosolic Ca2+. *Toxicol. Appl. Pharmacol.* 100(3): 485-97.

Eggert, M., C.C. Mows, D. Tripier et al. 1995. A fraction enriched in a novel glucocorticoid receptor-interacting protein stimulates receptor-dependent transcription in vitro. *J. Biol. Chem.* 270(51):30755-59.

Egloff, M.P., P.T. Cohen, P. Reinemer et al. 1995. Crystal structure of the catalytic subunit of human protein phosphatase 1 and its complex with tungstate. *J. Mol. Biol.* 254(5):942-59.

Elferink, J.G. & M. Deierkauf. 1985. Involvement of intracellular Ca2+ in chemotaxis and metabolic burst by neutrophils: the use of antagonists of intracellular Ca2+. *Res. Comm. Chem. Pathol. Pharmacol.* 50(1): 67-81.

Engfred, K., M. Kjaer, N.H. Secher et al. 1994. Hypoxia and training-induced adaptation of hormonal responses to exercise in humans. *Eur. J. Appl. Physiol.* 68:303-09.

Evans, A.A.L., S. Khan & M.E. Smith. 1997. Evidence for a hormonal action of β-endorphin to increase glucose uptake in resting and contracting skeletal muscle. *J. Endocrinol.* 155:387-92.

Evans, C.J., E. Erdelyi, E. Weber et al. 1983. Identification of pro-opiomelanocortin-derived peptides in the human adrenal medulla. *Science* 221:957-60.

Evans, V.R., A.B. Manning, L.H. Bernard et al. 1994. α-melanocyte-stimulating hormone and N-acetyl-β-endorphin immunoreactivities are localized in the human pituitary but are not restricted to the zona intermedia. *Endocrinology* 134:97-106.

Fagan, K.A., R. Mahey & D.M. Cooper. 1996. Functional co-localization of transfected Ca(2+)-stimulable adenylyl cyclases with Ca2+ entry. *J. Biol. Chem.* 271(21):12438-44.

Farrell, P.A., M. Kjaer, F.W. Bach et al. 1987. β-endorphin and adrenocorticotropin response to supramaximal treadmill exercise in trained and untrained males. *Acta Physiol. Scand.* 130:619-25.

Fauman, E.B. & M.A. Saper. 1996. Structure and function of the protein tyrosine phosphatases. *Trends Biochem. Sci.* 21(11):413-17.

Febbraio, M.A. & B.K. Pedersen. 2002. Muscle-derived interleukin-6: mechanisms for activation and possible biological roles. *FASEB J.* 16(11):1335-47.

Fell, R.D., F.H. Lizzo, P. Cervoni et al. 1985. Effect of contractile activity on rat skeletal muscle beta-adrenoceptor properties. *Proc. Soc. Exp. Biol. Med.* 180(3):527-32.

Fiers, W. 1991. Tumor necrosis factor. Characterization at the molecular, cellular and in vivo level. *FEBS Lett.* 285(2):199-212.

Flescher, E., H. Tripoli, K. Salnikow et al. 1998. Oxidative stress suppresses transcription factor activities in stimulated lymphocytes. *Clin. Exp. Immunol.* 112(2): 242-47.

Francis, S.H. & J.D. Corbin. 1994. Structure and function of cyclic nucleotide-dependent protein kinases. *Annu. Rev. Physiol.* 56:237-72.

Fricchione, G.L. & G.B. Stefano. 1994. The stress response and autoimmunoregulation. *Adv. Neuroimmunol.* 4: 13-27.

Fujii, N., H. Miyazaki, S. Homma et al. 1993. Dynamic exercise induces translocation of beta-adrenergic receptors in human lymphocytes. *Acta Physiol. Scand.* 148(4):463-64.

Fujii, N., T. Shibata, S. Homma et al. 1997. Exercise-induced changes in beta-adrenergic-receptor mRNA level measured by competitive RT-PCR. *J. Appl. Physiol.* 82(6):1926-31.

Fujii, N., T. Shibata, F. Yamazaki et al. 1996. Exercise-induced change in beta-adrenergic receptor number in lymphocytes from trained and untrained men. *Jap. J. Physiol.* 46(5):389-95.

Gadient, R.A. & P.H. Patterson. 1999. Leukemia inhibitory factor, interleukin 6, and other cytokines using the GP130 transducing receptor: roles in inflammation and injury. *Stem cells* 17(3):127-37.

Gamberucci, A., R. Fulceri, F.L. Bygrave et al. 1997. Unsaturated fatty acids mobilize intracellular calcium independent of IP_3 generation and VIA insertion at the plasma membrane. *Biochem. Biophys. Res. Com.* 241:312-16.

Ganong, W.F., M.F. Dallman & J.L. Roberts, eds. 1987. *Ann. N.Y. Acad. Sci. 512: The Hypothalamic-Pituitary-Adrenal Axis Revisited.* New York: New York Academy of Sciences.

Garbers, D.L. 1992. Guanylyl cyclase receptors and their endocrine, paracrine and autocrine ligands. *Cell* 71:1-4.

Genot, E.M., P.J. Parker & D.A. Cantrell. 1995. Analysis of the role of protein kinase C-alpha, -epsilon, and -zeta in T cell activation. *J. Biol. Chem.* 270(17):9833-39.

Geny, B., R. Richard, B. Mettauer et al. 2001. Cardiac natriuretic peptides during exercise and training after heart transplantation. *Cardiovasc. Res.* 3:521-28.

Ginn-Pease, M.E. & R.L. Whisler. 1998. Redox signals and NF-kappaB activation in T cells. *Free Radic. Biol. Med.* 25(3):346-61.

Gissel, H. & T. Clausen. 2000. Excitation-induced Ca(2+) influx in rat soleus and EDL muscle: mechanisms and effects on cellular integrity. *Am. J. Physiol.* 279: R917-24.

Giuliani, A. & B. Cestaro. 1997. Exercise, free radical generation and vitamins. *Eur. J. Cancer Prev.* 6(Suppl 1):S55-56.

Goalstone, M.L. & B. Draznin. 1999. Effect of insulin on farnesyltransferase gene transcription and mRNA stability. *Biochem. Biophys. Res. Com.* 8;254(1):243-47.

Goalstone, M.L., K. Wall, J.W. Leitner et al. 1999. Increased amounts of farnesylated p21Ras in tissues of hyperinsulinaemic animals. *Diabetologia* 42(3):310-16.

Goldfarb, A.H. & A.Z. Jamurtas. 1997. β-endorphin response to exercise. *Sports Med.* 24:8-16.

Goldfarb, A.H., A.Z. Jamurtas, G.H. Kamimori et al. 1998. Gender effect on β-endorphin response to exercise. *Med. Sci. Sports Exerc.* 30:1672-76.

Goldfarb, A.H., B.D. Hatfield, J. Potts et al. 1991. β-endorphin time course response to intensity of exercise: effect of training status. *Int. J. Sports Med.* 12:264-68.

Goldstone, S.D. & N.H. Hunt. 1997. Redox regulation of the mitogen-activated protein kinase pathway during lymphocyte activation. *Biochim. Biophys. Acta* 1355(3):353-60.

Goodman, O.B. Jr., J.G. Krupnick, F. Santini et al. 1996. Beta-arrestin acts as a clathrin adaptor in endocytosis of the beta2-adrenergic receptor. *Nature* 383(6599):447-50.

Graafsma, S.J., L.J. van Tits, P. van Heijst et al. 1989. Effects of isometric exercise on blood cell adrenoceptors in essential hypertension. *J. Cardiovasc. Pharmacol.* 14(4):598-602.

Graafsma, S.J., L.J. van Tits, P.H. Willems et al. 1990. Beta 2-adrenoceptor up-regulation in relation to cAMP production in human lymphocytes after physical exercise. *Brit. J. Clin. Pharmacol.* 30(Suppl 1):142S-144S.

Grady, E.F., S.K. Bohm & N.W. Bunnett. 1997. Turning off the signal: mechanisms that attenuate signaling by G protein-coupled receptors. *Am. J. Physiol.* 273(3 Pt 1):G586-601.

Guse, A.H. 1998. Ca2+ signaling in T-lymphocytes. *Crit. Rev. Immunol.* 18(5):419-48.

Guse, A.H. 2000. Cyclic ADP-ribose. *J. Mol. Med.* 78(1): 26-35.

Guse, A.H., I. Berg, C.P. da Silva et al. 1997. Ca2+ entry induced by cyclic ADP-ribose in intact T-lymphocytes. *J. Biol. Chem.* 272(13):8546-50.

Haahr, P.M., B.K. Pedersen, A. Fomsgaard et al. 1991. Effect of physical exercise on in vitro production of interleukin 1, interleukin 6, tumor necrosis factor-alpha, interleukin 2 and interferon-gamma. *Int. J. Sports Med.* 12(2):223-27.

Hack, V., C. Weiss, B. Friedmann et al. 1997. Decreased plasma glutamine level and CD4+ T cell number in response to 8 wk of anaerobic training. *Am. J. Physiol.* 272(5.1):E788-95.

Hall, R.A. & R.J. Lefkowitz. 2002. Regulation of G protein-coupled receptor signaling by scaffold proteins. *Circ. Res.* 91(8):672-80.

Haller, H., K. Jendroska, T. Lenz et al. 1996. Effect of strenuous exercise on agonist-induced platelet cytosolic calcium in man. *J. Hum. Hypert.* 10(2):99-104.

Hanoune, J. & N. Defer. 2001. Regulation and role of adenylyl cyclase isoforms. *Annu. Rev. Pharmacol. Toxicol.* 41:145-74.

Harbach, H., K. Hell, C. Gramsch et al. 2000. β-endorphin (1-31) in the plasma of male volunteers undergoing physical exercise. *Psychoneuroendocrinology* 25:551-62.

Hawkins, C.L. & M.J. Davies. 2001. Generation and propagation of radical reactions on proteins. *Biochim. Biophys. Acta* 1504(2-3):196-219.

Haynes, W.G., C.J. Ferro, K.P. O'Kane et al. 1996. Systemic endothelin receptor blockade decreases peripheral vascular resistance and blood pressure in humans. *Circulation* 10:1860-70.

Hegedus, Z., V. Chitu, G.K. Toth et al. 1999. Contribution of kinases and the CD45 phosphatase to the generation of tyrosine phosphorylation patterns in the T-cell receptor complex zeta chain. *Immunol. Lett.* 67(1):31-39.

Heitkamp, H.C., W. Huber & K. Scheib. 1996. β-endorphin and adrenocorticotrophin after incremental exercise and marathon running—female responses. *Eur. J. Appl. Physiol.* 72:417-24.

Heitkamp, H.C., K. Schmid & K. Scheib. 1993. β-endorphin and adrenocorticotropic hormone production during marathon and incremental exercise. *Eur. J. Appl. Physiol.* 66:269-74.

Heitkamp, H.C., H. Schulz, K. Röcker et al. 1998. Endurance training in females: changes in β-endorphin and ACTH. *Int. J. Sports Med.* 19:260-64.

Hellsten, Y., U. Frandsen, N. Orthenblad, N. Sjodin & E.A. Richter. 1997. Xanthine oxidase in human skeletal muscle following eccentric exercise: a role of inflammation. *J. Physiol.* 498:239-48.

Herman, J.P. & W.E. Cullinan. 1997. Neurocircuitry of stress: central control of the hypothalamo-pituitary-adrenocortical axis. *Trends Neurosci.* 20:78-84.

Hicke, L. 1997. Ubiquitin-dependent internalization and down-regulation of plasma membrane proteins. *FASEB J.* 11(14):1215-26.

Hicke, L. 1999. Gettin' down with ubiquitin: turning off cell-surface receptors, transporters and channels. *Trends Cell Biol.* 9(3):107-12.

Hinson, J.P., L.A. Cameron, A. Purbrick et al. 1994a. The role of neuropeptides in the regulation of adrenal zona glomerulosa function: effects of substance P, neuropeptide Y, neurotensin, Met-enkephalin, Leu-enkephalin and corticotropin releasing hormone on aldosterone secretion in the intact perfused rat adrenal. *J. Endocrinol.* 140:91-96.

Hinson, J.P., A. Purbrick, L.A. Cameron et al. 1994b. The role of neuropeptides in the regulation of adrenal zona fasciculata/reticularis function: effects of vasoactive intestinal polypeptide, substance P, neuropeptide Y, Met-enkephalin, Leu-enkephalin and neurotensin on corticosterone secretion in the intact perfused rat adrenal gland in situ. *Neuropeptides* 26:391-97.

Hirose, K., S. Kadowaki, M. Tanabe et al. 1999. Spatiotemporal dynamics of inositol 1,4,5-trisphosphate that underlies complex Ca2+ mobilization patterns. *Science* 28;284(5419):1527-30.

Hodgetts, V., S.W. Coppack, K.N. Frayn et al. 1991. Factors controlling fat mobilization from human subcutaneous adipose tissue during exercise. *J. Appl. Physiol.* 71(2):445-51.

Hoffmann, P., I.H. Jonsdottir & P. Thoren. 1996. Activation of different opioid systems by muscle activity and exercise. *News Physiol. Sci.* 11:223-28.

Höllt, V. 1993. Regulation of opioid peptide gene expression. In *Handbook of Experimental Pharmacology 104/I,* eds. A. Herz, H. Akil & E.J. Simon, 307-46. Berlin: Springer.

Horowitz, J.F. & S. Klein. 2000. Lipid metabolism during endurance exercise. *Am. J. Clin. Nutr.* 72(2 Suppl): S558-63.

Howlett, T.A., S. Tomlin, L. Ngahfoong et al. 1984. Release of β-endorphin and met-enkephalin during exercise in normal women: response to training. *Brit. Med. J.* 288:1950-52.

Hunter, T. 1991. Protein kinase classification. *Meth. Enzymol.* 200:3-37.

Hunter, T. 1995. Protein kinases and phosphatases: the yin and yang of protein phosphorylation and signaling. *Cell* 80(2):225-36.

Hurley, J.H. 1999. Structure, mechanism, and regulation of mammalian adenylyl cyclase. *J. Biol. Chem.* 274(12):7599-602.

Hutchison, K.A., K.D. Dittmar & W.B. Pratt. 1994. All of the factors required for assembly of the glucocorticoid receptor into a functional heterocomplex with heat shock protein 90 are preassociated in a self-sufficient protein folding structure, a "foldosome." *J. Biol. Chem.* 269(45):27894-99.

Ibfelt, T., E.W. Petersen, H. Bruunsgaard, M. Sandmand & B.K. Pedersen. 2002. Exercise-induced change in type 1 cytokine-producing CD8(+) T cells is related to a decrease in memory T cells. *J. Appl. Physiol.* 93(2):645-48.

Iemitsu, M., T. Miyauchi, S. Maeda et al. 2002. Effects of aging and subsequent exercise training on gene expression of endothelin-I in rat heart. *Clin. Sci.* 103(Suppl 48):S152-S157.

Inder, W.J., J. Hellemans, M.P. Swanney et al. 1998. Prolonged exercise increases peripheral plasma ACTH, CRH, and AVP in male athletes. *J. Appl. Physiol.* 85: 835-41.

Isakov, N., R.L. Wange & L.E. Samelson. 1994. The role of tyrosine kinases and phosphotyrosine-containing recognition motifs in regulation of the T cell-antigen receptor-mediated signal transduction pathway. *J. Leukoc. Biol.* 55(2):265-71.

Ishikawa, T., M. Yanagisawa, S. Kimura et al. 1988a. Positive chronotropic effects of endothelin, a novel endothelium-derived vasoconstrictor peptide. *Pflügers Arch.* 413:108-10.

Ishikawa, T., M. Yanagisawa, S. Kimura et al. 1988b. Positive inotropic action of novel vasoconstrictor peptide endothelin on guinea pig atria. *Am. J. Physiol.* 4(Pt 2):H970-73.

Ito, H., Y. Hirata, S. Adachi et al. 1993. Endothelin-1 is an autocrine/paracrine factor in the mechanism of angiotensin II-induced hypertrophy in cultured rat cardiomyocytes. *J. Clin. Invest.* 1:398-403.

Izawa, T., T. Komabayashi, K. Suda et al. 1989. An acute exercise-induced translocation of beta-adrenergic receptors in rat myocardium. *J. Biochem.* (Tokyo) 105(1):110-13.

Jacob, C., H. Zouhal, S. Vincent et al. 2002. Training status (endurance or sprint) and catecholamine response to the wingate-test in women. *Int. J. Sports Med.* 23:342-47.

James, S., J. Kelleher & L.K. Trejdosiewicz. 1990. Linoleic acid increases cytosolic Ca2+ in lymphocytes. *Biochem. Soc. Transact.* 18(5):903-04.

Jamieson, C.A. & K.R. Yamamoto. 2000. Crosstalk pathway for inhibition of glucocorticoid-induced apoptosis by T cell receptor signaling. *Proc. Natl. Acad. Sci.* 97(13):7319-24.

Jayaraman, T. & A.R. Marks. 1997. T cells deficient in inositol 1,4,5-trisphosphate receptor are resistant to apoptosis. *Mol. Cell Biol.* 17(6):3005-12.

Jen, C.J., H.P. Chan & H.I. Chen. 2002. Chronic exercise improves endothelial calcium signaling and vasodilatation in hypercholesterolemic rabbit femoral artery. *Arterioscl. Throm. Vasc. Biol.* 22:1219-24.

Jerzynska, J., I. Stelmach & P. Kuna. 2000. The role of heparin in allergic inflammation. *Pol. Merkuriusz Lek.* 8(47):341-46.

Jeukendrup, A.E., W.H. Saris & A.J. Wagenmakers. 1998a. Fat metabolism during exercise: a review. Part I: Fatty acid mobilization and muscle metabolism. *Int. J. Sports Med.* 19(4):231-44.

Jeukendrup, A.E., W.H. Saris & A.J. Wagenmakers. 1998b. Fat metabolism during exercise: a review—part II: regulation of metabolism and the effects of training. *Int. J. Sports Med.* 19(5):293-302.

Jonat, C., H.J. Rahmsdorf, K.K. Park et al. 1990. Antitumor promotion and antiinflammation: down-modulation of AP-1 (Fos/Jun) activity by glucocorticoid hormone. *Cell* 62(6):1189-204.

Jonsdottir, I.H., P. Hoffmann & P. Thoren. 1997. Physical exercise, endogenous opioids and immune function. *Acta Physiol. Scand. Suppl.* 640:47-50.

Joyeux, M., G.F. Baxter, D.L. Thomas et al. 1997. Protein kinase C is involved in resistance to myocardial infarction induced by heat stress. *J. Mol. Cell Cardiol.* 29(12):3311-19.

June, C.H., M.C. Fletcher, J.A. Ledbetter et al. 1990. Inhibition of tyrosine phosphorylation prevents T-cell receptor-mediated signal transduction. *Proc. Natl. Acad. Sci.* 87(19):7722-26.

Kambayashi, Y., K. Nakao, M. Mukoyama et al. 1990. Isolation and sequence determination of human brain natriuretic peptide in human atrium. *FEBS Lett.* 2:341-45.

Kangawa, K. & H. Matsuo. 1984. Purification and complete amino acid sequence of alpha-human atrial natriuretic polypeptide (alpha-hANP). *Biochem. Biophys. Res. Comm.* 1:131-39.

Karsten, S., G. Schäfer & P. Schauder. 1994. Cytokine production and DNA synthesis by human peripheral lymphocytes in response to palmitic, stearic, oleic, and linoleic acid. *J. Cell Physiol.* 161(1):15-22.

Kawakami, Y., J. Kitaura, S.E. Hartman et al. 2000. Regulation pf protein kinase Cbl by two protein-tyrosine kinases, Btk and Syk. *Proc. Natl. Acad. Sci.* 97(13): 7423-28.

Kazlauskas, A. 1994. Receptor tyrosine kinases and their targets. *Current Opin. Gen. Dev.* 1:5-14.

Keenan, C. & D. Kelleher. 1998. Protein kinase C and the cytoskeleton. *Cell Signal.* 10(4):225-32.

Keenan, C., A. Long & D. Kelleher. 1997. Protein kinase C and T cell function. *Biochim. Biophys. Acta* 1358(2): 113-26.

Kehlet, H. 1988. The modifying effect of anesthetic technique on the metabolic and endocrine responses to anesthesia and surgery. *Acta Anaesthesiol. Belgica* 39:143-46.

Keller, C., A. Steensberg, H. Pilegaard et al. 2001. Transcriptional activation of the IL-6 gene in human contracting skeletal muscle: influence of muscle glycogen content. *FASEB J.* 15(14):2748-50.

Khachaturian, H., M.K.H. Schaefer & M.E. Lewis. 1993. Anatomy and function of the endogenous opioid systems. In *Handbook of Experimental Pharmacology 104/I,* eds. A. Herz, H. Akil & E.J. Simon, 471-97. Berlin: Springer.

Khan, A.A., M.J. Soloski, A.H. Sharp et al. 1996. Lymphocyte apoptosis: mediation by increased type 3 inositol 1,4,5-trisphosphate receptor. *Science* 273(5274):503-07.

Kilpatrick, D.L. 1993. Opioid peptide expression in peripheral tissues and its functional implications. In *Handbook of Experimental Pharmacology 104/II,* eds. A. Herz, H. Akil & E.J. Simon, 551-70. Berlin: Springer.

Kjaer, M. 1998. Adrenal medulla and exercise training. *Eur. J. Appl. Physiol.* 77:195-99.

Kjaer, M., N.J. Christensen, B. Sonne et al. 1985. Effect of exercise on epinephrine turnover in trained and untrained male subjects. *J. Appl. Physiol.* 59:1061-67.

Kjaer, M. & H. Galbo. 1988. Effect of physical training on the capacity to secrete epinephrine. *J. Appl. Physiol.* 64:11-16.

Knudtzon, J. 1986. Effects of proopiomelanocortin-derived peptides on plasma levels of glucagon, insulin and glucose. *Horm. Metab. Res.* 18:579-83.

Koeler, H. 1987. Fluid metabolism in exercise. *Kidney Int.* 32(Suppl 21):S93-S96.

Korkushko, O.V., M.V. Frolkis & Y.T. Yaroshenko. 1988. Hormonal supply during physical exercise in elderly patients with ischemic heart disease. *Gerontology* 34: 88-94.

Korkushko, O.V., M.V. Frolkis, V.B. Shatilo et al. 1990. Hormonal and autonomic reactions to exercise in elderly healthy subjects and patients with ischemic heart disease. *Acta Clin. Belgica* 45:164-75.

Kraemer, R.R., S. Blair & G.R. Kraemer. 1989. Effects of treadmill running on plasma β-endorphin, corticotropin, and cortisol levels in male and female 10K runners. *Eur. J. Appl. Physiol.* 58:845-51.

Kraemer, W.J., J.E. Dziados, S.E. Gordon et al. 1990. The effects of graded exercise on plasma proenkephalin peptide F and catecholamine responses at sea level. *Eur. J. Appl. Physiol.* 61:214-17.

Kraemer, W.J., B. Noble, B. Culver et al. 1985. Changes in plasma proenkephalin peptide F and catecholamine levels during graded exercise in men. *Proc. Natl. Acad. Sci.* 82:6349-51.

Kraemer, W.J., J.F. Patton, H.G. Knuttgen et al. 1991. Effects of high-intensity cycle exercise on sympathoadrenal-medullary response patterns. *J. Appl. Physiol.* 70:8-14.

Kraemer, W.J., P.B. Rock, C.S. Fulco et al. 1988. Influence of altitude and caffeine during rest and exercise on plasma levels of proenkephalin peptide F. *Peptides* 9:1115-19.

Krause, K.H., N. Demaurex, M. Jaconi et al. 1993. Ion channels and receptor-mediated Ca^{2+} influx in neutrophil granulocytes. *Blood Cells* 9(1):165-75.

Krause, K.H., W. Schlegel, C.B. Wollheim et al. 1985. Chemotactic peptide activation of human neutrophils and HL-60 cells. Pertussis toxin reveals correlation between inositol trisphosphate generation, calcium ion transients, and cellular activation. *J. Clin. Invest.* 76(4):1348-54.

La Villa, G., L. Stefani, C. Lazzeri et al. 1995. Acute effects of physiological increments of brain natriuretic peptide in humans. *Hypertension* 4:628-33.

La Villa, G., S. Vena, A. Conti et al. 1993. Plasma levels of brain natriuretic peptide in healthy subjects and patients with essential hypertension: response to posture. *Clin. Sci.* 4:411-16.

Lang, C.C., A.M. Choy & A.D. Struthers. 1992. Atrial and brain natriuretic peptides: a dual natriuretic peptide system potentially involved in circulatory homeostasis. *Clin. Sci.* 5:519-27.

Leaf, D.A., M.T. Kleinman, M. Hamilton et al. 1997. The effect of exercise intensity on lipid peroxidation. *Med. Sci. Sports Exerc.* 29(8):1036-39.

Lee, H.C. 1997. Mechanisms of calcium signaling by cyclic ADP-ribose and NAADP. *Physiol. Rev.* 77(4): 1133-64.

Lee, H.C. 2000. Enzymatic functions and structures of CD38 and homologs. *Chem. Immunol.* 75:39-59.

Lee, J.S., C.R. Bruce, B.E. Spurrell et al. 2002. Effect of training on activation of extracellular signal-regulated kinase 1/2 and p38 mitogen-activated protein kinase pathways in rat soleus muscle. *Clin. Exp. Pharmacol. Physiol.* 29(8):655-60.

Legon, S., D.M. Glover, J. Hughes et al. 1982. The structure and expression of the preproenkephalin gene. *Nucleic Acids Res.* 10:7905-18.

Lehmann, M., K. Hasler, E. Bergdolt et al. 1986. Alpha-2-adrenoreceptor density on intact platelets and adrenaline-induced platelet aggregation in endurance- and nonendurance-trained subjects. *Int. J. Sports Med.* 7(3):172-76.

Lehmann, M., P. Schmid, H. Porzig et al. 1983. [Beta-adrenergic receptors and plasma catecholamine behavior in trained and untrained athletes]. *Klin. Wochenschr.* 61(17):865-71.

Leitenberg, D., Y. Boutin, D.D. Lu et al. 1999. Biochemical association of CD45 with the T cell receptor complex: regulation by CD45 isoform and during T cell activation. *Immunity* 10(6):701-11.

Lepple-Wienhues, A., I. Szabo, U. Wieland et al. 2000. Tyrosine kinases open lymphocyte chloride channels. *Cell. Physiol. Biochem.* 10(5-6):307-12.

Levkowitz, G., H. Waterman, S.A. Ettenberg et al. 1999. Ubiquitin ligase activity and tyrosine phosphorylation underlie suppression of growth factor signaling by c-Cbl/Sli. *Mol. Cell* 4(6):1029-40.

Lew, P.D. 1990. Receptors and intracellular signaling in human neutrophils. *Am. Rev. Resp. Dis.* 141: S127-31.

Lew, P.D., C.B. Wollheim, F.A. Waldvogel et al. 1984. Modulation of cytostolic-free calcium transients by changes in intracellular calcium-buffering capacity: correlation with exocytosis and O_2^- production in human neutrophils. *J. Cell Biol.* 99:1212-20.

Li, W. & H. She. 2000. The SH2 and SH3 adapter Nck: a two-gene family and a linker between tyrosine kinases and multiple signaling networks. *Histol. Histopathol.* (3):947-55.

Lijnen, P., P. Hespel, J.R. M'Buyamba-Kabangu et al. 1987. Plasma atrial natriuretic peptide and cyclic nucleotide levels before and after a marathon. *J. Appl. Physiol.* 3:1180-84.

Litt, M., N.E. Buroker, S. Kondoleon et al. 1988. Chromosomal localization of the human proenkephalin and prodynorphin genes. *Am. J. Hum. Gen.* 42:327-34.

Locke, M. 1997. The cellular stress response to exercise: role of stress proteins. *Exerc. Sport Sci. Rev.* 25:105-36.

Loh, Y.P. 1992. Molecular mechanisms of β-endorphin biosynthesis. *Biochem. Pharmacol.* 44:843-49.

Lowe, G.M., C.E. Hulley, E.S. Rhodes et al. 1998. Free radical stimulation of tyrosine kinase and phosphatase activity in human peripheral blood mononuclear cells. *Biochem. Biophys. Res. Com.* 245(1):17-22.

Lu, S.S., C.P. Lau, Y.F. Tung et al. 1997. Lactate and the effects of exercise on testosterone secretion: evidence for the involvement of a cAMP-mediated mechanism. *Med. Sci. Sports Exerc.* 29(8):1048-54.

Lucas, K.A., G.M. Pitari, S. Kazerounian et al. 2000. Guanylyl cyclases and signaling by cyclic GMP. *Pharmacol. Rev.* 52(3):375-414.

Luger, A., P.A. Deuster, S.B. Kyle et al. 1987. Acute hypothalamic-pituitary-adrenal response to stress of treadmill exercise. *N. Engl. J. Med.* 316:1309-15.

Luther, S.A. & J.G. Cyster. 2001. Chemokines as regulators of T cell differentiation. *Nat. Immunol.* 2(2): 102-07.

Lynch, G.S., C.J. Fary & D.A. Williams. 1997. Quantitative measurement of resting skeletal muscle [Ca2+]i following acute and long-term downhill running exercise in mice. *Cell Calcium* 22:373-83.

Maeda, S., T. Miyauchi, M. Iemitsu et al. 2002. Effects of exercise training on expression of endothelin-I mRNA in the aorta of aged rats. *Clin. Sci.* (Suppl 48): S118-S123.

Maeda, S., T. Miyauchi, T. Kakiyama et al. 2001. Effects of exercise training of 8 weeks and detraining on plasma levels of endothelium-derived factors, endothelin-1 and nitric oxide, in healthy young humans. *Life Sci.* 9:1005-16.

Maeda, S., T. Miyauchi, S. Sakai et al. 1998. Prolonged exercise causes an increase in endothelin-1 production in the heart in rats. *Am. J. Physiol.* 6 (Pt 2): H2105-12.

Maeda, S., T. Miyauchi, M. Sakane et al. 1997. Does endothelin-1 participate in the exercise-induced changes of blood flow distribution of muscles in humans? *J. Appl. Physiol.* 4:1107-11.

Maki, T. 1989. Density and functioning of human lymphocytic beta-adrenergic receptors during prolonged physical exercise. *Acta Physiol. Scand.* 136(4):569-74.

Maki, T., H. Leinonen, H. Naveri et al. 1989. Response of the beta-adrenergic system to maximal dynamic exercise in congestive heart failure secondary to idiopathic dilated cardiomyopathy. *Am. J. Cardiol.* 63(18):1348-53.

Marchant, B., V. Umachandran, P. Wilkinson et al. 1994. Reexamination of the role of endogenous opiates in silent myocardial ischemia. *J. Am. Coll. Cardiol.* 23: 645-51.

Marks, A.R. 1997. Intracellular calcium-release channels: regulators of cell life and death. *Am. J. Physiol.* 272(2 Pt 2):H597-605.

Marriott, I. & M.J. Mason. 1995. ATP depletion inhibits Ca^{2+} entry pathway in rat thymic lymphocytes. *Am. J. Physiol.* 269:C766-C774.

Marumoto, K., M. Hamada & K. Hiwada. 1995. Increased secretion of atrial and brain natriuretic peptides during acute myocardial ischaemia induced by dynamic exercise in patients with angina pectoris. *Clin. Sci.* 5:551-56.

Masaki, T., S. Kimura, M. Yanagisawa et al. 1991. Molecular and cellular mechanism of endothelin regulation. Implications for vascular function. *Circulation* 4: 1457-68.

Mastorakos, G., G.P. Chrousos & J.S. Weber. 1993. Recombinant interleukin-6 activates the hypothalamic-pituitary-adrenal axis in humans. *J. Clin. Endocrinol. Metab.* 77(6):1690-94.

Masuda, E.S., R. Imamura, Y. Amasaki et al. 1998. Signalling into the T-cell nucleus: NFAT regulation. *Cell Signal.* 10(9):599-611.

Matejec, R., R. Ruwoldt, R.H. Bödeker et al. 2003. Release of β-endorphin immunoreactive material under perioperative conditions into blood or cerebrospinal fluid: significance for postoperative pain? *Anesth. Analg.* 96:481-86.

Matsuda, S. & S. Koyasu. 2000. Mechanisms of action of cyclosporine. *Immunopharmacology* 47(2-3):119-25.

Matzen, S., C. Emmeluth, M.C. Milliken et al. 1992. Plasma endothelin-1 during central hypovolaemia in man. *Clin. Physiol.* 6:653-58.

Mazzeo, R.S. 1991. Catecholamine responses to acute and chronic exercise. *Med. Sci. Sports Exerc.* 23(7): 839-45.

McCarthy, D.A. & M.M. Dale. 1988. The leucocytosis of exercise. A review and model. *Sports Med.* 6(6): 333-63.

McLoughlin, L., S. Medback & A.B. Grossman. 1993. Circulating opioids in man. In *Handbook of Experimental Pharmacology 104/II,* eds. A. Herz, H. Akil & E.J. Simon, 673-96. Berlin: Springer.

McMurray, J.J., S.G. Ray, I. Abdullah et al. 1992. Plasma endothelin in chronic heart failure. *Circulation* 4: 1374-79.

Merkus, D., D.J. Duncker & W.M. Chilian. 2002. Metabolic regulation of coronary vascular tone: role of endothelin-1. *Am. J. Physiol.* 5:H1915-21.

Middeke, M., S. Reder & H. Holzgreve. 1994. Regulation of the beta-adrenoceptor-cAMP-system during dynamic exercise in patients with primary hypertension after acute beta-blockade. *Blood Press.* 3(3):189-92.

Mikoshiba, K., T. Furuichi & A. Miyawaki. 1994. Structure and function of IP3 receptors. *Sem. Cell Biol.* 5(4): 273-81.

Miller, W.E. & R.J. Lefkowitz. 2001. Expanding roles for beta-arrestins as scaffolds and adapters in GPCR signaling and trafficking. *Curr. Op. Cell Biol.* 13(2): 139-45.

Mitsuyama, T., K. Takeshige, T. Furuno, T. Tanaka et al. 1995. An inhibitor of cyclic AMP-dependent protein kinase enhances the superoxide production of human neutrophils stimulated by N-formyl-methionyl-leucyl-phenylalanine. *Mol. Cell. Biochem.* 145(1):19-24.

Mizuhara, H., E. O'Neill, N. Seki et al. 1994. T cell activation-associated hepatic injury: mediation by tumor necrosis factors and protection by interleukin 6. *J. Exp. Med.* 179(5):1529-37.

Mochly-Rosen, D. & L.M. Kauvar. 2000. Pharmacological regulation of network kinetics by protein kinase C localization. *Sem. Immunol.* 12:55-61.

Mochly-Rosen, D., H. Khaner, J. Lopez et al. 1991. Intracellular receptors for activated protein kinase C. Identification of a binding site for the enzyme. *J. Biol. Chem.* 266(23):14866-68.

Moldoveanu, A.I., R.J. Shephard & P.N. Shek. 2000. Exercise elevates plasma levels but not gene expression of IL-1beta, IL-6, and TNF-alpha in blood mononuclear cells. *J. Appl. Physiol.* 89(4):1499-1504.

Monteiro, H.P. & A. Stern. 1996. Redox modulation of tyrosine phosphorylation-dependent signal transduction pathways. *Free Radic. Biol. Med.* 21(3):323-33.

Moore, R.L., M. Riedy & P.D. Gollnick. 1982. Effect of training on beta-adrenergic receptor number in rat heart. *J. Appl. Physiol.* 1982 May;52(5):1133-37.

Mooren, F.C., A. Lechtermann, A. Fromme et al. 2001a. Alterations in intracellular calcium signaling of lymphocytes after exhaustive exercise. *Med. Sci. Sports Exerc.* 33(2):242-48.

Mooren, F.C., A. Lechtermann, A. Fromme et al. 2001b. Decoupling of intracellular calcium signalling in granulocytes after exhaustive exercise. *Int. J. Sports Med.* 22(5):323-28.

Mooren, F.C. & R.K. Kinne. 1998. Cellular calcium in health and disease. *Biochim. Biophys. Acta* 1406: 127-51.

Mortensen, L.H., C.M. Pawloski, N.L. Kanagy et al. 1990. Chronic hypertension produced by infusion of endothelin in rats. *Hypertension* 6(Pt 2):729-33.

Mougin, F., M.L. Simon-Rigaud, C. Mougin et al. 1992. Met-enkephalin, β-endorphin and cortisol responses to sub-maximal exercise after sleep disturbances. *Eur. J. Appl. Physiol.* 64: 37-76.

Mourey, R.J. & J.E. Dixon. 1994. Protein tyrosine phosphatases: characterization of extracellular and intracellular domains. *Curr. Opin. Gen. Dev.* 4(1):31-39.

Mukoyama, M., K. Nakao, K. Hosoda et al. 1991. Brain natriuretic peptide as a novel cardiac hormone in humans. Evidence for an exquisite dual natriuretic peptide system, atrial natriuretic peptide and brain natriuretic peptide. *J. Clin. Invest.* 4:1402-12.

Naesh, O., I. Hindberg, J. Trap-Jensen et al. 1990. Post-exercise platelet activation—aggregation and release in relation to dynamic exercise. *Clin. Physiol.* 10(3): 221-30.

Nahas, N., T.F. Molski, G.A. Fernandez et al. 1996. Tyrosine phosphorylation and activation of a new mitogen-activated protein (MAP)-kinase cascade in human neutrophils stimulated with various agonists. *Biochem. J.* 318 (Pt 1):247-53.

Neer, E.J. 1995. Heterotrimeric G proteins: organizers of transmembrane signals. *Cell* 80:249-57.

Neet, K. & T. Hunter. 1996. Vertebrate non-receptor protein-tyrosine kinase families. *Genes Cells* 1(2): 147-69.

Negulescu, P.A., N. Shastri & M.D. Cahalan. 1994. Intracellular calcium dependence of gene expression in single T lymphocytes. *Proc. Natl. Acad. Sci.* 91(7): 2873-77.

Nehlsen-Cannarella, S.L., O.R. Fagoaga, D.C. Nieman et al. 1997. Carbohydrate and the cytokine response to 2.5 h of running. *J. Appl. Physiol.* 82(5):1662-67.

Neubauer, S., G. Ertl, U. Haas et al. 1990. Effects of endothelin-1 in isolated perfused rat heart. *J. Cardiovasc. Pharmacol.* 1:1-8.

Nicholson, S., M. Richards, E. Espiner et al. 1993. Atrial and brain natriuretic peptide response to exercise in patients with ischaemic heart disease. *Clin. Exp. Pharmacol. Physiol.* 7-8:535-40.

Nielsen, H.B., N. Secher & B.K. Pedersen. 1996. Lymphocytes and NK cell activity during repeated bouts of maximal exercise. *Am. J. Physiol.* 271:R222-R227.

Nieman, D.C., O.R. Fagoaga, D.E. Butterworth et al. 1997a. Carbohydrate supplementation affects blood granulocyte and monocyte trafficking but not function after 2.5 hours of running. *J. Appl. Physiol.* 82(5):1385-94.

Nieman, D.C., D.A. Henson, E.B. Garner et al. 1997b. Carbohydrate affects natural killer cell redistribution but not activity after running. *Med. Sci. Sports Exerc.* 29(10):1318-24.

Nieman, D.C., S.L. Nehlsen-Canarella, O.R. Fagoaga et al. 1998a. Effects of mode and carbohydrate on the granulocyte and monocyte response to intensive prolonged exercise. *J. Appl. Physiol.* 84(4):1252-59.

Nieman, D.C., S.L. Nehlsen-Canarella, O.R. Fagoaga et al. 1998b. Influence of mode and carbohydrate on the cytokine response to heavy exertion. *Med. Sci. Sports Exerc.* 30(5):671-78.

Niess, A.M., M. Sommer & E. Schlotz. 2000. Expression of the inducible nitric oxide synthase (iNOS) in human leukocytes: responses to running exercise. *Med. Sci. Sports Exerc.* 32:1220-25.

Nieto, J., I. Diaz-Laviada, A. Guillen et al. 1996a. Effect of endurance physical training on rat liver adenlylyl cyclase system. *Cell Signal.* 8(4):317-22.

Nieto, J.L., I. Diaz-Laviada, A. Guillen et al. 1996b. Adenylyl cyclase system is affected differently by endurance physical training in heart and adipose tissue. *Biochem. Pharmacol.* 51(10):1321-29.

Nishizuka, Y. 1984. The role of protein kinase C in cell surface signal transduction and tumour promotion. *Nature* 308(5961):693-98.

Nishizuka, Y. 1992. Intracellular signaling by hydrolysis of phospholipids and activation of protein kinase C. *Science* 258(5082):607-14.

Noda, M., Y. Teranishi, H. Takahashi et al. 1982. Isolation and structural organisation of the human preproenkephalin gene. *Nature* 297:431-34.

Nomura, S., H. Kawanami, H. Ueda et al. 2002. Possible mechanisms by which adipocyte lipolysis is enhanced in exercise-trained rats. *Biochem. Biophys. Res. Com.* 12;295(2):236-42.

Nordstrom, T., C. Lindqvist, A. Stahls, T. Mustelin & L.C. Andersson. 1991. Inhibition of CD3-induced Ca2+ signals in Jurkat T-cells by myristic acid. *Cell Calcium* 12(7):449-55.

Northoff, H. & A. Berg. 1991. Immunologic mediators as parameters of the reaction to strenuous exercise. *Int. J. Sport Med.* 12(Suppl 1):S9-15.

Northoff, H., C. Weinstock & A. Berg. 1994. The cytokine response to strenuous exercise. *Int. J. Sports Med.* 15:S167-71.

Obba, H., H. Takada, H. Musha et al. 2001. Effects of prolonged strenuous exercise on plasma levels of atrial natriuretic peptide and brain natriuretic peptide in healthy men. *Am. Heart J.* 147:751-58.

Ogawa, Y., K. Nakao, M. Mukoyama et al. 1990. Rat brain natriuretic peptide—tissue distribution and molecular form. *Endocrinology* 4:2225-27.

Ohashi, M., N. Fujio, H. Nawata et al. 1987. Pharmacokinetics of synthetic alpha-human atrial natriuretic polypeptide in normal men; effect of aging. *Regul. Pept.* 5-6:265-72.

Oh-ishi, S., T. Kizaki, J. Nagasawa et al. 1997. Effects of endurance training on superoxide dismutase activity, content and mRNA expression in rat muscle. *Clin. Exp. Pharmacol. Physiol.* 24(5):326-32.

Ohman, E.M., J. Butler, J. Kelly et al. 1987. Beta-adrenoceptor adaptation to endurance training. *J. Cardiovasc. Pharmacol.* 10(6):728-31.

Oki, N., S.I. Takahashi, H. Hidaka et al. 2000. Short term feedback regulation of cAMP in FRTL-5 thyroid cells. Role of PDE4D3 phosphodiesteRase activation. *J. Biol. Chem.* 275(15):10831-37.

Oleshansky, M.A., J.M. Zoltick, R.H. Herman et al. 1990. The influence of fitness on neuroendocrine responses to exhaustive treadmill exercise. *Eur. J. Appl. Physiol.* 59:405-10.

Ostrowski, K., C. Hermann, A. Bangash et al. 1998a. A trauma-like elevation in plasma cytokines in humans in response to treadmill running. *J. Physiol.* 508:949-53.

Ostrowski, K., T. Rohde, S. Asp et al. 1999. The cytokine balance and strenuous exercise: TNF-alpha, IL-2beta, IL-6, IL-1ra, sTNF-r1, sTNF-r2, and IL-10. *J. Physiol.* 515(1):287-91.

Ostrowski, K., T. Rohde, S. Asp et al. 2001. Chemokines are elevated in plasma after strenuous exercise in humans. *Eur. J. Appl. Physiol.* 84:244-45.

Ostrowski, K., T. Rohde, M. Zacho et al. 1998b. Evidence that IL-6 is produced in skeletal muscle during intense long-term muscle activity. *J. Physiol.* 508(3):949-53.

Ostrowski, K., P. Schjerling & B.K. Pedersen. 2000. Physical activity and plasma interleukin-6 in humans—effect of intensity of exercise. *Eur. J. Appl. Physiol.* 83(6):512-15.

Owens, P.C. & R. Smith. 1987. Opioid peptides in blood and cerebrospinal fluid during acute stress. *Bailliere's Clin. Endocrinol. Metab.* 1:415-37.

Paradiso, K. & P. Brehm. 1998. Long-term desensitization of nicotinic acetylcholine receptors is regulated via protein kinase A-mediated phosphorylation. *J. Neurosci.* 18(22):9227-37.

Passene, J., M. Germain, A.M. Allevard et al. 1996. Water balance during and after marathon running. *Eur. J. Appl. Physiol.* 73:49-55.

Pearce, D. & K.R. Yamamoto. 1993. Mineralocorticoid and glucocorticoid receptor activities distinguished by nonreceptor factors at a composite response element. *Science* 59(5098):1161-65.

Pedersen, B.K. & L. Hoffman-Goetz. 2000. Exercise and the immune system: regulation, integration and adaption. *Physiol. Rev.* 80:1055-81.

Pedersen, B.K. & D.C. Nieman. 1998. Exercise and immunology: integration and regulation. *Immunol. Today* 19(5):204-06.

Pedersen, B.K., K. Ostrowski, T. Rohde et al. 1998. The cytokine response to strenuous exercise. *Can J. Physiol. Pharmacol.* 76:505-11.

Pedersen, B.K., A. Steensberg & P. Schjerling. 2001. Exercise and interleukin-6. *Curr. Op. Hematol.* 8(3):137-41.

Petersen, O.H., D. Burdakov & A.V. Tepikin. 1999. Polarity in intracellular calcium signaling. *Bioessays* 21(10):851-60.

Petraglia, F., C. Barletta, F. Facchinetti et al. 1988. Response of circulating adrenocorticotropin, β-endorphin, β-lipotropin, and cortisol to athletic competition. *Acta Endocrinol* 118:332-36.

Petrides, J.S., P.A. Deuster & G.P. Mueller. 1999. Lactic acid does not directly activate hypothalamic-pituitary corticotroph function. *Proc. Soc. Exp. Biol. Med.* 220:100-05.

Pettit, E.J. & M.B. Hallett. 1997. Pulsatile Ca2+ influx in human neutrophils undergoing CD11b/CD18 integrin engagement. *Biochem. Biophys. Res. Com.* 30(2):258-61.

Piacentini, M.P., E. Piatti, A. Bucchini et al. 1996. Modification of inositol 1,4,5-trisphosphate concentration of human erythrocytes under "in vivo" physiological conditions. *Biochem. Mol. Biol. Int.* 38(6):1265-69.

Pierce, E.F., N.W. Eastman, H.L. Tripathi et al. 1993. β-endorphin response to endurance exercise: relationship to exercise dependence. *Percept. Motor Skills* 77:767-70.

Plourde, G., S. Rousseau-Migneron & A. Nadeau. 1993. Effect of endurance training on beta-adrenergic system in three different skeletal muscles. *J. Appl. Physiol.* 74(4):1641-46.

Predel, H.G., H. Meyer-Lehnert, A. Backer et al. 1990. Plasma concentrations of endothelin in patients with abnormal vascular reactivity. Effects of ergometric exercise and acute saline loading. *Life Sci.* 20:1837-43.

Przewlocki, R. 1993. Opioid systems and stress. In *Handbook of Experimental Pharmacology 104/II,* eds. A. Herz, H. Akil & E.J. Simon, 293-324. Berlin: Springer.

Putney, J.W. Jr. & C.M. Ribeiro. 2000. Signaling pathways between the plasma membrane and endoplasmic reticulum calcium stores. *Cell. Mol. Life Sci.* 57(8-9):1272-86.

Qin, S., T. Inazu, M. Takata et al. 1996. Cooperation of tyrosine kinases p72syk and p53/56lyn regulates calcium mobilization in chicken B cell oxidant stress signaling. *Eur. J. Biochem.* 236(2):443-49.

Quick, M.W. & R.A. Lester. 2002. Desensitization of neuronal nicotinic receptors. *J. Neurobiol.* 53(4):457-78.

Rahkila, P., E. Hakala, M. Alen et al. 1988. β-endorphin and corticotropin release is dependent on a threshold intensity of running exercise in male endurance athletes. *Life Sci.* 43:551-58.

Ranallo, R.F. & E.C. Rhodes. 1998. Lipid metabolism during exercise. *Sports Med.* 26(1):29-42.

Rasmussen, L.B., B. Kiens, B.K. Pedersen et al. 1994. Effect of diet and plasma fatty acid composition on immune status in elderly men. *Am. J. Clin. Nutr.* 59(3):572-77.

Remes, J.J., U.E. Petaja-Repo, K.J. Tuukkanen et al. 1993. Significance of the extracellular domain and the carbohydrates of the human neutrophil N-formyl peptide chemotactic receptor for the signal transduction by the receptor. *Exp. Cell Res.* 209(1):26-32.

Rexin, M., W. Busch, B. Segnitz et al. 1992. Structure of the glucocorticoid receptor in intact cells in the absence of hormone. *J. Biol. Chem.* 267(14):9619-21.

Rhind, S.G., J.W. Castellani, I.K.M. Brenner et al. 2002. Intracellular monocyte and serum cytokine expression is modulated by exhausting exercise and cold exposure. *Am. J. Physiol.* 282.

Richards, C. & J. Gauldie. 1995. Role of cytokines in acute-phase response. In *Human Cytokines: Their Roles in Disease and Therapy,* eds. B.B. Aggarwal & R.K. Puri, 253-69. Cambridge, MA: Blackwell Science.

Richieri, G.V. & A.M. Kleinfeld. 1989. Free fatty acid perturbation of transmembrane signaling in cytotoxic T lymphocytes. *J. Immunol.* 143(7):2302-10.

Richter, E.A., C. Emmeluth, P. Bie et al. 1994. Biphasic response of plasma endothelin-1 concentration to exhausting submaximal exercise in man. *Clin. Physiol.* 4:379-84.

Rivier, C. & S. Rivest. 1991. Effect of stress on the activity of the hypothalamic-pituitary-gonadal axis: peripheral and central mechanisms. *Biol. Reproduct.* 45:523-32.

Roederer, M., F.J. Staal, H. Osada et al. 1991. CD4 and CD8 T cells with high intracellular glutathione levels are selectively lost as the HIV infection progresses. *Int. Immunol.* 3(9):933-37.

Rogers, P.J., G.M. Tyce, K.R. Bailey et al. 1991. Exercise-induced increases in atrial natriuretic factor are attenuated by endurance training. *J. Am. Coll. Cardiol.* 5:1236-41.

Rohde, T., D.A. MacLean, E.A. Richter et al. 1997. Prolonged submaximal eccentric exercise is associated with increased levels of plasma IL-6. *Am. J. Physiol.* 273(36):E85-E91.

Rossier, J. 1993. Biosynthesis of enkephalins and proenkephalin-derived peptides. In *Handbook of Experimental Pharmacology 104/I,* eds. A. Herz, H. Akil & E.J. Simon, 423-47. Berlin: Springer.

Saito, Y., K. Nakao, H. Itoh et al. 1989. Brain natriuretic peptide is a novel cardiac hormone. *Biochem. Biophys. Res. Comm.* 2:360-68.

Saito, Y., K. Nakao, A. Sugawara et al. 1987. Atrial natriuretic polypeptide during exercise in healthy man. *Acta Endocrinol.* 1:59-65.

Sbirrazzuoli, V. & P. Lapalus. 1989. Human lymphocyte and myocardial beta-adrenoceptors: up and down regulation. *Biomed. Pharmacotherapy* 43(5):369-74.

Schaller, K., D. Mechau, H.G. Scharmann et al. 1999. Increased training load and the beta-adrenergic-receptor system on human lymphocytes. *J. Appl. Physiol.* 87(1):317-24.

Schieven, G.L., J.M. Kirihara, D.L. Burg et al. 1993. p72syk tyrosine kinase is activated by oxidizing conditions that induce lymphocyte tyrosine phosphorylation and Ca2+ signals. *J. Biol. Chem.* 268(22):16688-92.

Schoonbroodt, S. & J. Piette. 2000. Oxidative stress interference with the nuclear factor-kappa B activation pathways. *Biochem. Pharmacol.* 60(8):1075-83.

Schulz, A., H. Harbach, N. Katz et al. 2000. β-endorphin immunoreactive material and authentic β-endorphin in the plasma of males undergoing anaerobic exercise on a rowing ergometer. *Int. J. Sports Med.* 21:513-17.

Schwartz, L. & W. Kindermann. 1990. β-endorphin, adrenocorticotropic hormone, cortisol and catecholamines during aerobic and anaerobic exercise. *Eur. J. Appl. Physiol.* 61:165-71.

Schwarz, L. & W. Kindermann. 1992. Changes in β-endorphin levels in response to aerobic and anaerobic exercise. *Sports Med.* 13:25-36.

Schwartz, M.W., S.C. Woods, D. Porte et al. 2000. Central nervous system control of food intake. *Nature* 404(6778):661-71.

Sforzo, G.A. 1988. Opioids and exercise. An update. *Sports Med.* 7:109-24.

Shubeita, H.E., P.M. McDonough, A.N. Harris et al. 1990. Endothelin induction of inositol phospholipid hydrolysis, sarcomere assembly, and cardiac gene expression in ventricular myocytes. A paracrine mechanism for myocardial cell hypertrophy. *J. Biol. Chem.* 33:20555-62.

Sibinga, A. & A. Goldstein. 1988. Opioid peptides and opioid receptors in cells of the immune system. *Annu. Rev. Immunol.* 6:219-49.

Siegel, J.N., R.D. Klausner, U.R. Rapp et al. 1990. T cell antigen receptor engagement stimulates c-raf phosphorylation and induces c-raf-associated kinase activity via a protein kinase C-dependent pathway. *J. Biol. Chem.* 265(30):18472-80.

Simonds, W.F. 1999. G protein regulation of adenylate cyclase. *Trends Pharmacol. Sci.* 20(2):66-73.

Sjödin, B., Y.H. Westing & F.S. Apple. 1990. Biochemical mechanisms for oxygen free radical formation during exercise. *Sports Med.* 10(4):236-54.

Soderling, S.H. & J.A. Beavo. 2000. Regulation of cAMP and cGMP signaling: new phosphodiesteRases and new functions. *Curr. Op. Cell Biol.* 12(2):174-79.

Solbrig, M.V., G.F. Koob & W.I. Lipkin. 2002. Key role for enkephalinergic tone in cortico-striatal-thalamic function. *Eur. J. Neurosci.* 16:1819-22.

Solomon, L.R., J.C. Atherton, H. Bobinski et al. 1986. Effect of posture on plasma immunoreactive atrial natriuretic peptide concentrations in man. *Clin. Sci.* 3:299-305.

Sperling, R.I., A.I. Benincaso, C.T. Knoell et al. 1993. Dietary omega-3 polyunsaturated fatty acids inhibit phosphoinositide formation and chemotaxis in neutrophils. *J. Clin. Invest.* 91(2):651-60.

Sprenger, H., C. Jacobs, M. Nain et al. 1992. Enhanced release of cytokines, interleukin-2 receptors, and neopterin after long-distance running. *Clin. Immunol. Immunopathol.* 63(2):188-95.

Staal, F.J., M.T. Anderson, M.T. Staal et al. 1994. Redox regulation of signal transduction: tyrosine phosphorylation and calcium influx *Proc. Natl. Acad. Sci.* 91:3619-22.

Stallknecht, B., M. Kjaer, T. Ploug et al. 1990. Diminished epinephrine response to hypoglycemia despite enlarged adrenal medulla in trained rats. *Am. J. Physiol.* 259:998-1003.

Starkie, R.L., D.J. Angus, J. Rolland et al. 2000. Effect of prolonged submaximal exercise and carbohydrate ingestion on monocyte intracellular cytokine production in humans. *J. Physiol.* 528(3):647-55.

Starkie, R.L., M.J. Arkinstall, I. Koukoulas et al. 2001a. Carbohydrate ingestion attenuates the increase in plasma interleukin-6, but not skeletal muscle interleukin-6 mRNA, during exercise in humans. *J. Physiol.* 533(Pt 2):585-91.

Starkie, R.L., J. Rolland, D.J. Angus, M.J. Anderson & M.A. Febbraio. 2001b. Circulating monocytes are not the source of elevations in plasma IL-6 and TNF-alpha levels after prolonged running. *Am. J. Physiol.* 280(4):C769-C774.

Steensberg, A., M.A. Febbraio, T. Osada et al. 2001a. Interleukin-6 production in contracting human skeletal muscle is influenced by pre-exercise muscle glycogen content. *J. Physiol.* 537(Pt 2):633-39.

Steensberg, A., A.D. Toft, H. Bruunsgaard et al. 2001b. Strenuous exercise decreases the percentage of type 1 T cells in the circulation. *J. Appl. Physiol.* 91(4): 1708-12.

Steensberg, A., A.D. Toft, J. Halkjaer-Kristensen et al. 2001c. Plasma interleukin-6 during strenuous exercise—role of adrenaline. *Am. J. Physiol.* 281(3): 1001-04.

Steensberg, A., C.P. Fischer, C. Keller et al. 2003. IL-6 enhances plasma IL-1ra, IL-10, and cortisol in humans. *Am. J. Physiol.* 285(2):E433-7.

Steensberg, A., G. van Hall, T. Osada et al. 2000. Production of IL-6 in contracting human skeletal muscles can account for the exercise-induced increase in plasma IL-6. *J. Physiol.* 529:237-42.

Steinberg, H. & E.A. Sykes. 1985. Introduction to symposium on endorphins and behavioural processes: review of literature on endorphins and exercise. *Pharmacol. Biochem. Behav.* 23:857-62.

Stephen, F.D., R.J. Kelleher Jr., M. Langner et al. 1997. Dietary fatty acids alter the adhesion properties of lymphocytes to extracellular matrix proteins. *Adv. Exp. Med. Biol.* 400B:775-88. 1997.

Stock, C., K. Schaller, M. Baum et al. 1995. Catecholamines, lymphocyte subsets, and cyclic adenosine monophosphate production in mononuclear cells and CD4+ cells in response to submaximal resistance exercise. *Eur. J. Appl. Physiol.* 71(2-3): 166-72.

Stone, R.L. & J.E. Dixon. 1994. Protein-tyrosine phosphatases. *J. Biol. Chem.* 269(50).31323-26.

Stouthard, J.M., R.P. Oude Elferink & H.P. Sauerwein. 1996. Interleukin-6 enhances glucose transport in 3T3-L1 adipocytes. *Biochem. Biophys. Res. Commun.* 220(2):241-45.

Strassman, R.J., O. Appenzeller, A.J. Lewy et al. 1989. Increase in plasma melatonin, β-endorphin, and cortisol after a 28.5-mile mountain race: relationship to performance and lack of effect of naltrexone. *J. Clin. Endocrinol. Metab.* 69:540-45.

Streb, H., R.F. Irvine, M.J. Berridge & I. Schulz. 1983. Release of Ca2+ from a nonmitochondrial intracellular store in pancreatic acinar cells by inositol-1,4,5-trisphosphate. *Nature* 3-9;306(5938):67-69.

Stulnig, T.M., M. Berger, M. Roden et al. 2000. Elevated serum free fatty acid concentrations inhibit T lymphocyte signaling. *FASEB J.* 14(7):939-47.

Sudoh, T., K. Kangawa, N. Minamino et al. 1988. A new natriuretic peptide in porcine brain. *Nature* 6159: 78-81.

Sugawara, A., K. Nakao, N. Morii et al. 1985. Alpha-human atrial natriuretic polypeptide is released from the heart and circulates in the body. *Biochem. Biophys. Res. Comm.* 2:439-46.

Sumpio, B.E. & M.D. Widmann. 1990. Enhanced production of endothelium-derived contracting factor by endothelial cells subjected to pulsatile stretch. *Surgery* 2:277-81; discussion 281-82.

Sun, H. & N.K. Tonks. 1994. The coordinated action of protein tyrosine phosphatases and kinases in cell signaling. *Trends Biochem. Sci.* 19(11):480-85.

Sun, Z., C.W. Arendt, W. Ellmeier et al. 2000. PKC-theta is required for TCR-induced NF-kappaB activation in mature but not immature T lymphocytes. *Nature* 404(6776):402-07.

Sundaresan, M., Z.X. Yu, V.J. Ferrans et al. 1995. Requirement for generation of H2O2 for platelet-derived growth factor signal transduction. *Science* 270(5234):296-99.

Suzuki, K., S. Nakaji, M. Yamada et al. 2002. Systemic inflammatory response to exhaustive exercise. Cytokine kinetics. *Exerc. Immunol. Rev.* 8:6-48.

Suzuki, K., M. Yamada, S. Kurakake et al. 2000. Circulating cytokines and hormones with immunosuppressive but neutrophil-priming potentials rise after endurance exercise in humans. *Eur. J. Appl. Physiol.* 81: 281-87.

Suzuki, T., T. Kumazaki & Y. Mitsui. 1993. Endothelin-1 is produced and secreted by neonatal rat cardiac myocytes in vitro. *Biochem. Biophys. Res. Comm.* 3: 823-30.

Sylvestre-Gervais, L., A. Nadeau, M.H. Nguyen et al. 1982. Effects of physical training on beta-adrenergic receptors in rat myocardial tissue. *Cardiovasc. Res.* 16(9):530-34.

Szamel, M., A. Appel, R. Schwinzer et al. 1998. Different protein kinase C isoenzymes regulate IL-2 receptor expression or Il-2 synthesis in human lymphocytes stimulated via the TCR. *J. Immunol.* 160(5):2207-14.

Tache, Y. & C. Rivier, eds. 1993. *Ann. N.Y. Acad. Sci. 697: Corticotropin-Releasing Factor and Cytokines: Role in the Stress Response.* New York: New York Academy of Sciences.

Tanabe, K., A. Yamamoto, N. Suzuki et al. 1999. Exercise-induced changes in plasma atrial natriuretic peptide and brain natriuretic peptide concentrations in healthy subjects with chronic sleep deprivation. *Jap. Circ. J.* 6:447-52.

Tanaka, M., Y. Ishizaka, Y. Ishiyama et al. 1995. Exercise-induced secretion of brain natriuretic peptide in essential hypertension and normal subjects. *Hypert. Res.* 2:159-66.

Tanaka, H., M. Shindo, J. Gutkowska et al. 1986. Effect of acute exercise on plasma immunoreactive-atrial natriuretic factor. *Life Sci.* 18:1685-93.

Tate, C.A., M.F. Hyek & G.E. Taffet. 1994. Mechanisms for the responses of cardiac muscle to physical activity in old age. *Med. Sci. Sports Exerc.* 26(5):561-67.

Taylor, D.V., J.G. Boyajian, N. James et al. 1994. Acidosis stimulates β-endorphin release during exercise. *J. Appl. Physiol.* 77:1913-18.

Tenbaum, S. & A. Baniahmad. 1997. Nuclear receptors: structure, function and involvement in disease. *Int. J. Biochem. Cell Biol.* 29(12):1325-41.

Teschemacher, H. 2003. Proopiomelanocortin—the precursor of ACTH and β-endorphin: functional significance for the adaptation to stress? *Arzneimittel, Therapie-Kritik.* 35:233-242.

Teschemacher, H., G. Koch, H. Scheffler et al. 1990. Opioid peptides: immunological significance? *Ann. N.Y. Acad. Sci.* 594:66-77.

Thamsborg, G., T. Storm, N. Keller et al. 1987. Changes in plasma atrial natriuretic peptide during exercise in healthy volunteers. *Acta Med. Scand.* 5:441-44.

Tholanikunnel, B.G. & C.C. Malbon. 1997. A 20-nucleotide (A + U)-rich element of beta2-adrenergic receptor (beta2AR) mRNA mediates binding to beta2AR-binding protein and is obligate for agonist-induced destabilization of receptor mRNA. *J. Biol. Chem.* 272(17):11471-78.

Tilg, H., C.A. Dinarello, & J.W. Mier. 1997. IL-6 and APPs: anti-inflammatory and immunosuppressive mediators. *Immunol Today.* 18(9):428-32.

Toft, A.D., Jensen, L.B., Bruunsgaard, H., Ibfelt, T., Halkjaer-Kristensen, J., Febbraio, M., and Pedersen, B.K. 2002. Cytokine response to eccentric exercise in young and elderly humans. *Am J Physiol* 283(1): C289-295.

Toft, A.D., K. Ostrowski, S. Asp et al. 2000. The effects of n-3 PUFA on the cytokine response to strenuous exercise. *J. Appl. Physiol.* 89:2401-05.

Tomiyama, H., T. Kushiro, S. Imai et al. 1995. Changes in the E/A ratio induced by handgrip-exercise are related to changes in the plasma atrial natriuretic peptide level, but not to changes in brain natriuretic peptide in mild essential hypertension. *Jap. Circ. J.* 9:617-23.

Trapp, T. & F. Holsboer. 1996. Heterodimerization between mineralocorticoid and glucocorticoid receptors increases the functional diversity of corticosteroid action. *Trends Pharmacol. Sci.* 17(4):145-49.

Triplett-McBride, N.T., A.M. Mastro, J.M. McBride et al. 1998. Plasma proenkephalin peptide F and human B cell responses to exercise stress in fit and unfit women. *Peptides* 19:731-38.

Tsao, P., T. Cao & M. von Zastrow. 2001. Role of endocytosis in mediating downregulation of G-protein-coupled receptors. *Trends Pharmacol. Sci.* 22(2):91-96.

Tsao, P.I. & M. von Zastrow. 2001. Diversity and specificity in the regulated endocytic membrane trafficking of G-protein-coupled receptors. *Pharmacol. Ther.* 89(2):139-47.

Turcotte, L.P. 1999. Role of fats in exercise. Types and quality. *Clin. Sports Med.* 18(3):485-98.

Ullrich, A. & J. Schlessinger. 1990. Signal transduction by receptors with tyrosine kinase activity. *Cell* 61(2): 203-12.

Ullum, H., P.M. Haahr, M. Diamant et al. 1994. Bicycle exercise enhances plasma IL-6 but does not change IL-1alpha, IL-1beta, IL-6, or TNF-alpha pre-mRNA in BMNC. *J. Appl. Physiol.* 77(1):93-97.

van Eeden, S.F. & T. Terashima. 2000. Interleukin 8 (IL-8) and the release of leukocytes from the bone marrow. *Leuk. Lymphoma* 37(3-4):259-71.

van Hall, G., A. Steensberg, M. Sacchetti et al. 2003. Interleukin-6 stimulates lipolysis and fat oxidation in humans. *J. Clin. Endocrinol. Metab.* 88(7): 3005-10.

Viru, A. & Z. Tendzegolskis. 1995. Plasma endorphin species during dynamic exercise in humans. *Clin. Physiol.* 15:73-79.

Wallenius, V., K. Wallenius, B. Ahren et al. 2002. Interleukin-6-deficient mice develop mature-onset obesity. *Nat. Med.* 8(1):75-79.

Walsh, D.A. & S.M. Van Patten. 1994. Multiple pathway signal transduction by the cAMP-dependent protein kinase. *FASEB J.* 8(15):1227-36. 1994.

Wang, H.Y., T.R. Bashore, Z.V. Tran et al. 2000. Age-related decreases in lymphocyte protein kinase C activity and translocation are reduced by aerobic fitness. *J. Gerontol. Series A Biol. Sci. Med. Sci.* 55(11):B545-51.

Wang, J.S. & L.J. Cheng. 1999. Effect of strenuous, acute exercise on alpha2-adrenergic agonist-potentiated platelet activation. *Arterioscler. Thromb. Vasc. Biol.* 19(6):1559-65.

Wang, J.S., C.J. Jen & H.I. Chen. 1997. Effects of chronic exercise and deconditioning on platelet function in women. *J. Appl. Physiol.* 83(6):2080-85.

Ward, S.G., A. Wilson, L. Turner et al. 1995. Inhibition of CD28-mediated T cell costimulation by the phosphoinositide 3-kinase wortmannin. *Eur. J. Immunol.* 25(2):526-32. 1995.

Weinstock, C., D. Konig, R. Harnischmacher et al. 1997. Effect of exhaustive exercise stress on the cytokine response. *Med. Sci. Sports. Exerc.* 29(3):345-54.

Weissman, C. 1990. The metabolic response to stress: an overview and update. *Anesthesiology* 73:308-27.

Werle, E.O., G. Strobel & H. Weicker. 1990. Decrease in rat cardiac beta 1- and beta 2-adrenoceptors by training and endurance exercise. *Life Sci.* 46(1):9-17.

Widegren, U., X.J. Jiang, A. Krook et al. 1998. Divergent effects of exercise on metabolic and mitogenic signaling pathways in human skeletal muscle. *FASEB J.* 12(13):1379-89.

Widegren, U., J.W. Ryder & J.R. Zierath. 2001. Mitogen-activated protein kinase signal transduction in skeletal muscle: effects of exercise and muscle contraction. *Acta Physiol. Scand.* 172(3):227-38.

Widegren, U., C. Wretman, A. Lionikas et al. 2000. Influence of exercise intensity on ERK/MAP kinase signalling in human skeletal muscle. *Pfluegers Arch.* 441(2-3):317-22.

Wiedemann, K. & H. Teschemacher. 1983. β-endorphin immunoreactive materials in human plasma determined by a multiple radioimmunoassay system. *Life Sci.* 33(Suppl I):89-92.

Wiley, H.S. & P.M. Burke. 2001. Regulation of receptor tyrosine kinase signaling by endocytic trafficking. *Traffic* 2(1):12-18.

Wilkinson, S.E. & J.S. Nixon. 1998. T-cell signal transduction and the role of protein kinase C. *Cell. Mol. Life Sci.* 54:1122-44.

Williamson, D., P. Gallagher, M. Harber et al. 2003. Mitogen-activated protein kinase (MAPK) pathway activation: effects of age and acute exercise on human skeletal muscle. *J. Physiol.* 15;547(Pt 3):977-87.

Winther, K. & J. Trap-Jensen. 1988. The effect of exercise on platelet beta-adrenoceptor function and platelet aggregation in healthy human volunteers. *Clin. Physiol.* 8(2):147-53.

Woodcock, E.A., J.F. Arthur & S.J. Matkovich. 2000. Inositol 1,4,5-trisphosphate and reperfusion arrhythmias. *Clin. Exp. Pharmacol. Physiol.* 27(9):734-37.

Woods, J.A., J.M. Davis, J.A. Smith et al. 1999. Exercise and cellular innate immune function. *Med. Sci. Sports Exerc.* 31(1):57-66.

Wretman, C., U. Widegren, A. Lionikas et al. 2000. Differential activation of mitogen-activated protein kinase signalling pathways by isometric contractions in isolated slow- and fast-twitch rat skeletal muscle. *Acta Physiol. Scand.* 170(1):45-49.

Wright, C.E. & J.R. Fozard. 1989. Regional vascular responses to endothelin in spontaneously hypertensive rats: comparison with Wistar-Kyoto controls. *Brit. J. Pharmacol.* 97:394P.

Wu, M.L., E.F. Kao, I.H. Liu et al. 1997. Capacitative Ca2+ influx in glial cells is inhibited by glycolytic inhibitors. *Glia* 21(3):315-26.

Xing, Z., J. Gauldie, G. Cox et al. 1998. IL-6 is an antiinflammatory cytokine required for controlling local or systemic acute inflammatory responses. *J. Clin. Invest.* 101(2):311-20.

Yamashita, N., G.F. Baxter & D.M. Yellon. 2001. Exercise directly enhances myocardial tolerance to ischaemia-reperfusion injury in the rat through a protein kinase C mediated mechanism. *Heart* 85(3): 331-36.

Yamazaki, T., I. Komuro, S. Kudoh et al. 1996. Endothelin-1 is involved in mechanical stress-induced cardiomyocyte hypertrophy. *J. Biol. Chem.* 6:3221-28.

Yanagisawa, M., H. Kurihara, S. Kimura et al. 1988. A novel potent vasoconstrictor peptide produced by vascular endothelial cells. *Nature* 6163:411-15.

Yang, Z.H., V. Richard, L. von Segesser et al. 1990. Threshold concentrations of endothelin-1 potentiate contractions to norepinephrine and serotonin in human arteries. A new mechanism of vasospasm? *Circulation* 1:188-95.

Yang-Yen, H.F., J.C. Chambard, Y.L. Sun et al. 1990. Transcriptional interference between c-Jun and the glucocorticoid receptor: mutual inhibition of DNA binding due to direct protein-protein interaction. *Cell* 62(6):1205-15.

Yokokawa, K., M. Kohno, K. Murakawa et al. 1989. Acute effects of endothelin on renal hemodynamics and blood pressure in anesthetized rats. *Am. J. Hypert.* 9:715-17.

Yoshizumi, M., H. Kurihara, T. Sugiyama et al. 1989. Hemodynamic shear stress stimulates endothelin production by cultured endothelial cells. *Biochem. Biophys. Res. Comm.* 2:859-64.

Young, E., D. Bronstein & H. Akil. 1993. Proopiomelanocortin biosynthesis, processing, and secretion: functional implications. In *Handbook of Experimental Pharmacology 104/I,* eds. A. Herz, H. Akil & E.J. Simon, 393-421. Berlin: Springer.

Yu, H.J., H. Ma & R.D. Green. 1993. Calcium entry via L-type calcium channels acts as a negative regulator of adenylyl cyclase activity and cyclic AMP levels in cardiac myocytes. *Mol. Pharmacol.* 44(4): 689-93.

Yu, M., E. Blomstrand, A.V. Chibalin et al. 2001. Marathon running increases ERK1/2 and p38 MAP kinase signalling to downstream targets in human skeletal muscle. *J. Physiol.* 1;536(Pt 1):273-82.

Zagon, I.S., M.F. Verderame, S.S. Allen et al. 1999. Cloning, sequencing, expression and function of a cDNA encoding a receptor for the opioid growth factor, [Met(5)]-enkephalin. *Brain Res.* 849:147-54.

Zagon, I.S., Y. Wu & P.J. McLaughlin. 1996. The opioid growth factor, [Met5]-enkephalin, and the zeta opioid receptor are present in human and mouse skin and tonically act to inhibit DNA synthesis in the epidermis. *J. Invest. Dermatol.* 106:490-97.

Zhang, L.Q., X.Q. Zhang, Y.C. Ng et al. 2000. Sprint training normalizes Ca(2+) transients and SR function in postinfarction rat myocytes. *J. Appl. Physiol.* 89(1):38-46.

Zhang, W. & L.E. Samelson. 2000. The role of membrane-associated adaptors in T cell receptor signalling. *Sem. Immunol.* 12(1):35-41.

Zu, Y.L., Y. Ai, A. Gilchrist et al. 1996. Activation of MAP kinase-activated protein kinase 2 in human neutrophils after phorbol ester or fMLP peptide stimulation. *Blood* 87(12):5287-96.

Zweifach, A. & R.S. Lewis. 1993. Mitogen-regulated Ca2+ current of T lymphocytes is activated by depletion of intracellular Ca2+ stores. *Proc. Natl. Acad. Sci.* 90(13):6295-99.

Chapter 8

Abu-Elheiga, L., M.M. Matzuk et al. 2001. Continuous fatty acid oxidation and reduced fat storage in mice lacking acetyl-CoA carboxylase 2. *Science* 291: 2613-16.

Abumrad, N.A., M.R. El-Maghrabi et al. 1993. Cloning of a rat adipocyte membrane protein implicated in binding or transport of long chain fatty acids that is induced during preadipocyte differentiation. Homology with human CD36. *J. Biol. Chem.* 268: 17665-68.

Ahlborg, G., P. Felig, L. Hagenfeldt et al. 1974. Substrate turnover during prolonged exercise in man. Splanchnic and leg metabolism of glucose, free fatty acids, and amino acids. *J. Clin. Invest.* 53: 1080-90.

Ahlborg, G.M. & Jensen-Urstad. 1991. Metabolism in exercising arm vs. leg muscle. *Clin. Physiol.* 11: 459-68.

Alam, N. & E.D. Saggerson. 1998. Malonyl-CoA and the regulation of fatty acid oxidation in soleus muscle. *Biochem. J.* 334: 233-41.

Angus, D.J., M.A. Febbraio & M. Hargreaves. 2002. Plasma glucose kinetics during prolonged exercise in trained humans when fed carbohydrate. *Am. J. Physiol.* 283: E573-E577.

Arkinstall, M.J., C.R. Bruce, V. Nikolopoulos et al. 2001. Effect of carbohydrate ingestion on metabolism during running and cycling. *J. Appl. Physiol.* 91: 2125-34.

Aronson, D., M.A. Violan, S.D. Dufresne et al. 1997. Exercise stimulates the mitogen-activated protein kinase pathway in human skeletal muscle. *J. Clin. Invest.* 99: 1251-57.

Azevedo, J.L. Jr., J.K. Linderman, S.L. Lehman et al. 1998. Training decreases muscle glycogen turnover during exercise. *Eur. J. Appl. Physiol.* 78: 479-86.

Baker, S.K., K.J.A. McCullagh & A. Bonen. 1998. Training intensity dependent and tissue specific increases in lactate uptake and MCT1 in heart and muscle. *J. Appl. Physiol.* 84: 987-94.

Baker, S.K., M.A. Tarnopolsky & A. Bonen. 2001. Expression of MCT1 and MCT4 in a patient with mitochondrial myopathy. *Muscle Nerve* 24: 394-98.

Bavenholm, P.N., J. Pigon et al. 2000. Fatty acid oxidation and the regulation of malonyl-CoA in human muscle. *Diabetes* 49: 1078-83.

Bergeron, R., S.F. Previs, G.W. Cline et al. 2001. Effect of 5-aminoimidazole-4-carboxamide-1-beta-D-ribofuranoside infusion on in vivo glucose and lipid metabolism in lean and obese Zucker rats. *Diabetes* 50: 1076-82.

Bergeron, R., R.R. Russell III, L.H. Young et al. 1999. Effect of AMPK activation on muscle glucose metabolism in conscious rats. *Am. J. Physiol.* 276: E938-E944.

Blomstrand, E. & B. Saltin. 1999. Effect of muscle glycogen on glucose, lactate and amino acid metabolism during exercise and recovery in human subjects. *J. Physiol.* 514: 293-302.

Boden, G. 1996. Perspectives in Diabetes. Role of fatty acids in the pathogenesis of insulin resistance and NIDDM. *Diabetes* 45: 3-10.

Boden, G., X. Chen et al. 1994. Mechanisms of fatty acid-induced inhibition of glucose uptake. *J. Clin. Invest.* 93: 2438-46.

Boden, G., F. Jadali et al. 1991. Effects of fat on insulin-stimulated carbohydrate metabolism in normal men. *J. Clin. Invest.* 88: 960-66.

Bolli, R., K.A. Nalecz & A. Azzi. 1989. Monocarboxylate and α-ketoglutarate carriers from bovine heart mitochondria. *J. Biol. Chem.* 264: 18024-30.

Bonen, A., C.J. Campbell, R.L. Kirby et al. 1979. A multiple regression model for blood lactate removal in man. *Pfluegers Arch.* 380: 205-10.

Bonen, A., D.J. Dyck et al. 1999a. Muscle contractile activity increases fatty acid metabolism and transport and FAT/CD36. *Am. J. Physiol.* 276: E642-E649.

Bonen, A., J.F. Glatz et al. 2002. Regulation of fatty acid transport and membrane transporters in health and disease. *Mol. Cell. Biochem.* 239: 181-92.

Bonen, A. & D. Homonko. 1994. Effects of exercise and glycogen depletion on glyconeogenesis in muscle. *J. Appl. Physiol.* 76: 1753-58.

Bonen, A., J.J.F.P. Luiken et al. 1998a. Palmitate transport and fatty acid transporters in red and white muscles. *Am. J. Physiol.* 275: E471-E478.

Bonen, A., J.J.F.P. Luiken et al. 2000a. Acute regulation of fatty acid uptake involves the cellular redistribution of fatty acid translocase. *J. Biol. Chem.* 275: 14501-08.

Bonen, A. & K.J.A. McCullagh. 1994. Effects of exercise on lactate transport into mouse skeletal muscles. *Can. J. Appl. Physiol.* 19: 275-85.

Bonen, A., K.J.A. McCullagh, C.T. Putman et al. 1998b. Short-term training increases human muscle MCT1 and femoral venous lactate in relation to muscle lactate. *Am. J. Physiol.* 274: E102-E107.

Bonen, A., J.C. McDermott & M.H. Tan. 1990. Glycogenesis and glyconeogenesis in skeletal muscle: effects of pH and hormones. *Am. J. Physiol.* 258: E693-E700.

Bonen, A., D. Miskovic et al. 1999b. Fatty acid transporters (FABPpm, FAT, FATP) in human muscle. *Can. J. Appl. Physiol.* 24: 515-23.

Bonen, A., D. Miskovic, M. Tonouchi et al. 2000b. Abundance and subcellular distribution of MCT1 and MCT4 in heart and fast-twitch skeletal muscles. *Am. J. Physiol.* 278: E1067-E1077.

Bonen, A., M. Tonouchi, D. Miskovic et al. 2000c. Isoform-specific regulation of the lactate transporters MCT1 and MCT4 by contractile activity. *Am. J. Physiol.* 279: E1131-E1138.

Boppart, M.D., S. Asp, J.F.P. Wojtaszewski et al. 2000. Marathon running transiently increases c-Jun NH_2-terminal kinase and p38 activities in human skeletal muscle. *J. Physiol.* 526. 3: 663-69.

Bosch, A.N., S.C. Dennis & T.D. Noakes. 1994. Influence of carbohydrate ingestion on fuel substrate turnover and oxidation during prolonged exercise. *J. Appl. Physiol.* 76: 2364-72.

Broer, S., A. Broer, H.P. Schneider et al. 1999. Characterization of the high-affinity monocarboxylate transporter MCT2 in Xenopus laevis oocytes. *Biochem. J.* 341: 529-35.

Broer, S., H.P. Schneider, A. Broer et al. 1998. Characterization of the monocarboxylate transporter 1 expressed in *Xenopus laevis* oocytes by changes in cytosolic pH. *Biochem. J.* 333: 167-74.

Brooks, G.A. 1986. Lactate production under fully aerobic conditions: the lactate shuttle during rest and exercise. *Federation Proc.* 45: 2924-29.

Brooks, G.A. 2002. Lactate shuttles in nature. *Biochem. Soc. Transact.* 30: 258-64.

Brooks, G.A., M.A. Brown, C.E. Butz et al. 1999a. Cardiac and skeletal muscle mitochondria have a monocarboxylate transporter MCT1. *J. Appl. Physiol.* 87: 1713-18.

Brooks, G.A., H. Dubouchaud, M. Brown et al. 1999b. Role of mitochondrial lactate dehydrogenase and lactate oxidation in the intracellular lactate shuttle. *Proc. Natl. Acad. Sci.* 96: 1129-34.

Brooks, G.A. & J. Mercier. 1994. Balance of carbohydrate and lipid utilization during exercise: the "crossover" concept. *J. Appl. Physiol.* 76: 2253-61.

Brooks, G.A. & J.K. Trimmer. 1996. Glucose kinetics during high-intensity exercise and the crossover concept. *J. Appl. Physiol.* 80(3): 1073-75.

Brown, M.A. & G.A. Brooks. 1994. Trans-acceleration of lactate transport from rat sarcolemmal membrane vesicles. *Arch. Biochem. Biophys.* 313: 22-28.

Burke, L.M. & J.A. Hawley. 1999. Carbohydrate and exercise. *Curr. Op. Clin. Nutr. Metab. Care* 2: 515-20.

Chasiotis, D. 1983. The regulation of glycogen phosphorylase and glycogen breakdown in human skeletal muscle. *Acta Physiol. Scand. Suppl.* 518: 1-68.

Chatham, J.C. & J.R. Forder. 1996. Metabolic compartmentation of lactate in the glucose-perfused rat heart. *Am. J. Physiol.* 270: H224-H229.

Chatham, J.C., Z.-P. Gao, A. Bonen et al. 1999. Preferential inhibition of lactate oxidation relative to glucose oxidation in the rat heart following diabetes. *Cardiovasc. Res.* 43: 96-106.

Chen, Z.-P., G.K. McConell, B.J. Michell et al. 2000. AMPK signaling in contracting human skeletal muscle: acetyl-CoA carboxylase and NO synthase phosphorylation. *Am. J. Physiol.* 279: E1202-06.

Chibalin, A.V., M. Yu, J.W. Ryder et al. 2000. Exercise-induced changes in expression and activity of proteins involved in insulin signal transduction in skeletal muscle: differential effects on insulin receptor substrates 1 and 2. *Proc. Natl. Acad. Sci.* 97: 38-43.

Clifford, G.M., C. Londos et al. 2000. Translocation of hormone-sensitive lipase and perilipin upon lipolytic stimulation of rat adipocytes. *J. Biol. Chem.* 275: 5011-15.

Cline, G.W., A.J. Vidal-Puig et al. 2001. In vivo effects of uncoupling protein-3 gene disruption on mitochondrial energy metabolism. *J. Biol. Chem.* 276: 20240-44.

Coburn, C.T., F.F. Knapp Jr. et al. 2000. Defective uptake and utilization of long chain fatty acids in muscle and adipose tissue of CD36 knockout mice. *J. Biol. Chem.* 275: 32523-29.

Coe, R., A. Johnston-Smith et al. 1999. The fatty acid transport protein (FATP1) is a very long chain acyl-CoA synthetase. *J. Biol. Chem.* 274: 36300-04.

Coggan, A.R. & E.F. Coyle. 1991. Carbohydrate ingestion during prolonged exercise: effects on metabolism and performance. *Exerc. Sport Sci. Rev.* 19. 1-40.

Coggan, A.R., W.M. Kohrt, R.J. Spina et al. 1990. Endurance training decreases plasma glucose turnover and oxidation during moderate-intensity exercise in men. *J. Appl. Physiol.* 68: 990-96.

Coggan, A.R., W.M. Kohrt, R.J. Spina et al. 1992. Plasma glucose kinetics during exercise in subjects with high and low lactate thresholds. *J. Appl. Physiol.* 73: 1873-80.

Coggan, A.R., S.C. Swanson, L.A. Mendenhall et al. 1995. Effect of endurance training on hepatic glycogenolysis and gluconeogenesis during prolonged exercise in men. *Am. J. Physiol.* 268: E375-E383.

Colberg, S.R., J.-A. Simoneau et al. 1995. Skeletal muscle utilization of free fatty acids in women with visceral obesity. *J. Clin. Invest.* 95: 1846-53.

Coleman, R.A., T.M. Lewin et al. 2000. Physiological and nutritional regulation of enzymes of triacylglycerol synthesis. *Annu. Rev. Nutr.* 20: 77-103.

Conlee, R.K., J.A. McLane, M.J. Rennie et al. 1979. Reversal of phosphorylase activation in muscle despite continued contractile activity. *Am. J. Physiol.* 237: R291-R296.

Connett, R.J., C.R. Honig, T.E. Gayeski et al. 1990. Defining hypoxia: a systems view of VO_2, glycolysis, energetics, and intracellular PO_2. *J. Appl. Physiol.* 68: 833-42.

Constable, S.H., R.J. Favier & J.O. Holloszy. 1986. Exercise and glycogen depletion: effects on ability to activate muscle phosphorylase. *J. Appl. Physiol.* 60: 1518-23.

Costill, D.L., E. Coyle et al. 1977. Effects of elevated plasma FFA and insulin on muscle glycogen usage during exercise. *J. Appl. Physiol.* 43: 695-99.

Coyle, E.F. 1997. Fuels for sports performance. *Perspectives in Exercise Science and Sports Medicine,* ed. D.R. Lamb & R. Murray, vol. 10, 95-129. Optimising sport performance. Cooper Publishing Group.

Coyle, E.F., A.R. Coggan, M.K. Hemmert et al. 1986. Muscle glycogen utilization during prolonged strenuous exercise when fed carbohydrate. *J. Appl. Physiol.* 61: 165-72.

Dean, D., J.R. Daugaard et al. 2000. Exercise diminishes the activity of acetyl-CoA carboxylase in human muscle. *Diabetes* 49: 1295-300.

DeFronzo, R.A., E. Jacot, E. Jequier et al. 1981. The effect of insulin on the disposal of intravenous glucose. Results from indirect calorimetry and hepatic and femoral venous catheterization. *Diabetes* 30: 1000-07.

Dela, F., J.J. Larsen, K.J. Mikines et al. 1995. Insulin-stimulated muscle glucose clearance in patients with NIDDM. Effects of one-legged physical training. *Diabetes* 44: 1010-20.

Derave, W., S. Lund, G.D. Holman et al. 1999. Contraction-stimulated muscle glucose transport and GLUT4 surface content are dependent on glycogen content. *Am. J. Physiol.* 277: E1103-E1110.

Derave, W., B.F. Hansen, S. Lund et al. 2000. Muscle glycogen content affects insulin-stimulated glucose transport and protein kinase B activity. *Am. J. Physiol.* 279: E947-55.

Dimmer, K-S., B. Friedrich, F. Lang et al. 2000. The low affinity monocarboxylate transporter MCT4 is adapted to the export of lactate in highly glycolytic cells. *Biochem. J.* 350: 219-27.

Douen, A.G., T. Ramlal, S. Rastogi et al. 1990. Exercise induces recruitment of the "insulin-responsive glucose transporter." Evidence for distinct intracellular insulin- and exercise-recruitable transporter pools in skeletal muscle. *J. Biochem.* 265: 13427-30.

Duan, C. & W.W. Winder. 1993. Control of malonyl-CoA by glucose and insulin in perfused skeletal muscle. *J. Appl. Physiol.* 74: 2543-47.

Dubouchaud, H., G.E. Butterfield, E.E. Wolfel et al. 2000. Endurance training, expression, and physiology of LDH, MCT1, and MCT4 in human skeletal muscle. *Am. J. Physiol.* 278: E571-E579.

Dubouchaud, H., N. Eydoux, P. Granier et al. 1999. Lactate transport activity in rat skeletal muscle sarcolemmal vesicles after acute exhaustive exercise. *J. Appl. Physiol.* 87: 955-61.

Dubouchaud, H., P. Granier, J. Mercier et al. 1996. Lactate uptake by skeletal muscle sarcolemmal vesicles decreases after 4 wk of hindlimb unweighting in rats. *J. Appl. Physiol.* 80: 416-21.

Durante, P.E., K.J. Mustard et al. 2002. Effects of endurance training on activity and expression of AMP-activated protein kinase isoforms in rat muscles. *Am. J. Physiol.* 283: E178-86.

Dyck, D.J. & A. Bonen. 1998. Muscle contraction increases palmitate esterification and oxidation, and triacylglycerol oxidation. *Am. J. Physiol.* 275: E888-E896.

Dyck, D.J., D. Miskovic et al. 2000. Endurance training increases FFA oxidation and reduces triacylglycerol utilization in contracting rat soleus. *Am. J. Physiol.* 278: E778-E785.

Dyck, D.J., S.J. Peters et al. 1996a. Effect of high FFA on glycogenolysis in oxidative rat hindlimb muscles during twitch stimulation. *Am. J. Physiol.* 270: R766-76.

Dyck, D.J., S.J. Peters et al. 1996b. Regulation of muscle glycogen phosphorylase activity during intense aerobic cycling with elevated FFA. *Am. J. Physiol.* 270: E116-25.

Dyck, D.J., S.J. Peters et al. 1997. Functional differences in lipid metabolism in resting skeletal muscle of various fiber types. *Am. J. Physiol.* 272: E340-E351.

Dyck, D.J., C.T. Putman et al. 1993. Regulation of fat-carbohydrate interaction in skeletal muscle during intense aerobic cycling. *Am. J. Physiol.* 265: E852-59.

Dyck, D.J. & L.L. Spriet. 1994. Elevated muscle citrate does not reduce carbohydrate utilization during tetanic stimulation. *Can. J. Physiol. Pharmacol.* 72: 117-25.

Dyck, D.J., G. Steinberg et al. 2001. Insulin increases FA uptake and esterification but reduces lipid utilization in isolated contracting muscle. *Am. J. Physiol.* 281: E600-07.

Elayan, I.M. & W.W. Winder. 1991. Effect of glucose infusion on muscle malonyl-CoA during exercise. *J. Appl. Physiol.* 70: 1495-99.

Enoksson, S., E. Degerman et al. 1998. Various phosphodiesterase subtypes mediate the in vivo antilipolytic effect of insulin on adipose tissue and skeletal muscle in man. *Diabetologia* 41: 560-68.

Entman, M.L., S.S. Keslensky, A. Chu et al. 1980. The sarcoplasmic reticulum-glycogenolytic complex in mammalian fast twitch skeletal muscle. Proposed in vitro counterpart of the contraction-activated glycogenolytic pool. *J. Biochem.* 255: 6245-52.

Eydoux, N., H. Dubouchaud, G. Py et al. 2000a. Lactate transport in rat sarcolemmal vesicles after a single bout of submaximal exercise. *Int. J. Sports Med.* 21: 393-99.

Eydoux, N., G. Py, K. Lambert et al. 2000b. Training does not protect against exhaustive exercise-induced lactate transport capacity. *Am. J. Physiol.* 278: E10454-52.

Febbraio, M., N.A. Abumrad et al. 1999. A null mutation in murine CD36 reveals an important role in fatty acid and lipoprotein metabolism. *J. Biol. Chem.* 274: 19055-62.

Fitts, R. & J. Holloszy. 1976. Lactate and contractile force in frog muscle during development of fatigue and recovery. *Am. J. Physiol.* 231: 430-33.

Foster, C., D.L. Costill & W.J. Fink. 1979. Effects of pre-exercise feedings on endurance performance. *Med. Sci. Sports* 11: 1-5.

Fujii, N., T. Hayashi, M.F. Hirshman et al. 2000. Exercise induces isoform-specific increase in 5′-AMP-activated protein kinase activity in human skeletal muscle. *Biochem. Biophys. Res. Com.* 273: 1150-55.

Furler, S.M., A.B. Jenkins, L.H. Storlien et al. 1991. In vivo location of the rate-limiting step of hexose uptake in muscle and brain tissue of rats. *Am. J. Physiol.* 261: E337-E347.

Garcia, C.K., M.S. Brown, R.K. Pathak et al. 1995. cDNA cloning of MCT2, a second monocarboxylate transporter expressed in different cells than MCT1. *J. Biol. Chem.* 270: 1843-49.

Garcia, C.K., J.L. Goldstein, R.K. Pathak et al. 1994a. Molecular characterization of a membrane transporter for lactate, pyruvate, and other monocarboxylates: implications for the Cori cycle. *Cell* 76: 865-73.

Garcia, C.K., X. Li, J. Luna et al. 1994b. cDNA cloning of the human monocarboxylate transporter 1 and chromosomal localization of the SLC16A1 locus to 1p13.2-p12. *Genomica* 23: 500-03.

Ghilardi, N., S. Ziegler et al. 1996. Defective STAT signaling by the leptin receptor in diabetic mice. *Proc. Natl. Acad. Sci.* 93: 6231-35.

Giacobino, J.P. 1999. Effects of dietary deprivation, obesity and exercise on UCP3 mRNA levels. *Int. J. Obes. Relat. Metab. Disord.* 23(Suppl 6): S60-63.

Gladden, L.B. 1996. Lactate transport and exchange during exercise. New York: Oxford University Press.

Goodyear, L.J. & B.B. Kahn. 1998. Exercise, glucose transport, and insulin sensitivity. *Annu. Rev. Med.* 49: 235-61.

Green, H.J., S. Jones, M. Ball-Burnett et al. 1995. Adaptations in muscle metabolism to prolonged voluntary exercise and training. *J. Appl. Physiol.* 78: 138-45.

Guo, Z., B. Burguera et al. 2000. Kinetics of intramuscular triglyceride fatty acids in exercising humans. *J. Appl. Physiol.* 89: 2057-64.

Hagenfeldt, L. 1979. Metabolism of free fatty acids and ketone bodies during exercise in normal and diabetic man. *Diabetes* 28(Suppl 1): 68-70.

Halestrap, A.P. & N.T. Price. 1999. The proton-linked monocarboxylate transporter family: structure, function and regulation. *Biochem. J.* 343: 281-99.

Halestrap, A.P., X. Wang, R.C. Poole et al. 1997. Lactate transport in heart in relation to myocardial ischemia. *Am. J. Cardiol.* 80: 17A-25A.

Hamilton, J.A. & F. Kamp. 1999. How are free fatty acids transported in membranes? Is it by proteins or by free diffusion through the lipids? *Diabetes* 48: 2255-69.

Hardie, D.G. & D. Carling. 1997. The AMP-activated protein kinase—fuel gauge of the mammalian cell? *Eur. J. Biochem.* 246: 259-73.

Hardie, D.G., D. Carling et al. 1998. The AMP-activated/SNF1 protein kinase subfamily: metabolic sensors of the eukaryotic cell? *Annu. Rev. Biochem.* 67: 821-55.

Hargreaves, M., G. McConell & J. Proietto. 1995. Influence of muscle glycogen on glycogenolysis and glucose uptake during exercise in humans. *J. Appl. Physiol.* 78: 288-92.

Haschke, R.H., L.M. Heilmeyer Jr., F. Meyer et al. 1970. Control of phosphorylase activity in a muscle glycogen particle. Regulation of phosphorylase phosphatase. *J. Biochem.* 245: 6657-63.

Hatta, H., M. Tonouchi & A. Bonen. 2001. Tissue-specific and isoform-specific changes in MCT1 and 4 in heart and soleus muscle during a 1-yr period. *Am. J. Physiol.* 281: E749-E756.

Hawley, J.A., A.N. Bosch, S.M. Weltan et al. 1994. Glucose kinetics during prolonged exercise in euglycaemic and hyperglycaemic subjects. *Pfluegers Arch.* 426: 378-86.

Hawley, J.A., S.C. Dennis & T.D. Noakes. 1995. Carbohydrate, fluid, and electrolyte requirements during prolonged exercise. In *Sports Nutrition Minerals and Electrolytes,* ed. C.V. Kies & J.A. Driskell, 235-65. Boca Raton, FL: CRC Press.

Hawley, J.A., E.J. Schabort, T.D. Noakes et al. 1997. Carbohydrate-loading and exercise performance: an update. *Sports Med.* 24: 73-81.

Hayashi, T., M.F. Hirshman, N. Fujii et al. 1999. Metabolic stress and altered glucose transport: activation of AMP-activated protein kinase as a unifying coupling mechanism. *Diabetes* 49: 527-31.

Hayashi, T., M.F. Hirshman, E.J. Kuth et al. 1998. Evidence for 5'-AMP-activated protein kinase mediation of the effect of muscle contraction on glucose transport. *Diabetes* 47: 1369-73.

Heilmeyer, L.M. Jr., F. Meyer, R.H. Haschke et al. 1970. Control of phosphorylase activity in a muscle glycogen particle. Activation by calcium. *J. Biochem.* 245: 6649-56.

Henin, N., M.F. Vincent, H.E. Gruber et al. 1995. Inhibition of fatty acid and cholesterol synthase by stimulation of AMP-activated protein kinase. *FASEB J.* 9: 541-46.

Henriksen, E.J., R.E. Bourey, K.J. Rodnick et al. 1990. Glucose transporter protein content and glucose transport capacity in rat skeletal muscles. *Am. J. Physiol.* 259: E593-E598.

Hespel, P. & E.A. Richter. 1990. Glucose uptake and transport in contracting, perfused rat muscle with different pre-contraction glycogen concentrations. *J. Physiol.* 427: 347-59.

Hirche, H.H., V. Hombach, D. Langohr et al. 1975. Lactic acid permeation rate in working gastrocnemii of dogs during metabolic alkalosis and acidosis. *Pfluegers Arch.* 356: 209-22.

Hirche, H.H., D. Langohr & U. Wacher. 1970. Lactic acid permeation from skeletal muscle. *Pfluegers Arch.* 319: R109.

Hirsch, D., A. Stahl et al. 1998. A family of fatty acid transporters conserved from mycobacterium to man. *Pro. Natl. Acad. Sci.* 95: 8625-29.

Holloszy, J.O. 2003. A forty-year memoir of research on the regulation of glucose transport into muscle. *Am. J. Physiol.* 284: E453-E467.

Holloszy, J.O., W.M. Kohrt & P.A. Hansen. 1998. The regulation of carbohydrate and fat metabolism during and after exercise. *Front. Biosci.* 15: D1011-D1027.

Holmes, B.F., E.J. Kurth-Kraczek & W.W. Winder. 1999. Chronic activation of 5'-AMP-activated protein kinase increases GLUT4, hexokinase and glycogen in muscle. *J. Appl. Physiol.* 87: 1990-95.

Hopp, J.F. & W.K. Palmer. 1990. Effect of electrical stimulation on intracellular triacyglycerol in isolated skeletal muscle. *J. Appl. Physiol.* 68: 348-54.

Hotta, K., T. Funahashi et al. 2001. Circulating concentrations of the adipocyte protein adiponectin are decreased in parallel with reduced insulin sensitivity during the progression to type 2 diabetes in rhesus monkeys. *Diabetes* 50: 1126-33.

Houmard, J.A., P.C. Egan, P.D. Neufer et al. 1991. Elevated skeletal muscle glucose transporter levels in exercise-trained middle-aged men. *Am. J. Physiol.* 261: E437-E443.

Houmard, J.A., C.D. Shaw, M.S. Hickey et al. 1999. Effect of short-term exercise training on insulin-stimulated PI 3-kinase activity in human skeletal muscle. *Am. J. Physiol. Endoc. Metab.* 277: E1055-E1060.

Hughes, V.A., M.A. Fiatarone, R.A. Fielding et al. 1993. Exercise increases muscle GLUT4-levels and insulin action in subjects with impaired glucose tolerance. *Am. J. Physiol.* 264: E855-E862.

Hurley, B.F., P.M. Nemeth, W.H. Martin 3rd et al. 1986. Muscle triglyceride utilization during exercise: effect of training. *J. Appl. Physiol.* 60: 562-67.

Ibrahimi, A., A. Bonen et al. 1999. Muscle-specific overexpression of FAT/CD36 enhances fatty acid oxidation by contracting muscles, reduces plasma triglycerides and fatty acids, and increases plasma glucose and insulin. *J. Biol. Chem.* 274: 26761-66.

Isola, L.M., S.L. Zhou et al. 1995. 3T3 fibroblasts transfected with a cDNA for mitochondrial aspartate aminotransferase express plasma membrane fatty acid-binding protein and saturable fatty acid uptake. *Pro. Natl. Acad. Sci.* 92: 9866-70.

Jackson, V.N., N.T. Price & A.P. Halestrap. 1997. Cloning of the monocarboxylate transporter isoform MCT2 from rat testis provides evidence that the expression is species specific and may involve post-transcriptional regulation. *Biochem. J.* 324: 447-53.

Jacob, S., B. Hauer et al. 1999. Lipolysis in skeletal muscle is rapidly regulated by low physiological doses of insulin. *Diabetologia* 42: 1171-74.

Jeukendrup, A.E., W.H. Saris et al. 1998a. Fat metabolism during exercise: a review. Part I: fatty acid mobilization and muscle metabolism. *Int. J. Sports Med.* 19: 231-44.

Jeukendrup, A.E., W.H. Saris et al. 1998b. Fat metabolism during exercise: a review—part II: regulation of metabolism and the effects of training. *Int. J. Sports Med.* 19: 293-302.

Jeukendrup, A.E., W.H. Saris et al. 1998c. Fat metabolism during exercise: a review—part III: effects of nutritional interventions. *Int. J. Sports Med.* 19: 371-79.

Jorfeldt, L., A. Juhlin-Dannfelt & J. Karlsson. 1978. Lactate release in relation to tissue lactate in human skeletal muscle during exercise. *J. Appl. Physiol.* 44: 350-52.

Jorfeldt, L. & J. Wahren. 1970. Human forearm muscle metabolism during exercise. Quantitative aspects of glucose uptake and lactate production during prolonged exercise. *Scand. J. Clin. Lab. Invest.* 26: 73-78.

Juel, C. 1991a. Muscle lactate transport studied in sarcolemmal giant vesicles. *Biochim. Biophys. Acta* 1065: 15-20.

Juel, C. 1991b. Human muscle lactate transport can be studied in sarcolemmal giant vesicles made from needle-biopsies. *Acta Physiol. Scand.* 142: 133-34.

Juel, C. 1995. Regulation of cellular pH in skeletal muscle fiber types studied with sarcolemmal giant vesicles obtained from rat muscles. *Biochim. Biophys. Acta* 1265: 127-32.

Juel, C. 1996a. Lactate/proton co-transport in skeletal muscle: regulation and importance for pH homeostasis. *Acta Physiol. Scand.* 156: 369-74.

Juel, C. 1996b. Symmetry and pH dependency of the lactate/proton carrier in skeletal muscle studied with rat sarcolemmal giant vesicles. *Biochim. Biophys. Acta* 1283: 106-10.

Juel, C. 1997. Lactate-proton cotransport in skeletal muscle. *Physiol. Rev.* 77: 1-37.

Juel, C., A. Honig & H. Pilegaard. 1991. Muscle lactate transport studied in sarcolemmal giant vesicles: dependence on fiber type and age. *Acta Physiol. Scand.* 143: 361-65.

Juel, C., S. Kristiansen, H. Pilegaard et al. 1994. Kinetics of lactate transport in sarcolemmal giant vesicles obtained from human skeletal muscle. *J. Appl. Physiol.* 76: 1031-36.

Juel, C. & F. Wibrand. 1989. Lactate transport in isolated mouse muscles studied with a tracer technique—kinetics, stereospecificity, pH dependency and maximal capacity. *Acta Physiol. Scand.* 137: 33-39.

Karlsson, J., S. Rosell & B. Saltin. 1972. Carbohydrate and fat metabolism in contracting canine muscle. *Pfluegers Arch. Eur. J. Physiol.* 331: 57-69.

Katz, A., S. Broberg, K. Sahlin et al. 1986. Leg glucose uptake during maximal dynamic exercise in humans. *Am. J. Physiol.* 251: E65-E70.

Kaushik, V.K., M.E. Young et al. 2001. Regulation of fatty acid oxidation and glucose metabolism in rat soleus muscle: effects of AICAR. *Am. J. Physiol.* 281(2): E335-40.

Kellerer, M., M. Koch et al. 1997. Leptin activates PI-3 kinase in C_2C_{12} myotubes via janus kinase-2 (JAK-2) and insulin receptor substrate-2 (IRS-2) dependent pathways. *Diabetologia* 40: 1358-62.

Kelley, D.E., B. Goodpaster et al. 1999. Skeletal muscle fatty acid metabolism in association with insulin resistance, obesity, and weight loss. *Am. J. Physiol.* 277: E1130-E1141.

Kelley, D.E., J.P. Reilly et al. 1990. Effects of insulin on skeletal muscle glucose storage, oxidation, and glycolysis in humans. *Am. J. Physiol.* 258: E923-29.

Kelley, D.E. & J.-A. Simoneau. 1994. Impaired free fatty acid utilization by skeletal muscle in non-insulin-dependent diabetes mellitus. *J. Clin. Invest.* 94: 2349-56.

Kelley, K.M., J.J. Hamann, C. Navarre et al. 2001. Lactate metabolism in resting and contracting skeletal muscle with elevated lactate concentrations. *J. Appl. Physiol.* 93: 865-72.

Kern, M., J.A. Wells, J.M. Stephens et al. 1990. Insulin responsiveness in skeletal muscle is determined by glucose transporter (GLUT4) protein level. *Biochem. J.* 270: 397-400.

Kiens, B., B. Essen-Gustavsson, N.J. Christensen et al. 1993. Skeletal muscle substrate utilization during submaximal exercise in man: effect of endurance training. *J. Physiol.* 469: 459-78.

Kiens, B., S. Kristiansen et al. 1997. Membrane associated fatty acid binding protein (FABPpm) in human skeletal muscle is increased by endurance training. *Biochem. Biophys. Res. Comm.* 231: 463-65.

Kim, J.-Y., R.C. Hickner et al. 2000. Lipid oxidation is reduced in obese human skeletal muscle. *Am. J. Physiol.* 279: E1039-19.

Kirk, P., M.C. Wilson, C. Heddle et al. 2000. CD 147 is tightly associated with lactate transporters MCT1 and MCT4 and facilitates their cell surface expression. *EMBO J.* 19: 3896-04.

Kirwan, J.P., L.F. Del Aguila, J.M. Hernandez et al. 2000. Regular exercise enhances activation of IRS-1-associated PI3-kinase in human skeletal muscle. *J. Appl. Physiol.* 88: 797-803.

Kjaer, M., K. Howlett et al. 2000. Adrenaline and glycogenolysis in skeletal muscle during exercise: a study in adrenalectomised humans. *J. Physiol.* 528: 371-78.

Klein, S., E.F. Coyle & R.R. Wolfe. 1994. Fat metabolism during low-intensity exercise in endurance-trained and untrained men. *Am. J. Physiol.* 267: E934-40.

Koehler-Stec, E.M., I.A. Simpson, S.J. Vannucci et al. 1998. Monocarboxylate transporter expression in mouse brain. *Am. J. Physiol.* 275: E516-24.

Krook, A., U. Widegren, X.J. Jiang et al. 2000. Effects of exercise on mitogen- and stress-activated kinase signal transduction in human skeletal muscle. *Am. J. Physiol.* 279: R1716-21.

Kurth-Kraczek, E., M.F. Hirshman, L.J. Goodyear et al. 1999. 5' AMP-activated protein kinase activation causes GLUT4 translocation in skeletal muscle. *Diabetes* 48: 1667-71.

Langfort, J., T. Ploug et al. 1999. Expression of hormone-sensitive lipase and its regulation by adrenaline in skeletal muscle. *Biochem. J.* 340(Pt 2): 459-65.

Lau, R., W.D. Blinn et al. 2001. Stimulatory effects of leptin and muscle contraction on fatty acid metabolism are not additive. *Am. J. Physiol.* 281: E122-29.

Lin, R.Y., J.C. Vera, R.S. Chaganti et al. 1998. Human monocarboxylate transporter 2 (MCT2) is a high affinity pyruvate transporter. *J. Biol. Chem.* 273: 28959-65.

Lindsay, R.S., T. Funahashi et al. 2002. Adiponectin and development of type 2 diabetes in the Pima Indian population. *Lancet* 360(9326): 57-58.

Lonnqvist, F., L. Nordfors et al. 1997. Leptin secretion from adipose tissue in women. Relationship to plasma levels and gene expression. *J. Clin. Invest.* 99(10): 2398-404.

Lopaschuk, G.D. & M. Saddik. 1992. The relative contribution of glucose and fatty acids to ATP production in hearts reperfused following ischemia. *Mol. Cell. Biochem.* 116: 111-16.

Luiken, J.J.F.P., Y. Arumugam et al. 2001. Increased rates of fatty acid uptake and plasmalemmal fatty acid transporters in obese Zucker rats. *J. Biol. Chem.* 276: 40567-73.

Luiken, J.J.F.P., Y. Arumugam et al. 2002a. Changes in fatty acid transport and transporters are related to the severity of insulin deficiency. *Am. J. Physiol.* 282: 612-21.

Luiken, J.J.F.P., D.J. Dyck et al. 2002b. Insulin induces the translocation of the fatty acid transporter FAT/CD36 to the plasma membrane. *Am. J. Physiol.* 282: E491-E495.

Luiken, J.J.F.P., D.P.Y. Koonen, J. Willems et al. 2002c. Insulin stimulates long-chain fatty acid utilization by rat cardiac myocytes through cellular redistribution of FAT/CD36. *Diabetes* 51: 3113-19.

Luiken, J.J.F.P., L.P. Turcotte et al. 1999. Protein-mediated palmitate uptake and expression of fatty acid transport proteins in heart giant vesicles. *J. Lipid Res.* 40: 1007-16.

Lund, S., G.D. Holman, O. Schmitz et al. 1995. Contraction stimulates translocation of glucose transporter GLUT4 in skeletal muscle through a mechanism distinct from that of insulin. *Proc. Natl. Acad. Sci.* 92: 5817-21.

MacRae, H.S., S.C. Dennis, A.N. Bosch et al. 1992. Effects of training on lactate production and removal during progressive exercise in humans. *J. Appl. Physiol.* 72: 1649-56.

Maehlum, S., A.T. Hostmark & L. Hermansen. 1977. Synthesis of muscle glycogen during recovery after prolonged severe exercise in diabetic and non-diabetic subjects. *Scand. J. Clin Lab. Invest.* 37: 309-16.

Maggs, D.G., R. Jacob et al. 1995. Interstitial fluid concentrations of glycerol, glucose, and amino acids in human quadricep muscle and adipose tissue. Evidence for significant lipolysis in skeletal muscle. *J. Clin. Invest.* 96: 370-77.

Manco, M., G. Mingrone et al. 2000. Insulin resistance directly correlates with increased saturated fatty acids in skeletal muscle triglycerides. *Metabolism* 49: 220-24.

Mandarino, L.J., A. Consoli et al. 1996. Interaction of carbohydrate and fat fuels in human skeletal muscle: impact of obesity and NIDDM. *Am. J. Physiol.* 270: E463-70.

Manning-Fox, J.E., D. Meredith, & A.P. Halestrap. 2000. Characterisation of human monocarboxylate transporter 4 substantiates its role in lactic acid efflux from skeletal muscle. *J. Physiol.* 529:285-93.

Matsubara, M., S. Maruoka et al. 2002. Inverse relationship between plasma adiponectin and leptin concentrations in normal-weight and obese women. *Eur. J. Endocrinol.* 147: 173-80.

McConell, G., S. Fabris, J. Proietto et al. 1994. Effect of carbohydrate ingestion on glucose kinetics during exercise. *J. Appl. Physiol.* 77: 1537-41.

McCullagh, K.J.A. & A. Bonen. 1995a. L(+)-lactate binding to a protein in rat skeletal muscle plasma membranes. *Can. J. Appl. Physiol.* 20: 112-24.

McCullagh, K.J.A. & A. Bonen. 1995b. Reduced lactate transport in denervated rat skeletal muscle. *Am. J. Physiol.* 268: R884-R888.

McCullagh, K.J.A., C. Juel, M. O'Brien et al. 1996a. Chronic muscle stimulation increases lactate transport in rat skeletal muscle. *Mol. Cell. Biochem.* 156: 51-57.

McCullagh, K.J.A., R.C. Poole, A.P. Halestrap et al. 1996b. Role of the lactate transporter (MCT1) in skeletal muscles. *Am. J. Physiol.* 271: E143-E150.

McCullagh, K.J.A., R.C. Poole, A.P. Halestrap et al. 1997. Chronic electrical stimulation increases MCT1 and lactate uptake in red and white skeletal muscle. *Am. J. Physiol.* 273: E 239-E246.

McDermott, J.C. & A. Bonen. 1992. Glyconeogenic and oxidative lactate utilization in skeletal muscle. *Can. J. Physiol. Pharmacol.* 70: 142-49.

McDermott, J.C. & A. Bonen. 1993a. Endurance training increases skeletal muscle lactate transport. *Acta Physiol. Scand.* 147: 323-27.

McDermott, J.C. & A. Bonen. 1993b. Lactate transport by skeletal muscle sarcolemmal vesicles. *Mol. Cell. Biochem.* 122: 113-21.

McDermott, J.C. & A. Bonen. 1994. Lactate transport in rat sarcolemmal vesicles and intact skeletal muscle, and after muscle contraction. *Acta Physiol. Scand.* 151: 17-28.

McGrail, J., A. Bonen & A. Belcastro. 1978. Dependence of lactate removal on muscle metabolism in man. *Eur. J. Appl. Physiol.* 39: 89-97.

McLane, J.A. & J.O. Holloszy. 1979. Glycogen synthesis from lactate in three types of skeletal muscle. *J. Biol. Chem.* 254: 6548-53.

Mendenhall, L.A., S.C. Swanson, D.L. Habash et al. 1994. Ten days of exercise training reduces glucose production and utilization during moderate-intensity exercise. *Am. J. Physiol.* 266: E136-E143.

Merezhinskaya, N., W. Fishbein, J.I. Davis et al. 2000. Mutations in MCT1 cDNA in patients with symptomatic deficiency in lactate transport. *Muscle Nerve* 23: 90-97.

Merrill, G.F., E.J. Kurth et al. 1997. AICA riboside increases AMP-activated protein kinase, fatty acid oxidation, and glucose uptake in rat muscle. *Am. J. Physiol.* 273: E1107-12.

Metzger, J. & R.H. Fitts. 1987. Role of intracellular pH in muscle fatigue. *J. Appl. Physiol.* 62: 1392-97.

Meyer, F., L.M. Heilmeyer Jr., R.H. Haschke et al. 1970. Control of phosphorylase activity in a muscle glycogen particle. I. Isolation and characterization of the protein-glycogen complex. *J. Biochem.* 245: 6642-48.

Miller, B.F., J.A. Fattor, K.A. Jacobs et al. 2002. Metabolic and cardiorespiratory responses to "the lactate clamp." *Am. J. Physiol.* 283: E889-E898.

Millet, L., H. Vidal et al. 1997. Increased uncoupling protein-2 and -3 mRNA expression during fasting in obese and lean humans. *J. Clin. Invest.* 100: 2665-70.

Minokoshi, Y., Y.B. Kim et al. 2002. Leptin stimulates fatty-acid oxidation by activating AMP-activated protein kinase. *Nature* 415(6869): 339-43.

Mu, J., J.T.J. Brozinick, O. Valladares et al. 2001. A role for AMP-activated protein kinase in contraction- and hypoxia-regulated glucose transport in skeletal muscle. *Mol. Cell* 7: 1085-94.

Muoio, D.M., G.L. Dohm et al. 1997. Leptin directly alters lipid partitioning in skeletal muscle. *Diabetes* 46: 1360-63.

Muoio, D.M., G.L. Dohm et al. 1999a. Leptin opposes insulin's effects on fatty acid partitioning in muscles isolated from obese ob/ob mice. *Am. J. Physiol.* 276: E913-21.

Muoio, D.M., K. Seefeld et al. 1999b. AMP-activated kinase reciprocally regulates triacylglycerol synthesis and fatty acid oxidation in liver and muscle: evidence that sn-glycerol-3-phosphate acyltransferase is a novel target. *Biochem. J.* 338(Pt 3): 783-91.

Nilsson, L.H., P. Furst & E. Hultman. 1973. Carbohydrate metabolism of the liver in normal man under varying dietary conditions. *Scand. J. Clin. Lab. Invest.* 32: 331-37.

Odland, L.M., G.J.F. Heigenhauser et al. 1996. Human skeletal muscle malonyl-CoA at rest and during prolonged submaximal exercise. *Am. J. Physiol.* 270: E541-E544.

Odland, L.M., G.J.F. Heigenhauser et al. 1998a. Effects of increased fat availability on fat-carbohydrate interaction during prolonged exercise in men. *Am. J. Physiol.* 274: R894-R902.

Odland, L.M., R.A. Howlett et al. 1998b. Skeletal muscle malonyl-CoA content at the onset of exercise at varying power outputs in humans. *Am. J. Physiol.* 274: E1080-85.

Ojuka, E.O., L.A. Nolte & J.O. Holloszy. 2000. Increased expression of GLUT4 and hexokinase in rat epitrochlearis muscles exposed to AICAR in vitro. *J. Appl. Physiol.* 88: 1072-75.

Owles, W.H. 1930. Alterations in the lactic acid content of the blood as a result of light exercise, and the associated changes in the CO2 combining power of the blood and in the alveolar CO2 pressure. *J. Physiol.* 69: 214-37.

Pagliassotti, M.J. & C.J. Donovan. 1990. Role of cell type in net lactate removal by skeletal muscle. *Am. J. Physiol.* 258: E635-E642.

Pan, D.A., S. Lillioja et al. 1997. Skeletal muscle triglyceride levels are inversely related to insulin action. *Diabetes* 46: 983-88.

Park, S.H., S.R. Gammon et al. 2002. Phosphorylation-activity relationships of AMPK and acetyl-CoA carboxylase in muscle. *J. Appl. Physiol.* 92: 2475-82.

Perseghin, G., P. Scifo et al. 1999. Intramyocellular triglyceride content is a determinant of in vivo insulin resistance in humans: a 1H-13C nuclear magnetic resonance spectroscopy assessment in offspring of type 2 diabetic parents. *Diabetes* 48: 1600-06.

Peters, S.J., D.J. Dyck et al. 1998. Effect of epinephrine on lipid metabolism in resting skeletal muscle. *Am. J. Physiol.* 275 : E300-E309.

Phillips, D.I., S. Caddy et al. 1996a. Intramuscular triglyceride and muscle insulin sensitivity: evidence for a relationship in nondiabetic subjects. *Metabolism* 45: 947-50.

Phillips, S.M., H.J. Green et al. 1996b. Progressive effect of endurance training on metabolic adaptations in working skeletal muscle. *Am. J. Physiol.* 270: E265-72.

Phillips, S.M., H.J. Green, M.A. Tarnopolsky et al. 1996c. Effects of training duration on substrate turnover and oxidation during exercise. *J. Appl. Physiol.* 81: 2182-91.

Pilegaard, H., J. Bangsbo, P. Henningsen et al. 1995. Effect of blood flow on muscle lactate release studied in pefused rat hindlimb muscle. *Am. J. Physiol.* 269: E1044-E1051.

Pilegaard, H., K. Domino, T. Noland et al. 1999a. Effect of high-intensity exercise training on lactate/H+ transport capacity in human skeletal muscle. *Am. J. Physiol.* 276: E255-61.

Pilegaard, H. & C. Juel. 1995. Lactate transport studied in sarcolemmal giant vesicles from rat skeletal muscle: effects of denervation. *Am. J. Physiol.* 269: E679-E682.

Pilegaard, H., C. Juel & F. Wibrand. 1993. Lactate transport studied in sarcolemmal giant vesicles from rats: effects of training. *Am. J. Physiol.* 264: E156-E160.

Pilegaard, H., G. Terzis, A. Halestrap et al. 1999b. Distribution of the lactate/H+ transporter isoforms MCT1 and MCT4 in human skeletal muscle. *Am. J. Physiol.* 276: E843-E848.

Poole, R.C. & A.P. Halestrap. 1993. Transport of lactate and other monocarboxylates across mammalian plasma membranes. *Am. J. Physiol.* 264: C761-C782.

Price, N.T., V.N. Jackson & A.P. Halestrap. 1998. Cloning and sequencing of four new mammalian monocarboxylate transporter (MCT) homologues confirms the existence of a transporter family with an ancient past. *Biochem. J.* 329: 321-28.

Py, G., N. Eydoux, A. Perez-Martin et al. 2001. Streptozotocin-induced diabetes decreases rat sarcolemmal lactate transport. *Metabolism* 50: 418-24.

Py, G., K. Lambert, O. Milhavet et al. 2002. Effects of streptozotocine-induced diabetes on markers of skeletal muscle metabolism and monocarboxylate transporter 1 to moncarboxylate transporters 4. *Metabolism* 51: 807-13.

Randle, P.J., C.N. Hales et al. 1963. The glucose fatty-acid cycle. Its role in insulin sensitivity and the metabolic disturbances of diabetes mellitus. *Lancet*: 785-89.

Rasmussen, B.B., U.C. Holmback et al. 2002a. Malonyl coenzyme A and the regulation of functional carnitine palmitoyltransferase-1 activity and fat oxidation in human skeletal muscle. *J. Clin. Invest.* 110: 1687-93.

Rasmussen, H.N., G. van Hall, U.F. Rasmussen et al. 2002b. Lactate dehydrogenase is not a mitochondrial enzyme in human and mouse vastus lateralis muscle. *J. Physiol.* 541: 575-80.

Rauch, L.H., A.N. Bosch, T.N. Noakes et al. 1995. Fuel utilisation during prolonged low-to-moderate intensity exercise when ingesting water or carbohydrate. *Pfluegers Arch.* 430: 971-77.

Ren, J.-M., C.F. Semenkovich, E.A. Gulve et al. 1994. Exercise induces rapid increases in GLUT4 expression, glucose transport capacity, and insulin-stimulated glycogen storage in muscle. *J. Biochem.* 269: 14396-14401.

Rennie, M.J. & J.O. Holloszy. 1977. Inhibition of glucose uptake and glycogenolysis by availability of oleate in well-oxygenated perfused skeletal muscle. *Biochem. J.* 168: 161-70.

Richter, E.A., W. Derave & J.F. Wojtaszewski. 2001. Glucose, exercise and insulin: emerging concepts. *J. Physiol.* 535: 313-22.

Richter, E.A. & H. Galbo. 1986. High glycogen levels enhance glycogen breakdown in isolated contracting skeletal muscle. *J. Appl. Physiol.* 61: 827-31.

Richter, E.A., P. Jensen, B. Kiens et al. 1998. Sarcolemmal glucose transport and GLUT-4 translocation during exercise are diminished by endurance training. *Am. J. Physiol.* 274: E89-E95.

Richter, E.A., N.B. Ruderman, H. Gavras et al. 1982. Muscle glycogenolysis during exercise: dual control by epinephrine and contractions. *Am. J. Physiol.* 242: E25-E32.

Romijn, J.A., E.F. Coyle et al. 1993. Regulation of endogenous fat and carbohydrate metabolism in relation to exercise intensity and duration. *Am. J. Physiol.* 265: E380-E391.

Roth, D.A. 1991. The sarcolemmal lactate transporter: transmembrane determinants of lactate flux. *Med. Sci. Sport Exerc.* 23: 925-34.

Roth, D.A. & G.A. Brooks. 1990a. Lactate and pyruvate transport is dominated by a pH gradient-sensitive carrier in rat skeletal muscle sarcolemmal vesicles. *Arch. Biochem. Biophys.* 279: 386-94.

Roth, D.A. & G.A. Brooks. 1990b. Lactate transport is mediated by a membrane-bound carrier in rat skeletal muscle sarcolemmal membranes. *Arch. Biochem. Biophys.* 279: 377-85.

Ruan, H. & H.J. Pownall. 2001. Overexpression of 1-acyl-glycerol-3-phosphate acyltransferase-alpha enhances lipid storage in cellular models of adipose tissue and skeletal muscle. *Diabetes* 50: 233-40.

Ruderman, N.B., A.K. Saha, D. Vavvas et al. 1999. Malonyl-CoA, fuel sensing, and insulin resistance. *Am. J. Physiol.* 276: E1-E18.

Ryder, J.W., R. Fahlman, H. Wallberg-Henriksson et al. 2000. Effect of contraction on mitogen-activated protein kinase signal transduction in skeletal muscle: involvement of the mitogen- and stress-activated protein kinase 1. *J. Biochem.* 275: 1457-62.

Sacchetti, M., B. Saltin et al. 2002. Intramuscular fatty acid metabolism in contracting and non-contracting human skeletal muscle. *J. Physiol.* 540: 387-95.

Saha, A.K., A.J. Schwarsin et al. 2000. Activation of malonyl-CoA decarboxylase in rat skeletal muscle by contraction and the AMP-activated protein kinase activator 5-aminoimidazole-4-carboxamide-1-beta-D-ribofuranoside. *J. Biol. Chem.* 275: 24279-83.

Sahlin, K., A. Katz & S. Broberg. 1990. Tricarboxylic acid cycle intermediates in human muscle during prolonged exercise. *Am. J. Physiol.* 259: C834-41.

Sahlin, K., M. Fernstrom & M. Tonkonogi. 2002. No evidence of an intracellular lactate shuttle in rat skeletal muscle. *J. Physiol.* 541: 569-74.

Salt, I., J.W. Celler, S.A. Hawley et al. 1998. AMP-activated protein kinase: greater AMP dependence and preferential nuclear localization, of complexes containing the α 2 isoform. *Biochem. J.* 334: 177-87.

Saltiel, A.R. & C.R. Kahn. 2002. Insulin signalling and the regulation of glucose and lipid metabolism. *Nature* 414: 799-806.

Santomauro, A.T.M.G., G. Boden et al. 1999. Overnight lowering of free fatty acids with acipimox improves insulin resistance and glucose tolerance in obese diabetic and nondiabetic subjects. *Diabetes* 48: 1836-41.

Schaffer, J.E. 2002. Fatty acid transport: the roads taken. *Am. J. Physiol.* 282: E239-E246.

Schaffer, J.E. & H.F. Lodish. 1994. Expression cloning and characterization of a novel adipocyte long chain fatty acid transport protein. *Cell* 79: 427-36.

Schrauwen, P., M.K. Hesselink et al. 2001a. Uncoupling protein 3 content is decreased in skeletal muscle of patients with type 2 diabetes. *Diabetes* 50: 2870-73.

Schrauwen, P., M.K. Hesselink et al. 2002. Effect of acute exercise on uncoupling protein 3 is a fat metabolism-mediated effect. *Am. J. Physiol.* 282: E11-E17.

Schrauwen, P., W.H. Saris et al. 2001b. An alternative function for human uncoupling protein 3: protection of mitochondria against accumulation of nonesterified fatty acids inside the mitochondrial matrix. *FASEB J.* 15: 2497-502.

Schrauwen, P., G. Schaart et al. 2000. The effect of weight reduction on skeletal muscle UCP2 and UCP3 mRNA expression and UCP3 protein content in type II diabetic subjects. *Diabetologia* 43: 1408-16.

Schrauwen, P., F.J. Troost et al. 1999. Skeletal muscle UCP2 and UCP3 expression in trained and untrained male subjects. *Int. J. Obes. Relat. Metab. Disord.* 23: 966-72.

Shimabukuro, M., K. Koyama et al. 1997. Direct antidiabetic effect of leptin through triglyceride depletion of tissues. *Proc. Natl. Acad. Sci.* 94: 4637-41.

Sidossis, L.S., C.A. Stuart et al. 1996. Glucose plus insulin regulate fat oxidation by controlling the rate of fatty acid entry into the mitochondria. *J. Clin. Invest.* 98: 2244-50.

Sidossis, L.S. & R.R. Wolfe. 1996. Glucose and insulin-induced inhibition of fatty acid oxidation: the glucose-fatty acid cycle reversed. *Am. J. Physiol.* 270: E733-E738.

Simoneau, J.-A. & D.A. Kelley. 1997. Altered glycolytic and oxidative capacities of skeletal muscle contribute to insulin resistance in NIDDM. *J. Appl. Physiol.* 83: 166-71.

Song, X.M., M. Fiedler, D. Galuska et al. 2002. 5-aminoimidazole-4-carboxamide ribonucleoside treatment improves glucose homeostasis in insulin-resistant diabetic (ob/ob) mice. *Diabetologia* 45: 56-65.

Spriet, L.L. & G.J.F. Heigenhasuer. 2002. Regulation of pyruvate dehydrogenase (PDH) activity in human skeletal muscle during exercise. *Exerc. Sport Sci. Rev.* 30: 91-95.

Starritt, E.C., R.A. Howlett et al. 2000. Sensitivity of CPT I to malonyl-CoA in trained and untrained human skeletal muscle. *Am. J. Physiol.* 278: E462-E468.

Steensberg, A., G. van Hall, C. Keller, T. Osada et al. 2002. Muscle glycogen content and glucose uptake during exercise in humans: influence of prior exercise and dietary manipulation. *J. Physiol.* 541: 273-81.

Steinberg, G.R., A. Bonen et al. 2002a. Fatty acid oxidation and triacylglycerol hydrolysis are enhanced after chronic leptin treatment in rats. *Am. J. Physiol.* 282: E593-E600.

Steinberg, G.R. & D.J. Dyck. 2000. Development of leptin resistance in rat soleus muscle in response to high-fat diets. *Am. J. Physiol.* 279: E1374-82.

Steinberg, G.R., M.L. Parolin et al. 2002b. Leptin increases FA oxidation in lean but not obese human skeletal muscle: evidence of peripheral leptin resistance. *Am. J. Physiol.* 283: E187-92.

Stephens, T.J., Z.P. Chen et al. 2002. Progressive increase in human skeletal muscle AMPKalpha2 activity and ACC phosphorylation during exercise. *Am. J. Physiol.* 282: E688-94.

Sullivan, J.E., K.J. Brocklehurst, A.E. Marley et al. 1994. Inhibition of lipolysis in isolated rat adipocytes with AICAR, a cell-permeable activator of AMP-activated protein kinase. *FEBS Lett.* 353: 33-36.

Szczesna-Kaczmarek, A. 1990. L-lactate oxidation by skeletal muscle mitochondria. *Int. J. Biochem.* 22: 617-20.

Tate, C.A. & R.W. Holtz. 1998. Gender and fat metabolism during exercise: a review. *Can. J. Appl. Physiol.* 23: 570-82.

Terjung, R.L., K.M. Baldwin, W.W. Winder et al. 1974. Glycogen repletion in different types of muscle and in liver after exhausting exercise. *Am. J. Physiol.* 226: 1387-91.

Thai, M., S. Guruswamy, K. Cao et al. 1998. Myocyte enhancer factor 2 (MEF2)-binding site is required for GLUT4 gene expression in transgenic mice. Regulation of MEF2 DNA binding activity in insulin-deficient diabetes. *J. Biochem.* 273: 14285-92.

Tonouchi, M., H. Hatta & A. Bonen. 2002. Muscle contraction increases lactate transport while reducing sarcolemmal MCT4, but not MCT1. *Am. J. Physiol.* 282: E1062-E1069.

Trimmer, J.K., J.M. Schwarz, G.A. Casazza et al. 2002. Measurement of gluconeogenesis in exercising men by mass isotopomer distribution analysis. *J. Appl. Physiol.* 93: 233-41.

Tunstall, R.J., K.A. Mehan et al. 2002. Exercise training increases lipid metabolism gene expression in human skeletal muscle. *Am. J. Physiol.* 283: E66-E72.

Turcotte, L.P. 2000. Muscle fatty acid uptake during exercise: possible mechanisms. *Exerc. Sport Sci. Rev.* 28: 4-9.

Turcotte, L.P., E.A. Richter & B. Kiens. 1992. Increased plasma FFA uptake and oxidation during prolonged exercise in trained vs. untrained humans. *Am. J. Physiol.* 262: E791-E799.

Van der Vusse, G.J. & R.S. Reneman. 1996. Exercise: regulation and integration of multiple systems. In: *Lipid Metabolism in Muscle,* ed. L.B. Rowell & J.T. Shepherd, 952-94. New York: Oxford University Press.

van Hall, G., G.J. van der Vusse, K. Soderlund et al. 1995. Deamination of amino acids as a source for ammonia production in human skeletal muscle during prolonged exercise. *J. Physiol.* 489: 251-61.

Vukovich, M.D., D.L. Costill et al. 1993. Effect of fat emulsion infusion and fat feeding on muscle glycogen utilization during cycle exercise. *J. Appl. Physiol.* 75: 1513-18.

Wahren, J., P. Felig, G. Ahlborg et al. 1971. Glucose metabolism during leg exercise in man. *J. Clin. Invest.* 50: 2715-25.

Wasserman, D.H. & A.D. Cherrington. 1991. Hepatic fuel metabolism during muscular work: role and regulation. *Am. J. Physiol.* 260: E811-E824.

Watkins, P.A., J.F. Lu et al. 1998. Disruption of the Saccharomyces cerevisiae FAT1 gene decreases very long-chain fatty acyl-CoA synthetase activity and elevates intracellular very long-chain fatty acid concentrations. *J. Biol. Chem.* 273: 18210-19.

Watt, M.J., G.J. Heigenhauser et al. 2002a. Intramuscular triacylglycerol, glycogen and acetyl group metabolism during 4 h of moderate exercise in man. *J. Physiol.* 541: 969-78.

Watt, M.J., G.J. Heigenhauser et al. 2002b. Intramuscular triacylglycerol utilization in human skeletal muscle during exercise: is there a controversy? *J. Appl. Physiol.* 93: 1185-95.

Weltan, S.M., A.N. Bosch, S.C. Dennis et al. 1998. Influence of muscle glycogen content on metabolic regulation. *Am. J. Physiol.* 274: E72-82.

Wendling, P.S., S.J. Peters et al. 1996. Variability of triacylglycerol content in human skeletal muscle biopsy samples. *J. Appl. Physiol.* 81: 1150-55.

Widegren, U., X.J. Jiang, A. Krook et al. 1998. Divergent effects of exercise on metabolic and mitogenic signaling pathways in human skeletal muscle. *FASEB J.* 12: 1379-89.

Widegren, U., J.W. Ryder & J.R. Zierath. 2001. Mitogen-activated protein kinase (MAPK) signal transduction in skeletal muscle: effects of exercise and muscle contraction. *Acta Physiol. Scand.* 172: 227-38.

Widegren, U., C. Wretman, A. Lionikas et al. 2000. Influence of exercise intensity on ERK/MAP kinase signalling in human skeletal muscle. *Pfluegers Arch.* 441: 317-22.

Wilson, M.C., V.N. Jackson, C. Hedle et al. 1998. Lactic acid efflux from white skeletal muscle is catalyzed by the monocarboxylate transporter MCT3. *J. Biol. Chem.* 273: 15920-26.

Winder, W.W. 2001. Energy-sensing and signaling by AMP-activated protein kinase in skeletal muscle. *J. Appl. Physiol.* 91: 1017-28.

Winder, W.W., J. Arogyasami et al. 1989. Muscle malonyl-CoA decreases during exercise. *J. Appl. Physiol.* 67: 2230-33.

Winder, W.W., J. Arogyasami et al. 1990. Time course of exercise-induced decline in malonyl-CoA in different muscle types. *Am. J. Physiol.* 259: E266-E271.

Winder, W.W. & B.F. Holmes. 2000. Insulin stimulation of glucose uptake fails to decrease palmitate oxidation in muscle if AMPK is activated. *J. Appl. Physiol.* 89: 2430-37.

Winder, W.W., B.F. Holmes, D.S. Rubink et al. 2000. Activation of AMP-activated protein kinase increases mitochondrial enzymes in skeletal muscle. *J. Appl. Physiol.* 88: 2219-26.

Wojtaszewski, J.F., B.F. Hansen, B. Kiens et al. 1997. Insulin signaling in human skeletal muscle. *Diabetes* 46: 1775-81.

Wojtaszewski, J.P.F., P. Nielsen, B.F. Hansen et al. 2000. Isoform-specific and exercise intensity-dependent activation of 5′-AMP-activated protein kinase in human skeletal muscle. *J. Physiol.* 528. 1: 221-26.

Woods, A., D. Azzout-Marniche, M. Foretz et al. 2000. Characterization of the role of AMP-activated protein kinase in the regulation of glucose-activated gene expression using constitutively active and dominant negative forms of the kinase. *Mol. Cell. Biol.* 20: 6704-11.

Woods, A., I. Salt, J. Scott et al. 1996. The α1 and α2 isoforms of the AMP-activated protein kinase have similar activities in rat liver but exhibit differences in substrate specificity in vitro. *FEBS Lett.* 397: 347-51.

Wu, Z., P. Puigserver, U. Andersson et al. 1999. Mechanisms controlling mitochondrial biogenesis and respiration through the thermogenic coactivator PGC-1. *Cell* 98: 115-24.

Yamauchi, T., J. Kamon et al. 2001. The fat-derived hormone adiponectin reverses insulin resistance associated with both lipoatrophy and obesity. *Nat. Med.* 7: 941-46.

Yamauchi, T., J. Kamon, Y. Minokoshi et al. 2002. Adiponectin stimulates glucose utilization and fatty-acid oxidation by activating AMP-activated protein kinase. *Nat. Med.* 8: 1288-95.

Yaspelkis, B.B., L. Ansari et al. 1999. Chronic leptin administration increases insulin-stimulated skeletal muscle glucose uptake and transport. *Metabolism* 48: 671-76.

Yee, A.J. & L.P. Turcotte. 2002. Insulin fails to alter plasma LCFA metabolism in muscle perfused at similar glucose uptake. *Am. J. Physiol.* 283: E73-77.

Yoon, H., A. Fanelli, E.F. Grollman et al. 1997. Identification of a unique monocarboxylate transporter (MCT3) in retinal pigment epithelium. *Biochem. Biophys. Res. Com.* 234: 90-94.

Yu, M., E. Blomstrand, A.V. Chibalin et al. 2001a. Exercise-associated differences in an array of proteins involved in signal transduction and glucose transport. *J. Appl. Physiol.* 90: 29-34.

Yu, M., E. Blomstrand, A.V. Chibalin et al. 2001b. Marathon running increases ERK1/2 and p38 MAP kinase signalling to downstream targets in human skeletal muscle. *J. Physiol.* 536: 273-82.

Yu, M., N.K. Stepto, L.D.G. Fryer et al. 2002. Metabolic and mitogenic signal transduction in skeletal muscle after intense cycling exercise. *J. Physiol.* 546. 2: 327-35.

Zhou, M., B.-Z. Lin, S. Coughlin et al. 2000. UCP-3 expression in skeletal muscle: effects of exercise, hypoxia, and AMP-activated protein kinase. *Am. J. Physiol.* 297: E622-E629.

Zhou, Y.T., M. Shimabukuro et al. 1997. Induction of leptin of uncoupling protein-2 and enzymes of fatty acid oxidation. *Proc. Natl. Acad. Sci.* 94: 6386-90.

Zierath, J.R., A. Krook & H. Wallberg-Henriksson. 2000. Insulin action and insulin resistance in human skeletal muscle. *Diabetologia* 43: 821-35.

Zong, H., J.M. Ren, L.H. Young et al. 2002. AMP kinase is required for mitochondrial biogenesis in skeletal muscle in response to chronic energy deprivation. *Proc. Natl. Acad. Sci.* 252625599.

Zou, Z., C. DiRusso et al. 2002. Fatty acid transport in *Sarccharomyces cerevisiae.* Directed mutagenesis of FAT1 distinguishes the biochemical activities associated with Fatp1. *J. Biol. Chem.* 277: 31062-71.

Chapter 9

Alessio, H.M. 1993. Exercise-induced oxidative stress. *Med. Sci. Sports Exerc.* 25(2): 218-24.

Alessio, H.M. & R.G. Cutler. 1990. Evidence that DNA damage-and-repair cycle activity increases following a marathon race [abstract]. *Med. Sci. Sports Exerc.* 22: 751.

Alessio, H.M., A.H. Goldfarb & R.G. Cutler. 1988. MDA content increases in fast- and slow-twitch skeletal muscle with intensity of exercise in a rat. *Am. J. Physiol.* 24: C874-C877.

Alessio, H.M., A.E. Hagerman, B.K. Fulkerson et al. 2000. Generation of reactive oxygen species after exhaustive aerobic and isometric exercise. *Med. Sci. Sports Exerc.* 32: 1576-81.

Anderson, R. 1982. Effects of ascorbate on normal and abnormal leukocyte functions. In *Vitamin C. New Clinical Applications in Immunology, Lipid Metabolism and Cancer,* ed. A. Hank, 23-52. Bern: Hans Huber.

Anggard, E. 1994. Nitric oxide: mediator, murderer, and medicine. *Lancet* 343: 1199-06.

Arner, E.S.J. & A. Holmgren. 2000. Physiological functions of thioredoxin and thioredoxin reductase. *Eur. J. Biochem.* 267: 6102-09.

Aronson, D., M.A. Violas, S.D. Dufresne et al. 1997. Exercise stimulates the mitogen-activated protein-kinase pathway in human skeletal muscle. *J. Clin. Invest.* 99: 1251-57.

Ashton, T., C.C. Rowlands, E. Jones et al. 1998. Electron spin resonance spectrometric detection of oxygen-centred radicals in human serum following exhaustive exercise. *Eur. J. Appl. Physiol.* 77: 498-502.

Atalay, M., P. Marnila, E.-M. Lilius et al. 1996. Glutathione-dependent modulation of exhausting exercise-induced changes in neutrophil function of rats. *Eur. J. Appl. Physiol.* 74: 342-47.

Baehner, R.L., L.A. Boxer, J.M. Allen et al. 1977. Autooxidation as a basis for altered function by polymorphonuclear leucocytes. *Blood* 50: 327-35.

Baeuerle, P.A. & T. Henkel. 1994. Function and activation of NF-kappaB. *Annu. Rev. Immunol.* 12: 141-79.

Baggiolini, M., P. Loetscher & B. Moser. 1995. Interleukin-8 and the chemokine family. *Int. J. Immunopharmacol.* 17: 103-08.

Balon, T.W. & J.L. Nadler. 1997. Evidence that nitric oxide increases glucose transport in skeletal muscle. *J. Appl. Physiol.* 82: 359-63.

Bauer, J.A., J.A. Wald, S. Doran et al. 1994. Endogenous nitric oxide in expired air: effects of acute exercise in humans. *Life Sci.* 55: 1903-09.

Becker, B.F. 1993. Towards the physiological function of uric acid. *Free Radic. Biol. Med.* 14: 615-31.

Beckman, K.B. & B.N. Ames. 2000. Oxidants and aging. In *Handbook of Oxidants and Antioxidants in Exercise,* ed. C.K. Sen, L. Packer & O. Hanninen, 755-96. Amsterdam: Elsevier.

Benzi, G. 1993. Aerobic performance and oxygen free radicals. *J. Sports Med. Phys. Fit.* 33: 205-22.

Bogdan, C. 2001. Nitric oxide and the immune response. *Nat. Immunol.* 2: 907-16.

Bogdan, C., M. Röllinghoff & A. Diefenbach. 2000. Reactive oxygen and reactive nitrogen intermediates in innate and specific immunity. *Curr. Opp. Immunol.* 12: 64-76.

Brouwer, S.E., C.G. Schnieders, H.P.F. Peters et al. 1997. Endogenous nitric oxide production after exercise [abstract]. *Med. Sci. Sports Exerc.* 29(Suppl): S72.

Buettner, G.R. & B.A. Jurkiewicz. 1993. Ascorbate free radical as a marker of oxidative stress: an EPR study. *Free Radic. Biol. Med.* 14: 49-55.

Bulkley, G.B. 1994. Reactive oxygen metabolites and reperfusion injury: aberrant triggering of reticuloendothelial function. *Lancet* 344: 934-36.

Camus, G., J. Pincemail, M. Ledent et al. 1992. Plasma levels of polymorphonuclear elastase and myeloperoxidase after uphill walking and downhill running at similar energy cost. *Int. J. Sports Med.* 13: 443-46.

Cannon, J.G., S.F. Orencole, R.A. Fielding et al. 1990. Acute phase response in exercise: interaction of age and vitamin E on neutrophils and muscle enzyme release. *Am. J. Physiol.* 259: R1214-R1219.

Chao, W.-H., E.W. Askew, D.E. Roberts et al. 1999. Oxidative stress in humans during work at moderate altitude. *J. Nutr.* 129: 2009-12.

Clarkson, P.M. & I. Tremblay. 1988. Exercise-induced muscle damage, repair, and adaptations in humans. *J. Appl. Physiol.* 65: 1-6.

Cochrane, C.G. 1992. Mechanisms of cell damage by oxidants. In *The Molecular Basis of Oxidative Damage by Leukocytes,* ed. A.J. Jesaitis & E.A. Dratz, 149-62. Boca Raton, Ann Arbor, London, Tokyo: CRC-Press.

Cooke, J.P., E. Rossitch, N.A. Andon et al. 1991. Flow activates an endothelial potassium channel to release an endogenous nitrovasodilatator. *J. Clin. Invest.* 88: 1663-67.

Coombes, J.S., S.K. Powers, B. Rowell et al. 2001. Effects of vitamin E and α lipoic acid on skeletal muscle contractile properties. *J. Appl. Physiol.* 90: 1424-30.

Croteau, D.L. & V.A. Bohr. 1997. Repair of oxidative damage to nuclear and mitochondrial DNA in mammalian cells. *J. Biol. Chem.* 41: 25409-12.

Davies, K.J.A., A.T. Quintanilha, G.A. Brooks et al. 1982. Free radicals and tissue damage produced by exercise. *Biochem. Biophys. Res. Com.* 107: 1198-1205.

Demirel, H.A., S.K. Powers, C. Caillaud et al. 1998. Exercise training reduces myocardial lipid peroxidation following short-term ischemia-reperfusion. *Med. Sci. Sports Exerc.* 30: 1211-16.

Desai, V.G., R. Weindruch, R.W. Hart et al. 1996. Influence of age and dietary restriction on gastrocnemius electron transport system activities in mice. *Arch. Biochem. Biophys.* 333: 145-51.

Dillard, C.J., R.E. Litov & A.L. Tappel. 1978. Effects of vitamine E, selenium and polysaturated fats on in vivo lipid peroxidation in the rat as measured by pentane production. *Lipids* 13: 396-402.

Dröge, W. 2002. Free radicals in the physiological control of cell function. *Physiol. Rev.* 82: 47-95.

Duarte, J.A., F. Carvalho, M.L. Bastos et al. 1994. Do invading leucocytes contribute to the decrease in glutathione concentrations indicating oxidative stress in exercised muscle, or are they important for its recovery? *Eur. J. Appl. Physiol.* 68: 48-53.

Emerit, I. 1994. Reactive oxygen species, chromosome mutation, and cancer: possible role of clastogenic factors in carcinogenesis. *Free Radic. Biol. Med.* 16: 99-109.

Ennezat, P.V., S.L. Malendowicz, M. Testa et al. 2001. Physical training in patients with chronic heart failure enhances the expression of genes encoding antioxidative enzymes. *J. Am. Coll. Cardiol.* 38: 194-98.

Essig, D.A. & D.R. Borger. 1997. Induction of heme oxygenase-1 (HSP32) mRNA in skeletal muscle following contractions. *Am. J. Physiol.* 272: C59-67.

Essig, D.A. & T.M. Nosek. 1997. Muscle fatigue and induction of stress protein genes: a dual function of reactive oxygen species? *Can. J. Appl. Physiol.* 22: 409-28.

Fehrenbach, E., A.M. Niess, F. Passek et al. 2003. Influence of different types of exercise on the expression of heme oxygenase-1 in leukocytes. *J. Sport Sci.* 21: 383-389.

Fehrenbach, E., A.M. Niess, E. Schlotz et al. 2000. Transcriptional and translational regulation of heat shock proteins in leukocytes of endurance runners. *J. Appl. Physiol.* 89: 704-10.

Fenton, H.J.H. 1894. Oxidation of tartaric acid in the presence of iron. *J. Chem. Soc.* 65: 899-910.

Fielding, R.A., T.J. Manfredi, W. Ding et al. 1993. Acute response to exercise III. Neutrophil and IL-1β accumulation in skeletal muscle. *Annu. Rev. Physiol.* 53: 201-06.

Forman, H.J. & M. Torres. 2001. Signaling by the respiratory burst in macrophages. *IUBMB Life* 51: 365-71.

Frandsen, U., L. Hoeffner, A. Betak et al. 2000. Endurance training does not alter the level of neuronal nitric oxide synthase in human skeletal muscle. *J. Appl. Physiol.* 89: 1033-38.

Galter, D., S. Mihm & W. Döge. 1994. Distinct effects of glutathione disulphide on the nuclear transcription factor-kappaB and the activator protein-1. *Eur. J. Biochem.* 221: 639-48.

Granger, D.N. 1988. Role of xanthine oxidase and granulocytes in ischemia reperfusion injury. *Am. J. Physiol.* 254: H1269-H1275.

Griendling, K.K., C.A. Minieri, J.D. Ollerenshaw et al. 1994. Angiotensin II stimulates NADH and NADPH oxidase activity in cultured vascular smooth muscle cells. *Circ. Res.* 74: 1141-48.

Guarnieri, C., G. Melandri, I. Caldarera et al. 1992. Spontaneous superoxide generation by polymorphonuclear leukocytes isolated from patients with stable angina after physical exercise. *Int. J. Cardiol.* 37: 301-07.

Haber, F. & J.J. Weiss. 1943. The catalytic decomposition of hydrogen peroxide by iron salts. *Proc. Royal Soc. London Series A* 147: 332-52.

Hack, V., C. Weiss, B. Friedmann et al. 1997. Decreased plasma glutamine level and CD4+ T cell number in response to 8 wk of anaerobic training. *Am. J. Physiol.* 272: E788-E795.

Halliwell, B. 1994a. Free radicals and antioxidants: a personal review. *Nutr. Rev.* 52: 253-65.

Halliwell, B. 1994b. Free radicals, antioxidants and human disease: curiosity, cause or consequence? *Lancet* 344: 721-24.

Halliwell, B. 1998. Free radicals and oxidative damage in biology and medicine: an introduction. In *Oxidative Stress in Skeletal Muscle,* ed. A.Z. Reznik, L. Packer, C.K. Sen, J.O. Holloszy & M.J. Jackson, 1-27. Basel: Birkhaeuser Verlag.

Hartmann, A. & A.M. Niess. 2000. DNA damage in exercise. In *Handbook of Oxidants and Antioxidants in Exercise,* ed. C.K. Sen, L. Packer & O. Hanninen, 195-217. Amsterdam: Elsevier.

Hartmann, A., A.M. Niess, M. Grünert-Fuchs et al. 1995. Vitamin E prevents exercise-induced DNA damage. *Mut. Res.* 346: 195-202.

Hartmann, A., S. Pfuhler, C. Dennog et al. 1998. Exercise-induced DNA effects in human leucocytes are not accompanied by increased formation of 8-hydroxy-2´-deoxyguanosine or induction of micronuclei. *Free Radic. Biol. Med.* 245-51.

Hartmann, A., U. Plappert, K. Raddatz et al. 1994. Does physical activity induce DNA damage? *Mutagenesis* 9(3): 269-72.

Hellsten-Westing, Y., A. Sollevi & B. Sjödin. 1991. Plasma accumulation of hypoxanthine, uric acid and creatine kinase following exhaustive runs of differing durations in man. *Eur. J. Appl. Physiol.* 62: 380-84.

Henschke, P.N. & S.J. Elliott. 1995. Oxidized glutathione decreases luminal Ca2+ content of the endothelial cell ins(1,4,5)P3-sensitive Ca2+ store. *Biochem. J.* 312: 485-89.

Hoffman-Goetz, L. & B.K. Pedersen. 1994. Exercise and the immune system: a model of the stress response? *Immunol. Today* 15: 382-87.

Hollander, J., R. Fiebig, M. Gore et al. 1999. Superoxide dismutase gene expression in skeletal muscle: fiber-specific adaptation to endurance training. *Am. J. Physiol.* 277: R856-R862.

Hollander, J., R. Fiebig, M. Gore et al. 2001. Superoxide dismutase gene expression is activated by a single bout of exercise in rat skeletal muscle. *Pfluegers Arch.* 442: 426-34.

Iemitsu, M., T. Miyauchi, S. Maeda et al. 2000. Intense exercise causes decrease in expression of both endothelial NO synthase and tissue NOx level in hearts. *Am. J. Physiol.* 279: R951-R959.

Inoue, T., Z. Mu, K. Sumikawa et al. 1993. Effect of physical exercise on the content of 8-hydroxydeoxyguanosine in nuclear DNA prepared from human lymphocytes. *Jap. J. Canc. Res.* 84: 720-25.

Itoh, H., T. Ohkuwa, T. Yamamoto et al. 1998. Effects of endurance physical training on hydroxyl radical generation in rat tissues. *Life Sci.* 63: 1921-29.

Jackson, M.J. 1998. Free radical mechanisms in exercise-related muscle damage. In *Oxidative Stress in Skeletal Muscle,* ed. A.Z. Reznik, L. Packer, C.K. Sen, J.O. Holloszy & M.J. Jackson, 75-86. Basel: Birkhaeuser Verlag.

Jackson, M.J., R.H.T. Edwards & M.C.R. Symons. 1985. Electron spin resonance of intact mammalian skeletal muscle. *Biochim. Biophys. Acta* 847: 185-90.

Javesghani, D., S.A. Magder, E. Barreiro et al. 2002. Molecular characterization of a superoxide-generating NAD(P)H oxidase in the ventilatory muscles. *Am. J. Resp. Crit. Care Med.* 165: 412-18.

Jenkins, R.R., K. Krause & L.S. Schofield. 1993. Influence of exercise on clearance of oxidant stress products and loosely bound iron. *Med. Sci. Sports Exerc.* 25: 213-17.

Ji, L.L. 1992. Responses of glutathione system and antioxidant enzymes to exhaustive exercise and hydroperoxides. *J. Appl. Physiol.* 72: 549-54.

Ji, L.L. 1998. Antioxidant enzyme response to exercise and training in the skeletal muscle. In *Oxidative Stress in Skeletal Muscle,* ed. A.Z. Reznik, L. Packer, C.K. Sen, J.O. Holloszy & M.J. Jackson, 103-25. Basel: Birkhaeuser Verlag.

Ji, L.L. 2002. Exercise-induced modulation of antioxidant defense. *Ann. N.Y. Acad. Sci.* 959: 82-92.

Johns, D.R. 1995. Mitochondrial DNA and disease. *New Engl. J. Med.* 333: 638.

Jones, S.A., V.B. O'Donnell, J.D. Wood et al. 1996. Expression of phagocyte NADPH oxidase components in human endothelial cells *Am. J. Physiol.* 271: C626-C634.

Jungersten, L., A. Ambring, B. Wall et al. 1997. Both physical fitness and acute exercise regulate nitric oxide formation in healthy humans. *J. Appl. Physiol.* 82: 760-64.

Kayatekin, B.M., S. Gönenc, O. Acikgöz et al. 2002. Effects of sprint exercise on oxidative stress in skeletal muscle and liver. *Eur. J. Appl. Physiol.* 87: 141-44.

Kelly, D.A., P.M. Tiidus, M.E. Houston et al. 1996. Effect of vitamin E deprivation and exercise training on induction of HSP70. *J. Appl. Physiol.* 81: 2379-85.

Khanna, S., M. Atalay, D.E. Laaksonen et al. 1999. Alpha-lipoic acid supplementation: tissue glutathione homeostasis at rest and after exercise. *J. Appl. Physiol.* 86: 1191-96.

Kobzik, L., M.B. Reid, D.S. Bredt et al. 1994. Nitric oxide in skeletal muscle. *Nature* 372: 546-48.

Koh, T.J. 2002. Do small heat shock proteins protect skeletal muscle from injury? *Exerc. Sport Sci. Rev.* 30: 117-121.

Laughlin, M.H., J.S. Pollock, J.F. Amann et al. 2001. Training induces nonuniform increases in eNOS content along the coronary arterial tree. *J. Appl. Physiol.* 90: 501-10.

Leaf, C.D., J.S. Wishnok, J.P. Hurley et al. 1990. Nitrate biosynthesis in rats, ferrets and humans: precursor studies with L-arginine. *Carcinogenesis* 11: 855-58.

Levine, R.L. 1983. Oxidative modification of glutamine synthetase. II. characterization of the ascorbate model system. *J. Biol. Chem.* 258: 11828-33.

Lewicki, R., H. Tchorzewski, A. Denys et al. 1987. Effect of physical exercise on some parameters of immunity in conditioned sportsmen. *Int. J. Sports Med.* 8: 309-314.

Lincoln, J., C.H.V. Hoyle & G. Burnstock. 1997. *Nitric Oxide in Health and Disease.* Cambridge, UK: Cambridge University Press.

Lindsay, M.A. & M.A. Giembycz. 1997. Signal transduction and activation of the NADPH oxidase in eosinophils. *Memorias do Instituto Oswaldo Cruz* 92(Suppl 2): 115-23.

Loft, S. & H.E. Poulsen. 1999. Markers of oxidative damage to DNA: antioxidants and molecular damage. *Meth. Enzymol.* 300: 166-84.

Los, M., H. Schenk, K. Hexel et al. 1995. IL-2 gene expression and NFkappaB activation through CD28 requires reactive oxygen production by 5-lipooxygenase. *EMBO J.* 14: 3731-40.

Lovlin, R., W. Cottle, I. Pyke et al. 1987. Are indices of free radical damage related to exercise intensity? *Eur. J. Appl. Physiol.* 56: 313-16.

Marini, M., F. Frabetti, D. Musiani et al. 1996. Oxygen radicals induce stress proteins and tolerance to oxidative stress in human lymphocytes. *Int. J. Rad. Biol.* 70: 337-50.

Martindale, J.L. & N.J. Holbrook. 2002. Cellular response to oxidative stress: signal for suicide and survival. *J. Cell Physiol.* 192: 1-15.

Maughan, R.J., A.E. Donnelly, M. Gleeson et al. 1989. Delayed-onset muscle damage and lipid peroxidation in man after a downhill run. *Muscle Nerve* 12: 332-336.

McKelvey-Martin, V.J., H.M.L. Green, P. Schmezer et al. 1993. The single gel electrophoresis assay (comet assay): a European review. *Mut. Res.* 288: 7-63.

Mena, P., M. Maynar, J.M. Gutierrez et al. 1991. Erythrocyte free radical scavenger enzymes in bicycle professional racers, adaption to training. *Int. J. Sports Med.* 12: 563-66.

Meydani, M. 1995. Vitamin E. *Lancet* 345: 170-75.

Mikami, T., K. Kita, S. Tomita et al. 2000. Is allantoin in serum and urine a useful indicator of exercise-induced oxidative stress in humans? *Free Radic. Res.* 32: 235-44.

Miyauchi, T., S. Maeda, M. Iemitsu et al. 2003. Exercise causes a tissue-specific change of NO production in the kidney and lung. *J. Appl. Physiol.* 94: 60-68.

Miyazaki, H., S. Oh-ishi, T. Ookawara et al. 2001. Strenuous endurance training in humans reduces oxidative stress following exhaustive exercise. *Eur. J. Appl. Physiol.* 84: 1-6.

Mohr, S., J.S. Stamler & B. Brune. 1994. Mechanism of covalent modification of glyceraldehyde-3-phosphate dehydrogenase at its active site thiol by nitric oxide, peroxynitrite and related nitrolsylating agents. *FEBS Lett.* 348: 223-27.

Moncada, S. & A. Higgs. 1993. The L-arginine-nitric oxide pathway. *N. Engl. J. Med.* 329: 2002-11.

Morse, D. & A.M.K. Choi. 2002. Heme oxygenase-1—the "emerging molecule" has arrived. *Am. J. Resp. Cell Mol. Biol.* 27: 8-16.

Nielsen, H.B., B. Hanel, S. Loft et al. 1995. Restricted pulmonary diffusion capacity after exercise is not an ARDS-like injury. *J. Sports Sci.* 84: 720-25.

Nieman, D.C., S. Simandle, D.A. Henson et al. 1995. Lymphocyte proliferative response to 2.5 hours of running. *Int. J. Sports Med.* 16: 404-08.

Niess, A.M., M. Baumann, K. Roecker et al. 1998. Effects of intensive endurance exercise on DNA damage in leucocytes. *J. Sports Med. Phys. Fit.* 38: 111-15.

Niess, A.M., H.-H. Dickhuth, H. Northoff et al. 1999a. Free radicals and oxidative stress in exercise—immunological aspects. *Exerc. Immunol. Rev.* 5: 22-56.

Niess, A.M., E. Fehrenbach, P. Krejzek et al. 1999b. Influence of various types of running exercise on DNA effects in leukocytes—a comparative comet assay study [abstract]. *J. Sports Sci.* 17: 584.

Niess, A.M., E. Fehrenbach, E. Schlotz et al. 2002a. Basal expression of leukocyte iNOS-mRNA is attenuated in moderately endurance trained subjects. *Eur. J. Appl. Physiol.* 87: 93-95.

Niess, A.M., E. Fehrenbach, E. Schlotz et al. 2002b. Effects of RRR-a-tocopherol on leukocyte expression of HSP72 in response to exhaustive treadmill exercise. *Int. J. Sports Med.* 23: 445-52.

Niess, A.M., E. Fehrenbach, M. Vogel et al. 2002c. Impact of endurance training on the regulation of the inducible nitric oxide synthase (iNOS) in human blood mononuclear cells [abstract]. *Pfluegers Arch.* 443: S362.

Niess, A.M., A. Hartmann, M. Grünert-Fuchs et al. 1996. DNA damage after exhaustive treadmill running in trained and untrained men. *Int. J. Sports Med.* 17: 397-403.

Niess, A.M., F. Passek, I. Lorenz et al. 1999c. Expression of the antioxidant stress protein heme oxygenase-1 (HO-1) in human leukocytes—acute and adaptional responses to endurance exercise. *Free Radic. Biol. Med.* 26: 184-92.

Niess, A.M., M. Sommer, E. Schlotz et al. 2000a. Expression of the inducible NO-synthase (iNOS) in human leukocytes—responses to running exercise. *Med. Sci. Sports Exerc.* 32: 1220-25.

Niess, A.M., M. Sommer, M. Schneider et al. 2000b. Physical exercise-induced expression of inducible nitric oxide synthase and heme oxygenase-1 in human leukocytes: effects of RRR-alpha-tocopherol supplementation. *Antiox. Redox Signal.* 2: 113-26.

Northoff, H. & A. Berg. 1991. Immunologic mediators as parameters of the reaction to strenuous exercise. *Int. J. Sports Med.* 12(Suppl): 9-15.

Ohno, H., H. Yamashita, T. Ookawara et al. 1992. Training effects on concentrations of immunoreactive superoxide dismutase iso-enzymes in human plasma. *Acta Physiol. Scand.* 146: 291-292.

Okamura, K., T. Doi, K. Hamada et al. 1997. Effect of repeated exercise on urinary 8-hydroxy-deoxyguanosine excretion in humans. *Free Radic. Res.* 26: 507-14.

Ortenblad, N., K. Madsen & M.S. Djurhuus. 1997. Antioxidant status and lipid peroxidation after short-term maximal exercise in trained and untrained humans. *Am. J. Physiol.* 272: 1258-63.

O'Neill, C.A., C.L. Stebbins, S. Bonigut et al. 1996. Production of hydroxyl radicals in contracting skeletal muscle of cats. *J. Appl. Physiol.* 81: 1197-1206.

Pansarasa, O., G. D'Antona, M.R. Gualea et al. 2002. "Oxidative stress": effects of mild endurance training and testosterone treatment on rat gastrocnemius muscle. *Eur. J. Appl. Physiol.* 87: 550-55.

Parks, D.A. & D.N. Granger. 1986. Xanthine oxidase: biochemistry, distribution and physiology. *Acta Physiol. Scand. Suppl.* 548: 87-99.

Paroo, Z., M.J. Meredith, M. Locke et al. 2002. Redox signaling of cardiac HSF1 binding. *Am. J. Physiol.* 283: C404-C411.

Paroo, Z., P.M. Tiidus & E.G. Noble. 1999. Estrogen attenuates HSP72 expression in acutely exercised male rodents. *Eur. J. Appl. Physiol.* 80: 180-84.

Peake, J.M. 2002. Exercise-induced alterations in neutrophil degranulation and respiratory burst activity: possible mechanisms of action. *Exerc. Immunol. Rev.* 8: 49-100.

Phillips, C.R., G.D. Giraud & W.E. Holden. 1996. Exhaled nitric oxide during exercise: site of release and modulation by ventilation and blood flow. *J. Appl. Physiol.* 80: 1865-71.

Pilegaard, H., G.A. Ordway, B. Saltin et al. 2000. Transcriptional regulation of gene expression in human skeletal muscle during recovery from exercise. *Am. J. Physiol.* 279: E806-E814.

Poulsen, H.E., S. Loft & K. Vistisen. 1996. Extreme exercise and oxidative DNA modification. *J. Sport Sci.* 14: 343-46.

Powers, S.K. & K. Hamilton. 1999. Antioxidants and exercise. *Clin. Sports Med.* 18: 525-36.

Powers, S.K. & C.K. Sen. 2000. Physiological antioxidants and exercise training. In *Handbook of Oxidants and Antioxidants in Exercise,* ed. C.K. Sen, L. Packer & O. Hanninen, 221-42. Amsterdam: Elsevier.

Preville, X., S. Salvemini, S. Giraud et al. 1999. Mammalian small stress proteins protect against oxidative stress through their ability to increase glucose-6-phosphatase dehydrogenase activity and by maintaining optimal cellular detoxifying machinery. *Exp. Cell Res.* 247: 61-78.

Pyne, D.B., M.S. Baker, P.A. Fricker et al. 1995. Effects of an intensive 12-wk training program by elite swimmers on neutrophil oxidative activity. *Med. Sci. Sports Exerc.* 27: 536-42.

Radák, Z., K. Asano, K.C. Lee et al. 1997. High altitude increases reactive carbonyl derivates but not lipid peroxidation in skeletal muscle of rats. *Free Radic. Biol. Med.* 22: 1109-14.

Radák, Z., T. Kaneko, S. Tahara et al. 1999a. The effect of exercise training on oxidative damage of lipids, proteins, and DNA in rat skeletal muscle: evidence for beneficial outcomes. *Free Radic. Biol. Med.* 27: 69-74.

Radák, Z., H. Naito, T. Kaneko et al. 2002. Exercise training decreases DNA damage and increases DNA repair and resistance to oxidative stress of proteins in aged rat skeletal muscle. *Pfluegers Arch.* 445: 273-78.

Radák, Z., A. Nakamura, H. Nakamoto et al. 1998. A period of anaerobic exercise increases the accumulation of reactive carbonyl derivates in the lungs of rats. *Pfluegers Arch.* 435: 439-41.

Radák, Z., J. Pucsok, S. Mecseki et al. 1999b. Muscle soreness-induced reduction in force generation is accompanied by increased nitric oxide content and DNA damage in human skeletal muscle. *Free Radic. Biol. Med.* 26: 1059-63.

Radák, Z., J. Pucsuk, S. Boros et al. 2000. Changes in urine 8-hydroxygeoxyguanosine levels of super-marathon runners during a four-day race period. *Life Sci.* 66: 1763-67.

Reid, M.B. 2001. Nitric oxide, reactive oxygen species, and skeletal muscle contraction. *Med. Sci. Sports Exerc.* 33: 371-76.

Reid, M.B. 2001. Redox modulation of skeletal muscle contraction: what we know and what we don't. *Eur. J. Appl. Physiol.* 90: 724-31.

Reid, M.B., S. Dobrivoje, S. Stokic et al. 1994. N-acetylcysteine inhibits muscle fatigue in humans. *J. Clin. Invest.* 94: 2468-74.

Reid, M.B. & W.J. Durham. 2002. Generation of reactive oxygen and nitrogen species in contracting skeletal muscle. *Ann. N.Y. Acad. Sci.* 959: 108-16.

Reznick, A.Z., E. Witt, M. Matsumoto et al. 1992. Vitamin E inhibits protein oxidation in skeletal muscle of resting and exercising rats. *Biochem. Biophys. Res. Comm.* 189: 801-06.

Rice-Evans, C., J. Sampson, P.M. Bramley et al. 1997. Why do we expect carotenoids to be antioxidants in vivo? *Free Radic. Res.* 26: 381-98.

Roy, R.S. & J.M. McCord. 1983. Superoxide and ischemia: conversion of xanthine dehydrogenase to xanthine oxidase. In *Oxy Radicals and Their Scavenger Systems, Vol. II: Cellular and Medical Aspects,* ed. R.A. Greenwald & G. Cohen, 145-53. Amsterdam: Elsevier.

Sakurai, T., J. Hollander, T. Izawa, H. Ohno, L.L. Ji & T. Best. 2001. Inducible nitric oxide synthase content following acute muscle stretch injury in rabbits [abstract]. *Med. Sci. Sports Exerc.* 33: S41.

Saxton, J.M., A.E. Donnelly & H.P. Roper. 1994. Indices of free-radical-mediated damage following maximum voluntary eccentric and concentric muscular work. *Eur. J. Appl. Physiol.* 68: 189-93.

Schulze, P.C., S. Gielen, G. Schuler & R. Hambrecht. 2002. Chronic heart failure and skeletal muscle catabolism: effects of exercise training. *Int. J. Cardiol.* 85: 141-49.

Sen, C.K. 2000. Biological thiols and redox regulation of cellular signal transduction pathways. In *Handbook of Oxidants and Antioxidants in Exercise,* ed. C.K. Sen, L. Packer & O. Hanninen, 375-401. Amsterdam: Elsevier.

Sen, C.K. 1998. Glutathione: a key role in skeletal muscle metabolism. In *Oxidative Stress in Skeletal Muscle,* ed. A.Z. Reznik, L. Packer, C.K. Sen, J.O. Holloszy & M.J. Jackson, 127-39. Basel: Birkhaeuser Verlag.

Sen, C.K., M. Atalay, J. Agren et al. 1997. Fish oil and vitamin E supplementation in oxidative stress at rest and after physical exercise. *Eur. J. Appl. Physiol.* 83: 189-95.

Sen, C.K., M. Atalay & O. Hänninen. 1994. Exercise-induced oxidative stress: glutathione supplementation and deficiency. *Eur. J. Appl. Physiol.* 77: 2177-87.

Sen, C.K., T. Rankinen, S. Väisänen et al. 1994. Oxidative stress after human exercise: effect of N-acetylcysteine supplementation. *J. Appl. Physiol.* 76: 2570-77.

Shen, W., X. Zhang, G. Zhao et al. 1995. Nitric oxide production and NO synthase gene expression contribute to vascular regulation during exercise. *Med. Sci. Sports Exerc.* 27: 1125-34.

Sjödin, B., Y. Hellsten-Westing & F.S. Apple. 1990. Biochemical mechanisms for oxygen free radical formation during exercise. *Sports Med.* 10(4): 236-54.

Smith, J.A., S.J. McKenzie, R.D. Telford et al. 1992. Why does moderate exercise enhance, but intense training depress immunity? In *Behaviour and Immunity,* ed. A.J. Husband. Boca Raton, FL: CRC Press.

Smith, J.A., R.D. Telford, I.B. Mason et al. 1990. Exercise, training and neutrophil microbicidal activity. *Int. J. Sports Med.* 11: 179-87.

Smolka, M.B., C.C. Zoppi, A.A. Alves et al. 2000. HSP72 as a complementary protection against oxidative stress induced by exercise in the soleus muscle of rats. *Am. J. Physiol.* 279: R1539-R1545.

Stadtman, E.R. 2001. Protein oxidation in aging and age-related diseases. *Ann. N.Y. Acad. Sci.* 928: 22-38.

Stamler, J.S. & G. Meissner. 2001. Physiology of nitric oxide in skeletal muscle. *Physiol. Rev.* 81: 209-237.

Steensberg, A., J. Morrow, A.D. Toft et al. 2002. Prolonged exercise, lymphocyte apoptosis and F2-isoprostanes. *Eur. J. Appl. Physiol.* 87: 38-42.

Stefanelli, C., C. Pignatti, B. Tantini et al. 1999. Nitric oxide can function as either a killer molecule or an antiapoptotic effector in cardiomyocytes. *Biochim. Biophys. Acta* 1450: 406-413.

Stuehr, D.J. & C.F. Nathan. 1989. Nitric oxide. A macrophage product responsible for cytostasis and respiratory inhibition in tumor cells. *J. Exp. Med.* 169:

Sumida, S., K. Tanaka, H. Kitao & F. Nakadomo. 1988. Exercise-induced lipid peroxidation and leakage of enzymes before and after vitamine E supplementation. *Int. J. Biochem.* 21: 835-38.

Sutton, J.R., C.J. Toews, J.R. Ward et al. 1980. Purine metabolism during strenuous muscular exercise in man. *Metabolism* 29: 254-60.

Suzuki, K., H. Ohno, S. Oh-ishi et al. 2000. Superoxide dismutases in exercise and disease. In *Handbook of Oxidants and Antioxidants in Exercise,* ed. C.K. Sen, L. Packer & O. Hanninen, 243-95. Amsterdam: Elsevier.

Suzuki, K., M. Yamada, S. Kurakake et al. 2000. Circulating cytokines and hormones with immunosuppressive but neutrophil-priming potentials rise after endurance exercise in humans. *Eur. J. Appl. Physiol.* 81: 281-87.

Thompson, H.S., S.P. Scordilis, P.M. Clarkson et al. 2001. A single bout of eccentric exercise increases HSP27 and HSC/HSP70 in human skeletal muscle. *Acta Physiol. Scand.* 171: 187-93.

Tiidus, P.M. 1998. Radical species in inflammation and overtraining. *Can. J. Physiol. Pharmacol.* 76: 533-38.

Tiidus, P.M., J. Pushkarenko & M.E. Houston. 1996. Lack of antioxidant adaptation to short-term aerobic training in human muscle. *Am. J. Physiol.* 271: R832-R836.

Tirosh, O. & A.Z. Reznick. 2000. Chemical bases and biological relevance of protein oxidation. In *Handbook of Oxidants and Antioxidants in Exercise,* ed. C.K. Sen, L. Packer & O. Hanninen, 89-114. Amsterdam: Elsevier.

Tonkonogi, M., B. Walsh, M. Svensson et al. 2000. Mitochondrial function and antioxidative defense in human muscle: effects of endurance training and oxidative stress. *J. Physiol.* 528: 379-88.

Tsai, K., T.G. Hsu, K.M. Hsu et al. 2001. Oxidative DNA damage in human peripheral leukocytes induced by massive aerobic exercise. *Free Radic. Biol. Med.* 31: 1465-72.

Vasankari, T.J., U.M. Kujala, T.M. Vasankari et al. 1998. Reduced oxidized LDL levels after a 10-month exercise program. *Med. Sci. Sports Exerc.* 30: 1496-1501.

Venkatraman, J.T., X. Feng & D. Pendergast. 2001. Effects of dietary fat and endurance exercise on plasma cortisol, prostaglandin E2, interferon-gamma and lipid peroxides in runners. *J. Am. Coll. Nutr.* 20: 529-36.

Vider, J., D.E. Laaksonen, A. Kilk et al. 2001. Physical exercise induced activation of NF-kappaB in human peripheral blood lymphocytes. *Antiox. Redox Signal.* 3: 1131-37.

Vina, J., A. Gimeno, J. Sastre et al. 2000. Mechanism of free radical production in exhaustive exercise in humans and rats: role of xanthine oxidase and protection by allopurinol. *IUBMB Life* 49: 539-44.

Vincent, H.K., S.K. Powers, H.A. Demirel et al. 1999. Exercise training protects against contraction-induced lipid peroxidation in the diaphragm. *Eur. J. Appl. Physiol.* 79: 268-73.

Vincent, K.R., H.K. Vincent, R.W. Braith et al. 2002. Resistance exercise training attenuates exercise-induced lipid peroxidation in the elderly. *Eur. J. Appl. Physiol.* 87: 416-23.

Warren, J.A., R.R. Jenkins, L. Packer et al. 1992. Elevated muscle vitamin E does not attenuate eccentric exercise-induced muscle injury. *J. Appl. Physiol.* 72: 2168-75.

Weiss, C., A. Bierhaus, R. Kinscherf et al. 2002. Tissue factor-dependent pathway is not involved in exercise-induced formation of thrombin and fibrin. *J. Appl. Physiol.* 92: 211-18.

Weiss, S.J. 1989. Tissue destruction by neutrophil. *N. Engl. J. Med.* 320: 365-76.

Wenger, R.H. 2000. Mammalian oxygen sensing, signalling and gene regulation. *J. Exp. Biol.* 203: 1253-63.

Widegren, C., X.J. Jiang, A. Krook et al. 1998. Divergent effects of exercise on metabolic and mitogenic signaling pathways in human skeletal muscle. *FASEB J.* 12: 1379-89.

Wink, D., I. Hanbauer, M. Krishna et al. 1993. Nitric oxide protects against cellular damage and cytotoxicity from reactive oxygen species. *Proc. Natl. Acad. Sci.* 90: 9813-17.

Witt, E.H., A.Z. Reznick, C.A. Viguie et al. 1992. Exercise, oxidative damage and effects of antioxidant manipulation. *J. Nutr.* 122: 766-73.

Woods, J.A., Q. Lu, M.A. Ceddia & T. Lowder. 2000. Exercise-induced modulation of macrophage function. *Immunol. Cell Biol.* 78: 545-53.

Wretman, C., A. Lionikas, U. Widegren et al. 2001. Effects of concentric and eccentric contractions on phosphorylation of $MAPK^{erk1/2}$ and $MAPK^{p38}$ in isolated rat skeletal muscle. *J. Physiol.* 535: 155-64.

Yang, A.L., S.J. Tsai, M.J. Jiang, C.J. Jen & H.I. Chen. 2002. Chronic exercise increases both inducible and endothelial nitric oxide synthase gene expression in endothelial cells of rat aorta. *J. Biomedic. Sci.* 9: 149-55.

Chapter 10

Abraham, N.G., R.D. Levere, J.H. Lin et al. 1991. Co-regulation of heme oxygenase and erythropoietin genes. *J. Cell Biochem.* 47: 43-48.

Abraham, N.G., J.H. Lin, M.L. Schwartzman et al. 1988. The physiological significance of heme oxygenase. *Int. J. Biochem.* 20: 543-58.

Abravaya, K., B. Phillips, R.I. Morimoto et al. 1991. Attenuation of the heat shock response in HeLa cells is mediated by the release of bound heat shock transcription factor and is modulated by changes in growth and in heat shock temperatures. *Genes Dev.* 5: 2117-27.

Amin, J., J. Ananthan, R. Voellmym et al. 1988. Key features of heat shock regulatory elements. *Mol. Cell Biol.* 8: 3761-69.

Angelidis, C.E., I. Lazaridis & G.N. Pagoulatos. 1991. Constitutive expression of heat-shock protein 70 in mammalian cells confers thermoresistance. *Eur. J. Biochem.* 199: 35-39.

Applegate, L.A., P. Luscher & R.M. Tyrrell. 1991. Induction of heme oxygenase: a general response to oxidant stress in cultured mammalian cells. *Cancer Res.* 51: 974-78.

Arcasoy, M.O., B.A. Degar, K.W. Harris et al. 1997. Familial erythrocytosis associated with a short deletion in the erythropoietin receptor gene. *Blood* 89: 4628-35.

Arrigo, A.P. 2000. sHsp as novel regulators of programmed cell death and tumorigenicity. *Pathol. Biol. Paris* 48: 280-88.

Arrigo, A.P., J.P. Suhan & W.J. Welch. 1988. Dynamic changes in the structure and intracellular locale of the mammalian low-molecular-weight heat shock protein. *Mol. Cell Biol.* 8: 5059-71.

Arrigo, A.P. & W.J. Welch. 1987. Characterization and purification of the small 28,000-dalton mammalian heat shock protein. *J. Biol. Chem.* 262: 15359-69.

Asami, S., T. Hirano, R. Yamaguchi et al. 1998. Reduction of 8-hydroxyguanine in human leukocyte DNA by physical exercise. *Free Radic. Res.* 29: 581-84.

Asano, M., K. Kaneoka, T. Nomura et al. 1998. Increase in serum vascular endothelial growth factor levels during altitude training. *Acta Physiol. Scand.* 162: 455-59.

Asea, A., S.-K. Kraft, E.A. Kurt-Jones et al. 2000. HSP70 stimulates cytokine production through a CD14-dependent pathway, demonstrating its dual role as a chaperone and cytokine. *Nat. Med.* 6: 435-42.

Ashenden, M.J., C.J. Gore, G.P. Dobson et al. 1999. "Live high, train low" does not change the total haemoglobin mass of male endurance athletes sleeping at a simulated altitude of 3000 m for 23 nights. *Eur. J. Appl.* 80: 479-84.

Bailey, D.M. & B. Davies. 1997. Physiological implications of altitude training for endurance performance at sea level: a review. *Brit. J. Sports Med.* 31: 183-90.

Baler, R., G. Dahl & R. Voellmy. 1993. Activation of human heat shock genes is accompanied by oligomerization, modification, and rapid translocation of heat shock transcription factor HSF1. *Mol. Cell Biol.* 13: 2486-96.

Baler, R., W.J. Welch & R. Voellmy. 1992. Heat shock gene regulation by nascent polypeptides and denatured proteins: hsp70 as a potential autoregulatory factor. *J. Cell Biol.* 117: 1151-59.

Bansal, G.S., P.M. Norton & D.S. Latchman. 1991. The 90-kDa heat shock protein protects mammalian cells from thermal stress but not from viral infection. *Exp. Cell Res.* 195: 303-06.

Bar, P.R. & G.J. Amelink. 1997. Protection against muscle damage exerted by oestrogen: hormonal or antioxidant action? *Biochem Soc. Trans.* 25: 50-54.

Basu, S., R.J. Binder, R. Suto et al. 2000. Necrotic but not apoptotic cell death releases heat shock proteins, which deliver a partial maturation signal to dendritic cells and activate the NF-kappa B pathway. *Int. Immunol.* 12: 1539-46.

Beckmann, R.P., L.E. Mizzen & W.J. Welch. 1990. Interaction of Hsp 70 with newly synthesized proteins: implications for protein folding and assembly. *Science* 248: 850-54.

Benjamin, I.J., S. Horie, M.L. Greenberg et al. 1992. Induction of stress proteins in cultured myogenic cells. Molecular signals for the activation of heat shock transcription factor during ischemia. *J. Clin. Invest.* 89: 1685-89.

Benoni, G., P. Bellavite, A. Adami et al. 1995. Effect of acute exercise on some haematological parameters and neutrophil functions in active and inactive subjects. *Eur. J. Appl. Physiol.* 70: 187-91.

Bienz, M. & H.R. Pelham. 1986. Heat shock regulatory elements function as an inducible enhancer in the Xenopus hsp70 gene and when linked to a heterologous promoter. *Cell* 45: 753-60.

Binder, R.J., N.E. Blachere & P.K. Srivastava. 2001. Heat shock protein-chaperoned peptides but not free peptides introduced into the cytosol are presented efficiently by major histocompatibility complex I molecules. *J. Biol. Chem.* 276: 17163-71.

Boning, D. 1997. Altitude and hypoxia training—a short review. *Int. J. Sports Med.* 18: 565-70.

Borger, D.R. & D.A. Essig. 1998. Induction of HSP 32 gene in hypoxic cardiomyocytes is attenuated by treatment with N-acetyl-L-cysteine. *Am. J. Physiol.* 274: H965-73.

Boshoff, T., F. Lombard, R. Eiselen et al. 2000. Differential basal synthesis of Hsp70/Hsc70 contributes to interindividual variation in Hsp70/Hsc70 inducibility. *Cell Mol. Life Sci.* 57: 1317-25.

Bouissou, P., J. Fiet, C.Y. Guezennec et al. 1988. Plasma adrenocorticotrophin and cortisol responses to acute hypoxia at rest and during exercise. *Eur. J. Appl. Physiol.* 57: 110-13.

Breen, E.C., E.C. Johnson, H. Wagner et al. 1996. Angiogenic growth factor mRNA responses in muscle to a single bout of exercise. *J. Appl. Physiol.* 81: 355-61.

Breymann, C., C. Bauer, A. Major et al. 1996. Optimal timing of repeated rh-erythropoietin administration improves its effectiveness in stimulating erythropoiesis in healthy volunteers. *Brit. J. Haematol.* 92: 295-01.

Bunn, H.F. & R.O. Poyton. 1996. Oxygen sensing and molecular adaptation to hypoxia. *Physiol. Rev.* 76: 839-85.

Bury, T.B. & F. Pirnay. 1995. Effect of prolonged exercise on neutrophil myeloperoxidase secretion. *Int. J. Sports Med.* 16: 410-12.

Camhi, S.L., P. Lee & A.M. Choi. 1995. The oxidative stress response. *New Horiz.* 3: 170-82.

Cartee, G.D., A.G. Douen, T. Ramlal et al. 1991. Stimulation of glucose transport in skeletal muscle by hypoxia. *J. Appl. Physiol.* 70: 1593-1600.

Chapman, R.F., J. Stray-Gundersen & B.D. Levine. 1998. Individual variation in response to altitude training. *J. Appl. Physiol.* 85: 1448-56.

Charron, M.J., F.C. Brosius III, S.L. Alper et al. 1989. A glucose transport protein expressed predominately in insulin-responsive tissues. *Proc. Natl. Acad. Sci.* 86: 2535-39.

Chen, H.W., H.L. Yang, H.H. Jing et al. 1995. Synthesis of Hsp72 induced by exercise in high temperature. *Chin. J. Physiol.* 38: 241-46.

Child, D.F., C.P. Williams, R.P. Jones et al. 1995. Heat shock protein studies in type 1 and type 2 diabetes and human islet cell culture. *Diabet. Med.* 12: 595-99.

Ciechanover, A., A. Orian & A.L. Schwartz. 2000. Ubiquitin-mediated proteolysis: biological regulation via destruction. *Bioessays* 22: 442-51.

Clifford, S.C. & E.R. Maher. 2001. Von Hippel-Lindau disease: clinical and molecular perspectives. *Adv. Cancer Res.* 82: 85-105.

Craig, E.A. & C.A. Gross. 1991. Is hsp70 the cellular thermometer? *Trends Biochem. Sci.* 16: 135-40.

Creagh, E.M., D. Sheehan & T.G. Cotter. 2000. Heat shock proteins—modulators of apoptosis in tumour cells. *Leukemia* 14: 1161-73.

Criswell, D., S. Powers, S. Dodd et al. 1993. High intensity training-induced changes in skeletal muscle antioxidant enzyme activity. *Med. Sci. Sports Exerc.* 25: 1135-40.

Dalman, F.C., L.C. Scherrer, L.P. Taylor et al. 1991. Localization of the 90-kDa heat shock protein-binding site within the hormone-binding domain of the glucocorticoid receptor by peptide competition. *J. Biol. Chem.* 266: 3482-90.

Davies, K.J., L. Packer & G.A. Brooks. 1981. Biochemical adaptation of mitochondria, muscle, and whole-animal respiration to endurance training. *Arch. Biochem. Biophys.* 209: 539-54.

Davies, K.J., A.T. Quintanilha, G.A. Brooks et al. 1982. Free radicals and tissue damage produced by exercise. *Biochem. Biophys. Res. Commun.* 107: 1198-1205.

Delcayre, C., J.L. Samuel, F. Marotte et al. 1988. Synthesis of stress proteins in rat cardiac myocytes 2-4 days after imposition of hemodynamic overload. *J. Clin. Invest.* 82: 460-68.

Demirel, H.A., S.K. Powers, C. Caillaud et al. 1998. Exercise training reduces myocardial lipid peroxidation following short-term ischemia-reperfusion. *Med. Sci. Sports Exerc.* 30: 1211-16.

Deshaies, R.J., B.D. Koch, M. Werner Washburne et al. 1988. A subfamily of stress proteins facilitates translocation of secretory and mitochondrial precursor polypeptides. *Nature* 332: 800-05.

Desplanches, D., H. Hoppeler, M.T. Linossier et al. 1993. Effects of training in normoxia and normobaric hypoxia on human muscle ultrastructure. *Pfluegers Arch.* 425: 263-67.

DiDomenico, B.J., G.E. Bugaisky & S. Lindquist. 1982. Heat shock and recovery are mediated by different translational mechanisms. *Proc. Natl. Acad. Sci.* 79: 6181-85.

Dillmann, W.H. 1999. Small heat shock proteins and protection against injury. *Ann. N.Y. Acad. Sci.* 874: 66-68.

Ebert, B.L. & H.F. Bunn. 1998. Regulation of transcription by hypoxia requires a multiprotein complex that includes hypoxia-inducible factor 1, an adjacent transcription factor, and p300/CREB binding protein. *Mol. Cell Biol.* 18: 4089-96.

Eckardt, K.U., J. Dittmer, R. Neumann et al. 1990. Decline of erythropoietin formation at continuous hypoxia is not due to feedback inhibition. *Am. J. Physiol.* 258: F1432-F1437.

Ellis, J. 1987. Proteins as molecular chaperones [news]. *Nature* 328: 378-79.

Erkeller, Y.F.M., D.A. Isenberg, V.B. Dhillon et al. 1992. Surface expression of heat shock protein 90 by blood mononuclear cells from patients with systemic lupus erythematosus. *J. Autoimmun.* 5: 803-14.

Essig, D.A., D.R. Borger & D.A. Jackson. 1997. Induction of heme oxygenase-1 (HSP32) mRNA in skeletal muscle following contractions. *Am. J. Physiol.* 272: C59-67.

Essig, D.A. & T.M. Nosek. 1997. Muscle fatigue and induction of stress protein genes: a dual function of reactive oxygen species? *Can. J. Appl. Physiol.* 22: 409-28.

Febbraio, M.A., A. Steensberg, R. Walsh et al. 2002. Reduced glycogen availability is associated with an elevation in HSP72 in contracting human skeletal muscle. *J. Physiol.* 538: 911-17.

Fehrenbach, E., A.M. Niess, F. Passek et al. 2002. Influence of different types of exercise on the expression of heme oxygenase-1 in leukocytes. *J. Sport. Sci.* 21: 383-389.

Fehrenbach, E., A.M. Niess, E. Schlotz et al. 2000a. Transcriptional and translational regulation of heat shock proteins (HSP27, HSP70) in leukocytes of endurance runners. *J. Appl. Physiol.* 89: 704-10.

Fehrenbach, E., A.M. Niess, R. Veith et al. 2001. Changes of HSP72-expression in leukocytes are associated with adaptation to exercise under conditions of high environmental temperature. *J. Leuk. Biol.* 69: 747-54.

Fehrenbach, E. & H. Northoff. 2001. Free radicals, exercise, apoptosis, and heat shock proteins. *Exerc. Immunol. Rev.* 7: 66-89.

Fehrenbach, E., F. Passek, A.M. Niess et al. 2000b. HSP expression in human leucocytes is modulated by endurance exercise. *Med. Sci. Sports Exerc.* 32: 592-600.

Fehrenbach, E., R. Veith, M. Schmid et al. 2003. Inverse response of leukocyte heat shock proteins and DNA damage to exercise and heat. *Free Radic. Res.* 37: 975-982.

Ferm, M.T., K. Soderstrom, S. Jindal et al. 1992. Induction of human hsp60 expression in monocytic cell lines. *Int. Immunol.* 4: 305-11.

Ferrara, N. 2000. Vascular endothelial growth factor and the regulation of angiogenesis. *Recent Prog. Horm. Res.* 55: 15-35.

Fisher, J.W. 1993. Recent advances in erythropoietin research. *Prog. Drug Res.* 41: 293-311.

Fisher, J.W. 2003. Erythropoietin: physiology and pharmacology update. *Exp. Biol. Med.* (Maywood) 228: 1-14.

Flanagan, S.W., A.J. Ryan, C.V. Gisolfi et al. 1995. Tissue-specific HSP70 response in animals undergoing heat stress. *Am. J. Physiol.* 268: R28-32.

Freeman, M.L., M.J. Borrelli, M.J. Meredith et al. 1999. On the path to the heat shock response: destabilization and formation of partially folded protein intermediates, a consequence of protein thiol modification. *Free Radic. Biol. Med.* 26: 737-45.

Fuller, K.J., R.D. Issels, D.O. Slosman et al. 1994. Cancer and the heat shock response. *Eur. J. Cancer* 30: 1884-91.

Gabriel, H., T. Kullmer, L. Schwarz et al. 1993. Circulating leukocyte subpopulations in sedentary subjects following graded maximal exercise with hypoxia. *Eur. J. Appl. Physiol.* 67: 348-53.

Gabriel, H., H.J. Muller, A. Urhausen et al. 1994. Suppressed PMA-induced oxidative burst and unimpaired phagocytosis of circulating granulocytes one week after a long endurance exercise. *Int. J. Sports Med.* 15: 441-45.

Garrido, C., J.M. Bruey, A. Fromentin et al. 1999. HSP27 inhibits cytochrome c-dependent activation of procaspase-9. *FASEB J.* 13: 2061-70.

Giaccia, A.J., E.A. Auger, A. Koong et al. 1992. Activation of the heat shock transcription factor by hypoxia in normal and tumor cell lines in vivo and in vitro. *Int. J. Radiat. Oncol. Biol. Phys.* 23: 891-97.

Giuliani, A. & B. Cestaro. 1997. Exercise, free radical generation and vitamins. *Eur. J. Cancer Prev.* 6(Suppl 1): S55-S67.

Gonzalez, B., R. Hernando & R. Manso. 2000. Stress proteins of 70 kDa in chronically exercised skeletal muscle. *Pfluegers Arch.* 440: 42-49.

Goodson, M.L. & K.D. Sarge. 1995. Regulated expression of heat shock factor 1 isoforms with distinct leucine zipper arrays via tissue-dependent alternative splicing. *Biochem. Biophys. Res. Commun.* 211: 943-49.

Green, H., J. MacDougall, M. Tarnopolsky et al. 1999. Downregulation of Na+-K+-ATPase pumps in skeletal muscle with training in normobaric hypoxia. *J. Appl. Physiol.* 86: 1745-48.

Gu, Y.Z., J.B. Hogenesch & C.A. Bradfield. 2000. The PAS superfamily: sensors of environmental and developmental signals. *Annu. Rev. Pharmacol. Toxicol.* 40: 519-61.

Guesdon, F., N. Freshney, R.J. Waller et al. 1993. Interleukin 1 and tumor necrosis factor stimulate two novel protein kinases that phosphorylate the heat shock protein hsp27 and beta-casein. *J. Biol. Chem.* 268: 4236-43.

Gunther, E. & L. Walter. 1994. Genetic aspects of the hsp70 multigene family in vertebrates. *Experientia* 50: 987-1001.

Gustafsson, T., A. Puntschart, L. Kaijser et al. 1999. Exercise-induced expression of angiogenesis-related transcription and growth factors in human skeletal muscle. *Am. J. Physiol.* 276: H679-H685.

Guzhova, I.V., A.C. Arnholdt, Z.A. Darieva et al. 1998. Effects of exogenous stress protein 70 on the functional properties of human promonocytes through binding to cell surface and internalization. *Cell Stress Chaperones* 3: 67-77.

Hahn, A.G., C.J. Gore, D.T. Martin et al. 2001. An evaluation of the concept of living at moderate altitude and training at sea level. *Comp. Biochem. Physiol. A Mol. Integr. Physiol.* 128: 777-89.

Hall, T.J. 1994. Role of hsp70 in cytokine production. *Experientia* 50: 1048-53.

Hammond, G.L., Y.K. Lai & C.L. Markert. 1982. Diverse forms of stress lead to new patterns of gene expression through a common and essential metabolic pathway. *Proc. Natl. Acad. Sci.* 79: 3485-88.

Harris, M.B. & J.W. Starnes. 2001. Effects of body temperature during exercise training on myocardial adaptations. *Am. J. Physiol.* 280: H2271-H2280.

Hellwig-Burgel, T., K. Rutkowski, E. Metzen et al. 1999. Interleukin-1beta and tumor necrosis factor-alpha stimulate DNA binding of hypoxia-inducible factor-1. *Blood* 94: 1561-67.

Hernando, R. & R. Manso. 1997. Muscle fibre stress in response to exercise: synthesis, accumulation and isoform transitions of 70-kDa heat-shock proteins. *Eur. J. Biochem.* 243: 460-67.

Hightower, L.E. & P.T. Guidon, Jr. 1989. Selective release from cultured mammalian cells of heat-shock (stress) proteins that resemble glia-axon transfer proteins. *J. Cell Physiol.* 138: 257-66.

Hoffman, E.C., H. Reyes, F.F. Chu et al. 1991. Cloning of a factor required for activity of the Ah (dioxin) receptor. *Science* 252: 954-58.

Hoppeler, H. & M. Vogt. 2001. Hypoxia training for sea-level performance. Training high-living low. *Adv. Exp. Med. Biol.* 502: 61-73.

Hoppeler, H., M. Vogt & E.R. Weibel. 2003. Response of skeletal muscle mitochondria to hypoxia. *Exp. Physiol.* 88: 109-19.

Huang, L.E., J. Gu, M. Schau & H.F. Bunn. 1998. Regulation of hypoxia-inducible factor 1alpha is mediated by an O2-dependent degradation domain via the ubiquitin-proteasome pathway. *Proc. Natl. Acad. Sci.* 95: 7987-92.

Iwaki, K., S.H. Chi, W.H. Dillmann et al. 1993. Induction of HSP70 in cultured rat neonatal cardiomyocytes by hypoxia and metabolic stress. *Circulation* 87: 2023-32.

Jaakkola, P., D.R. Mole, Y.M. Tian et al. 2001. Targeting of HIF-alpha to the von Hippel-Lindau ubiquitylation complex by O2-regulated prolyl hydroxylation. *Science* 292: 468-72.

Jaattela, M. 1993. Overexpression of major heat shock protein hsp70 inhibits tumor necrosis factor-induced activation of phospholipase A2. *J. Immunol.* 151: 4286-94.

Janssen, Y.M., B. Van Houten, P.J. Borm et al. 1993. Cell and tissue responses to oxidative damage. *Lab. Invest.* 69: 261-74.

Jelkmann, W. & T. Hellwig-Burgel. 2001. Biology of erythropoietin. *Adv. Exp. Med. Biol.* 502: 169-87.

Jenkins, R.R. & J. Beard. 2000. Metal binding agents: possible role in exercise. In: Sen, C.K., Packer, L. & O. Hanninen, eds. *Handbook of Oxidants and Antioxidants in Exercise.* 1st ed. Amsterdam: Elsevier, 129-52.

Jewell, U.R., I. Kvietikova, A. Scheid et al. 2001. Induction of HIF-1alpha in response to hypoxia is instantaneous. *FASEB J.* 15: 1312-14.

Jiang, B.H., E. Rue, G.L. Wang et al. 1996. Dimerization, DNA binding, and transactivation properties of hypoxia-inducible factor 1. *J. Biol. Chem.* 271: 17771-78.

Johnston, R.N. & B.L. Kucey. 1988. Competitive inhibition of hsp70 gene expression causes thermosensitivity. *Science* 242: 1551-54.

Kallio, P.J., K. Okamoto, S. O'Brien et al. 1998. Signal transduction in hypoxic cells: inducible nuclear translocation and recruitment of the CBP/p300 coactivator by the hypoxia-inducible factor-1alpha. *EMBO J.* 17: 6573-86.

Katschinski, D.M., L. Le, D. Heinrich, K.F. Wagner et al. 2002. Heat induction of the unphosphorylated form of hypoxia-inducible factor-1alpha is dependent on heat shock protein-90 activity. *J. Biol. Chem.* 277: 9262-67.

Kaufmann, S.H. 1990. Heat shock proteins and the immune response. *Immunol. Today* 11: 129-36.

Kedziora, J., A. Buczynski & K.K. Kedziora. 1995. Effect of physical exercise on antioxidative enzymatic defense in blood platelets from healthy men. *Int. J. Occup. Med. Environ. Health* 8: 33-39.

Kee, A.J., A.J. Taylor, A.R. Carlsson et al. 2002. IGF-I has no effect on postexercise suppression of the ubiquitin-proteasome system in rat skeletal muscle. *J. Appl. Physiol.* 92: 2277-84.

Kelly, D.A., P.M. Tiidus, M.E. Houston et al. 1996. Effect of vitamin E deprivation and exercise training on induction of HSP70. *J. Appl. Physiol.* 81: 2379-85.

Khassaf, M., R.B. Child, A. McArdle et al. 2001. Time course of responses of human skeletal muscle to oxidative stress induced by nondamaging exercise. *J. Appl. Physiol.* 90: 1031-35.

Kilgore, J.L., T.I. Musch & C.R. Ross. 1998. Physical activity, muscle, and the HSP70 response. *Can. J. Appl. Physiol.* 23: 245-60.

Kim, D., H. Ouyang & G.C. Li. 1995. Heat shock protein hsp70 accelerates the recovery of heat-shocked mammalian cells through its modulation of heat shock transcription factor HSF1. *Proc. Natl. Acad. Sci.* 92: 2126-30.

Kleinau, S., K. Soderstrom, R. Kiessling et al. 1991. A monoclonal antibody to the mycobacterial 65 kDa heat shock protein (ML 30) binds to cells in normal and arthritic joints of rats. *Scand. J. Immunol.* 33: 195-202.

Kline, M.P. & R.I. Morimoto. 1997. Repression of the heat shock factor 1 transcriptional activation domain is modulated by constitutive phosphorylation. *Mol. Cell Biol.* 17: 2107-15.

Knauf, U., E.M. Newton, J. Kyriakis et al. 1996. Repression of human heat shock factor 1 activity at control temperature by phosphorylation. *Genes Dev.* 10: 2782-93.

Koistinen, P.O., H. Rusko, K. Irjala et al. 2000. EPO, red cells, and serum transferrin receptor in continuous and intermittent hypoxia. *Med. Sci. Sports Exerc.* 32: 800-04.

Kondo, K. & W.G. Kaelin Jr. 2001. The von Hippel-Lindau tumor suppressor gene. *Exp. Cell Res.* 264: 117-25.

Kregel, K.C. & P.L. Moseley. 1996. Differential effects of exercise and heat stress on liver HSP70 accumulation with aging. *J. Appl. Physiol.* 80: 547-51.

Kretzschmar, M., D. Muller, J. Hubscher et al. 1991. Influence of aging, training and acute physical exercise on plasma glutathione and lipid peroxides in man. *Int. J. Sports Med.* 12: 218-22.

Kukreja, R.C., M.C. Kontos, K.E. Loesser et al. 1994. Oxidant stress increases heat shock protein 70 mRNA in isolated perfused rat heart. *Am. J. Physiol.* 267: H2213-19.

Kumae, T., K. Yamasaki, K. Ishizaki et al. 1999. Effects of summer camp endurance training on non-specific immunity in long-distance runners. *Int. J. Sports Med.* 20: 390-95.

Kuppner, M.C., R. Gastpar, S. Gelwer et al. 2001. The role of heat shock protein (hsp70) in dendritic cell maturation: hsp70 induces the maturation of immature dendritic cells but reduces DC differentiation from monocyte precursors. *Eur. J. Immunol.* 31: 1602-09.

La Ferla, K., C. Reimann, W. Jelkmann et al. 2002. Inhibition of erythropoietin gene expression signaling involves the transcription factors GATA-2 and NF-kappaB. *FASEB J.* 16: 1811-13.

Landry, J., P. Chretien, H. Lambert et al. 1989. Heat shock resistance conferred by expression of the human HSP27 gene in rodent cells. *J. Cell Biol.* 109: 7-15.

Landry, J., P. Chretien, A. Laszlo et al. 1991. Phosphorylation of HSP27 during development and decay of thermotolerance in Chinese hamster cells. *J. Cell Physiol.* 147: 93-101.

Laroia, G., R. Cuesta, G. Brewer et al. 1999. Control of mRNA decay by heat shock-ubiquitin-proteasome pathway. *Science* 284: 499-502.

Levine, B.D. & J. Stray-Gundersen. 1997. "Living high-training low": effect of moderate-altitude acclimatization with low-altitude training on performance. *J. Appl. Physiol.* 83: 102-12.

Levine, B.D. & J. Stray-Gundersen. 2001. The effects of altitude training are mediated primarily by acclimatization, rather than by hypoxic exercise. *Adv. Exp. Med. Biol.* 502: 75-88.

Li, G.C. 1983. Induction of thermotolerance and enhanced heat shock protein synthesis in Chinese hamster fibroblasts by sodium arsenite and by ethanol. *J. Cell Physiol.* 115: 116-22.

Li, G.C. 1985. Elevated levels of 70,000 dalton heat shock protein in transiently thermotolerant Chinese hamster fibroblasts and in their stable heat resistant variants. *Int. J. Radiat. Oncol. Biol. Phys.* 11: 165-77.

Lindebro, M.C., L. Poellinger & M.L. Whitelaw. 1995. Protein-protein interaction via PAS domains: role of the PAS domain in positive and negative regulation of the bHLH/PAS dioxin receptor-Arnt transcription factor complex. *EMBO J.* 14: 3528-39.

Lindquist, S. & E.A. Craig. 1988. The heat-shock proteins. *Annu. Rev. Genet.* 22: 631-77.

Liu, Y., S. Mayr, A. Opitz-Gress et al. 1999. Human skeletal muscle HSP70 response to training in highly trained rowers. *J. Appl. Physiol.* 86: 101-04.

Locke, M. 1997. The cellular stress response to exercise: role of stress proteins. *Exerc. Sport Sci. Rev.* 25: 105-36.

Locke, M. & E.G. Noble. 1995. Stress proteins: the exercise response. *Can. J. Appl. Physiol.* 20: 155-67.

Locke, M., E.G. Noble & B.G. Atkinson. 1990. Exercising mammals synthesize stress proteins. *Am. J. Physiol.* 258: C723-29.

Locke, M., E.G. Noble, R.M. Tanguay et al. 1995a. Activation of heat-shock transcription factor in rat heart after heat shock and exercise. *Am. J. Physiol.* 268: C1387-94.

Locke, M., R.M. Tanguay, R.E. Klabunde et al. 1995b. Enhanced postischemic myocardial recovery following exercise induction of HSP 72. *Am. J. Physiol.* 269: H320-25.

Lutton, J.D., J.L. da Silva, S. Moqattash et al. 1992. Differential induction of heme oxygenase in the hepatocarcinoma cell line (Hep3B) by environmental agents [published erratum appears in *J. Cell Biochem.* 1992 Nov 50(3):336]. *J. Cell Biochem.* 49: 259-65.

Maher, J.T., L.G. Jones & L.H. Hartley. 1974. Effects of high-altitude exposure on submaximal endurance capacity of men. *J. Appl. Physiol.* 37: 895-98.

Maines, M.D. 1988. Heme oxygenase: function, multiplicity, regulatory mechanisms, and clinical applications. *FASEB J.* 2: 2557-68.

Maines, M.D. 1997. The heme oxygenase system: a regulator of second messenger gases. *Annu. Rev. Pharmacol. Toxicol.* 37: 517-54.

Major, A., C. Bauer, C. Breymann et al. 1994. rh-erythropoietin stimulates immature reticulocyte release in man. *Brit. J. Haematol.* 87: 605-08.

Maloney, J., D. Wang, T. Duncan et al. 2000. Plasma vascular endothelial growth factor in acute mountain sickness. *Chest* 118: 47-52.

Manjili, M.H., X.Y. Wang, J. Park et al. 2002. Immunotherapy of cancer using heat shock proteins. *Front. Biosci.* 7: d43-d52.

Martel, E., E. Fehrenbach, A.M. Niess et al. 2000. CD45 isoforms appear to be a major target for oxidative stress in mononuclear cells of trained athletes [abstract]. Deut. Zeitschr. Sportmed. 51, 9.

Martin, J., A.L. Horwich & F.U. Hartl. 1992. Prevention of protein denaturation under heat stress by the chaperonin Hsp60. *Science* 258: 995-98.

Matzinger, P. 2002. An innate sense of danger. *Ann. N.Y. Acad. Sci.* 961: 341-42.

Maxwell, P.H., M.S. Wiesener, G.W. Chang et al. 1999. The tumour suppressor protein VHL targets hypoxia-inducible factors for oxygen-dependent proteolysis. *Nature* 399: 271-75.

Mazzeo, R.S., D. Donovan, M. Fleshner et al. 2001. Interleukin-6 response to exercise and high-altitude exposure: influence of alpha-adrenergic blockade. *J. Appl. Physiol.* 91: 2143-49.

McCoubrey, W.K. Jr., T.J. Huang & M.D. Maines. 1997. Isolation and characterization of a cDNA from the rat brain that encodes hemoprotein heme oxygenase-3. *Eur. J. Biochem.* 247: 725-32.

McTiernan, A., C. Ulrich, S. Slate et al. 1998. Physical activity and cancer etiology: associations and mechanisms. *Cancer Causes Control* 9: 487-509.

Mehlen, P., C. Kretz-Remy, X. Preville et al. 1996. Human hsp27, Drosophila hsp27 and human alphaB-crystallin expression-mediated increase in glutathione is essential for the protective activity of these proteins against TNFalpha-induced cell death. *EMBO J.* 15: 2695-2706.

Melissa, L., J.D. MacDougall, M.A. Tarnopolsky et al. 1997. Skeletal muscle adaptations to training under normobaric hypoxic versus normoxic conditions. *Med. Sci. Sports Exerc.* 29: 238-43.

Minchenko, A., T. Bauer, S. Salceda et al. 1994. Hypoxic stimulation of vascular endothelial growth factor expression in vitro and in vivo. *Lab. Invest.* 71: 374-79.

Mitani, K., H. Fujita, A. Kappas et al. 1992. Heme oxygenase is a positive acute-phase reactant in human Hep3B hepatoma cells. *Blood* 79: 1255-59.

Mizzen, C.A. & C.D. Allis. 2000. Transcription. New insights into an old modification. *Science* 289: 2290-91.

Mizzen, L.A., C. Chang, J.I. Garrels et al. 1989. Identification, characterization, and purification of two mammalian stress proteins present in mitochondria, grp 75, a member of the hsp 70 family and hsp 58, a homolog of the bacterial groEL protein. *J. Biol. Chem.* 264: 20664-75.

Mizzen, L.A. & W.J. Welch. 1988. Characterization of the thermotolerant cell. I. Effects on protein synthesis activity and the regulation of heat-shock protein 70 expression. *J. Cell Biol.* 106: 1105-16.

Morimoto, R.I. 1998. Regulation of the heat shock transcriptional response: cross talk between a family of heat shock factors, molecular chaperones, and negative regulators. *Genes Dev.* 12: 3788-96.

Morimoto, R.I., A. Tissieres & C. Georgopoulos. 1994. *The Biology of Heat Shock Proteins and Molecular Chaperones.* New York: Cold Spring Harbor Laboratory Press.

Moseley, P. 2000b. Stress proteins and the immune response. *Immunopharmacology* 48: 299-302.

Moseley, P.L. 2000a. Exercise, stress, and the immune conversation. *Exerc. Sport Sci. Rev.* 28: 128-32.

Moseley, P.L., E.S. Wallen, J.D. McCafferty et al. 1993. Heat stress regulates the human 70-kDa heat-shock gene through the 3'-untranslated region. *Am. J. Physiol.* 264: L533-37.

Mosser, D.D., J. Duchaine & B. Massie. 1993. The DNA-binding activity of the human heat shock transcription factor is regulated in vivo by hsp70. *Mol. Cell Biol.* 13: 5427-38.

Multhoff, G. & C. Botzler. 1998. Heat-shock proteins and the immune response. *Ann. N.Y. Acad. Sci.* 851: 86-93.

Nakai, A. & R.I. Morimoto. 1993. Characterization of a novel chicken heat shock transcription factor, heat shock factor 3, suggests a new regulatory pathway. *Mol. Cell Biol.* 13: 1983-97.

Nakai, A., M. Tanabe, Y. Kawazoe et al. 1997. HSF4, a new member of the human heat shock factor family which lacks properties of a transcriptional activator. *Mol. Cell Biol.* 17: 469-81.

Neckers, L. 2002. Hsp90 inhibitors as novel cancer chemotherapeutic agents. *Trends Mol. Med.* 8: S55-S61.

Neufer, P.D., G.A. Ordway & R.S. Williams. 1998. Transient regulation of c-fos, alpha B-crystallin, and hsp70 in muscle during recovery from contractile activity. *Am. J. Physiol.* 274: C341-46.

Newton, E.M., U. Knauf, M. Green et al. 1996. The regulatory domain of human heat shock factor 1 is sufficient to sense heat stress. *Mol. Cell Biol.* 16: 839-46.

Nieman, D.C. & B.K. Pedersen. 1999. Exercise and immune function. Recent developments. *Sports Med.* 27: 73-80.

Niess, A.M., M. Baumann, K. Roecker et al. 1997. Effects of intensive endurance exercise on DNA damage in leukocytes. *J. Sports Med. Phys. Fit.* 38: 111-15.

Niess, A.M., E. Fehrenbach, E. Schlotz et al. 2002. Effects of RRR-alpha-tocopherol on leukocyte expression of HSP72 in response to exhaustive treadmill exercise. *Int. J. Sports Med.* 23: 445-52.

Niess, A.M., E. Fehrenbach, G. Strobel et al. 2003. Evaluation of stress responses to interval training at low and moderate altitude. *Med. Sci. Sports Exerc.* 35: 263-69.

Niess, A.M., F. Passek, I. Lorenz et al. 1999. Expression of the antioxidant stress protein heme oxygenase-1 (HO-1) in human leukocytes—acute and adaptational responses to endurance exercise. *Free Radic. Biol. Med.* 26: 184-92.

Niess, A.M., M. Sommer, M. Schneider et al. 2000. Physical exercise-induced expression of inducible nitric oxide and heme oxygenase-1 in human leukocytes: effects of RRR-alpha-tocopherol supplementation. *Antiox. Redox Signal.* 2: 113-26.

Northoff, H., S. Enkel & C. Weinstock. 1995. Exercise, injury, and immune function. *Exerc. Immunol. Rev.* 1: 1-25.

Okinaga, S., K. Takahashi, K. Takeda et al. 1996. Regulation of human heme oxygenase-1 gene expression under thermal stress. *Blood* 87: 5074-84.

Ornatsky, O.I., M.K. Connor & D.A. Hood. 1995. Expression of stress proteins and mitochondrial chaperonins in chronically stimulated skeletal muscle. *Biochem. J.* 311: 119-23.

Palmer, L.A., B. Gaston & R.A. Johns. 2000. Normoxic stabilization of hypoxia-inducible factor-1 expression and activity: redox-dependent effect of nitrogen oxides. *Mol. Pharmacol.* 58: 1197-1203.

Parmiani, G., C. Castelli, P. Dalerba et al. 2002. Cancer immunotherapy with peptide-based vaccines: what have we achieved? Where are we going? *J. Natl. Cancer Inst.* 94: 805-18.

Paroo, Z., E.S. Dipchand & E.G. Noble. 2002. Estrogen attenuates postexercise HSP70 expression in skeletal muscle. *Am. J. Physiol.* 282: C245-C251.

Paroo, Z. & E.G. Noble. 2002. Gender-specific regulation of HSP70: mechanisms and consequences. In: Locke, M., Noble, E.G., eds. *Exercise and Stress Response.* 1st ed. Boca Raton, FL: CRC Press, 163-78.

Pedersen, B.K. 1996. Immune response to acute exercise. In: Hoffmann-Goetz, L., ed. *Exercise and Immune Function.* Boca Raton, FL: CRC Press, 79-92.

Pedersen, B.K. & A. Steensberg. 2002. Exercise and hypoxia: effects on leukocytes and interleukin-6-shared mechanisms? *Med. Sci. Sports Exerc.* 34: 2004-13.

Pelham, H.R. 1986. Speculations on the functions of the major heat shock and glucose-regulated proteins. *Cell* 46: 959-61.

Percy, M.J., M.F. McMullin, A.W. Roques et al. 1998. Erythrocytosis due to a mutation in the erythropoietin receptor gene. *Brit. J. Haematol.* 100: 407-10.

Petersen, R. & S. Lindquist. 1988. The Drosophila hsp70 message is rapidly degraded at normal temperatures and stabilized by heat shock. *Gene* 72: 161-68.

Petronini, P.G., R. Alfieri, C. Campanini et al. 1995. Effect of an alkaline shift on induction of the heat shock response in human fibroblasts. *J. Cell Physiol.* 162: 322-29.

Ploug, T., H. Galbo & E.A. Richter. 1984. Increased muscle glucose uptake during contractions: no need for insulin. *Am. J. Physiol.* 247: E726-E731.

Plumier, J.C., B.M. Ross, R.W. Currie et al. 1995. Transgenic mice expressing the human heat shock protein 70 have improved post-ischemic myocardial recovery. *J. Clin. Invest.* 95: 1854-60.

Podhorska-Okolow, M., M. Sandri, S. Zampieri et al. 1998. Apoptosis of myofibres and satellite cells: exercise-induced damage in skeletal muscle of the mouse. *Neuropathol. Appl. Neurobiol.* 24: 518-31.

Powers, S.K., L.L. Ji & C. Leeuwenburgh. 1999. Exercise training-induced alterations in skeletal muscle antioxidant capacity: a brief review. *Med. Sci. Sports Exerc.* 31: 987-97.

Powers, S.K. & S.L. Lennon. 1999. Analysis of cellular responses to free radicals: focus on exercise and skeletal muscle. *Proc. Nutr. Soc.* 58: 1025-33.

Pratt, W.B. 1987. Transformation of glucocorticoid and progesterone receptors to the DNA-binding state. *J. Cell Biochem.* 35: 51-68.

Puntschart, A., M. Vogt, H.R. Widmer et al. 1996. Hsp70 expression in human skeletal muscle after exercise. *Acta Physiol. Scand.* 157: 411-17.

Pyne, D.B. 1994. Regulation of neutrophil function during exercise. *Sports Med.* 17: 245-58.

Radak, Z., A.W. Taylor, H. Ohno et al. 2001. Adaptation to exercise-induced oxidative stress: from muscle to brain. *Exerc. Immunol. Rev.* 7: 90-107.

Rammensee, H.G. 1996. Antigen presentation—recent developments. *Int. Arch. Allergy Immunol.* 110: 299-307.

Rankinen, T., L. Perusse, R. Rauramaa et al. 2002. The human gene map for performance and health-related fitness phenotypes: the 2001 update. *Med. Sci. Sports Exerc.* 34: 1219-33.

Ren, J.M., C.F. Semenkovich, E.A. Gulve et al. 1994. Exercise induces rapid increases in GLUT4 expression, glucose transport capacity, and insulin-stimulated glycogen storage in muscle. *J. Biol. Chem.* 269: 14396-14401.

Renner, W., S. Kotschan, C. Hoffmann et al. 2000. A common 936 C/T mutation in the gene for vascular endothelial growth factor is associated with vascular endothelial growth factor plasma levels. *J. Vasc. Res* 37: 443-48.

Richardson, R.S., E.A. Noyszewski, K.F. Kendrick et al. 1995. Myoglobin O2 desaturation during exercise. Evidence of limited O2 transport. *J. Clin. Invest.* 96: 1916-26.

Richter, K. & J. Buchner. 2001. Hsp90: chaperoning signal transduction. *J. Cell Physiol.* 188: 281-90.

Ritossa, F. 1962. A new puffing pattern induced by temperature shock and DNP in Drosophila. *Experientia* 13: 571-73.

Ritz, M.F., A. Masmoudi, N. Matter et al. 1993. Heat stressing stimulates nuclear protein kinase C raising diacylglycerol levels. Nuclear protein kinase C activation precedes Hsp70 mRNA expression. *Receptor* 3: 311-24.

Rizzardini, M., M. Terao, F. Falciani et al. 1993. Cytokine induction of haem oxygenase mRNA in mouse liver. Interleukin 1 transcriptionally activates the haem oxygenase gene. *Biochem. J.* 290: 343-47.

Rolfs, A., I. Kvietikova, M. Gassmann et al. 1997. Oxygen-regulated transferrin expression is mediated by hypoxia-inducible factor-1. *J. Biol. Chem.* 272: 20055-62.

Roskamm, H., F. Landry, L. Samek et al. 1969. Effects of a standardized ergometer training program at three different altitudes. *J. Appl. Physiol.* 27: 840-47.

Ryan, A.J., C.V. Gisolfi & P.L. Moseley. 1991. Synthesis of 70K stress protein by human leukocytes: effect of exercise in the heat. *J. Appl. Physiol.* 70: 466-71.

Salceda, S. & J. Caro. 1997. Hypoxia-inducible factor 1alpha (HIF-1alpha) protein is rapidly degraded by the ubiquitin-proteasome system under normoxic conditions. Its stabilization by hypoxia depends on redox-induced changes. *J. Biol. Chem.* 272: 22642-47.

Salo, D.C., C.M. Donovan & K.J. Davies. 1991. HSP70 and other possible heat shock or oxidative stress proteins are induced in skeletal muscle, heart, and liver during exercise. *Free Radic. Biol. Med.* 11: 239-46.

Saltin, B. 1996. Exercise and the environment: focus on altitude. *Res. Q. Exerc. Sport* 67: S1-10.

Sarge, K.D., S.P. Murphy & R.I. Morimoto. 1993. Activation of heat shock gene transcription by heat shock factor 1 involves oligomerization, acquisition of DNA-binding activity, and nuclear localization and can occur in the absence of stress [published errata appear in *Mol. Cell Biol.* 1993 May 13(5):3122-3 and 1993 Jun 13(6):3838-9]. *Mol. Cell Biol.* 13: 1392-1407.

Sarge, K.D., V. Zimarino, K. Holm et al. 1991. Cloning and characterization of two mouse heat shock factors with distinct inducible and constitutive DNA-binding ability. *Genes Dev.* 5: 1902-11.

Schmidt, W., K.U. Eckardt, A. Hilgendorf et al. 1991. Effects of maximal and submaximal exercise under normoxic and hypoxic conditions on serum erythropoietin level. *Int. J. Sports Med.* 12: 457-61.

Schneider, E.M., A.M. Niess, I. Lorenz et al. 2002. Inducible HSP70 expression analysis after heat and physical exercise—transcriptional, protein expression and subcellular localization. *Ann. N.Y. Acad. Sci.* 973: 8-12.

Schobersberger, W., P. Hobisch-Hagen, D. Fries et al. 2000. Increase in immune activation, vascular endothelial growth factor and erythropoietin after an ultramarathon run at moderate altitude. *Immunobiology* 201: 611-20.

Schuetz, T.J., G.J. Gallo, L. Sheldon et al. 1991. Isolation of a cDNA for HSF2: evidence for two heat shock factor genes in humans. *Proc. Natl. Acad. Sci.* 88: 6911-15.

Schwane, J.A. & R.B. Armstrong. 1983. Effect of training on skeletal muscle injury from downhill running in rats. *J. Appl. Physiol.* 55: 969-75.

Semenza, G.L. 2000. HIF-1: mediator of physiological and pathophysiological responses to hypoxia. *J. Appl. Physiol.* 88: 1474-80.

Shastry, S., D.O. Toft & M.J. Joyner. 2002. HSP70 and HSP90 expression in leucocytes after exercise in moderately trained humans. *Acta Physiol. Scand.* 175: 139-46.

Shi, Y., D.D. Mosser & R.I. Morimoto. 1998. Molecular chaperones as HSF1-specific transcriptional repressors. *Genes Dev.* 12: 654-66.

Shibahara, S., M. Sato, R.M. Muller et al. 1989. Structural organization of the human heme oxygenase gene and the function of its promoter. *Eur. J. Biochem.* 179: 557-63.

Singh, J.H., R.E. Toes, P. Spee et al. 2000. Cross-presentation of glycoprotein 96-associated antigens on major histocompatibility complex class I molecules requires receptor-mediated endocytosis. *J. Exp. Med.* 191: 1965-74.

Siren, A.L., F. Knerlich, W. Poser et al. 2001. Erythropoietin and erythropoietin receptor in human ischemic/hypoxic brain. *Acta Neuropathol.* (Berlin) 101: 271-76.

Sistonen, P., A.L. Traskelin, H. Lehvaslaiho et al. 1993. Genetic mapping of the erythropoietin receptor gene. *Hum. Genet.* 92: 299-301.

Skidmore, R., J.A. Gutierrez, V. Guerriero et al. 1995. HSP70 induction during exercise and heat stress in rats: role of internal temperature. *Am. J. Physiol.* 268: R92-97.

Smith, J.A., R.D. Telford, I.B. Mason et al. 1990. Exercise, training and neutrophil microbicidal activity. *Int. J. Sports Med.* 11: 179-87.

Sokol, L. & J.T. Prchal. 1994. Two microsatellite repeat polymorphisms in the EPO gene. *Hum. Mol. Genet.* 3: 219.

Somani, S.M., S. Frank & L.P. Rybak. 1995. Responses of antioxidant system to acute and trained exercise in rat heart subcellular fractions. *Pharmacol. Biochem. Behav.* 51: 627-34.

Sondermann, H., T. Becker, M. Mayhew et al. 2000. Characterization of a receptor for heat shock protein 70 on macrophages and monocytes. *Biol. Chem.* 381: 1165-74.

Sorger, P.K. & H.R. Pelham. 1988. Yeast heat shock factor is an essential DNA-binding protein that exhibits temperature-dependent phosphorylation. *Cell* 54: 855-64.

Srivastava, P. 2002. Roles of heat-shock proteins in innate and adaptive immunity. *Nat. Rev. Immunol.* 2: 185-94.

Srivastava, P.K. 1994. Heat shock proteins in immune response to cancer: the fourth paradigm. *Experientia* 50: 1054-60.

Stocker, R. 1990. Induction of heme oxygenase as a defense against oxidative stress. *Free Radic. Res. Commun.* 9: 101-12.

Subjeck, J.R. & T.T. Shyy. 1986. Stress protein systems of mammalian cells. *Am. J. Physiol.* 250: C1-17.

Tacchini, L., L. Bianchi, A. Bernelli-Zazzera et al. 1999. Transferrin receptor induction by hypoxia. HIF-1-mediated transcriptional activation and cell-specific post-transcriptional regulation. *J. Biol. Chem.* 274: 24142-46.

Tanabe, M., N. Sasai, K. Nagata et al. 1999. The mammalian HSF4 gene generates both an activator and a repressor of heat shock genes by alternative splicing. *J. Biol. Chem.* 274: 27845-56.

Tenhunen, R., H.S. Marver & R. Schmid. 1970. The enzymatic catabolism of hemoglobin: stimulation of microsomal heme oxygenase by hemin. *J. Lab. Clin. Med.* 75: 410-21.

Terrados, N., E. Jansson, C. Sylven et al. 1990. Is hypoxia a stimulus for synthesis of oxidative enzymes and myoglobin? *J. Appl. Physiol.* 68: 2369-72.

Terrados, N., J. Melichna, C. Sylven et al. 1988. Effects of training at simulated altitude on performance and muscle metabolic capacity in competitive road cyclists. *Eur. J. Appl. Physiol.* 57: 203-09.

Thompson, H.S., P.M. Clarkson & S.P. Scordilis. 2002. The repeated bout effect and heat shock proteins: intramuscular HSP27 and HSP70 expression following two bouts of eccentric exercise in humans. *Acta Physiol. Scand.* 174: 47-56.

Thompson, H.S. & S.P. Scordilis. 1994. Ubiquitin changes in human biceps muscle following exercise-induced damage. *Biochem. Biophys. Res. Commun.* 204: 1193-98.

Torok, Z., I. Horvath, P. Goloubinoff et al. 1997. Evidence for a lipochaperonin: association of active protein-folding GroESL oligomers with lipids can stabilize membranes under heat shock conditions. *Proc. Natl. Acad. Sci.* 94: 2192-97.

Trudell, J.R., W.Q. Lin, D.A. Chrystof et al. 1995. Induction of HSP72 in rat liver by chronic ethanol consumption combined with exercise: association with the prevention of ethanol-induced fatty liver by exercise. *Alcohol Clin. Exp. Res.* 19: 753-58.

Vayssier, M., N. Banzet, D. Francois et al. 1998. Tobacco smoke induces both apoptosis and necrosis in mammalian cells: differential effects of HSP70. *Am. J. Physiol.* 275: L771-L779.

Vesely, M.J., D.J. Exon, J.E. Clark et al. 1998. Heme oxygenase-1 induction in skeletal muscle cells: hemin and sodium nitroprusside are regulators in vitro. *Am. J. Physiol.* 275: C1087-94.

Vigh, L., B. Maresca & J.L. Harwood. 1998. Does the membrane's physical state control the expression of heat shock and other genes? *Trends Biochem. Sci.* 23: 369-74.

Vogt, B.A., J. Alam, A.J. Croatt et al. 1995. Acquired resistance to acute oxidative stress—possible role of heme oxygenase and ferritin. *Lab. Invest.* 72: 474-83.

Vogt, M., A. Puntschart, J. Geiser et al. 2001. Molecular adaptations in human skeletal muscle to endurance training under simulated hypoxic conditions. *J. Appl. Physiol.* 91: 173-82.

Wagner, K.F., D.M. Katschinski, J. Hasegawa et al. 2001. Chronic inborn erythrocytosis leads to cardiac dysfunction and premature death in mice overexpressing erythropoietin. *Blood* 97: 536-42.

Walsh, R.C., I. Koukoulas, A. Garnham et al. 2001. Exercise increases serum Hsp72 in humans. *Cell Stress Chaperones* 6: 386-93.

Walters, T.J., K.L. Ryan, M.R. Tehrany et al. 1998. HSP70 expression in the CNS in response to exercise and heat stress in rats. *J. Appl. Physiol.* 84: 1269-77.

Wang, G.L. & G.L. Semenza. 1995. Purification and characterization of hypoxia-inducible factor 1. *J. Biol. Chem.* 270: 1230-37.

Wang, X.Y., Y. Kaneko, E. Repasky et al. 2000. Heat shock proteins and cancer immunotherapy. *Immunol. Invest.* 29: 131-37.

Welch, W.J., H.S. Kang, R.P. Beckmann et al. 1991. Response of mammalian cells to metabolic stress; changes in cell physiology and structure/function of stress proteins. *Curr. Top. Microbiol. Immunol.* 167: 31-55.

Welch, W.J. & J.P. Suhan. 1986. Cellular and biochemical events in mammalian cells during and after recovery from physiological stress. *J. Cell Biol.* 103: 2035-52.

Welsh, M.J. & M. Gaestel. 1998. Small heat-shock protein family: function in health and disease. *Ann. N.Y. Acad. Sci.* 851: 28-35.

Wenger, R.H. 2002. Cellular adaptation to hypoxia: O2-sensing protein hydroxylases, hypoxia-inducible transcription factors, and O2-regulated gene expression. *FASEB J.* 16: 1151-62.

Willoughby, D.S., J.W. Priest & M. Nelson. 2002. Expression of the stress proteins, ubiquitin, heat shock protein 72, and myofibrillar protein content after 12 weeks of leg cycling in persons with spinal cord injury. *Arch. Phys. Med. Rehabil.* 83: 649-54.

Wiseman, H., P. Quinn & B. Halliwell. 1993. Tamoxifen and related compounds decrease membrane fluidity in liposomes. Mechanism for the antioxidant action of tamoxifen and relevance to its anticancer and cardioprotective actions? *FEBS Lett.* 330: 53-56.

Wu, C. 1995. Heat shock transcription factors: structure and regulation. *Annu. Rev. Cell Dev. Biol.* 11: 441-69.

Yoshida, T., P. Biro, T. Cohen et al. 1988. Human heme oxygenase cDNA and induction of its mRNA by hemin. *Eur. J. Biochem.* 171: 457-61.

Young, J.C., I. Moarefi & F.U. Hartl. 2001. Hsp90: a specialized but essential protein-folding tool. *J. Cell Biol.* 154: 267-73.

Zelzer, E., P. Wappner & B.Z. Shilo. 1997. The PAS domain confers target gene specificity of Drosophila bHLH/PAS proteins. *Genes Dev.* 11: 2079-89.

Zeng, S.M., J. Yankowitz, J.A. Widness et al. 2001. Etiology of differences in hematocrit between males and females: sequence-based polymorphisms in erythropoietin and its receptor. *J. Gend. Specif. Med.* 4: 35-40.

Zierath, J.R., T.S. Tsao, A.E. Stenbit et al. 1998. Restoration of hypoxia-stimulated glucose uptake in GLUT4-deficient muscles by muscle-specific GLUT4 transgenic complementation. *J. Biol. Chem.* 273: 20910-15.

Zou, J., Y. Guo, T. Guettouche et al. 1998. Repression of heat shock transcription factor HSF1 activation by HSP90 (HSP90 complex) that forms a stress-sensitive complex with HSF1. *Cell* 94: 471-80.

Chapter 11

Albert, C.M., M.A. Mittleman, C.U. Chae et al. 2000. Triggering of sudden death from cardiac causes by vigorous exertion. *New Engl. J. Med.* 343: 1355-61.

Allen, D.G. & J.C. Kentish. 1985. The cellular basis of the length-tension relation in cardiac muscle. *J. Mol. Cell. Cardiol.* 17: 821-40.

Allen, D.L., B.C. Harrison, A. Maass et al. 2001. Cardiac and skeletal muscle adaptations to voluntary wheel running in the mouse. *J. Appl. Physiol.* 90: 1900-08.

Anderson, P.A., A. Greig, T.M. Mark et al. 1995. Molecular basis of human cardiac troponin T isoforms expressed in the developing, adult, and failing heart. *Circ. Res.* 76: 681-86.

Antos, C.L., T.A. McKinsey, N. Frey et al. 2002. Activated glycogen synthase-3 beta suppresses cardiac hypertrophy in vivo. *Proc. Natl. Acad. Sci.* 99: 907-12.

Baar, K., E. Blough, B. Dineen et al. 1999. Transcriptional regulation in response to exercise. *Exerc. Sport Sci. Rev.* 27: 333-79.

Belardinelli, R., D. Georgiou, G. Cianci et al. 1999. Randomized, controlled trial of long-term moderate exercise training in chronic heart failure: effects on functional capacity, quality of life, and clinical outcome. *Circulation* 99: 1173-82.

Bers, D.M. 2001. *Excitation-Contraction Coupling and Cardiac Contractile Force.* London: Kluwer Academic.

Bers, D.M. 2002. Cardiac excitation-contraction coupling. *Nature* 415: 198-205.

Beyer, R.E., P.G. Morales-Corral, B.J. Ramp et al. 1984. Elevation of tissue coenzyme Q (ubiquinone) and cytochrome c concentrations by endurance exercise in the rat. *Arch. Biochem. Biophys.* 234: 323-29.

Bing, O.H., W.W. Brooks, C.H. Conrad et al. 1991. Intracellular calcium transients in myocardium from spontaneously hypertensive rats during the transition to heart failure. *Circ. Res.* 68: 1390-1400.

Boluyt, M.O., A.M. Loyd, M.H. Roth et al. 2003. Activation of JNK in rat heart by exercise: effect of training. *Am. J. Physiol. Heart. Circ. Physiol.* 285:H2639-47.

Braunwald, E. 1997. Pathophysiology of heart failure. In *Heart Disease. A Textbook of Cardiovascular Medicine,* ed. E. Braunwald, 394-418. Philadelphia: Saunders.

Breisch, E.A., F.C. White, L.E. Nimmo et al. 1986. Exercise-induced cardiac hypertrophy: a correlation of blood flow and microvasculature. *J. Appl. Physiol.* 60: 1259-67.

Brittsan, A.G. & E.G. Kranias. 2000. Phospholamban and cardiac contractile function. *J. Mol. Cell. Cardiol.* 32: 2131-39.

Bupha-Intr, T., & J. Wattanapermpool. 2004. Cardioprotective effects of exercise training on myofilament calcium activation in ovariectomized rats. *J. Appl. Physiol.* 96:1755-60.

Buttrick, P.M., M. Kaplan, L.A. Leinwand et al. 1994. Alterations in gene expression in the rat heart after chronic pathological and physiological loads. *J. Mol. Cellular Cardiol.* 26: 61-67.

Calderone, A., R.J. Murphy, J. Lavoie et al. 2001. TGF-beta(1) and prepro-ANP mRNAs are differentially regulated in exercise-induced cardiac hypertrophy. *J. Appl. Physiol.* 91: 771-76.

Cheung, J.Y., J.M. Constantine & J.V. Bonventre. 1986. Regulation of cytosolic free calcium concentration in cultured renal epithelial cells. *Am. J. Physiol.* 251: F690-F701.

Chien, K.R. 2004. *Molecular Basis of Cardiovascular Disease. 2nd ed.* Philadelphia: Saunders.

Chien, K.R., K.U. Knowlton, H. Zhu et al. 1991. Regulation of cardiac gene expression during myocardial growth and hypertrophy: molecular studies of an adaptive physiologic response. *Fed. Am. Soc. Exp. Biol.* 5: 3037-46.

Cittadini, A., Y. Ishiguro, H. Stromer et al. 1998. Insulin-like growth factor-1 but not growth hormone augments mammalian myocardial contractility by sensitizing the myofilament to Ca2+ through a wortmannin-sensitive pathway: studies in rat and ferret isolated muscles. *Circ. Res.* 83: 50-59.

Consitt, L.A., J.L. Copeland & M.S. Tremblay. 2001. Hormone responses to resistance vs. endurance exercise in premenopausal females. *Can. J. Appl. Physiol.* 26: 574-87.

Copeland, J.L., L.A. Consitt & M.S. Tremblay. 2002. Hormonal responses to endurance and resistance exercise in females aged 19-69 years. *J. Gerontol. Series A: Biol. Sci. Med. Sci.* 57: B158-B165.

Crabtree, G.R. & E.N. Olson. 2002. NFAT signaling: choreographing the social lives of cells. *Cell* 109(Suppl): S67-S79.

Currie, S. & G.L. Smith. 1999. Enhanced phosphorylation of phospholamban and downregulation of sarco/endoplasmic reticulum Ca2+ ATPase type 2 (SERCA 2) in cardiac sarcoplasmic reticulum from rabbits with heart failure. *Cardiovasc. Res.* 41: 135-46.

D'Agnolo, A., G.B. Luciani, A. Mazzucco et al. 1992. Contractile properties and Ca2+ release activity of the sarcoplasmic reticulum in dilated cardiomyopathy. *Circulation* 85: 518-25.

Davies, C.H., K. Davia, J.G. Bennett, et al. 1995. Reduced contraction and altered frequency response of isolated ventricular myocytes from patients with heart failure. *Circulation* 92: 2540-49.

del Monte, F., S.E. Harding, U. Schmidt et al. 1999. Restoration of contractile function in isolated cardiomyocytes from failing human hearts by gene transfer of SERCA2a. *Circulation* 100: 2308-11.

Demirel, H.A., S.K. Powers, C. Caillaud et al. 1998. Exercise training reduces myocardial lipid peroxidation following short-term ischemia-reperfusion. *Med. Sci. Sports Exerc.* 30: 1211-16.

Diffee, G.M. & E. Chung. 2003. Altered single cell force-velocity and power properties in exercise-trained rat myocardium. *J. Appl. Physiol.* 94: 1941-48.

Diffee, G.M. & D.F. Nagle. 2003a. Exercise training alters length dependence of contractile properties in rat myocardium. *J. Appl. Physiol.* 94: 1137-44.

Diffee, G.M. & D. Nagle. 2003b. Regional differences in effects of exercise training on contractile and biochemical properties of rat cardiac myocytes. *J. Appl. Physiol.* 95:35-42.

Diffee, G.M., E.A. Seversen, T.D. Stein et al. 2003. Microarray expression analysis of effects of exercise training: increase in atrial MLC-1 in rat ventricles. *Am. J. Physiol. Heart Circ. Physiol.* 284: H830-H837.

Diffee, G.M., E.A. Seversen & M.M. Titus. 2001. Exercise training increases the Ca(2+) sensitivity of tension in rat cardiac myocytes. *J. Appl. Physiol.* 91: 309-15.

Ellingsen, O., O.A. Vengen, S.E. Kjeldsen et al. 1987. Myocardial potassium uptake and catecholamine release during cardiac sympathetic nerve stimulation. *Cardiovasc. Res.* 21: 892-901.

Ferguson, S., N. Gledhill, V.K. Jamnik et al. 2001. Cardiac performance in endurance trained and moderately active young adult females. *Med. Sci. Sports Exerc.* 33: 1114-19.

Ferrier, G.R. & S.E. Howlett. 2001. Cardiac excitation-contraction coupling: role of membrane potential in regulation of contraction. *Am. J. Physiol.* 280: H1928-H1944.

Frey, N., T.A. McKinsey & E.N. Olson. 2000. Decoding calcium signals involved in cardiac growth and function. *Nature Med.* 6: 1221-27.

Giannuzzi, P., P.L. Temporelli, U. Corra et al. 1997. Attenuation of unfavorable remodeling by exercise training in postinfarction patients with left ventricular dysfunction: results of the Exercise in Left Ventricular Dysfunction (ELVD) trial. *Circulation* 96: 1790-97.

Gomez, A.M., H. Cheng, W.J. Lederer et al. 1996. Ca2+ diffusion and sarcoplasmic reticulum transport both contribute to [Ca2+]i decline during Ca2+ sparks in rat ventricular myocytes. *J. Physiol.* 496: 575-81.

Gomez, A.M., S. Guatimosim, K.W. Dilly et al. 2001. Heart failure after myocardial infarction: altered excitation-contraction coupling. *Circulation* 104: 688-93.

Gomez, A.M., B.G. Kerfant & G. Vassort. 2000. Microtubule disruption modulates Ca(2+) signaling in rat cardiac myocytes. *Circ. Res.* 86: 30-36.

Gwathmey, J.K., L. Copelas, R. MacKinnon et al. 1987. Abnormal intracellular calcium handling in myocardium from patients with end-stage heart failure. *Circ. Res.* 61: 70-76.

Hajjar, R.J., J.K. Gwathmey, G.M. Briggs et al. 1988. Differential effect of DPI 201-106 on the sensitivity of the myofilaments to Ca2+ in intact and skinned trabeculae from control and myopathic human hearts. *J. Clin. Invest.* 82: 1578-84.

Hajjar, R.J., R.H. Schwinger, U. Schmidt et al. 2000. Myofilament calcium regulation in human myocardium. *Circulation* 101: 1679-85.

Hambrecht, R., S. Gielen, A. Linke et al. 2000. Effects of exercise training on left ventricular function and peripheral resistance in patients with chronic heart failure: a randomized trial. *J. Am. Med. Assoc.* 283: 3095-3101.

Hardt, S.E. & J. Sadoshima. 2002. Glycogen synthase kinase-3beta: a novel regulator of cardiac hypertrophy and development. *Circ. Res.* 90: 1055-63.

Hasenfuss, G., H. Reinecke, R. Studer et al. 1994. Relation between myocardial function and expression of sarcoplasmic reticulum Ca(2+)-ATPase in failing and nonfailing human myocardium. *Circ. Res.* 75: 434-42.

Heller, L.J. 1979. Augmented aftercontractions in papillary muscles from rats with cardiac hypertrophy. *Am. J. Physiol.* 237: H649-H654.

Hoshijima, M. & K.R. Chien. 2002. Mixed signals in heart failure: cancer rules. *J. Clin. Invest.* 109: 849-55.

Hudlicka, O. 1991. What makes blood vessels grow? *J. Physiol.* 444: 1-24.

Iemitsu, M., T. Miyauchi, S. Maeda et al. 2001. Physiological and pathological cardiac hypertrophy induce different molecular phenotypes in the rat. *Am. J. Physiol.* 281: R2029-R2036.

Iemitsu, M., T. Miyauchi, S. Maeda et al. 2002a. Effects of aging and subsequent exercise training on gene expression of endothelin-1 in rat heart. *Clin. Sci.* (Lond) 103(Suppl 48): 152S-157S.

Iemitsu, M., T. Miyauchi, S. Maeda et al. 2002b. Aging-induced decrease in the PPAR-alpha level in hearts is improved by exercise training. *Am. J. Physiol. Heart. Circ. Physiol.* 283:H1750-60.

Iemitsu, M., T. Miyauchi, S. Maeda et al. 2003. Cardiac hypertrophy by hypertension and exercise training exhibits different gene expression of enzymes in energy metabolism. *Hypertens. Res.* 26:829-37.

Iemitsu, M., T. Miyauchi, S. Maeda et al. 2004. Exercise training improves cardiac function-related gene levels through thyroid hormone receptor signaling in aged rats. *Am. J. Physiol. Heart. Circ. Physiol.* 286: H1696-705.

Jain, M., R. Liao, S. Ngoy et al. 2000. Angiotensin II receptor blockade attenuates the deleterious effects of exercise training on post-MI ventricular remodelling in rats. *Cardiovasc. Res.* 46: 66-72.

Jew, K.N. & R.L. Moore. 2001. Glibenclamide improves postischemic recovery of myocardial contractile function in trained and sedentary rats. *J. Appl. Physiol.* 91: 1545-54.

Jew, K.N. & R.L. Moore. 2002. Exercise training alters an anoxia-induced, glibenclamide-sensitive current in rat ventricular cardiocytes. *J. Appl. Physiol.* 92: 1473-79.

Jin, H., R. Yang, W. Li et al. 2000. Effects of exercise training on cardiac function, gene expression, and apoptosis in rats. *Am. J. Physiol.* 279: H2994-H3002.

Joyeux, M., D. Godin-Ribuot & C. Ribuot. 1998. Resistance to myocardial infarction induced by heat stress and the effect of ATP-sensitive potassium channel blockade in the rat isolated heart. *Brit. J. Pharmacol.* 123: 1085-88.

Katz, A.M. 2001. *Physiology of the Heart.* Philadelphia: Lippincott Williams & Wilkins.

Kaye, D., D. Pimental, S. Prasad et al. 1996. Role of transiently altered sarcolemmal membrane permeability and basic fibroblast growth factor release in the hypertrophic response of adult rat ventricular myocytes to increased mechanical activity in vitro. *J. Clin. Invest.* 97: 281-91.

Kemi, O.J., P.M. Haram, U. Wisløff, et al. 2004. Aerobic fitness is associated with cardiomyocyte contractile capacity and endothelial function in exercise training and detraining. *Circulation.* 109: 2897-904.

Kerfant, B.G., G. Vassort & A.M. Gomez. 2001. Microtubule disruption by colchicine reversibly enhances calcium signaling in intact rat cardiac myocytes. *Circ. Res.* 88: E59-E65.

Kleiman, R.B. & S.R. Houser. 1988. Calcium currents in normal and hypertrophied isolated feline ventricular myocytes. *Am. J. Physiol.* 255: H1434-H1442.

Lajoie, C., A. Calderone, F. Trudeau et al. 2004. Exercise training attenuated the PKB and GSK-3 dephosphorylation in the myocardium of ZDF rats. *J. Appl. Physiol.* 96:1606-12.

Lajoie, C., A. Calderone, & L. Béliveau. 2002. Exercise enhanced PKB phosphorylation and HSP72 expression in the myocardium of spontaneously hypertensive rats (SHR). *Can. J. Appl. Physiol*, 27 (Suppl.): 29.

Latronico, M.V.G., S. Costinean, M. L. Lavitrano et al. 2004. Regulation of cell size and contractile function by AKT in cardiomyocytes. *Ann. N.Y. Acad. Sci.* 1015: 1-11.

Laughlin, M.H., M.E. Schaefer & M. Sturek. 1992. Effect of exercise training on intracellular free Ca2+ transients in ventricular myocytes of rats. *J. Appl. Physiol.* 73: 1441-48.

Lederer, W.J., C.G. Nichols & G.L. Smith. 1989. The mechanism of early contractile failure of isolated rat ventricular myocytes subjected to complete metabolic inhibition. *J. Physiol.* 413: 329-49.

Libonati, J.R., J.P. Gaughan, C.A. Hefner et al. 1997. Reduced ischemia and reperfusion injury following exercise training. *Med. Sci. Sports Exerc.* 29: 509-16.

Loennechen, J.P., A. Stoylen, V. Beisvag et al. 2001. Regional expression of endothelin-1, ANP, IGF-1, and LV wall stress in the infarcted rat heart. *Am. J. Physiol.* 280: H2902-H2910.

Loennechen, J.P. 2002. Heart failure after myocardial infarction. Regional differences, myocyte function, gene expression, and response to Cariporide, Losartan, and exercise training. PhD Thesis. Tapir Press, Tirondheim, Norway, 1-40.

Loennechen, J.P., U. Wisloff, G. Falck et al. 2002. Cardiomyocyte contractility and calcium handling partially recover after early deterioration during post-infarction failure in rat. *Acta Physiol. Scand.* 176: 17-26.

Loennechen, J.P., U. Wisloff, G. Falck et al. 2003. Effects of cariporide and losartan on hypertrophy, calcium transients, contractility, and gene expression in congestive heart failure. *Circulation* 105: 1380-86.

Loennechen, J.P. 2003. Heart failure after myocardial infarction. Regional differences, myocyte function, gene expression, and response to cariporide, losartan, and exercise training. *Ph.D. Thesis*. Norwegian council on cardiovascular diseases, Oslo, Norway. Hjerteforum suppl 1 vol 16, 1-40.

Lu, L., D.F. Mei, A.G. Gu et al. 2002. Exercise training normalizes altered calcium-handling proteins during development of heart failure. *J. Appl. Physiol.* 92: 1524-30.

Lue, W.M. & P.A. Boyden. 1992. Abnormal electrical properties of myocytes from chronically infarcted canine heart. Alterations in Vmax and the transient outward current. *Circulation* 85: 1175-88.

Maeda S., T. Miyauchi, S. Sakai et al. 1998. Prolonged exercise causes an increase in endothelin-1 production in the heart in rats. *Am J Physiol.* 275:H2105-12.

McIntosh, M.A., S.M. Cobbe & G.L. Smith. 2000. Heterogeneous changes in action potential and intracellular Ca2+ in left ventricular myocyte sub-types from rabbits with heart failure. *Cardiovasc. Res.* 45: 397-409.

McMullen, J.R., T. Shioi, L. Zhang et al. 2003. Phosphoinositide 3-kinase(p110α) plays a critical role for the induction of physiological, but not pathological, cardiac hypertrophy. *Proc. Acad. Nat. Sci.* 100:12355-60.

Milne, K.J. & E.G. Noble. 2002. Exercise-induced elevation of HSP70 is intensity dependent. *J. Appl. Physiol.* 93: 561-68.

Mole, P.A. 1978. Increased contractile potential of papillary muscles from exercise-trained rat hearts. *Am. J. of Physiol.* 234: H421-H425.

Moore, R.L. & D.H. Korzick. 1995. Cellular adaptations of the myocardium to chronic exercise. *Prog. Cardiovasc. Dis.* 37: 371-96.

Moore, R.L., T.I. Musch, R.V. Yelamarty et al. 1993. Chronic exercise alters contractility and morphology of isolated rat cardiac myocytes. *Am. J. Physiol.* 264: C1180-C1189.

Moran, M., A. Saborido, & A. Megias. 2003. Ca2+ regulatory systems in rat myocardium are altered by 24 weeks treadmill training. *Pflugers Arch.*446:161-8.

Myers, J., M. Prakash, V. Froelicher et al. 2002. Exercise capacity and mortality among men referred for exercise testing. *New Engl. J. Med.* 346: 793-801.

Nakao, K., W. Minobe, R. Roden et al. 1997. Myosin heavy chain gene expression in human heart failure. *J. Clin. Invest.* 100: 2362-70.

Natali, A.J., L.A. Wilson, M. Peckham et al. 2002. Different regional effects of voluntary exercise on the mechanical and electrical properties of rat ventricular myocytes. *J. Physiol.* 541: 863-75.

Neary, P., S.M. Cobbe & G.L. Smith. 1998. Reduced sarcoplasmic reticulum Ca2+ release in rabbits with left ventricular dysfunction. *Ann. N.Y. Acad. Sci.* 853: 338-40.

Neri Serneri, G.G., M. Boddi, P.A. Modesti et al. 2001. Increased cardiac sympathetic activity and insulin-like growth factor-I formation are associated with physiological hypertrophy in athletes. *Circ Res 89:* 977-82.

Nishizawa, J., A. Nakai, M. Komeda et al. 2002. Increased preload directly induces the activation of heat shock transcription factor 1 in the left ventricular overloaded heart. *Cardiovasc. Res.* 55: 341-48.

O'Connell, T.D., S. Ishizaka, A. Nakamura et al. 2003. The $\alpha_{1A/C}$– and α_{1B}–adrenergic receptors are required for physiological cardiac hypertrophy in the double-knckout mouse. *J. Clin. Invest.* 111: 1783-91.

Orenstein, T.L., T.G. Parker, J.W. Butany et al. 1995. Favorable left ventricular remodeling following large myocardial infarction by exercise training. Effect on ventricular morphology and gene expression. *J. Clin. Invest.* 96: 858-66.

Palmer, B.M., A.M. Thayer, S.M. Snyder et al. 1998. Shortening and [Ca2+] dynamics of left ventricular myocytes isolated from exercise-trained rats. *J. Appl. Physiol.* 85: 2159-68.

Paroo, Z., J.V. Haist, M. Karmazyn et al. 2002. Exercise improves postischemic cardiac function in males but not females: consequences of a novel sex-specific heat shock protein 70 response. *Circ. Res.* 90: 911-17.

Penpargkul, S., D.I. Repke, A.M. Katz et al. 1977. Effect of physical training on calcium transport by rat cardiac sarcoplasmic reticulum. *Circ. Res.* 40: 134-38.

Perez, N.G., K. Hashimoto, S. McCune et al. 1999. Origin of contractile dysfunction in heart failure: calcium cycling versus myofilaments. *Circulation* 99: 1077-83.

Pierce, G.N., P.S. Sekhon, H.P. Meng et al. 1989. Effects of chronic swimming training on cardiac sarcolemmal function and composition. *J. Appl. Physiol.* 66: 1715-21.

Qin, D., Z.H. Zhang, E.B. Caref et al. 1996. Cellular and ionic basis of arrhythmias in postinfarction remodeled ventricular myocardium. *Circ. Res.* 79: 461-73.

Richardson, R.S., B. Grassi, T.P. Gavin et al. 1999. Evidence of O2 supply-dependent VO2 max in the exercise-trained human quadriceps. *J. Appl. Physiol.* 86: 1048-53.

Ritter, O., H.P. Luther, H. Haase et al. 1999. Expression of atrial myosin light chains but not alpha-myosin heavy chains is correlated in vivo with increased ventricular function in patients with hypertrophic obstructive cardiomyopathy [see comments]. *J. Mol. Med.* 77: 677-85.

Rozanski, G.J. & Z. Xu. 2002. Glutathione and K(+) channel remodeling in postinfarction rat heart. *Am. J. Physiol.* 282: H2346-H2355.

Ruwhof, C. & A. van der Laarse. 2000. Mechanical stress-induced cardiac hypertrophy: mechanisms and signal transduction pathways. *Cardiovasc. Res.* 47: 23-37.

Sadoshima, J. & S. Izumo. 1997. The cellular and molecular response of cardiac myocytes to mechanical stress. *Annu. Rev. Physiol.* 59: 551-71.

Schaible, T.F. & J. Scheuer. 1981. Cardiac function in hypertrophied hearts from chronically exercised female rats. *J. Appl. Physiol.* 50: 1140-45.

Scheuer, J. & C.M. Tipton. 1977. Cardiovascular adaptations to physical training. *Annu. Rev. Physiol.* 39: 221-51.

Schwinger, R.H., M. Bohm, A. Koch et al. 1994. The failing human heart is unable to use the Frank-Starling mechanism. *Circ. Res.* 74: 959-69.

Sipido, K.R. & W.G. Wier. 1991. Flux of Ca2+ across the sarcoplasmic reticulum of guinea-pig cardiac cells during excitation-contraction coupling. *J. Physiol.* (Lond) 435: 605-30.

Snoeckx, L.H., R.N. Cornelussen, F.A. Van Nieuwenhoven et al. 2001. Heat shock proteins and cardiovascular pathophysiology. *Physiol. Rev.* 81: 1461-97.

Song, J., X.Q. Zhang, & J. Wang. 2004. Sprint Training Improves Contractility in Postinfarction Rat Myocytes: Role of Na+/Ca2+ Exchange. *J. Appl. Physiol.* [Epub ahead of print].

Spencer, R.G., P.M. Buttrick & J.S. Ingwall. 1997. Function and bioenergetics in isolated perfused trained rat hearts. *Am. J. Physiol.* 272: H409-H417.

Stern, M.D., H.S. Silverman, S.R. Houser et al. 1988. Anoxic contractile failure in rat heart myocytes is caused by failure of intracellular calcium release due to alteration of the action potential. *Proc. Natl. Acad. Sci.* 85: 6954-58.

Sugden, P.H. 2001. Signalling pathways in cardiac myocyte hypertrophy. *Ann. Med.* 33: 611-22.

Sugden, P.H. & A. Clerk. 1998. Cellular mechanisms of cardiac hypertrophy. *J. Mol. Med.* 76: 725-46.

Tagawa, H., M. Koide, H. Sato et al. 1998. Cytoskeletal role in the transition from compensated to decompensated hypertrophy during adult canine left ventricular pressure overloading. *Circ. Res.* 82: 751-61.

Tajima, M., E.O. Weinberg, J. Bartunek et al. 1999. Treatment with growth hormone enhances contractile reserve and intracellular calcium transients in myocytes from rats with postinfarction heart failure. *Circulation* 99: 127-34.

Tate, C.A., T. Helgason, M.F. Hyek et al. 1996. SERCA2a and mitochondrial cytochrome oxidase expression are increased in hearts of exercise-trained old rats. *Am. J. Physiol.* 71: H68-H72.

Tate, C.A., G.E. Taffet, E.K. Hudson et al. 1990. Enhanced calcium uptake of cardiac sarcoplasmic reticulum in exercise-trained old rats. *Am. J. Physiol.* 258: H431-H435.

Terracciano, C.M., A.I. Souza, K.D. Philipson et al. 1998. Na+-Ca2+ exchange and sarcoplasmic reticular Ca2+ regulation in ventricular myocytes from transgenic mice overexpressing the Na+-Ca2+ exchanger. *J. Physiol.* 512: 651-67.

Tibbits, G.F., R.J. Barnard, K.M. Baldwin et al. 1981a. Influence of exercise on excitation-contraction coupling in rat myocardium. *Am. J. Physiol.* 240: H472-H480.

Tibbits, G.F., H. Kashihara & K. O'Reilly. 1989. Na+-Ca2+ exchange in cardiac sarcolemma: modulation of Ca2+ affinity by exercise. *Am. J. Physiol.* 256: C638-C643.

Tibbits, G., B.J. Koziol, N.K. Roberts et al. 1978. Adaptation of the rat myocardium to endurance training. *J. Appl. Physiol.* 44: 85-89.

Ueki, K., D.A. Fruman, S.M. Brachmann, Y.H. Tseng, L.C. Cantley & C.R. Kahn. 2002. Molecular balance between the regulatory and catalytic subunits of phosphoinositide 3-kinase regulates cell signaling and survival. *Mol. Cell. Biol.* 2: 965-77.

Vahl, C.F., A. Bonz, C. Hagl et al. 1994. Reversible desensitization of the myocardial contractile apparatus for calcium. A new concept for improving tolerance to cold ischemia in human myocardium? *Eur. J. Cardio-Thorac. Surg.* 8: 370-78.

Wagner, P.D. 2000. New ideas on limitation to VO2max. *Exerc. Sport Sci. Rev.* 28: 10-14.

Wamel, A.J., C. Ruwhof, P.I. Schrier et al. 2002. Stretch-induced paracrine hypertrophic stimuli increase TGF-beta1 expression in cardiomyocytes. *Mol. Cell. Biochem.* 236: 147-53.

Wattanapermpool, J., P.J. Reiser & R.J. Solaro. 1995. Troponin I isoforms and differential effects of acidic pH on soleus and cardiac myofilaments. *Am. J. Physiol.* 268: C323-C330.

Wiebe, C.G., N. Gledhill, V.K. Jamnik et al. 1999. Exercise cardiac function in young through elderly endurance trained women. *Med. Sci. Sports Exerc.* 31: 684-91.

Wiebe, C.G., N. Gledhill, D.E. Warburton et al. 1998. Exercise cardiac function in endurance-trained males versus females. *Clin. J. Sports Med.* 8: 272-79.

Wielenga, R.P., A.J. Coats, W.L. Mosterd et al. 1997. The role of exercise training in chronic heart failure. *Heart* 78: 431-36.

Wilkins, B.J. & J.D. Molkentin. 2002. Calcineurin and cardiac hypertrophy: where have we been? Where are we going? *J. Physiol.* 541: 1-8.

Wilkins, B.J., Y.-S. Dai, O.F. Bueno et al. 2004. Calcineurin/NFAT coupling participates in pathological, but not physiological, cardiac hypertrophy. *Circ. Res.* 94: 110-8.

Wisløff, U. 2001. Cardiac effects of aerobic endurance training. Hypertrophy, contractility and calcium handling in normal and failing heart. *Ph.D. Thesis.* Tapir Press, Trondheim, Norway, 1-49.

Wisløff, U., J.P. Loennechen, S. Currie et al. 2002. Aerobic exercise reduces cardiomyocyte hypertrophy and increases contractility, Ca2+ sensitivity and SERCA-2 in rat after myocardial infarction. *Cardiovasc. Res.* 54: 162-74.

Wisløff, U., J.P. Loennechen, G. Falck et al. 2001. Increased contractility and calcium sensitivity in cardiac myocytes isolated from endurance trained rats. *Cardiovasc. Res.* 50: 495-508.

Yasukawa, H., M. Hoshijima, Y. Gu et al. 2001. Suppressor of cytokine signaling-3 is a biomechanical stress-inducible gene that suppresses gp130-mediated cardiac myocyte hypertrophy and survival pathways. *J. Clin. Invest.* 108:1459-67.

Yatani, A., K. Okabe, J. Codina et al. 1990. Heart rate regulation by G proteins acting on the cardiac pacemaker channel. *Science* 249: 1163-66.

Zhang, X.Q., T.I. Musch, R. Zelis et al. 1999. Effects of impaired Ca2+ homeostasis on contraction in postinfarction myocytes. *J. Appl. Physiol.* 86: 943-50.

Zhang, X.Q., Y.C. Ng, T.I. Musch et al. 1998. Sprint training attenuates myocyte hypertrophy and improves Ca2+ homeostasis in postinfarction myocytes. *J. Appl. Physiol.* 84: 544-52.

Zhang, L.Q., X.Q. Zhang, T.I. Musch et al. 2000a. Sprint training restores normal contractility in postinfarction rat myocytes. *J. Appl. Physiol.* 89: 1099-1105.

Zhang, L.Q., X.Q. Zhang, Y.C. Ng et al. 2000b. Sprint training normalizes Ca(2+) transients and SR function in postinfarction rat myocytes *J. Appl. Physiol.* 89: 38-46.

Zhang, X.Q., L.Q. Zhang, B.M. Palmer et al. 2001. Sprint training shortens prolonged action potential duration in postinfarction rat myocyte: mechanisms. *J. Appl. Physiol.* 90: 1720-28.

Zhang, X.Q., J. Song, L.L. Carl et al. 2002. Effects of sprint training on contractility and [Ca(2+)](i) transients in adult rat myocytes. *J. Appl. Physiol.* 93:1310-7.

Zile, M.R., M. Koide, H. Sato et al. 1999. Role of microtubules in the contractile dysfunction of hypertrophied myocardium. *J. Am. Coll. Cardiol.* 33: 250-60.

Chapter 12

Augustin, H.G., D.H. Kozian & R.C. Johnson. 1994. Differentiation of endothelial cells: analysis of the constitutive and activated endothelial cell phenotypes. *Bioessays* 16(12): 901-06.

Awolesi, M.A., M.D. Widmann, W.C. Sessa et al. 1994. Cyclic strain increases endothelial nitric oxide synthase activity. *Surgery* 116(2): 439-44; discussion 444-45.

Ayajiki, K., M. Kindermann, M. Hecker et al. 1996. Intracellular pH and tyrosine phosphorylation but not calcium determine shear stress-induced nitric oxide production in native endothelial cells. *Circ. Res.* 78(5): 750-58.

Bergholm, R., S. Makimattila, M. Valkonen et al. 1999. Intense physical training decreases circulating antioxidants and endothelium-dependent vasodilatation in vivo. *Atherosclerosis* 145(2): 341-49.

Bolotina, V.M., S. Najibi, J.J. Palacino et al. 1994. Nitric oxide directly activates calcium-dependent potassium channels in vascular smooth muscle. *Nature* 368(6474): 850-53.

Bowles, D.K., M.H. Laughlin & M. Sturek. 1998. Exercise training increases K+-channel contribution to regulation of coronary arterial tone. *J. Appl. Physiol.* 84(4): 1225-33.

Brayden, J.E. & M.T. Nelson. 1992. Regulation of arterial tone by activation of calcium-dependent potassium channels. *Science* 256(5056): 532-35.

Brewster, J.L., T. de Valoir, N.D. Dwyer et al. 1993. An osmosensing signal transduction pathway in yeast. *Science* 259(5102): 1760-63.

Celermajer, D.S., K.E. Sorensen, D.J. Spiegelhalter et al. 1994. Aging is associated with endothelial dysfunction in healthy men years before the age-related decline in women. *J. Am. Coll. Cardiol.* 24(2): 471-76.

Chauhan, A., R.S. More, P.A. Mullins et al. 1996. Aging-associated endothelial dysfunction in humans is reversed by L-arginine. *J. Am. Coll. Cardiol.* 28(7): 1796-804.

Chen, S.J., C.C. Wu & M.H. Yen. 2001. Exercise training activates large-conductance calcium-activated K(+) channels and enhances nitric oxide production in rat mesenteric artery and thoracic aorta. *J. Biomed. Sci.* 8(3): 248-55.

Cheng, L., C. Yang, L. Hsu et al. 1999. Acute exercise enhances receptor-mediated endothelium-dependent vasodilation by receptor upregulation. *J. Biomed. Sci.* 6(1): 22-27.

Chu, T.F., T.Y. Huang, C.J. Jen et al. 2000. Effects of chronic exercise on calcium signaling in rat vascular endothelium. *Am. J. Physiol.* 279(4): H1441-46.

Cines, D.B., E.S. Pollak, C.A. Buck et al. 1998. Endothelial cells in physiology and in the pathophysiology of vascular disorders. *Blood* 91(10): 3527-61.

Clarkson, P., H.E. Montgomery, M.J. Mullen et al. 1999. Exercise training enhances endothelial function in young men. *J. Am. Coll. Cardiol.* 33(5): 1379-85.

Cooke, J.P., J. Stamler, N. Andon et al. 1990. Flow stimulates endothelial cells to release a nitrovasodilator that is potentiated by reduced thiol. *Am. J. Physiol.* 259(3 Pt 2): H804-12.

Davis, M.J., J.A. Donovitz & J.D. Hood. 1992. Stretch-activated single-channel and whole cell currents in vascular smooth muscle cells. *Am. J. Physiol.* 262(4 Pt 1): C1083-88.

De Caterina, R., P. Libby, H.B. Peng et al. 1995. Nitric oxide decreases cytokine-induced endothelial activation. Nitric oxide selectively reduces endothelial expression of adhesion molecules and proinflammatory cytokines. *J. Clin. Invest.* 96(1): 60-68.

Delp, M.D. & M.H. Laughlin. 1998. Regulation of skeletal muscle perfusion during exercise. *Acta Physiol. Scand.* 162(3): 411-19.

DeSouza, C.A., L.F. Shapiro, C.M. Clevenger et al. 2000. Regular aerobic exercise prevents and restores age-related declines in endothelium-dependent vasodilation in healthy men. *Circulation* 102(12): 1351-57.

Diamond, S.L., S.G. Eskin & L.V. McIntire. 1989. Fluid flow stimulates tissue plasminogen activator secretion by cultured human endothelial cells. *Science* 243(4897): 1483-85.

Dubin, D., R.E. Pratt, J.P. Cooke et al. 1989. Endothelin, a potent vasoconstrictor, is a vascular smooth muscle mitogen. *J. Vasc. Med. Biol.* (1): 150-54.

Dull, R.O. & P.F. Davies. 1991. Flow modulation of agonist (ATP)-response (Ca2+) coupling in vascular endothelial cells. *Am. J. Physiol.* 261(1 Pt 2): H149-54.

Dyke, C.K., D.N. Proctor, N.M. Dietz et al. 1995. Role of nitric oxide in exercise hyperaemia during prolonged rhythmic handgripping in humans. *J. Physiol.* 488 (Pt 1): 259-65.

Ekelund, U., J. Bjornberg, P.O. Grande et al. 1992. Myogenic vascular regulation in skeletal muscle in vivo is not dependent on endothelium-derived nitric oxide. *Acta Physiol. Scand.* 144(2): 199-207.

Ekelund, U. & S. Mellander. 1990. Role of endothelium-derived nitric oxide in the regulation of tonus in large-bore arterial resistance vessels, arterioles and veins in cat skeletal muscle. *Acta Physiol. Scand.* 140(3): 301-09.

Endo, T., T. Imaizumi, T. Tagawa et al. 1994. Role of nitric oxide in exercise-induced vasodilation of the forearm. *Circulation* 90(6): 2886-90.

Feletou, M. & P.M. Vanhoutte. 1988. Endothelium-dependent hyperpolarization of canine coronary smooth muscle. *Brit. J. Pharmacol.* 93(3): 515-24.

Fishman, A.P. 1982. Endothelium: a distributed organ of diverse capabilities. *Ann. N.Y. Acad. Sci.* 401: 1-8.

Frangos, J.A., S.G. Eskin, L.V. McIntire et al. 1985. Flow effects on prostacyclin production by cultured human endothelial cells. *Science* 227(4693): 1477-79.

Franke, W.D., G.M. Stephens & P.G. Schmid 3rd. 1998. Effects of intense exercise training on endothelium-dependent exercise-induced vasodilatation. *Clin. Physiol.* 18(6): 521-28.

Furchgott, R.F. & J.V. Zawadzki. 1980. The obligatory role of endothelial cells in the relaxation of arterial smooth muscle by acetylcholine. *Nature* 288(5789): 373-76.

Geiger, R.V., B.C. Berk, R.W. Alexander et al. 1992. Flow-induced calcium transients in single endothelial cells: spatial and temporal analysis. *Am. J. Physiol.* 262(6 Pt 1): C1411-17.

Gilligan, D.M., J.A. Panza, C.M. Kilcoyne et al. 1994. Contribution of endothelium-derived nitric oxide to exercise-induced vasodilation. *Circulation* 90(6): 2853-58.

Goligorsky, M.S., H. Tsukahara, H. Magazine et al. 1994. Termination of endothelin signaling: role of nitric oxide. *J. Cell Physiol.* 158(3): 485-94.

Green, D.J., N.T. Cable, C. Fox et al. 1994. Modification of forearm resistance vessels by exercise training in young men. *J. Appl. Physiol.* 77(4): 1829-33.

Groves, P., S. Kurz, H. Just et al. 1995. Role of endogenous bradykinin in human coronary vasomotor control. *Circulation* 92(12): 3424-30.

Haskell, W.L., E.L. Alderman, J.M. Fair et al. 1994. Effects of intensive multiple risk factor reduction on coronary atherosclerosis and clinical cardiac events in men and women with coronary artery disease. The Stanford Coronary Risk Intervention Project (SCRIP). *Circulation* 89(3): 975-90.

Haskell, W.L., C. Sims, J. Myll et al. 1993. Coronary artery size and dilating capacity in ultradistance runners. *Circulation* 87(4): 1076-82.

Hecker, M., A. Mulsch, E. Bassenge et al. 1993. Vasoconstriction and increased flow: two principal mechanisms of shear stress-dependent endothelial autacoid release. *Am. J. Physiol.* 265(3 Pt 2): H828-33.

Hester, R.L., A. Eraslan & Y. Saito. 1993. Differences in EDNO contribution to arteriolar diameters at rest and during functional dilation in striated muscle. *Am. J. Physiol.* 265(1 Pt 2): H146-51.

Hilton, S.M. 1959. A peripheral arterial conduction mechanism underlying dilation of the femoral artery and concerned in functional vasodilation in the skeletal muscle. *J. Physiol.* 149: 93-111.

Hilton, S.M., M.G. Jefferies & G. Vrboba. 1970. Functional specialization of the vascular bed of soleus. *J. Physiol.* 206: 13147.

Hirata, Y., Y. Takagi, Y. Fukuda et al. 1989. Endothelin is a potent mitogen for rat vascular smooth muscle cells. *Atherosclerosis* 78(2-3): 225-28.

Hornig, B., V. Maier & H. Drexler. 1996. Physical training improves endothelial function in patients with chronic heart failure. *Circulation* 93(2): 210-14.

Jen, C.J., H.P. Chan & H.I. Chen. 2002. Acute exercise enhances vasorelaxation by modulating endothelial calcium signaling in rat aortas. *Am. J. Physiol.* 282(3): H977-82.

Kessler, P., R. Popp, R. Busse et al. 1999. Proinflammatory mediators chronically downregulate the formation of the endothelium-derived hyperpolarizing factor in arteries via a nitric oxide/cyclic GMP-dependent mechanism. *Circulation* 99(14): 1878-84.

Kingwell, B.A., B. Sherrard, G.L. Jennings et al. 1997. Four weeks of cycle training increases basal production of nitric oxide from the forearm. *Am. J. Physiol.* 272(3 Pt 2): H1070-77.

Koller, A., A. Huang, D. Sun et al. 1995. Exercise training augments flow-dependent dilation in rat skeletal muscle arterioles. Role of endothelial nitric oxide and prostaglandins. *Circ. Res.* 76(4): 544-50.

Kramsch, D.M., A.J. Aspen, B.M. Abramowitz et al. 1981. Reduction of coronary atherosclerosis by moderate conditioning exercise in monkeys on an atherogenic diet. *N. Engl. J. Med.* 305(25): 1483-89.

Kubes, P., M. Suzuki & D.N. Granger. 1991. Nitric oxide: an endogenous modulator of leukocyte adhesion. *Proc. Natl. Acad. Sci.* 88(11): 4651-55.

Kuchan, M.J. & J.A. Frangos. 1993. Shear stress regulates endothelin-1 release via protein kinase C and cGMP in cultured endothelial cells. *Am. J. Physiol.* 264(1 Pt 2): H150-56.

Langille, B.L. & F. O'Donnell. 1986. Reductions in arterial diameter produced by chronic decreases in blood flow are endothelium-dependent. *Science* 231(4736): 405-07.

Laughlin, M.H. 1995. Endothelium-mediated control of coronary vascular tone after chronic exercise training. *Med. Sci. Sports Exerc.* 27(8): 1135-44.

Laughlin, M.H. & R.M. McAllister. 1992. Exercise training-induced coronary vascular adaptation. *J. Appl. Physiol.* 73(6): 2209-25.

Laughlin, M.H., J.S. Pollock, J.F. Amann et al. 2001. Training induces nonuniform increases in eNOS content along the coronary arterial tree. *J. Appl. Physiol.* 90(2): 501-10.

Laughlin, M.H., L.J. Rubin, J.W. Rush et al. 2003. Short-term training enhances endothelium-dependent dilation of coronary arteries, not arterioles. *J. Appl. Physiol.* 94(1): 234-44.

Leon, A.S. & C.M. Bloor. 1968. Effects of exercise and its cessation on the heart and its blood supply. *J. Appl. Physiol.* 24(4): 485-90.

Lie, M., O.M. Sejersted & F. Kiil. 1970. Local regulation of vascular cross section during changes in femoral arterial blood flow in dogs. *Circ. Res.* 27(5): 727-37.

Linz, W., P. Wohlfart, B.A. Scholkens et al. 1999. Interactions among ACE, kinins and NO. *Cardiovasc. Res.* 43(3): 549-61.

Maeda, S., T. Miyauchi, T. Kakiyama et al. 2001. Effects of exercise training of 8 weeks and detraining on plasma levels of endothelium-derived factors, endothelin-1 and nitric oxide, in healthy young humans. *Life Sci.* 69(9): 1005-16.

Maiorana, A., G. O'Driscoll, C. Cheetham et al. 2001. The effect of combined aerobic and resistance exercise training on vascular function in type 2 diabetes. *J. Am. Coll. Cardiol.* 38(3): 860-66.

Maiorana, A., G. O'Driscoll, L. Dembo et al. 2000a. Effect of aerobic and resistance exercise training on vascular function in heart failure. *Am J. Physiol.* 279(4): H1999-2005.

Maiorana, A., G. O'Driscoll, L. Dembo et al. 2000b. Exercise training, vascular function, and functional capacity in middle-aged subjects. *Med. Sci. Sports Exerc.* 33(12): 2022-28.

Malek, A.M., A.L. Greene & S. Izumo. 1993. Regulation of endothelin 1 gene by fluid shear stress is transcriptionally mediated and independent of protein kinase C and cAMP. *Proc. Natl. Acad. Sci.* 90(13): 5999-6003.

Martin, W.H. 3rd, W.M. Kohrt, M.T. Malley et al. 1990. Exercise training enhances leg vasodilatory capacity of 65-yr-old men and women. *J. Appl. Physiol.* 69(5): 1804-09.

Meininger, G.A. & M.J. Davis. 1992. Cellular mechanisms involved in the vascular myogenic response. *Am. J. Physiol.* 263(3 Pt 2): H647-59.

Mellander, S. & J. Bjornberg. 1992. Regulation of vascular smooth muscle tone and capillary pressure. *News Physiol. Sci.* (7): 113-19.

Mendelsohn, M.E., S. O'Neill, D. George et al. 1990. Inhibition of fibrinogen binding to human platelets by S-nitroso-N-acetylcysteine. *J. Biol. Chem.* 265(31): 19028-34.

Michelson, A.D., S.E. Benoit, M.I. Furman et al. 1996. Effects of nitric oxide/EDRF on platelet surface glycoproteins. *Am. J. Physiol.* 270(5 Pt 2): H1640-48.

Mitsumata, M., R.S. Fishel, R.M. Nerem et al. 1993. Fluid shear stress stimulates platelet-derived growth factor expression in endothelial cells. *Am J. Physiol.* 265(1 Pt 2): H3-8.

Mo, M., S.G. Eskin & W.P. Schilling. 1991. Flow-induced changes in Ca2+ signaling of vascular endothelial cells: effect of shear stress and ATP. *Am. J. Physiol.* 260(5 Pt 2): H1698-707.

Muller, J.M., P.R. Myers & M.H. Laughlin. 1994. Vasodilator responses of coronary resistance arteries of exercise-trained pigs. *Circulation* 89(5): 2308-14.

Nakamura, S., M. Naruse, K. Naruse et al. 1990. Immunocytochemical localization of endothelin in cultured bovine endothelial cells. *Histochemistry* 94(5): 475-77.

Nelson, M.T., H. Cheng, M. Rubart et al. 1995. Relaxation of arterial smooth muscle by calcium sparks. *Science* 270(5236): 633-37.

Nelson, M.T. & J.M. Quayle. 1995. Physiological roles and properties of potassium channels in arterial smooth muscle. *Am. J. Physiol.* 268(4 Pt 1): C799-822.

Niebauer, J. & J.P. Cooke. 1996. Cardiovascular effects of exercise: role of endothelial shear stress. *J. Am. Coll. Cardiol.* 28(7): 1652-60.

Noris, M., M. Morigi, R. Donadelli et al. 1995. Nitric oxide synthesis by cultured endothelial cells is modulated by flow conditions. *Circ. Res.* 76(4): 536-43.

O'Driscoll, G., D. Green, A. Maiorana et al. 1999. Improvement in endothelial function by angiotensin-converting enzyme inhibition in non-insulin-dependent diabetes mellitus. *J. Am. Coll. Cardiol.* 33(6): 1506-11.

Ohno, M., J.P. Cooke, V.J. Dzau et al. 1995. Fluid shear stress induces endothelial transforming growth factor beta-1 transcription and production. Modulation by potassium channel blockade. *J. Clin. Invest.* 95(3): 1363-69.

Olesen, S.P., D.E. Clapham & P.F. Davies. 1988. Haemodynamic shear stress activates a K+ current in vascular endothelial cells. *Nature* 331(6152): 168-70.

Oltman, C.L., J.L. Parker, H.R. Adams et al. 1992. Effects of exercise training on vasomotor reactivity of porcine coronary arteries. *Am. Coll. Physiol.* 263(2 Pt 2): H372-82.

Oltman, C.L., J.L. Parker & M.H. Laughlin. 1995. Endothelium-dependent vasodilation of proximal coronary arteries from exercise-trained pigs. *J. Appl. Physiol.* 79(1): 33-40.

Parker, J.L., C.L. Oltman, J.M. Muller et al. 1994. Effects of exercise training on regulation of tone in coronary arteries and arterioles. *Med. Sci. Sports Exerc.* 26(10): 1252-61.

Persson, M.G., L.E. Gustafsson, N.P. Wiklund et al. 1990. Endogenous nitric oxide as a modulator of rabbit skeletal muscle microcirculation in vivo. *Brit. J. Pharmacol.* 100(3): 463-66.

Powell, K.E., P.D. Thompson, C.J. Caspersien et al. 1987. Physical activity and the incidence of coronary heart disease. *Ann. Rev. Public Health* (8): 253-87.

Radomski, M.W., R.M. Palmer & S. Moncada. 1987. Comparative pharmacology of endothelium-derived relaxing factor, nitric oxide and prostacyclin in platelets. *Brit. J. Pharmacol.* 92(1): 181-87.

Resnick, N., T. Collins, W. Atkinson et al. 1993. Platelet-derived growth factor B chain promoter contains a cis-acting fluid shear-stress-responsive element. *Proc. Natl. Acad. Sci.* 90(10): 4591-95.

Resnick, N. & M.A. Gimbrone Jr. 1995. Hemodynamic forces are complex regulators of endothelial gene expression. *FASEB J.* 9(10): 874-82.

Rinder, M.R., R.J. Spina & A.A. Ehsani. 2000. Enhanced endothelium-dependent vasodilation in older endurance-trained men. *J. Appl. Physiol.* 88(2): 761-66.

Rogers, P.J., T.D. Miller, B.A. Bauer et al. 1991. Exercise training and responsiveness of isolated coronary arteries. *J. Appl. Physiol.* 71(6): 2346-51.

Rywik, T.M., M.R. Blackman, A.R. Yataco et al. 1999. Enhanced endothelial vasoreactivity in endurance-trained older men. *J. Appl. Physiol.* 87(6): 2136-42.

Sadoshima, J. & S. Izumo. 1993. Signal transduction pathways of angiotensin II-induced c-fos gene expression in cardiac myocytes in vitro. Roles of phospholipid-derived second messengers. *Circ. Res.* 73(3): 424-38.

Sessa, W.C., J.K. Harrison, C.M. Barber et al. 1992. Molecular cloning and expression of a cDNA encoding endothelial cell nitric oxide synthase. *J. Biol. Chem.* 267(22): 15274-76.

Sessa, W.C., K. Pritchard, N. Seyedi et al. 1994. Chronic exercise in dogs increases coronary vascular nitric oxide production and endothelial cell nitric oxide synthase gene expression. *Circ. Res.* 74(2): 349-53.

Shen, J., F.W. Luscinskas, A. Connolly et al. 1992. Fluid shear stress modulates cytosolic free calcium in vascular endothelial cells. *Am. J. Physiol.* 262(2 Pt 1): C384-90.

Simonson, M.S. & M.J. Dunn. 1990. Cellular signalling by peptides of the endothelin gene family. *FASEB* (4): 2989.

Smith, D.T., G.L. Hoetzer, J.J. Greiner et al. 2003. Effects of ageing and regular aerobic exercise on endothelial fibrinolytic capacity in humans. *J. Physiol.* 546(Pt 1): 289-98.

Snell, P.G., W.H. Martin, J.C. Buckey et al. 1987. Maximal vascular leg conductance in trained and untrained men. *J. Appl. Physiol.* 62(2): 606-10.

Stamler, J.S., D.E. Vaughan & J. Loscalzo. 1989. Synergistic disaggregation of platelets by tissue-type plasminogen activator, prostaglandin E1, and nitroglycerin. *Circ. Res.* 65(3): 796-804.

Taddei, S., A. Virdis, P. Mattei et al. 1995. Aging and endothelial function in normotensive subjects and patients with essential hypertension. *Circulation* 91(7): 1981-87.

Topper, J.N., J. Cai, D. Falb et al. 1996. Identification of vascular endothelial genes differentially responsive to fluid mechanical stimuli: cyclooxygenase-2, manganese superoxide dismutase, and endothelial cell nitric oxide synthase are selectively up-regulated by steady laminar shear stress. *Proc. Natl. Acad. Sci.* 93(19): 10417-22.

Venema, R.C., H.S. Sayegh & J.F. Arnal. 1995. Role of the enzyme calmodulin-binding domain in membrane association and phospholipid inhibition of endothelial nitric oxide synthase. *J. Biol. Chem.* 270(24): 14705-11.

Wang, J., M.S. Wolin & T.H. Hintze. 1993. Chronic exercise enhances endothelium-mediated dilation of epicardial coronary artery in conscious dogs. *Circ. Res.* 73(5): 829-38.

Wellner, M.-C. & G. Isenberg. 1994. Stretch effects on whole-cell currents of guinea-pig urinary bladder monocytes. *J. Physiol.* 480: 439-48.

Woodman, C.R., J.M. Muller, M.H. Laughlin et al. 1997. Induction of nitric oxide synthase mRNA in coronary resistance arteries isolated from exercise-trained pigs. *Am. J. Physiol.* 2: 73(6 Pt 2): H2575-79.

Yang, Z.H., V. Richard, L. von Segesser et al. 1990. Threshold concentrations of endothelin-1 potentiate contractions to norepinephrine and serotonin in human arteries. A new mechanism of vasospasm? *Circulation* 82(1): 188-95.

Yoshizumi, M., H. Kurihara, T. Sugiyama et al. 1989. Hemodynamic shear stress stimulates endothelin production by cultured endothelial cells. *Biochem. Biophys. Res. Comm.* 161(2): 859-64.

Chapter 13

Allen, D.L. & L.A. Leinwand. 2002. Intracellular calcium and myosin isoform transitions. Calcineurin and calcium-calmodulin kinase pathways regulate preferential activation of the IIa myosin heavy chain promoter. *J. Biol. Chem.* 277: 45323-30.

Allen, D.L., C.A. Sartorius, L.K. Sycuro et al. 2001. Different pathways regulate expression of the skeletal myosin heavy chain genes. *J. Biol. Chem.* 276: 43524-33.

Andersen, J.L., T. Gruschy-Knudsen, C. Sandri at al. 1999. Bed rest increases the amount of mismatched fibers in human skeletal muscle. *J. Appl. Physiol.* 86: 455-60.

Andersen, J.L., H. Klitgaard & B. Saltin. 1994. Myosin heavy chain isoforms in single fibres from m. vastus lateralis of sprinters: influence of training. *Acta Physiol. Scand.* 151: 135-42.

Andersen, J.L., T. Mohr, F. Biering-Sörensen et al. 1996. Myosin heavy chain isoform transformation in single fibres from m. vastus lateralis in spinal cord injured individuals: effects of long-term functional electrical stimulation (FES). *Pfluegers Arch.* 431: 513-18.

Arnold, H.H. & B. Winter. 1998. Muscle differentiation: more complexity to the network of myogenic regulators. *Curr. Opin. Gen. Dev.* 8: 539-44.

Ausoni, S., L. Gorza, S. Schiaffino et al. 1990. Expression of myosin heavy chain isoforms in stimulated fast and slow rat muscles. *J. Neurosci.* 10: 153-60.

Baldwin, K.M. & F. Haddad. 2001. Effects of different activity and inactivity paradigms on myosin heavy chain gene expression in striated muscle. *J. Appl. Physiol.* 90: 345-57.

Barash, I.A., L. Mathew, A.F. Ryan et al. 2003. Rapid muscle-specific gene expression changes after a single bout of eccentric contractions in the mouse. *Am. J. Physiol.* 286: C355-C364.

Bassel-Duby, R. & E.N. Olson. 2003. Role of calcineurinnin striated muscle: development, adaptation, and disease. *Biochem. Biophys. Res. Com.* 311: 1133-41.

Beals, C.R., N.A. Clipstone, S.N. Ho et al. 2002. Nuclear localization of NF-ATc by a calcineurin-dependent, cyclosporine-sensitive intramolecular interaction. *Gen. Dev.* 11: 824-34.

Bergeron, R., J.M. Ren, K.S. Cadman et al. 2001. Chronic activation of AMP kinase results in NRF-1 activation and mitochondrial biogenesis. *Am. J. Physiol.* 281: E1340-E1346.

Bigard, X., H. Sanchez, J. Zoll et al. 2000. Calcineurin co-regulates contractile and metabolic components of slow muscle phenotype. *J. Biol. Chem.* 275: 19653-60.

Bonen, A., D.J. Dyck, A. Ibrahimi et al. 1999. Muscle-contractile activity increases fatty acid metabolism and transport and FAT/CD36. *Am. J. Physiol.* 276: E642-E649.

Bonen, A., J.J.F.P. Luiken, Y. Arumugam et al. 2001. Acute regulation of fatty acid uptake involves the cellular redistribution of fatty acid translocase. *J. Biol. Chem.* 275: 14501-08.

Booth, F.W. & K.M. Baldwin. 1996. Muscle plasticity: energy demanding and supply processes. In *Handbook of Physiology, Exercise: Regulation and Integration of Multiple Systems*, sect. 12, ed. L.B. Rowell & J.T. Shepherd, 1075-1123. New York: Oxford University Press.

Bottinelli, R., R. Betto, S. Schiaffino et al. 1994a. Unloaded shortening velocity and myosin heavy chain and alkali light chain isoform composition in rat skeletal muscle fibres. *J. Physiol.* 478: 341-49.

Bottinelli, R., M. Canepari, C. Reggiani et al. 1994b. Myofibrillar ATPase activity during isometric contraction and isomyosin composition in rat single skinned muscle fibres. *J. Physiol.* 481: 663-75.

Bottinelli, R. & C. Reggiani. 1995. Force-velocity properties and myosin light chain isoform composition of an identified type of skinned fibres from rat skeletal muscle. *Pfluegers Arch.* 429: 592-94.

Bottinelli, R. & C. Reggiani. 2000. Human skeletal muscle fibres: molecular and functional diversity. *Prog. Biophys. Mol. Biol.* 73: 195-262.

Bottinelli, R., S. Schiaffino & C. Reggiani. 1991. Force-velocity relationship and myosin heavy chain isoform compositions of skinned fibres from rat skeletal muscle. *J. Physiol.* 437: 655-72.

Brownson, C., P. Little, J.C. Jarvis et al. 1992. Reciprocal changes in myosin isoform messenger RNAs of rabbit skeletal muscle in response to the initiation and cessation of chronic electrical stimulation. *Muscle Nerve* 15: 694-700.

Burnham, R., T. Martin, R. Stein et al. 1997. Skeletal muscle fibre type transformation following spinal cord injury. *Spinal Cord* 35: 86-91.

Caiozzo, V.J., M.J. Baker & K.M. Baldwin. 1998. Novel transitions in MHC isoforms: separate and combined effects of thyroid hormone and mechanical unloading. *J. Appl. Physiol.* 85: 2237-48.

Caiozzo, V.J., F. Haddad, M.J. Baker et al. 1996. Microgravity-induced transformations of myosin isoforms and contractile properties of skeletal muscle. *J. Appl. Physiol.* 81: 123-32.

Carroll, S., P. Nicotera & D. Pette. 1999. Calcium transients in single fibers of low-frequency stimulated fast-twitch muscle of rat. *Am. J. Physiol.* 277: C1122-C1129.

Chin, E.R., E.N. Olson, J.A. Richardson et al. 1998. A calcineurin-dependent transcriptional pathway controls skeletal muscle fiber type. *Gen. Dev.* 12: 2499-2509.

Conjard, A., H. Peuker & D. Pette. 1998. Energy state and myosin isoforms in single fibers of normal and transforming rabbit muscles. *Pfluegers Arch.* 436: 962-69.

DiMario, J.X. 2001. Protein kinase C signaling controls skeletal muscle fiber types. *Exp. Cell Res.* 263: 23-32.

Dohm, G.L. 2002. Invited review: regulation of skeletal muscle GLUT-4 expression by exercise. *J. Appl. Physiol.* 93: 782-87.

Dolmetsch, R.E., R.S. Lewis, C.C. Goodnow et al. 1997. Differential activation of transcriptional factors induced by Ca^{2+} response amplitude and duration. *Nature* 386: 855-58.

Edgerton, V.R. & R.R. Roy. 1996. Neuromuscular adaptations to actual and simulated spaceflight. In *Handbook of Physiology, Environmental Physiology*, sect. 4, vol. II, chapt. 32, 721-63. Bethesda, MD: American Physiological Society.

Egginton, S. & O. Hudlická. 1999. Early changes in performance, blood flow and capillary fine structure in rat fast muscles induced by electrical stimulation. *J. Physiol.* 515: 265-75.

Eisenberg, B.R. & S. Salmons. 1981. The reorganization of subcellular structure in muscle undergoing fast-to-slow type transformation. A stereological study. *Cell Tiss. Res.* 220: 449-71.

Ennion, S., J. S'Antana Pereira, A.J. Sargeant et al. 1995. Characterization of human skeletal muscle fibres according to the myosin heavy chains they express. *J. Muscle Res. Cell Mot.* 16: 35-43.

Etgen, G.J., J.T. Brozinick, H.Y. Kang et al. 1993. Effects of exercise training on skeletal muscle glucose uptake and transport. *Am. J. Physiol.* 264: C727-C733.

Fields, R.D., F. Eshete, B. Stevens et al. 1997. Action potential-dependent regulation of gene expression: temporal specificity in Ca^{2+}, cAMP-responsive element binding proteins, and mitogen-activated protein kinase signaling. *J. Neurosci.* 17: 7252-66.

Finkbeiner, S. & M.E. Greenberg. 1996. Ca^{2+}-dependent routes to Ras: mechanisms for neuronal survival, differentiation, and plasticity? *Neuron* 16: 233-36.

Fitts, R.H., D.R. Riley & J.J. Widrick. 2000. Physiology of a microgravity environment. Invited review: microgravity and skeletal muscle. *J. Appl. Physiol.* 89: 823-39.

Freyssenet, D., M. DiCarlo & D.A. Hood. 1999. Calcium-dependent regulation of cytochrome c gene expression in skeletal muscle cells—identification of a protein kinase C-dependent pathway. *J. Biol. Chem.* 274: 9305-11.

Fujii, N., T. Hayashi, M.F. Hirshman et al. 2000. Exercise induces isoform-specific increase in 5'AMP-activated protein kinase activity in human skeletal muscle. *Biochem. Biophys. Res. Com.* 273: 1150-55.

Galler, S., E. Puchert, B. Gohlsch et al. 2002. Kinetic properties of cardiac myosin heavy chain isoforms in rat. *Pfluegers Arch.* 445: 218-223.

Galler, S., T. Schmitt & D. Pette. 1994. Stretch activation, unloaded shortening velocity, and myosin heavy chain isoforms of rat skeletal muscle fibres. *J. Physiol.* 478: 523-31.

Goldspink, G., A. Scutt, P.T. Loughna et al. 1992. Gene expression in skeletal muscle in response to stretch and force generation. *Am. J. Physiol.* 262: R356-R363.

Gollnick, P.D. & D.W. King. 1969. Effect of exercise and training on mitochondria of rat skeletal muscle. *Am. J. Physiol.* 216: 1502-09.

Goodyear, L.J. 2000. AMP-activated protein kinase: a critical signaling intermediary for exercise-stimulated glucose transport? *Exerc. Sport Sci. Rev.* 28: 113-16.

Green, H.J., S. Düsterhöft, L. Dux et al. 1992a. Metabolite patterns related to exhaustion, recovery, and transformation of chronically stimulated rabbit fast-twitch muscle. *Pfluegers Arch.* 420: 359-66.

Green, H.J., R. Helyar, M. Ball-Burnett et al. 1992b. Metabolic adaptations to training precede changes in muscle mitochondrial capacity. *J. Appl. Physiol.* 72: 484-91.

Green, H.J., G.A. Klug, H. Reichmann et al. 1984. Exercise-induced fibre type transitions with regard to myosin, parvalbumin, and sarcoplasmic reticulum in muscles of the rat. *Pfluegers Arch.* 400: 432-38.

Gundersen, K., E. Leberer, T. Lömo et al. 1988. Fibre types, calcium-sequestering proteins and metabolic enzymes in denervated and chronically stimulated muscles of the rat. *J. Physiol.* 398: 177-89.

Hayashi, T., M.F. Hirshman, S.D. Dufresne et al. 1999. Skeletal muscle contractile activity in vitro stimulates mitogen-activated protein kinase signaling. *Am. J. Physiol.* 277: C701-C707.

Hämäläinen, N. & D. Pette. 1997. Coordinated fast-to-slow transitions of myosin and SERCA isoforms in chronically stimulated fast-twitch muscles of euthyroid and hyperthyroid rabbits. *J. Muscle Res. Cell Mot.* 18: 545-54.

Hardie, D.G. 2004. AMP-activated protein kinase: a key system mediating metabolic responses to exercise. *Med. Sci. Sports. Exerc.* 36: 28-34.

Heizmann, C.W., M.W. Berchtold & A.M. Rowlerson. 1982. Correlation of parvalbumin concentration with relaxation speed in mammalian muscles. *Proc. Natl. Acad. Sci.* 79: 7243-47.

Hennig, R. & T. Lömo. 1985. Firing patterns of motor units in normal rats. *Nature* 314: 164-66.

Henriksson, J. 1992. Effects of physical training on the metabolism of skeletal muscle. *Diabet. Care* 15: 1701-11.

Hicks, A., K. Ohlendieck, S.O. Göpel et al. 1997. Early functional and biochemical adaptations to low-frequency stimulation of rabbit fast-twitch muscle. *Am. J. Physiol.* 273: C297-C305.

Hilber, K., S. Galler, B. Gohlsch et al. 1999. Kinetic properties of myosin heavy chain isoforms in single fibers from human skeletal muscle. *FEBS Lett.* 455: 267-70.

Hofmann, S. & D. Pette. 1994. Low-frequency stimulation of rat fast-twitch muscle enhances the expression of hexokinase II and both the translocation and expression of glucose transporter 4 (GLUT-4). *Eur. J. Biochem.* 219: 307-15.

Holloszy, J.O. 1967. Biochemical adaptations in muscle. Effects of exercise on mitochondrial oxygen uptake and respiratory enzyme activity in skeletal muscle. *J. Biol. Chem.* 242: 2278-82.

Holloszy, J.O. & E.F. Coyle. 1984. Adaptations of skeletal muscle to endurance exercise and their metabolic consequences. *J. Appl. Physiol.* 56: 831-38.

Holmes, B.F., E.J. Kurth-Kraczek & W.W. Winder. 1999. Chronic activation of 5 '-AMP-activated protein kinase increases GLUT-4, hexokinase, and glycogen in muscle. *J. Appl. Physiol.* 87: 1990-95.

Hood, D.A. 2001. Invited review: contractile activity-induced mitochondrial biogenesis in skeletal muscle. *J. Appl. Physiol.* 90: 1137-57.

Hood, D.A. & D. Pette. 1989. Chronic long-term stimulation creates a unique metabolic enzyme profile in rabbit fast-twitch muscle. *FEBS Lett.* 247: 471-74.

Horton, M.J., C.A. Brandon, T.J. Morris et al. 2001. Abundant expression of myosin heavy-chain IIB RNA in a subset of human masseter muscle fibres. *Arch. Oral Biol.* 46: 1039-50.

Howald, H., H. Hoppeler, H. Claassen et al. 1985. Influence of endurance training on the ultrastructural composition of the different muscle fiber types in humans. *Pfluegers Arch.* 403: 369-76.

Hudlická, O., M. Brown, M. Cotter et al. 1977. The effect of long-term stimulation of fast muscles on their blood flow, metabolism and ability to withstand fatigue. *Pfluegers Arch.* 369: 141-49.

Hughes, S.M., J.M. Taylor, S.J. Tapscott et al. 1993. Selective accumulation of MyoD and myogenin messenger RNAs in fast and slow adult skeletal muscle is controlled by innervation and hormones. *Development* 118: 1137-47.

Hutber, C.A., D.G. Hardie & W.W. Winder. 1997. Electrical stimulation inactivates muscle acetyl-CoA carboxylase and increases AMP-activated protein kinase. *Am. J. Physiol.* 272: E262-E266.

Jagoe, R.T. & A.L. Goldberg. 2001. What do we really know about the ubiquitin-proteasome pathway in muscle atrophy? *Curr. Op. Clin. Nutr. Metab. Care* 4: 183-90.

Jansson, E., M. Esbjörnsson & I. Jacobs. 1990. Increase in the proportion of fast-twitch muscle fibres by sprint training in males. *Acta Physiol. Scand.* 140: 359-63.

Jansson, E. & L. Kaijser. 1977. Muscle adaptation to extreme endurance training in man. *Acta Physiol. Scand.* 100: 315-24.

Jaschinski, F., M. Schuler, H. Peuker et al. 1998. Transitions in myosin heavy chain mRNA and protein isoforms of rat muscle during forced contractile activity. *Am. J. Physiol.* 274: C365-C371.

Jorgensen, A.O. & L.R. Jones. 1986. Localization of phospholamban in slow but not fast canine skeletal muscle fibers. An immunocytochemical and biochemical study. *J. Biol. Chem.* 261: 3775-81.

Karasseva, N., G. Tsika, J. Ji et al. 2003. Transcription enhancer factor 1 binds multiple muscle MEF2 and A/T-rich elements during fast-to-slow skeletal muscle fiber type transitions. *Mol. Cell Biol.* 23: 5143-64.

Kirschbaum, B.J., H.-B. Kucher, A. Termin et al. 1990a. Antagonistic effects of chronic low frequency stimulation and thyroid hormone on myosin expression in rat fast-twitch muscle. *J. Biol. Chem.* 265: 13974-80.

Kirschbaum, B.J., S. Schneider, S. Izumo et al. 1990b. Rapid and reversible changes in myosin heavy chain expression in response to increased neuromuscular activity of rat fast-twitch muscle. *FEBS Lett.* 268: 75-78.

Klitgaard, H., O. Bergman, R. Betto et al. 1990. Co-existence of myosin heavy chain I and IIA isoforms in human skeletal muscle fibres with endurance training. *Pfluegers Arch.* 416: 470-72.

Krenács, T., E. Molnar, E. Dobo et al. 1989. Fibre typing using sarcoplasmic reticulum Ca^{2+}-ATPase and myoglobin immunohistochemistry in rat gastrocnemius muscle. *Histochem. J.* 21: 145-55.

LaFramboise, W.A., M.J. Daood, R.D. Guthrie et al. 1990. Electrophoretic separation and immunological identification of type 2X myosin heavy chain in rat skeletal muscle. *Biochim. Biophys. Acta* 1035: 109-112.

Larsson, L. & R.L. Moss. 1993. Maximum velocity of shortening in relation to myosin isoform composition in single fibres from human skeletal muscles. *J. Physiol.* 472: 595-614.

Läuger, P. 1991. *Electrogenic Ion Pumps.* Sunderland, MA: Sinauer Associates.

Leeuw, T. & D. Pette. 1993. Coordinate changes in the expression of troponin subunit and myosin heavy chain isoforms during fast-to-slow transition of low-frequency stimulated rabbit muscle. *Eur. J. Biochem.* 213: 1039-46.

Leberer, E. & D. Pette 1986. Immunochemical quantification of sarcoplasmic reticulum Ca-ATPase, of calsequestrin and of parvalbumin in rabbit skeletal muscles of defined fiber composition. *Eur. J. Biochem.* 156: 489-96.

Li, X.P. & L. Larsson. 1997. Contractility and myosin isoform compositions of skeletal muscles and muscle cells from rats treated with thyroid hormone for 0, 4 and 8 weeks. *J. Muscle Res. Cell Mot.* 18: 335-44.

Lin, J., H. Wu, P.T. Tarr et al. 2002. Transcriptional co-activator PGC-1^{N}_{U}alpha drives the formation of slow-twitch muscle fibres. *Nature* 418: 797-801.

Liu, Y., Z. Cseresnyés, W.R. Randall et al. 2001. Activity-dependent nuclear translocation and intranuclear distribution of NFATc in adult skeletal muscle fibers. *J. Cell Biol.* 155: 27-39.

Lompré, A.-M., B. Nadal-Ginard & V. Mahdavi. 1984. Expression of the cardiac ventricular $^{D}{}_{T}$- and β-myosin heavy chain genes is developmentally and hormonally regulated. *J. Biol. Chem.* 259: 6437-46.

Loughna, P.T., S. Izumo, G. Goldspink et al. 1990. Disuse and passive stretch cause rapid alterations in expression of developmental and adult contractile protein genes in skeletal muscle. *Development* 109: 217-23.

Lowey, S., G.S. Waller & K.M. Trybus. 1993. Function of skeletal muscle myosin heavy and light chain isoforms by an in vitro motility assay. *J. Biol. Chem.* 268: 20414-18.

Luginbuhl, A.J. & G.A. Dudley. 1984. Fiber type changes in rat skeletal muscle after intense interval training. *Histochemistry* 81: 55-58.

Mathieu-Costello, O., P.J. Agey, L. Wu et al. 1996. Capillary-to-fiber surface ratio in rat fast-twitch hindlimb muscles after chronic electrical stimulation. *J. Appl. Physiol.* 80: 904-09.

Meissner, J.D., G. Gros, R.J. Scheibe et al. 2001. Calcineurin regulates slow myosin, but not fast myosin or metabolic enzymes, during fast-to-slow transformation in rabbit skeletal muscle cell culture. *J. Physiol.* 533: 215-26.

Meissner, J.D., H.P. Kubis, R.J. Scheibe et al. 2000. Reversible Ca^{2+}-induced fast-to-slow transition in primary skeletal muscle culture cells at the mRNA level. *J. Physiol.* 523: 19-28.

Michel, J.B., G.A. Ordway, J.A. Richardson et al. 1994. Biphasic induction of immediate early gene expression accompanies activity-dependent angiogenesis and myofiber remodeling of rabbit skeletal muscle. *J. Clin. Invest.* 94: 277-85.

Murgia, M., A.L. Serrano, E. Calabria et al. 2000. Ras is involved in nerve-activity-dependent regulation of muscle genes. *Nat. Cell Biol.* 2: 142-47.

Naya, F.J., B. Mercer, J. Shelton et al. 2000. Stimulation of slow skeletal muscle fiber gene expression by calcineurin in vivo. *J. Biol. Chem.* 275: 4545-48.

Odoherty, R.M., D.P. Bracy, D.K. Granner et al. 1996. Transcription of the rat skeletal muscle hexokinase II gene is increased by acute exercise. *J. Appl. Physiol.* 81: 789-93.

Ohlendieck, K. 2000. Changes in Ca^{2+}-regulatory muscle membrane proteins during the chronic low-frequency stimulation induced fast-to-slow transition process. *Basic Appl. Myol.* 10: 99-106.

Ohlendieck, K., B.E. Murray, G.R. Froemming et al. 1999. Effects of chronic low-frequency stimulation on Ca^{2+}-regulatory membrane proteins in rabbit fast muscle. *Pfluegers Arch.* 438: 700-08.

Olson, E.N. & R.S. Williams. 2000a. Calcineurin signaling and muscle remodeling. *Cell* 101: 689-92.

Olson, E.N. & R.S. Williams. 2000b. Remodeling muscles with calcineurin. *BioEssays* 22: 510-19.

Parra, J., D. Brdiczka, R. Cusso et al. 1997. Enhanced catalytic activity of hexokinase by work-induced mitochondrial binding in fast-twitch muscle of rat. *FEBS Lett.* 403: 279-82.

Pattullo, M.C., M.A. Cotter, N.E. Cameron et al. 1992. Effects of lengthened immobilization on functional and histochemical properties of rabbit tibialis anterior muscle. *Exp. Physiol.* 77: 433-42.

Pedersen, B.K., A. Steensberg & P. Schjerling. 2001. Muscle-derived interleukin-6: possible biological effects. *J. Physiol.* 536: 329-37.

Pellegrino, M.A., M. Canepari, R. Rossi et al. 2003. Orthologous myosin isoforms and scaling of shortening velocity with body size in mouse, rat, rabbit and human muscles. *J. Physiol.* 546: 677-89.

Peter, J.B., R.J. Barnard, V.R. Edgerton et al. 1972. Metabolic profiles of three fiber types of skeletal muscle in guinea pigs and rabbits. *Biochemistry* 11: 2627-33.

Pette, D. 2001. Historical perspectives: plasticity of mammalian skeletal muscle. *J. Appl. Physiol.* 90: 1119-24.

Pette, D. 2002. The adaptive potential of skeletal muscle fibers. *Can. J. Appl. Physiol.* 27: 423-48.

Pette, D., B.U. Ramirez, W. Müller et al. 1975. Influence of intermittent long-term stimulation on contractile, histochemical and metabolic properties of fibre populations in fast and slow rabbit muscles. *Pfluegers Arch.* 361: 1-7.

Pette, D. & R.S. Staron. 1990. Cellular and molecular diversities of mammalian skeletal muscle fibers. *Rev. Physiol. Biochem. Pharmacol.* 116: 1-76.

Pette, D. & R.S. Staron. 1997. Mammalian skeletal muscle fiber type transitions. *Int. Rev. Cytol.* 170: 143-223.

Pette, D. & R.S. Staron. 2000. Myosin isoforms, muscle fiber types, and transitions. *Microsc. Res. Tech.* 50: 500-09.

Pette, D. & G. Vrbová. 1992. Adaptation of mammalian skeletal muscle fibers to chronic electrical stimulation. *Rev. Physiol. Biochem. Pharmacol.* 120: 116-202.

Pette, D. & G. Vrbová. 1999. Invited review: what does chronic electrical stimulation teach us about muscle plasticity? *Muscle Nerve* 22: 666-77.

Peuker, H., A. Conjard & D. Pette. 1998. $^{D}{}_{T}$-cardiac-like myosin heavy chain as an intermediate between MHCIIa and MHCIβ in transforming rabbit muscle. *Am. J. Physiol.* 274: C595-C602.

Phillips, S.M., H.J. Green, M.A. Tarnopolsky et al. 1996. Progressive effect of endurance training on metabolic adaptations in working skeletal muscle. *Am. J. Physiol.* 270: E265-E272.

Pilegaard, H., G.A. Ordway, B. Saltin et al. 2000. Transcriptional regulation of gene expression in human skeletal muscle during recovery from exercise. *Am. J. Physiol.* 279: E806-E814.

Putman, C.T., M. Kiricsi, J. Pearcey et al. 2003. AMPK activation increases uncoupling protein-3 expression and mitochondrial and enzyme activities in rat muscle without fiber type transitions. *J. Physiol.* 551: 169-78.

Reichmann, H., H. Hoppeler, O. Mathieu-Costello et al. 1985. Biochemical and ultrastructural changes of skeletal muscle mitochondria after chronic electrical stimulation in rabbits. *Pfluegers Arch.* 404: 1-9.

Roy, D., E. Johannsson, A. Bonen et al. 1997. Electrical stimulation induces fiber type-specific translocation of GLUT-4 to T tubules in skeletal muscle. *Am. J. Physiol.* 36: E688-E694.

Salmons, S. 1994. Exercise, stimulation and type transformation of skeletal muscle. *Int. J. Sports Med.* 15: 136-41.

Salmons, S. & G. Vrbová. 1969. The influence of activity on some contractile characteristics of mammalian fast and slow muscles. *J. Physiol.* 201: 535-49.

Saltin, B. & P.D. Gollnick. 1983. Skeletal muscle adaptability: significance for metabolism and performance. In *Handbook of Physiology, Skeletal Muscle*, sect. 10, ed. L.D. Peachey, R.H. Adrian & S.R. Geiger, 555-631. Baltimore: Williams & Wilkins.

Schantz, P., R. Billeter, J. Henriksson et al. 1982. Training-induced increase in myofibrillar ATPase intermediate fibres in human skeletal muscle. *Muscle Nerve* 5: 628-36.

Schantz, P. & J. Henriksson. 1983. Increases in myofibrillar ATPase intermediate human skeletal muscle fibers in response to endurance training. *Muscle Nerve* 6: 553-56.

Schiaffino, S. & C. Reggiani. 1996. Molecular diversity of myofibrillar proteins: gene regulation and functional significance. *Physiol. Rev.* 76: 371-423.

Schmitt, T. & D. Pette. 1991. Fiber type-specific distribution of parvalbumin in rabbit skeletal muscle—a quantitative immunohistochemical and microbiochemical study. *Histochemistry* 96: 459-65.

Serrano, A.L., M. Murgia, G. Pallafacchina et al. 2001. Calcineurin controls nerve activity-dependent specification of slow skeletal muscle fibers but not muscle growth. *Proc. Natl. Acad. Sci.* 98: 13108-13.

Simoneau, J.A., G. Lortie, M.R. Boulay et al. 1985. Human skeletal muscle fiber type alteration with high-intensity intermittent training. *Eur. J. Appl. Physiol.* 54: 250-53.

Škorjanc, D., F. Jaschinski, G. Heine et al. 1998. Sequential increases in capillarization and mitochondrial enzymes in low-frequency stimulated rabbit muscle. *Am. J. Physiol.* 274: C810-C818.

Smerdu, V., I. Karsch-Mizrachi, M. Campione et al. 1994. Type IIx myosin heavy chain transcripts are expressed in type IIb fibers of human skeletal muscle. *Am. J. Physiol.* 267: C1723-C1728.

Sréter, F.A., J.R. Lopez, L. Alamo et al. 1987. Changes in intracellular ionized Ca concentration associated with muscle fiber type transformation. *Am. J. Physiol.* 253: C296-C300.

Staron, R.S., B. Gohlsch & D. Pette. 1987. Myosin polymorphism in single fibers of chronically stimulated rabbit fast-twitch muscle. *Pfluegers Arch.* 408: 444-50.

Staron, R.S. & R.S. Hikida. 1992. Histochemical, biochemical, and ultrastructural analyses of single human muscle fibers, with special reference to the C-fiber population. *J. Histochem. Cytochem.* 40: 563-68.

Staron, R.S., W.J. Kraemer, R.S. Hikida et al. 1998. Comparison of soleus muscles from rats exposed to microgravity for 10 versus 14 days. *Histochem. Cell. Biol.* 110: 73-80.

Stephens, T.J., Z.P. Chen, B.J. Canny et al. 2002. Progressive increase in human skeletal muscle AMPK alpha 2 activity and ACC phosphorylation during exercise. *Am. J. Physiol.* 282: E688-E694.

Stevens, L., B. Bastide, P. Kischel et al. 2002. Time-dependent changes in expression of troponin subunit isoforms in unloaded rat soleus muscle. *Am. J. Physiol.* 285: C1025-C1030.

Stevens, L., B. Gohlsch, Y. Mounier et al. 1999a. Changes in myosin heavy chain mRNA and protein isoforms in single fibers of unloaded rat soleus muscle. *FEBS Lett.* 463: 15-18.

Stevens, L., K.R. Sultan, H. Peukerm et al. 1999b. Time-dependent changes in myosin heavy chain mRNA and protein isoforms in unloaded soleus muscle of rat. *Am. J. Physiol.* 277: C1044-C1049.

Sutherland, H., J.C. Jarvis, M.M.N. Kwende et al. 1998. The dose-related response of rabbit fast muscle to long-term low-frequency stimulation. *Muscle Nerve* 21: 1632-46.

Talmadge, R.J. 2000. Myosin heavy chain isoform expression following reduced neuromuscular activity: potential regulatory mechanisms. *Muscle Nerve* 23: 661-79.

Turner, D.L., H. Hoppeler, H. Claassen et al. 1997. Effects of endurance training on oxidative capacity and structural composition of human arm and leg muscles. *Acta Physiol. Scand.* 161: 459-64.

van Houten, D.R., J.M. Davis, D.M. Meyers et al. 1992. Altered cellular distribution of hexokinase in skeletal muscle after exercise. *Int. J. Sports Med.* 13: 436-38.

Vikstrom, K.L., S.H. Seiler, R.L. Sohn et al. 1997. The vertebrate myosin heavy chain: genetics and assembly properties. *Cell Struct. Funct.* 22: 123-29.

Weiss, A. & L.A. Leinwand. 1996. The mammalian myosin heavy chain gene family. *Annu. Rev. Cell Dev. Biol.* 12: 417-39.

Weiss, A., S. Schiaffino & L.A. Leinwand. 1999. Comparative sequence analysis of the complete human sarcomeric myosin heavy chain family: implications for functional diversity. *J. Mol. Biol.* 290: 61-75.

Westerblad, H. & D.G. Allen. 1991. Changes of myoplasmic calcium concentration during fatigue in single mouse muscle fibers. *J. Gen. Physiol.* 98: 615-35.

Widegren, U., X.J. Jiang, A. Krook et al. 1998. Divergent effects of exercise on metabolic and mitogenic signaling pathways in human skeletal muscle. *FASEB J.* 12: 1379-89.

Williams, R.S. & P.D. Neufer. 1996. Regulation of gene expression in skeletal muscle by contractile activity, In *Handbook of Physiology, Exercise: Regulation and Integration of Multiple Systems*, sect. 12, ed. L.B. Rowell & J.T. Shepherd, 1124-50. New York: Oxford University Press.

Winder, W.W. 2001. Invited review: energy-sensing and signaling by AMP-activated protein kinase in skeletal muscle. *J. Appl. Physiol.* 91: 1017-28.

Winder, W.W. & D.G. Hardie. 1996. Inactivation of acetyl-CoA carboxylase and activation of AMP-activated protein kinase in muscle during exercise. *Am. J. Physiol.* 270: E299-E304.

Winder, W.W., B.F. Holmes, D.S. Rubink et al. 2000. Activation of AMP-activated protein kinase increases mitochondrial enzymes in skeletal muscle. *J. Appl. Physiol.* 88: 2219-26.

Wojtaszewski, J.F.P., P. Nielsen, B.F. Hansen et al. 2000. Isoform-specific and exercise intensity-dependent activation of 5'-AMP-activated protein kinase in human skeletal muscle. *J. Physiol.* 528: 221-26.

Wright, D.C., K.A. Hucker, J.O. Holloszy et al. 2004. Ca^{2+} and AMPK both mediate stimulation of glucose transport by muscle contractions. *Diabetes* 53: 330-35.

Wu, H., S.B. Kanatous, F.A. Thurmond et al. 2002. Regulation of mitochondrial biogenesis in skeletal muscle by CaMK. *Science* 296: 349-52.

Xu, Q., L. Yu, L. Liu et al. 2002. p38 Mitogen-activated protein kinase-, calcium-calmodulin-dependent protein kinase-, and calcineurin-mediated signaling pathways transcriptionally regulate myogenin expression. *Mol. Biol. Cell.* 13: 1940-52.

Zheng, D.H., P.S. Maclean, S.C. Pohnert et al. 2001. Regulation of muscle GLUT-4 transcription by AMP-activated protein kinase. *J. Appl. Physiol.* 91: 1073-83.

Zoll, J., H. Sanchez, B. N'Guessan et al. 2002. Physical activity changes the regulation of mitochondrial respiration in human skeletal muscle. *J. Physiol.* 543: 191-200.

Zong, H., J.M. Ren, L.H. Young et al. 2002. AMP kinase is required for mitochondrial biogenesis in skeletal muscle in response to chronic energy deprivation. *Proc. Natl. Acad, Sci.* 99: 15983-87.

Chapter 14

Andersen, S.D. & E. Daviskas. 2000. The mechanism of exercise-induced asthma is. . . . *J. Allergy Clin. Immunol.* 106:453-59.

Andersen, S.D. & K. Holzer. 2000. Exercise-induced asthma: is it the right diagnosis in elite athletes? *J. Allerg. Clin. Immunol.* 106:419-28.

Bärtsch, P. 1997. High altitude pulmonary edema. *Respiration* 64:435-43.

Bonsignore, M.R., G. Morici, A.M. Vignola et al. 2003. Increased airway inflammatory cells in endurance athletes: what do they mean? *Clin. Exp. Allergy* 33:14-21.

Clerici, C. 1998. Sodium transport in alveolar epithelial cells: modulation by O_2 tension. *Kidney Int.* 65:S79-S83.

Costello, M.L., O. Mathieu-Costello & J.B. West. 1992. Stress failure of alveolar epithelial cells studied by scanning electron microscopy. *Am. Rev. Resp. Dis.* 145:1446-55.

Doyle, I.R., S. Morton, A.J. Crockett et al. 2000. Composition of alveolar surfactant changes with training in humans. *Respirology* 5:211-20.

Factor, P., F. Saldias, K. Ridge et al. 1998. Augmentation of lung liquid clearance via adenovirus-mediated transfer of a Na,K-ATPase beta(1) subunit gene. *J. Clin. Invest.* 102:1421-30.

Gabriel, S.E., L.L. Clarke, R.C. Boucher et al. 1993. CFTR and outward rectifying chloride channels are distinct proteins with a regulatory relationship. *Nature* 363: 263-66.

Goodman, B.E. & E.D. Crandall. 1982. Dome formation in primary cultured monolayers of alveolar epithelial cells. *Am. J. Physiol.* 243:C96-C100.

Guggino, W.B. 1999. Cystic fibrosis and the salt controversy. *Cell* 96:607-10.

Guggino, W.B. & S.E. Guggino. 2000. Amiloride-sensitive sodium channels contribute to the woes of the flu. *Proc. Natl. Acad. Sci.* 97:9827-29.

Haas, M. & B. Forbush. 2000. The Na-K-Cl cotransporter of secretory epithelia. *Annu. Rev. Physiol.* 62:515-34.

Haas, M. & D.G. McBrayer. 1994. Na-K-Cl cotransport in nystatin-treated tracheal cells: regulation by isoproterenol, apical UTP, and [Cl](i). *Am. J. Physiol.* 266: C1440-C1452.

Haas, M., D.G. McBrayer & J.R. Yankaskas. 1993. Dual mechanisms for Na-K-Cl cotransport regulation in airway epithelial cells. *Am. J. Physiol.* 264:C189-C200.

Hebestreit, A., U. Kersting, B. Basler et al. 2001. Exercise inhibits epithelial sodium channels in patients with cystic fibrosis. *Am. J. Resp. Crit. Care Med.* 164: 443-46.

Höschele, S. & H. Mairbäurl. 2003. Alveolar flooding at high altitude: failure of reabsorption? *News Physiol. Sci.* 18:55-59.

Hummler, E., P. Barker, J. Gatzy et al. 1996. Early death due to defective neonatal lung liquid clearance in α-ENaC-deficient mice. *Nature Genet.* 13:325-28.

Knowles, M.R., J.L. Carson, J.T. Collier et al. 1981. Measurements of nasal transepithelial electric potential differences in normal human subjects in vivo. *Am. Rev. Resp. Dis.* 124:484-90.

Ko, Y.H. & P.L. Pedersen. 1997. Frontiers in research on cystic fibrosis: understanding its molecular and chemical basis and relationship to the pathogenesis of the disease. *J. Bioenerg. Biomembr.* 29:417-27.

König, J., R. Schreiber, T. Voelcker et al. 2001. The cystic fibrosis transmembrane conductance regulator (CFTR) inhibits ENaC through an increase in the intracellular Cl-concentration. *EMBO Rep.* 2: 1047-51.

Lingrel, J.B. 1992. Na,K-ATPase: isoform structure, function, and expression. *J. Bioenerg. Biomembr.* 24:263-70.

Mairbäurl, H., K. Mayer, K.J. Kim et al. 2002. Hypoxia decreases active Na transport across primary rat alveolar epithelial cell monolayers. *Am. J. Physiol.* 282:L659-L665.

Mason, R.J., K. Greene & D.R. Voelker. 1998. Surfactant protein A and surfactant protein D in health and disease. *Am. J. Physiol.* 19:L1-L13.

Matthay, M.A., H.G. Folkesson & C. Clerici. 2002. Lung epithelial fluid transport and the resolution of pulmonary edema. *Physiol. Rev.* 82:569-600.

Morrison, K.E., R.F. Slocombe, S.A. McKane et al. 1999. Functional and compositional changes in pulmonary surfactant in response to exercise. *Equine Vet. J. Suppl.* 30:62-66.

O'Brodovich, H. 1991. Epithelial ion transport in the fetal and perinatal lung. *Am. J. Physiol.* 261:C555-C564.

O'Grady, S.M., X.P. Jiang & D.H. Ingbar. 2000. Cl-channel activation is necessary for stimulation of Na transport in adult alveolar epithelial cells. *Am. J. Physiol.* 278:L239-L244.

Parker, J.C., A.C. Guyton & A.E. Taylor. 1979. Pulmonary transcapillary exchange and pulmonary edema. *Int. Rev. Physiol.* 18:261-315.

Planes, C., B. Escoubet, M. Blot-Chabaud et al. 1997. Hypoxia downregulates expression and activity of epithelial sodium channels in rat alveolar epithelial cells. *Am. J. Resp. Cell Mol. Biol.* 17:508-18.

Rafii, B., C. Coutinho, G. Otulakowski et al. 2000. Oxygen induction of epithelial Na^+ transport requires heme proteins. *Am. J. Physiol.* 278:L399-L406.

Rafii, B., A.K. Tanswell, O. Pitkanen et al. 1999. Induction of epithelial sodium channel (ENaC) expression and sodium transport in distal lung epithelia by oxygen. *Curr. Top. Membr.* 47:239-54.

Robbins, R.A. & S.I. Rennard. 1997. Biology of airway epithelial cells. In *The Lung,* R.G. Crystal, J.B. West, E.R. Weibel & P.J. Barnes, eds. Philadelphia: Lippincott-Raven, 445-57.

Rooney, S.A., S.L. Young & C.R. Mendelson. 1994. Molecular and cellular processing of lung surfactant. *FASEB J.* 8:957-67.

Rossier, B.C., S. Pradervand, L. Schild et al. 2002. Epithelial sodium channel and the control of sodium balance: interaction between genetic and environmental factors. *Annu. Rev. Physiol.* 64:877-97.

Sartori, C., Y. Allemann, H. Duplain et al. 2002. Salmeterol for the prevention of high-altitude pulmonary edema. *N. Engl. J. Med.* 346:1631-36.

Schoene, R.B. 1987. High-altitude pulmonary edema: pathophysiology and clinical review. *Ann. Emerg. Med.* 16:987-92.

Stutts, M.J., C.M. Canessa, J.C. Olsen et al. 1995. CFTR as a cAMP-dependent regulator of sodium channels. *Science* 269:847-50.

Swenson, E.R., M. Maggiorini, S. Mongovin et al. 2002. Pathogenesis of high-altitude pulmonary edema: inflammation is not an etiologic factor. *JAMA* 287: 2228-35.

Verkman, A.S., M.A. Matthay & Y. Song. 2000. Aquaporin water channels and lung physiology. *Am. J. Physiol.* 278:L867-L879.

Voilley, N., A. Galibert, F. Bassilana et al. 1997. The amiloride-sensitive Na+ channel: from primary structure to function. *Comp. Biochem. Physiol. A* 118:193-200.

Weibel, E.R. 1984. *The Pathway for Oxygen.* Cambridge: Harvard University Press.

Welsh, M.J. 1986. The respiratory epithelium. In *Membrane Transport Processes in Organized Systems,* T.E.

Andreoli, J.F. Hoffman, D.D. Fanestil & S.G. Schultz, eds. New York: Plenum Medical Book, 367-74.

West, J.B. 2000. Invited review: pulmonary capillary stress failure. *J. Appl. Physiol.* 89:2483-89.

West, J.B. & O. Mathieu-Costello. 1995. Stress failure of pulmonary capillaries as a limiting factor for maximal exercise. *Eur. J. Appl. Physiol.* 70:99-108.

Widdicombe, J.H. 1997. Ion transport by airway epithelia. In *The Lung,* R.G. Crystal, J.B. West, E.R. Weibel & P.J. Barnes, eds. Philadelphia: Lippincott-Raven, 573-84.

Wodopia, R., H.S. Ko, J. Billian et al. 2000. Hypoxia decreases proteins involved in transepithelial electrolyte transport of A549 cells and rat lung. *Am. J. Physiol.* 279:L1110-L1119.

Chapter 15

Anthony, T.G., J.C. Anthony, M.S. Lewitt et al. 2001. Time course changes in IGFBP-1 after treadmill exercise and postexercise food intake in rats. *Am. J. Physiol.* 280: E650-E656.

Ardies, C.M., G.S. Morris, C.K. Erickson et al. 1989. Both acute and chronic exercise enhance in vivo ethanol clearance in rats. *J. Appl. Physiol.* 66: 555-60.

Barzilai, N., S. Li, B.-Q. Liu et al. 1999. Surgical removal of visceral fat reverses hepatic insulin resistance. *Diabetes* 48: 94-98.

Beattie, M.A. & W.W. Winder. 1985. Attenuation of postexercise ketosis in fasted endurance-trained rats. *Am. J. Physiol.* 248: R63-R67.

Bélanger, P., F. Désy, M.-S. Gauthier et al. 2003. Effects of endurance training on the gluconeogenic capacity of periportal and perivenous hepatocytes. *Physiol. Behav.* 78: 27-32.

Bjorntorp, P. 1990. Portal adipose tissue as a generator of risk factors for cardiovascular disease and diabetes. *Arteriosclerosis* 10: 493-96.

Bonen, A., P.A. Clune & M.H. Tan. 1986. Chronic exercise increases insulin binding in muscles but not in liver. *Am. J. Physiol.* 251: E196-E203.

Brosnan, J.T., H.S. Ewart & S.A. Squires. 1995. Hormonal control of hepatic glutaminase. *Adv. Enzyme Reg.* 35: 131-46.

Burcelin, R., W. Dolci & B. Thorens. 2000. Portal glucose infusion in the mouse induces hypoglycemia. Evidence that the hepatoportal glucose sensor stimulates glucose utilization. *Diabetes* 49: 1635-42.

Burelle, Y., C. Fillipi, F. Péronnet et al. 2000. Mechanisms of increased gluconeogenesis from alanine in rat isolated hepatocytes after endurance training. *Am. J. Physiol.* 278: E35-E42.

Carlson, C.L. & W.W. Winder. 1999. Liver AMP-activated protein kinase and acetyl-CoA carboxylase during and after exercise. *J. Appl. Physiol.* 86: 669-74.

Coggan, A.R., S.C. Swanson, L.A. Mendenhall et al. 1995. Effect of endurance training on hepatic glycogenolysis and gluconeogenesis during prolonged exercise in men. *Am. J. Physiol.* 268: E375-E383.

Coker, R.H., Y. Koyama, J.C. Denny et al. 2002. Prevention of overt hypoglycemia during exercise. Stimulation of endogenous glucose production independent of hepatic catecholamine action and changes in pancreatic hormone concentration. *Diabetes* 51: 1310-18.

Désy, F., Y. Burelle, P. Bélanger et al. 2001. Effects of acute exercise on the gluconeogenic capacity of periportal and perivenous hepatocytes. *J. Appl. Physiol.* 91: 1099-1104.

Dohm, G.L. & E.A. Newsholme. 1983. Metabolic control of hepatic gluconeogenesis during exercise. *Biochem. J.* 212: 633-39.

Dohm, G.L., G.J. Kasparek & H.A. Barakat. 1985. Time course of changes in gluconeogenic enzyme activities during exercise and recovery. *Am. J. Physiol.* 249: E6-E11.

Donovan, C.M. & K.D. Sumida. 1990. Training improves glucose homeostasis in rats during exercise via glucose production. *Am. J. Physiol.* 258: R770-R776.

Drouin, R., C. Lavoie, J. Bourque et al. 1998. Increased hepatic glucose production response to glucagon in trained subjects. *Am. J. Physiol.* 274: E23-E28.

Ferreira, E.B., R.B. Ceddia, R. Curi et al. 1998. Swimming-exercise increases the capacity of perfused rat liver to produce urea from ammonia and L-glutamine. *Res. Com. Mol. Pathol. Pharmacol.* 102: 289-303.

Fiebig, R., M.A. Griffiths, M.T. Gore et al. 1998. Exercise training down-regulates hepatic lipogenic enzymes in meal-fed rats: fructose versus complex-carbohydrate diets. *J. Nutr.* 128: 810-17.

Fiebig, R., M.T. Gore & L.L. Ji. 1999. Exercise attenuates nuclear protein binding to gene regulatory sequences of hepatic fatty acid synthetase. *J. Appl. Physiol.* 87: 1009-15.

Fiebig, R.G., J.M. Hollander & L.L. Ji. 2001. Exercise down-regulates hepatic fatty acid synthetase in streptozotocin-treated rats. *J. Nutr.* 131: 2252-59.

Friedman, J.E. 1994. Role of glucocorticoids in activation of hepatic PEPCK gene transcription during exercise. *Am. J. Physiol.* 266: E560-E566.

Friedman, M.I. & M.G. Tordoff. 1996. Hepatic metabolic signal for control of food intake: stimulus generation, signal transduction and neural transmission. In *Liver Innervation and the Neural Control of Hepatic Function,* ed. T. Shimazu, 373-80. London: John Libbey.

Galassetti, P., Y. Koyama, R.H. Coker et al. 1999. Role of a negative arterial-portal venous glucose gradient in the postexercise state. *Am. J. Physiol.* 277: E1038-E1045.

Galassetti, P., M. Shiota, B.A. Zinker et al. 1998. A negative arterial-portal venous glucose gradient decreases skeletal muscle glucose uptake. *Am. J. Physiol.* 275: E101-E111.

Galassetti, P., R.H. Coker, D.B. Lacy et al. 1999. Prior exercise increases net hepatic glucose uptake during a glucose load. *Am. J. Physiol.* 276: E1022-E1029.

Gauthier, M.-S., K. Couturier, J.-G. Latour et al. 2003. Concurrent exercise prevents high-fat diet-induced macrovesicular hepatic steatosis *J. Appl. Physiol.* 94(6): 2127-34.

Ghanbari-Niaki, A., R. Bergeron, M.G. Latour et al. 1999. Effects of physical exercise on liver ATP levels in fasted and phosphate-injected rats. *Arch. Physiol. Biochem.* 107: 393-402.

Griffiths, M.A., D.H. Baker, X.-X. Yu et al. 1995. Effects of acute exercise on hepatic lipogenic enzymes in fasted and refed rats. *J. Appl. Physiol.* 79: 879-85.

Griffiths, M.A., R. Fiebig, M.T. Gore et al. 1996. Exercise down-regulates hepatic lipogenic enzymes in food-deprived and refed rats. *J. Nutr.* 126: 1959-71.

Halseth, A.E., N. Rhéaume, A.B. Messina et al. 1998. Regulation of hepatic glutamine metabolism during exercise in the dog. *Am. J. Physiol.* 275: E655-E664.

Hamada, K., K. Matsumoto, K. Minehira et al. 1998. Effect of glucose on ureagenesis during exercise in amino acid-infused dogs. *Metabolism* 47: 1303-07.

Hardie, D.G. & D. Carling. 1997. The AMP-activated protein kinase: fuel gauge of the mammalian cell? *Eur. J. Biochem.* 246: 259-73.

Horn, D.B., D.A. Podolin, J.E. Friedman et al. 1997. Alterations in key gluconeogenic regulators with age and endurance training. *Metabolism* 46: 1-7.

Johnson, J.L. & G.J. Bagby. 1988. Gluconeogenic pathway in liver and muscle glycogen synthesis after exercise. *J. Appl. Physiol.* 64: 1591-99.

Jungermann, K. & T. Kietzmann. 1996. Zonaton of parenchymal and nonparenchymal metabolism in the liver. *Annu. Rev. Nutr.* 16: 179-203.

Jungermann, K. & R.G. Thurman. 1992. Hepatocyte heterogeneity in the metabolism of carbohydrates. *Enzyme* 46: 33-58.

Keller, C., A. Steensberg, H. Pilegaard et al. 2001. Transcriptional activation of the IL-6 gene in human contracting skeletal muscle: influence of muscle glycogen content. *FASEB J.* 15: 2748-50.

Kjaer, M. 1992. Regulation of hormonal and metabolic responses during exercise in humans. *Exerc. Sport Sci. Rev.* 20: 161-84.

Kjaer, M. 1998. Hepatic glucose production during exercise. *Adv. Exp. Med. Biol.* 441: 117-27.

Krishna, M.G., R.H. Coker, D. Brooks et al. 2000. Glucagon response to exercise is critical for accelerated hepatic glutamine metabolism and nitrogen disposal. *Am. J. Physiol.* 279: E638-E645.

Lavoie, J.-M. 2002. Afferent contribution of the liver to metabolic regulation of physical exercise. *Can. J. Physiol. Pharmacol.* 80: 1035-44.

Lavoie, J.-M., S. Cardin & B. Doiron. 1989. Influence of hepatic vagus nerve on pancreatic hormone secretion during exercise. *Am. J. Physiol.* 257: E855-E859.

Lavoie, J.-M., Y. Fillion, K. Couturier et al. 2002. Selected contribution. Evidence that the decrease in liver glycogen is associated with the exercise-induced increase in IGFBP-1. *J. Appl. Physiol.* 93: 798-804.

Lee, P.D.K., C.A. Conover & D.R. Powell. 1993. Regulation and function of insulin-like growth factor binding protein-1. *Proc. Soc. Exp. Biol. Med.* 204: 4-29.

Légaré, A., R. Drouin, M. Milot et al. 2001. Increased density of glucagon receptors in liver from endurance trained rats. *Am. J. Physiol.* 280: E193-E196.

Matas Bonjorn, V., M.G. Latour, P. Bélanger et al. 2002. Influence of prior exercise and liver glycogen content on the sensitivity of the liver to glucagon. *J. Appl. Physiol.* 92: 188-94.

Matsuhisa, M., H. Nishizawa, M. Ikeda et al. 1998. Prior muscular exercise contraction enhances disposal of glucose analog in the liver and muscle. *Metabolism* 47: 44-49.

McGarry, J.D. & N.F. Brown. 1997. The mitochondrial carnitine palmitoyltransferase system: from concept to molecular analysis. *Eur. J. Biochem.* 244: 1-14.

Müller, M.J. 1995. Hepatic fuel selection. *Proc. Nutr. Soc.* 54: 139-50.

Myers, S.R., O.P. McGuinness, D.W. Neal et al. 1991. Intraportal glucose delivery alters the relationship between net hepatic glucose uptake and the insulin concentration. *J. Clin. Invest.* 87: 930-39.

Nieto, J.L., I. Diaz-Laviada, A. Guillén et al. 1996. Effect of endurance physical training on rat liver adenyl cyclase system. *Cell Signal.* 8: 317-22.

Nizielski, S.E., C. Arizmendi, A.R. Shteyngarts et al. 1996. Involvement of transcription factor C/EBP-β in stimulation of PEPCK gene expression during exercise. *Am. J. Physiol.* 270: R1005-R1012.

O'Brien, R.M., E.L. Noisin, A. Suwanichkul et al. 1995. Hepatic nuclear factor 3- and hormone-regulated expression of the phosphoenolpyruvate carboxykinase and insulin-like growth factor-binding protein 1 genes. *Mol. Cell. Biol.* 15: 1747-58.

Park, H., V.K. Kaushik, S. Constant et al. 2002. Coordinate regulation of malonyl-CoA decarboxylase, *sn*-glycerol-3-phosphate acyltransferase and acetyl-CoA carboxylase by AMP-activated protein kinase in rat tissues in response to exercise. *J. Biol. Chem.* 277: 32571-77.

Patel, Y.M., J.S. Yun, J. Liu et al. 1994. An analysis of regulatory elements in the PEPCK gene which are responsible for its tissue-specific expression and metabolic control in transgenic mice. *J. Biol. Chem.* 269: 5619-28.

Paulauskis, J.D. & H.S. Sul. 1989. Hormonal regulation of mouse fatty acid synthetase gene transcription in liver. *J. Biol. Chem.* 264: 574-77.

Pessayre, D., A. Mansouri & B. Fromenty. 2002. Nonalcoholic steatosis and steatohepatitis V. Mitochondrial dysfunction in steatohepatitis. *Am. J. Physiol.* 282: G193-G199.

Podolin, D.A., B.K. Wills, I.O. Wood et al. 2001. Attenuation of age-related declines in glucagon-mediated signal transduction in rat liver by exercise training. *Am. J. Physiol.* 281: E516-E523.

Podolin, D.A., M.J. Pagliassotti, T.T. Gleeson et al. 1994. Influence of endurance training on the age-related decline in hepatic glyconeogenesis. *Mech. Ageing Dev.* 75: 81-93.

Podolin, D.A., T.T. Gleeson & R.S. Mazzeo. 1994. Role of epinephrine in hepatic gluconeogenesis: evidence of aging and training effects. *Am. J. Physiol.* 267: E680-E686.

Remesy, C. & C. Demigne. 1983. Changes in availability of glucogenic and ketogenic substrates and liver metabolism in fed or starved rats. *Ann. Nutr. Metab.* 27: 57-70.

Roder, K., S.S. Wolf, K.F. Beck et al. 1997. NF-Y binds to the inverted CCAAT box, as essential element for cAMP-dependent regulation of the fatty acid synthetase (FAS) gene. *Gene* 184: 21-26.

Saltiel, A.R. & R. Kahn. 2001. Insulin signalling and the regulation of glucose and lipid metabolism. *Nature* 414: 799-806.

Sasse, D., N. Katz & K. Jungermann. 1975. Functional heterogeneity of rat liver parenchyma and of isolated hepatocytes. *Fed. Eur. Biochem. Soc.* 57: 83-88.

Straczkowski, M., I. Kowalska, S. Dzienis-Straczkowska et al. 2001. The effect of exercise training on glucose tolerance and skeletal muscle triacylglycerol content in rats fed with a high-fat diet. *Diab. Metab.* (Paris) 27: 19-23.

Sumida, K.D. & C.M. Donovan. 1996. Enhanced hepatic gluconeogenesis from selected precursors following endurance training. *J. Appl. Physiol.* 79: 1883-88.

Teli, M.R., O.F. James, A.D. Burt et al. 1995. The natural history of non-alcoholic fatty liver: a follow-up study. *Hepatology* 22: 1714-19.

Trudell, J.R., W.-Q. Lin, D.A. Chrystof et al. 1995. Induction of HSP72 in rat liver by chronic ethanol consumption combined with exercise: association with the prevention of ethanol-induced fatty liver by exercise. *Alcoholism: Clin. Exp. Res.* 19: 753-58.

Ueno, T.H. Sugawara, K. Sujaku et al. 1997. Therapeutic effects of restricted diet and exercise in obese patients with fatty liver. *J. Hepatol.* 27: 103-07.

Ulbright, C. & P.J. Snodgrass. 1993. Coordinate induction of the urea cycle enzymes by glucagon and dexamethasone is accomplished by three different mechanisms. *Arch. Biochem. Biophys.* 301: 237-43.

Wahren, J., L. Hagenfeldt & P. Felig. 1975. Splanchnic and leg exchange of glucose, amino acids and free fatty acids during exercise in diabetes mellitus. *J. Clin. Invest.* 55: 1303-14.

Wasserman, D.H. 1995. Regulation of glucose fluxes during exercise in the postabsorptive state. *Annu. Rev. Physiol.* 57: 191-218.

Wasserman, D.H. & A.D. Cherrington. 1991. Hepatic fuel metabolism during exercise: role and regulation. *Am. J. Physiol.* 260: E811-E824.

Wasserman, D.H., D.B. Lacy, D.R. Green et al. 1987. Dynamics of hepatic lactate and glucose balances during prolonged exercise and recovery in the dog. *J. Appl. Physiol.* 63: 2411-17.

Wasserman, D.H., R.J. Geer, P.E. Williams et al. 1991. Interaction of gut and liver in nitrogen metabolism during exercise. *Metabolism* 40: 307-14.

Wasserman, D.H., R.M. O' Doherty & B.A. Zingler. 1995. Role of the endocrine pancreas in control of fuel metabolism by the liver during exercise. *Int. J. of Obes. Metab. Dis.* 19(Suppl 4): 320-29.

Wegner, M., Z. Cao & M.G. Rosenfeld. 1992. Calcium-regulated phosphorylation within the leucine zipper of C/EBP-β. *Science* 256: 370-73.

Winder, W.W. & D.G. Hardie. 1999. AMP-activated protein kinase, a metabolic master switch: possible role in type 2 diabetes. *Am. J. Physiol.* 277: E1-E10.

Yamatani, K., Z. Qing Shi, A. Giacca et al. 1992. Role of FFA-glucose cycle in glucoregulation during exercise in total absence of insulin. *Am. J. Physiol.* 263: E646-E653.

Yang, S.Q., H. Zhu, Y. Li et al. 2000. Mitochondrial adaptations to obesity-related oxidant stress. *Arch. Biochem. Biophys.* 378: 259-68.

Chapter 16

Allen, D.O. & R.R. Beck. 1986. Role of calcium ion in hormone-stimulated lipolysis. *Biochem. Pharmacol.* 35:767-72.

Arner, P. & R.H. Eckel. 1998. Adipose tissue as a storage organ. In *Handbook of Obesity,* ed. G.A. Bray, C. Bouchard & W.P.T. James, 379-95. New York: Dekker.

Askew, E.W. & A.L. Hecker. 1976. Adipose tissue cell size and lipolysis in the rat: response to exercise intensity and food restriction. *J. Nutr.* 106:1351-60.

Askew, E.W., A.L. Hecker, V.G. Coppes et al. 1978. Cyclic AMP metabolism in adipose tissue of exercise-trained rats. *J. Lipid Res.* 19:729-36.

Baba, T., T. Kanda, A. Yoshida et al. 2000. Reciprocal changes in leptin and tumor necrosis factor-alpha with exercise in insulin resistant rats. *Res. Com. Mol. Pathol. Pharmacol.* 108:133-43.

Bai, Y., S. Zhang, K.S. Kim et al. 1996. Obese gene expression alters the ability of 30A5 preadipocytes to respond to lipogenic hormones. *J. Biol. Chem.* 271:13939-42.

Bouchard, C., J.-P. Despres & P. Mauriege. 1993. Genetic and nongenetic determinants of regional fat distribution. *Endoc. Rev.* 14:72-93.

Brasaemle, D.L., D.M. Levin, D.C. Adler-Wailes et al. 2000. The lipolytic stimulation of 3T3-L1 adipocytes promotes the translocation of hormone-sensitive lipase to the surface of lipid storage droplets. *Biochim. Biophys. Acta* 1483:251-62.

Bryant, N.J., R. Govers & D.E. James. 2002. Regulated transport of the glucose transporter GLUT4. *Nature Rev. Mol. Cell Biol.* 3:267-77.

Bukowiecki, L., J. Lupien, N. Follea et al. 1980. Mechanism of enhanced lipolysis in adipose tissue of exercise-trained rats. *Am. J. Physiol.* 239:E422-E429.

Carey, G.B. 1998. Mechanisms regulating adipocyte lipolysis. In *Advances in Experimental Medicine and Biology,* ed. E.A. Richter, B. Kiens, H. Galbo & B. Saltin, 157-70. New York: Plenum Press.

Carey, G.B. 2000. Cellular adaptations in fat tissue of exercise-trained miniature swine: role of excess energy intake. *J. Appl. Physiol.* 88:881-87.

Carey, G.B. & K.A. Sidmore. 1994. Exercise attenuates the antilipolytic effect of adenosine in adipocytes isolated from miniature swine. *Int. J. Obes.* 18:155-60.

Cinti, S. 2001. The adipose organ: morphological perspectives of adipose tissues. *Proc. Nutr. Soc.* 60: 319-328.

Crampes, F., M. Beauville, D. Riviere et al. 1986. Effect of physical training in humans on the response of isolated fat cells to epinephrine. *J. Appl. Physiol.* 61: 25-29.

Crampes, F., D. Riviere, M. Beauville et al. 1989. Lipolytic response of adipocytes to epinephrine in sedentary and exercise-trained subjects: sex-related differences. *Eur. J. Appl. Physiol.* 59:249-55.

de Glisezinski, I., F. Crampes, I. Harant et al. 1998. Endurance training changes in lipolytic responsiveness of obese adipose tissue. *Am. J. Physiol.* 275:E951-E956.

Dong, Q., J. Schuchman & G.B. Carey. 1994. Characterization of the swine adipocyte A1 adenosine receptor using an optimized assay system. *Comp. Biochem. Physiol.* 108C:269-80.

Egan, J.J., A.S. Greenberg, M.K. Chang et al. 1992. Mechanism of hormone-stimulated lipolysis in adipocytes: translocation of hormone-sensitive lipase to the lipid storage droplet. *Proc. Natl. Acad. Sci.* 89:8537-41.

Fan, G., E. Shumay, H. Wang et al. 2001. The scaffold protein gravin (cAMP-dependent protein kinase-anchoring protein 250) binds the β_2-adrenergic receptor via the receptor cytoplasmic arg-329 to leu-413 domain and provides a mobile scaffold during desensitization. *J. Biol. Chem.* 276:24005-14.

Fried, S.K. & C.D. Russell. 1998. Diverse roles of adipose tissue in the regulation of systemic metabolism and energy balance. In *Handbook of Obesity,* G.A. Bray, C. Bouchard & W.P.T. James, 397-413. New York: Dekker.

Friedman, J.E., C.M. Ferrara, K.S. Aulak et al. 1997. Exercise training down-regulates ob gene expression in the genetically obese SHHF/Mcc-fa(cp) rat. *Horm. Metab. Res.* 29:214-19.

Fruhbeck, G., S.A. Jebb & A.M. Prentice. 1998. Leptin: physiology and pathophysiology. *Clin. Physiol.* 18: 399-419.

Fruhbeck, G. 2001. A heliocentric view of leptin. *Proc. Nutr. Soc.* 60:301-08.

Fruhbeck, G., J. Gomez-Ambrosi, F.J. Muruzabal et al. 2001. The adipocyte: a model for integration of endocrine and metabolic signaling in energy metabolism regulation. *Am. J. Physiol.* 280:E827-E847.

Goodyear, L.J. & B.B. Kahn. 1998. Exercise, glucose transport, and insulin sensitivity. *Annu. Rev. Med.* 49:235-61.

Greenberg, A.S., J.J. Egan, S.A. Wek et al. 1991. Perilipin, a major hormonally regulated adipocyte-specific phosphoprotein associated with the periphery of lipid storage droplets. *J. Biol. Chem.* 266:11341-46.

Guan, H.-P., Y. Li, M.V. Jensen et al. 2002. A futile metabolic cycle activated in adipocytes by antidiabetic agents. *Nat. Med.* 8:1122-28.

Halle, M., A. Berg, H. Northoff et al. 1998. Importance of TNF-alpha and leptin in obesity and insulin resistance: a hypothesis on the impact of physical exercise. *Exerc. Immunol. Rev.* 4:77-94.

Hayase, H., S. Nomura, T. Abe et al. 2002. Relation between fat distributions and several plasma adipocytokines after exercise training in premenopausal and postmenopausal women. *J. Physiol. Anthropol. Appl. Hum. Sci.* 21:105-13.

Hellstrom, L., E. Blaak & E. Hagstrom-Toft. 1996. Gender differences in adrenergic regulation of lipid mobilization during exercise. *Intern. J. Sports Med.* 17: 439-47.

Hickey, M.S. & D.J. Calsbeek. 2001. Plasma leptin and exercise: recent findings. *Sports Med.* 31:583-89.

Hirshman, M.F., L.J. Wardzala, L.J. Goodyear et al. 1989. Exercise training increases the number of glucose transporters in rat adipose cells. *Am. J. Physiol.* 257: E520-E530.

Honnor, R.C., G.S. Dhillon & C. Londos. 1985. cAMP-dependent protein kinase and lipolysis in rat adipocytes. II. Definition of steady-state relationship with lipolytic and antilipolytic modulators. *J. Biol. Chem.* 260:15130-38.

Ishii, T., T. Yamakita, K. Yamagami et al. 2001. Effect of exercise training on serum leptin levels in type 2 diabetic patients. *Metabolism* 50:1136-40.

Izawa, T. & T. Komabayashi. 1994. Ca2+ and lipolysis in adipocytes from exercise-trained rats. *J. Appl. Physiol.* 77:2618-24.

Izawa, T., T. Komabayashi, T. Mochizuki et al. 1991. Enhanced coupling of adenylate cyclase to lipolysis in permeabilized adipocytes from trained rats. *J. Appl. Physiol.* 71:23-29.

Izawa, T., T. Komabayashi, S. Shinoda et al. 1988. Possible mechanism of regulating adenylate cyclase activity in adipocyte membranes from exercise-trained male rats. *Biochem. Biophys. Res. Com.* 151: 1262-68.

Kenno, K.A., J.L. Durstine & R.E. Shepherd. 1986. Distribution of cyclic AMP phosphodiesterase in adipose tissue from trained rats. *J. Appl. Physiol.* 61:1546-51.

Lafontan, M. & M. Berlan. 1993. Fat cell adrenergic receptors and the control of white and brown fat cell function. *J. Lipid Res.* 34:1057-91.

Londos, C., D.L. Brasaemle, C.J. Schultz et al. 1999. On the control of lipolysis in adipocytes. In *The Metabolic Syndrome X. Convergence of Insulin Resistance, Glucose Intolerance, Hypertension, Obesity, and Dyslipidemias: Searching for the Underlying Defects,* ed. B.C. Hansen, J. Saye & L.P. Wennogle, 155-68. New York: New York Academy of Sciences.

Londos, C., J. Gruia-Gray, D.L. Brasaemle et al. 1996. Perilipin: possible roles in structure and metabolism of intracellular neutral lipids in adipocytes and steroidogenic cells. *Int. J. Obes. Relat. Metab. Disord.* 20:S97-S101.

Mauriege, P., D. Prud'homme, M. Marcotte et al. 1997. Regional differences in adipose tissue metabolism between sedentary and endurance-trained women. *Am. J. Physiol.* 273:E497-E506.

Michel, J.J. & J.D. Scott. 2002. AKAP mediated signal transduction. *Annu. Rev. Pharmacol. Toxicol.* 42: 235-57.

Mohamed-Ali, V., J.H. Pinkney & S.W. Coppack. 1998. Adipose tissue as an endocrine and paracrine organ. *Int. J. Obes.* 22:1145-58.

Nieto, J.L., I.D. Laviada, A. Guillen et al. 1996. Adenylyl cyclase system is affected differently by endurance physical training in heart and adipose tissue. *Biochem. Pharmacol.* 51:1321-29.

Nomura, S., H. Kawanami, H. Ueda et al. 2002. Possible mechanisms by which adipocyte lipolysis is enhanced in exercise-trained rats. *Biochem. Biophys. Res. Com.* 295:236-42.

Okuda, H., C. Morimoto & T. Tsujita. 1992. Relationship between cyclic AMP production and lipolysis induced by forskolin in rat fat cells. *J. Lipid Res.* 33: 225-31.

Ong, J.M., R.B. Simsolo, M. Saghizadeh et al. 1995. Effects of exercise training and feeding on lipoprotein lipase gene expression in adipose tissue, heart, and skeletal muscle of the rat. *Metabolism* 44:1596-1605.

Petridou, A. & V. Mougios. 2002. Acute changes in triacylglycerol lipase activity of human adipose tissue during exercise. *J. Lipid Res.* 43:1331-34.

Ploug, T. & E. Ralston. 2002. Exploring the whereabouts of GLUT4 in skeletal muscle. *Mol. Membr. Biol.* 19: 39-49.

Pond, C.M. 1992. An evolutionary and functional view of mammalian adipose tissue. *Proc. Nutr. Soc.* 51:367-77.

Pond, C.M. & C.A. Mattacks. 1998. In vitro evidence for the involvement of the adipose tissue surrounding lymph nodes in immune responses. *Immunol. Lett.* 63:159-67.

Romijn, J.A., S. Klein, E.F. Coyle et al. 1993. Strenuous endurance training increases lipolysis and triglyceride-fatty acid cycling at rest. *J. Appl. Physiol.* 75: 108-13.

Savard, R. & C. Bouchard. 1990. Genetic effects in the response of adipose tissue lipoprotein lipase activity to prolonged exercise. A twin study. *Int. J. Obes.* 14:771-77.

Seip, R.L., T.J. Angelopoulos & C.F. Semenkovich. 1995. Exercise induces human lipoprotein lipase gene expression in skeletal muscle but not adipose tissue. *Am. J. Physiol.* 268:E229-E236.

Sengenes, C., M. Berlan, I. de Glisezinski et al. 2000. Natriuretic peptides: a new lipolytic pathway in human adipocytes. *FASEB J.* 14:1345-51.

Shepherd, R.E., M.D. Bah & K.M. Nelson. 1986. Enhanced lipolysis is not evident in adipocytes from exercise-trained SHR. *J. Appl. Physiol.* 61:1301-08.

Shepherd, R.E., E.G. Noble, G.A. Klug et al. 1981. Lipolysis and cAMP accumulation in adipocytes in response to physical training. *J. Appl. Physiol.* 50:143-48.

Shinoda, S., T. Izawa, T. Komabayashi et al. 1989. Effects of adenosine and pertussis toxin on lipolysis in adipocytes from exercise-trained male rats. *Res. Commun. Chem. Pathol. Pharmacol.* 66:397-410.

Simsolo, R.B., J.M. Ong & P.A. Kern. 1993. The regulation of adipose tissue and muscle lipoprotein lipase in runners by detraining. *J. Clin. Invest.* 92:2124-30.

Stich, V., I. DeGlisezinski, F. Crampes et al. 1999. Activation of antilipolytic α_2-adrenergic receptors by epinephrine during exercise in human adipose tissue. *Am. J. Physiol.* 277:R1076-R1083.

Stich, V., I. de Glisezinski, F. Crampes et al. 2000. Activation of α_2-adrenergic receptors impairs exercise-induced lipolysis in SCAT of obese subjects. *Am. J. Physiol.* 279:R499-R504.

Suda, K., T. Izawa, T. Komabayashi et al. 1993. Effect of insulin on adipocyte lipolysis in exercise-trained rats. *J. Appl. Physiol.* 74:2935-39.

Syu, L.-J. & A.R. Saltiel. 1999. Lipotransin: a novel docking protein for hormone-sensitive lipase. *Mol. Cell* 4:109-15.

Tansey, J.T., C. Sztalryd, J. Gruia-Gray et al. 2001. Perilipin ablation results in a lean mouse with aberrant adipocyte lipolysis, enhanced leptin production, and resistance to diet-induced obesity. *Proc. Natl. Acad. Sci.* 98:6494-99.

Trayhurn, P. & J.H. Beattie. 2001. Physiological role of adipose tissue: white adipose tissue as an endocrine and secretory organ. *Proc. Nutr. Soc.* 60:329-39.

Vidal, H. 2001. Gene expression in visceral and subcutaneous adipose tissues. *Ann. Med.* 33:547-55.

Wahrenberg, H., J. Bolinder & P. Arner. 1991. Adrenergic regulation of lipolysis in human fat cells during exercise. *Eur. J. Clin. Invest.* 21:534-41.

Wahrenberg, H., P. Engfeldt, J. Bolinder et al. 1987. Acute adaptation in adrenergic control of lipolysis during physical exercise in humans. *Am. J. Physiol.* 253:E383-E390.

Williams, R.S. & T. Bishop. 1982. Enhanced receptor-cyclase coupling and augmented catecholamine-stimulated lipolysis in exercising rats. *Am. J. Physiol.* 243:E345-E351.

Wolfe, R.R., S. Klein, F. Carraro et al. 1990. Role of triglyceride-fatty acid cycle in controlling fat metabolism in humans during and after exercise. *Am. J. Physiol.* 258:E382-E389.

Xue, B., A.G. Greenberg, F.B. Kraemer et al. 2001. Mechanism of intracellular calcium ([Ca2+]i) inhibition of lipolysis in human adipocytes. *FASEB J.* 15:2527-29.

Zhang, Y., R. Proenca, M. Maffei et al. 1994. Positional cloning of the mouse obese gene and its human homologue. *Nature* 372:425-32.

Chapter 17

Baumann, R., C. Bauer & H. Bartels. 1984. Blood oxygen transport. In *Handbook of Physiology, Respiration* section, ed. M. Tenny & F. Farhi. Bethesda, MD: American Physiological Society.

Bellingham, A.J., J.C. Detter & C. Lenfant. 1971. Regulatory mechanisms of haemoglobin oxygen affinity in acidosis and alkalosis. *J. Clin. Invest.* 50: 700-06.

Bencowitz, H.Z., P.D. Wagner & J. West. 1982. Effect of change in p50 on exercise tolerance at high altitude: theoretical study. *J. Appl. Physiol.* 53: 1487-95.

Böning, D. 2001. Differences between whole blood and plasma lactate concentration have to be considered when comparing various studies. Letter to the editor-in-chief. *Med. Sci. Sports Exerc.* 33: 1411-12.

Böning, D., W. Draude, F. Trost et al. 1978. Interrelation between Bohr and temperature effect on the oxygen dissociation curve in men and women. *Resp. Physiol.* 34: 195-200.

Böning, D., T. Fetzer & R. Beneke. 2000. Influences of temperature and lactic acid on oxylabile CO_2 in human blood. *Med. Sci. Sports Exerc.* 32: 65.

Böning, D., C. Hollnagel, A. Boecker et al. 1991. Bohr shift by lactic acid and the supply of the O_2 to skeletal muscle. *Resp. Physiol.* 85: 231-43.

Böning, D. & N. Maassen. 1983. Blood osmolality in vitro: dependence on PCO_2, lactic acid concentration, and O_2 saturation. *J. Appl. Physiol.* 54: 118-22.

Böning, D., N. Maassen, J.M. Steinacker et al. 1999. Carbon dioxide storage and nonbicarbonate buffering in the human body before and after an Himalayan expedition. *Eur. J. Appl. Physiol.* 79: 457-66.

Böning, D., J. Rojas & M. Serrato. 2001. Hemoglobin mass and peak oxygen uptake in untrained and trained residents of moderate altitude. *Int. J. Sports Med.* 22: 572-78.

Böning, D., J. Rojas, M. Serrato et al. 2002. Extracellular pH defense against lactic acid in untrained and trained altitude residents. *Int. J. Sports Med.* 23(Suppl 2): 110.

Böning, D., H.J. Schünemann, N. Maassen et al. 1993. Reduction of oxylabile CO_2 in human blood by lactate. *J. Appl. Physiol.* 74: 710-14.

Böning, D., U. Schweigart, U. Tibes et al. 1975. Influences of exercise and endurance training on the oxygen dissociation curve of blood under "in vivo" and "in vitro" conditions. *Eur. J. Appl. Physiol.* 34: 1-10.

Böning, D., F. Trost, K.-M. Braumann et al. 1982. The Bohr effect in blood of physically trained subjects. In *Metabolic and Functional Changes During Exercise,* ed. B. Semiginovsky & S. Tucek, 107-11. Prague: Charles University.

Brahm, J. 1982. Diffusional water permeability of human erythrocytes and their ghosts *J. Gen. Physiol.* 79: 791-819.

Braumann, K.M., D. Böning & F. Trost. 1979. Oxygen dissociation curves in trained and untrained subjects. *Eur. J. Appl. Physiol.* 42: 51-60.

Braumann, K.M., D. Böning & F. Trost. 1982. Bohr effect and slope of the oxygen dissociation curve after physical training. *J. Appl. Physiol.* 52: 1524-29.

Cooper, G.J. & W.F. Boron. 1998. Effect of PCMBS on CO_2 permeability of Xenopus oocytes expressing aquaporin 1 or its C189S mutant. *Am. J. Physiol.* 275: C1481-C1486.

Eckardt, K.U., U. Bouttellier, A. Kurtz et al. 1989. Rate of erythropoietin formation in humans in response to acute hypobaric hypoxia. J. Appl. Physiol. 66: 1785-88.

Gary-Bobo, C.M. & A.K. Solomon. 1968. Properties of hemoglobin solutions in red cells. *J. Gen. Physiol.* 52: 825-53.

Geers, C. & G. Gros. 2000. Carbon dioxide transport and carbonic anhydrase in blood and muscle. *Physiol. Rev.* 80: 681-715.

Greger, R. & U. Windhorst, eds. 1996. *Comprehensive Human Physiology. From Cellular Mechanisms to Integration.* Berlin, Heidelberg, New York: Springer.

Heinicke, K., B. Wolfarth, P. Winchenbach et al. 2001. Blood volume and hemoglobin mass in elite athletes of different disciplines. *Int. J. Sports Med.* 22: 504-12.

Hester, R.L. & J. Choi. 2002. Blood flow control during exercise: role for the venular endothelium? *Exerc. Sport Sci. Rev.* 30: 147-51.

Hütler, M., C. Pollmann, R. Beneke et al. 2000. Measurable amount of glutathione in blood is influenced by oxygen saturation of hemoglobin. *Clin. Chim. Acta* 301: 213-17.

Hütler, M., S. Woweries, R. Leithäuser et al. 2001. Exercise-induced changes in blood levels of alpha-tocopherol. *Eur. J. Appl. Physiol.* 85: 151-56.

Jaakkola, P., D. Mole, Y.M. Tian et al. 2001. Targeting of HIFα to the von Hippel-Lindau ubiquitylation complex by O_2-regulated prolyl hydroxylation. Science 292: 468-72.

Juel, C. 1997. Lactate-proton cotransport in skeletal muscle *Physiol. Rev.* 77: 321-58.

Juel, C., J. Bangsbo, T. Graham et al. 1990. Lactate and potassium fluxes from human skeletal muscle during and after intense, dynamic, knee extensor exercise. *Acta Physiol. Scand.* 140: 147-59.

Juel, C., Y. Hellsten, B. Saltin et al. 1999. Potassium fluxes in contracting human skeletal muscle and red blood cells. *Am. J. Physiol.* 276: R184-R188.

Kjellberg, S.R., U. Rudhe & T. Sjöstrand. 1949. Increase of the amount of hemoglobin and blood volume in connection with physical training. *Acta Physiol. Scand.* 19: 146-52.

Kostka, T., J. Drai, S.E. Berthouze et al. 2000. Physical activity, aerobic capacity and selected markers of oxidative stress and the anti-oxidant defense system in healthy elderly men. *Clin. Physiol.* 20: 185-90.

Lee, M.D., L.S. King & P. Agre. 1997. The aquaporin family of water channel proteins in clinical medicine. *Medicine* 76: 141-56.

Lindinger, M.I., P.L. Horn & S.P. Grudzien. 1999. Exercise-induced stimulation of K(+) transport in human erythrocytes *J. Appl. Physiol.* 87: 2157-67.

Lipton, A.J., M.A. Johnson, T. Macdonald et al. 2001. S-nitrosothiols signal the ventilatory response to hypoxia. *Nature* 413: 171-74.

Maassen, N. 1984. Die Abhängigkeit kurzfristiger Schwankungen der Blutosmolalität von Säuren-Basen-Gleichgewicht und Stoffwechselintensität. PhD diss., Medical School Hannover.

Maassen, N. & D. Böning. 1984. Arbeitsbedingte Hämokonzentration und Osmolalität. In *Stellenwert der Sportmedizin,* ed. D. Jeschke, 93-99. Berlin, New York, Heidelberg: Springer Verlag.

Maassen, N. & D. Böning. 1987. Blood osmolality in vitro: dependence on base addition, buffer value, and temperature. *J. Appl. Physiol.* 62: 2174-79.

Maassen, N., M. Foerster & H. Mairbaurl. 1998. Red blood cells do not contribute to removal of K^+ released from exhaustively working forearm muscle. *J. Appl. Physiol.* 85: 326-32.

Mairbaurl, H., S. Schulz & J.F. Hoffman. 2000. Cation transport and cell volume changes in maturing rat reticulocytes. *Am. J. Physiol.* 279: C1621-C1630.

Mathai, J.C., S. Mori, B.L. Smith et al. 1996. Functional analysis of aquaporin-1 deficient red cells. The Colton-null phenotype. *J. Biol. Chem.* 271: 1309-13.

McMahon, T., R. Moon, B.P. Luschinger et al. 2002. Nitric oxide in the human respiratory cycle. *Nat. Med.* 8: 711-17.

Mena, P., M. Maynar, J.M. Guitierrez et al. 1991. Erythrocyte free radical scavenger enzymes in bicycle professional racers. Adaptation to training. *Int. J. Sports Med.* 12: 563-66.

Moura, T.F., R.I. Macey, D.Y. Chien et al. 1984. Thermodynamics of all-or-none water channel closure in red cells. *J. Membr. Biol.* 81: 105-11.

Ohno, H., T. Yahata, Y. Sato et al. 1988. Physical training and fasting erythrocyte activities of free radical scavenging enzyme systems in sedentary men. *Eur. J. Appl. Physiol.* 57: 173-76.

Parker, J.C., P.B. Dunham & A.P. Minton. 1995. Effects of ionic strength on the regulation of Na/H exchange and K-Cl cotransport in dog red blood cells. *J. Gen. Physiol.* 105: 677-99.

Preston, G.M., B.L. Smith, M.L. Zeidel et al. 1994. Mutations in aquaporin-1 in phenotypically normal humans without functional CHIP water channels. *Science* 265: 1585-87.

Qian, Z.M., D.S. Xiao, P.L. Tang et al. 1999. Increased expression of transferrin receptor on membrane of erythroblasts in strenuously exercised rats. *J. Appl. Physiol.* 87: 523-29.

Reefsum, H.E., G. Jordfald & S.B. Strömme. 1976. Hematological changes following prolonged heavy exercise. *Medicine Sport* 9 *(Advances in Exercise Physiology).* Basel: Karger.

Sachs, J.R. 1970. Sodium movements in the human red blood cell. *J. Gen. Physiol.* 56: 322-41.

Sachs, J.R. & D.W. Martin. 1999. Role of polyamine structure in inhibition of K^+-Cl^- cotransport in human red cell ghosts. *J. Physiol.* 520: 723-35.

Schmidt, W., D. Böning & K.M. Braumann. 1987. Red cell age effects on metabolism and oxygen affinity in humans. *Resp. Physiol.* 68: 215-25.

Schmidt, W., H.W. Dahners, R. Correa et al. 1990. Blood gas transport properties in endurance trained athletes living at different altitudes. *Int. J. Sports Med.* 11: 15-21.

Schmidt, W., K. Eckardt, A. Hilgendorf et al. 1991. Effects of maximal and submaximal exercise under normoxic and hypoxic conditions on serum erythropoietin level. Int. J. Sports Med. 12: 457-61.

Schmidt, W., K. Heinicke, J. Rojas et al. 2002. Blood volume and hemoglobin mass in endurance athletes from moderate altitude. *Med. Sci. Sports Exerc.* 34: 1934-40.

Schmidt, W., N. Maassen, F. Trost et al. 1988. Training induced effects on blood volume, erythrocyte turnover and haemoglobin oxygen binding properties. *Eur. J. Appl. Physiol.* 57: 490-98.

Semenza, G.L. 2000. HIF-1: mediator of physiological and pathophysiological responses to hypoxia. J. Appl. Physiol. 88: 1474-80.

Sejersted, O.M., J.I. Medbö & L. Hermansen. 1982. Metabolic acidosis and changes in water and electrolyte balance after maximal exercise. *Ciba Found. Symp.* 87: 153-67.

Sentürk, Ü.K., F. Gündüz, O. Kuru et al. 2001. Exercise-induced oxidative stress affects erythrocytes in sedentary rats but not exercise-trained rats. *J. Appl. Physiol.* 91: 1999-2004.

Shoemaker, J.K., H.J. Green, J. Coates et al. 1996. Failure of prolonged exercise training to increase red cell mass in humans. *Am. J. Physiol.* 270: H121-26.

Stager, J.M., A. Tucker, L. Cordian, B.J. Engebretsen, W.F. Brechue & C.C. Matulich. 1990. Normoxic and acute hypoxic exercise tolerance in man following acetazolamide. *Med. Sci. Sports Exerc.* 22: 178-84.

Stroka, D.M., T. Burkhardt, I. Desbaillets et al. 2001. HIF-1 is expressed in normoxic tissue and displays an organ specific regulation under systemic hypoxia. FASEB J. 15: 2445-53.

Stryer, L. 1995. Biochemistry, 4th ed., 567-69. New York: Freeman.

Swenson, E.R. & T.H. Maren. 1978. A quantitative analysis of CO_2 transport at rest and during maximal exercise. *Resp. Physiol.* 35: 129-59.

Tauler, P., I. Gimeno, A. Aguilo et al. 1999. Regulation of erythrocyte antioxidant enzyme activities in athletes during competition and short term recovery. *Pfluegers Arch.* 438: 782-87.

West, J.B., P. Hackett, K. Maret et al. 1983. Pulmonary gas exchange on the summit of Mount Everest. *J. Appl. Physiol.* 55: 678-87.

Wittmann, B. & G. Gros. 1981. The osmotic properties of haemoglobin under physiological conditions—implications for the osmotic behaviour of red cells. Proceedings of a working conference at Girton College, Cambridge, 121-24.

Wozniak, A., G. Drewa, G. Chesy et al. 2001. Effect of altitude training on the peroxidation and antioxidant enzymes in sportsmen. *Med. Sci. Sports Exerc.* 33: 1109-13.

Chapter 18

Baj, Z., J. Kantorski, E. Majewska et al. 1994. Immunological status of competitive cyclists before and after the training season. *Int. J. Sports Med.* 15(6):319-24.

Belcastro, A.N., G.D. Arthur, T.A. Albisser et al. 1996. Heart, liver, and skeletal muscle myeloperoxidase activity during exercise. *J. Appl. Physiol.* 80(4): 1331-35.

Bell, E.B., S. Spartshott & C. Bunce. 1998. CD4+ T cell memory, CD45R subsets and the persistence of antigen—a unifying concept. *Immunol. Today* 19:60-64.

Brahmi, Z., J.E. Thomas, M. Park et al. 1985. The effect of acute exercise on natural killer-cell activity of trained and sedentary human subjects. *J. Clin. Immunol.* 5(5): 321-28.

Brines, R., L. Hoffman-Goetz & B.K. Pedersen. 1996. Can you exercise to make your immune system fitter? *Immunol. Today* 17(6):252-54.

Brunda, M.J., R.B. Herberman & H.T. Holden. 1980. Inhibition of murine natural killer cell activity by prostaglandins. *J. Immunol.* 124(6):2682-87.

Bruunsgaard, H., A. Hartkopp, T. Mohr et al. 1997. In vivo cell mediated immunity and vaccination response following prolonged, intense exercise. *Med. Sci. Sports Exerc.* 29(9):1176-81.

Bruunsgaard, H., M.S. Jensen, P. Schjerling et al. 1999. Exercise induces recruitment of lymphocytes with an activated phenotype and short telomere lengths in young and elderly humans. *Life Sci.* 65(24):2623-33.

Cannon, J.G. & M.J. Kluger. 1984. Exercise enhances survival rate in mice infected with Salmonella typhimurium. *Proc. Soc. Exp. Biol. Med.* 175(4): 518-21.

Eskola, J., O. Ruuskanen, E. Soppi et al. 1978. Effect of sport stress on lymphocyte transformation and antibody formation. *Clin. Exp. Immunol.* 32(2):339-45.

Essen, P., J. Wernerman, T. Sonnenfeld et al. 1992. Amino acids in plasma and muscle during 24 hours post-operatively—a descriptive study. *Clin. Physiol.* 12(2):163-77.

Field, C.J., R. Gougeon & E.B. Marliss. 1991. Circulating mononuclear cell numbers and function during intense exercise and recovery. *J. Appl. Physiol.* 71(3):1089-97.

Fry, R.W., A.R. Morton, G.P. Crawford et al. 1992. Cell numbers and in vitro responses of leukocytes and lymphocyte subpopulations following maximal exercise and interval training sessions of different intensities. *Eur. J. Appl. Physiol.* 64(3):218-27.

Gabriel, H., B. Schmitt, A. Urhausen et al. 1993. Increased CD45RA+CD45R0+ cells indicate activated T cells after endurance exercise. *Med. Sci. Sports Exerc.* 25(12):1352-57.

Galbo, H. 1983. *Hormonal and Metabolic Adaption to Exercise.* New York: Thieme Verlag.

Gleeson, M., W.A. McDonald, A.W. Cripps et al. 1995. The effect on immunity of long-term intensive training in elite swimmers. *Clin. Exp. Immunol.* 102(1):210-16.

Green, R.L., S.S. Kaplan, B.S. Rabin et al. 1981. Immune function in marathon runners. *Ann. Allergy* 47(2): 73-75.

Hack, V., G. Strobel, J.P. Rau et al. 1992. The effect of maximal exercise on the activity of neutrophil granulocytes in highly trained athletes in a moderate training period. *Eur. J. Appl. Physiol.* 65(6):520-24.

Hack, V., G. Strobel, M. Weiss et al. 1994. PMN cell counts and phagocytic activity of highly trained athletes depend on training period. *J. Appl. Physiol.* 77(4):1731-35.

Heath, G.W., E.S. Ford, T.E. Craven et al. 1991. Exercise and the incidence of upper respiratory tract infections. *Med. Sci. Sports Exerc.* 23(2):152-57.

Hinton, J.R., D.G. Rowbottom, D. Keast et al. 1997. Acute intensive interval training and in vitro T-lymphocyte function. *Int. J. Sports Med.* 18(2):130-35.

Hiscock, N. & B.K. Pedersen. 2002. Exercise-induced immunodepression—plasma glutamine is not the link. *J. Appl. Physiol.* 93(3):813-22.

Hoffman-Goetz, L. 1996. Exercise, immunity, and colon cancer. In Hoffman-Goetz, L. & Husted, J., eds. *Exercise and Immune Function.* Boca Raton, FL: CRC Press, 179-98.

Hoffman-Goetz, L., Y. Arumugam & L. Sweeny. 1994. Lymphokine activated killer cell activity following voluntary physical activity in mice. *J. Sports. Med. Phys. Fit.* 34:83-90.

Hoffman-Goetz, L., B. MacNeil, Y. Arumugam et al. 1992. Differential effects of exercise and housing condition on murine natural killer cell activity and tumor growth. *Int. J. Sports Med.* 13(2):167-71.

Hoffman-Goetz, L. & B.K. Pedersen. 1994. Exercise and the immune system: a model of the stress response? *Immunol. Today* 15(8):382-87.

Hoffman-Goetz, L., J.R. Simpson, N. Cipp et al. 1990. Lymphocyte subset responses to repeated submaximal exercise in men. *J. Appl. Physiol.* 68(3):1069-74.

Housh, T.J., G.O. Johnson, D.J. Housh et al. 1991. The effect of exercise at various temperatures on salivary levels of immunoglobulin A. *Int. J. Sports Med.* 12(5): 498-500.

Ibfelt, T., E.W. Petersen, H. Bruunsgaard, M. Sandmand & B.K. Pedersen. 2002. Exercise-induced change in type 1 cytokine-producing CD8(+) T cells is related to a decrease in memory T cells. *J. Appl. Physiol.* 93(2):645-48.

Ilback, N.G., G. Friman, D.J. Crawford et al. 1991. Effects of training on metabolic responses and performance capacity in Streptococcus pneumoniae infected rats. *Med. Sci. Sports Exerc.* 23(4):422-27.

Jonsdottir, I.H., P. Hoffman & P. Thoren. 1997. Physical exercise, endogenous opioids and immune function. *Acta Physiol. Scand.* 640:47-50.

Kappel, M., M.B. Hansen, M. Diamant et al. 1993. Effects of an acute bolus growth hormone infusion on the human immune system. *Horm. Metab. Res.* 25(11): 579-85.

Kappel, M., N. Tvede, H. Galboet al. 1991. Evidence that the effect of physical exercise on NK cell activity is mediated by epinephrine. *J. Appl. Physiol.* 70(6): 2530-34.

Keast, D., D. Arstein, W. Harper et al. 1995. Depression of plasma glutamine concentration after exercise stress and its possible influence on the immune system. *Med. J. Aust.* 162:15-18.

Keast, D., K. Cameron & A.R. Morton. 1988. Exercise and the immune response. *Sports Med.* 5(4):248-67.

Keast, D. & A.R. Morton. 1992. Long-term exercise and immune functions. In Watson, R.R. & Eisinger, M., eds. *Exercise and Disease.* Boca Raton, FL: CRC Press, 89-120.

Kendall, A., L. Hoffman-Goetz, M. Houston et al. 1990. Exercise and blood lymphocyte subset responses: intensity, duration, and subject fitness effects. *J. Appl. Physiol.* 69(1):251-60.

Khan, M.M., P. Sansoni, E.D. Silverman et al. 1986. Beta-adrenergic receptors on human suppressor, helper, and cytolytic lymphocytes. *Biochem. Pharmacol.* 35(7):1137-42.

Kjaer, M. & F. Dela. 1996. Endocrine responses to exercise. In Hoffman-Goetz, L., ed. *Exercise and Immune Function.* Boca Raton, New York, London, Tokyo: CRC Press, 1-20.

Krzywkowski, K., E.W. Petersen, K. Ostrowski et al. 2001a. Effect of glutamine supplementation on exercise-induced changes in lymphocyte function. *Am. J. Physiol.* 281(4):C1259-C1265.

Krzywkowski, K., E.W. Petersen, K. Ostrowski et al. 2001b. Effect of glutamine and protein supplementation on exercise-induced decreases in salivary IgA. *J. Appl. Physiol.* 91(2):832-38.

Kurokawa, Y., S. Shinkai, J. Torii et al. 1995. Exercise-induced changes in the expression of surface adhesion molecules on circulating granulocytes and lymphocytes subpopulations. *Eur. J. Appl. Physiol.* 71(2-3):245-52.

Lehmann, M., M. Huonker, F. Dimeo et al. 1993. Serum amino acid concentrations in nine athletes before and after the 1993 Colmar ultra triathlon. *Int. J. Sports Med.* 16(3):155-59.

Lewicki, R., H. Tchorzewski, E. Majewska et al. 1988. Effect of maximal physical exercise on T-lymphocyte subpopulations and on interleukin 1 (IL 1) and interleukin 2 (IL2) production in vitro. *Int. J. Sports Med.* 9:114-17.

Liew, F.Y., S.M. Russell, G. Appleyard et al. 1984. Cross-protection in mice infected with influenza A virus by the respiratory route is correlated with local IgA antibody rather than serum antibody or cytotoxic T cell reactivity. *Eur. J. Immunol.* 14(4):350-56.

Mackinnon, L.T. 1989. Exercise and natural killer cells: what is the relationship? *Sports Med.* 7:141-49.

Mackinnon, L.T. 1994. Current challenges and future expectations in exercise immunology: back to the future. *Med. Sci. Sports Exerc.* 26(2):191-94.

Mackinnon, L.T., T.W. Chick, A. van As et al. 1987. The effect of exercise on secretory and natural immunity. *Adv. Exp. Med. Biol.* 216A:869-76.

Mackinnon, L.T. & S. Hooper. 1994. Mucosal (secretory) immune system responses to exercise of varying intensity and during overtraining. *Int. J. Sports Med.* 15:S179-S183.

MacNeil, B. & L. Hoffman-Goetz. 1993a. Effect of exercise on natural cytotoxicity and pulmonary tumor metastases in mice. *Med. Sci. Sports Exerc.* 25(8):922-28.

MacNeil, B. & L. Hoffman-Goetz. 1993b. Exercise training and tumor metastasis in mice: influence of time of exercise onset. *Anticancer Res.* 13(6A):2085-88.

Madden, K. & D.L. Felten. 1995. Experimental basis for neural-immune interactions. *Physiol. Rev.* 75(1): 77-106.

Maisel, A.S., C. Harris, C.A. Rearden et al. 1990. Beta-adrenergic receptors in lymphocyte subsets after exercise. Alterations in normal individuals and patients with congestive heart failure. *Circulation* 82:2003-10.

McCarthy, D.A. & M.M. Dale. 1988. The leucocytosis of exercise. A review and model. *Sports Med.* 6(6): 333-63.

McDowell, S.L., R.A. Hughes, R.J. Hughes et al. 1992. The effect of exhaustive exercise on salivary immunoglobulin A. *J. Sports Med. Phys. Fit.* 32(4):412-15.

Nehlsen-Canarella, S.L., O.R. Fagoaga & D.C. Nieman. 1997. Carbohydrate and the cytokine response to 2.5 hours of running. *J. Appl. Physiol.* 82:1662-67.

Newsholme, E.A. 1990. Psychoimmunology and cellular nutrition: an alternative hypothesis [editorial]. *Biol. Psychiatry* 27(1):1-3.

Newsholme, E.A. 1994. Biochemical mechanisms to explain immunosuppression in well-trained and overtrained athletes. *Int. J. Sports Med.* 15:S142-47.

Nicholls, E.E. & R.A. Spaeth. 1922. The relation between fatigue and the susceptibility of guinea pigs to infections of type I pneumococcus. *Am. J. Hygiene* 2:527-35.

Nielsen, H.B. & B.K. Pedersen. 1997. Lymphocyte proliferation in response to exercise. *Eur. J. Appl. Physiol.* 75:375-79.

Nielsen, H.B., N. Secher & B.K. Pedersen. 1996. Lymphocytes and NK cell activity during repeated bouts of maximal exercise. *Am. J. Physiol.* 271:R222-R227.

Nielsen, H.B., N.H. Secher, M. Kappel et al. 1996. Lymphocyte, NK, and LAK cell responses to maximal exercise. *Int. J. Sports Med.* 17:60-65.

Nieman, D.C. 1994. Exercise, upper respiratory tract infection, and the immune system. *Med. Sci. Sports Exerc.* 26(2):128-39.

Nieman, D.C., D. Brendle, D.A. Henson et al. 1995. Immune function in athletes versus nonathletes. *Int. J. Sports Med.* 16(5):329-33.

Nieman, D.C., K.S. Buckley, D.A. Henson et al. 1995. Immune function in marathon runners versus sedentary controls. *Med. Sci. Sports Exerc.* 27(7):986-92.

Nieman, D.C., O.R. Fagoaga, D.E. Butterworth et al. 1997. Carbohydrate supplementation affects blood granulocyte and monocyte trafficking but not function after 2.5 hours of running. *J. Appl. Physiol.* 82(5):1385-94.

Nieman, D.C., D.A. Henson, E.B. Garner et al. 1997. Carbohydrate affects natural killer cell redistribution but not activity after running. *Med. Sci. Sports Exerc.* 29(10):1318-24.

Nieman, D.C., D.A. Henson, G. Gusewitch et al. 1993. Physical activity and immune function in elderly women. *Med. Sci. Sports Exerc.* 25(7):823-31.

Nieman, D.C., L.M. Johanssen & J.W. Lee. 1989. Infectious episodes in runners before and after a roadrace. *J. Sports Med. Phys. Fit.* 29(3):289-96.

Nieman, D.C., L.M. Johanssen, J.W. Lee et al. 1990. Infectious episodes in runners before and after the Los Angeles marathon. *J. Sports Med. Phys. Fit.* 30(3):316-28.

Nieman, D.C., S.L. Nehlsen-Cannarella, P.A. Markoff et al. 1990. The effects of moderate exercise training on natural killer cells and acute upper respiratory tract infections. *Int. J. Sports Med.* 11(6):467-73.

O'Shea, J. & J.R. Ortaldo. 1992. The biology of natural killer cells: insights into the molecular basis of function. In Lewis, C.E. & McGee J.O., eds. *The Natural Killer Cell.* Oxford: Oxford University Press, 1-40.

Ortaldo, J.R., A. Mantovani, D. Hobbs et al. 1983. Effects of several species of human leukocyte interferon on cytotoxic activity of NK cells and monocytes. *Int. J. Cancer* 31(3):285-89.

Ortega, E., M.E. Collazos, M. Maynar et al. 1993. Stimulation of the phagocytic function of neutrophils in sedentary men after acute moderate exercise. *Eur. J. Appl. Physiol.* 66(1):60-64.

Oshida, Y., K. Yamanouchi & S. Hayamizu. 1988. Effect of acute physical exercise on lymphocyte subpopulations in trained and untrained subjects. *Int. J. Sports Med.* 9(2):137-40.

Papa, S., M. Vitale, G. Mazzotti et al. 1989. Impaired lymphocyte stimulation induced by long-term training. *Immunol. Lett.* 22:29-33.

Parry Billings, M., R. Budgett, Y. Koutedakis et al. 1992. Plasma amino acid concentrations in the overtraining syndrome: possible effects on the immune system. *Med. Sci. Sports Exerc.* 24(12):1353-58.

Peake, J.M. 2002. Exercise-induced alterations in neutrophil degranulation and respiratory burst activity: possible mechanisms of action. *Exerc. Immunol. Rev.* 8:49-100.

Pedersen, B.K. 1997. In Pedersen B.K., ed. *Exercise Immunology.* Austin, TX: Landes BioSci, 1-206.

Pedersen, B.K. & L. Hoffman-Goetz. 2000. Exercise and the immune system: regulation, integration and adaption. *Physiol. Rev.* 80:1055-81.

Pedersen, B.K. & D.C. Nieman. 1998. Exercise and immunology: integration and regulation. *Immunol. Today* 19(5):204-06.

Pedersen, B.K., K. Ostrowski, T. Rohde et al. 1998. Nutrition, exercise and the immune system. *Proc. Nutr. Soc.* 57:43-47.

Pedersen, B.K., B.S. Thomsen & H. Nielsen. 1986. Inhibition of natural killer cell activity by antigen-antibody complexes. *Allergy* 41(8):568-74.

Pedersen, B.K., N. Tvede, L.D. Christensen et al. 1989. Natural killer cell activity in peripheral blood of highly trained and untrained persons. *Int. J. Sports Med.* 10(2):129-31.

Pedersen, B.K., N. Tvede, K. Klarlund et al. 1990. Indomethacin in vitro and in vivo abolishes post-exercise suppression of natural killer cell activity in peripheral blood. *Int. J. Sports Med.* 11(2):127-31.

Pedersen, B.K. & H. Ullum. 1994. NK cell response to physical activity: possible mechanisms of action. *Med. Sci. Sports Exerc.* 26(2):140-46.

Pyne, D.B. 1994. Regulation of neutrophil function during exercise. *Sports Med.* 17(4):245-58.

Rabin, B.S., M.N. Moyna, A. Kusnecov et al. 1996. Neuroendocrine effects of immunity. In Hoffman-Goetz, L., ed. *Exercise and Immune Function.* Boca Raton, New York, London, Tokyo: CRC Press, 21-38.

Rohde, T., S. Asp, D.A. MacLean & B.K. Pedersen. 1998. Competitive sustained exercise in humans, lymphokine activated killer cell activity, and glutamine—an intervention study. *Eur. J. Appl. Physiol.* 78:448-53.

Rohde, T., D.A. MacLean, A. Hartkopp et al. 1996. The immune system and serum glutamine during a triathlon. *Eur. J. Appl. Physiol.* 74(5):428-34.

Rohde, T., D. MacLean & B.K. Pedersen. 1998. Effect of glutamine on changes in the immune system induced by repeated exercise. *Med. Sci. Sports Exerc.* 30(6): 856-62.

Ronsen, O., B.K. Pedersen, T.R. Oritsland et al. 2001. Leukocyte counts and lymphocyte responsiveness associated with repeated bouts of strenuous endurance exercise. *J. Appl. Physiol.* 91(1):425-34.

Smith, J.A., A.B. Gray, D.B. Pyne et al. 1996. Moderate exercise triggers both priming and activation of neutrophil subpopulations. *Am. J. Physiol.* 270(4): R838-R845.

Smith, J.A., S.J. McKenzie, R.D. Telford et al. 1992. Why does moderate exercise enhance, but intense training depress, immunity? In Husband A.J., ed. *Behavior and Immunity.* Boca Raton, FL: CRC Press, 155-68.

Smith, J.A., R.D. Telford, I.B. Mason et al. 1990. Exercise, training and neutrophil microbicidal activity. *Int. J. Sports Med.* 11:179-87.

Steensberg, A., A.D. Toft, J. Halkjaer-Kristensen & B.K. Pedersen. 2001. Plasma interleukin-6 during strenuous exercise—role of adrenaline. *Am. J. Physiol.* 281(3):1001-04.

Steensberg, A., A.D. Toft, H. Bruunsgaard et al. 2001. Strenuous exercise decreases the percentage of type 1 T cells in the circulation. *J. Appl. Physiol.* 91(4): 1708-12.

Steerenberg, P.A., I.A. van-Aspersen, A. van-Nieuw-Amerongen et al. 1997. Salivary levels of immunoglobulin A in triathletes. *Eur. J. Oral Sci.* 105(4): 305-09.

Tharp, G.D. & M.W. Barnes. 1990. Reduction of saliva immunoglobulin levels by swim training. *Eur. J. Appl. Physiol.* 60(1):61-64.

Tiidus, P.M. & E. Bombardier. 1999. Oestrogen attenuates post-exercise myeloperoxidase activity in skeletal muscle of male rats. *Acta Physiol. Scand.* 166(2):85-90.

Tomasi, T.B., F.B. Trudeau, D. Czerwinski et al. 1982. Immune parameters in athletes before and after strenuous exercise. *J. Clin. Immunol.* 2(3):173-78.

Tonnesen, E., N.J. Christensen & M.M. Brinklov. 1987. Natural killer cell activity during cortisol and adrenaline infusion in healthy volunteers. *Eur. J. Clin. Invest.* 17(6):497-503.

Tvede, N., C. Heilmann, J. Halkjaer Kristensen et al. 1989. Mechanisms of B-lymphocyte suppression induced by acute physical exercise. *J. Clin. Lab. Immunol.* 30(4):169-73.

Tvede, N., M. Kappel, K. Klarlund et al. 1994. Evidence that the effect of bicycle exercise on blood mononuclear cell proliferative responses and subsets is mediated by epinephrine. *Int. J. Sports Med.* 15(2):100-04.

Tvede, N., B.K. Pedersen, F.R. Hansen et al. 1989. Effect of physical exercise on blood mononuclear cell subpopulations and in vitro proliferative responses. *Scand. J. Immunol.* 29(3):383-89.

Tvede, N., J. Steensberg, B. Baslund et al. 1991. Cellular immunity in highly trained elite racing cyclists during periods of training with high and low intensity. *Scand. J. Med. Sci. Sports* 1:163-66.

van Tits, L.J. & S.J. Graafsma. 1991. Stress influences CD4+ lymphocyte counts [letter]. *Immunol. Lett.* 30(1):141-42.

Volek, J.S., W.J. Kraemer, J.A. Bush et al. 1997. Testosterone and cortisol in relationship to dietary nutrients and resistance exercise. *J. Appl. Physiol.* 82(1):49-54.

Welsh, R.M. & R.M. Vargas-Cortes. 1992. Natural killer cells in viral infection. In Lewis, C.E. & McGee, J.O., eds. *The Natural Killer Cell.* Oxford: Oxford University Press, 108-50.

Chapter 19

Adams, J., Y. Collaco-Moraes et al. 1996. Cyclooxygenase-2 induction in cerebral cortex: an intracellular response to synaptic excitation. *J. Neurochem.* 66(1): 6-13.

Adlard, P.A. and C.W. Cotman. 2004. Voluntary exercise protects against stress-induced decreases in brain-derived neurotrophic factor protein expression. *Neurosci,* 124: 985-992.

Adlard, P.A., V.M. Perreau et al. 2004. The timecourse of induction of brain-derived neurotrophic factor mRNA and protein in the rat hippocampus following voluntary exercise. *Neurosci. Lut.* 363: 43-48.

Alonso, M., M.R. Vianna et al. 2002. Signaling mechanisms mediating BDNF modulation of memory formation in vivo in the hippocampus. *Cell. Mol. Neurobiol.* 22(5-6): 663-74.

Altar, C.A. 1999. Neurotrophins and depression. *Trends Pharmacol. Sci.* 20(2): 59-61.

Anderson, B.J., D.N. Rapp et al. 2000. Exercise influences spatial learning in the radial arm maze. *Physiol. and Behav.* 70(5): 425-29.

Barde, Y.A. 1994. Neurotrophins: a family of proteins supporting the survival of neurons. *Prog. Clin. Biol. Res.* 390: 45-56.

Berchtold, N.C., J.P. Kesslak et al. 2001. Estrogen and exercise interact to regulate brain-derived neurotrophic factor mRNA and protein expression in the hippocampus. *Eur. J. Neurol.* 14(12): 1992-2002.

Berchtold, N.C., J.P. Kesslak et al. 2002. Hippocampal brain-derived neurotrophic factor gene regulation by exercise and the medial septum. *J. Neurosci. Res.* 68(5): 511-21.

Berchtold, N.C., J.P. Kesslak et al. Regulation and maintenance of brain-derived neurotrophic factor (BDNF) protein in the rat hippocampus by continuous vs intermittent activity. Submitted.

Berkman, L.F., T.E. Seeman et al. 1993. High, usual and impaired functioning in community-dwelling older men and women: findings from the MacArthur Foundation Research Network on Successful Aging. *J. Clin. Epidemiol.* 46(10): 1129-40.

Blomquist, K.B. & F. Danner. 1987. Effects of physical conditioning on information-processing efficiency. *Percept. Mot. Skills* 65(1): 175-86.

Blumenthal, J.A., M.A. Babyak et al. 1999. Effects of exercise training on older patients with major depression. *Arch. Intern. Med.* 159(19): 2349-56.

Bowman, R.E., M.C. Zrull et al. 2001. Chronic restraint stress enhances radial arm maze performance in female rats. *Brain Res.* 904(2): 279-89.

Busiguina, S., A.M. Fernandez et al. 2000. Neurodegeneration is associated to changes in serum insulin-like growth factors. *Neurobiol. Dis.* 7(6 Pt B): 657-65.

Byrne, A. & D.G. Byrne. 1993. The effect of exercise on depression, anxiety and other mood states: a review. *J. Psychosomatic Res.* 37(6): 565-74.

Carro, E., A. Nunez et al. 2000. Circulating insulin-like growth factor I mediates effects of exercise on the brain. *J. Neurosci.* 20(8): 2926-33.

Carro, E., J.L. Trejo et al. 2001. Circulating insulin-like growth factor I mediates the protective effects of physical exercise against brain insults of different etiology and anatomy. *J. Neurosci.* 21(15): 5678-84.

Carro, E., J.L. Trejo et al. 2002. Serum insulin-like growth factor I regulates brain amyloid-beta levels. *Nat. Med.* 8(12): 1390-97.

Carter, C.S., M.M. Ramsey et al. 2002. A critical analysis of the role of growth hormone and IGF-1 in aging and lifespan. *Trends Genet.* 18(6): 295-301.

Chen, C., J.C. Magee et al. 2002. Cyclooxygenase-2 regulates prostaglandin E2 signaling in hippocampal long-term synaptic plasticity. *J. Neurophysiol.* 87(6): 2851-57.

Chen, Y.C., Q.S. Chen et al. 1998. Physical training modifies the age-related decrease of GAP-43 and synaptophysin in the hippocampal formation in C57BL/6J mouse. *Brain Res.* 806(2): 238-45.

Colcombe, S.J., K.I. Erickson et al. 2003. Aerobic fitness reduces brain tissue loss in aging humans. *J. Gerontol. Series A, Biolog. Sci. Med. Sci.* 58(2): 176-80.

Colcombe, S. & A.F. Kramer. 2003. Fitness effects on the cognitive function of older adults: a meta-analytic study. *Psychol. Sci.* 14(2): 125-30.

Cotman, C.W. & N.C. Berchtold. 2002. Exercise: a behavioral intervention to enhance brain health and plasticity. *Trends Neurosci.* 25(6): 292-98.

Diamond, M.C. 2001. Response of the brain to enrichment. *Anais de academia brasileira de ciencas* 73(2): 211-20.

DiLorenzo, T.M., E.P. Bargman et al. 1999. Long-term effects of aerobic exercise on psychological outcomes. *Prev. Med.* 28(1): 75-85.

Dimeo, F., M. Bauer et al. 2001. Benefits from aerobic exercise in patients with major depression: a pilot study. *Brit. J. Sports Med.* 35(2): 114-17.

Dishman, R.K., K.J. Renner et al. 1997. Activity wheel running reduces escape latency and alters brain monoamine levels after footshock. *Brain Res. Bull.* 42(5): 399-406.

Dudar, J.D., I.Q. Whishaw et al. 1979. Release of acetylcholine from the hippocampus of freely moving rats during sensory stimulation and running. *Neuropharmacology* 18(8-9): 673-78.

Duman, R.S., G.R. Heninger et al. 1997. A molecular and cellular theory of depression [see comments]. *Arch. Gen. Psychiatry* 54(7): 597-606.

Dunn, A.L., M.H. Trivedi et al. 2002. The DOSE study: a clinical trial to examine efficacy and dose response of exercise as treatment for depression. *Control. Clin. Trials* 23(5): 584-603.

Egan, M.F., M. Kojima et al. 2003. The BDNF val66met polymorphism affects activity-dependent secretion of BDNF and human memory and hippocampal function. *Cell* 112(2): 257-69.

Ferencz, I., M. Kokaia et al. 1997. Effects of cholinergic denervation on seizure development and neurotrophin messenger RNA regulation in rapid hippocampal kindling. *Neuroscience* 80(2): 389-99.

Figurov, A., M.L. Pozzo et al. 1996. Regulation of synaptic responses to high-frequency stimulation and LTP by neurotrophins in the hippocampus. *Nature* 381(6584): 706-09.

Friedland, R.P., T. Fritsch et al. 2001. Patients with Alzheimer's disease have reduced activities in midlife compared with healthy control-group members. *Proc. Natl. Acad. Sci.* 98: 3440-45.

Fujimaki, K., S. Morinobu et al. 2000. Administration of a cAMP phosphodiesterase 4 inhibitor enhances antidepressant-induction of BDNF mRNA in rat hippocampus. *Neuropsychopharmacology* 22(1): 42-51.

Garcia, C., M.J. Chen et al. In press. The influence of specific noradrenergic and serotonergic lesions on the expression of hippocampal BDNF transcripts following voluntary physical activity. *Neuroscience.*

Gomez-Pinilla, F., V. So et al. 1998. Spatial learning and physical activity contribute to the induction of fibroblast growth factor: neural substrates for increased cognition associated with exercise. *Neuroscience* 85(1): 53-61.

Gomez-Pinilla, F., Z. Ying et al. 2001. Differential regulation by exercise of BDNF and NT-3 in rat spinal cord and skeletal muscle. *Eur. J. Neurol.* 13(6): 1078-84.

Greenwood, B.N., T.E. Foley et al. 2003. Freewheel running prevents learned helplessness/behavioral depression: role of dorsal raphe serotonergic neurons. *J. Neurosci.* 23(7): 2889-98.

Hassmen, P., N. Koivula et al. 2000. Physical exercise and psychological well-being: a population study in Finland. *Prev. Med.* 30(1): 17-25.

Hill, R.D., M. Storandt et al. 1993. The impact of long-term exercise training on psychological function in older adults. *J. Gerontol.* 48(1): P12-17.

Holscher, C. 1999. Stress impairs performance in spatial water maze learning tasks. *Behav. Brain Res.* 100(1-2): 225-35.

Holzenberger, M., J. Dupont et al. 2003. IGF-1 receptor regulates lifespan and resistance to oxidative stress in mice. *Nature* 421(6919): 182-87.

Ivy, A.S., F.G. Rodriguez et al. 2003. Noradrenergic and serotonergic blockade inhibits BDNF mRNA activation following exercise and antidepressant. *Pharmacol Biochem Behav.* 2003. 75(1):81-8.

Karege, F., G. Perret et al. 2002. Decreased serum brain-derived neurotrophic factor levels in major depressed patients. *Psychiatry Res.* 109(2): 143-48.

Kaufmann, W.E., P.F. Worley et al. 1996. COX-2, a synaptically induced enzyme, is expressed by excitatory neurons at postsynaptic sites in rat cerebral cortex. *Proc. Natl. Acad. Sci.* 93(6): 2317-21.

Knipper, M. et al. 1994. Positive feedback between acetylcholine and the neurotrophins nerve growth factor and brain-derived neurotrophic factor in the rat hippocampus. *Eur. J. Neurol.* 6(4): 668-71.

Korte, M., O. Griesbeck et al. 1996. Virus-mediated gene transfer into hippocampal CA1 region restores long-term potentiation in brain-derived neurotrophic factor mutant mice. *Proc. Natl. Acad. Sci.* 93(22): 12547-52.

Kunugi, H., A. Ueki et al. 2001. A novel polymorphism of the brain-derived neurotrophic factor (BDNF) gene associated with late-onset Alzheimer's disease. *Mol. Psychiatry* 6(1): 83-86.

Lampinen, P., R.L. Heikkinen et al. 2000. Changes in intensity of physical exercise as predictors of depressive symptoms among older adults: an eight-year follow-up. *Prev. Med.* 30(5): 371-80.

Lapchak, P.A., D.M. Araujo et al. 1993. Cholinergic regulation of hippocampal brain-derived neurotrophic factor mRNA expression: evidence from lesion and chronic cholinergic drug treatment studies. *Neuroscience* 52(3): 575-85.

Laurin, D., R. Verreault et al. 2001. Physical activity and risk of cognitive impairment and dementia in elderly persons. *Arch. Neurol.* 58(3): 498-504.

Lawlor, D.A. & S.W. Hopker. 2001. The effectiveness of exercise as an intervention in the management of depression: systematic review and meta-regression analysis of randomised controlled trials. *Brit. Med. J.* 322(7289): 763-67.

Lawson, V.H. & B.H. Bland. 1993. The role of the septohippocampal pathway in the regulation of hippocampal field activity and behavior: analysis by the intraseptal microinfusion of carbachol, atropine, and procaine. *Exp. Neurol.* 120(1): 132-44.

Lee, M.G., J.J. Chrobak et al. 1994. Hippocampal theta activity following selective lesion of the septal cholinergic system. *Neuroscience* 62(4): 1033-47.

Levine, E.S., C.F. Dreyfus et al. 1995. Brain-derived neurotrophic factor rapidly enhances synaptic transmission in hippocampal neurons via postsynaptic tyrosine kinase receptors. *Proc. Natl. Acad. Sci.* 92(17): 8074-77.

Lichtenwalner, R.J., M.E. Forbes et al. 2001. Intracerebroventricular infusion of insulin-like growth factor-I ameliorates the age-related decline in hippocampal neurogenesis. *Neuroscience* 107(4): 603-13.

Lu, B. & A. Chow. 1999. Neurotrophins and hippocampal synaptic transmission and plasticity. *J. Neurosci. Res.* 58(1): 76-87.

Luine, V., C. Martinez et al. 1996. Restraint stress reversibly enhances spatial memory performance. *Physiol. Behav.* 59(1): 27-32.

Marcheselli, V.L. & N.G. Bazan. 1996. Sustained induction of prostaglandin endoperoxide synthase-2 by seizures in hippocampus. Inhibition by a platelet-activating factor antagonist. *J. Biol. Chem.* 271(40): 24794-99.

Markowska, A.L., M. Mooney et al. 1998. Insulin-like growth factor-1 ameliorates age-related behavioral deficits. *Neuroscience* 87(3): 559-69.

Martinez, A., S. Alcantara et al. 1998. TrkB and TrkC signaling are required for maturation and synaptogenesis of hippocampal connections. *J. Neurosci.* 18(18): 7336-50.

Matthews, K.A., L.H. Kuller et al. 1996. Prior to use of estrogen replacement therapy, are users healthier than nonusers? [see comments]. *Am. J. Epidemiol.* 143(10): 971-78.

McAllister, A.K., L.C. Katz et al. 1999. Neurotrophins and synaptic plasticity. *Annu. Rev. Neurosci.* 22: 295-318.

McIntosh, L.J. & R.M. Sapolsky. 1996. Glucocorticoids increase the accumulation of reactive oxygen species and enhance adriamycin-induced toxicity in neuronal culture. *Exp. Neurol.* 141(2): 201-06.

Mizuno, T., Y. Endo et al. 1991. Acetylcholine release in the rat hippocampus as measured by the microdialysis method correlates with motor activity and exhibits a diurnal variation. *Neuroscience* 44(3): 607-12.

Molteni, R., Z. Ying et al. 2002. Differential effects of acute and chronic exercise on plasticity-related genes in the rat hippocampus revealed by microarray. *Eur. J. Neurol.* 16(6): 1107-16.

Moraska, A. & M. Fleshner. 2001. Voluntary physical activity prevents stress-induced behavioral depression and anti-KLH antibody suppression. *Am. J. Physiol.* 281(2): R484-89.

Neeper, S.A., P.F. Gomez et al. 1996. Physical activity increases mRNA for brain-derived neurotrophic factor and nerve growth factor in rat brain. *Brain Res.* 726(1-2): 49-56.

Neeper, S.A., F. Gómez-Pinilla et al. 1995. Exercise and brain neurotrophins. *Nature* 373: 109.

Nibuya, M., S. Morinobu et al. 1995. Regulation of BDNF and trkB mRNA in rat brain by chronic electroconvulsive seizure and antidepressant drug treatments. *J. Neurosci.* 15(11): 7539-47.

Nilsson, O.G., P. Kalen et al. 1990. Acetylcholine release in the rat hippocampus as studied by microdialysis is dependent on axonal impulse flow and increases during behavioural activation. *Neuroscience* 36(2): 325-38.

Nitta, A., M. Ohmiya et al. 1999. Brain-derived neurotrophic factor prevents neuronal cell death induced by corticosterone. *J. Neurosci. Res.* 57(2): 227-35.

O'Brien, R., D. Xu et al. 2002. Synaptically targeted narp plays an essential role in the aggregation of AMPA receptors at excitatory synapses in cultured spinal neurons. *J. Neurosci.* 22(11): 4487-98.

Oliff, H.S., N.C. Berchtold et al. 1998. Exercise-induced regulation of BDNF transcripts in the rat hippocampus. *Mol. Brain Res.* 61: 147-53.

Patterson, S.L., T. Abel et al. 1996. Recombinant BDNF rescues deficits in basal synaptic transmission and hippocampal LTP in BDNF knockout mice. *Neuron* 16(6): 1137-45.

Persson, I., L. Bergkvist et al. 1997. Hormone replacement therapy and major risk factors for reproductive cancers, osteoporosis, and cardiovascular diseases: evidence of confounding by exposure characteristics. *J. Clin. Epidemiol.* 50(5): 611-18.

Pollock, K.M. 2001. Exercise in treating depression: broadening the psychotherapist's role. *J. Clin. Psychol.* 57(11): 1289-300.

Pozzo-Miller, L.D., W. Gottschalk et al. 1999. Impairments in high-frequency transmission, synaptic vesicle docking, and synaptic protein distribution in the hippocampus of BDNF knockout mice. *J. Neurosci.* 19: 4972-83.

Radak, Z., T. Kaneko et al. 2001. Regular exercise improves cognitive function and decreases oxidative damage in rat brain. *Neurochem. Int.* 38(1): 17-23.

Rogers, R.L., J.S. Meyer et al. 1990. After reaching retirement age physical activity sustains cerebral perfusion and cognition. *J. Am. Geriatr. Soc.* 38(2): 123-28.

Russo-Neustadt, A., R.C. Beard et al. 1999. Exercise, antidepressant medications, and enhanced brain derived neurotrophic factor expression. *Neuropsychopharmacology* 21(5): 679-82.

Russo-Neustadt, A., T. Ha et al. 2001. Physical activity-antidepressant treatment combination: impact on brain-derived neurotrophic factor and behavior in an animal model. *Behav. Brain Res.* 120(1): 87-95.

Sapolsky, R.M. 1999. Glucocorticoids, stress, and their adverse neurological effects: relevance to aging. *Exp. Gerontol.* 34(6): 721-32.

Schaaf, M.J., J. de Jong et al. 1998. Downregulation of BDNF mRNA and protein in the rat hippocampus by corticosterone. *Brain Res.* 813(1): 112-20.

Shirayama, Y., A.C. Chen et al. 2002. Brain-derived neurotrophic factor produces antidepressant effects in behavioral models of depression. *J. Neurosci.* 22(8): 3251-61.

Singh, M., E.M. Meyer et al. 1995. The effect of ovariectomy and estradiol replacement on brain-derived neurotrophic factor messenger ribonucleic acid expression in cortical and hippocampal brain regions of female Sprague-Dawley rats. *Endocrinology* 136(5): 2320-24.

Smith, M.A. & G. Cizza. 1996. Stress-induced changes in brain-derived neurotrophic factor expression are attenuated in aged fischer 344/N rats. *Neurobiol. Aging* 17(6): 859-64.

Sonntag, W.E., J.K. Brunso-Bechtold et al. 2001. Age-related decreases in growth hormone and insulin-like growth factor (IGF)-1: implications for brain aging. *J. Anti-Aging Med.* 4(4): 311-29.

Strawbridge, W.J., S. Deleger et al. 2002. Physical activity reduces the risk of subsequent depression for older adults. *Am. J. Epidemiol.* 156(4): 328-34.

Stummer, W., K. Weber et al. 1994. Reduced mortality and brain damage after locomotor activity in gerbil forebrain ischemia [see comments]. *Stroke* 25(9): 1862-69.

Tong, L., H. Shen et al. 2001. Effects of exercise on gene-expression profile in the rat hippocampus. *Neurobiol. Dis.* 8(6): 1046-56.

Tsui, C.C., N.G. Copeland et al. 1996. Narp, a novel member of the pentraxin family, promotes neurite outgrowth and is dynamically regulated by neuronal activity. *J. Neurosci.* 16(8): 2463-78.

Tu, B. & N.G. Bazan. 2003. Hippocampal kindling epileptogenesis upregulates neuronal cyclooxygenase-2 expression in neocortex. *Exp. Neurol.* 179(2): 167-75.

Tu, J.C., B. Xiao et al. 1998. Homer binds a novel proline-rich motif and links group 1 metabotropic glutamate receptors with IP3 receptors. *Neuron* 21(4): 717-26.

Tyler, W.J., M. Alonso et al. 2002. From acquisition to consolidation: on the role of brain-derived neurotrophic factor signaling in hippocampal-dependent learning. *Learn. Mem.* 9(5): 224-37.

van Praag, H., B.R. Christie et al. 1999. Running enhances neurogenesis, learning, and long-term potentiation in mice. *Proc. Natl. Acad. Sci.* 96(23): 13427-31.

van Praag, H., G. Kempermann et al. 1999. Running increases cell proliferation and neurogenesis in the adult mouse dentate gyrus. *Nat. Neurosci.* 2(3): 266-70.

van Praag, H., G. Kempermann et al. 2000. Neural consequences of environmental enrichment. *Nature Rev. Neurosci.* 1(3): 191-98.

Vanderwolf, C.H. 1969. Hippocampal electrical activity and voluntary movement in the rat. *Electroencephalogr. Clin. Neurophysiol.* 26(4): 407-18.

Ventriglia, M., L. Bocchio Chiavetto et al. 2002. Association between the BDNF 196 A/G polymorphism and sporadic Alzheimer's disease. *Mol. Psychiat.* 7(2): 136-37.

Vicario-Abejon, C., D. Owens et al. 2002. Role of neurotrophins in central synapse formation and stabilization. *Nat. Neurosci.* 3(12): 965-74.

Yamagata, K., K.I. Andreasson et al. 1993. Expression of a mitogen-inducible cyclooxygenase in brain neurons: regulation by synaptic activity and glucocorticoids. *Neuron* 11(2): 371-86.

Zhou, J., F. Zhang et al. 2000. Corticosterone inhibits generation of long-term potentiation in rat hippocampal slice: involvement of brain-derived neurotrophic factor. *Brain Res.* 885(2): 182-91.

Index

The italicized *f* stands for figure.

H

I

N

U

V

X

About the Editors

Frank C. Mooren, MD, is associate professor at the Institute of Sports Medicine at University Hospital Muenster in Germany, where he leads a molecular exercise physiology research group. Drawing from an academic background in medicine and chemistry, he worked as a researcher at the Max Planck Institute for Sytemphysiology in Dortmund. He has also been a consultant in internal medicine and sports medicine at the University Hospital Muenster.

Dr. Mooren is a member of the American College of Sports Medicine and the International Society of Exercise Immunology. He received the Heinz Zumkley Prize from the German Society on Minerals and Trace Elements in 1997 and the Arno Arnold Prize from the German Society of Sports Medicine and Prevention in 2001.

Klaus Völker, MD, is director of the Institute of Sports Medicine at University Hospital Muenster in Germany, where he established a group for molecular research. He has devoted his career to the study of both sports science and medicine.

He is vice president of the German Society of Sports Medicine and Prevention and chairman of its Commission of Sports for Youth, Leisure Time and Seniors. He is vice chairman of the Commission of Health of the Landessportbund NRW.

He is a member of the American College of Sports Medicine, European College of Sports Medicine, and European Hypertension League.